Computational modeling and visualization of physical systems with Python

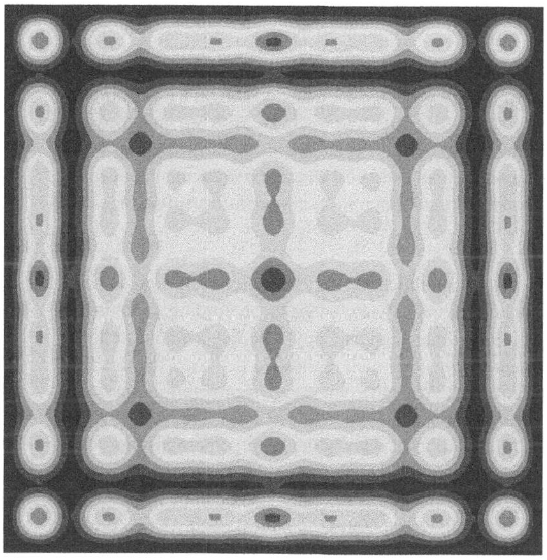

J. Wang

WILEY

VICE PRESIDENT & DIRECTOR	Petra Recter
EXECUTIVE EDITOR	Jessica Fiorillo
DEVELOPMENT EDITOR	Jennifer Yee
ASSISTANT	Kathryn Hancox
SENIOR DIRECTOR	Don Fowley
PROJECT MANAGER	Gladys Soto
PROJECT SPECIALIST	Nichole Urban
PROJECT ASSISTANT	Anna Melhorn
ASSISTANT MARKETING MANAGER	Puja Katariwala
ASSOCIATE DIRECTOR	Kevin Holm
SENIOR CONTENT SPECIALIST	Nicole Repasky
PRODUCTION EDITOR	Ezhilan Vikraman
PHOTO RESEARCHER	Nicholas Olin

This book was set in 10/12 Times LT Std by SPi Global and printed and bound by Strategic Content Imaging.

Founded in 1807, John Wiley & Sons, Inc. has been a valued source of knowledge and understanding for more than 200 years, helping people around the world meet their needs and fulfill their aspirations. Our company is built on a foundation of principles that include responsibility to the communities we serve and where we live and work. In 2008, we launched a Corporate Citizenship Initiative, a global effort to address the environmental, social, economic, and ethical challenges we face in our business. Among the issues we are addressing are carbon impact, paper specifications and procurement, ethical conduct within our business and among our vendors, and community and charitable support. For more information, please visit our website: www.wiley.com/go/citizenship.

ISBN: 978-1-119-23988-8 (PBK)
ISBN: 978-1-119-22418-1 (EVALC)

Library of Congress Cataloging in Publication Data:

Names: Wang, J. (Jianyi), 1962- author.
Title: Computational modeling and visualization of physical systems with
Python / J. Wang.
Description: Hoboken, NJ : John Wiley & Sons, 2015. | Includes
bibliographical references and index.
Identifiers: LCCN 2015029743 | ISBN 978-1-119-23988-8 (pbk.)
Subjects: LCSH: Physics–Computer simulation. | Physics–Graphic methods. |
Python (Computer program language)
Classification: LCC QC30 .W33 2015 | DDC 530.0285/5133–dc23 LC record available at http://lccn.loc.gov/2015029743

Printing identification and country of origin will either be included on this page and/or the end of the book. In addition, if the ISBN on this page and the back cover do not match, the ISBN on the back cover should be considered the correct ISBN.

Printed in the United States of America
10 9 8 7 6 5 4 3 2 1

To my Father and Mother,

who taught me that knowledge is to be revered.

Contents

4 Planetary Motion and Few-Body Problems 92

5 Nonlinear Dynamics and Chaos 144

6 Oscillations and Waves 184

7 Electromagnetic Fields 226

Preface

Computer modeling has had a significant impact on the way we do science and physics, both in research and in teaching. We are able to study problems of all scales from atoms to galaxies, and of complexities that would be impossible without computer modeling. Computation has even led to an entirely new field of science, chaos. In many ways, computational modeling has become the third pillar of physics alongside experimentation and theory.

In this book, we introduce computational modeling and visualization of physical systems that are commonly found in physics and related areas. Our first and foremost goal is to introduce a framework that integrates model building, algorithm development, and data visualization for problem solving via scientific computing. Through carefully selected problems, methods, and projects, the student is guided to learning and discovery by actively doing rather than just knowing physics. By constructing models and algorithms, programming and testing them, and analyzing the results, we will gain insight, ask new questions, tweak the model as necessary, and change the parameters to test what-if scenarios analogous to turning knobs in a virtual experiment. Another goal is to broaden the scope and depth of problems that may be studied with computational modeling. Many fundamental physical systems, despite their apparent simplicity, are beyond reach without computer simulation. Take projectile motion with air resistance and quantum free fall, for example. Though the problems can be expressed in basic calculus, they are unsolvable analytically in closed-form solutions. Computer modeling enables us to not only solve these problems, but also comparatively study the differences and similarities of their behavior in classical and quantum mechanical realms. We also aim to integrate applied computational tools and methods with effective visualization techniques in the simulations, including in situ data graphing and real-time animation within standard and open-source frameworks.

We take a problem-centric approach to the presentation of the material. For most problems, we include sufficient background information or essential relations to make them self-contained as much as possible, as well as to indicate their general importance and relevance. To streamline the text, many of the details are given in appendices or exercises, allowing readers of diverse backgrounds to skip ahead or study the detail at their own pace. We have also created a Digital Online Companion (DOC) for in-depth and advanced topics. Following the description of a problem, we discuss model building, the appropriate computational methods and tools, and visualization techniques to the simulations. Most results are represented in graphical form directly from the simulation programs for analysis and discussion.

Graphical representation of data and effective visualization techniques, including real time animations of three-dimensional objects, are standard parts of our simulations because they are essential to help interpret and analyze the results, especially in time-dependent studies. They also bring the simulations alive, making them more dynamic, instructive, and exciting.

The tight integration of advanced graphics and visualization into the simulations is made possible with standard, easy-to-use, open-source packages such as Matplotlib, SciPy, and VPython. These packages only require us to manipulate an object's attributes such as the physical position of a particle, so we can avoid low-level graphics programming and focus on modeling and proper techniques in using these tools. Except for schematic illustrations, most graphical representations and animations are created using techniques in the actual codes contained in the book. However, many programs are written such that the integrated graphics can be decoupled (disabled), replaced with either Matplotlib output or results written to a file, without affecting the calculations.

Because coding is essential to understanding an algorithm or to gaining insight to a physical process, and the most effective way to learn to code is by studying examples, the reader is guided in the process of building over ninety fully working sample programs, with detailed explanations for most of them. Many of the coded algorithms form part of basic building blocks in scientific computing. The associated tools and methods, whether well-known or otherwise still being currently researched, share common elements found in computational research of more complex problems. Through building fully working programs complete with graphical output, the student not only will learn how to write codes for computation, but their finished product is suitable for demonstration as well.

We use Python as the default programming language to show concrete, working examples and to take several advantages it offers: being easy to learn and use, readable, flexible, and powerful. One can think of Python as the modern equivalent of BASIC, but with a large and growing body of open source libraries for common tasks in scientific computing such as those mentioned earlier. Even newcomers to Python can learn from examples and grow quickly, writing functional programs and being productive in about two to three weeks. However, the choice of a programming language can be highly personal. Readers with programming background in other languages should find most Python codes to be expressive, pseudocode-like, and readily adaptable.

Sample systems are drawn from across the fundamental areas of physical science including mechanical, electromagnetic, quantum, and statistical systems. Representative problems within a given topic or theme are organized by chapter. Most chapters start with an animated simulation related to the central topic and end with a brief summary. The chapters and topics are organized as follows. After a brief introduction in Chapter 1 to our approach to modeling and basic programming elements, we discuss methods of solving ordinary differential equations (ODEs) in Chapter 2 as it constitutes the beginning of our framework. Thematic studies begin with the familiar example of projectile motion with air resistance (Chapter 3), followed by planetary and few-body motion including exoplanet modeling (Chapter 4), and nonlinear dynamics and classical chaos (Chapter 5). We study waves and oscillations in Chapter 6 and electromagnetic fields in Chapter 7. In Chapters 8 and 9, we present simulations of time-dependent and time-independent quantum mechanics, respectively. We consider simple random systems in Chapter 10 and statistical and thermal dynamics in Chapter 11. Finally, we investigate classical and quantum scattering in Chapter 12.

To effectively study these topics, the student should be familiar with basic calculus and introductory mechanics for mechanical systems (Chapters 2 through 6). For quantum and thermal systems (Chapters 8, 9, and 11), some familiarity with concepts typically covered in a modern physics course such as the wave function, expectation value, eigenstates, and entropy, will be helpful. We also assume the student to have some programming background, as familiarity with Python is a plus but not required. Past experience has shown that students unfamiliar with Python can successfully learn from examples and become productive quickly. By the time Chapter 3 is finished, most students will have grasped the essential programming elements needed to carry on.

Most of the introductory topics are aimed at the undergraduate level (up to Chapter 7). Starting from Chapter 5 and increasingly toward Chapter 12, we include advanced and more in-depth topics in the DOC content. These topics are geared at upper undergraduate or graduate levels, including quantum chaos, particle transport, Bose-Einstein condensation, quantum transitions, inelastic scattering and atomic reactions.

Moreover, we put a special emphasis on the simulation of quantum systems, expanding the coverage over three chapters (Chapters 8, 9, and 12). Ideally, these topics are better studied after a first course in quantum mechanics. However, some sections should be accessible even without a quantum course, such as Sections 8.1 through 8.3 and Section 9.1, for which a basic understanding of wave motion should be sufficient. Our motivation for including these topics is due to the fact that, unlike classical mechanics, quantum mechanics has far fewer analytically solvable problems and is less intuitive and less visual compared to classical motion. These factors cause considerable difficulties to students' understanding in introductory quantum mechanics. By simulating and visualizing the behavior of quantum systems such as the quantum oscillator or free fall, we are in a position to help effectively address some of the difficulties, pre- or postquantum class. We regularly use computer simulations to augment and enhance traditional physics classes such as quantum mechanics.

Although some chapters are interrelated, most from Chapter 3 on can be independently studied. Figure 1 shows the major dependencies of the chapters where the primary numerical methods (under the chapter number) are discussed, which are used in subsequent chapters. As the book progresses, most chapters are dependent on two previous chapters or less, except Chapters 9 and 12, which are dependent on five. There are several methods spanning multiple chapters, including ODE solvers (discussed in Chapter 2) for most time-dependent simulations in Chapters 3 through 9 and Chapter 11; the finite element method introduced in Chapter 6 and extended in Chapters 7 and 9; and other methods such as root finding (Chapter 3). In many cases, the reader may skip the methodology section without much disruption. For instance, the fast Fourier transform (FFT) discussed in Chapter 5 of DOC is lightly used in Chapter 6 to analyze fundamental oscillation frequencies, but heavily used in Chapter 8 for evolving quantum systems and for extracting momentum distributions. In both instances, the reader initially unfamiliar with FFT can still work through the simulations without worrying about the technical details of FFT, using it as a supplied library. The reader may come back and review it later for a fuller understanding if desired.

The instructor can pick what sections to cover within a given chapter in most cases. For example, the instructor may skip some sections on ODE solvers to save time (see Introduction in Chapter 2). However, this may be limited due to progression in methodology or conceptual

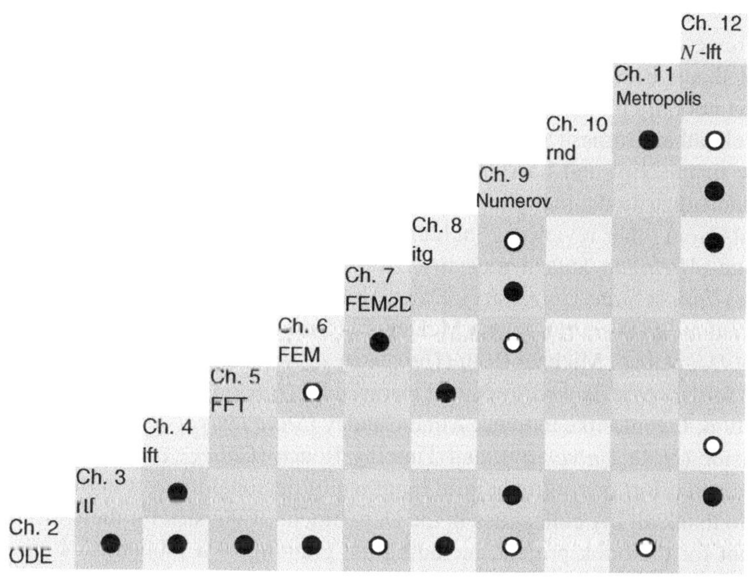

FIGURE 1 Major chapter dependencies at a glance. Open circles indicate minor or Digital Online Companion only dependencies. The abbreviations are: FEM—finite element method; FFT—fast Fourier transform; itg—numerical integration; lft—leapfrog with time-transformation; ODE—ordinary differential equations; rnd—random numbers; and rtf—root finding.

buildup. In such situations, indicated in the narrative and aided by Figure 1, the instructor may choose to discuss the prerequisite section lightly and leave further investigation to a project. Carefully constructed and tested projects given at the end of each chapter are an integral resource enabling the students to extend the models, make predictions, test them, analyze and present the results in project reports. Many in-depth projects, some found in the DOC, are designed for two-person teams to encourage peer learning and collaboration. There are also numerous exercises for the purpose of analytical derivations and proofs or simple programming tasks.

Throughout, references to the DOC content are indicated by the "S:" prefix, including the completely merged index. The DOC content and other electronic resources such as programs, data files, solution manual to select exercises, feedback and contact information, and installation instruction and files can be found at:

`http://www.wiley.com/college/wang` and
`https://github.com/com-py/compy/`

ACKNOWLEDGMENTS

I am indebted to Joachim Burgdörfer from whom I learned much of the craft of the trade during my PhD study. I also benefited from professional and personal interactions with Wolfgang Christian (*Davidson College*), Steven Gottlieb (*Indiana University, Bloomington*), Harvey Gould (*Clark University*), and Jan Tobochnik (*Kalamazoo College*), who have been pioneers in computational physics and education. I am grateful to Bruce Sherwood, the original developer of VPython and GlowScript, for being a constant resource and teacher. I am thankful to Steve Spicklemire (*University of Indianapolis*) for reviewing the manuscript, testing the programs, and making suggestions that substantially changed the structure of the codes on vector operations.

I owe a debt of deep gratitude to Michael Carlozzi (Writing and Reading Center, UMass Dartmouth) for his great generosity in time and effort to offer expert critique of the manuscript over two semesters that helped to significantly improve the writing, organization, and presentation of the material.

Many reviewers have provided critical and constructive feedback that improved the manuscript. I thank Brad Marston (*Brown University*) for extensive comments and valuable suggestions on not only modeling, but the validation and interpretation of computational results. I also thank Johnathan Barnes (*Salt Lake Community College*) for careful reading and annotation of part of the manuscript and Larry Engelhardt (*Francis Marion University*) for many useful discussions and suggestions, including code optimization. I sincerely acknowledge useful input and viewpoints from other reviewers, including Eric Ayers (*California State University, Chico*), Simon Billinge (*Michigan State University*), Alexander Godunov (*Old Dominion University*), Rainer Grobe (*Illinois State University*), Dawn Hollenbeck (*Rochester Institute of Technology*), Lev Kaplan (*Tulane University*), James McDonald (*University of Delaware*), Brianislav Nikolic (*University of Delaware*), Michael Roth (*University of Northern Iowa*), Steffen Vojta (*Missouri University of Science and Technology*), and Haowen Xi (*Bowling Green State University*).

In addition, I would like to thank Anne Caraley (*State University of New York at Oswego*), Henry Greenside (*Duke University*), and Timothy Roach (*College of the Holy Cross*) for their generous time and effort providing helpful input out of personal interest in the manuscript.

I wish to thank all my colleagues in the Department of Physics for their strong support and encouragement for the book project, including the granting of a sabbatical. I am very grateful to Robert Fisher and Gaurav Khanna who gave specific comments and valuable suggestions, and to Glenn Volkema, who tested and advised on the installation procedures. I also wish to thank Adam Hausknecht and Alfa Heryudono in the Department of Mathematics for helpful ideas and feedback, Andreas Ernst for discussions on fewbody problems, and Amit Tandon for source references on fluid motion. I thank my own students in the computational physics course over the years who left their mark on the project.

Furthermore, I would like to express my thanks to the Wiley team for their able support and guidance: to editors Stuart Johnson for initiating and developing the project and to Jessica Fiorillo for expertly and energetically steering it to completion, to Mary O'Sullivan, Amanda Rillo, Gladys Soto, Marcus Van Harpen, Ezhilan Vikraman and Karen Slaght for production and logistics support.

Finally, I wish to thank my family for their unconditional love and support. I want to say thanks to my sons, Gregory and Adam, for giving advice and opinion whenever I asked, for helping with literature research, and for trying out some topics and projects in the book. Last but not least, a heartfelt thank-you to my wife, Huihua, for showing great understanding and patience throughout the long project, without which this project would not have been possible.

J Wang
Dartmouth, Massachusetts
www.faculty.umassd.edu/j.wang/

Introduction

CHAPTER 1

In this book, we will be discussing computational modeling of physical systems ranging from classical to quantum systems. The primary factor that makes this possible is the speed of computation offered by digital computers. In this sense, computer modeling is simply a tool to get the job done, just like the abacus or the calculator of the past. It is an important tool, enabling us to simulate a range of systems from the simple projectile motion but with added realism such as drag and spin effects to the complex behaviors of many-body problems or time-dependent quantum systems. Computational modeling is important and often indispensable in the study of these systems.

In the following, we outline a three-step strategy in computational modeling. We also briefly discuss sources of error and introduce the basic elements of Python programming with our first program as well as the required packages for computation and visualization. Installation of these packages and numerical array operations are discussed in the Appendices.

1.1 COMPUTATIONAL MODELING AND VISUALIZATION

Before computational modeling can begin, there are several required steps to set it up. We generally follow a three-step approach for a given problem:

model building → algorithm construction → implementation

First, we build a model based on physical laws governing the problem. We usually aim to keep the model as simple as possible but still contains the essential physics. On the one hand, key physical concepts are best understood from the simplest of the models. On the other hand, many problems are sufficiently complex that, without simplifying approximations, they cannot be solved even with the most powerful computers available.

Second, we construct a simulation method, or an algorithm, to solve the problem based on the mathematical description of the physical model. Consideration of the operational behavior of an algorithm can often lead to physical insight about how the physical processes unfold. There is typically a trade-off in the selection of algorithms. The main criteria affecting our choice include accuracy, simplicity, stability, and robustness.

Last, we implement the algorithm in a programming language and present the results accordingly. Here again, we face several possible choices. We choose Python because, among other reasons, it is easy to write and read codes, and is open source. Python is flexible and is made more attractive and powerful for scientific computing due to the availability of external libraries for numerical simulations. For our purpose, four open-source packages are especially useful: SciPy and NumPy, an essential core base for numerical and scientific computing in Python; Matplotlib, a library for data plotting and graphical presentation in 2D and 3D; and VPython, a 3D interactive animation and graphics library.

The NumPy library adds advanced multidimensional array capability that is missing in native Python. The SciPy library provides many fundamental numerical analysis tools, among which the linear algebra routines and special functions are especially useful to us. These routines not only extend the functionalities of Python, but many of them also allow vectorized operations (implicit looping) over whole arrays. We gain the added benefits of execution speeds of compiled codes, as well as simpler and more elegant programs.

The graphics libraries enable us to seamlessly integrate visualization into our Python simulations. Data visualization is becoming an essential part of computer modeling that is increasingly data-centric. We will encounter simulations that yield a large amount of spatial and temporal data. With Matplotlib, we can produce with ease graphical representation of simulation results as simple line graphs, contour and surface plots, and vector fields. For many systems, we also use VPython to animate the process or data sets as dynamic movies.[1] All these techniques help us to more effectively analyze and interpret the simulation results. We can do all this with these packages within four or five dozen lines of code in most cases (see Section 1.B for installation of required packages).

We make every effort to present the physical systems as self-contained models. For mechanical systems, the student should have a good grasp of physics at the introductory level or above. For quantum and thermal systems, some familiarity with the relevant concepts such as the wave function, expectation value, entropy, and so on, will be helpful. We also assume the student to have some programming background. Familiarity with Python is helpful but not required. Past experience has shown that students unfamiliar with Python can successfully learn from the examples and become productive quickly.

1.2 THE SCIENCE AND ART OF NUMERICS

Scientific computing naturally involves numerical computation (number crunching). It uses integers and real values, or floating point numbers. Integer arithmetic on the computer is exact, but its range is limited. It is unsuitable for scientific computing because our physical world encompasses a vast scale over many orders of magnitude. Floating point numbers have a much expanded range, but their operations may not be exact in the mathematical sense, so exactness is sacrificed for enlarged range. Here we briefly discuss floating point operations so we are aware of some of the limitations.

1.2.1 FLOATING POINT OPERATIONS

To check floating point operations, consider $a = \sqrt{1-x^2}$ for some real value $|x| < 1$. We will use IPython (other interactive shells such as IDLE work as well; see Section 1.B on installing and running programs). At the prompt `In [1]`, enter the following (in notebook mode, press `Shift Enter` at the end of each input cell to get output):

```
In [1]: %precision 17
        x=0.6
        a=(1-x*x)**(0.5)
        a
Out[1]: 0.80000000000000004
In [2]: 1-a
Out[2]: 0.19999999999999996
```

The "%precision 17" keyword (also called a magic function) tells IPython to print numbers to 17 decimal places. If nothing follows the keyword, default behavior is restored (see below).

Mathematically, we should have exact values $a = 0.8$ and $1 - a = 0.2$. But the numerical values are not exact though very close, agreeing to 16 digits. We see right away that mathematical rules are bent a bit in floating point operations. This is because of the way decimal numbers (floats) are represented in the computer. A float f is internally represented by

$$f = (-1)^S \times M \times 2^E, \tag{1.1}$$

[1] See http://www.faculty.umassd.edu/j.wang/vp/movie.htm or the Digital Online Companion site for sample movies.

where S is the sign, M a fractional number between $\frac{1}{2} \leq M < 1$ (except 0), known as the mantissa, and E an integer exponent. Python provides an easy way to find M and E for a given f as follows,

```
In [3]:  import math as ma        # import math library
         ma.frexp(0.2)
Out[3]:  (0.8000000000000004, -2)
```

In the preceding code, we first import the mathematics library, rename it to `ma`, and use the Python function `frexp()` to find the mantissa and the exponent of 0.2, which are $M = 0.8$, and $E = -2$. Indeed, we have $0.2 = 0.8/4 = 0.8 \times 2^{-2}$, according to Eq. (1.1). Note the mantissa is not exactly 0.8 as seen by the computer. Why?

The reason is due to the representation of a float by a finite number of bits (see Appendix Section 1.A). According to Eq. (1.7), the mantissa is a sum of powers $(\frac{1}{2})^i, i = 1, 2, ..., n$, similar to a geometric series. Not all real values can fit exactly in a finite n-bit range (see Table 1.2). Python uses double-precision floats, which has an effective $n = 53$ bits. For a fixed exponent, the smallest difference between a pair of mantissas representable on the computer is $\sim 2^{-53} \approx 10^{-16}$. For single-precision floats, the difference is larger. Double precision, the default in Python, is most common in scientific computing, though higher precisions such as quadruple precision are used for special purposes. Regardless, a measure of precision is given by the machine accuracy ϵ, defined as the largest ϵ value that the sum $1 + \epsilon$ is still 1 in floating point arithmetic (not mathematical arithmetic),

$$1 + \epsilon \doteq 1. \quad \text{(machine accuracy)} \tag{1.2}$$

Let us see what the machine thinks of this:

```
In [4]:  1. + 1.e-15 == 1.
Out[4]:  False
In [5]:  1. + 1.e-16 == 1.
Out[5]:  True
```

The operator "==" tests whether a condition is true or false. Evidently, the machine accuracy ϵ is below 10^{-15} but above 10^{-16}.

1.2.2 NUMERICAL ERROR

When doing numerical calculations, we should be aware that numerical results are not always mathematically sensical. For instance, using the same example earlier with $x = 10^{-7}$, we obtain the following,

```
In [6]:  x=1.e-7
         a=(1-x*x)**(0.5)
         a
Out[6]:  0.99999999999999500
In [7]:  %precision
         1-a
Out[7]:  4.9960036108132044e-15
```

Although the value a is as expected, the accuracy of $1 - a$ is significantly reduced compared to the approximate value $1 - a \sim \frac{1}{2}x^2 = 5 \times 10^{-15}$, a loss of more than 10 significant digits in double precision. Further reducing $x = 10^{-8}$, the numerical answers become $a = 1$ and $1 - a = 0$. The latter is far worse than a loss of accuracy because there is a qualitative difference between a zero value and a finite but nonzero value, albeit a small one ($\sim 5 \times 10^{-17}$). This occurs because x^2 is less than the machine accuracy in this case, so that $a = \sqrt{1 - x^2} = 1$ in floating point operations, and we have a total cancellation error when evaluating $1 - a$.

Because floating point operations are inexact, mathematically equivalent expressions can yield different numerical results. This often leads to various tricks (on the level of art) in the quest for better numerical results. One trick to remedy the preceding situation is to rewrite it equivalently: let $b = \sqrt{(1 - x)/(1 + x)}$, so that $\sqrt{1 - x^2} = b + bx$, and $1 - a = 1 - b - bx$. Numerically, we have

```
In [8]:   x=1.e-8
          b=((1-x)/(1+x))**0.5
          1-b-b*x
Out[8]:   3.922529011343002e-17
```

The preceding step improves $1 - a$ somewhat, at least it is nonzero, and on the same order of magnitude as the mathematical answer. A better way is to write $1 - a = 1 - \sqrt{1 - x^2} = x^2/(1 + \sqrt{1 - x^2})$, so that continuing from the preceding, we have

```
In [9]:   x*x/(1+(1-x*x)**0.5)
Out[9]:   5.0000000000000005e-17
```

There is no cancellation involved. This result is as close as we can get to the mathematical answer within double precision.

The preceding tricks are mathematically equivalent, but numerically different. This simple example illustrates that numerics involves both science and art. Real-world examples are not as simple, but the underlying idea is valid: we need mathematical thinking as well as numerical thinking in scientific computing.

Round-Off Error

Round-off error can occur in arithmetic operations of floating point numbers for several reasons (see Section 1.A). The floats may not be exactly representable, or some significant bits are lost during shifting. Even if the floats are exactly representable or no shifting is involved, the intermediate results could require more mantissa bits to hold the full accuracy than the working precision could provide, resulting in rounding off.

The severity of round-off error varies. The preceding example is a case of round-off errors due to cancellation of two floats ($1 - a$) within close proximity of each other, leading to a severe loss of accuracy. We cannot eliminate round-off error. But with proper choice of methods (algorithms), we may be able to control or minimize its impact, as seen earlier.

Truncation Error

To produce a concrete result, we normally use a practical method, or an algorithm, for the calculation. There is usually a difference between the obtained result and the exact result. This is known as the truncation error. The size of the error should be dependent on the method, and unlike round-off error, it is entirely under our control by the choice of the method.

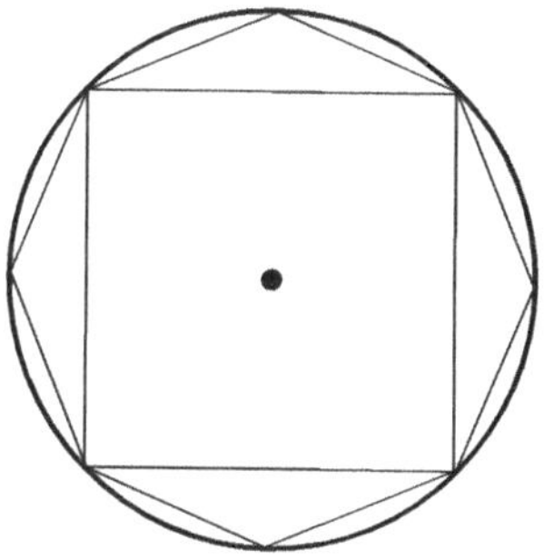

FIGURE 1.1 Approximating the circumference of a circle by polygons.

Table 1.1 The Ratio of an n-Sided Polygon Perimeter to the Circumference of a Unit Circle.

n	4	8	16	32	64	128
$\mathcal{P}/2\pi$	0.900316	0.974495	0.993586	0.998394	0.999598	0.999899

As an illustrative example, let us calculate the circumference of a unit circle approximated by the perimeter of a regular polygon inside the circle shown in Figure 1.1. For an n-sided regular polygon, adding up all the sides yields the perimeter

$$\mathcal{P} = 2n \sin\left(\frac{\pi}{n}\right). \tag{1.3}$$

Figure 1.1 shows two polygons, a square, and an octagon. We can see error cumulating as the arc length is replaced by straight-line segments. With increasing n, we expect the approximation to be ever more accurate.

Table 1.1 lists the ratio of the polygon perimeter to the exact circumference of the unit circle for several n values. Roughly with every quadrupling of n, the numerical result gains one significant digit. The method is stable and allows systematic improvement of accuracy. Most algorithms in our simulations are chosen to minimize or control truncation error.

Stability

Often, stability is a factor in controlling truncation error. Consider the golden mean $\phi = \frac{1}{2}(1+\sqrt{5}) \sim 1.618034$. It is a root to the quadratic equation[2]

$$\phi^2 = \phi + 1. \tag{1.4}$$

We can multiply both sides of Eq. (1.4) by ϕ^{n-1} to get $\phi^{n+1} = \phi^n + \phi^{n-1}$. This indicates a recursion relation

$$y_{n+1} = y_n + y_{n-1}, \quad \text{with } y_n = \phi^n. \tag{1.5}$$

Recursion relations are very useful because they are simple, are fast, and can generate a series of values on the fly. Suppose we want to compute the series y_n for a range of n values by the recursion method. We set the seeds $y_0 = 1$ and $y_1 = \phi$ in Eq. (1.5) and obtain y_2, then y_3, and so on. To check the accuracy against the exact results, we calculate the relative error $|y_n - \phi^n|/|\phi^n|$ and plot the results in Figure 1.2. Over the n range shown, the error for the ϕ series is on the order

[2]Another form for Eq. (1.4) is $\phi/1 = (\phi+1)/\phi$, meaning that the ratio ϕ to 1 is the same as $\phi+1$ to ϕ. This "perfect" ratio has inspired artists and architects in history.

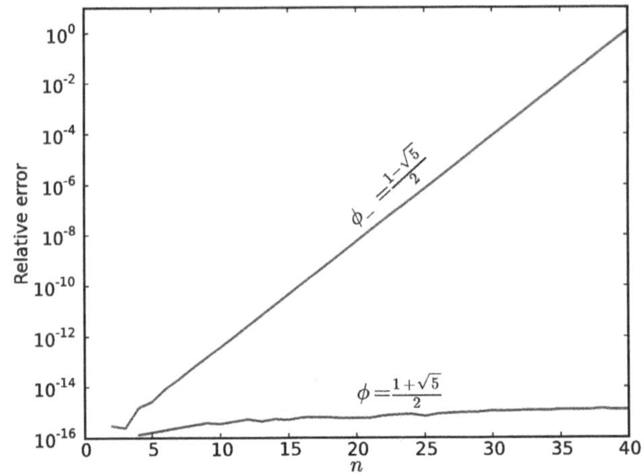

FIGURE 1.2 The relative error of ϕ^n and ϕ_-^n by recursion.

of $\sim 10^{-15}$, so small that it is practically at the machine accuracy in double precision. Though there is a slight rise in the error with increasing n due to inevitable round-off, the recursion method works well in this case.

There is a second root to Eq. (1.4), $\phi_- = -\phi^{-1} = \frac{1}{2}(1 - \sqrt{5}) \sim -0.618034$. Like the golden mean, the negative root ϕ_- also satisfies the recursion relation (1.5), so we could follow the same way as given earlier to generate a series of values for ϕ_-^n via recursion, with the initial seeds set as $y_0 = 1$ and $y_1 = \phi_-$. Repeating the calculation, we find quite a surprise: the relative error growing exponentially, and without bound. This behavior is easy to spot because we plot the results on the semilog scale, which shows the trend (a straight line) and, at the same time, clearly displays data differing over many orders of magnitude. The exponential explosion of the error cannot be attributed to round-off error. Rather, it must be due to the inherent ill-behavior of the algorithm (recursion). A significant digit is lost in roughly every 2 to 3 steps, rendering recursion unstable and useless for the ϕ_- series.

We may wonder why the method works well for one series (ϕ) but fails terribly for another (ϕ_-). The reason is that both ϕ and ϕ_- are present (even a small amount caused by round-off error) in the recursion because both are solutions of Eq. (1.4). The growing ϕ series completely overwhelms the vanishing ϕ_- series. A small error is quickly amplified and washes out the latter in a few steps.[3] The behavior is not related to, and cannot be solved by, higher float precisions. It is the property of the recursion (1.5). There is a silver lining, though: if the upward recursion is unstable, the reverse direction must be stable. We can use this fact to accurately obtain the ϕ_- series by downward recursion (Project P1.3). Stability conditions of some common recursions can be found in Ref. [1].

Through these examples, we demonstrate that proper choice of algorithms is important, and algorithm-induced error (truncation error) is entirely under our control.

1.3 FUNDAMENTALS OF PROGRAMMING AND VISUALIZATION

We are concerned with applying appropriate programming techniques to computational simulations but not with fundamental programming itself. Ideally, the reader should already have some computer programming background. Knowledge of the Python programming language is preferred but not required. If you are using Python for the first time, I recommend that you start with the tutorial at https://docs.python.org/2/tutorial/. You should find that Python is a

[3] A quantitative von Neumann stability analysis is discussed later (Section 6.5.4).

straightforward and forgiving language and is easy to learn from examples [54]. Many examples are found in the included programs in the chapters ahead. Next, we discuss the basic elements of Python and introduce the relevant packages to be used: Matplotlib, NumPy and SciPy, and VPython. We assume Python 2.7x is installed first, along with the latest versions of the packages just mentioned (Section 1.B).

1.3.1 PYTHON

Like driving, the only way to learn programming is to just do it. Let us write a Python program to compute and graph one-dimensional motion with constant acceleration a. Taking the initial value to be zero, the position x as a function of time t is $x = v_0 t + \frac{1}{2}at^2$, where v_0 is the initial velocity. The program is given here and results in Figure 1.3 (see Section 1.B on how to run programs).

Program listing 1.1 Motion with constant acceleration (`motion.py`)

```
   import matplotlib.pyplot as plt # get matplotlib plot functions
 2
   a, v0 = input('enter a, v0 (eg 1.0, 0.0) : ') # read input
 4 t, h, n = 0.0, 0.1, 20            # init time, step size, number of steps
   ta, xa = [], []                   # time and position arrays for plotting
 6
   for i in range(n):                # loop for n steps
 8     ta.append(t)                  # record time and position
       xa.append(v0*t + a*t*t/2.0)
10     t = t + h                     # update time

12 plt.figure()                      # start a figure; no indent->end of loop
   plt.plot(ta, xa, '-o')            # plot data
14 plt.xlabel('t (s)')               # add labels
   plt.ylabel('x (m)')
16 plt.show()                        # show figure
```

As our first full example, Program 1.1 may feel longer than necessary. It is okay if you do not understand fully what is going on here. We will explain the program later, but after working through the rest of this chapter to the middle of Chapter 2, it will become clear. The program has three basic blocks: initialization (lines 1 to 5), computation (lines 7 to 10), and data presentation (lines 12 to end). Most programs generally follow this template.

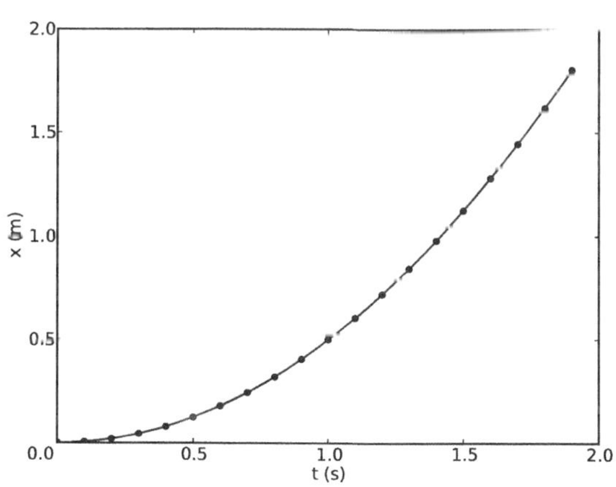

FIGURE 1.3 The position-time curve from Program 1.1.

Program 1.1 starts by importing from Matplotlib the `pyplot` library, which is the main workhorse of 2D plotting and is short-named to `plt`. Any external library must be imported before its use.[4] To keep separate namespaces, we will follow this convention from now on using short descriptive aliases for several libraries, including VPython as `vp` and NumPy as `np`. The program then reads (line 3, blank lines are ignored) the values a and v_0, assumed to be in MKS units, from the keyboard via the `input()` function, which can parse multiple parameters from user input. In this case, we should enter two numbers separated by a comma (also see what happens without the comma). The next line initializes the time, step size, and number of steps, the first two being floats, and the last one being an integer. We can assign values to multiple variables on a single line, provided the number of variables on the left is equal to the number of values on the right (called sequence unpacking). This saves print space without sacrificing clarity. In Python, variables need not be declared before assignment. The type of a variable is automatically derived from the assigned object. The same variable could be assigned to a different object later, changing its type. The last line before the loop creates two empty lists to store time and position values, rounding up the initialization block. For readability, it is usually a good idea to group similar constants and parameters together.

After the initialization statements, the `for` loop starts at line 7 and terminates just before the blank line. The `range()` function specifies the number of iterations. Unlike other languages where the programmer controls the appearance of logical blocks for easy identification, Python *requires* vertical alignment (indentation) for block control, including `while` loops, `if` conditions, function declarations `def()`, and other logical blocks. The next three lines, indented by the same amount, belong to the `for` loop. Inside the loop, the current time and position are appended to the respective lists `ta` and `xa`,[5] and time is incremented. Like the append method, there is a host of methods for manipulating lists, including insert, remove, pop, and so on.

The rest of the program graphs the results (Figure 1.3, $a = 1$ m/s^2, $v_0 = 0$). We first open a figure (line 12, it may be omitted if there is only one figure in the program) and plot the data with `plot()` from the Matplotlib library. The plot function accepts the required pair of (x, y) data arrays, plus an optional string specifying color, line style, or symbol types (see Section 1.C). After adding axis labels, the plot is displayed on the screen, ending the program. We can manipulate the figure, zoom in, drag around, or save to a file.

Conditional Controls

Besides the "for" loop above, Python has a conditional "while" loop (see Program 2.1)

```
while (condition):
    ......
```

The conditional `if` statement takes the form (also see Program 2.1)

```
if (condition):
    ......
elif (condition):
    ......
else:
    ......
```

[4]There seems to be a specialized library for any task. Want to fly? Try `import antigravity`.

[5]Some may prefer to use more descriptive names. We try to use short names if they do not lead to confusion. Here, `ta` and `xa` designate time and position arrays.

Both the `elif` and `else` are optional, and there can be multiple `elif` statements. There is also an inline `if-else` expression that yields a value depending on the condition:

```
x = a if condition else b
```

In the preceding statement, `x` will be equal to `a` if the condition is true and equal to `b` otherwise (see Program 5.1 for an example).

Function Definitions and Variable Scope

We can define functions with the `def` keyword as (see Program 2.5)

```
def func(x1, x2=1.0):
    ......
    return y1, y2
```

The function may accept input parameters with default values (`x2 = 1.0`). Unlike traditional programming languages, Python functions can return multiple values or none at all. All variables defined inside the function are local and inaccessible outside. The function can access any global variables defined in the main program, but cannot modify them. If the latter is necessary (should be rarely used and very well controlled, see Program 9.2), the variables to be modified must be declared as `global` inside.

Procedural and Object-Oriented Programming

With few exceptions, we choose procedural programming for its direct approach to problem solving. We can set up a problem quickly and solve it by building a series of compact functions. There are cases when it is more convenient to use object-oriented programming, such as Brownian motion (Program 10.2), Einstein solids (Program 11.1), and VPython modules (Program S:6.4).

Object-oriented programming combines both methods (i.e., functions) and data into stand-alone entities called objects. It may be more logical and suitable for building models that are expandable (see Program 11.1 and Project P11.2). For example, it would be natural to define particle objects with properties such as position and velocity. When combined with an integration method (Chapter 2), one could instantiate one particle object for projectile motion (Chapter 3), or another one for planetary motion (Chapter 4), all within a unified approach. One could even extend it to N-body problems. The interested reader may wish to explore this object-oriented approach.

File Handling

Among Python file handlers, the easiest way may be to read from and write data to files via the `pickle` module. Structured data records can be loaded or dumped easily (see Program 9.7 for an example). A quick way to handle text files can be found in Program 4.4.

Python 2.7x and 3.xx Compatibility

The programs presented assume Python version 2.7x, though they should run without problems in version 3.xx, provided the required packages are 3.xx specific. For our purpose, there are two incompatibilities between the two versions that affect us: print and division operations. First, the `print` statement in 2.7x becomes a function in 3.xx (see Program 3.5), and second, divisions are carried out as floats in 3.xx. To retain forward compatibility, the first line of a 2.7x program should be

```
from __future__ import print_function, division
```

Python 2.7x treats divisions according to operand types, but 3.xx defaults divisions to float operations. For example, in 2.7x, we have (using IPython)

```
In [1]:  1/2
Out[1]:  0
In [2]:  2.0**(1/2)
Out[2]:  1.0
```

The last result was probably unintended even in 2.7x and a source of unintentional human error. To always use float division like in 3.xx, add the compatibility statement as

```
In [3]:  from __future__ import print_function, division
         1/2
Out[3]:  0.5
In [4]:  2.0**(1/2)
Out[4]:  1.4142135623730951
```

We believe the programmer should be explicit about whether integer or float divisions are intended. For instance, the following works in both versions.

```
In [5]:  1//2      # integer division
Out[5]:  0
In [6]:  2.**(1./2.) # float division
Out[6]:  1.4142135623730951
```

We follow the preceding practice in our programs. The included programs run correctly under 3.xx without relying on the division compatibility.

Built-In Help

Python has an online help system that can provide information on keywords, modules, and so on. It is invoked by

```
In [7]:  help()
help>
```

Enter terms at the `help>` prompt, for example, `range` or `numpy`, to get help on them. Type `q` to return to the prompt or exit help.

1.3.2 NumPy AND SciPy

At its core, the NumPy library offers advanced multidimensional array creation and manipulation functions. In concert with Python, they operate at a high and often abstract level and execute at the speed of compiled codes. They form a unique and powerful combination for scientific computing. Many other libraries require NumPy, including Matplotlib and VPython. Appendix 1.D introduces the basic operations on NumPy arrays. We also use elementary mathematical functions in NumPy, many of which are the so-called universal functions (ufunc), functions that operate on scalar and NumPy array inputs transparently (see Program 4.6 for examples).

We will mainly use the SciPy library for solving linear systems and eigenvalue problems. It has highly specialized routines for linear algebraic problems, including efficient banded matrix solvers (see Program 8.3). SciPy also offers many special functions useful to us, such as Legendre and Hermite polynomials, spherical Bessel functions, and so on. Like NumPy functions, many SciPy routines are universal functions.

Speed Boost with Language Extensions

Though easy to learn and use, flexible yet powerful, Python by itself was not designed for sheer number crunching, a core task in scientific computing. Being an interpretive language, Python parses and translates each statement it encounters, one at a time, into machine instructions. Therefore, explicit, nested looping is slow and expensive, making the use of NumPy array operations practically indispensable in many simulations. Programs 6.7 and 8.2 are such examples where the proper use of NumPy arrays speeds up the code by 50 times or more over explicit looping (see also Programs 6.8 and 7.2 and Section 8.2.2).

Sometimes the speed problem cannot be solved so easily with NumPy alone. In most cases, a small part of the code hogs most of the computing time. Code profiling (Section S:8.B) can tell us where the bottleneck is. Are there ways to alleviate, even eliminate, the bottleneck without losing the benefits of Python? The answer is yes: with language extensions. There are several such easy-to-use extensions. Numba is the easiest package that can compile a standard Python function at runtime. It does not require us to change the code or use a different language. We just put the so-called decorator symbol `@jit` in front of a function, and almost by magic, it runs faster, sometimes by a lot. So Numba is nearly effortless to use. This is shown in a simple example in Program 5.7 for fractals. Cython is a package providing a simple interface so Python can easily call compiled C/C++ codes. Another package is Weave, part of the SciPy library, that can turn inline expressions into C/C++ codes. For Fortran, one can use F2Py, which is part of NumPy. F2Py can convert Fortran codes into Python callable binary modules.

For example, molecular dynamics (Section 11.4) is a typical case where the bottleneck is localized in the computation of forces among the particles (Program 11.7). When we use F2Py extensions to convert that part of the code to Fortran, we get a handsome payoff: a 100x speed gain (see Programs S:11.3 and S:11.4).

When we have a critical speed bottleneck, one of the aforementioned extensions should solve the problem. With just a little effort, we can have the best of both worlds.

1.3.3 Matplotlib

Matplotlib is a Python library for data plotting and graphical presentation in 2D and 3D. It is easy to make quality graphs with Matplotlib, interact with them like zooming, and save to standard file formats. At the same time, almost every aspect of a Matplotlib plot is customizable, making it possible to do hard or odd things, such as twin axes with different scales or even data animation (see Figure 3.16, Programs 8.1 and 10.2, and Matplotlib gallery for examples).

Almost all graphs of simulation results in the book are made in situ with Matplotlib. Source codes for a graph can be used as a template for similar types (e.g., Program 4.6 for surface plots), with the user substituting data only. This saves us time as we can simply copy and paste that part of the code. Furthermore, Matplotlib functions should be familiar to Matlab users.

Matplotlib also works in IPython notebooks, so graphs can be embedded in an interactive document. This is illustrated in Figure 1.4. The statement "`%matplotlib inline`" in input cell 1 tells IPython to show Matplotlib output within the notebook. The first line of input cell 2 imports `pylab`, which combines both `pyplot` and NumPy into a single entity (namespace), effectively importing two packages at once. The next line divides the interval $[-10, 10]$ into 100 points using the NumPy function `linspace`, which returns the points in a NumPy array assigned to x (see Program 4.6). The last line plots the function $\sin(x)/x$, which is shown immediately below. Here, the plotted function accepts an array input automatically, an example of a ufunc. We can also graph functions symbolically with SymPy (see Section 1.B).

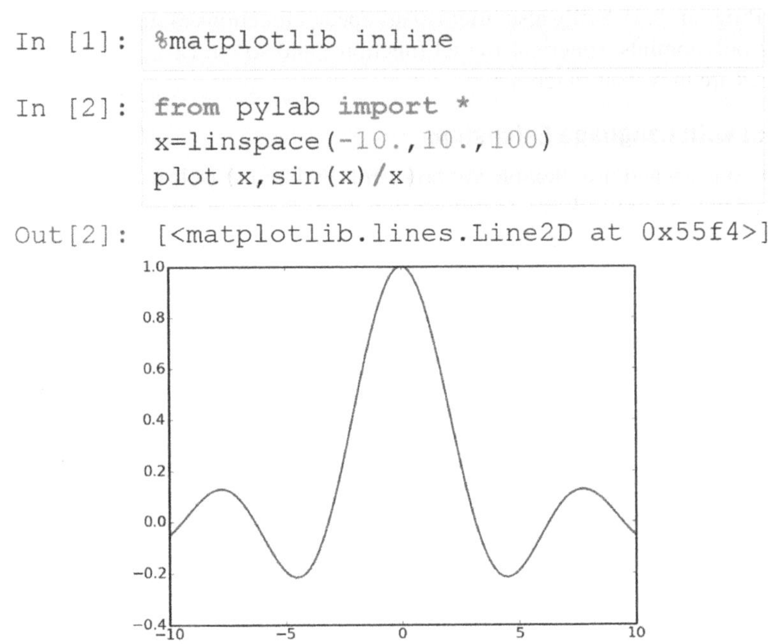

FIGURE 1.4 Matplotlib and IPython notebook.

It is also possible to load an existing program such as Program 1.1 in input cell 2. This way, we have a flexible, interactive environment within which both simulation and visualization can be done.

1.3.4 VPython

VPython is a graphics programming library for Python. It makes manipulation and dynamic animation of illuminated three-dimensional objects easy and accessible to the end user. It offers a rich and interactive environment to render the simulation process or data sets as dynamic movies on the screen. Program 2.1 shows a simple example of a bouncing ball (Figure 2.1). We use VPython extensively. In nearly all cases, animation can be turned off (commented out) without affecting the nonvisual aspect of the simulations, or replaced with either Matplotlib output (see Exercise E2.1) or data files for separate analysis.

Starting from VPython 6, a frame-limiting function `rate()` is required in the animation loop for it to function correctly across different computer platforms due to technical reasons. This normally does not pose a problem because without it, most animations would run too fast given the speed of today's computers. In cases where we need all the speed we can get (e.g., restricted three-body motion, Program 4.7), we can set the limit very high.

VPython programs are most robustly run by themselves (see Section 1.B on ways to run programs). Very recently, however, it has become possible to run VPython animation inline within IPython notebooks, as is the case with inline Matplotlib. To do so, just replace VPython with a wrapper called IVisual (see Section 1.B) and create VPython objects as usual. Figure 1.5 shows a spinning cube in the IPython notebook.[6] The cube is created using the VPython `box` object with

[6]If the animation does not show in the notebook, restarting the IPython kernel usually fixes it.

```
In [1]: import ivisual as vp
```

```
In [*]: cube = vp.box()
        while True:
            vp.rate(20)
            cube.rotate(angle=0.1, axis=(1,1,1))
```

FIGURE 1.5 Spinning cube in the IPython notebook.

default attributes because none is specified in the argument, so it appears as a cube with equal length, height, and width (see Program 2.1). Inside the `while` loop, the `rate` statement limits the animation to about 20 loops per second. We make the cube spin using the `rotate` method, which rotates an object for the specified angle (radians) about an axis vector (see Program 3.8 for a spinning baseball).

Run the example, and interact with the animation. You can view the scene from different perspectives by right-button drag (or Ctrl-drag) to change the camera angle. You can also zoom in or out by middle-button drag (Alt-drag, or left+right on a two-button mouse).

VPython also defines vector objects and operations, including cross and dot products. We make use of these capabilities (see Programs 3.8, S:6.2, and 7.5).

1.3.5 DEBUGGING AND OPTIMIZATION

It is rare not to make programming errors when writing functional programs of more than a dozen or so lines. Debugging can be challenging and time-consuming. Syntax errors are easy to find, but logical errors can be hard to trace. A few things can make debugging easier. Write modular and readable codes, test them often and in isolation if possible, and comment out, rather than delete, the changed lines (delete only when finished). Python is forgiving because it will print out the lines where errors occurred, but the chain-error messages can be bewildering. Try to find the error from the earliest line in the error chain, and print the values of the variables in question. For really hard-to-find bugs, try the integrated debugger built in the Python shell, IDLE. It allows stepping through the code and examination of variables.

Optimization, if necessary, should be done once a working program is at hand. It should keep the loss of clarity and readability to a minimum. Before we start optimizing, find the bottleneck by profiling first (Section S:8.B).

CHAPTER SUMMARY

We outlined the three-step strategy in computational modeling, namely: model building, algorithm construction, and implementation. We discussed floating point numerics and its limitations, sources of error, including round-off and truncation errors. We very briefly described the basic Python elements, and the required libraries: Matplotlib, NumPy and SciPy, and VPython. Familiarity with these libraries comes from using them, which has plenty of help in the chapters ahead.

GENERAL REFERENCES

We list below several general references on computational physics. A comprehensive bibliography is included at the end of the book.

- P. DeVries and J. Hasbun, *A first course in computational physics*, 2nd ed. (Jones & Bartlett, 2010) [24]. An introductory computational physics text that emphasized graphical representation of data even in the early days of the first edition, with the graphics-challenged Fortran language. This edition uses the MATLAB environment.
- N. Giordano and H. Nakanishi, *Computational physics*, 2nd ed. (Benjamin Cummings, 2005) [34]. A general computational physics book that uses both Fortran and C as the programming languages. Most examples are available online only.
- H. Gould, J. Tobochnik, and W. Christian, *An introduction to computer simulation methods: Applications to physical systems*, 3rd ed. (Addison-Wesley, 2007) [42]. A classic text since its first edition using the True Basic language, which had integrated graphical elements. This book contains a wide breadth of topics. The third edition uses Java and Open Source Physics library for graphical output.
- R. Landau, M. Páez, and C. Bordeianu, *Computational physics: Problem solving with computers*, 2nd ed. (Wiley, 2007) [53]. A comprehensive how-to book on numerical methods as applied to computational physics using the Java programming language. The third edition (2011) eBook uses Python.
- M. Newman, *Computational physics*. (CreateSpace Independent Publishing) [68]. A computational physics text using the Python programming language and the graphics libraries, including VPython.
- T. Pang, *An introduction to computational physics*, 2nd ed. (Cambridge University Press, 2006) [70]. A text on traditional computational physics topics as well as special topics such as generic algorithm. Sample programs are given in Java.
- J. Thijssen, *Computational physics*, 2nd ed. (Cambridge University Press, 2007) [89]. A text aimed at graduate level with advanced topics such as quantum physics. Examples in Fortran or C are available online only.

1.4 EXERCISES AND PROJECTS

Exercises

E1.1 Find the exponent and mantissa of the following floating point numbers: 0.5, 1.0, 5.0, 10^{10}, and 10^{-10}.

E1.2 Write down the bit patterns of the following fractions in their fixed-point format (1.6), keeping only the first 4 most-significant bits.
(a) 1/4; (b) 4/5; (c) 5/8; (d) 3/7.

E1.3 Write a program to find the machine accuracy ϵ on the computer you are using, to within a factor of 2.

E1.4 (a) Use the sample code with `struct` (Section 1.A) to find the mantissa bits in double precision (53 bits including the phantom bit) for the floats in Table 1.2. Replace `'f'` and `'i'` by `'d'` and `'q'`, respectively.

(b)* Find the exponent of each float. The machine typically includes a bias to the exponent. What is the bias?

E1.5 Derive an expression for the area of a regular n-sided polygon inside a unit circle. Approximate the area of the unit circle by the area of the polygon, and calculate the truncation error for the n-values listed in Table 1.1.

Projects

P1.1 Write a program to convert a floating point number to its bit pattern. The program should take the input from the user and print out the bit pattern (sign, and the rest in groups of 4 bits) for the float, assuming single precision. Test your program with the results shown in Table 1.2. See if you can break the program into logical modules.

P1.2 Consider the roots of the quadratic equation $ax^2 + bx + c = 0$,

$$x_{1,2} = \frac{-b \pm \sqrt{b^2 - 4ac}}{2a} = \frac{2c}{-b \mp \sqrt{b^2 - 4ac}}.$$

(a) Write a program to calculate the roots, taking a, b, c as input from the user, and printing out the answer to the screen. Choose any of the preceding equivalent expressions. Test it on the following numbers:

$$a = 3.0 \times 10^{-3},\ b = -2.0,\ c = 10^{-4},$$
$$a = 1.2 \times 10^{-17},\ b = 5.5,\ c = 10^{-13}.$$

Are the results correct for all cases? What kind of numerical error is responsible?

(b) To control the error, modify your program so that it correctly computes the roots to within machine precision, no matter the values of a, b, and c.

P1.3 We showed that the recursion relation (1.5) is unstable in the upward direction for $\phi_- = (1 - \sqrt{5})/2$. Thus, it should be stable in the downward direction, $y_{n-1} = -y_n + y_{n+1}$.

Iterate the downward recursion for N times, say $N = 50$, starting with arbitrary seeds, for example, $y_{50} = 0$ and $y_{49} = 1.0$. Normalize your series so $y_0 = \phi_-^0 = 1$ because the seeds are arbitrary. Compare with the exact results (ϕ_-^n). Also plot the relative error as in Figure 1.2. Discuss why it works.

Because the downward recursion is equally valid for the golden mean $\phi = (1 + \sqrt{5})/2$, the series obtained earlier should also be applicable to ϕ^n. Calculate and plot the relative error on the same graph. Why does it not work? Given that the seeds are the same, how does the algorithm "know" what series to converge to?

1.A FLOATING POINT REPRESENTATION

Fixed Point Numbers

We are familiar with bit patterns representing integers. They can also be used to represent fractional numbers, if we imagine a fixed point placed in front and interpret the n-th bit as a fraction 2^{-n}.

For example, consider the n-bit binary pattern X_{bin} given by

$$X_{bin} = .b_1 b_2 \cdots b_n, \tag{1.6}$$

where b_i is either 0 or 1. Interpreted as above, the decimal fraction number X_{dec} can be written as

$$X_{dec} = \sum_{i=1}^{n} b_i 2^{-i}. \tag{1.7}$$

The range of X_{dec} is $0 \leq X_{dec} \leq 1 - 2^{-n}$, and the smallest increment representable is 2^{-n}. By increasing the number of bits n, the increment will be reduced, and finer fractional values, or greater precision, can be achieved. It is also possible to shift the fixed point to different positions to represent a mixture of whole numbers plus fractions. In addition, by adding a sign bit, negative values can be included using two's complementary binary representation.

Floating Point Numbers

Floating point numbers are natural extensions of fixed point numbers. As Eq. (1.1) shows, a nonzero real number f can be decomposed as $f = M \times 2^E$ where the mantissa M is a fraction between $\frac{1}{2} \leq |M| < 1$, and the exponent $E = \text{ceiling}(\log_2 |f|)$, the smallest integer above $\log_2 |f|$. For example, the number 0.5 has a mantissa $M = 0.5$ and exponent 0, and 1.0 has the same mantissa but a different exponent of 1.

Typically, a floating point number is stored in a finite number of bits. For example, single-precision floats use 32 bits (4 bytes) as

$$\underbrace{\,|\,}_{S}\cdot\underbrace{||||\,||||}_{E}\cdot\underbrace{||||\,||||\,||||\,||||\,||||\,|||}_{M}, \tag{1.8}$$

including 1 bit for the sign, 8 bits for the exponent interpreted as a signed integer between $[-128, 127]$, and 23 bits for the mantissa as a fraction according to Eq. (1.6). Because the mantissa has a lower bound $M \geq \frac{1}{2}$, the first bit (most significant) of the mantissa in Eq. (1.8) is always 1, and by convention, it is normally not stored explicitly (so-called phantom bit). The mantissa actually represents the next 23 bits, for an effective 24-bit length. This means that single-precision floats are accurate to $\epsilon = 2^{-24} \sim 6 \times 10^{-8}$, or about 7 digits. The value ϵ is called the machine accuracy in Eq. (1.2). The range of single-precision floats is $[2^{-128}, 2^{127}]$, or roughly 10^{-38} to 10^{38}.

Double-precision floats, the default in Python, use 64 bits (8 bytes): 1 bit for the sign, 11 bits for the exponent, and 52 bits for the mantissa (53 effective bits). The machine accuracy for double precision is $\epsilon \sim 10^{-16}$ (15 digits), and the range is about 10^{-308} to 10^{308}. Results exceeding the range will cause under- or overflow.

We can write a program to calculate the mantissa binary pattern for a given float (Project P1.1). However, like the earlier case about the exponent and mantissa (Section 1.2.1), there is an easier way provided by Python via the `struct` conversion module. The following code demonstrates this.

```
In [1]:  import struct as st
         fpack=st.pack('f',6.e-8)        # pack float to bytes
         fint=st.unpack('i',fpack)[0]    # unpack to int
         bin(fint)[-23:]                 # mantissa bits
Out[1]:  '00000001101100101011001'
```

This code converts a float (6×10^{-8} in the example) to a byte pattern (string) in memory, which is subsequently converted to an integer. The mantissa is contained in the last 23 bits of the integer in binary representation, excluding the phantom bit. Several other examples including the

Table 1.2 The Exponent E and Mantissa M of Single-Precision Floats.

float	E	M (decimal)	M (binary including phantom bit)
10.0	4	0.625	1010 0000 0000 0000 0000 0000
1.0	1	0.5	1000 0000 0000 0000 0000 0000
0.1	−3	0.8	1100 1100 1100 1100 1100 1101
6×10^{-8}	−23	0.50331648	1000 0000 1101 1001 0101 1001

phantom bit are listed in Table 1.2. The decimal exponent and the mantissa are obtained from `frexp()` explained earlier (Section 1.2.1).

When two floats are to be added, the exponent of the smaller number is adjusted upward so it is equal to the exponent of the larger number, and the remainders are added. For example, suppose we wish to add $1 + 6 \times 10^{-8}$. Using the E and M values from Table 1.2, we have

$$1 + 6 \times 10^{-8} = 0.5 \times 2^1 + 0.50331648 \times 2^{-23} = 2^1(0.5 + 0.50331648/2^{24}). \qquad (1.9)$$

In machine operation, the second mantissa in the parenthesis in Eq. (1.9) will be shifted to the right 24 times,[7] with the most significant bit (left most) replaced by 0 each time. In single precision, the effective mantissa is zero after the shifts because all the bits are flushed to the right. Consequently, the result from Eq. (1.9) is (numerical, not mathematical)

$$1 + 6 \times 10^{-8} = 2^1 \times 0.5 = 1. \quad \text{(single precision)} \qquad (1.10)$$

We see from the preceding example that, in general, addition of two floats whose relative magnitudes differ by more than the machine accuracy will cause the smaller float to be neglected, and the result is equal to the larger float. Even when the floats are within the machine accuracy, there is usually a loss of accuracy due to the flushed bits. This is the round-off error. Round-off errors also occur in other arithmetic operations even if no shifting is involved because the finite mantissa may not be sufficient to hold the resultant answer. In particular, cancellation error from the subtraction of two floats within close proximity of each other could lead to a complete loss of significant digits.

1.B PYTHON INSTALLATION

In addition to Python 2.7x, we need the following four packages: NumPy, SciPy, Matplotlib, and VPython. Optional packages include IPython, SymPy (both recommended), and IVisual. IPython is a user-friendly Python shell (see following). SymPy, a symbolic mathematics library for Python, is not used in our numerical computation, but is handy nonetheless for many tasks such as algebra, exercises (see Exercise E2.7 and Exercise E3.10), and so on.

Python Distributions

The easiest way to get a full installation up and running quickly is through all-in-one distributions such as:

Anaconda,	http://continuum.io/downloads;
Canopy,	https://www.enthought.com/products/canopy/;
Python(x,y),	https://code.google.com/p/pythonxy/ (Windows only);
WinPython,	http://sourceforge.net/projects/winpython/.

[7] In binary arithmetic, dividing by 2 is equivalent to shifting the bits to the right by one position.

Each distribution contains all the required packages (and more) listed here, except Anaconda and Canopy, which do not include VPython as of this writing. To add VPython to Anaconda, enter the following conda command

```
$ conda install -c mwcraig vpython
```

This will install VPython built for Anaconda by Matt Craig on both Mac OS and Windows. If you are using Canopy on the Mac OS, there is a ready-made binary installer at vpython.org.

The Python(x,y) and WinPython distributions do include VPython but are for Windows only. WinPython is also portable.

Custom Installation

In general, use binary installers if possible, and stick with one version (32- or 64-bit) consistently. All required packages have binary installers at their web sites built for the official python.org Python (Table 1.3). Binaries built for other Python installations including all-in-one distributions are usually not compatible. The online resource of the book has updated installation instructions and the needed binaries for Windows at:

`http://www.wiley.com/college/wang`

You can also use the `pip` tool to install packages from Python Package Index, https://pypi.python.org/, for any platform including Linux. The typical command line is `$ pip install xyz` for installing package xyz. Note that `pip` (included starting Python 2.7.9) is in the `Scripts` directory of a Python installation, so make sure it is in the command path (C:\Python27\Scripts\ on Windows).

The `pip` tool can also install packages in the "wheel" format, for example, `$ pip install xyz.whl`. For Windows, many binaries built by Christoph Gohlke for the official Python including Matplotlib dependencies can be found at www.lfd.uci.edu/~gohlke/pythonlibs/ (referred to as UCI).

Table 1.3 lists the recommended order to install only the needed packages individually (file names are for Windows or Mac at the time of writing, always use the most current versions). Check each package's installation instructions, and test the installed package before moving on to the next one. When running into a brick wall, ask the instructor or colleagues for help.

To check the installed packages, issue this from a command terminal,

```
$ pip list
```

or start IDLE and enter the following

```
>>> help('modules')
```

This will list the locally installed packages. Use `pip` to install missing dependencies (if any) or other packages.

Running Programs

If you have installed the required packages successfully, congratulations! You are about to enter the exciting world of simulations, and you can now run all the programs discussed in the book.

Table 1.3 Package Installation Guide (W=Windows, M=Mac OS).

Package	Files to download or method
Python python.org	W: `python-2.7.8.msi` M: `python-2.7.8-macosx10.6.dmg`
`pip` pip.pypa.io	Download `get-pip.py` from the site, then run it: `$ python get-pip.py` (from a command terminal)
VPython vpython.org	W: `VPython-Win-32-Py2.7-6.10.exe` M: `VPython-Mac-Py2.7-6.10.dmg`
NumPy numpy.org	W: `numpy-1.8.2-win32-superpack-python2.7.exe` M: `$ pip install numpy` This can be skipped because VPython includes NumPy.
SciPy scipy.org	W: `scipy-0.14.0-win32-superpack-python2.7.exe` M: `scipy-0.14.0-py2.7-python.org-macosx10.6.dmg`
Matplotlib matplotlib.org	W: `matplotlib-1.4.0.win32-py2.7.exe`; requires `python-dateutil`, `pyparsing` and `six` (binaries at UCI) M: `matplotlib-1.3.1-py2.7-python.org-macosx10.6.dmg`
IPython ipython.org	W: `ipython-2.4.0-py27-none-any.whl` (from UCI) W or M: `$ pip install ipython`
SymPy sympy.org	W: `sympy-0.7.5.win32.exe` M: `$ pip install sympy`
IVisual python.org/ pypi/IVisual	Install from the command terminal `$ pip install ivisual` Optional. Some VPython codes may not run properly.

The most direct way to run a program is from a command terminal. For instance, the following will run Program 1.1, `motion.py`[8]

```
$ python motion.py
```

One can also navigate to the folder of source programs and click on them. Programs can be prepared with a text editor and be executed this way.

To run Python interactively, we can use IPython, which is an interactive shell for Python. It can act like an interactive calculator and also integrate codes and other media like graphics in the notebook mode running in a browser (see Figure 1.4). IPython has many other useful features, including online help and tab completion. Start IPython (in `Scripts` path) in either the standard line mode, or the notebook mode (without []),

```
$ ipython [notebook]
```

[8]On Windows, the keyword "python" can be omitted. Any program entered at the command line (cmd.exe) with `.py` extension should automatically invoke Python.

Once in IPython, enter statements line by line (in notebook mode, press `Shift Enter` at the end of each input cell to execute the cell). For example, you can test SymPy with

```
In [1]:  %matplotlib inline
In [2]:  from sympy import *
         x = symbols('x')
         plot(sin(x)/x)
```

and you should see the same output as Figure 1.4.

We can also load an existing program in IPython

```
In [1]:  load motion
```

This will load `motion.py`, Program 1.1, to be edited or executed.

An alternative to IPython is the standard Python integrated development environment (IDLE), with an editor and a debugger. From IDLE, we can write a new program, open an existing one, and run it from a menu-driven system.

VPython also comes with its own shell called VIDLE. It offers several improvements over IDLE, including more graceful handling of exceptions when Matplotlib and VPython functions are used simultaneously. For maximum reliability, run VPython programs from the command line or from IDLE or VIDLE.[9]

1.C THE Matplotlib PLOT FUNCTION

The `plot()` function from Matplotlib produces line graphs. It is invoked by

```
plt.plot(x, y, string)
```

where `x` and `y` are the data set, and `string` is an optional string specifying the color, line style, or marker type of the curve. There are also many other optional arguments. The string is normally a three-character field, `'CMS'`, for color, marker, and line style, respectively. Refer to Table 1.4 for common codes. The order of the character in the string is unimportant. For instance, the string `'g-o'` means a green solid line with filled circle markers.

Axis labels as well as text can be added. For example,

```
plt.text(xloc, yloc, 'message', rotation=45)
```

Table 1.4 Common Line Properties for the Matplotlib `Plot()` Function.

color	b blue	g green	r red	m magenta	c cyan	y yellow	k black	w white
marker	o ●	s ■	x ×	+ +	D ◆	* ★	^ ▲	, ·
style	- solid	-- dashed	: dots	-. dash-dot				

[9] One can also run most VPython programs without VPython installation in a web browser via GlowScript (glowscript.org) or capture output to a movie (see vpython.org).

would put the "message" at (x_{loc}, y_{loc}), and rotated 45° counterclockwise. Often, the font size looks fine on the screen, but may be too small for print after resizing. Such parameters can be specified via the optional arguments, or via configuration records. For example, the following code would change font size, axis and line thickness, frame spacing, and so on.

```
import matplotlib.pyplot as plt
plt.rc('lines',linewidth=2) # change line style, fonts
plt.rc('font',size=16)        # one group at a time
plt.rc('axes',linewidth=2)
plt.rc('figure.subplot',right=0.96)   # frame spacing
plt.rc('figure.subplot',top=0.96)
plt.rc('figure.subplot',left=0.1)
```

Many figures in the book are produced with these parameters. For convenience, the code can be included in a file (say `rcpara.py`) and imported before use.

1.D BASIC NumPy ARRAY OPERATIONS

We give a very brief introduction to the usage of NumPy arrays (also called ndarrays). There are many other powerful features that can be understood through use, examples (see programs mentioned at the end of this section), and the NumPy documentation.[10] Remember, import NumPy before use, and online help is available in IPython or IDLE on any installed packages. To get help on `np.array` for instance, issue

```
In [1]:   import numpy as np
In [2]:   help(np.array)
```

Array Creation, Indexing, and Slicing

A NumPy array can be created from any array-like objects:

```
In [3]    np.array([1.,2.,3.])
Out[3]:   array([1.,  2.,  3.])
In [4]:   a=np.arange(5)
          a
Out[4]:   array([0, 1, 2, 3, 4])
```

The NumPy `arange` function is just like the Python `range` except it returns an ndarray. The data type (e.g., float or integer) is derived from the assigned objects or specified with `dtype`. NumPy has other functions for initializing arrays, for example, `zeros(5)` and `ones(5)` will create two, 5-element arrays filled with zeros and ones, respectively, with float as the default type. The array a will be continuously updated and used in the following examples using IPython notebook mode.

[10]There are also online tutorials, for example, scipy-lectures.github.io.

Array elements can be accessed by an index, either positive or negative,

```
In [5]:   a[2]
Out[5]:   2
In [6]:   a[-2]
Out[6]:   3
```

A positive index counts from the beginning of the array. Remember that array indices start from zero, thus a[0] is the first element (0), and a[2] the third element (2). A negative index is allowed, which counts backward from the end. So a[-1] refers to the last element (4), a[-2] the second last element (3), and so on.

Slicing is triggered by the colon [:] symbol, and is an important part of using NumPy arrays. It normally takes the form a[start:end:step], producing an array from start up to but not including end, in steps step. If start or end are absent, the first or the last elements (inclusive), respectively, are assumed. If step is omitted, it defaults to 1. A negative step means going in the reverse direction of the array. Some basic slicing examples are the following:

```
In [7]:   a[:]
Out[7]:   array([0, 1, 2, 3, 4])
In [8]:   a[1:3]
Out[8]:   array([1, 2])
In [9]:   a[:3]
Out[9]:   array([0, 1, 2])
In [10]:  a[2:]
Out[10]:  array([2, 3, 4])
In [11]:  a[::-1]
Out[11]:  array([4, 3, 2, 1, 0])
```

Note that a[:] returns the whole array, but a[] would be illegal.

Array Manipulations

Elements can be inserted or removed from ndarrays. The following example inserts a "9" in front of the original index "1" of a and deletes the element at index "2":

```
In [12]:  np.insert(a,1,9)
Out[12]:  array([0, 9, 1, 2, 3, 4])
In [13]:  np.delete(a,2)
Out[13]:  array([0, 1, 3, 4])
```

Each time, a new array is created, and the original array a is not modified.

Operations on an array occur in an element-wise, vectorized fashion. For example:

```
In [14]:  2*a
Out[14]:  array([0, 2, 4, 6, 8])
In [15]:  a+3
Out[15]:  array([3, 4, 5, 6, 7])
```

When the scalar "3" is added to the array, it is cast into an array of the same dimension (called broadcasting) internally, and then the arrays are added element by element. Because no explicit looping is invoked in Python, the speed gain is considerable for larger arrays. For explicit looping, Python lists are faster.

Elements can also be selectively changed via slicing. The following assigns two elements to new values:

```
In [16]:   a[1:3]=[-1,-2]
           a
Out[16]:   array([0, -1, -2,  3,  4])
```

Basic slicing like the preceding returns a "view" of the array, that is, it points to the elements within the same array. In contrast, slicing of a Python list returns a copy.

If we wish to work with a copy of an ndarray, we can use the `copy()` function. The following copies part of the array and changes the copy without affecting the original array:

```
In [17]:   b=np.copy(a[1:3])
           b
Out[17]:   array([-1, -2])
In [18]:   b[:]=[-3,-4]
           b
Out[18]:   array([-3, -4])
In [19]:   a
Out[19]:   array([0, -1, -2,  3,  4])
```

Advanced Indexing

Copies of ndarrays can also be obtained with advanced indexing. This occurs when the index is not simply an integer. The following code obtains a copy with a list index, which triggers advanced indexing. Again, the source array is not affected:

```
In [20]:   c=a[[1,2]]
           c
Out[20]:   array([-1, -2])
In [21]:   c[:]=[-5,-6]
           c
Out[21]:   array([-5, -6])
In [22]:   a
Out[22]:   array([ 0, -1, -2,  3,  4])
```

It is also possible to select elements with a boolean or truth array. A truth array can be created by logical relations such as <, >, ==, and so on. For example,

```
In [23]:   a>0
Out[23]:   array([False, False, False,  True,  True], dtype=bool)
In [24]:   a==0
Out[24]:   array([True, False, False, False, False], dtype=bool)
```

The statement a>0 tests every element of a against 0 and sets the corresponding indices to `True` if they are greater than zero, and `False` otherwise.

When a truth array is an index to another array, only the true elements are taken. This is sometimes called fancy indexing in NumPy. The following are some examples of truth array indexing:

```
In [25]:   a[a>0]
Out[25]:   array([3, 4])
In [26]:   a[a<0]
Out[26]:   array([-1, -2])
In [27]:   a[a==0]
Out[27]:   array([0])
```

The positive elements are selected in the first instance, negative ones in the second instance. Note that the third instance has only a single element, but it is still returned as a one-element array.

Like assignment with slicing, we can also selectively assign elements with a truth array:

```
In [28]:   a[a<0]=0
           a
Out[28]:   array([0, 0, 0, 3, 4])
```

All elements of the array less than zero are set to 0 in the preceding example.

Advanced indexing can be very powerful but also convoluted at the same time, making the code harder to understand. Unless necessary, we prefer the direct approach, for example, to obtain array copies with the `copy()` function.

Multidimensional Arrays

Multidimensional arrays can be created and handled similarly. For instance, `zeros((2,5))` will create a 2-row by 5-column array filled with zeros. Each dimension is called an axis, so axis 0 refers to going down the rows and axis 1 to going across the columns.

Next, we create a two-dimensional array and find its dimensions via the `shape` attribute:

```
In [29]:   d=np.array([[11,12,13],[21,22,23],[31,32,33]])
           d
Out[29]:   array([[11, 12, 13],
                  [21, 22, 23],
                  [31, 32, 33]])
In [30]:   d.shape
Out[30]:   (3, 3)
```

The array can be accessed with two indices and sliced like before, but the results may be scalars, one-dimensional arrays (row or column vectors), or two-dimensional subarrays:

```
In [31]:   d[1,2]
Out[31]:   23
In [32]:   d[0]
Out[32]:   array([11, 12, 13])
In [33]:   d[:,1]
Out[33]:   array([12, 22, 32])
In [34]:   d[1:,:2]
Out[34]:   array([[21, 22],
                  [31, 32]])
```

In this example, d[1,2] refers to the element at the second row and third column, d[0] the first (entire) row, and d[:,1] the second column, both 1D arrays. But d[1:,:2] refers to the second row and above, and all columns up to, but not including, the third column. The result is a 2D subarray. See Program 7.2 for actual uses.

Rows or columns can be inserted or deleted as well. The following deletes the second row along axis 0 and the third column along axis 1, respectively:

```
In [35]:   np.delete(d,1,axis=0)
Out[35]:   array([[11, 12, 13],
                  [31, 32, 33]])
In [36]:   np.delete(d,2,axis=1)
Out[36]:   array([[11, 12],
                  [21, 22],
                  [31, 32]])
```

Operations on ndarrays with advanced indexing can be very efficient and elegant. For instance, we can swap the first and third columns simply by:

```
In [37]:   d[:,[0,2]]=d[:,[2,0]]
           d
Out[37]:   array([[13, 12, 11],
                  [23, 22, 21],
                  [33, 32, 31]])
```

The target d[:,[0,2]] refers to the first and third columns of all rows, and the source d[:,[2,0]] specifies the third and first columns of all rows. Slicing and advanced indexing work together to enable us to swap the columns in a single statement. See Program 9.9 for examples in actual codes.

Universal Functions

NumPy arrays of the same shape can be added, subtracted, multiplied, or divided, all element-wise. Python operators like +, −, *, /, **, and so on, are context-aware. When acting on ndarrays, operations are carried out element-wise transparently and automatically, irrespective of the type of operands as long as they can be broadcast into an ndarray.

```
In [38]:   x=np.array([1.,2.,3.,4])
           x-2
Out[38]:   array([-1.,  0.,  1.,  2.])
In [39]:   x/2.
Out[39]:   array([0.5,  1. ,  1.5,  2. ])
In [40]:   x*x
Out[40]:   array([1.,  4.,  9.,  16.])
```

In the first two cases, the second operand is broadcast to an ndarray [2., 2., 2., 2.] before element-wise subtraction/division is carried out. In the third case, no broadcasting is necessary, and the elements are multiplied one to one, amounting to x**2.

In NumPy the concept of element-wise operations is also extended to functions, called universal functions (ufunc). They operate on scalar and ndarray inputs transparently. For example, with the same x as used previously,

```
In [41]:   np.sqrt(x)
Out[41]:   array([1.        , 1.41421356, 1.73205081, 2.        ])
In [42]:   np.maximum(x,x[::-1])
Out[42]:   array([4., 3., 3., 4.])
```

In the first instance, the square root of every element is taken. Next, each element of x is compared to the element of the reversed array, and the greater one is picked. Using ufuncs with ndarrays involves implicit looping. It is not only simple and elegant, but also efficient: the actual coding or computation is often done in compiled languages like `CPython` or `C/C++` for increased speed.

Like `sqrt` and `maximum`, most NumPy functions are ufuncs. Thus, any function we define using NumPy versions of mathematical functions such as `sqrt` is also a universal function.[11] See `potential()` in Program 4.6 for an example.

We have described very briefly some features of the NumPy library. There are many other operations and advanced features on NumPy arrays, including concatenation, stacking, sorting, and logical operations, and so on. Numerous examples are found and explained in the accompanying programs, particularly starting from Chapter 4 such as Programs 4.6, 6.8, 7.3, 8.2, 9.7, and S:6.1, to name but a few.

[11] If necessary, we can also make a non-ufunc into a unfunc by vectorizing it with a NumPy module named, well, you guessed it: `vectorize()`.

Free Fall and Ordinary Differential Equations

CHAPTER 2

Many problems in physics and engineering are expressed in the form of either ordinary or partial differential equations (denoted ODEs, or PDEs). To a large part, this seems to be due to the fact that mathematical descriptions of problems of interest, especially the complex ones, simplify significantly when we consider only infinitesimal changes of the variables involved. The result is the emergence of differential equations of one form or another. Take Newton's second law $F = ma$, for example. Because the acceleration a is the second derivative of position with respect to time, Newtonian mechanics always involves differential equations, be it free fall on earth, or star motion in the sky. Another example is wave motion such as a vibrating string. To describe waves, we must treat the space and time variables independently. This leads to wave equations in the form of PDEs. Therefore, solving differential equations is a major undertaking in scientific computing.

Some ODEs, like those for free fall, yield analytic solutions in closed-form, known functions. Others do not, so numerical solutions are needed. Because ODEs are among the single, most common class of mathematical relations used across science and engineering, including this book, we will devote this chapter to the discussion of methods of numerical solutions suitable to the problems we study (PDEs will be discussed separately starting in Chapter 6). The ODE solvers developed here will be used throughout this book. We will first study one ODE at a time and generalize it to a set of coupled ODEs. The collection of these methods serves as a toolbox that will be applied to computer simulations in subsequent chapters where analytic solutions are rarely known.

If you are familiar with ODE solvers or wish to save time, you can skip Sections 2.1 to 2.3. Also see Program 2.10 (Section 2.B), which serves as a blackbox ODE wrapper for a general integrator from SciPy.

2.1 FREE FALL WITH EULER'S METHOD

Let us begin with a very familiar problem, free fall. We will use it as the main example in this chapter. This is useful because it allows us to test the methods against exact solutions.

2.1.1 A FIRST EXAMPLE: FREE FALL

To get a feel for our approach to modeling and visualization first, we start with the example of a bouncing ball: A ball is let go from a certain height above the floor, and on reaching the floor it reverses its velocity and bounces right back up without loss of speed. Thus, between the contacts with the floor, the ball is in constant free fall motion. The following program simulates this motion.

Program listing 2.1 A bouncing ball (bouncing_ball.py)

```
import visual as vp     # get VPython modules for animation

# draw the ball at (0,5,0) and a thin floor at (0,−5,0)
ball = vp.sphere(pos=(0,5,0), radius=1, color=vp.color.yellow) # ball
```

```
floor = vp.box(pos=(0,−5,0), length=8, height=0.2, width=4)  # floor
6
h = 0.01            # time step size
8 v = 0.0           # initial velocity
while True:         # loop forever
10   vp.rate(400)   # limit animation rate to 400 loops/sec
     ball.pos.y = ball.pos.y + v*h       # update y position
12   if ball.pos.y > floor.y + ball.radius:
        v = v − 9.8*h         # above floor, update velocity
14   else:
        v = − v               # below floor, reverse velocity
```

If the program has been entered correctly and everything goes smoothly (see Section 1.B on running programs), we should see the free fall animation, with the ball bouncing up and down continuously. See the motion from different angles by dragging with the right mouse button (camera) or both buttons (zoom). Several snapshots of the motion are shown in Figure 2.1.

Program 2.1 consists of about a dozen working lines, with brief comments after each line. Next, we explain the program flow. The first line of the program imports the VPython package. The next two lines (4 and 5) create the ball and the floor using the "sphere" and "box" objects of VPython, respectively. Each object accepts a set of parameters (or attributes) such as the position, radius, color, and so on. For instance, the position vector is specified by a point (x, y, z). The objects are assigned names, `ball` and `floor` in this case, for later reference. VPython objects can have more attributes than specified. Those unspecified will be assigned default values automatically. Once created, the objects can be modified by changing the attributes. For example, we can "move" the ball to a new position (a, b, c) via `ball.pos=(a,b,c)`, or `ball.pos.x=a` if we only want to move in the x direction.

The rest of the program sets the step size and the initial velocity (lines 7 and 8), then enters the `while` loop. The "rate" function (line 10) inside the loop is another VPython module that limits the rate of animation so it will not run too fast on fast machines. The position of the ball is updated by changing the variable `ball.pos.y` by an amount `v*h` each time (line 11). Because it is an attribute of the sphere object assigned to `ball`, VPython, running in the background and keeping track of all attributes of objects created, will automatically redraw the ball at the new position, creating the animated effect. The last if-else statement checks to see if the ball is above the floor: if it is, the velocity is updated according to free fall with constant acceleration -9.8 m/s^2 (line 13); otherwise, it is reversed (line 15).

After understanding the program flow, you should experiment a little and play with different parameters to see how they affect the animation. For example, change the ball to your favorite color (mine is green), give it a nonzero initial velocity (positive and negative), a slower or faster frame rate, or even a different value of g. In the problems ahead, the programs may grow to

FIGURE 2.1 Snapshots of a bouncing ball from Program 2.1 using VPython.

more lines (not too many more in most cases, I assure you), and the structure may be a bit more complex, but the basic elements and approach will remain the same.

This object-oriented approach of VPython makes it easy for us to create 3D animation and data visualization effects in our programs. All we need to do is to change the attributes such as the positions of the objects in motion. We do not have to worry about drawing or moving the objects. It allows us to focus on the task of modeling. We will use this to our fullest advantage throughout. If you do not use VPython, you can use Matplotlib output (see Exercise E2.1 and Program 2.2) or save the position and time data to a file (Section 1.3.1) for separate analysis.

2.1.2 EULER'S METHOD

The free fall example given earlier calculates the changes in position and velocity of the ball with a method as $\Delta y = v\Delta t$ and $\Delta v = a\Delta t$, respectively, where $v = dy/dt$ is the velocity, $a = dv/dt = -g$ the acceleration, Δt the step size (h), and $g = 9.8$ m/s^2 the gravitational acceleration. It is convenient to describe free fall in terms of differential equations as

$$\frac{dy}{dt} = v, \quad \frac{dv}{dt} = -g, \tag{2.1}$$

with the analytic solutions given by

$$y(t) = v_0 t - \frac{1}{2}gt^2, \quad v(t) = v_0 - gt, \tag{2.2}$$

where the initial position and velocity are assumed to be $y(0) = 0$ and $v(0) = v_0$, respectively. Accordingly, the method of calculating the changes in Program 2.1 is first order in Δt, also known as Euler's method.

To better understand Euler's method, including an estimation of the error term, we work with a single ordinary differential equation for now. This simplifies the discussion without loss of generality. The general form of an ODE is

$$\frac{dy}{dt} = f(y, t), \tag{2.3}$$

where y is the solution we seek, and $f(y, t)$ is a well-behaved but otherwise arbitrary function.

The problem at hand now becomes an initial value problem, that is, given the initial condition $y(t = 0) = y_0$, find y at all later times. More generally, assuming we know $y(t)$, we could obtain $y(t + \Delta t)$ from the Taylor series

$$y(t + \Delta t) = y(t) + \frac{dy}{dt}\Delta t + \frac{d^2y}{dt^2}\frac{(\Delta t)^2}{2!} + \frac{d^3y}{dt^3}\frac{(\Delta t)^3}{3!} + \ldots \tag{2.4}$$

It is understood that in Eq. (2.4) the derivatives are carried out at t. Looking at Eq. (2.4), the easiest thing to do is to drop all orders higher than Δt, to yield a first order solution as

$$y(t + \Delta t) \approx y(t) + \frac{dy}{dt}\Delta t + O[(\Delta t)^2]. \tag{2.5}$$

Here $O[(\Delta t)^2] = [\frac{1}{2}(\Delta t)^2]\, d^2y/dt^2$ is basically the leading-order error correction that we throw away.[1] This is also called the local truncation error. A method is called order n if it is accurate to order $(\Delta t)^n$ with an error of order $O[(\Delta t)^{n+1}]$. Hence Eq. (2.5) is order 1, or first order. For linear

[1]We may also turn Eq. (2.5) around to find the derivative by two-point numerical differentiation, $dy/dt \approx [y(t + \Delta t) - y(t)]/\Delta t$. The error term is $O(\Delta t)$. However, we can do better with two points using the midpoint method, Eq. (2.13).

functions, $d^2y/dt^2 = 0$, and the error term vanishes. Therefore we expect this first-order method to be exact, numerically speaking, for linear solutions.

Given its first order accuracy and dropping the approximation sign, we can write down the classic Euler's method as

$$y(t + \Delta t) = y(t) + \frac{dy}{dt}\Delta t = y(t) + f(y, t)\Delta t. \tag{2.6}$$

Note that we have replaced dy/dt with $f(y, t)$ by virtue of Eq. (2.3). It is to be evaluated at t. Therefore Euler's method is said to be explicit and self-starting, that is, given $y(0)$, we know $y(\Delta t)$, $y(2\Delta t)$, $y(3\Delta t)$, and so on.

So here is Euler's algorithm, given $y(t = 0)$ and $h = \Delta t$ as the time step:

$$\begin{aligned}
\text{Let } t_n = nh,\ y(t_n) \equiv y_n, n = 0, 1, 2, ..., \\
y_1 = y_0 + f(y_0, t)h, \\
y_2 = y_1 + f(y_1, t)h, \\
\cdots\cdots \\
y_{n+1} = y_n + f(y_n, t)h.
\end{aligned} \tag{2.7}$$

How accurate is Euler's method as applied to the free fall problem? Because we know the exact solutions, we can readily answer this question by comparing the numerical and analytic results for the position y. From Eqs. (2.1) and (2.2), we have

$$f(y, t) \equiv v(t) = v_0 - gt, \quad \text{so that } dy/dt = f(y, t). \tag{2.8}$$

Here is the program that solves the preceding equation with Euler's method (2.7) and plots the results.

Program listing 2.2 Free fall with Euler's method (`freefall_plot.py`)

```python
import matplotlib.pyplot as plt  # get matplotlib plot functions

g, h = 9.8, 0.02         # g, step size. Multiple assignments in one line
y, v0 = 0.0, 5.0         # initial position, velocity
ta, ya = [], []          # time and position arrays for plotting
t, yb = 0.0, []          # yb[] holds analytic solution

while t<1.0:             # loop for one second
    ta.append(t)         # record time and position
    ya.append(y)
    yb.append(v0*t-g*t*t/2.0) # analytic solution
    v = v0 - g*t              # calc v(t)
    y = y + v*h               # Euler's method
    t = t + h

plt.figure()                  # start a figure
plt.plot(ta,ya, ta,yb,'--')   # draw 2nd curve as dashed
plt.xlabel('t (s)')           # add labels
plt.ylabel('y (m)')
plt.show()                    # show figure
```

The program flow is very similar to Program 1.1. It first imports the Matplotlib library, then sets parameters and initial conditions. As before, empty lists such as `ta`, `ya`, and `yb` are created to hold intermediate values for later plotting. Inside the loop, elements are added to lists via the append method in the first three statements. The rest of the loop updates the position with Euler's method.

When presenting many data points, instead of printing numbers, graphs are preferred. This is done after the main loop. We plot two curves with the `plot` function (Section 1.C): the first pair `ta,ya` plots the numerical (Euler) y-t results, and the second pair `ta,yb` plots the analytic results. An optional string argument immediately after the data arrays specifies the line style, and a dashed line is drawn with '--' in this case (see Table 1.4). The plots are shown in Figure 2.2.

Starting with the same value $y = 0$ at $t = 0$, the numerical and analytic results stay close to each other in the beginning (Figure 2.2). But as time increases, the difference between them grows linearly. The absolute error, which measures the global error at a given t, is doubled from $t \sim 0.5$ to $t \sim 1$ s. Though the exact global error is unknown, we can roughly estimate it to be the cumulation of local error. As stated earlier, the local truncation error of Euler's method is second order, $O(h^2)$. The number of steps to reach t is proportional to $1/h$, so we expect the global error to be $\sim O(h^2) \times h^{-1} \propto O(h)$. This means that if h is halved, the global error should be reduced by half. Project P2.3 shows that this is indeed the case.

Aside from the error, the peak position (around $t \sim 0.5$ s) is shifted as well, with the numerical peak occurring at a slightly later time. Because Euler's method is exact for linear solutions, error in the numerical result is expected because the analytic solution (2.2) is quadratic. Nonetheless, Euler's method does yield an overall correct parabolic shape for free fall.

In addition to accuracy, another important factor in the consideration of an algorithm is its efficiency. The efficiency of an ODE solver is usually measured in terms of the number of function calls made to $f(y, t)$ per step. The cost of computation of Euler's method is one function evaluation per time step. This is the minimum number we would expect, so it is apparently very efficient, 100% in fact.

Euler's method is an instructive example to introduce us to numerical solutions of ODEs, but is very limited for practical use. For instance, to try to reduce the error, we might use a smaller step size h. This partially works, but is very limited. To reach a certain time t, we need

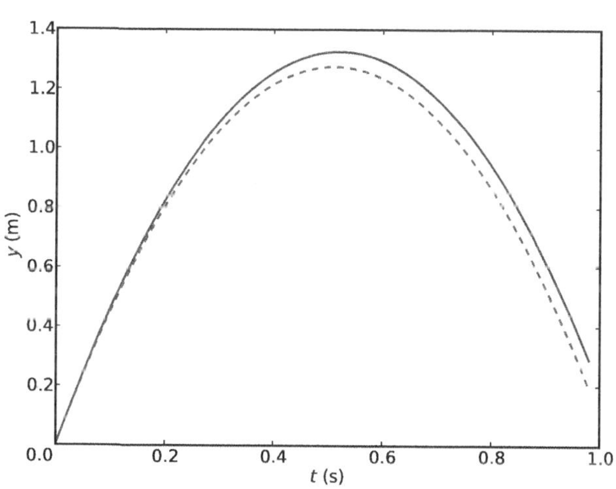

FIGURE 2.2 The position-time curves of free fall with Euler's method (solid/blue) and the analytic result (dashed/green).

more steps, using more computing time and becoming less efficient, and partially defeating our purpose. Furthermore, there is a fundamental limit to the lower bound of h due to machine accuracy. More accurate methods with reasonable efficiency would be needed for problems whose solutions may not be as smooth as free fall. We examine those methods next.

2.2 THE RUNGE-KUTTA (RK) METHODS

In this section, we introduce a family of ODE solvers known as the Runge-Kutta methods to improve accuracy over Euler's method. We will discuss a second-order method and a fourth-order one.

2.2.1 SECOND-ORDER RK METHOD: RK2

Looking at Euler's method (2.6), we notice that the use of the derivative $f(y, t)$ is not quite symmetrical, that is, at each step, only information at the beginning of the step is used, not at the end. This asymmetry is illustrated graphically in Figure 2.3(a). To arrive at the second step (t_1), only the derivative at t_0 is used (similarly for subsequent steps).

Here is where we could make an improvement: to make a more symmetrical use of the information at the beginning and at the end of each step. The idea is shown schematically in Figure 2.3(b). If we used an average derivative at the middle point (dash circles), which is halfway between t_0 and t_1, we just might get a better approximation of the solution at t_1. This is exactly the rationale behind the second-order Runge-Kutta method, which we will call RK2 from now on.

To see how this works out, let us use the Taylor series (2.4) again, first at the beginning of the step as (using same notation as in last section)

$$y_0 = y(0) = y\left(\frac{h}{2}-\frac{h}{2}\right) = y\left(\frac{h}{2}\right) + \frac{dy}{dt}\left[-\frac{h}{2}\right] + \frac{d^2y}{dt^2}\frac{[-\frac{h}{2}]^2}{2!} + \frac{d^3y}{dt^3}\frac{[-\frac{h}{2}]^3}{3!}\dots$$
$$= y_{\frac{1}{2}} - y'_{\frac{1}{2}}\frac{h}{2} + y''_{\frac{1}{2}}\frac{[\frac{h}{2}]^2}{2!} - y'''_{\frac{1}{2}}\frac{[\frac{h}{2}]^3}{3!}\dots, \quad (2.9)$$

and then at the end of the step as

$$y_1 = y(h) = y\left(\frac{h}{2}+\frac{h}{2}\right) = y_{\frac{1}{2}} + y'_{\frac{1}{2}}\frac{h}{2} + y''_{\frac{1}{2}}\frac{[\frac{h}{2}]^2}{2!} + y'''_{\frac{1}{2}}\frac{[\frac{h}{2}]^3}{3!}\dots \quad (2.10)$$

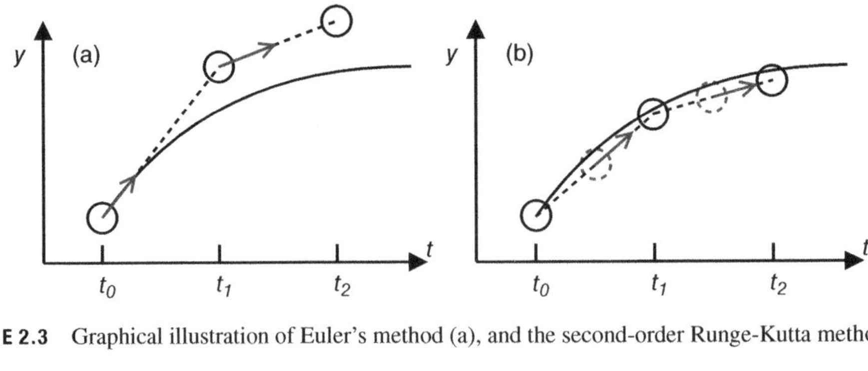

FIGURE 2.3 Graphical illustration of Euler's method (a), and the second-order Runge-Kutta method (b).

In Eqs. (2.9) and (2.10), $y_{\frac{1}{2}} \equiv y(\frac{h}{2})$, and it is understood that the derivatives are also evaluated at $h/2$, the midpoint.

Because all we know from Eq. (2.3) is the first derivative $y' = f(y, t)$, we really don't know the second-(y'') or the third-(y''') order derivatives in Eqs. (2.9) and (2.10). Fortunately we can eliminate y'' by subtracting both sides of Eqs. (2.9) and (2.10),

$$y_1 - y_0 = y'_{\frac{1}{2}} h + 2 \times y'''_{\frac{1}{2}} \frac{[\frac{h}{2}]^3}{3!}. \tag{2.11}$$

Cancellation helped us to get rid of the second derivative, but the third (and other odd) derivative is still present. Because our goal is to obtain second-order accuracy, we can safely throw away the term with the third derivative because it is a third-order correction h^3. Doing so, we have

$$y_1 = y_0 + y'_{\frac{1}{2}} h + O(h^3). \tag{2.12}$$

This is the RK2 method. Comparing Eq. (2.6) with (2.12), we see they are very similar in form, except that for RK2 the derivative is evaluated at $\frac{h}{2}$, the midpoint. Thus RK2 has another name, the midpoint method. It is locally accurate to second order. The global error is expected to scale as $O(h^2)$, so it would decrease by a factor of 4 if h is halved (Project P2.3).

Incidentally, Eq. (2.12) leads to the midpoint method for numerical differentiation,

$$y'_{\frac{1}{2}} = \frac{y_1 - y_0}{h} + O(h^2), \quad \text{or} \quad y'_1 = \frac{y_2 - y_0}{2h} + O(h^2). \tag{2.13}$$

This is a commonly used method for finding derivatives between two points.

Now only one question remains: How do we get $y'_{\frac{1}{2}}$? We cannot use Eq. (2.13) because we do not have y_1 yet. According to Eq. (2.3), we could also calculate it as follows

$$y'_{\frac{1}{2}} = \left.\frac{dy}{dt}\right|_{h/2} = f\left(y\left(\frac{h}{2}\right), \frac{h}{2}\right). \tag{2.14}$$

However, at this point, we only have the value $y_0 = y(0)$, not $y(\frac{h}{2})$.

Consistent with second-order accuracy, we only need $y'_{\frac{1}{2}}$ to be accurate to first order because of the additional h in the term $y'_{\frac{1}{2}} h$ in Eq. (2.12). One way is to obtain $y(\frac{h}{2})$ from Euler's method, Eq. (2.6), and put it in Eq. (2.14) to get $y'_{\frac{1}{2}}$ as

$$y(h/2) \simeq y_0 + f(y_0, 0)\frac{h}{2}, \text{ and } y'_{\frac{1}{2}} \simeq f\left(y_0 + f(y_0, 0)\frac{h}{2}, \frac{h}{2}\right). \tag{2.15}$$

Combining Eqs. (2.12) and (2.15), we have the algorithm for the RK2 method, given y_n and $t_n = nh$, to advance to t_{n+1} as

$$\begin{aligned} k_1 &= hf(y_n, t_n), \\ k_2 &= hf\left(y_n + k_1/2, t_n + \frac{h}{2}\right), \\ y_{n+1} &= y_n + k_2. \end{aligned} \tag{2.16}$$

In each step, the RK2 method makes two function evaluations, one for k_1 and another one for k_2. Because it is second order, two calls are the minimum expected, so it is still very efficient and self-starting.

As it turns out, the RK2 method as expressed in Eq. (2.16) is not unique. The RK2 family has an infinite number of ways to achieve second order accuracy with the same efficiency. For example, we could approximate $y'_{\frac{1}{2}}$ as the average of $f(y,t)$ at the two end points, in which case the RK2 method reads

$$\begin{aligned} k_1 &= hf(y_n, t_n), \\ k_2 &= hf(y_n + k_1, t_n + h), \\ y_{n+1} &= y_n + \frac{1}{2}(k_1 + k_2). \end{aligned} \tag{2.17}$$

We leave the proof of Eq. (2.17) and another popular variation to Exercise E2.2. We will use Eq. (2.16) as the RK2 method of choice in our study.

2.2.2 FOURTH-ORDER RK METHOD: RK4

The RK4 method is of fourth-order accuracy, two orders higher than RK2, but the basic idea is the same, that is, using more of available information at the end points of the step, plus two in the middle. With proper weighting factors, we get higher-order cancellations (see Sec. 3.4 of Ref. [56]).

As with RK2, the RK4 method is not unique. Two common forms are illustrated in Figure 2.4. In the Runge method (Figure 2.4(a)), the derivative is evaluated four times in each step, twice at the end points, and twice in the middle. The mathematical proof that it is indeed of fourth-order accuracy is given elsewhere [78]. The final results – the Runge formula – for the RK4 algorithm are

$$\begin{aligned} k_1 &= hf(y_n, t_n), \\ k_2 &= hf\left(y_n + k_1/2, t_n + \frac{h}{2}\right), \\ k_3 &= hf\left(y_n + k_2/2, t_n + \frac{h}{2}\right), \\ k_4 &= hf(y_n + k_3, t_n + h), \\ y_{n+1} &= y_n + \frac{k_1}{6} + \frac{k_2}{3} + \frac{k_3}{3} + \frac{k_4}{6} + O(h^5). \end{aligned} \tag{2.18}$$

The notations used in Eq. (2.18) are the same as in Eq. (2.16).

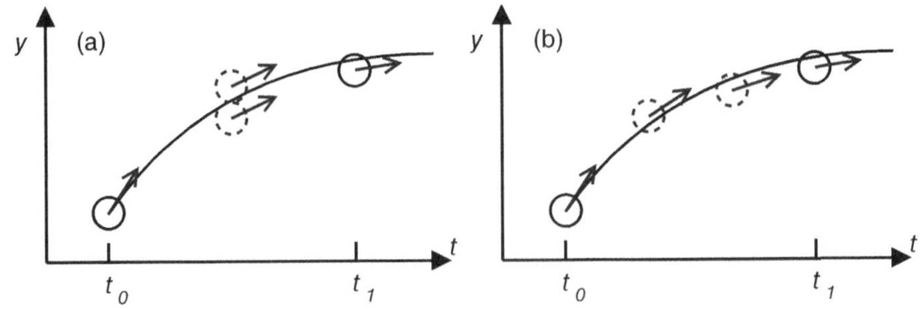

FIGURE 2.4 Illustration of the fourth-order Runge method (a) and the Kutta method (b).

Alternatively, we can divide the step size h into three intervals and evaluate the derivative at 0, $h/3$, $2h/3$, and h (Figure 2.4(b)). This leads to the Kutta formula

$$
\begin{aligned}
k_1 &= hf(y_n, t_n), \\
k_2 &= hf(y_n + k_1/3, t_n + h/3), \\
k_3 &= hf(y_n - k_1/3 + k_2, t_n + 2h/3), \\
k_4 &= hf(y_n + k_1 - k_2 + k_3, t_n + h), \\
y_{n+1} &= y_n + \frac{k_1}{8} + \frac{3k_2}{8} + \frac{3k_3}{8} + \frac{k_4}{8} + O(h^5).
\end{aligned}
\tag{2.19}
$$

Both formulas work fine. We will stick with the simpler Runge formula (2.18). The RK4 method is self-starting and efficient. There are more elaborate methods around, but a basic RK4 with some step size adjustment as necessary will serve us well in most situations. Except for certain cases such as Hamiltonian systems to be discussed later, RK4 is the recommended method as the general ODE solver in our simulations. Of course, we will develop appropriate methods for special problems we encounter going forward.

We might well ask, why stop at the fourth-order RK4? Can we not continue to higher orders? As it turns out, higher than fourth order will require more function calls to $f(y, t)$ than the order number. For example, a fifth-order Runge-Kutta method would need six function calls (see Runge-Kutta-Fehlberg method, Section 2.3.2). So the RK4 is a natural place to stop.

2.2.3 APPLICATION TO FREE FALL

Now let us apply the Runge-Kutta methods to free fall in the same manner as Euler's method in Program 2.2. Starting with RK2 first and according to Eq. (2.16), we need only to change two lines of that program inside the `while` loop, from

```
v = v0 - g*t     # calc v(t)
y = y + v*h      # Euler's method
```

to

```
......
k1 = h*(v0 - g*t)  # k1, for clarity only, not needed
k2 = h*(v0 - g*(t+h/2.0))  # k2=f(y+k1/2, t+h/2)
y = y + k2                  # RK2 method
......
```

The modified code works, but the way of putting everything in one chunk is not a good idea for organizing programs in terms of well-laid structure and clarity, or for reusing certain parts. Following the modular revolution, we break the code down into self-contained functions, or callable subroutines. For RK2, we would like the function to take as input the initial value y_0 at t and advance it one time step to y_1 at $t + h$. We code it as follows.

Program listing 2.3 RK2 method (`rk2.py`)

```
def RK2(diffeq, y0, t, h):
    """ RK2 method for ODEs.
        Given y0 at t, returns y1 at t+h """
    k1 = h*diffeq(y0, t)                  # get dy/dt at t first
    k2 = h*diffeq(y0+0.5*k1, t + h/2.)    # get dy/dt at t+h/2,
    return y0 + k2                        # calc. y1 = y(t+h)
```

The function `diffeq(y,t)` is a generic name for the user supplied function that returns the derivative, that is, the right-hand side (RHS) of Eq. (2.3), $dy/dt = f(y,t)$. It looks like an argument, but in Python it is the name of a user-defined function (in other languages such as C/C++, it might be a pointer to a function). For free fall, this function is, according to Eq. (2.8),

```
def velocity(y, t):         # returns the RHS of Eq. (2.8)
    v0 = 0                  # locally defined init value
    v = v0 - 9.8*t
    return v
```

Note that in the preceding function, the initial value `v0` is defined locally. Of course, one could also define it in a scope accessible to the function, as a global variable for instance. But the use of global variables that can be modified by other functions should not be too casual in larger codes because it is hard to track down potential bugs caused by them.

With `RK2()` and `velocity()` thus defined, the modularized code snippet looks like

```
    ......
    y = RK2(velocity, y, t, h)    # call RK2
    ......
```

The plots generated from running this program should be similar to Figure 2.2, but the numerical and analytic results should show no difference because the RK2 method is exact for quadratic solutions. We will explore this in Project P2.1.

When `RK2()` is called with the function name `velocity`, the program will match the generic function `diffeq()` inside to `velocity()`, and any reference to `diffeq()` is redirected to `velocity()` instead. This way, the program is much clearer structurally. This is helpful in understanding what we are doing, much easier, in fact, than to hardwire the RK2 method in the main body of codes. Furthermore, the RK2 method is totally encapsulated (hidden) from the main code, so it can be tested separately and called from anywhere.

In addition, the user-supplied function can be changed to suit the problem of the day. We could use the same RK2 module to solve any problem that is a single, first-order ODE, by supplying a different function `diffeq(y,t)`.

In an analogous way, the RK4 method, Eq. (2.18), can be implemented as a stand-alone function that uses the same `diffeq(y,t)`, as given next.

Program listing 2.4 RK4 method (`rk4.py`)

```
def RK4(diffeq, y0, t, h):
    """ RK4 method for ODEs:
        Given y0 at t, returns y1 at t+h
        For a system of ODEs, diffeq must return a NumPy array"""
    k1 = h*diffeq(y0, t)                      # dy/dt at t
    k2 = h*diffeq(y0+0.5*k1, t + h/2.)        # dy/dt at t+h/2
    k3 = h*diffeq(y0+0.5*k2, t + h/2.)        # dy/dt at t+h/2
    k4 = h*diffeq(y0+k3, t + h)               # dy/dt at t+h
    return y0 + (k1+k4)/6.0 + (k2+k3)/3.0
```

For example, we can solve the nuclear decay problem this way (Project P2.3). Though the RK2 and RK4 codes are developed with a single ODE, they can be used for multiple ODEs without change using vector NumPy arrays described next.

2.3 SYSTEM OF FIRST-ORDER ODEs

With what we have now, we can solve any problem expressible as a single, first-order ordinary differential equation with Euler's method, RK2, or RK4. What about multiple, or even higher order differential equations? As the main example, we have treated free fall as a single ODE so far, assuming the analytic formula for velocity for simplicity. But free fall actually consists of two ODEs, Eq. (2.1), which come from this second-order equation

$$\frac{d^2y}{dt^2} = -g. \tag{2.20}$$

As we shall see, finding solutions to higher-order ODEs is not any more difficult than dealing with a single ODE. There are just two steps involved, each is straightforward. First, convert one (or more) second- or higher-order ODE into a set of first-order ODEs. Second, solve them using the methods we have already discussed in this chapter.

2.3.1 CONVERSION FROM THE nTH TO THE FIRST ORDER

Consider an ODE of the general form

$$\frac{d^ny}{dt^n} = f\left(y, \frac{dy}{dt}, \frac{d^2y}{dt^2}, \ldots, \frac{d^{n-1}y}{dt^{n-1}}, t\right). \tag{2.21}$$

Here we assume the highest order being n (do not confuse derivative order with the order of accuracy of algorithms discussed earlier), and the RHS function f contains all orders up to $n - 1$. This is not too restrictive, as most problems can be formulated this way.

To convert Eq. (2.21) into a set of first-order ODEs, we define some new variables as

$$\text{Let } x_1 = y,\ x_2 = \frac{dy}{dt}, \ldots, x_n = \frac{d^{n-1}y}{dt^{n-1}}. \tag{2.22}$$

We can express Eq. (2.22) as a recursive relationship between the first derivative dx_i/dt and x_{i+1} as

$$\frac{dx_i}{dt} = \frac{d\frac{d^{i-1}y}{dt^{i-1}}}{dt} = \frac{d^iy}{dt^i} = x_{i+1}. \tag{2.23}$$

Running through the indices fully, we have

$$\begin{aligned}
\frac{dx_1}{dt} &= x_2,\\
\frac{dx_2}{dt} &= x_3,\\
&\ldots\\
\frac{dx_n}{dt} &= x_{n+1} = \frac{d^ny}{dt^n} = f(x_1, x_2, \ldots, x_n, t).
\end{aligned} \tag{2.24}$$

Equation (2.24) is what we wanted: a set of n, first-order ODEs in terms of x_i. The RHS of the first $n - 1$ equations is trivial, it is just a simple assignment. The last equation is where the real work is done in function f.

The preceding can be summarized by the following statement: Any ODE higher than first order can be converted into a system of n first-order ODEs, where n is the highest-order derivative in Eq. (2.21).

The surrogate variables x_i in Eqs. (2.22) and (2.23) are used not just for convenience, they often have real physical meaning as well. Again using the free fall example (2.20), we associate

$$x_1 = y = \text{ position } y, \; x_2 = \frac{dy}{dt} = \text{ velocity } v, \tag{2.25}$$

and the set of ODEs is

$$\begin{aligned} \frac{dx_1}{dt} &= x_2, \text{ this is just } \frac{dy}{dt} = v, \\ \frac{dx_2}{dt} &= -g, \text{ which is } \frac{dv}{dt} = -g. \end{aligned} \tag{2.26}$$

The original Eq. (2.21) is of nth order. The solution will be uniquely determined given n initial conditions, $y, dy/dt, ...,$ and so on. In terms of the converted Eq. (2.24), it means knowing the values of x_i, $i = 1, 2, ..., n$. For free fall, $n = 2$, and this translates into x_1 for the initial position, and x_2 for the initial velocity.

How do we deal with arbitrary n? The answer in one word: vectors.

2.3.2 SOLUTION OF COUPLED ODEs

We know now how to: (a) solve one single, first-order ODE and (b) reduce one or more ODEs of second or higher orders to a set of first-order ones. We just need to find a way to solve (b) in a way similar to solving (a).

As it turns out, all the methods we have discussed so far about (a) could be applied straightforwardly to (b), if we generalize these methods using vector representation.

First, let us say we have a vector y that has n components $y[1]$ to $y[n]$. We denote it by the short-hand $\{y\}$, which means a collection of all its components as

$$\{y\} \equiv y[1], y[2], ..., y[n]. \tag{2.27}$$

Next, we consider a system of n ODEs similar to the set (2.24) but more general in form as

$$\frac{dy[i]}{dt} = f_i(\{y\}, t), \; i = 1, 2, ..., n. \tag{2.28}$$

In each of the n equations (2.28), the left-hand side (LHS) is the derivative of one component $y[i]$. But on the RHS, the function f_i is understood to depend on possibly all other components of y, that is, with the shorthand notation (2.27), $f_i(\{y\}, t) = f_i(y[1], y[2], ..., y[n], t)$. Thus, the value of one component $y[i]$ depends on the values of all other n components. This interdependency is the reason that Eq. (2.28) is called a set of coupled equations. The coupled equations must be solved together, meaning all the components of $\{y\}$ need to be updated at every time step, even though in the end perhaps only a subset of the components are desired or actually useful.

If we compare Eq. (2.28) with (2.3), $dy/dt = f(y, t)$, we see that they are formally equivalent. The only difference is that the first argument to the RHS function $f(y, t)$ of Eq. (2.3) is replaced by a vector $f_i(\{y\}, t)$ in Eq. (2.28). This is purely cosmetic, as we know how to deal with that, just plugging in more arguments. What this means is that all the algorithms we discussed, be it Euler's, RK2, or RK4 methods, can be used for a system of ODEs without modification. We only require that `diffeq` return the derivatives in a NumPy (vector) array. In fact, we can broaden the scope to include an array of vector objects (2D or multidimensional NumPy arrays) returned by `diffeq` (e.g., see Program 3.8, most codes in Chapter 4, and Program 6.8).

Euler Method

To illustrate this point, let us port Euler's method, Eq. (2.7), for a single ODE to a system of coupled ODEs, Eq. (2.28). Instead of using y_n, y_{n+1} for the start and end values in each step, we adopt a slightly different notation as

$$y0 = \text{ the value of } y \text{ at } t, \; y1 = \text{ the value of } y \text{ at } t+h. \tag{2.29}$$

Then, using vector notation, Euler's method is simply

$$\begin{aligned} y1 &= y0 + f(y,t) * h, \text{ for a single ODE,} \\ \{y1\} &= \{y0\} + \{f(\{y0\},t)\} * h, \text{ for a system of ODEs.} \end{aligned} \tag{2.30}$$

Note we had generalized $\{f\}$ to mean a function vector. Expanded out, Eq. (2.30) reads

$$\begin{aligned} y1[1] &= y0[1] + f_1(y0[1], y0[2], ..., y0[n], t) * h, \\ y1[2] &= y0[2] + f_2(y0[1], y0[2], ..., y0[n], t) * h, \\ &\ldots \\ y1[n] &= y0[n] + f_n(y0[1], y0[2], ..., y0[n], t) * h. \end{aligned} \tag{2.31}$$

This is the explicit formula for Euler's method for a system of ODEs. It is essentially one Euler's method repeated n times for all the components.

As an example that expressly ties it all together, consider the set of equations (2.26) resulting from free fall (2.20),

$$\text{Let } y[1] \equiv \text{ position, } y[2] \equiv \text{ velocity,}$$

$$\frac{dy[1]}{dt} = f_1(y[1], y[2], t) = y[2], \tag{2.32}$$

$$\frac{dy[2]}{dt} = f_2(y[1], y[2], t) = -g. \tag{2.33}$$

We would write a subroutine implementing Eq. (2.31) called `Euler`. To make it general, we would want to use a user-supplied subroutine that returns the derivatives $f_i(\{y\}, t)$ in Eq. (2.31). Similar to Program 2.3, we shall call it generically `diffeq`. The standalone `Euler`, the user-supplied subroutine `freefall`, and the main code go are listed here.

Program listing 2.5 Free fall with modular Euler (`freefall_euler.py`)

```
import numpy as np                # get numpy functions
import matplotlib.pyplot as plt # get matplotlib plot functions
g = 9.8                           # gravitational constant

def Euler(diffeq, y0, t, h): # uses docstring """ """
    """ Euler's method for n ODEs:
    Given y0 at t, returns y1 at t+h """
    dydt = diffeq(y0, t)      # get {dy/dt} at t
    return y0 + h*dydt        # Euler method on a vector

```

```
11 def freefall(y, t):            # returns {dy/dt}, the RHS of ODEs
12     dydt = np.zeros(2)         # initialize 2-element numpy array
13     dydt[0] = y[1]             # f1(), Eq. (2.32)
14     dydt[1] = -g               # f2(), Eq. (2.33)
15     return dydt                # note: returns whole array dydt[]
16 
17 def go(v0):                    # main program, v0=initial velocity
18     y0 = [0.0, v0]             # initial values
19     t, h = 0.0, 0.02           # init time, step size
20     ta,ya,yb = [], [], []      # declare arrays for plotting
21     while t<1.0:               # loop for one second
22         ta.append(t)           # record time and position
23         ya.append(y0[0])
24         yb.append(v0*t-g*t*t/2.0)
25         y1 = Euler(freefall, y0, t, h)   # Euler's method
26         for i in range(len(y0)):         # reseed y0
27             y0[i] = y1[i]
28         t = t + h
29 
30     plt.figure()                    # start a figure
31     plt.plot(ta,ya, ta,yb,'--')     # draw 2nd curve as dashed
32     plt.xlabel('t (s)')             # add labels
33     plt.ylabel('y (m)')
34     plt.show()                      # show figure
35 
36 go(5.0)                        # run the program
```

The program defines the global variable g so it is accessible to all modules within the program. The vector operations in Eq. (2.31) are effected by element-wise operations with NumPy arrays (ndarrays, see Section 1.D) at line 9 in `Euler`: `y0+h*dydt`, with `dydt` returned from `diffeq`. The actual form of `diffeq` for free fall, Eqs. (2.32) and (2.33), is contained in `freeall`, which defines the derivatives as an ndarray (line 12). Note that we made a name switch in the main code from the generic name `diffeq` to `freefall` when calling `Euler`. As was pointed out earlier (Section 2.2.3), the advantage of doing things this way is that we can supply another function to calculate something entirely different without having to change `Euler`. But if a programming language does not support function names as arguments, you may have to rename `freefall` to its generic name, `diffeq`.

The main code is similar to Program 2.2, but is modularly contained in the function `go()`. Here we used another variable, `y1`, to store the returned array from `Euler`. We could have saved space without it, using `y0` instead, by changing to `y0=Euler(...)`. We introduced `y1` mainly for clarity here, but there may be other reasons. For example, if in situ analysis requires the data between time steps (e.g., calculating the difference), it is necessary to use another variable. The program should produce plots exactly like Figure 2.2.

Before going further, it is instructive to walk through the vector method step-by-step for one cycle. To do so, let us denote the position (pos) and velocity (vel) pair by a 2×1 matrix and represent the equations of motion (2.32) and (2.33) as

$$\begin{bmatrix} \text{pos} \\ \text{vel} \end{bmatrix} \stackrel{inp}{\rightarrow} \begin{bmatrix} f_1() \\ f_2() \end{bmatrix} = \begin{bmatrix} \text{vel} \\ -g \end{bmatrix}. \tag{2.34}$$

Equation (2.34) means that the derivative function just transforms the input vector into another vector (where position is discarded). Then the vectorized Euler's algorithm, Eq. (2.30), may be schematically represented as (assuming initial position and velocity are 0, v_0)

$$\{y0\} = \begin{bmatrix} 0 \\ v_0 \end{bmatrix} \overset{inp}{\rightarrow} \begin{bmatrix} f_1() \\ f_2() \end{bmatrix} = \begin{bmatrix} v_0 \\ -g \end{bmatrix},$$

$$\{y1\} = \{y0\} + \begin{bmatrix} v_0 \\ -g \end{bmatrix} * h = \begin{bmatrix} v_0 h \\ v_0 - gh \end{bmatrix}. \tag{2.35}$$

With input $\{y0\}$, the derivative function generates a vector that, when multiplied by h and added to $\{y0\}$, gives new values $\{y1\}$. As we can see, the new position, $v_0 h$, is linear in h. Exactly, it should be quadratic, $v_0 h - gh^2/2$. So Euler's method, accurate to first order, misses the second-order correction as expected. For velocity, it is exact, $v_0 - gh$, also as expected because it is linear.

RK2 Method

Higher-order methods like RK2 can be ported over in exactly the same way. We vectorize the single ODE version (2.16) as

$$\{k_1\} = \{f(\{y0\}, t)\} * h,$$
$$\{k_2\} = \{f(\{y0\} + \{k_1\}/2, t + h/2)\} * h,$$
$$\{y1\} = \{y0\} + \{k_2\}. \tag{2.36}$$

The actual code given in Program 2.3 works for a single ODE as well as a system of ODEs, provided the derivatives are returned as vector ndarrays.

We can use `freefall` as defined in Program 2.5 and only change one line in the main code (`go`) to use `RK2` instead of `Euler`,

```
def go(v0):
    ......
    y1 = RK2(freefall, y0, t, h)
    ......
```

It would be helpful if you took time to follow through one working cycle of `RK2` for the free fall problem to really understand what is going on (see Exercise E2.3). And you will convince yourself of not only the vectorization process (especially if you are new to it), but also that `RK2` is exact for free fall, which we know is a second-order process (t^2).

RK4 Method

By now we know the routine pretty well. We can vectorize RK4 (2.18) so it becomes applicable to a system of ODEs. This is shown here.

$$\{k_1\} = h * \{f(\{y0\}, t)\},$$
$$\{k_2\} = h * \{f(\{y0\} + \frac{\{k_1\}}{2}, t + \frac{h}{2})\},$$
$$\{k_3\} = h * \{f(\{y0\} + \frac{\{k_2\}}{2}, t + \frac{h}{2})\},$$
$$\{k_4\} = h * \{f(\{y0\} + \{k_3\}, t + h)\},$$
$$\{y1\} = \{y0\} + \frac{\{k_1\}}{6} + \frac{\{k_2\}}{3} + \frac{\{k_3\}}{3} + \frac{\{k_4\}}{6}. \tag{2.37}$$

Like RK2, Program 2.4 can be used for a system of ODEs as well, as long as the derivatives are returned in ndarrays.

Nonvectorized RK4 Method

The vectorized ODE solvers in Programs 2.3 and 2.4 are simple and elegant and should be suitable for most problems. However, vectorized operations involving ndarrays can be much slower than Python lists in some cases. This typically happens if the number of ODEs is small ($\lesssim 10$) and the ODEs are relatively simple to evaluate. Then, the overhead caused by creating ndarrays can be a relatively large portion of the load, yielding significant speed loss.[2] The difference in running time could be minutes instead of seconds.

For problems that call simple ODEs repeatedly and where the need for speed is critical, we should use nonvectorized methods suitable for lists. One could also use the ODE solvers in the SciPy library `scipy.integrate` as blackbox routines (see Program 2.10). We give a nonvectorized RK4 using lists and explicit looping in Program 2.8 (see Section 2.B), called `RK4n`.

To make it easy to reuse the ODE solvers `Euler, RK2, RK4/n`, we will put them in one file, say `ode.py`. It will be our standalone library of ODE solvers and should be placed in the Python installation folder. To use the solvers, just import the ODE library as

```
import ode      # get ODE solvers
```

Step Size and Characteristic Time

We will use `RK4` as the general-purpose ODE solver in our simulations. To be accurate, the step size h should be small compared to the characteristic time scale of the problem. What is the characteristic time, τ, you may wonder? It depends on the problem under consideration. For Earth moving around the Sun, τ is on the order of one year. For tossing a ball across the room, τ would be the time it takes the ball to hit the wall. So a step size of one hour would be small for planet motion (see Section 4.2.4), but it would be too large for a ball toss. In general, we can estimate τ as $\tau \sim L/v$, where L is the typical size of the problem, and v the typical speed. To ensure accuracy, we should set $h/\tau \ll 1$ (say 0.01). But, it is important to always keep the characteristic time in mind when choosing a step size.

Sometimes we need to balance between speed and accuracy. To increase speed, we could implement a variable step size h between calls to RK4. One simple way to adjust the step size is to monitor the differences $\Delta_{32} = |k_3 - k_2|$ relative to $\Delta_{21} = |k_2 - k_1|$. Because k_2 and k_3 are refinements of the derivative at the midpoint, Δ_{32} should be small compared to Δ_{21}. Generally, if $\Delta_{32}/\Delta_{21} \sim 0.01$, h should be adequate. If Δ_{32}/Δ_{21} is several times larger (say ~ 0.05), the step size h may be too large, and it should be reduced by half and the integration restarted from the beginning of the step. On the other hand, if $\Delta_{32}/\Delta_{21} < 0.01$, h may be too small. To avoid overkill, the step size h should be doubled. With some type of automatic step size control, we would be able to take baby steps around rough patches and giant leaps over smooth terrains.

The Runge-Kutta-Fehlberg Method

The Runge-Kutta-Fehlberg method is a fifth-order method that requires six function calls, two more than RK4. The method achieves fifth-order accuracy by a combination of weighting factors and proper choice of abscissa points. It has some advantages over RK4 other than the higher accuracy. By a different combination of the (same) function values, the method gives a fourth-order result at the same time, hence it is also called the embedded Runge-Kutta method. This is useful for step size adjustment because the difference between the fourth- and the fifth-order results tells us about the error at a given step size h. Therefore, adjustment of h can be made to achieve a specified tolerance.

---

[2]Tests show that creating 10^5 two-element ndarrays is about 30 times slower than comparable lists.

We include a fifth-order embedded Runge-Kutta program in Section 2.B (Program 2.9) named `RK45n`. The function `RK45n` is nonvectorized like `RK4n` and can be used in place of the latter. It should be included in our standalone ODE library.

Implicit Methods

All the methods we have discussed so far are explicit: Given the initial conditions, we march forward for solutions at later times. There are cases where explicit methods are not suitable and even fail to produce acceptable solutions. For instance, stiff differential equations are known to be ill-suited for explicit methods. To obtain accurate solutions to these equations, the step size h needed with explicit methods is often much smaller than that for nonstiff differential equations. The reason is usually due to terms varying over very different time scales. Implicit methods such as the backward Euler method are often used for such cases. We will encounter one such case in Section 8.3.1.

2.4 THE LEAPFROG METHOD

So far we have discussed solutions of ODEs purely from the mathematical point of view. However, there are physical systems possessing special properties that should be respected in the numerical solution. Hamiltonian systems fall into this category. Take the simple harmonic oscillator as an example. The motion is oscillatory and time reversible, that is, if the velocity at a certain point in its path is reversed, the trajectory will move backward over the original path, as if time is reversed, $t \to -t$. The motion is also area preserving in phase space, which is a multidimensional space consisting of all coordinates and momenta [31]. Two phase space trajectories for the harmonic oscillator are shown in Figure 2.5.

Their shape is an ellipse, controlled by the energy

$$E = \frac{1}{2}mv^2 + \frac{1}{2}m\omega^2 x^2, \quad m = \text{mass}, \ \omega = \text{frequency}. \tag{2.38}$$

For a given energy E, the area within each ellipse is proportional to E and is constant because energy is conserved. The Euler and the Runge-Kutta methods do not preserve these properties. Next we discuss the leapfrog method that does (also known as the Verlet method [92]).

Let us start with the equations of motion of a one-dimensional Hamiltonian system,

$$\frac{dx}{dt} = v, \tag{2.39}$$

$$\frac{dv}{dt} = a(x), \tag{2.40}$$

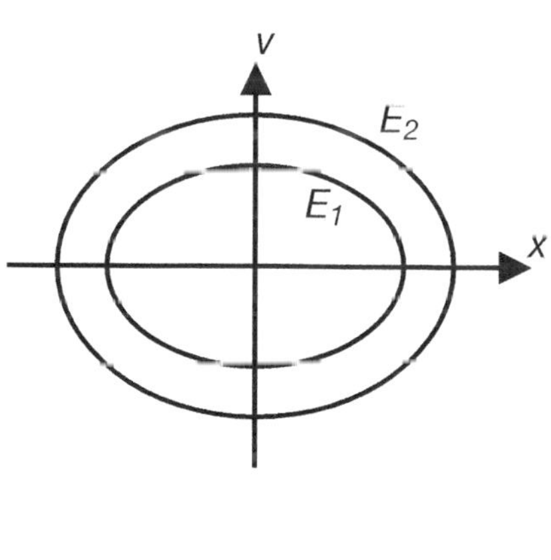

FIGURE 2.5 Phase space plot of the harmonic oscillator for two energies, $E_2 > E_1$.

where $a(x) = F(x)/m$ is the acceleration. The force $F(x)$ depends on the coordinate x only. For the harmonic oscillator, $a(x) = -\omega^2 x$. The important thing is that the RHS of the first equation (2.39) depends on the velocity v only, and the second equation (2.40) on the coordinate x only. This separation of coordinates and velocity is required for the leapfrog method (Appendix 2.A).

Given the values x_0 and v_0 at t, the leapfrog method calculates the values x_1 and v_1 at $t + h$ in three steps,

$$x_{1/2} = x_0 + v_0 \frac{h}{2}, \tag{2.41}$$

$$v_1 = v_0 + a(x_{1/2})h, \tag{2.42}$$

$$x_1 = x_{1/2} + v_1 \frac{h}{2}. \tag{2.43}$$

First, a half step is taken to obtain the position $x_{1/2}$ at the midpoint. Then a full step is taken to calculate the new velocity v_1 using acceleration evaluated at $x_{1/2}$. Finally, the new position x_1 is found by taking another half step, using the new velocity. Each step uses the newest values as they become available.

If the leapfrog method looks a lot like the midpoint (RK2) method, it partially is, at least in the order of accuracy. The leapfrog method is second order, as is the midpoint method. But there is a big difference: the leapfrog method is area preserving. To understand this property, let us think of the algorithm as a transformation in phase space that maps the points (x, v) at t to the new points (q, p) at $t + h$. Figure 2.6 illustrates such a transformation.

The corners of the rectangular area element at t are transformed to new points at $t + h$. When h is small enough, the transformed points form an approximate parallelogram, whose area dA' is given by the cross product of the vectors $d\vec{u}$ and $d\vec{w}$. We can associate a transformation with each step in the leapfrog method represented by Eqs. (2.41) to (2.43). Using these transformations, it can be shown that the area is preserved, $dA' = dA$. Details are given in Section 2.A. The leapfrog method is also time reversible. We leave its proof to Exercise E2.4.

We wish to write a standalone leapfrog subroutine for a system of ODEs more general than Eqs. (2.39) and (2.40). Similar to Eq. (2.27), let us introduce the coordinate and velocity vectors as $\{r\} = r_1, r_2, ..., r_n$ and $\{v\} = v_1, v_2, ..., v_n$, respectively. The system of ODEs can be written as two sets, with $i = 1, 2, ..., n$,

$$\frac{dr_i}{dt} = f_i(\{v\}, t), \tag{2.44}$$

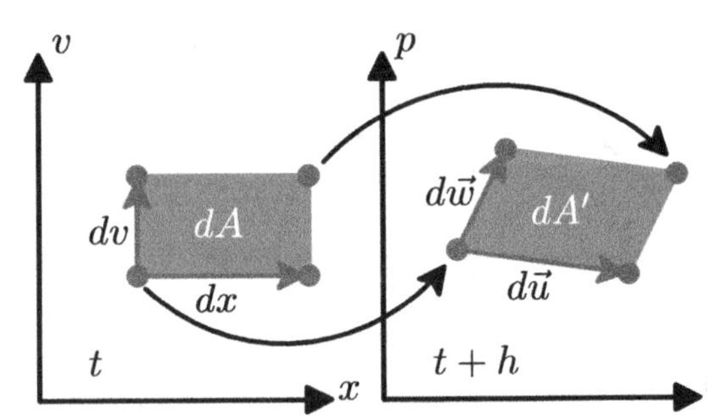

FIGURE 2.6 Schematic of area transformation. The transformation maps the points in phase space from $[x, v]$ at t to $[q, p]$ at $t + h$, e.g., $[x, v] \rightarrow [q(x, v),\ p(x, v)]$ for the lower-left corner, similarly for other points. In the leapfrog method, the area is preserved, $dA' = dA$.

$$\frac{dv_i}{dt} = g_i(\{r\}, t). \tag{2.45}$$

The function $f_i(\{v\}, t)$ depends on the velocity $\{v\}$ only, and $g_i(\{r\}, t)$ on the coordinate $\{r\}$ only. Again we need the separation of $\{r\}$ and $\{v\}$ for the leapfrog method to be applicable. We can think of $f_i(\{v\}, t)$ as the generalized velocity, and $g_i(\{r\}, t)$ the generalized acceleration. For Hamiltonian systems, they become the actual velocity and acceleration. For others, such as time-dependent quantum systems discussed in Chapter 8, they are pseudovelocity and pseudoacceleration.

By vectorizing the leapfrog method (2.41) to (2.43) in a way analogous to Eq. (2.16), we can write a leapfrog solver for Eqs. (2.44) and (2.45) as follows.

Program listing 2.6 The leapfrog method (`leapfrog.py`)

```
def leapfrog( lfdiffeq,  r0,  v0,  t,  h):          # vectorized leapfrog
    """ vector leapfrog method using numpy arrays.
        It solves general (r,v) ODEs as:
        dr[i]/dt = f[i](v), and dv[i]/dt = g[i](r).
        User supplied lfdiffeq (id, r, v, t) returns
        f[i](r) if id=0, or g[i](v) if id=1.
        It must return a numpy array if i>1 """
    hh = h/2.0
    r1 = r0 + hh*lfdiffeq(0, r0, v0, t)        # 1st: r at h/2 using v0
    v1 = v0 + h* lfdiffeq (1, r1, v0, t+hh) # 2nd: v1 using a(r) at h/2
    r1 = r1 + hh*lfdiffeq(0, r0, v1, t+h)   # 3rd: r1 at h using v1
    return r1, v1
```

Like RK2/RK4 requiring `diffeq`, `leapfrog` needs a user-supplied function that returns the RHS of Eqs. (2.44) and (2.45). This function is defined as `lfdiffeq(id, r, v, t)`. Note that by our convention, the leapfrog method assumes a different interface in the derivative function (`lfdiffeq`) than the Runge-Kutta methods. See Program 2.7 for an example.

Here, `lfdiffeq` is basically the same as `diffeq` except for the `id` flag: when `id=0`, `lfdiffeq` returns $f_i(\{v\}, t)$, and when `id=1`, it returns $g_i(\{r\}, t)$. We could have split it into two functions, but it is generally more convenient to have just one function. As before, we put `leapfrog` in the file `ode.py`, which should be imported before any ODE solver is used.

The Harmonic Oscillator with the Leapfrog Method

As an example, let us apply the leapfrog method to a Hamiltonian system, the simple harmonic oscillator. The equations of motion analogous to Eqs. (2.39) and (2.40) are

$$\frac{dx}{dt} = v, \quad \frac{dv}{dt} = -\omega^2 x. \tag{2.46}$$

The solution is a perfect harmonic,

$$x = A\cos(\omega t + \varphi), \quad v = -\omega A \sin(\omega t + \varphi). \tag{2.47}$$

The constants A and φ are the amplitude and initial phase, respectively. We see that $x^2/A^2 + v^2/(\omega A)^2 = 1$, indeed an ellipse in phase space.

The program to solve Eq. (2.46) numerically is as follows ($\omega = 1$ rad/s).

Program listing 2.7 Oscillator with leapfrog method (`sho_lf.py`)

```
import ode                          # get ODE solvers
import math as ma                   # get math functions
import matplotlib.pyplot as plt # get matplotlib plot functions

def oscillator(id, x, v, t):     # return dx/dt or dv/dt
    if (id==0):                  # calc velocity
        return v
    else:                        # calc acceleration
        return -x                # -ω²x, ω = 1,

def go():                        # main program
    x, v = 1.0, 0.0              # initial values
    t, h = 0.0, 0.1              # init time, step size
    xa, va = [], []              # declare arrays for plotting
    while t<4*ma.pi:             # loop for two periods
        xa.append(x)             # record position and velocity
        va.append(v)
        x, v = ode.leapfrog( oscillator, x, v, t, h) # solve it
        t = t + h

    plt.figure()                 # start a figure
    plt.plot(xa,va)              # plot vel-pos
    plt.xlabel('x (m)')          # add labels
    plt.ylabel('v (m/s)')
    plt.show()                   # show figure

go()                             # run the program
```

The program first imports several libraries: our ODE solvers in `ode.py`, standard mathematical functions, as well as Matplotlib. According to the leapfrog scheme (2.44) and (2.45), we have $n = 1$ in Eq. (2.46) (with r replaced by x), and $f_1(v,t) = v$, $g_1(x,t) = -x$. Therefore, the function `oscillator` returns v if `id=0`, and $-x$ otherwise. The main program `go()` is very similar to the one in Program 2.5. It sets up the initial values and enters the `while` loop, which runs for two periods, storing the position and velocity at each time step before advancing to the next step with `leapfrog`. The velocity-position curve in phase space is plotted after the loop.

Figure 2.7 (top) shows a sample plot from the preceding program. For comparison, results using the Euler and RK2 methods are also shown (generated separately, see Project P2.4). Starting from the initial point $(1, 0)$, the path by Euler's method spirals outward noticeably. The RK2 and leapfrog methods produce ellipses similar to what is expected from Figure 2.5. Although hard to see on the scale shown, the path of the RK2 method is not closed and the area is not preserved. If we wait long enough, it would spiral outward eventually, but at a slower pace than Euler's method. In contrast, the leapfrog method has a closed path, thus preserving the area.

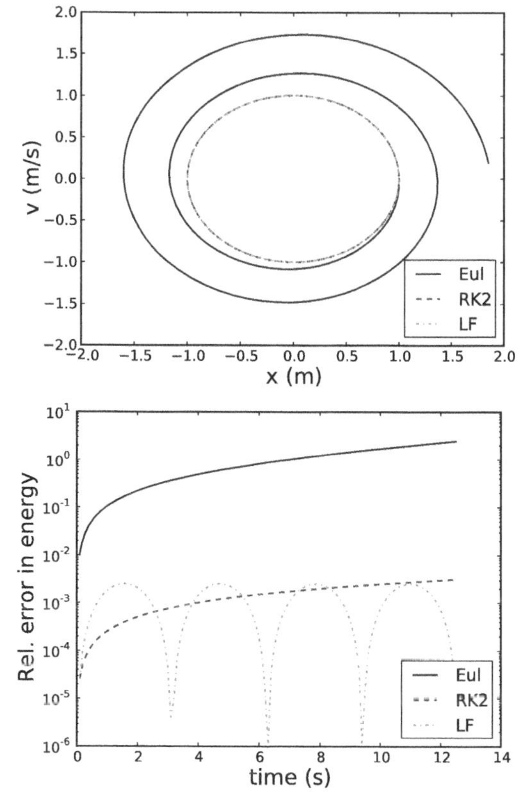

FIGURE 2.7 Phase space plot of the harmonic oscillator (top), and the relative error in energy (bottom) for $\omega = 1$ rad/s. The results are calculated with the Euler (Eul), RK2, and the leapfrog (LF) methods.

We can get a clearer picture by considering the relative error in energy defined as $|E_{numeric} - E_{exact}|/E_{exact}$.[3] The numerical energy $E_{numeric}$ can be obtained by slightly modifying Program 2.7. The results are shown in Figure 2.7 (bottom) as a semilog plot. When plotting results that vary over orders of magnitude, the logarithmic scale (semilog or log-log) should be used to show clearly the difference. The energy by Euler's method increases with time. The RK2 method shows a similar trend, albeit with an offset and at a slower rate because it is second order. The energy in both methods increases monotonically and shows no upper bound. In the leapfrog method it fluctuates around the exact energy and does not diverge without bound. Being of finite order of accuracy, the leapfrog method does not yield the exact energy, but it does preserve the proper structure in phase space of a Hamiltonian system, including a closed path as required. Accordingly, area-preserving methods (also known as symplectic methods) such as the leapfrog method are preferred when integrating systems whose motion is bounded, including periodic and oscillatory motion. Later, we discuss a modified leapfrog method with step size control in Section 4.3.3.[4]

[3]To validate a program, it is always important to check the results against exact solutions, if available, or against conservation laws such as energy and angular momentum, and so on.

[4]The use of symplectic methods has completely transformed our understanding of the stability of the solar system by extending accurate time integrations from millions of years prior to 1990s to timescales comparable to the age of the solar system (see Ref. [65] and Chapter 4).

CHAPTER SUMMARY

We have discussed several methods of numerical solutions, namely, Euler, RK2, RK4, and leapfrog methods. Using free fall as the main example, they have been applied to a single ODE at first and generalized to a system of coupled ODEs later. Along the way we have showed animation and data plotting techniques using the simple yet powerful and flexible Matplotlib, NumPy, and VPython functions.

We have constructed stand-alone ODE solvers with these methods, including `Euler`, `RK2`, `RK4/n`, `RK45n`, and `leapfrog`. They have been included in one file, `ode.py`, which will be our toolbox for ODE solvers. To use it, import `ode.py` first, and supply an appropriate `diffeq` or `lfdiffeq` function for the problem being solved. In subsequent chapters, except for specific cases requiring special methods, we will regard `RK4` as the general-purpose ODE solver, and `leapfrog` for Hamiltonian problems. The vectorized solvers can be applied to a single or a system of ODEs. For the latter, `diffeq` must return a NumPy array. The nonvectorized routines `RK4n` and `RK45n` use explicit looping through lists or NumPy arrays and are intended for systems of a handful ODEs where speed is important.

2.5 EXERCISES AND PROJECTS

Exercises

E2.1 Modify Program 2.1 to replace VPython animation with Matplotlib output. Use elements from Program 1.1, and initialize three lists for time, position, and velocity before the loop. Terminate the loop after 10s, and plot position and velocity vs. time, respectively.

E2.2 (a) Show that the alternative RK2 algorithm (2.17) is second order in accuracy and is equivalent to Eq. (2.16).

(b) Among other possibilities of the RK2 family is the following

$$k_1 = hf(y_n, t_n),\ k_2 = hf(y_n + 2k_1/3, t_n + 2h/3),$$

$$y_{n+1} = y_n + \frac{1}{4}(k_1 + 3k_2),$$

where the second point is chosen at $2h/3$. Show that this is equivalent to Eq. (2.16) to second order. Hint: Expand k_1 and k_2 around the midpoint. For simplicity, assume $f(y, t) = f(t)$, that is, no explicit y dependence.

E2.3 Follow through one working cycle of the RK2 method for free fall similar to Eq. (2.35) of Euler's method. Show that in RK2 (2.16)

$$\{k_1\} = \begin{bmatrix} v_0 h \\ -gh \end{bmatrix},\ \{k_2\} = \begin{bmatrix} v_0 h - \frac{1}{2}gh^2 \\ -gh \end{bmatrix},$$

proving that RK2 is exact for free fall, no matter what step size h is.

E2.4 Prove that the leapfrog method, Eqs. (2.41) to (2.43), is time reversible. Use primed variables to represent reversed motion, and let $x_0' = x_1$, $v_0' = -v_1$. Show that after one step, the position is restored but the velocity is reversed, that is, $x_1' = x_0$, $v_1' = -v_0$.

E2.5 Use an approach analogous to Exercise E2.4, show that Euler's method when applied to Eqs. (2.39) and (2.40) is not time reversible unless $a(x) = 0$. In other words, Euler's method (as well as RK2 and RK4) is time reversible only for a free particle (straight-line motion).

E2.6 The Euler-Cromer algorithm is a first-order symplectic method. It solves Eqs. (2.39) and (2.40) by the following,

$$x_1 = x_0 + v_0 h, \tag{2.48}$$

$$v_1 = v_0 + a(x_1)h, \tag{2.49}$$

which would be the standard Euler's method except for the acceleration, $a(x_1)$, evaluated at x_1 rather than at x_0 in Eq. (2.49). Show that the Euler-Cromer method is time reversible and area preserving by their Jacobians.

E2.7 Nuclear decay is a random process. But given sufficient number of nuclei N, it may be modeled as a deterministic process,

$$\frac{dN}{dt} = -\lambda N,$$

where λ is the decay constant.

(a) One interpretation of the lifetime of a single nucleus is the average time of their existence before decay, that is, $\tau = \int t(-dN)/\int(-dN)$. Based on this interpretation and given $N = N_0 \exp(-\lambda t)$, show analytically that the lifetime of an unstable nucleus is $\tau = 1/\lambda$.

(b) If you have not used SymPy before, this might be a good place to start (see Section 1.B). Try the following in IPython

```
In [1]:  from sympy import *      # import SymPy
In [2]:  t, lamb = symbols('t lamb',positive=True) # declare vars
In [3]:  integrate(exp(-lamb*t),(t,0,oo)) # integrate from [0,∞]
Out[3]:  1/lamb
```

Now compute τ with SymPy.

Projects

P2.1 Simulate free fall using Euler's method and VPython.

(a) Do this by combining and modifying Programs 2.1 and 2.5. Delete `Euler` from Program 2.5 because we will be using `ode.py`, but keep `freefall` as is. Optionally, you may also delete the plotting functions below the "while" loop. Your program blocks should look like

```
from visual import *
from ode import *
import numpy as np
# draw ball and floor
# set initial conditions
......
while True:
    ......
    y = Euler (...)

    # update ball position
    # reverse velocity if below floor
```

Note that the position and velocity of the ball are computed by the standalone `Euler` and contained in the array `y` and should not be computed separately as in Program 2.1. Reversing the ball's velocity on hitting the floor amounts to changing the initial condition (or seed value for next step). Once the program is running, what do you see? Notice how the "camera" keeps zooming out to keep the ball in view. Why? Does the ball bounce higher and higher? Is it numerical or physical? Try a few different initial velocities, and see if anything changes.

(b) Replace Euler's method with the RK2 method. Change just one line as

```
#y = Euler(...)
y = RK2(...)
```

It is helpful that in testing out different things such as we are doing here, rather than deleting the original code, simply comment out the lines we do not need. It makes it easier to go back to an earlier state by uncommenting them. You should now see the ball bouncing up and down, but not gaining height after each bounce.

(c) Compare your observations about the different results with Euler's and the RK2 methods. Discuss the reasons. Refer to Figure 2.2 for hints.

(d) Using energy conservation as a further check, add two statements in the loop that records time and energy (kinetic and potential, setting mass to one). Break the loop when time reaches a certain value (say 10s), and plot energy vs. time. Compare results obtained with Euler's and the RK2 methods. What went "wrong" with Euler's method?

P2.2 Let us consider free fall in two dimensions, that is, projectile motion. The equations of motion are

$$\frac{dx}{dt} = v_x, \quad \frac{dy}{dt} = v_y,$$
$$\frac{dv_x}{dt} = 0, \quad \frac{dv_y}{dt} = -g.$$

(a) Solve these equations with the Euler and RK2 (or RK4) methods. Your program should be very similar to Program 2.5. The key part is to write a derivative function that plays the role of the generic `diffeq()`. Let us name it `projectile(y,t)`, where `y[]` should be a 4-element array. The first two elements should contain the positions (x, y) and the last two the velocities (v_x, v_y) (see Program 3.5). Assume initial conditions as: $x = y = 0$, launching speed and angle at 10 m/s and 30°, respectively. Follow the motion until the projectile hits the ground. Plot the x-t and y-t curves, and compare with analytic results.

(b) Animate projectile motion using VPython. As in Program 2.1, you only need to update the position of the projectile after each step. The following line will change both the x and y positions (assuming they are stored in `y[0]`,`y[1]` as suggested earlier)

```
ball.pos=(y[0],y[1])
```

P2.3 (a) Write a program to investigate nuclear decay (Exercise E2.7) using several methods: `Euler`, `RK2`, `RK4`, and `RK45n`. Assume $\lambda = 0.5$ min^{-1} and $N_0 = 1000$. Choose your time step h judiciously and integrate to 8 min. Plot the results as N-t curves for each method on one plot. Discuss their differences.

Next, plot the absolute errors of each method on a semilog plot. Compare the magnitudes of the error to each other. Then halve h and plot the errors again. Compare the reduction in error for each method. Does the global error for a given method scale with its order?

(b) Calculate numerically the lifetime τ according to the interpretation in Exercise E2.7. This can be done by first computing the cumulative lifetime as $\sum t(-\Delta N)$ in the main loop of (a) because

"$-\Delta N$" is the number of nuclei that "lived" to time t but decayed within the next time step, and then averaging over all the nuclei that decayed, that is, $N_0 - N$, where N is taken after the loop. Compare the numerical results for each method with the exact answer. Discuss the discrepancies, the reason, and how they may be reduced.

P2.4 (a) Modify Program 2.7 to plot the phase space trajectories with different methods, including `Euler`, `RK2`, and `leapfrog`. Of course, you can use the derivative function `oscillator()` as is with the `leapfrog()` function. However, with the Euler and RK2 methods, you need a slightly different derivative function, call it `oscillator_rk(y,t)`. It is similar to `freefall()` in Program 2.5 but returns the RHS of the oscillator equations (2.46) based on the input of the 2-element array `y[]`, with `y[0]` = position and `y[1]` = velocity. Run the program with different initial conditions: $x_0 = 1$ and 1.5, and $v_0 = 0$.

(b) Calculate the relative error in the energy of the harmonic oscillator with the same methods. Plot the results on a semilog scale as in Figure 2.7 with `plt.semilogy()` after the `plot()` function. All other plotting commands are the same.

(c) Reduce the step size h by half and repeat (b). Discuss the changes and trends in the results.

(d) Redo the calculation in part (b) using a higher order method, `RK4` or `RK45n`. Comment on your results.

P2.5 Consider the motion of an electron ($m = 1$) in a one-dimensional Coulomb field $V(x) = 1/x$ (all in atomic units (a.u.), see Table 8.1).

(a) Using conservation of energy E, obtain the relationship between velocity and position, $v(x)$, for a given E.

(b) Convert $d^2x/dt^2 = F/m = -\partial V/\partial x$ into two first-order ODEs.

(c) Use `RK2` to simulate the motion of the electron with initial conditions as: position $x = 5$ a.u. and velocity $v = -1$ a.u. You need a proper derivative function based on part (b). Plot position-time and velocity-time curves from the beginning to when it returns to its original position. Also plot the phase space trajectories, the v-x curves, one of the numerical results, and another of the analytic result, part (a). How do they compare?

(d) Repeat (c) with the `leapfrog` method. You would need a slightly different derivative function than the preceding one.

(e) Do the same using `RK4` and `RK45n`, but with different initial conditions of $x = 5$ and $v = -4$. You can use the same derivative function as in (c). Discuss and compare the performance of the two methods relative to each other.

Observation: As you work on this project, think about the following questions. What would be the "characteristic" time scale? How did you choose the step size h relative to this time scale? Be conservative with h because the potential is singular at the origin. Because the motion is unbound (the electron can escape to $x = \infty$), is there still any advantage of the leapfrog method compared to the RK2 method?

P2.6 Let us play a game popular on many campuses: Humans vs. Zombies! The students join one of two groups, designated as Humans and Zombies, respectively. Members of each group are armed with toy dart guns. To make it interesting, let us assume that the firing rates, or "kill" rates, between the groups are different. Let H be the number of Humans with a kill rate λ_H, and Z the number of Zombies with a kill rate λ_Z. In the simplest model, we can assume that the number of Humans decreases at a rate proportional to the kill rate and the number of Zombies, and vice versa. This model can be written as

$$\frac{dH}{dt} = -\lambda_Z Z, \quad \frac{dZ}{dt} = -\lambda_H H.$$

The game ends when one side reaches zero, and the other side is the winner.

(a) Say you are the last one to join, and you are free to choose either group. The current situation is as follows: 69 Humans with a kill rate 2, and 99 Zombies with a kill rate 1. There are fewer Humans, but they have twice the kill rate. If your goal is to win, which group would you join? Choose wisely (and write it down).

(b) It is war now! Simulate the game by solving the model with the initial condition as: $\lambda_H = 2.0, \lambda_Z = 1.0$; if you decided to join the Humans, then $H = 70, Z = 99$; otherwise $H = 69, Z = 100$. Plot the H-t and Z-t curves when the game ends. You may use either RK2/RK4 or the leapfrog method. Are you victorious? Why?

(c) Switch sides and repeat (b).

(d)* Run the program with different initial H, say 80, 100, 120, and so on, and for each H, find the upper limit Z such that the Humans barely win. Plot H vs. Z, and you should discover a scaling law, that is, $H/Z = C$. Find the constant C. As it turns out, the model yields analytic solutions. Find them, and discuss how C is related to λ_H and λ_Z.

2.A AREA PRESERVATION OF THE LEAPFROG METHOD

If we view a numerical method that advances the solution (x, v) at t to (q, p) at $t + h$ as a transformation in phase space, we have

$$q = q(x, v), \; p = p(x, v). \tag{2.50}$$

The functional forms of $q(x, v)$ and $p(x, v)$ will depend on the numerical method, or the algorithm. As we will see shortly, each of the three steps (2.41) to (2.43) in the leapfrog method represents one transformation.

Let us first consider the transformation of area elements by Eq. (2.50), schematically shown in Figure 2.6. Specifically, let us consider small changes in dx and dv, respectively. There are four points associated with these changes: (x, v), $(x + dx, v)$, $(x + dx, v + dv)$, and $(x, v + dv)$. They are among the possible points in phase space at t.

The corner points of the rectangle with sides $dx\,\hat{\imath}$ and $dv\,\hat{\jmath}$ will be mapped to new locations at $t + h$ according to Eq. (2.50). The change dx, while holding v constant, will transform the vector $dx\,\hat{\imath}$ to a new vector $d\vec{u}$ (Figure 2.6). Similarly, the change dv, while holding x constant, will transform the vector $dv\,\hat{\jmath}$ to a new vector $d\vec{w}$. The two vectors $d\vec{u}$ and $d\vec{w}$ will not be perpendicular to each other in general. Provided the changes are small, the transformed points form an approximate parallelogram with $d\vec{u}$ and $d\vec{w}$ as the sides.

The vectors $d\vec{u}$ and $d\vec{w}$ can be found by

$$\begin{aligned} d\vec{u} &= \frac{\partial q}{\partial x} dx\,\hat{\imath} + \frac{\partial p}{\partial x} dx\,\hat{\jmath}, \quad \text{(holding } v \text{ constant)} \\ d\vec{w} &= \frac{\partial q}{\partial v} dv\,\hat{\imath} + \frac{\partial p}{\partial v} dv\,\hat{\jmath}. \quad \text{(holding } x \text{ constant)} \end{aligned} \tag{2.51}$$

The area element of the transformed parallelogram is $dA' = du\,dw \sin\theta$, where θ is the angle between $d\vec{u}$ and $d\vec{w}$ (Figure 2.6). This is just the magnitude of the cross product of the vectors $d\vec{u}$ and $d\vec{w}$, that is, $dA' = |d\vec{u} \times d\vec{w}|$. Using Eq. (2.51), we find the area element to be

$$dA' = \left(\frac{\partial q}{\partial x}\frac{\partial p}{\partial v} - \frac{\partial q}{\partial v}\frac{\partial p}{\partial x} \right) dxdv.$$

Introducing the Jacobian J as

$$J = \begin{vmatrix} \dfrac{\partial q}{\partial x} & \dfrac{\partial q}{\partial v} \\ \dfrac{\partial p}{\partial x} & \dfrac{\partial p}{\partial v} \end{vmatrix}, \tag{2.52}$$

we can express the area element as

$$dA' = JdA, \tag{2.53}$$

where $dA = dxdv$ is the area element of the original rectangle.

Now we can construct the three transformations represented by the three steps in the leapfrog method, Eqs. (2.41) to (2.43). We may consider the first step (2.41) as a transformation that changes only the coordinate while keeping the velocity unchanged. In terms of pretransformed variables x_0, v_0, the transformation q_1, p_1 is

$$q_1(x_0, v_0) = x_0 + v_0\frac{h}{2}, \; p_1(x_0, v_0) = v_0.$$

The associated Jacobian is

$$J_1 = \begin{vmatrix} \dfrac{\partial q_1}{\partial x_0} & \dfrac{\partial q_1}{\partial v_0} \\ \dfrac{\partial p_1}{\partial x_0} & \dfrac{\partial p_1}{\partial v_0} \end{vmatrix} = \begin{vmatrix} 1 & h/2 \\ 0 & 1 \end{vmatrix}. \tag{2.54}$$

Similarly, the transformations and the Jacobians for the next two steps, respectively, are

$$q_2(x_{1/2}, v_0) = x_{1/2}, \; p_2(x_{1/2}, v_0) = v_0 + a(x_{1/2})h,$$

$$J_2 = \begin{vmatrix} \dfrac{\partial q_2}{\partial x_{1/2}} & \dfrac{\partial q_2}{\partial v_0} \\ \dfrac{\partial p_2}{\partial x_{1/2}} & \dfrac{\partial p_2}{\partial v_0} \end{vmatrix} = \begin{vmatrix} 1 & 0 \\ a'(x_{1/2})h & 1 \end{vmatrix}, \tag{2.55}$$

and

$$q_3(x_{1/2}, v_1) = x_{1/2} + v_1\frac{h}{2}, \; p_3(x_{1/2}, v_1) = v_1,$$

$$J_3 = \begin{vmatrix} \dfrac{\partial q_3}{\partial x_{1/2}} & \dfrac{\partial q_3}{\partial v_1} \\ \dfrac{\partial p_3}{\partial x_{1/2}} & \dfrac{\partial p_3}{\partial v_1} \end{vmatrix} = \begin{vmatrix} 1 & h/2 \\ 0 & 1 \end{vmatrix}. \tag{2.56}$$

Note that the pretransformed variables for the second and the third steps are $(x_{1/2}, v_0)$ and $(x_{1/2}, v_1)$, respectively.

The cumulative transformation of the area element is the product of the three successive transformations, $dA' = J_3 J_2 J_1 dA$. We see from Eqs. (2.54) to (2.56) that each Jacobian (determinant) is exactly 1, $J_1 = J_2 = J_3 = 1$, such that

$$dA' = dA. \quad \text{(leapfrog method)} \tag{2.57}$$

We have shown that one area element enclosed by the four points at the corners of the rectangle is preserved (Figure 2.6). To consider the total area of the accessible phase space, we would divide it into many small (infinitesimal) rectangular elements. Because each area element is preserved by the transformation, the total area, representing all possible points in phase space, must also be preserved.

Hamiltonian systems preserve the phase space area. This is known as the Liouville theorem [40]. Area-preserving methods are said to be symplectic, so the leapfrog method is symplectic. The leapfrog method (or other symplectic methods) is generally preferred when integrating Hamiltonian systems. However, if the force is velocity dependent, the leapfrog method as presented by Eqs. (2.54) to (2.56) is ambiguous with respect to velocity and is not area preserving because the Jacobian J_2 would not be 1. A modified scheme would be necessary, but we will not pursue it here. Speaking of modifications, it is also possible to start the leapfrog method (2.54) to (2.56) with velocity $v_{1/2}$, basically swapping x and v. There is no advantage to this approach because it is equivalent to the original method, but is less efficient because two calls to $a(x)$ are required per step, instead of just one.

It should be noted that the leapfrog method, being of finite order, may not yield the exact area in phase space. But, however inexact that area may be, it is preserved exactly.

2.B PROGRAM LISTINGS AND DESCRIPTIONS

Program listing 2.8 RK4 nonvector method (`rk4n.py`)

```
def RK4n(diffeq, y0, t, h):       # non-vectorized with lists
    """ RK4 method for n ODEs:
        Given y0 at t, returns y1 at t+h """
    n, y1 = len(y0), [0.0]*len(y0)
    k1 = diffeq(y0, t)                  # dy/dt at t
    for i in range(n):                  # loop thru n ODEs
        y1[i] = y0[i] + 0.5*h*k1[i]     # prep for k2[]
    k2 = diffeq(y1, t + h/2.)           # dy/dt at t+h/2
    for i in range(n):
        y1[i] = y0[i] + 0.5*h*k2[i]
    k3 = diffeq(y1, t + h/2.)           # dy/dt at t+h/2
    for i in range(n):
        y1[i] = y0[i] + h*k3[i]
    k4 = diffeq(y1, t + h)              # dy/dt at t+h
    for i in range(n):
        y1[i] = y0[i] + h*(k1[i]+2*k2[i]+2*k3[i]+k4[i])/6.0
    return y1
```

Program 2.8 works with both lists and NumPy arrays returned by `diffeq`. It should be applicable for problems fitting the discussed criteria (Section 2.3.2).

Program listing 2.9 RK45 nonvector method (`rk45n.py`)

```python
def RK45n(diffeq, y0, t, h):            # RK45 method
    a2, a3, a4, a5, a6 = 0.2, 0.3, 0.6, 1.0, 0.875
    b21, b31, b32, b41, b42, b43 = 0.2, 3./40., 9./40., 0.3, -0.9, 1.2
    b51, b52, b53, b54 = -11./54., 2.5, -70./27., 35./27.
    b61, b62, b63, b64, b65 = [1631./55296., 175./512., 575./13824.,
                               44275./110592., 253./4096.]
    c1, c3, c4, c6 = 37./378., 250./621., 125./594., 512./1771.

    n, y1 = len(y0), [0.0]*len(y0)
    k1 = diffeq(y0, t)
    for i in range(n):
        y1[i] = y0[i] + h*b21*k1[i]
    k2 = diffeq(y1, t + a2*h)
    for i in range(n):
        y1[i] = y0[i] + h*(b31*k1[i] + b32*k2[i])
    k3 = diffeq(y1, t + a3*h)
    for i in range(n):
        y1[i] = y0[i] + h*(b41*k1[i] + b42*k2[i] + b43*k3[i])
    k4 = diffeq(y1, t + a4*h)
    for i in range(n):
        y1[i] = y0[i] + h*(b51*k1[i] + b52*k2[i] + b53*k3[i] + b54*k4[i])
    k5 = diffeq(y1, t + a5*h)
    for i in range(n):
        y1[i] = y0[i] + h*(b61*k1[i] + b62*k2[i] + b63*k3[i] + b64*k4[i]
                           + b65*k5[i])
    k6 = diffeq(y1, t + a6*h)
    for i in range(n):
        y1[i] = y0[i] + h*(c1*k1[i] + c3*k3[i] + c4*k4[i] + c6*k6[i])
    return y1
```

The preceding program (nonvectorized) is based on the Cash-Karp coefficients for the embedded Runge-Kutta formula [71]. We omitted calculation of error estimate in the code. It seems from our experience that for some conservative oscillating systems, the error estimate substantially underestimates the true error. However, this does not impact the accuracy of the method (fifth order).

Program listing 2.10 SciPy ODE wrapper (`odewrap.py`)

```python
from scipy.integrate import odeint       # SciPy integrator

def odewrapper(diffeq, y0, t, h):         # ode wrapper
    y = odeint(diffeq, y0, [t, t+h])
    return y[1]

Euler = RK2 = RK4 = RK45 = RK4n = RK45n = odewrapper # alias
```

This is just a wrapper for the general ODE integrator `odeint` from the SciPy library `scipy.integrate`. We call `odeint` with two points in time $(t, t+h)$ and return the solutions at $t+h$. It is functionally (but not accuracy wise) equivalent to the Runge-Kutta methods, all assigned to the same alias on the last line.

The wrapper may be useful to those who wish to skip the discussion of the RK methods or to use it as a blackbox. To use the wrapper, import it as

```
import odewrap as ode      # use wrapper as RK substitutes
```

All programs using the RK family of methods should work the same way as with our own ODE library. However, the wrapper is not a substitute for the leapfrog method.

Realistic Projectile Motion with Air Resistance

CHAPTER 3

One of the most classic and well-known problems in physics is projectile motion. Tossing a ball in the air, we expect to see it move in a familiar arched path. Under ideal conditions assuming constant gravitational acceleration and negligible air resistance, projectile motion is analytically solvable, that is, its solutions are expressible in closed-form, known functions. Its properties such as the parabolic path are well explained from introductory physics. Free fall discussed in Chapter 2 is a special case of one-dimensional ideal projectile motion, whereas the general case is three dimensional.

To describe realistic projectile motion, we need to consider the effects of air resistance, or drag, which can be significant and interesting. However, the inclusion of these effects renders the problem analytically nonsolvable, and no closed-form solutions are known except under limited conditions. Numerically, this presents no particular difficulty for us, given the toolbox and ODE solvers we just developed in Chapter 2. In fact, realistic projectile motion is an ideal case study for us to begin application of these numerical and visualization techniques to this classic problem in this chapter, for it is relatively simple and intuitive, and its basic features are already familiar to us. We will learn to construct models of appropriate degree of complexity to reflect the effects of drag and spin. Furthermore, we will also discuss the application of a recently discovered Lambert W function to projectile motion, as well as methods of finding roots.

3.1 VISUALIZATION OF IDEAL PROJECTILE MOTION

In his work *Two New Sciences* in 1638, Galileo theorized that a free body in motion was "compounded from equable horizontal and naturally accelerated downward motion, which I call 'projection'." Of course we recognize in today's nomenclature that Galileo was talking about ideal projectile motion, which, if air resistance is ignored, moves with constant horizontal velocity and constant vertical acceleration, that is, $\vec{v} = d\vec{r}/dt$ and $\vec{a} = d\vec{v}/dt = -g\hat{j}$. In 2D, this can be expressed in terms of differential equations similar to Eq. (2.1) as

$$\frac{dx}{dt} = v_x, \quad \frac{dy}{dt} = v_y, \quad \frac{dv_x}{dt} = 0, \quad \frac{dv_y}{dt} = -g, \tag{3.1}$$

where x, y and v_x, v_y are the components of position and velocity, respectively, and g the gravitational acceleration as usual. The analytical solutions to Eq. (3.1) are well known from introductory physics. For example, the projectile range is given simply by

$$R_0 = \frac{v_0^2 \sin 2\theta_0}{g}, \tag{3.2}$$

where v_0 is the initial speed and θ_0 the initial projection angle. It is worth noting that the angle that maximizes the range is 45°, independent of speed in ideal projectile motion. We shall see that this is no longer true if air resistance is included.

Visualizing projectile motion is also straightforward. The following program displays perpetual projectile motion of a ball, captured in Figure 3.1.

FIGURE 3.1 Snapshots of a ball toss from Program 3.1.

Program listing 3.1 Ball toss (`balltoss.py`)

```
1  import visual as vp                   # get VPython modules for animation

3  vp.display(background=(.2,.5,1))      # make scene, ball, floor, and path
   ball  = vp.sphere(pos=(-4,-4,0), radius=1, color=vp.color.yellow)
5  floor = vp.box(pos=(0,-5,0), length=12, height=0.2, width=4)
   path  = vp.points(pos=ball.pos, size=4)      # use make_trail in newer VP
7  h, g, vx, vy = 0.01, 9.8, 4.0, 9.8           # Δt, g, and initial velocity
   while True:
9      vp.rate(400)                              # limit animation rate
       ball.pos.x += vx*h                        # update position
11     ball.pos.y += vy*h
       path.append(pos=ball.pos, retain=300)     # draw path, keep 300 pts
13     if ball.pos.y > floor.y + ball.radius:
           vy = vy - g*h                         # above floor, update vy
15     else: vx, vy = - vx, - vy                 # below floor, reverse vel.
```

The program is very similar to Program 2.1 in both structure and flow. Note that the position attributes of objects such as the sphere include now both x and y components in 2D motion. We have added the path of the ball made with `points`, which is continually appended to in the loop (lines 6 and 12). The `retain=n` attribute keeps only the last n points, useful for ensuring smooth animation where a large number of points are generated. The position and the vertical velocity v_y are updated according to Eq. (3.1) by first-order Euler's method. When the ball would fall below the floor, its velocity vector is reversed, so the ball moves back and forth perpetually. The path traces out the familiar parabolic shape, with front-back symmetry. Experiment with different parameters such as g, and initial v_x, v_y, and so on. What happens if only v_y is reversed on collision with the floor?

3.2 MODELING AIR RESISTANCE

Anyone riding a bicycle can feel the rush of oncoming air and the resistive force that grows with increasing speed. As a fact of life, air resistance is easy to understand but hard to quantify because the underlying physics involves the interactions of a very large number of particles, and there is no exact form for that force. Nonetheless, we observe that, among other things, the force is dependent on the velocity of relative motion $\vec{v}$. In particular, it is zero if $\vec{v} = 0$. This leads us to describe the resistive force phenomenologically by a power series as

$$\vec{F}_d = -b_1\vec{v} - b_2 v\vec{v} + \cdots \tag{3.3}$$

Here $\vec{F}_d$ is the force due to air resistance (or drag), $\vec{v}$ is the velocity of the projectile relative to the medium (air), and $v = |\vec{v}|$ is the speed. The constants b_1 and b_2 are called linear and quadratic coefficients. The negative sign in front of b_1 and b_2 indicates that the drag force is opposite the direction of velocity.

Linear Drag

Physically, the first term in Eq. (3.3) comes from viscosity of the medium. The linear coefficient b_1 can be calculated from Stokes' law applicable to laminar (smooth) flow. We can understand qualitatively how b_1 comes about from Newtonian physics.

Let us consider a fluid between two large parallel plates shown in Figure 3.2. The top plate is moving with velocity v to the right, and the bottom one is stationary. We assume v is small, so the flow is smooth and separates into thin layers from top to bottom. The topmost layer moves at speed v, and the bottom one at rest because they "stick" to the top and bottom plates, respectively. In between, we expect the fluid velocity to decrease uniformly, depicted as shrinking arrows in the figure.

To keep the top plate moving, a force to the right has to be applied on it to counter the drag from the layer below. Newton's conjecture is that the force should be proportional to the velocity v, the area of contact A between the plate and the fluid, and inversely proportional to the plate separation d, that is, $F_1 \propto Av/d$. Proportionality to v and A can be understood as follows: a larger v means a higher rate of momentum change for the fluid,[1] and a larger A means a greater amount of fluid being pulled, so both factors lead to a greater force. As to the inverse proportionality to d, we imagine trying to push a piece of Jell-O from the top, and the taller the Jell-O, the easier to deform (less shear force).

Let the proportionality constant be μ, then

$$F_1 = -\mu A \frac{v}{d}. \quad \text{(plates)} \tag{3.4}$$

The coefficient μ is the viscosity. Its unit is kg/s/m, or Pa·s. The negative sign is added because the force is exerted on the plate by the fluid (Newton's third law).

For a sphere moving in a fluid, Eq. (3.4) does not apply because the velocity gradient (v/d) around it is not constant. Nevertheless, we expect Eq. (3.4) to give at least a rough estimate. Replacing $A = 4\pi r^2$ (surface of sphere) and $d = 2r$, with r being the radius of the sphere, we have $F_1 = -2\pi\mu r v$. The exact expression is

$$F_1 = -6\pi\mu r v. \quad \text{(sphere)} \tag{3.5}$$

This is Stokes' law. Our estimate gives the correct dependence on the relevant parameters, but the constant factor is wrong (2 instead of 6), which is to be expected given the crudeness of our model.

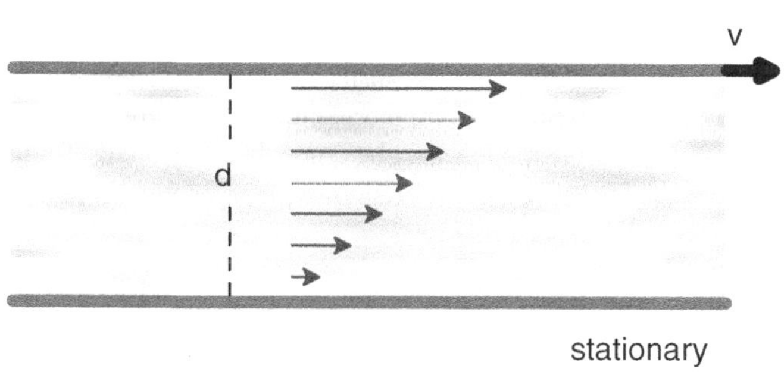

FIGURE 3.2 Viscous fluid flow between two moving plates.

[1] We should not confuse the velocity dependence with kinetic friction, which is independent of velocity. For kinetic friction, the contact surface does not move with the body.

If the velocity v is very small, linear drag is the dominant force. But as it turns out, the speed where the linear term dominates is usually quite low (see Exercise E3.3). Therefore, for all practical purposes, the second-order term is the important one.

Quadratic Drag

The second term in Eq. (3.3) arises when the flow is turbulent, meaning that it is no longer laminar in nature. The dimensionless Reynolds number defined as

$$Re = \rho v L/\mu, \tag{3.6}$$

is the determining factor of whether the flow is laminar or turbulent.[2] Here, ρ is the fluid density, μ the fluid viscosity defined in Eq. (3.4), L the characteristic length of the object, and v the velocity. It is generally noted that the transition from laminar to turbulent flow for circular objects occurs at $Re \simeq 2300$, that is, the flow is laminar below 2300 and turbulent above it. For example, the transition occurs at about 0.5 m/s for a baseball, so the quadratic term plays the dominant role in practice.

We can obtain the quadratic coefficient from elementary considerations. Let us assume an object (say a disc) with an effective cross-sectional area A. As it moves through air with speed v in time interval Δt, it will displace all the air mass in its path (Figure 3.3). The amount of air displaced is equal to the cylindrical volume $Av\Delta t$ swept by the disc, multiplied by air density ρ, or $\Delta m = \rho A v \Delta t$. If, on average, the air mass Δm is accelerated to the speed v, the net gain of impulse is $\Delta p = \Delta m v = \rho A v^2 \Delta t$. The force necessary to exert this impulse is given by Newton's second law, $F = \Delta p/\Delta t = \rho A v^2$. By the third law, the object (disc) experiences an equal magnitude but opposite force, $-\rho A v^2$. By this estimate, the coefficient b_2 in Eq. (3.3) is identified to be $b_2 \approx \rho A$.

To be more precise, we let $b_2 = C_d \rho A/2$ and express the second term in Eq. (3.3) as

$$\vec{F}_2 = -b_2 v \vec{v} = -\frac{1}{2} C_d \rho A v \vec{v}. \tag{3.7}$$

Now C_d is dimensionless and is called the (quadratic) drag coefficient. To the extent that the linear drag force is negligible, then $F_2 \simeq F_d$, and F_2 is usually referred to as the drag force. It is the dominant force from air resistance. This expression (3.7) is by no means fully accurate, but contains the essential physics.

Model Building, Realism, and Complexity

Numerically, the nominal drag coefficient for spherical shapes is $C_d \sim 0.5$. According to our earlier estimation, it should be about 2. The difference reflects the crudeness of our model, assuming uniform acceleration of displaced air mass, flat surface, and so on. Here a word is in order about model building in computer modeling. A computer simulation is only as accurate as the ingredients contained in the model. Simple models may be crude, but they are useful for bringing out the

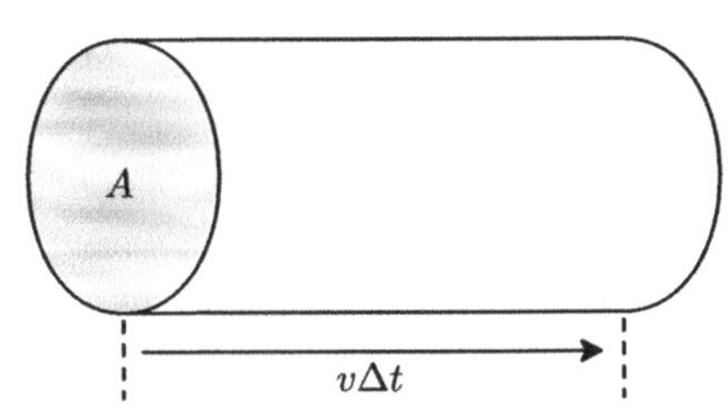

FIGURE 3.3 An object of cross-section area A moving at speed v. In time interval Δt, the air mass in its path is displaced and accelerated.

[2]Laminar and turbulent flows can be observed in a column of rising smoke, say from a burning incense. At first, the column is smooth as silk and is laminar. As the column rises, it spreads and starts to develop twirls and twists and becomes turbulent.

essential physical features for a better understanding of the problem. Complex models, though more complete and accurate, may give results that are difficult to interpret. One can increase the accuracy by increasing sophistication of the model. There is clearly a trade-off. In practice, it is usually a good idea keep the model as simple as possible within required realism.

For instance, the drag force of the form (3.7) with constant C_d is adequate for most projectile problems. We could certainly fine-tune the model, for example, by including some velocity dependence in the coefficient C_d. In fact, experimental evidence shows that C_d is velocity dependent and not a constant over all the range of v. This dependence is displayed in Figure 3.4.

Figure 3.4 presents a rather interesting feature: a sharp drop in C_d at the critical speed ~34 m/s for baseball. A similar drop occurs for smooth balls at a larger speed. Before and after the sudden drop, the drag coefficient remains relatively flat. The difference in C_d between smooth and rough balls is very interesting and is mainly due to the boundary layer surrounding the ball. A rough ball (stitches on the baseball) creates a more turbulent layer than a smooth ball, so the onset of the drop occurs earlier. It is thought that at the critical speed, the turbulence produces a smaller wake, thus less air resistance and the precipitous drop in C_d.

It is convenient to fit the baseball data in Figure 3.4 to a simple form for use in simulations. We choose the following empirical formula

$$C_d = a + \frac{b}{1+\exp(\chi)} - c \times \begin{cases} \exp(-\chi^2), & \chi < 0, \\ \exp(-\chi^2/4), & \chi > 0, \end{cases} \tag{3.8}$$

$$\chi = (v - v_c)/4, \; v_c = 34, \; a = 0.36, \; b = 0.14, \; c = 0.27.$$

Here, speed v is in m/s, and v_c denotes the critical speed. The second term in Eq. (3.8) mainly causes the steep drop near v_c, and the third term, an asymmetric Gaussian, yields the steep rise after v_c.

From the preceding discussions on the velocity dependence of the drag coefficient, we see that the power series (3.3) is effectively of higher order than quadratic. In other words, an adequate representation of air resistance should include orders v^3, v^4, and so on. Then, how meaningful is it to speak of quadratic air resistance? Because the drag coefficient C_d is rather flat outside the critical speed region, we take the view that Eq. (3.7) captures the essential quadratic nature of the drag force. However, we should keep in mind that if the velocity-dependent coefficient (3.8) is used, Eq. (3.7) effectively takes into account all orders within the specified range shown in Figure 3.4.

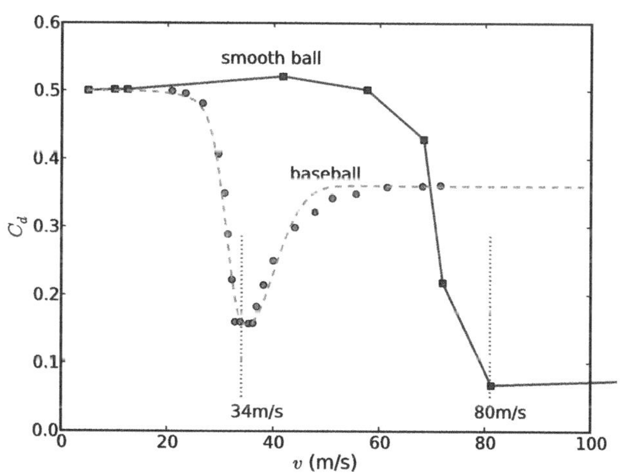

FIGURE 3.4 The drag coefficient as a function of velocity for smooth balls and baseball [32]. The data for baseball were derived from actual games at 1996 Atlanta Olympics [81]. The dashed line is a fit to the data.

3.3 LINEAR AIR RESISTANCE

Though for practical purposes quadratic air resistance dominates for all but extremely small speeds, linear air resistance is nonetheless very insightful from the viewpoint of modeling because it is simpler and, surprisingly, even admits analytic solutions in terms of special functions. It is an ideal, first case study to begin our exploration into simulations.[3]

In this section, we shall discuss linear air resistance only. Ignoring the quadratic term in Eq. (3.3), we take the drag force to be linear

$$\vec{F}_1 = -b_1 \vec{v}. \tag{3.9}$$

The Newtonian equations of motion with the net force of gravity plus linear drag $\vec{F} = -mg\hat{j} + \vec{F}_1$ are

$$\frac{dx}{dt} = v_x, \quad \frac{dy}{dt} = v_y,$$
$$\frac{dv_x}{dt} = -\frac{b_1}{m}v_x, \quad \frac{dv_y}{dt} = -g - \frac{b_1}{m}v_y, \tag{3.10}$$

where m is the mass of the projectile, and x, y, v_x, v_y are the same as in Eq. (3.1). These four first-order ODEs (3.10) are slightly modified versions from Eq. (3.1).

3.3.1 NUMERICAL SIMULATION

The equations of motion (3.10) are tailor made for a numerical integrator like Euler's or Runge-Kutta methods discussed in Chapter 2. Similar to Program 2.5, we only need to supply a subroutine `diffeq`, which returns the RHS of Eq. (3.10). So here it is, named `Lin_drag`:

Program listing 3.2 Linear drag (`projectile.py`)

```python
import ode, numpy as np             # ODE solvers, numpy
import matplotlib.pyplot as plt     # plot functions

def Lin_drag(Y, t):             # projectile motion with linear drag
    # returns RHS of Eq. (3.10) : [vx, vy, ax, ay]
    return np.array([Y[2], Y[3], -b1*Y[2]/m, -g - b1*Y[3]/m])

def go(vx=5., vy=5.):                   # default velocity =(5,5)
    Y = [0., 0., vx, vy]                # initial values, [x,y,vx,vy]
    t, h, x, y = 0., .01, [], []        # time, step size, temp arrays
    while Y[1] >= 0.0:                  # loop as long as y>=0
        x.append(Y[0]), y.append(Y[1])              # record pos.
        Y, t = ode.RK4(Lin_drag, Y, t, h), t+h     # update Y, t

    plt.figure()                        # open fig, plot, label, show fig
    plt.plot(x, y), plt.xlabel('x (m)'), plt.ylabel('y (m)')
    plt.show()

g, b1, m = 9.8, 0.2, 0.5                # g, linear coeff., mass
go(vx=10.0, vy=10.0)
```

[3] To quote the 5th-century BC philosopher Laozi: The longest journey starts with a single step.

In Lin_drag, the input array Y[0..3] is assumed to contain x, y, v_x, v_y in order. It returns a NumPy array converted from a list to use vectorized RK4. The array just contains the respective derivatives on the RHS of Eq. (3.10). It also requires the parameters b1 and m, which are chosen (somewhat arbitrarily) at the end of the code. The main function go() initializes the variables, calls RK4 in the main loop until the y-position becomes negative and plots the results (similar to Figure 3.5). Note the front-back asymmetry in the ascending and descending parts of the trajectory and the sharp falloff toward the end. This is a signature of air resistance breaking the symmetry in ideal projectile motion.

3.3.2 ANALYTIC SOLUTIONS

Analytic solutions can be obtained in nearly the same straightforward manner as for numerical solutions. To make things a bit tidy, we redefine the linear drag coefficient by including the mass m of the projectile as

$$b = b_1/m. \tag{3.11}$$

We begin with the x component of the velocity of Eq. (3.10), $dv_x/dt = -bv_x$, whose solution is

$$v_x = v_{0x} \exp(-bt). \tag{3.12}$$

It shows that v_x declines exponentially from v_{0x}, the initial x velocity, when air resistance is included. Integrating v_x to find x gives

$$x = \frac{v_{0x}}{b}[1 - \exp(-bt)], \tag{3.13}$$

where a starting position at the origin is assumed.

For the y component of the velocity, $dv_y/dt = -bv_y - g$, and the solutions are

$$v_y = (v_{0y} + g/b) \exp(-bt) - g/b, \tag{3.14}$$

$$y = \frac{v_{0y} + g/b}{b}[1 - \exp(-bt)] - gt/b. \tag{3.15}$$

Note that the terminal y velocity as $t \to \infty$ is $-g/b$.

Time t may be eliminated from x and y to get the trajectory as

$$y = \frac{x}{v_{0x}}\left(v_{0y} + \frac{g}{b}\right) + \frac{g}{b^2} \ln\left(1 - \frac{bx}{v_{0x}}\right). \tag{3.16}$$

Several sample trajectories are shown in Figure 3.5. Unlike ideal projectile motion, the trajectories with air resistance are no longer front-back symmetric. Note the steep drop-off after the peak at larger angles. It also shows that the range increases with decreasing angle. In addition, the maximum range is not at 45°. It is around 20° in this case (see Exercise E3.9).

Explicitly, the range, R, is defined as the x coordinate where $y = 0$, so

$$\frac{R}{v_{0x}}\left(v_{0y} + \frac{g}{b}\right) + \frac{g}{b^2} \ln\left(1 - \frac{bR}{v_{0x}}\right) = 0. \tag{3.17}$$

The range R is a root to Eq. (3.17), a transcendental equation. There are no known solutions in terms of elementary functions, though analytic solutions do exist, albeit in terms of a special function (see Section 3.4). To find the root without resorting to special functions, we need a numerical root finder. We take up this subject next.

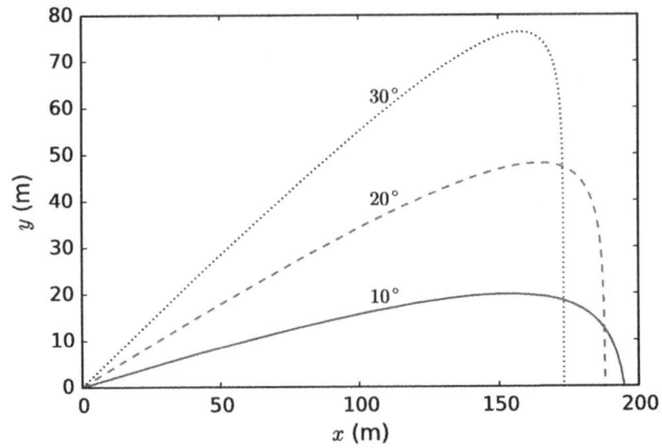

FIGURE 3.5 The trajectories for projectile motion with linear air resistance at a large initial speed ($b = 1\ s^{-1}$, $v_0 = 200$ m/s) for several firing angles.

3.3.3 ROOT FINDERS

In scientific computing, we frequently encounter situations just described earlier where we need to find the roots of equations. Root finding is an important operation that we will be using in several chapters ahead. The basic task of root finding may be stated as follows: Given a function $f(x)$, find a root x^* such that $f(x^*) = 0$. Equation (3.17) is such an example with $f(R)$ equal to the LHS of the equation. We shall limit our discussion to two common, practical methods.[4]

Bisection Method

The bisection method is a simple and sure way of finding a root, provided we know in advance that there is at least one root within an interval (bracket). It relies on the fact that the function $f(x)$ changes sign when passing through a root. The basic idea is shown graphically in Figure 3.6.

Suppose a root is known to lie within an initial bracket $[a, b]$, that is to say, $f(a) \times f(b) < 0$. The bisection method works simply as 1-2-3 (see Figure 3.6): (1) cutting the bracket in half; (2) determining which new bracket contains the root; and (3) repeating steps 1–2 until the bracket is sufficiently small.

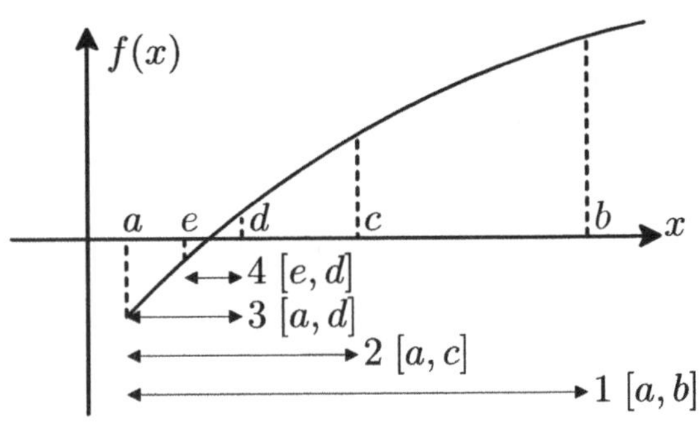

FIGURE 3.6 Bisection root finding method. Starting with an initial bracket $[a, b]$ in which a root lies, the interval containing the root is successively halved (iterations marked 1 to 4), until the required accuracy is reached.

---

[4]SciPy also has a general root solver, `scipy.optimize.fsolve`. See Exercise E3.8.

Program 3.6 implementing these steps is given in Appendix 3.A. The function has the form `bisect(f, a, b, eps)`. It requires as input a user-defined function `f(x)`, initial bracket `[a,b]`, and an optional error tolerance `eps`.

Newton's Method

Compared to the bisection method, Newton's method for root finding requires not only the function $f(x)$ itself, but its derivative $f'(x)$ as well. It can be a lot faster and more efficient when it works, but may also be unstable and less robust. Nonetheless, there are situations when this method is useful and appropriate, or even preferred.

The idea for Newton's method is illustrated in Figure 3.7. Starting with the initial guess x_1, Newton's method uses the local slope to find the x-intercept of the tangent, x_2 on the x-axis. Presumably, the point x_2 is closer to the actual root, and it is used to find the next approximation, x_3, and so on. Various stages are related by

$$\begin{aligned} x_2 &= x_1 - \frac{f(x_1)}{f'(x_1)}, \\ &\cdots \\ x_{n+1} &= x_n - \frac{f(x_n)}{f'(x_n)}. \end{aligned} \tag{3.18}$$

Unless the derivative f' is readily available, Newton's method is not suitable, especially if f' is expensive to evaluate.

The mathematical basis of Newton's method is easy to understand. Suppose we expand $f(x+\delta)$ with our trusted Taylor series,

$$f(x+\delta) = f(x) + f'(x)\delta + O(\delta^2). \tag{3.19}$$

If we wish to find the correction δ, which hopefully brings us closer to the root where $f(x+\delta) \sim 0$, then from Eq. (3.19) we have

$$0 = f(x) + f'(x)\delta, \quad \text{or } \delta = -\frac{f(x)}{f'(x)}. \tag{3.20}$$

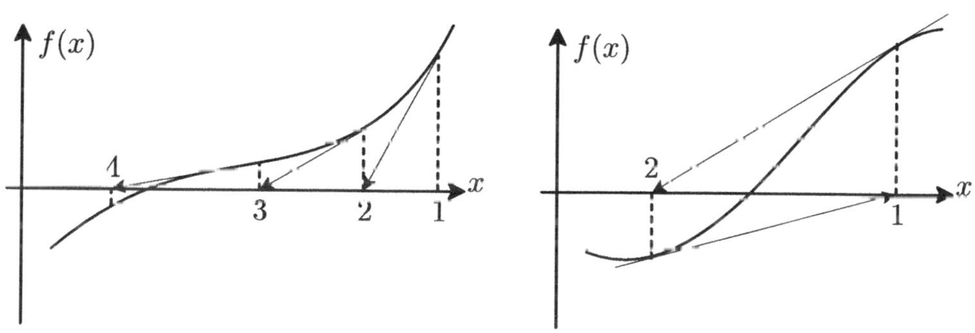

FIGURE 3.7 Newton's method of root finding. The left panel shows the successive steps of the process. In each step, the x-intercept of the tangent line is found, which is used as the next estimate for the root. The process is repeated until the change of the intercept is within allowed accuracy. The right panel shows the unstable nature of this method. In this case, the intercept cycles between two points endlessly.

This leads precisely to the correction terms in Eq. (3.18). The root solver, `newton(f, df, x, eps)`, is given in Program 3.7 (Section 3.A). We supply the function and its derivative `f(x)` and `df(x)`, the initial guess `x`, and the error tolerance `eps` as before.

Root Finders Library

As with the library for ODE solvers we develop, we will put both Programs 3.6 and 3.7 in a standalone library called `rootfinder.py` and import it before use. Practice the use of `bisect()` and `newton()` with the next example and Exercise E3.8. The bisection method is robust, that is, given a root known to exist within a bracket, it will find the root without fail. As such, the bisection method should be the first method to consider in a practical problem.

Numerical Solution of Range by Root-Finding

As an example, we solve Eq. (3.17) to obtain the dependence of range R on launching angle θ_0. We use the following program, which yields the results shown in Figure 3.8.

Program listing 3.3 Range vs. angle (`range.py`)

```
import rootfinder as rtf, math as ma # get root finders, math
import matplotlib.pyplot as plt # get matplotlib plot functions

def f(R):                          # range function, Eq. (3.17)
    return R*(vy+g/b)/vx + g*ma.log(1.-b*R/vx)/(b*b)

g, b, v0 = 9.8, 1.0, 100.          # g, linear coeff., firing speed
x, y, dtr = [], [], ma.pi/180.  # temp arrays, deg-to-rad conv

for theta in range(1,90):
    vx, vy = v0*ma.cos(theta*dtr), v0*ma.sin(theta*dtr) # init vel
    R = rtf.bisect(f, 0.01, vx/b-1.e-5, 1.e-6)        # solve
    if (R != None): x.append(theta), y.append(R)

plt.figure()                       # open fig, plot, label, show fig
plt.plot(x, y), plt.xlabel('angle (deg)'), plt.ylabel('R (m)')
plt.show()
```

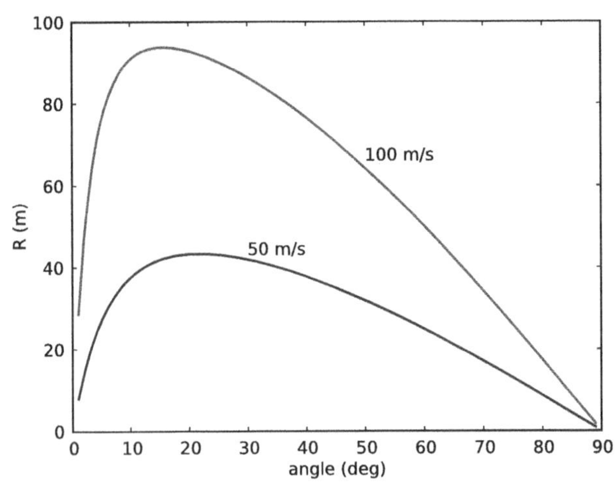

FIGURE 3.8 The range as a function of launching angle for projectile motion with linear air resistance ($b = 1\ s^{-1}$, $v_0 = 50$ and 100 m/s).

The function `f(R)` represents the range equation (3.17) to be solved. It uses several global variables defined in the main program, including v_x and v_y, components of the initial velocity that depend on the launching angle θ_0. The function is supplied to the bisection root finder (line 12). Though the root lies between zero and infinity, the lower end of the initial bracket should be set to just above zero because it is a trivial solution. The upper end should be a large number, but not so large that the argument to the logarithm term in Eq. (3.17) becomes negative. This yields a maximum of v_x/b for the upper end. Therefore, we set the bracket to $[\epsilon_1, v_x/b - \epsilon_2]$, $\epsilon_{1,2} > 0$, on line 12. The next line checks to make sure a valid root is found (not `None`) before storing the results.

The results (Figure 3.8) for two initial speeds show a steep rise from small angles to the maximum and a shallower falloff toward larger angles. The maxima occur at angles much earlier than 45°, and there is no symmetry as there would be for ideal projectile motion. Achieving the maximum range requires optimal balancing between the time of flight and the horizontal velocity. With air resistance, it is more important to have a larger horizontal velocity and a smaller time of flight because the former decreases exponentially (3.12), leading to a smaller angle θ_{max} that maximizes the range.

We note that θ_{max} decreases as the speed increases ($\sim 22°$ vs. 15°). Qualitatively, this can be understood as the balancing act discussed earlier. However, it begs an interesting question: How fast does θ_{max} decrease? How does it scale with the initial speed? We turn to a special function next for answers.

3.4 THE LAMBERT *W* FUNCTION

As mentioned earlier, Eq. (3.17) does not possess analytic solutions in terms of *elementary* functions, so numerical solutions have traditionally been obtained by means of root finding. However, it turns out that it does have analytically exact solutions in terms of a *special* function, the Lambert *W* function. Though discovered (or rediscovered) only recently, the Lambert *W* function has been linked to the solutions of a growing number of problems where only numerical solutions were thought possible previously. These problems include projectile motion with drag, Wien's displacement law in blackbody radiation, and the Dirac δ-molecule (see Section 9.3.2 and Ref. [90, 94] for more examples). Because it is important to compare numerical and analytic solutions where possible for testing numerical methods, we will briefly introduce this special function and its application to projectile motion with linear drag.

The Lambert *W* function is defined as a function satisfying the following inverse relation for a given z,

$$W(z)\exp(W(z)) = z. \tag{3.21}$$

This definition is similar to stating that the square root function f is defined such that $f(x) \times f(x) = x$ for a given value of $x \geq 0$. Some special values can be inferred from Eq. (3.21):

$$W(0) = 0,\ \ W(e) = 1,\ \ W(-e^{-1}) = -1. \tag{3.22}$$

The Lambert *W* function is generally complex-valued (see Ref. [20, 46] for general properties). For our purpose we are interested in the real-valued $W(x)$, shown in Figure 3.9. There are two branches, the principal branch W_0 and the secondary branch W_{-1}.

The power series expansion for W is

$$W_0(x) = x - x^2 + \frac{3}{2}x^3 - \frac{8}{3}x^4 + \cdots \tag{3.23}$$

Asymptotically, $W(x)$ behaves largely like the logarithm function,

$$W_{-1}(x \to 0^-) \simeq \ln(x/\ln|x|),\ \ W_0(x \to \infty) \simeq \ln(x/\ln x). \tag{3.24}$$

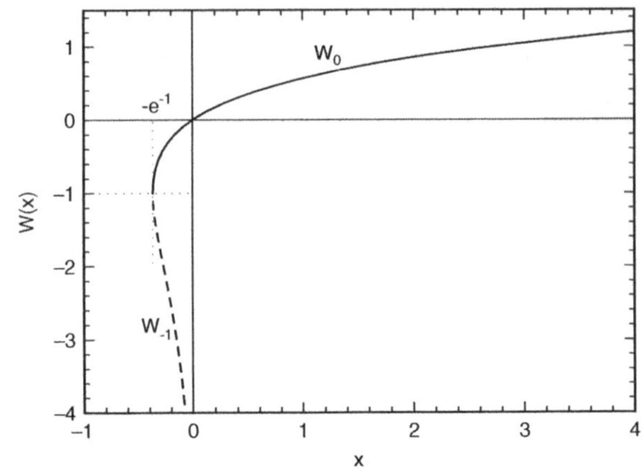

FIGURE 3.9 The Lambert W function for real argument x. The two real branches are the principal branch W_0 (solid line) and the secondary branch W_{-1} (dashed line). The branch point is at $x = -e^{-1}$ and $W = -1$.

Without the prefactor, Eq. (3.21) would be the inverse of an exponential function, that is, an exact logarithm.

Evaluation of *W*

Unlike elementary functions, the Lambert W function has yet to make its way on a handheld calculator with a dedicated button. Fortunately, it can be computed efficiently and accurately using the root finders we have discussed.[5] For a given x, the value of the W function is the root w to

$$f(w) = w\exp(w) - x, \tag{3.25}$$

according to Eq. (3.21). Whereas the bisection method would certainly work, Newton's method is preferred here for its rapid convergence. Making use of the derivative of $f(w)$ in Eq. (3.25) as

$$f'(w) = (1 + w)\exp(w), \tag{3.26}$$

Newton's method for the W works as follows (see Eq. (3.18)): Starting with an initial guess w_1 and assuming local stability, successive iterations w_n approach rapidly to the true value as

$$w_{n+1} = \frac{w_n^2 + x\exp(-w_n)}{1 + w_n}, \quad n = 1, 2, 3, \ldots \tag{3.27}$$

It only remains to determine the initial guess w_1. We choose w_1 to be

$$w_1 = \begin{cases} \left.\begin{array}{ll} x, & \text{if } -e^{-1} < x \le 1.5 \\ \ln(x), & \text{if } 1.5 < x < \infty \end{array}\right\} & \text{for } W_0, \\ \begin{array}{ll} \ln(-x), & \text{if } -e^{-1} < x < 0 \end{array} & \text{for } W_{-1}. \end{cases} \tag{3.28}$$

Equation (3.27) with the seed from (3.28) always converges to the correct branch W_0 or W_{-1}.

The actual implementation for the Lambert W function is given next.

[5]The Lambert W function is also available in the SciPy special function library, `lambertw`, and in SymPy as `LambertW`.

Program listing 3.4 Lambert *W* function (`lambertw.py`)

```
import math as ma                    # import math library
def lambertw(x, branch = 0):         # branch = 0 or -1
    if (x<-1.0/ma.e or branch<0 and x>=0.0): return None # bad arg
    w, nmax = x, 20
    if (x > 1.5 or branch == -1): w = ma.log(abs(x))

    for i in range(nmax):
        w, v = (w*w + x*ma.exp(-w))/(1.0+w), w # Eq. (3.27)
        if (abs(w-v) <= 1.e-15*abs(w)): break # to double precision
    return w
```

The module `lambertw()` accepts an argument x and an optional branch switch. If the argument is invalid, a type `None` is returned. Otherwise, it computes the value by Eq. (3.27) and checks for convergence in the loop. The convergence is fast, usually to machine accuracy in just a few iterations in nearly all cases.

There is one exception, however. It occurs near $x \sim -1/e$ for both branches, where $f(w) \sim -1$ and the derivative vanishes, $f'(w) \sim 0$. This causes slow convergence in Newton's method in the immediate neighborhood around $-1/e$. If accurate values are required in this narrow region, we may switch to the bisection method, which does not suffer from the vanishing derivative problem. Alternatively, we may also use the rational polynomial approximation (Section S:3.A). Otherwise, Program 3.4 works very well for computing the Lambert *W* function on the real branches, W_0 and W_{-1}. We will use it next.

Analytic Solution of Range by Lambert *W* Function

Until recently it was thought that the transcendental range equation (3.17) admitted only numerical solutions. Having introduced the Lambert *W* function, we are now ready to discuss how it can be used to solve Eq. (3.17) analytically.[6] To begin, Eq. (3.17) may be rewritten in exponential form as

$$\left[R - \frac{v_{0x}}{b}\right] \exp(cR) = -\frac{v_{0x}}{b}, \qquad (3.29)$$
$$c = b^2 v_{0y}/g v_{0x} + b/v_{0x}.$$

Our strategy is to cast the preceding equation into the same form as the defining equation for the Lambert *W* function (3.21), where the multiplicative factor in front of the exponential is the same as the exponent. This can be accomplished by multiplying both sides of (3.29) by $c\exp(-cv_{0x}/b)$ to give

$$\left[cR - \frac{cv_{0x}}{b}\right] \exp\left(cR - \frac{cv_{0x}}{b}\right) = -\frac{cv_{0x}}{b} \exp\left(-\frac{cv_{0x}}{b}\right). \qquad (3.30)$$

All R-dependence has been moved to the LHS of Eq. (3.30), and the prefactor is the same as the exponent. Comparing Eq. (3.30) with (3.21), we identify the following,

$$W(z) = cR - \frac{cv_{0x}}{b}, \quad z = -\frac{cv_{0x}}{b} \exp\left(-\frac{cv_{0x}}{b}\right). \qquad (3.31)$$

[6]It can also be solved more quickly (and opaquely) with SymPy, see Exercise E3.10.

Expressing R in terms of $W(z)$ and explicitly substituting in c, we have for the range

$$R = \frac{v_{0x}}{b} + \frac{W(z)}{c} = \frac{v_{0x}}{b}\left[1 + \frac{W(z)}{1 + bv_{0y}/g}\right], \tag{3.32}$$

$$z = -\left[1 + \frac{bv_{0y}}{g}\right] \exp[-(1 + bv_{0y}/g)].$$

This is the analytic solution for the range in terms of the special Lambert W function. Because the argument z in Eq. (3.32) is negative, and $W(z)$ is multivalued with two branches (Figure 3.9), how do we decide which one to use, W_0 or W_{-1}? Mathematically, both are correct solutions. Physically, the valid choice is the primary branch W_0 because W_{-1} would produce a negative range. It is understood that $W_0(z)$ must be used in Eq. (3.32).

With the use of the special W function, projectile motion with linear air resistance is now completely and analytically solvable. The analytic solutions can help us determine the accuracy of the numerical methods and to validate the program (see Project P3.2). Furthermore, it can also help us answer several interesting questions, including one posed earlier about the scaling law, namely: How does the range-maximizing angle, θ_{max}, scale with the initial speed? It turns out the scaling law is rather simple but powerful,

$$\theta_{max} \simeq 2 \ln \beta / \beta, \qquad \text{as } \beta = bv_0/g \gg 1. \tag{3.33}$$

It predicts that, for increasing speed v_0 (or equivalently drag b), there exists a small but finite angle that maximizes the range. Proof of the scaling law and its further exploration are left to Exercise E3.12.[7]

3.5 QUADRATIC AIR RESISTANCE AND SPIN

In the preceding sections we had a thorough airing-out of projectile motion with linear resistance. But in most cases, the quadratic term in Eq. (3.3) dominates. No known analytic solutions exist with this type of resistive force, except for one special case where we restrict the projectile to a straight up-down motion.

3.5.1 QUADRATIC AIR RESISTANCE

For the general case involving quadratic drag forces, we will have to rely on an ODE solver to get numerical solutions. The equations of motion similar to Eq. (3.10) but for quadratic drag forces are

$$\frac{dx}{dt} = v_x, \quad \frac{dy}{dt} = v_y,$$
$$\frac{dv_x}{dt} = -\frac{b_2}{m} v v_x, \quad \frac{dv_y}{dt} = -g - \frac{b_2}{m} v v_y, \tag{3.34}$$

where $v = \sqrt{v_x^2 + v_y^2}$ is the projectile speed. The appearance of the square root in Eq. (3.34) makes the problem unsolvable analytically.

Numerically, we need a `diffeq` that returns the RHS of Eq. (3.34) for quadratic drag. We call it `Quad_drag` given here.

[7]Numerical extraction of the scaling law is rather tricky because of the slow-varying logarithmic term $\ln \beta$ in Eq. (3.33) (see Project P3.2).

```
def Quad_drag(Y, t): # returns RHS of Eq. (3.34): [vx, vy, ax, ay]
  v = np.sqrt(Y[2]*Y[2] + Y[3]*Y[3])
  return np.array([Y[2], Y[3], -b2*v*Y[2]/m, -g - b2*v*Y[3]/m])
```

Being equivalent to `Lin_drag` in Program 3.2, `Quad_drag` assumes the array `Y[]` holding x, y, v_x, v_y in that order. We can plug `Quad_drag` in an ODE solver just like in Program 3.2 and obtain the desired results.

One complication with quadratic air resistance is that the drag coefficient C_d changes with speed and has a dramatic drop near critical speeds (Figure 3.4). The effect can be simulated with `Quad_drag`, and sample results are shown in Figure 3.10 for both constant C_d and changing C_d from Eq. (3.8).

With realistic drag coefficient used, we can see significant effects of air resistance. As expected, a constant $C_d = 0.5$ produces the shortest range. A changing C_d from Eq. (3.8) increases the range by about 20%, mainly because of reduced drag at the critical speed. Of course, both are significantly smaller than the ideal range without air resistance.

It is interesting to note that greater dynamical influence happens sooner than the closeness of the trajectories in Figure 3.10 seems to suggest because the points in the trajectories correspond to different moments in time. In fact, the greatest difference in acceleration, particularly the horizontal component, occurs at the beginning where the speed is largest. To a significant degree, this initial phase determines subsequent motion, including the difference in range.

Even with air resistance, projectile motion remains in a vertical plane. Sideway (transverse) motion can occur if airflow such as wind is introduced. In that case, the motion becomes three dimensional, and 3D simulation is necessary. We can extend the model because the drag force depends on the relative velocity, $\vec{v}_r$, between the projectile and air (wind). If air is flowing with a wind velocity $\vec{v}_{wind}$, the drag force becomes

$$\vec{F}_2 = -b_2 v_r \vec{v}_r, \quad \vec{v}_r = \vec{v} - \vec{v}_{wind}. \tag{3.35}$$

The full 3D equations of motion are (see effect in Project P3.8)

$$\frac{d\vec{r}}{dt} = \vec{v}, \quad \frac{d\vec{v}}{dt} = -\frac{b_2}{m} v_r(\vec{v} - \vec{v}_{wind}) - g\hat{j}, \quad v_r = |\vec{v} - \vec{v}_{wind}|. \tag{3.36}$$

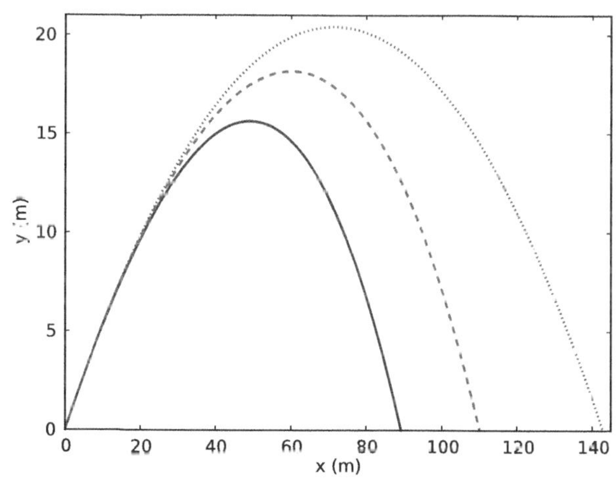

FIGURE 3.10 The trajectories for projectile motion (baseball) with quadratic air resistance with constant C_d (solid), changing C_d (dashed), and zero C_d (dotted).

3.5.2 EFFECTS OF SPIN

Transverse motion can also occur if airflow is deflected off the direction of motion. Many an example can be found in everyday life, including in sports such as baseball and Ping-Pong where the ball can move in curving, lateral directions.

Redirection of airflow can be accomplished by moving objects having uneven shapes or spin. Figure 3.11 shows a sketch of airflow around an airfoil and a spinning ball in a wind tunnel.

It is clear from the sketch that as air flows past the airfoil, it is bent downward. There must be a net downward force exerted on the airflow by the airfoil. By Newton's third law, the airfoil experiences an equal magnitude but opposite force from the airflow, that is, an upward lift.[8]

Similar effect exists for a spinning ball, as the wind-tunnel experiment shows. The ball is spinning counterclockwise, and the wind is blowing to the left. We can see a puff of smoke downstream which comes off the spinning ball and is deflected downward. The airflow over the top is "carried" around the ball more due to rotation. Again, by Newton's third law, this means a net downward force is acting on the air mass due to the spinning ball. The ball must experience a reciprocal upward force. This effect is known as the Magnus effect (force). It is schematically shown in Figure 3.12.

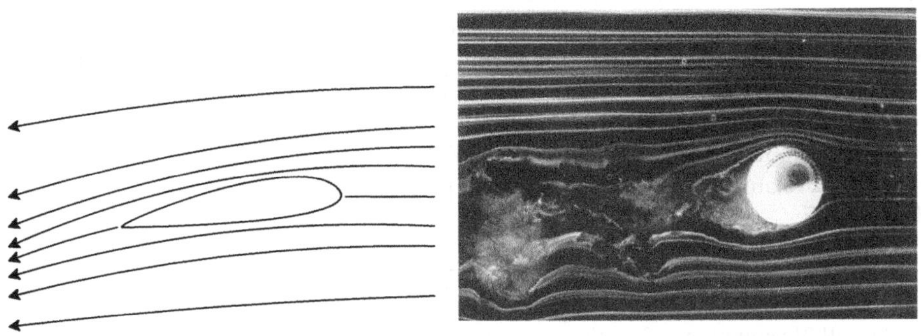

FIGURE 3.11 Schematic airflow around an airfoil (left) and measurement of a suspended ball spinning in a wind tunnel (right). The ball is spinning counterclockwise (Photo credit: F. N. M. Brown, Univ. of Notre Dame).

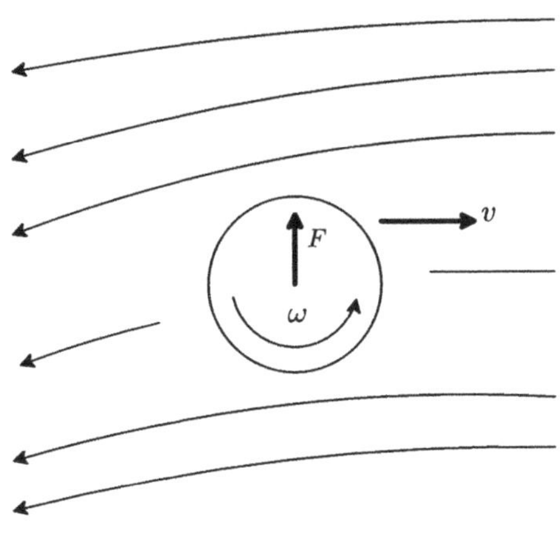

FIGURE 3.12 The Magnus effect on a spinning ball. Airflow past the spinning ball (counterclockwise) is bent downward. The direction of the force $\vec{F}$ on the ball is given by the vector product $\vec{\omega} \times \vec{v}$, upward here.

[8]Another common explanation invokes Bernoulli's principle. The flow velocity is greater on the top than on the bottom, resulting in a net upward pressure difference, hence the lift. But we think the third-law argument is more direct.

The magnitude of the Magnus force is found to be proportional to the angular velocity $\vec{\omega}$ as well as to the linear velocity $\vec{v}$ of the ball. Its direction is given by the cross product $\vec{\omega} \times \vec{v}$. The final result is

$$\vec{F}_{magnus} = \alpha\, \vec{\omega} \times \vec{v}, \tag{3.37}$$

where α is a constant with the dimensions of mass. Because the direction of the Magnus force is perpendicular to the velocity, the ball will always curve, but will not change speed by this force alone (no work). Thus, if a ball travels forward spinning around its vertical axis, the Magnus force will be lateral, inducing a sideways motion to either the left or the right.

Analogous to Eq. (3.7), the Magnus force for a sphere is often written as

$$\vec{F}_{magnus} = \frac{1}{2} C_L \rho A \frac{v}{\omega}\, \vec{\omega} \times \vec{v}, \tag{3.38}$$

where C_L is called the lift coefficient. The virtue of writing it this way is that when $\vec{\omega} \perp \vec{v}$, the magnitude is simply $F_{magnus} = C_L \rho A v^2/2$.

Undoubtedly, we expect C_L to depend on quite a few things such as the roughness of the ball. Given the roughness, C_L is experimentally found to be mainly a function of the so-called spin parameter, $S = r\omega/v$. Putting all together, we have

$$\vec{F}_{magnus} = \alpha\, \vec{\omega} \times \vec{v} = \frac{1}{2} C_L \frac{\rho A r}{S}\, \vec{\omega} \times \vec{v}, \tag{3.39}$$

$$\alpha = \frac{1}{2} C_L \frac{\rho A r}{S}, \quad S = \frac{r\omega}{v}.$$

Here, ρ and A are the air density and cross section, respectively (same as in (3.7)), and r is the radius of the ball. Effects of both drag and Magnus forces should be included in more realistic simulations of projectile motion and are considered next.

3.6 PHYSICS OF BALL SPORTS

Sports games such as baseball, Ping-Pong (table tennis), and soccer bring out the fans to share the excitement. To a physicist, the games are fascinating because they bring out the interesting effects due to drag and spin [22]. At times it seems the ball is doing nearly impossible tricks as it moves from point A to point B, be it a curveball, a back chop, or a corner kick. We have discussed the basic physics needed to understand the complexity in the ball movement. We are interested in the combined effects of drag and spin on a ball in flight.

3.6.1 PHYSICS OF BASEBALL

Baseball may be low in scoring (normally), but very high in physics content [2, 3]. We lead off with baseball in our discussion.

Lift Coefficient for a Baseball

Like the drag coefficient, accurate lift coefficient C_L in Eq. (3.39) needs to be determined experimentally. Figure 3.13 shows experimental data of C_L as a function of the spin parameter S for a baseball.

The data are an aggregate of several measurements using a variety of techniques in wind tunnels and open labs. The value of C_L increases with S. Within the data scatter, it is approximately linear in S, though there is noticeable curvature near small values of S. As we might expect from the parameterization of C_d (3.7), C_L must have a more complex dependence than on S alone.

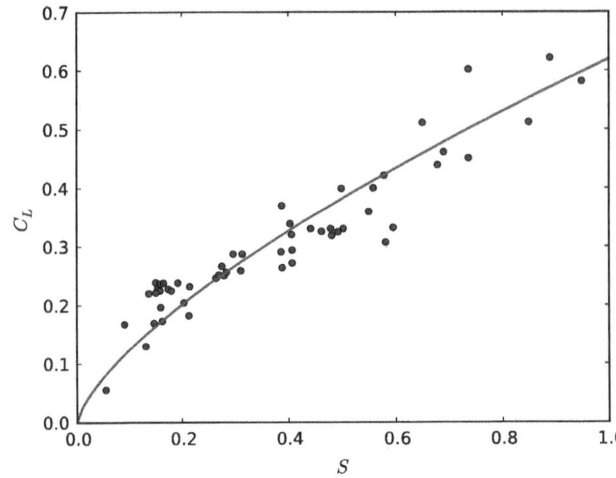

FIGURE 3.13 The lift coefficient as a function of spin parameter for a baseball. The line is a fit to the data [67] shown as symbols.

Experimental evidence shows that C_L is also dependent on the Reynolds number, although it is relatively weak. Nonetheless, it is useful to have a simple form similar to Eq. (3.8) for use in simulations. We choose a power law as

$$C_L = 0.62 \times S^{0.7}. \tag{3.40}$$

This relationship is displayed in Figure 3.13.

Forces on a Baseball

With the drag and lift coefficients determined, we can examine and compare quantitatively the forces acting on a baseball in flight. They are shown in Figure 3.14.

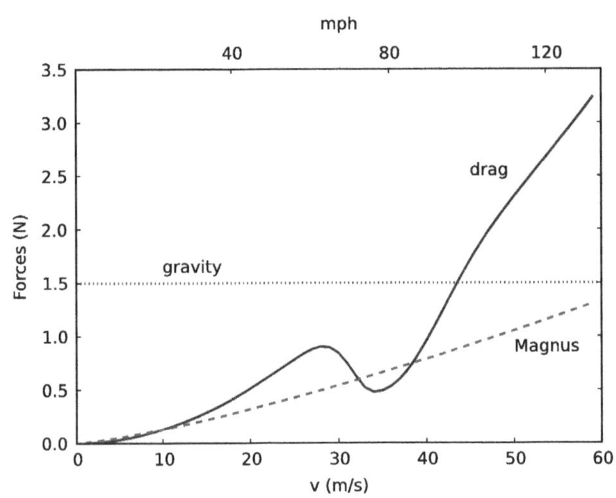

FIGURE 3.14 The forces acting on a baseball as a function of speed. The spin of the baseball is 200 rad/s (1900 rpm).

The quadratic drag force grows with v^2 until around 30 m/s, where it starts to decrease due to the drop-off in the drag coefficient near the critical speed (Figure 3.4). It rebounds afterward, is equal to the weight of the baseball (gravity) at about 43 m/s (97 mph), and exceeds gravity eventually. Therefore, quadratic drag has considerable effect on the path and range of a baseball, as Figure 3.10 depicts.

The Magnus force, on the other hand, grows monotonically as a power law, slightly faster than linearly in v. Of course, it would be exactly linear if the exponent in the scaling law (3.40) were 1, which is reasonably valid within the scatter of Figure 3.13 for larger spin parameters S. In the latter case, it is simpler to use Eq. (3.37) with a constant α for the Magnus force.

Because the Magnus force is perpendicular to the velocity, it always deflects the moving body, making it appear to "curve" and more perceptive to visual detection by the naked eye. This is especially true in sideways curves where gravity has zero effect.

The equations of motion, including gravity, drag $\vec{F}_2$, and the Magnus force $\vec{F}_{magnus}$, are

$$\frac{d\vec{r}}{dt} = \vec{v}, \quad \frac{d\vec{v}}{dt} = -g\hat{j} - \frac{b_2}{m}v\vec{v} + \frac{\alpha}{m}\vec{\omega} \times \vec{v}. \tag{3.41}$$

Here $\vec{r} = (x, y, z)$ is the position vector, with x referring to the forward motion, y the vertical motion, and z the lateral (sideways) motion. This is a set of 6 ODEs to be solved.

Before doing a full-scale simulation, it is useful to have an estimate of the effects. Let us estimate the magnitude of the Magnus force at speed $v = 30$ m/s (~70 mph) and angular velocity $\omega = 100$ rad/s (950 rpm). By scaling from Figure 3.14, we obtain a Magnus force that is roughly 1/6 of gravity, which is equivalent to an acceleration of $g/6$. If the ball is thrown from the pitcher's mound, it takes about 0.6 seconds for it to travel to home plate (a distance of 18.4 m). Over this period, the Magnus force causes a lateral displacement of 0.3 meters (nearly one foot), a substantial amount of adjustment for the hitter. For comparison, a ball would drop about 1.8 m in the same period by gravity.

Visualization of Realistic Flight of a Baseball

We solve Eq. (3.41) and display the motion with VPython in Program 3.8 listed in Section 3.B. The trajectory is shown in Figure 3.15.

The essential physics is contained in the function `baseball()`.

```
def baseball(Y, t):  # Y = [r, v] assumed
    v = Y[1]  # name v for clarity
    fm = alpha*vp.cross(omega, v) # Magnus force
    a = (fm - b2*vp.mag(v)*v)/mass - [0,g,0] # minus g-vec
    return np.array([v, a])  # np array
```

Because the motion becomes three dimensional when spin is involved, it is much easier to use vector objects. The program uses vectors as the basic building blocks instead of individual vector components like x, v_x, and so on. Vector operations can be simulated using NumPy arrays, which can be added or multiplied like true vectors (see Program 2.5 and Section 1.D). As stated in Chapter 2, our vectorized ODE solvers like RK4 works transparently on vector objects as on scalar objects without any change.

The input `Y` is assumed to be a 2×3 NumPy array containing the position and velocity vectors as $Y = [\vec{r}, \vec{v}]$. The output is the RHS of Eq. (3.41) in a NumPy array consisting of velocity and acceleration vectors $[\vec{v}, \vec{a}]$. We calculate the Magnus and drag forces using VPython's vector functions such as the magnitude `vp.mag()` and the cross product `vp.cross()`. This makes our

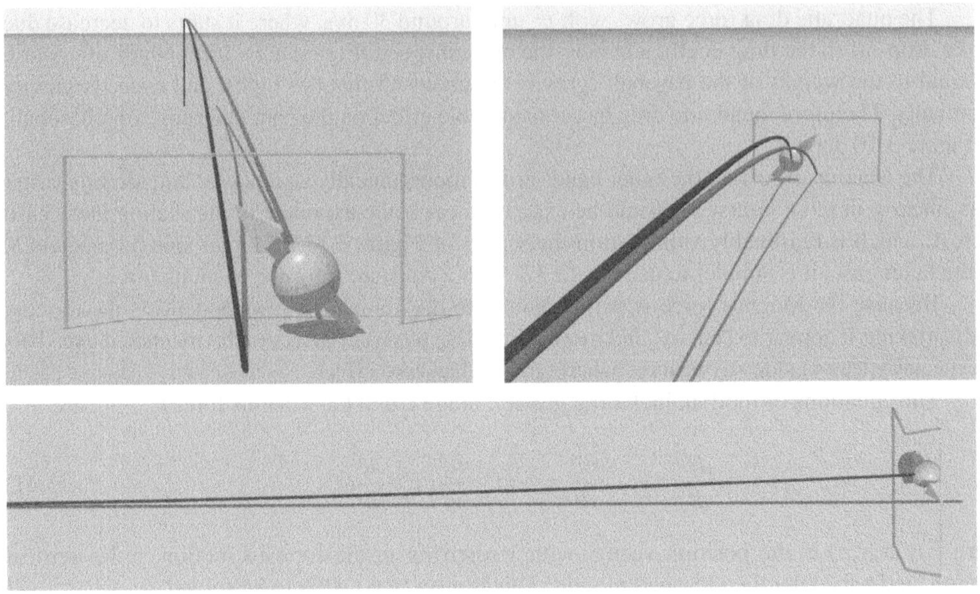

FIGURE 3.15 Views of a curveball from the front, behind, and above (bird's-eye). Also shown is the ideal projectile motion (darker/green, without the ball). The coordinates are such that from the bird's-eye, x-axis points to the right, y-axis out of page, and z-axis down.

life easier and the code more compact. Note that in the calculation of acceleration, the gravity vector $-g\hat{j}$ is simply represented by a list $-[0, g, 0]$, not a NumPy array. This is possible because when a list is added to a NumPy array, it is automatically broadcast to an array type before adding.

The actual trajectory can be seen to rise above and curve to the left relative to ideal projectile motion (Figure 3.15). This is due to the ball rotating with a back and side spin, that is, $\omega_z > 0$ (back) and $\omega_y > 0$ (side).

Quantitative results are shown in Figure 3.16. Both the lateral shift (z, curving) and the vertical rise (Δy) represent the difference between two cases: with or without spin (drag is included in both cases). The side spin—rotation about the vertical axis—is responsible for curving to the left (negative z), and the back spin for the rise (lift). The amount of curving and rise is nearly

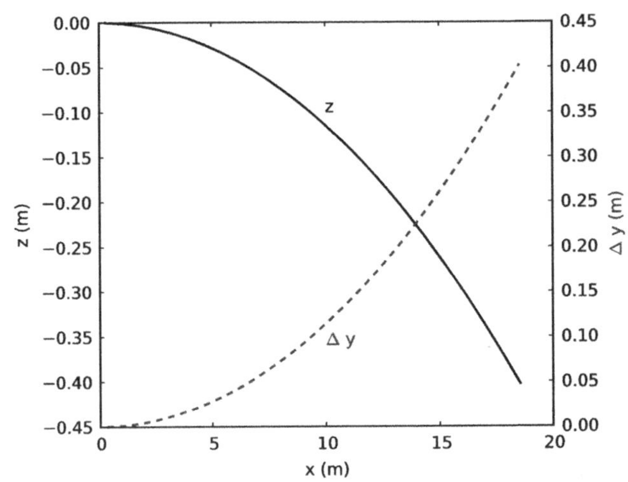

FIGURE 3.16 The lateral shift (z) and vertical rise (Δy) of a baseball with spin relative to no spin. The parameters are the same as in Program 3.8.

identical because the angular velocity components are the same. We note that the lift produced by the back spin actually increases the range of motion. This fact turns out to be very important in golf for long drives, as we will see shortly. Conversely, we could reverse it to a top spin (as if rolling in the forward direction). Then the ball would appear to sink faster.

The program uses constant drag and lift coefficients. It can be improved by using variable C_d and C_L. Then, the simulation could be considered fully realistic. Even then, we must remember that there are uncertainties and fluctuations in actual games, so our results should not be taken as perfectly accurate. However, the trend and conclusions we draw from these effects are valid.

A final caveat: In all likelihood, the ball's spin rate (ω) is not constant, and there is little data at present on how it precisely changes. If we assume the torque is proportional to angular velocity, we expect an exponential decay (see Eq. (3.12)) as

$$\omega = \omega_0 \exp(-t/\tau), \tag{3.42}$$

where ω_0 is the initial angular velocity and τ the characteristic decay time.

3.6.2 GOLF, PING-PONG, AND SOCCER

We briefly discuss key elements of physics involved in golf, Ping-Pong, and soccer games. We mainly present the results and leave the details to be investigated in projects.

Long Golf Drives

Golf balls have a roughened surface with small depressions, or dimples. Coupled with its smaller size (diameter 4.3 cm) and mass (46 g) and high spin rate (up to 10,000 rpm), drag and Magnus forces can have much larger effects on the range and direction of golf drives.

The drag coefficient for a golf ball is qualitatively similar to data for smooth balls shown in Figure 3.4. Quantitatively, however, the critical speed happens at a much smaller value, and the dip, present in baseball data, appears to be much smoothened [32]. As a result, we can approximate the drag coefficient by dropping the third term from Eq. (3.8) as

$$C_d = \frac{1}{4}\left(1 + \frac{1}{1 + \exp(2v - 12)}\right), \tag{3.43}$$

with v in m/s.

The lift coefficient is also qualitatively similar to that for baseball (see Figure 3.13), at least at higher Reynolds numbers ($Re \geq 10^5$), and can be roughly described similarly to Eq. (3.40)

$$C_L = 0.55 \times S^{0.4}. \quad \text{(golf)} \tag{3.44}$$

Sample results are shown in Figure 3.17 for a firing angle of 20° and speed of 60 m/s. For simplicity, we assume constant backspin only. The spin rates are none, normal (~8000 rpm), and double spin. Compared to the case of no spin, the range is increased by roughly one-third with normal spin, and nearly one-half with double spin. The increase in range is due to lift, which primarily increases the height and, hence, the time of flight. The change in the maximum height is even more dramatic, approximately doubling in each step. It shows the strong influence of the Magnus force that can be greater than gravity and drag combined. We can see this in the double spin case where the beginning part of the trajectory actually curves upward. Another notable feature is that the front-back asymmetry is less pronounced for golf than for baseball when spin is included. The back end of the trajectory falls off less steeply, due to the relatively strong Magnus force that "stretches" the peak out on both sides.

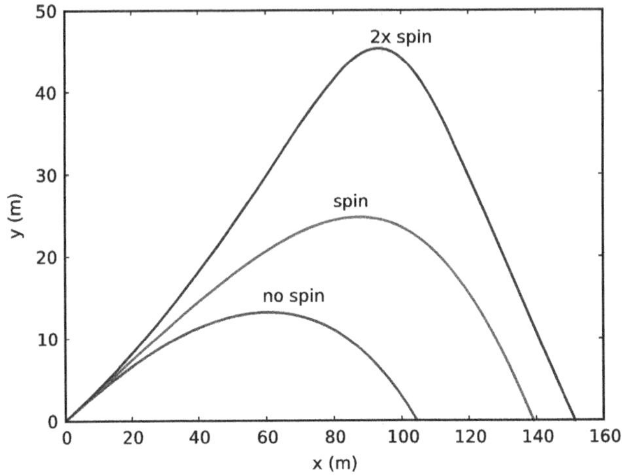

FIGURE 3.17 Trajectories of golf with different amounts of spin.

If the rotation is not purely backspin, curving can also occur. This may be used to counter such factors as wind or other obstacles. Of course, given how strongly spin affects the trajectory of a golf ball, controlling the flight by imparting a proper spin on the ball can be tricky.

Ping-Pong Smash

Ping-Pong games are played with a ball that has a 4.0-cm diameter and 2.7-g mass. Unlike other ball sports, Ping-Pong balls have a rather smooth surface. It is played over a table that is 2.72 m long and 1.53 m wide. In the middle of the table is a net, 15.3 cm high.

Ping-Pong is rather unique due to its small mass and the compact playing area. Of all ball sports, spin is most important in Ping-Pong games. Top spin rates can easily reach 6000 to 8000 rpm, and top speeds can be more than 40 m/s. These factors make Ping-Pong one of the fastest sports where quick reaction and accurate anticipation of ball movement is crucial [98].

Figure 3.18 shows part of a rally in a singles Ping-Pong game at the 2012 London Olympics. The plot shows the last two plays by each of the players (not shown). It starts with the player to the right hitting the ball from the upper-right corner and ends with the player to the left, who returns the ball with sufficiently large side-spin that the ball curves and bounces off the table near the edge, out of the opponent's reach. The right panel shows the last play, with a dotted line superimposed to illustrate the amount of curving.

Because Ping-Pong balls are smooth, we expect the drag coefficient shown in Figure 3.4 should work well. Furthermore, because the critical speed for smooth balls occurs at a rather large value relative to the speeds Ping-Pong is played at, we can safely set the drag coefficient to $\frac{1}{2}$. For the lift coefficient, available data suggests that a linear function of the spin parameter should be

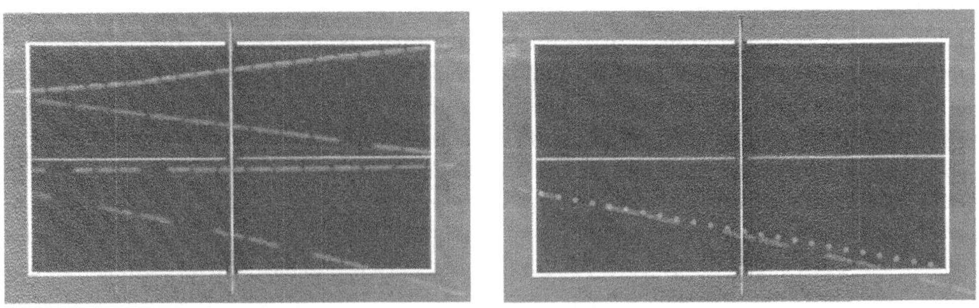

FIGURE 3.18 Overhead view of trajectories of a Ping-Pong ball in a game at the 2012 London Olympics. Left: The rally starts from upper-right corner and ends at the lower-right corner, when the ball bounces off the table. Right: The last return in the rally showing curved path; the dots are in a straight line to guide the eye.

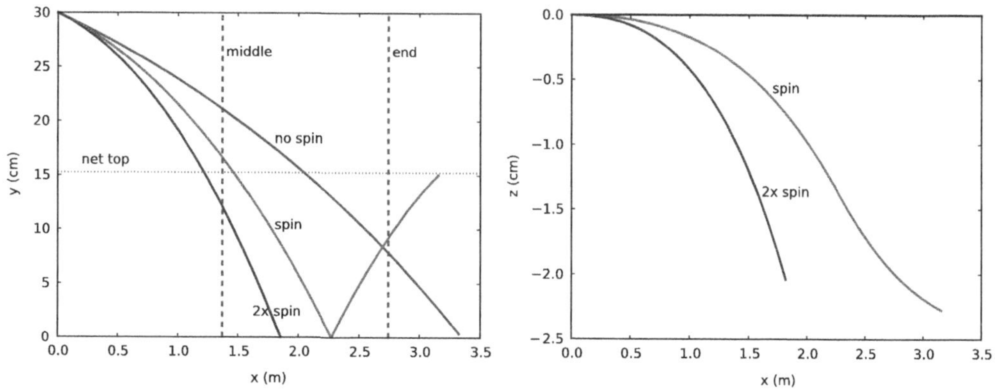

FIGURE 3.19 Trajectories of Ping-Pong with different amount of spin. Left: side view; Right: vertical view.

adequate [83]. Therefore, we can use the following in our Ping-Pong simulation,

$$C_d = \frac{1}{2}, \quad C_L = \frac{1}{2}S. \quad \text{(Ping-Pong)} \tag{3.45}$$

Representative results are shown in Figure 3.19. Parameters are used to mimic the conditions of an overhand smash: an initial speed of 25 m/s, and a spin of 60 rev/s (3600 rpm) in the direction $(1, 0, -1)$, that is, roll-spin ($\omega_x > 0$) and topspin ($\omega_z < 0$). The coordinates are such that the x-axis runs along the length of the table, the z-axis along the width, and the y-axis is vertical as usual. The left plot again shows three cases, no spin, normal spin, and double spin. Without spin, the ball flies past the far end of the table, and the player loses a point. But with normal topspin, the Magnus force is downward (negative lift). The ball sinks, just clears the net, and lands in play. This is a good smash. In this case, we also allow the ball to bounce back up with a loss of energy, that is, reversing $v_y' = 0.85v_y$. Too much spin, the ball accelerates downward too fast and is blocked by the net (but allowed to continue in the simulation).

In the overhand smash, we also assume a roll-spin in the positive x-direction, a clockwise rotation from the attacking player's perspective. This causes a slight curving motion to the left (Figure 3.19, right panel). The amount is small because in the cross product $\vec{\omega} \times \vec{v}$, the z component of the Magnus force is proportional to $\omega_x v_y$. The initial value of v_y is small, and $|v_y|$ increases due to gravity and the negative lift as the ball travels across. Hence we see increased curving later on. After the bounce (normal spin), the vertical velocity is reversed, and that slows down the rate of curving.

In comparison to baseball, the effect of spin is about 5% to 10% for Ping-Pong judging from Figure 3.19 (left), and 1% to 2% for baseball from Figure 3.16. Spin plays a more important role in Ping-Pong games.

Soccer Corners

Last but not least, we turn our attention to the most popular sport the world over, soccer. The soccer ball is considerably bigger and heavier than the cases discussed so far, but the effects of drag and spin are no less important [97]. Casual spectators can easily spot the curving of the ball in long passes. Curving is such an important art in soccer that set plays such as direct kicks or corner kicks are major scoring opportunities. In direct kicks, for example, players form a human wall to block the direct path of the ball to the goal. However, a skilled player could curve the ball around the wall and score. Direct scoring is also possible, albeit more difficult, in corner kicks.

Typical soccer speeds are below 30 m/s, and spin rates are roughly under 160 rad/s (25 rev/s). The drag and lift coefficients for soccer are shown in Figure 3.20. The drag coefficient

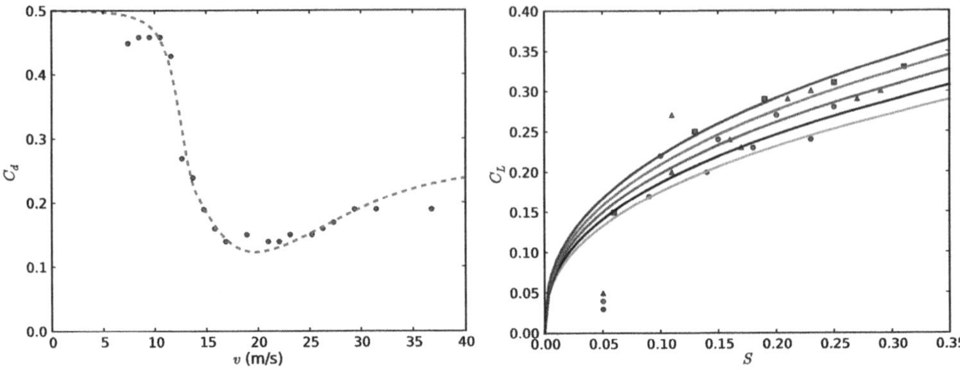

FIGURE 3.20 Drag and lift coefficients for soccer [5, 36]. The data for the lift coefficient correspond to different Reynolds numbers from 3.3 to 4.5×10^5.

again exhibits the critical speed phenomenon, but the dip is relatively shallow because of low roughness of soccer balls. We see weak dependence of the lift coefficient on the Reynolds number *Re*: A higher *Re* leads to a slightly higher C_L.

The parameterized expression similar to Eq. (3.8) for the drag coefficient can be written as

$$C_d = a + \frac{b}{1+\exp(\chi)} - c \times \begin{cases} \exp(-\chi^2), & \chi < 0, \\ \exp(-\chi^2/64), & \chi > 0, \end{cases} \quad \text{(soccer)} \tag{3.46}$$

$$\chi = (v - v_c)/2,\ v_c = 14,\ a = 0.25,\ b = 0.25,\ c = 0.16.$$

This relation is shown in Figure 3.20 as the dashed line. The average lift coefficient is modeled as

$$C_L = \frac{1}{2} S^{0.4}. \quad \text{(soccer)} \tag{3.47}$$

Except for the smallest spin parameter (S) values, we may assume C_L as a linear function in S, with appropriate adjustment of the constant factor. In this case, we can use the constant α defined in Eq. (3.39).

Using these coefficients, we have the necessary physics to simulate realistically the flight of soccer balls with a code similar to Program 3.8. Figure 3.21 displays a corner kick, which, absent a virtual goalkeeper, would result in a direct, curling goal.

3.7 SHOOTING METHODS

We have dealt with initial value problems, that is, given initial conditions, we integrate the equations of motion to find out where the trajectory ends up. There is another scenario in which we want the ball to end at some specified position, and we wish to know what initial conditions will lead it there. For instance, if we want to kick a ball so that it clears a bar (or hits a spot) at certain distance, what kicking angle should we aim initially? Problems of this kind are called two-point boundary value problems.

One way to solve such a problem seems straightforward, trial and error: Pick a value, run the simulation, check the result, adjust, and repeat. As illustrated in Figure 3.22, we want to find the launching angle leading the projectile to the boundary point at (x_b, y_b). The first pass overshoots, the second one undershoots, and the third try is the correct angle. This is the basic idea of the shooting method.

Manual adjustment can work, but is inefficient if high accuracy is required. In that case, it is better to cast the problem into a form suitable for automatic adjustment. Next we discuss an example using root finders.

FIGURE 3.21 Trajectories of a soccer ball in a corner kick with (curving) and without spin, from forward (top) and reverse (bottom) angles.

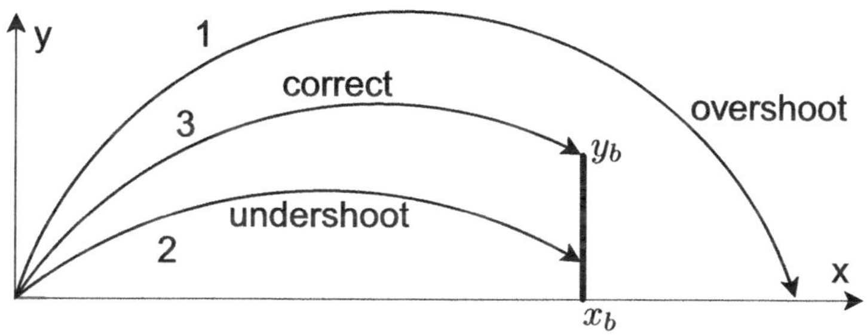

FIGURE 3.22 The shooting method. Trial 3 is the correct solution.

We assume a projectile is launched from the origin with a given speed v_0. Our goal is to find the firing angle θ such that the projectile hits the point at (x_b, y_b) (Figure 3.22). For a given θ, let the projectile reach x_b in time t, where its vertical position is $y(t)$. The condition for the correct θ satisfying the boundary values is

$$x(t) - x_b = 0, \quad y(t) - y_b = 0. \tag{3.48}$$

The first equation in (3.48) defines time t, which is an implicit function of θ. We rewrite Eq. (3.48) as a pair of functions

$$f_x(t) = x(t) - x_b, \tag{3.49a}$$

$$f_y(\theta) = y(t(\theta)) - y_b. \tag{3.49b}$$

We have treated θ explicitly as the independent variable in Eq. (3.49b). We can solve the two equations by finding the roots to $f_x(t) = 0$ for t first, then $f_y(\theta) = 0$ for θ, the desired solution.

The shooting method can be implemented as follows.

1. For a given θ, solve (3.49a) for t, by integrating the equation of motion such as (3.1) or (3.34) to obtain $x(t)$ and by root-finding as necessary.
2. Find $y(t)$ by integrating the same equations of motion to t. Return $y(t) - y_b$ as $f_y(\theta)$, (3.49b).
3. Feed $f_y(\theta)$ to a root finder to obtain the desired θ, using the bisection method to be safe.

This strategy is used in sample Program 3.5.

Program listing 3.5 Shooting method (`shoot.py`)

```
 1 from __future__ import print_function        # use print() as function
   import ode, rootfinder as rtf, numpy as np  # ode, root solvers, numpy
 3
   def proj(Y, t):                      # ideal projectile motion
 5     return np.array([Y[2],Y[3], 0.0,-g])   # [vx,vy,ax,ay]

 7 def fy(theta):                       # return f as a func of theta
       Y = [0., 0., v0*np.cos(theta), v0*np.sin(theta)]    # [x,y,vx,vy]
 9     t = xb / Y[2]                    # Step 1: time to xb
       h = t / nsteps
11     for i in range(nsteps):
           Y = ode.RK2(proj, Y, t, h)          # no need to update t
13
       return Y[1] - yb                        # Step 2: y(θ) − y_b
15
   # number of steps, g, init speed, x and y boundary values
17 nsteps, g, v0, xb, yb = 100, 9.8, 22., 40., 3.   # para.

19 theta = rtf.bisect(fy, 0.6, 1.2, 1.e-6)      # Step 3: shoot for θ
   if (theta != None): print('theta(deg)=',theta*180/np.pi) # result
```

To be concrete and avoid unnecessary clutter, we neglect air resistance, that is, assume ideal projectile motion. This simplifies the first step (line 9) because we can find the time simply as $t = x_b/v_{0x}$ without integrating the equations of motion or using a root finder.

In the function `fy(theta)`, $y(t)$ is calculated using the RK2 ODE solver. Though it is unnecessary because analytic solutions are available, we do it this way so it can be easily generalized to cases involving air resistance, for example. By the same token, the value of `nsteps` does not matter in the ideal case because RK2 yields exact solutions for ideal projectile motion regardless of the step size. After setting the initial speed and boundary point (line 17), the bisection method is called to find the solution. The key to be successful lies in choosing the initial bracket. There are usually two solutions if the initial speed is sufficiently large. In the present case, the solutions are 33.1° (ascending) and 61.2° (descending).

CHAPTER SUMMARY

The focus of this chapter is on one of the most classic and celebrated problems in physics, realistic projectile motion. We have discussed the effects of air resistance and spin. The drag force is broken down into linear and quadratic powers in speed. We modeled them separately. In the case of linear drag, we have presented a new method of solution in terms of the Lambert W function. We briefly described the key properties of W and several ways of evaluating it. With the help of the W function, the closed-form solutions enable us to compare numerical and analytic solutions and to gain more insight than possible with only numerical solutions, such as the asymptotic scaling law for the angle that maximizes the range.

For the quadratic drag that dominates in practice, we discussed the dependence of the drag coefficient on speed, models to parameterize it, and its effect on projectile motion. We also introduced the Magnus force and discussed its effect, as well as modeling of the lift coefficient. Finally, simulations combining drag and spin are applied to the study of physics of sports, including baseball, golf, Ping-Pong, and soccer. Both qualitative and quantitative results have been presented to help explain the observed effects.

Numerical methods for root finding have been discussed, including bisection and Newton's methods. They are included in one of the libraries that we will use frequently in the chapters ahead. We introduced the shooting method for two-point boundary value problems. We have also introduced visualization techniques using 3D capabilities of VPython and applied them to illustrate realistic projectile motion effectively and from unique perspectives.

3.8 EXERCISES AND PROJECTS

Exercises

E3.1 Modify Program 3.1 to use the RK2 ODE solver.

E3.2 Use dimensional analysis to show that the linear drag force must be of the form given by Eq. (3.5) without the explicit factor 6π.

E3.3 Calculate the speed at which the transition from laminar to turbulent flow occurs for a baseball. The viscosity and density of air are $\mu = 1.8 10^{-5}$ kg/m/s and $\rho = 1.2$ kg/m^3 under standard atmospheric conditions. The mass of a baseball is 0.15 kg, and the diameter is 7.4 cm

E3.4 Estimate the size of falling rain drops if their speed ranges typically from 10 to 30 km/h.

E3.5 Find the speed at which the linear and quadratic drag forces are equal for a baseball. Do the same for a Ping-Pong ball.

E3.6 Verify that Eqs. (3.13), (3.14), and (3.15) are solutions to (3.10). Also derive the range (3.16).

E3.7 In the limit $b \to 0$, show that Eqs. (3.16) and (3.17) reduce to the expressions for ideal projectile motion, with the former to

$$y = \tan\theta_0\, x - \frac{g}{2v_0^2\cos^2\theta_0}\, x^2, \tag{3.50}$$

and the latter to Eq. (3.2), respectively.

E3.8 (a) Use both the bisection and Newton's methods to find the roots of $f(x) = 5x - x^3$. Investigate what happens when the initial guess is at ± 1 in Newton's method.

(b) Alternatively, find the roots using SciPy's equation solver, `fsolve`, which requires a (vector) function and an initial guess, as shown here:

```
In [1]: from scipy.optimize import fsolve
In [2]: def f(x): return x*(5.-x*x)
In [3]: x = fsolve(f, -2.)
```

E3.9 Solve Eq. (3.17) to find R as a function of θ with both the bisection and Newton's methods. Take $b = 1$ s^{-1}, and do this for several initial speeds: 10, 20, 50, 100, 200, and 500 m/s. Note the position of the range-maximizing angle in each case. Compare with the scaling law (3.33). Be warned that it is challenging to calculate R by root-finding at large angles and large speeds.

E3.10 (a) Show that $W(x \ln x) = \ln x$. (b) Solve $x^x = a$, and express x in terms of W. Verify your answer for $a = 4$ and 27. Here you can also use SymPy's solve,

```
In [1]: from sympy import *
In [2]: x, a = symbols('x a')
In [3]: solve(x**x-a, x) # solve x^x - a = 0
Out[3]: [exp(LambertW(log(a)))]
```

(c)* Solve Eq. (3.29) with SymPy to obtain Eq. (3.32). (d) Verify the first two terms in the series expansion of W, Eq. (3.23). Note that $W' = W/[x(1+W)]$.

E3.11 Consider projectile motion with linear air resistance. (a) Find the maximum height. (b) Calculate the time it takes to reach the top, to fall back to the ground, and the total time of flight, respectively. (c) Give the numerical values for each if $b = 1\ \mathrm{s}^{-1}$, $v_0 = 50$ m/s, and $\theta_0 = 20°$.

E3.12 Prove the scaling law (3.33). Hint: Simplify the range (3.32) with the assumption $\beta \sin\theta \gg 1$ using (3.23), and solve $dR/d\theta = 0$ in terms of the Lambert W function. See Project P3.4 for a numerical analysis.

Projects

P3.1 (a) Integrate animation into Program 3.2 so that the projectile and its trail are displayed during flight as in Program 3.1. Place the camera to capture the scene from different angles, for example, forward ($+\hat{x}$), reverse ($-\hat{x}$), top ($-\hat{y}$), or other arbitrary angles. Also add a keyboard check to the main loop so the program can be paused or resumed with key presses.

(b) Modify the program and add two arrows to the projectile indicating the vector forces of gravity and drag, respectively. The lengths of the arrows should be proportional to the magnitudes of the forces. Scale them appropriately so they are clearly visible but not overwhelming the scene. Note that the drag force (both direction and magnitude) changes with time.

Refer to Program 3.8 on useful VPython techniques.

P3.2 Numerically investigate projectile motion with linear air resistance.

(a) Add necessary elements to Program 3.2 so the position-time and velocity-time curves (x-t, y-t, v_x-t, and v_y-t) are graphed. In each case, add the corresponding curve for the analytic results (Eqs. (3.12) to (3.15)) and for ideal projectile motion.

Do the numerical and analytic results agree? The curves may be too close to see any difference. Instead, plot the differences, $\Delta x = x_{num} - x_{ana}$ and $\Delta y = y_{num} - y_{ana}$ as a function of time. Comment on the error.

Now turn off drag ($b_1 = 0$) and run the program again. Do the results converge to the ideal case?

Note: We have assumed the step size h to be small compared to the characteristic time scale τ_s. Estimate τ_s in this case. How do we check whether h is adequate?

(b) Calculate and plot the range as a function of the launching angle like Figure 3.8 for the same parameters. This is best done automatically with a double loop: the outer one over the angle (say in 5-degree intervals), and the inner one over time. The range for a given angle is the last value of x before y turns negative. Your results should agree with Figure 3.8 to within the accuracy of the step size.

(c) On the same graph as part (b), plot the analytic range using the Lambert W function. To do so, you should write a function that accepts as input the launching speed and angle and returns the range given by Eq. (3.32). The function can access other parameters such as b and g as global variables. In the main code, add a loop iterating through the angles (for a given speed) and append the results to a list. How do the numerical and analytic results compare? Are the differences as expected for the step size h used? Halve h and repeat the comparison.

As an optional challenge, find the angle θ_{max} for which the range is maximum, from both the numerical and analytic results. Study the asymptotic behavior as the speed (or β in Eq. (3.33)) becomes large. Plot θ_{max} vs. $1/\beta$ for $1 \leq \beta \leq 10^3$. Compare with the prediction of the scaling law (3.33). Is it easy to extract the $\ln \beta$ dependence from the graph? Why? See also Project P3.4.

(d) Plot as a function of the launching angle the time of flight, τ, and the mean horizontal velocity, $\langle v_x \rangle$. Compute the latter as $\langle v_x \rangle = \sum v_x/N$ where the sum runs over the number of time steps N. Discuss your results. Plot $\langle v_x \rangle \times \tau$ as a function of the launching angle, and compare how they relate to the range calculated earlier.

(e)* When recording the range or time of flight, it is unlikely that the projectile would hit the ground exactly at $y = 0$ because of the finite step size h. More often than not, we will have the projectile above ground ($y > 0$) at t and below it ($y < 0$) at $t + h$. To pinpoint the moment more exactly without too small a step size, we can linearly interpolate between the two points just above and just below.

Let (x_1, y_1) and (x_2, y_2) be the positions of the projectile at t_1 and t_2, respectively. Connecting the points by a straight line, the equation of the line is

$$\frac{x - x_1}{x_2 - x_1} = \frac{y - y_1}{y_2 - y_1}. \quad \text{(linear interpolation)} \tag{3.51}$$

Hence, if the projectile crosses $y = 0$ between t_1 and t_2, we can approximate the range and time of flight as

$$R \simeq x_1 - \frac{x_2 - x_1}{y_2 - y_1} y_1, \quad \tau \simeq t_1 - \frac{t_2 - t_1}{y_2 - y_1} y_1. \tag{3.52}$$

Refine the calculation of R and τ accordingly. Change the main loop in the program to keep the positions of the last two iterations. Apply Eq. (3.52) after the loop. Compare your results with part (d). Optionally, do the comparison for a larger step size h and comment on your observations.

P3.3 Investigate one-dimensional free fall with quadratic air resistance.

(a) Consider only vertical (y) motion in Eq. (3.34) (i.e., set $x = 0, v_x = 0$). Calculate the terminal speed, assuming $b_2/m = 10^{-2}$ m^{-1}.

(b) Predict how position, velocity, and acceleration will change with time if motion starts from rest. Sketch them. With initial conditions $y = 200$ m and $v_y = 0$, simulate the motion until it hits the ground ($y = 0$). Plot the y-t, v_y-t, and a_y-t curves. Discuss your results and predictions. At what time and position does the velocity reach 95% of terminal velocity?

(c) Animate the motion with VPython. Mark the position of the projectile at regular time intervals. Make sure to space out the marks so they are not too close (say every 1/4 seconds or so). What do you conclude from the trail of the position marks?

(d) We have used a constant b_2. However, air density is not constant, which means that b_2 is a function of y. Let us model the effect of changing air density. Assuming constant temperature, thermodynamics predicts that air density decreases exponentially with height,

$$\rho(y) = \rho_0 \exp(-y/y_0). \tag{3.53}$$

Here, ρ_0 is the density at $y = 0$, and $y_0 \simeq 9000$ m. This model is not terribly accurate, but not grossly inaccurate either.

Repeat parts (a) to (c), assuming $b_2/m = 10^{-3} \times \exp(-y/y_0)$ (m^{-1}). You will have to adjust the initial position so it reaches at least within 5% of terminal speed before hitting the ground. Analyze your results and observations.

P3.4 The project concerns numerical exploration of the scaling law (3.33).

(a) Differentiate Eq. (3.32) with respect to θ, set $dR/d\theta = 0$, and show that the angle maximizing the range satisfies

$$\sin^2\theta_{max} + \sin\theta_{max}\,(1 + W)/\beta + W = 0. \qquad (3.54)$$

Consult Exercise E3.10 on the chain derivative of W. Also see Exercise E3.12.

(b) Solve Eq. (3.54) for θ_{max} as a function of β from 10 to 1000 using a root finder. Space out the data points by increasing β by a constant factor, for example, doubling it each time. Plot your exact results along with the scaling law (3.33) on a semilog scale. Discuss your findings.

P3.5 Consider a batted baseball hit near the ground level at an angle $\theta = 32°$ above the horizontal with speed $v_0 = 150$ km/h. Assume air resistance in the form of Eq. (3.7).

(a) Is the use of quadratic air resistance justified? Support your answer by computing the Reynolds number for a typical size and speed of the baseball. Obtain a reasonable value for the constant b_2 (including units). The relevant parameters for a baseball is given in Exercise E3.3.

(b) Modify Program 3.8 to simulate the motion of a baseball. As input, the program should ask the user for v_0 and θ. Verify the program works correctly.

(c) Plot the x-t, y-t, and y-x curves calculated with the drag force and compare each with the respective curve in the ideal case. Find the range and maximum height of the baseball.

(d) Let us assume that when the baseball hits the ground, it rebounds with 60% of speed (coefficient of restitution), but keeps the same angle of "incidence." Plot the trajectory up to the third bounce. (Fun exploration: Let the bounce continue until it does not move appreciably. Find the maximum horizontal distance traveled.)

(e)* Track the work done by the drag force, and verify that the work-energy theorem is satisfied by testing the relation $W_{drag} \approx E - E_i$, where $W_{drag} = \int \vec{F}_{drag} \cdot d\vec{r} \approx \sum \vec{F}_{drag} \cdot \Delta\vec{r}$ is the work done up to time t, E the energy at t (kinetic and potential), and E_i the initial energy. Plot E, W_{drag}, and $E - W_{drag}$ as a function of time. This relation, another form of energy conservation, may be used to validate the program and the step size h.

P3.6 Let us continue the investigation of the batted ball with regard to the angle maximizing the range and a scaling law.

(a) With the same initial condition as in Project P3.5, vary the launching angle θ and locate the optimum θ_{max} to within 0.1° for which the range is maximum. Use linear interpolation (3.52) for improved accuracy. Compare with the theoretical maximum range without air resistance.

(b) How does θ_{max} scale with large v_0? First, make a qualitative prediction. Now, calculate θ_{max} for different values of v_0, say between 10 and 1000 m/s. Double the speed each time to space out the data points. Plot the results on a semilog scale for v_0. You should automate the process of finding θ_{max}. Keep θ_{max} to at least two significant digits. Do your prediction and the results agree? Why or why not?

(c)* Assume a power scaling law $\theta_{max} = cv_0^{\gamma}$. Visually fit it to the data, and extract γ. Comment on how the results compare with the results of Project P3.4 if available.

P3.7 Like other ball sports, the American football is also full of physics [33].[9] It is typically thrown with a speed of 80 to 90 km/h. Use the optimum launch angle from Project P3.6 if available, or else $\theta = 30°$.

---

[9] See the so-called Deflategate in which the New England Patriots team was alleged of deflating or underinflating the football to gain an unfair advantage in a playoff game, and the science behind it in The *New York Times* Sports, "Upon scientific review," January 30, 2015. The team won that year's Super Bowl. See also Section 11.4.4 on pressure, temperature and the ideal gas law.

(a) Given that the mass of the football is about 400 g and the radius at the center of the football is ~9 cm, estimate the b_2 value for quadratic drag (ignore the fact that the football is prolate, and assume a reasonable frontal cross section). Write a program, or use one you developed, to calculate the range of the football. Should we worry about any forward spin (zip)? Plot the trajectory and compare with no drag. Are they reasonable?

(b) Suppose the b_2 above is for standard atmospheric conditions. Let us assume that air density is inversely proportional to temperature T (kelvin), so that the temperature-dependent coefficient is $b_2(T) = b_2 T_0/T$. Assuming $T_0 = 300$ K, repeat the range calculation as a function of T from T_0 to $T_0 - 30$ K in steps of 2 K. Plot and discuss the results. What is the difference in range between T_0 and freezing (273 K)? What is the range reduction per 5 K (~9 F) drop in temperature?

P3.8 Study the effects of constant versus variable drag and lift coefficients.

(a) Calculate the trajectories depicted in Figure 3.10 of a baseball with constant $C_d = 0.5$ and variable C_d given by Eq. (3.8). See Exercise E3.3 for relevant parameters of a baseball. Use initial values $v_x = 30$ m/s and $v_y = 20$ m/s, for example.

(b) Add spin to the calculation. Let us give it a side-spin $\vec{\omega} = 200\hat{j}$ rad/s. First, assume a constant ratio $C_L/S = 1.0$ so α is a constant in Eq. (3.39). Second, use a variable C_L from (3.40). Compare the sideways motion curves (z-t), and discuss the difference. Note the net displacement in z direction at the end.

(c) Examine the equivalency between wind and spin. Assume cross wind in the $-\hat{z}$ direction. Use Eq. (3.36) in your simulation. Find the wind speed v_{wind} that gives the same net displacement as part (b). Convert it into km/h or mph, and comment on your answer. This may be done by trial and error, or by the shooting method. Try the latter for added challenge!

P3.9 Let us investigate the necessary parameters to hit a home run or a treetop clearing golf drive.

(a) If the baseball leaves the bat at 35° above the horizontal and with a backspin of 2000 rpm, what should be the initial speed to make a 420-foot home run? Use realistic drag and lift coefficients.

(b) Assume you want to hit a golf ball directly over a treetop that stands at 10 m high and 120 m away. With an initial speed of 60 m/s and a backspin of 8000 rpm, find the angle (at least two significant figures) so your shot just clears the treetop. Compare the cases of constant and changing drag and lift coefficients.

P3.10 Pick your favorite sport, and simulate an event or play that interests you, for example, a corner kick in soccer or an overhand smash in Ping-Pong. Show your results both quantitatively by graphs and visually via animation. Additionally, study the effect of changing spin rate during the flight according to Eq. (3.42). You will need to use a reasonable τ.

3.A BISECTION AND NEWTON'S ROOT FINDERS

In this section we implement the root-finding methods outlined in Section 3.3.3 and discuss their properties, including applicability and convergence.

Bisection Root Finder

The key step in the implementation of the bisection method is to determine which new bracket contains the root (Section 3.3.3). It may be accomplished by first finding the midpoint x_{mid} and then deciding which end point to replace. To determine which side the root is on, the function value $f(x_{mid})$ at $x_{mid} = (b + a)/2$ (the first being c in Figure 3.6) is compared to an end point, say $f(a)$. If they are of the same sign, then a is replaced by x_{mid}, and the new bracket for the root is $[x_{mid}, b]$. Otherwise, b is replaced, and the bracket is set to $[a, x_{mid}]$. This strategy may be expressed as

$$
\begin{aligned}
&\text{Let } x_{mid} = \frac{1}{2}(a+b), \\
&\text{If } f(a)f(x_{mid}) > 0 \quad \text{then new bracket is } [x_{mid}, b], \\
&\qquad\qquad\qquad\qquad\quad \text{else new bracket is } [a, x_{mid}].
\end{aligned} \tag{3.55}
$$

The code below implements the preceding strategy.

Program listing 3.6 Bisection root finder (`bisect.py`)

```python
1  def bisect(f, a, b, eps=1.e-6):  # user-defined f(x) and bracket [a,b]
2      fa, fb, gap = f(a), f(b), abs(b-a)  # end points and initial gap
3      if (fa*fb > 0.0):                   # no root in bracket
4          print('Bisection error: no root bracketed')
5          return None
6      elif fa == 0.0:   return a
7      elif fb == 0.0:   return b
8
9      while (True):
10         xmid = 0.5*(a+b)
11         fmid = f(xmid)
12         if (fa*fmid > 0.0):          # root in [xmid, b]
13             a, fa = xmid, fmid       # set a=xmid and save a function call
14         else: b=xmid                 # root in [a, xmid]
15         if (fmid == 0.0 or abs(b-a) < eps*gap): break # root found
16
17     return xmid
```

This code needs a user-supplied function `f(x)`, an initial bracket `[a, b]`, and an optional relative accuracy `eps` (default: 10^{-6}). It starts by checking easy cases first: If no root is bracketed, it returns a type `None`;[10] or if either end happens to be a root, it just returns as such. The main loop computes the midpoint and decides which end point to replace according to Eq. (3.55). If the midpoint satisfies $f(x) = 0$, or the relative accuracy is satisfied (line 15), a root is presumed found. The program exits the loop and returns the root.

The bisection method never fails, provided the initial bracket $[a, b]$ contains at least one root. Although for an arbitrary function it is usually not trivial to locate the initial bracket, it is relatively easy to do so based on physics because physical considerations should lead to a reasonable guess. For example, for the projectile motion problem (3.17), we know the range must be between $(0, R_0]$, where R_0 is given by Eq. (3.2), the range for ideal projectile motion.

The cost of the bisection method amounts to one function call per iteration. The rate of convergence is linear because iteration reduces the interval by half. Let δ be the bracket width after n iterations, then $\delta \sim |b-a|/2^n$, or $n \sim -\log_2 \epsilon$ with $\epsilon = \delta/|b-a|$ being the relative error. For example, if $\epsilon = 10^{-6}$, roughly 20 iterations are required. It is not the fastest, but fast enough for most cases. Coupled with its robustness, it should be the first root finding method to consider for a given problem.

---

[10] Having no root bracketed is not an uncommon error, so type `None` should be inspected by the calling program to make sure a legitimate root is found.

Newton's Root Finder

In implementing Newton's method, we need to be careful about potential pitfalls lurking behind it. Figure 3.7 illustrates one of those, where Newton's method happily jumps in a never-ending loop. Other pitfalls might occur such that the algorithm marches to $\pm\infty$ without bounds, unless safeguards are built into the implementation of the method. We implement Newton's method (3.18) in Program 3.7.

Program listing 3.7 Newton's method (`newton.py`)

```
1  def newton(f, df, x, eps=1.e-6):  # user-defined f(x),df/dx, init root
2      nmax, fx = 20, f(x)            # max number of iterations
3      if (fx == 0.0): return x
4  
5      for i in range(nmax):
6          delta = fx/df(x)
7          if (i == 0): gap = abs(delta)  # save initial gap
8          x = x - delta                  # 'improved' root
9          fx = f(x)                      # prep for next round
10         if (fx == 0.0 or abs(delta) < eps*gap): break    # root found
11 
12     return x
```

To use `newton()`, the user supplies both the function `f(x)` and its derivative `df(x)`, the initial guess `x`, and the optional relative error `eps`. The choice of the starting value can be critical; it could mean a successful run or a not-so-successful run into the abyss. We safeguard against the latter by setting a maximum number of iterations `nmax`. However, no check is made on whether the loop exited normally due to accuracy satisfied (line 10) or just reached the maximum iterations.

The rate of convergence of Newton's method turns out to be quadratic for well-behaved functions in the neighborhood of a root. This means that if the initial error is ϵ_0, the error after one iteration is $\sim\epsilon_0^2$. After n iterations, the error roughly scales as $\epsilon \sim \epsilon_0^{2^n}$, or $n \sim \log_2(\ln\epsilon/\ln\epsilon_0)$. For instance, if $\epsilon_0 = 10^{-2}$ and $\epsilon = 10^{-16}$, then $n \sim 3$, that is, each iteration roughly doubles the number of significant digits. It is very efficient and usually converges to machine accuracy in a few iterations. When we are certain that Newton's method is locally stable in the proximity of a root, it can be a powerful tool, and we should use it.

Program listing 3.8 Motion of a baseball (`baseball.py`)

```
import ode, visual as vp, numpy as np   # get ODE, VPython, numpy

def baseball(Y, t):                     # Y = [r, v] assumed
    v = Y[1]
    fm = alpha*vp.cross(omega, v)                    # Magnus force
    a = (fm - b2*vp.mag(v)*v)/mass - [0,g,0]   # minus g-vec
    return np.array([v, a])                          # np array

def set_scene(R):              # draw scene, ball, trails, spin, info box
    scene = vp.display(background=(.2,.5,1), forward=(-1,-.1,-.1),
                       center=(.5*R,1,0), ambient=.4, fullscreen=1)
    floor  = vp.box(pos=(R/2,0,0), length=1.1*R, height=.1, width=8,
                    color=vp.color.orange, opacity=0.7)  # transparent
```

```
14      zone = vp.curve(pos=[(R,0,1),(R,1,1),(R,1,-1),(R,0,-1)], radius=.02)
15      ball = vp.sphere(pos=(0,0,0), radius=.2, material=vp.materials.rough)
16      trail = vp.curve(pos=(0,0,0), radius=0.04)
17      ideal = vp.curve(pos=(0,0,0), radius=0.04, color=vp.color.green)
18      spin = vp.arrow(axis=omega,pos=(0,0,0),length=1) # omega dir
19      info = vp.label(pos=(1.1*R,2,-2),text='Any key=repeat')
20      return scene, ball, trail, ideal, spin
21
22  def go(x, y, vx, vy):          # motion with full drag and spin effects
23      h, t, Y = 0.01, 0., np.array([[x,y,0.], [vx,vy,0.]])   # initialize
24      while (Y[0,0]<R and Y[0,1]>0.2):  # before homeplate&above ground
25          vp.rate(40)
26          t, Y = t+h, ode.RK4(baseball, Y, t, h)  # integrate
27          ball.pos, spin.pos = Y[0], Y[0]-offset  # move ball, arrow
28          spin.rotate(angle=phi), ball.rotate(angle=phi,axis=omega) #spin
29          trail.append(pos=ball.pos)
30          ideal.append(pos=(x+vx*t, y+vy*t-0.5*g*t*t, 0.)) # ideal case
31      while (not scene.kb.keys):                 # check for key press
32          vp.rate(40)
33          spin.rotate(angle=phi), ball.rotate(angle=phi,axis=omega)
34      scene.kb.getkey()                          # clear key
35      trail.append(pos=(0,0,0), retain=0)        # reset trails
36      ideal.append(pos=(0,0,0), retain=0)
37
38  g, b2, alpha, mass = 9.8, .0013, 5e-5, .15   # parameters
39  R, omega = 18.4, 200.*np.array([0,1,1])       # range, angular velocity
40  phi, offset = np.pi/16., 0.4*omega/vp.mag(omega)
41
42  scene, ball, trail, ideal, spin = set_scene(R)
43  while (1):
44      go(x=0., y=2., vx=30., vy=0.)             # initially z=0, vz=0
```

As discussed in the main text, `baseball()` returns the equations of motion, RHS of Eq. (3.41), in a NumPy array in order to use a vectorized ODE solver. Subroutine `set_scene()` does what it says: It orients the camera to the scene containing various objects: the ground (floor), strike-zone, ball, trails (`vp.curve()`, one for the actual motion and another for ideal projectile motion), angular velocity arrow, and an informational text box, `vp.label()`. The floor is drawn as a transparent box via the `opacity` parameter (line 13), so that even if the camera is below the ground while being swung around, the baseball remains visible. The lower the opacity parameter, the higher the transparency.

The function `go()` takes input for initial conditions, and integrates the motion until the ball reaches the strike-zone or hits the ground. The array `Y` holds the position and velocity vectors as $[\vec{r}, \vec{v}]$, which are represented as NumPy arrays. At each step, the positions of the ball and of the arrow representing spin are updated. The rotation is effected via the `rotate` method of the object (line 28), which rotates a given angle counterclockwise about the object's axis. For objects that do not have an axis such as the sphere, we must specify the axis as a direction vector. After the numerical integration ends, the second loop keeps rotating the ball and the arrow until a key has been pressed (so we can admire the scenery, explore it from different angles, or take a snapshot). This is done by monitoring whether a key has been pressed (line 31). If so, `scene.kb.keys` will be set to True in VPython, and the loop exits. The function `scene.kb.getkey()` is called to clear the keyboard buffer (line 34). Finally, the trails are cleared by setting `retain=0` (line 35) so fresh ones can be drawn in the next go-around.

The main body of the program sets the necessary parameters. The angular velocity vector $\vec{\omega}$ is set (line 39) such that the ball has a back and side spin. The purpose of the variable `offset`, a vector along the direction of $\vec{\omega}$, is to center the spin arrow at the ball (line 27). Once the scene is set, `go()` is called repeatedly after each key press until the window is closed.

CHAPTER 4

Planetary Motion and Few-Body Problems

In projectile motion discussed in Chapter 3, gravity is considered to be constant. This is an approximation, valid only for distances small compared to the size of Earth. Now imagine we stood atop Mt. Everest and fired a cannon horizontally. At low speed, the cannonball would travel just like projectile motion. With greater firing power and ever-increasing speed, the cannonball would travel further and curve less. At some critical speed, it would fly all the way around Earth and come back from behind (ignoring drag, of course). This is exactly how the Moon moves around Earth, or the planets around the Sun. In a sense, planetary motion is akin to projectile motion in that a planet falls continuously toward the Sun. We can no longer treat gravity as a constant force, however.

The motion of planets, or the heavenly bodies as they were known, is a fascinating subject. Observations of planetary motion led to the discovery of Kepler's laws, which, strictly speaking, are valid only for a two-body system (the planet and the Sun). The fact that these empirical laws could be successfully explained by Newton's theory of gravity confirmed the validity and universality of the classical theory of gravity. However, beyond two-body systems, no analytic solutions are known for the general configuration, even though simple three-body systems are still at the heart of physics at all scales, such as helium (two electrons plus the nucleus), water (one oxygen and two hydrogen atoms), or celestial systems. These systems, part of the so-called few-body problems, prove to be a fertile ground for numerical investigation.

We discuss several interesting cases in celestial mechanics ranging from differential precession of Mercury including correction from Einstein's theory of relativity, to exoplanets and three-body problems.

4.1 MOTION OF A PLANET

Planetary motion is by and large governed by Newton's theory of gravity. The force F obeys the inverse-square law

$$F(r) = -\frac{k}{r^2} \text{ (force)}, \quad V(r) = -\frac{k}{r} \text{ (potential)}, \tag{4.1}$$

where r is the radial distance, and the constant k is

$$k = GMm, \; G = 6.67 \times 10^{-11} \text{ Nm}^2/\text{kg}^2. \tag{4.2}$$

Here, M and m are the masses.

From Newton's second law $\vec{a} = \vec{F}/m$, the equations of motion are

$$\frac{d\vec{r}}{dt} = \vec{v}, \quad \frac{d\vec{v}}{dt} = -GM\frac{\vec{r}}{r^3}. \tag{4.3}$$

Note that the mass of the planet, m, is canceled out, and the force components have $1/r^3$ dependence because of the $\vec{r}$ vector in the numerator. Assuming the motion is confined in the xy plane, the vectors have only x and y components.

To see what a typical motion is like, run the following program.

Program listing 4.1 Planetary motion (`earth.py`)

```
import ode, numpy as np  # get ODE solvers, numpy
import visual as vp       # get VPython modules for animation

def earth(id, r, v, t):             # return the eqns of motion
    if (id == 0): return v          # velocity, dr/dt
    s = vp.mag(r)                   # s = |r|
    return -GM*r/(s*s*s)            # accel dv/dt, faster than s**3

def go():
    r = np.array([1.017, 0.0])      # initial x,y position for earth
    v = np.array([0.0, 6.179])      # initial vx, vy

    # draw the scene, planet earth/path, sun/sunlight
    scene = vp.display(title='Planetary motion',        # scene start
                       background=(.2,.5,1), forward=(0,2,-1))
    planet= vp.sphere(pos=r, radius=0.1, make_trail=True,
                      material=vp.materials.earth, up=(0,0,1))
    sun   = vp.sphere(pos=(0,0), radius=0.2, color=vp.color.yellow,
                      material=vp.materials.emissive)
    sunlight = vp.local_light(pos=(0,0), color=vp.color.yellow) #scn end

    t, h = 0.0, 0.001
    while True:
        vp.rate(200)   # limit animation speed
        r, v = ode.leapfrog(earth, r, v, t, h)  # integrate
        planet.pos = r                          # move planet
        if (scene.kb.keys): scene.kb.getkey(), scene.kb.getkey() #pause

GM = 4*np.pi*np.pi          # G*Msun
go()
```

We should see planet Earth orbiting the Sun as shown in Figure 4.1.

Do not be too concerned if Program 4.1 seems to have increased complexity in a hurry, for we have yet to explain key elements such as the model or the units in it. By the time we finish Section 4.3, you should understand everything about this program. We put the program up front because it serves as a central problem around which most problems are built subsequently. We note, however, that the program follows the basic templates and structure of programs we have seen so far, especially Programs 2.7, 3.1, and 3.8. For instance, Program 3.1 first sets up the visualization scene, then goes into the "while" loop for computation. This is basically what the

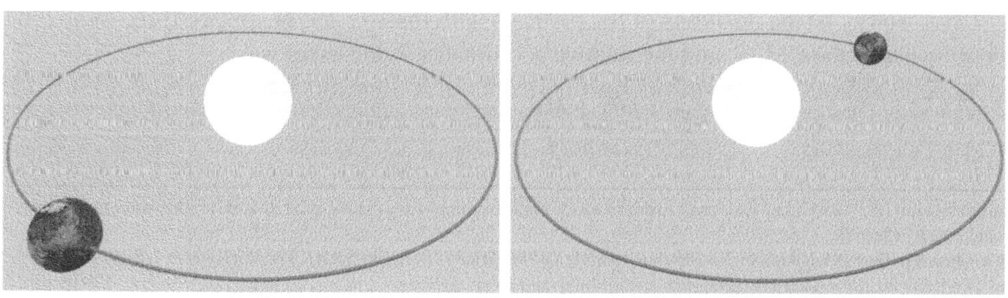

FIGURE 4.1 Motion of Earth around the Sun.

main function go() does here. The difference is that the former uses Euler's method within the loop, whereas the latter calls the leapfrog method separately. As stated earlier, most programs generally follow this template, with problem-specific differences.

In Program 4.1, the equations of motion (4.3) are computed by `earth()` in a form suitable to leapfrog integration (see Chapter 2, Section 2.4).[1] It returns either the velocity or the acceleration depending on the flag `id`, which is requested by the leapfrog integrator. The function `earth()` is similar to `oscillator()` in Program 2.7 but is two dimensional, so it uses two-element NumPy arrays to represent the position, velocity, and acceleration vectors (lines 10, 11, and 7, respectively). As a result, the acceleration is computed vectorially in identical expression as is written in Eq. (4.3). Besides simplifying the computation, it is also conceptually clearer with vector operations. The advantage of this approach, last used in Program 3.8, will be seen frequently in subsequent problems (see Programs 4.3, 4.5, and 4.7).

In the main code go(), the initial position and velocity vectors are given as two-component vectors (lines 10 and 11). These initial values represent the motion of Earth at the point farthest away from the Sun (aphelion, see Table 4.1). The next few lines (14 to 20) add visualization effects much like those in Program 3.1. In order, they set the display window and viewing direction, draw an Earth-like planet via `material.earth` attribute, put a glowing (emissive) Sun at origin, and add a light source representing sunlight (when attributes such as position are given as (x, y) pairs, the z value defaults to zero). Only the line for the "planet" is required, the rest are just bells and whistles, if you will. The `make_trail` attribute for the planet will leave a trail marking the path of VPython objects.

After setting the time and step size, the program enters the main loop, which is functionally similar to the loop in Program 2.7. Because this is a Hamiltonian system, the leapfrog integrator is called to advance the solution in time (line 25). The position of the planet is updated (line 26) after every step, leaving a trail automatically. The program pauses if a key has been pressed (line 27). It then calls `scene.kb.getkey()` twice, first to read the key and then to wait for the second key press before continuing.[2] Finally, we define a constant GM=$4\pi^2$, which sets the unit of mass (to be discussed shortly) before executing the main function. Earth should move around the Sun repeatedly over the same path. In other words, the orbit is closed.

I encourage you to play around with Program 4.1, for example, change the initial position or velocity slightly, turn off (commenting out) the sunlight, disable the trail, or add spin to the planet (see Program 3.8), and so on.

4.2 PROPERTIES OF PLANETARY MOTION

Because planetary motion falls under the broader scope of two-body central field problems, we will briefly review some helpful properties of motion with central forces for easy reference [40].

4.2.1 REDUCTION TO AN EFFECTIVE ONE-BODY PROBLEM

A general two-body problem can be reduced to an equivalent one-body problem in the center of mass (CM) coordinates. Figure 4.2 illustrates this.

Let $\vec{r}_1$ and $\vec{r}_2$ be the positions of the particles with mass m and M, respectively. We define the coordinates of the CM, $\vec{R}$, and the relative coordinates, $\vec{r}$, the usual way

$$\vec{R} = \frac{m\vec{r}_1 + M\vec{r}_2}{m+M}, \qquad \vec{r} = \vec{r}_1 - \vec{r}_2. \tag{4.4}$$

[1] In the calculation of s^3, instead of the power operator `s**3`, multiplications (`s*s*s`) are used. It is faster this way (up to power ~ 5), and equally readable.

[2] Such keyboard polling is included in the VPython module (VPM) library (Program S:6.4). The library also defines several classes of objects for use in later chapters, such as the slinky and mesh surfaces.

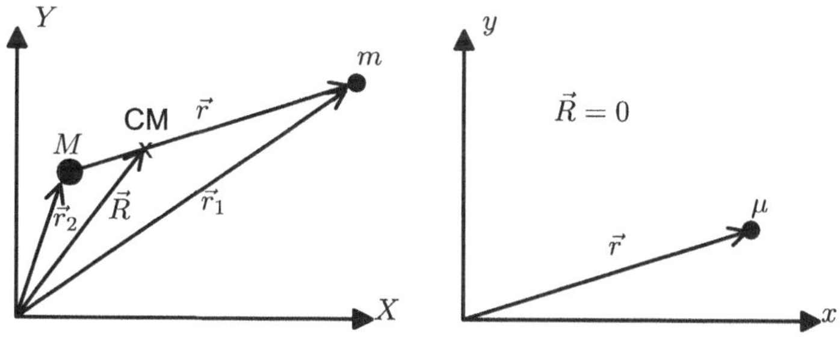

FIGURE 4.2 The reduction of a two-body problem to a one-body problem in the center of mass coordinates. Left: The space-fixed coordinate system; Right: The center of mass coordinate system. In either case, the relative coordinate $\vec{r} = \vec{r}_1 - \vec{r}_2$ is the same.

It can be shown (see Exercise E4.1) that in the CM coordinate system, the equations of motion may be written as

$$\ddot{\vec{R}} = 0, \quad \mu\ddot{\vec{r}} = F\hat{r}, \quad \text{with } \mu = \frac{mM}{m+M}, \tag{4.5}$$

where the force F is given by Eq. (4.1). The center of mass moves at constant velocity (can be set to zero for simplicity), and the relative motion is a one-body problem with the reduced mass, μ. If one mass (say M) is large compared to the other, that is, $M/m \gg 1$ as in the case of Sun/planet mass ratio, the reduced mass takes on the value of the smaller mass, $\mu = m$. In this case the bigger mass (Sun) can be regarded as infinitely heavy and frozen in space. Only the smaller mass (planet) moves. It is for this reason that the Earth–Sun system in Program 4.1 is effectively a one-body problem. Hereafter, we will use μ and m interchangeably except where confusion might occur.

4.2.2 ANGULAR MOMENTUM AND EFFECTIVE POTENTIAL

Let $\dot{r}$ and $\dot{\theta}$ be the radial and angular velocities, respectively. The angular momentum $\vec{L} = \vec{r} \times (m\vec{v})$ will be in the z direction as

$$\vec{L} = L\hat{k}, \quad L = mr(r\dot{\theta}) = mr^2\dot{\theta} = \text{ constant.} \tag{4.6}$$

Because the torque is zero for a central force, $\vec{L}$ is a constant of motion. This is the basis of Kepler's second law, which states that the radial vector sweeps an equal area in equal time.

The second constant of motion is energy, of course. From Eq. (4.5), we have

$$E = \frac{1}{2}mv^2 + V(r) = \frac{1}{2}m(\dot{r}^2 + r^2\dot{\theta}^2) + V(r) \tag{4.7}$$

We can eliminate $\dot{\theta}$ in E by substituting $\dot{\theta} = L/mr^2$ into Eq. (4.7) as

$$E = \frac{1}{2}m\dot{r}^2 + V_{eff}(r), \quad V_{eff}(r) = \frac{L^2}{2mr^2} + V(r) = \text{ Effective potential.} \tag{4.8}$$

The term $L^2/2mr^2$ is the centrifugal potential, and it keeps the particle from falling into the origin. The effective potential V_{eff} is the sum of the real potential plus the centrifugal potential. Equation (4.8) describes the motion as if it is one dimensional along the radial direction only in the potential V_{eff}. For bound motion in an attractive potential, the radius r usually oscillates between $r_{min} \leq r \leq r_{max}$. The values r_{min} and r_{max} are called turning points because at those points, the radial velocity $\dot{r} = 0$.

4.2.3 KEPLER ORBITS

Of course, the actual motion is two dimensional, say in radius r and angle θ. The orbit is most conveniently expressed as $r(\theta)$. For the inverse-square law (4.1), $r(\theta)$ is

$$r = \frac{L^2/mk}{1 + e\cos(\theta + \theta_0)}, \tag{4.9}$$

$$e = \sqrt{1 + \frac{2EL^2}{mk^2}}. \tag{4.10}$$

The constant θ_0 determines the orientation of the orbit and can be chosen for convenience. The other constant e is called the eccentricity. Geometrically, Eq. (4.9) is the equation of a conic section (slice of a cone) with the focus at the origin.

The shape of the orbit is determined by the value of e as follows:

$$\begin{array}{lll} E > 0, & e > 1; & \text{hyperbola (unbound)}, \\ E = 0, & e = 1; & \text{parabola (borderline)}, \\ E < 0, & e < 1; & \text{ellipse (bound, Kepler's first law)}, \\ E = -mk^2/2L^2, & e = 0; & \text{circle (bound).} \end{array} \tag{4.11}$$

The unbound, hyperbolic motion corresponds to scattering of particles and will be discussed later (Chapter 12). For bound, elliptic motion, it is useful to introduce a, the semimajor axis, and rewrite e and L^2/mk as

$$a = -\frac{k}{2E}, \Rightarrow e = \sqrt{1 - \frac{L^2}{mka}}, \Rightarrow \frac{L^2}{mk} = a(1 - e^2). \tag{4.12}$$

To be more definite, let us choose $\theta_0 = \pi$ in Eq. (4.9), and with the equalities (4.12), Eq. (4.9) becomes

$$r = \frac{a(1 - e^2)}{1 - e\cos\theta}. \tag{4.13}$$

With this choice of θ_0,[3] it follows that r is bounded by

$$r_{max} = a(1 + e)\ (\text{when } \theta = 0), \quad r_{min} = a(1 - e)\ (\text{when } \theta = \pi). \tag{4.14}$$

Figure 4.3 illustrates the parameters of an elliptic orbit. The particle moves around the center of force at the origin, which is also one focus of the ellipse. For planetary motion, the force center is at (very nearly) the Sun. The radial velocity ($\dot{r}$) is zero at the maximum distance (aphelion) and also at the minimum distance (perihelion) from the force center. The velocity at these two positions is perpendicular to the x-axis and may be obtained from energy conservation (4.7) and (4.12) as

$$\begin{aligned} v_{ap} &= \left[\frac{k}{ma}\frac{1-e}{(1+e)}\right]^{1/2} \quad (\text{at aphelion, } r = r_{max}), \\ v_{peri} &= \left[\frac{k}{ma}\frac{1+e}{(1-e)}\right]^{1/2} \quad (\text{at perihelion, } r = r_{min}). \end{aligned} \tag{4.15}$$

We list in Table 4.1 some properties of Kepler's orbits of the planets.[4]

---

[3] Another common choice is $\theta_0 = 0$. In that case, r_{min} and r_{max} would be swapped.
[4] Though Pluto was reclassified as a minor planet recently, it does not affect our treatment.

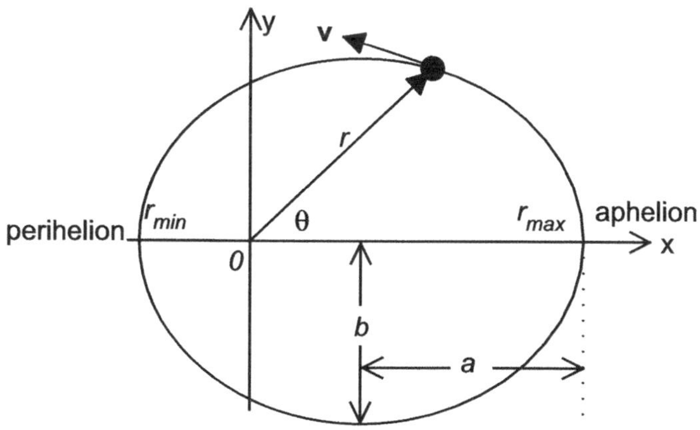

FIGURE 4.3 An elliptic orbit ($e = 0.4$) of a particle moving about the origin O (focus) with semimajor and semiminor axes a and b $\left(b = a\sqrt{1-e^2}\right)$. The orbit is oriented such that $r = r_{max}$ at $\theta = 0$ (aphelion) and $r = r_{min}$ at $\theta = \pi$ (perihelion). Note $r_{max} + r_{min} = 2a$ by Eq. (4.14).

4.2.4 UNITS FOR PLANETARY MOTION

As in Table 4.1, we will use the astronomical units (AU) for distance, 1 AU=1.496 $\times 10^{11}$ m. This is convenient for planetary motion in our solar system. It is equal to the semimajor axis of Earth's orbit and represents the average distance between the Earth and the Sun. The natural unit of time is chosen to be one year (365.25 days), or 3.156 $\times 10^7$ s. This is the characteristic time scale for planet motion. The unit of speed is 1 AU/year = 4740 m/s.

We now have two of the three basic units. The third one should be about mass. Often we need the ratio $k/m = GM$ with M being the mass of the Sun, as in Eq. (4.3). If we assume circular motion such as the Earth around the Sun, then

$$\mu \frac{v^2}{r} = |F(r)| = \frac{GMm}{r^2}. \tag{4.16}$$

With a mean radius $r = 1$ AU, the period is 1 year, and the speed is $v = 2\pi$ AU/year (29.77 km/s). Using the reduced mass given by (4.2), we get

$$G(M+m) \simeq GM = v^2 r = 4\pi^2 \text{ AU}^3/\text{year}^2, \tag{4.17}$$

Table 4.1 Properties of Kepler's Orbits of the Planets.

Planet Name	e	a (AU)	Period (year)	r_{max} (AU)	v_{ap} (AU/year)	Mass ($M_\odot$)
Mercury	0.2056	0.3871	0.2408	0.4667	8.198	1.660×10^{-7}
Venus	0.0068	0.7233	0.6152	0.7282	7.338	2.448×10^{-6}
Earth	0.0167	1.0000	1.0000	1.0167	6.179	3.004×10^{-6}
Mars	0.0934	1.5237	1.8808	1.6660	4.635	3.227×10^{-7}
Jupiter	0.0484	5.2029	11.863	5.4547	2.624	9.551×10^{-4}
Saturn	0.0539	9.5367	29.447	10.051	1.928	2.859×10^{-4}
Uranus	0.0473	19.189	84.017	20.097	1.368	4.355×10^{-5}
Neptune	0.0086	30.070	164.79	30.329	1.136	5.178×10^{-5}
Pluto	0.2488	39.482	247.92	49.305	0.776	6.582×10^{-9}

where we have used the condition that $m/M \ll 1$. In effect, Eq. (4.17) defines the mass of the Sun $M_\odot$ as the unit of mass. We will use $GM = 4\pi^2$ for planetary motion in our units system, as in Program 4.1. In absolute units (rarely needed), $M_\odot = 1.99 \times 10^{30}$ kg, and $GM = 1.327 \times 10^{20}$ m^3/s^2.

Note that Eq. (4.17) is a special case of Kepler's third law

$$T^2 = \frac{4\pi^2}{G(m+M)} a^3 \simeq \frac{4\pi^2}{GM} a^3, \tag{4.18}$$

which states that the square of period T is proportional to the cube of semimajor axis a. For the purpose of setting the value of GM in our unit system, Eq. (4.18) is equivalent to Eq. (4.17).

Finally, in the unit system thus defined, everything becomes dimensionless. Distance and time are to be compared with 1, and the time step must satisfy $h \ll 1$ in numerical integration, as was the case in Program 4.1.

4.2.5 OPEN AND CLOSED ORBITS

We have seen from running Program 4.1 and Eq. (4.13) that Earth moves in a fixed orbit, retracing itself over the same path periodically (Figure 4.1). The orbit is said to be closed.

Closed orbits may seem ordinary at first, but they are extraordinary. They are the exception rather than the rule. Gravity is one of only two forces supporting them. To wit, slightly modify Program 4.1 by changing the force to $1/r^{2+\delta}$ with $\delta \sim 0.1$, set the initial values for Mercury (so the effect is clearer), run it, and we should see an open orbit as shown in Figure 4.4. The small $\delta = 0.1$ is such that the force deviates from the inverse-square law (4.1) only slightly. But significant differences are seen in the orbit after about two revolutions.

We can understand qualitatively the nature of open and closed orbits by considering the stability of circular motion.[5] Circular motion is possible for any attractive potential. One just has to adjust the energy such that Eq. (4.16) holds. But circular motion may be stable or unstable. If, by increasing the energy slightly from the condition (4.16), the orbit deviates only a small amount from the circular orbit, the motion is said to be stable (with a slightly deformed orbit). Because the full motion consists of radial $r(t)$ and angular $\theta(t)$ motion, we can think of a deformed orbit as composed of radial oscillations and angular sweeps, illustrated in Figure 4.5.

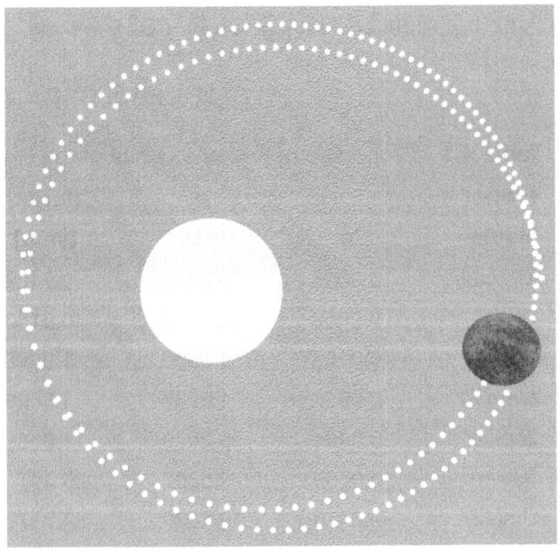

FIGURE 4.4 Open orbit of Mercury with a presumed force $\propto 1/r^{2+\delta}$, $\delta = 0.1$. The initial condition is $[x, y] = [0.4667, 0.0]$, $[v_x, v_y] = [0.0, 8.198]$.

[5]For readers interested in technical details, see Sec. 3.6 of Ref. [40].

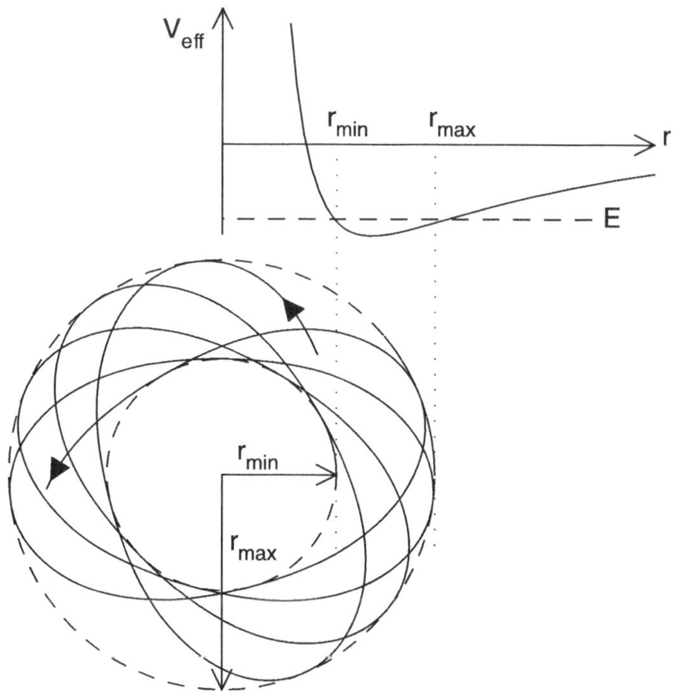

FIGURE 4.5 An open orbit for bound motion ($E < 0$) in an effective potential.

The radial distance oscillates between the turning points $r_{min} \leq r \leq r_{max}$, so the motion is restricted to the region between the concentric circles shown in Figure 4.5 (bottom). If the radius completes n_r integer oscillations and the angle θ makes n_θ integer sweeps (revolutions) in the same time interval, then the orbit is closed, since n_r/n_θ is rational (commensurate). As it turns out, of all the possible forces in the universe, only $1/r$ (gravitational and Coulombic) and r^2 (harmonic, i.e., Hooke's law) potentials can produce closed orbits.

Other forms of potentials do not support stable circular motion, and an integer pair of (n_r, n_θ) cannot be found, that is, n_r/n_θ is irrational. The orbit is open, as it does not repeat itself within a finite period of time. Furthermore, the orbit orientation rotates a bit with each revolution. This rotation is known as orbital precession, to be discussed next.

4.3 PRECESSION OF MERCURY

Astronomers have long known that the orbits of planets precess with time [25]. Because observations are carried out on Earth, which wobbles like a gyroscope (called precession of equinoxes, not our concern here), the data must be analyzed to separate the kinematic effects from true dynamic effects. This is equivalent to going to the fixed space frame where the Sun is at rest. Reduction of measurements of the innermost planet, Mercury, indicates a precession rate of 574 seconds of arc per century in the space frame (1 arcsecond, or $''$, is 1/3600°). In other words, the perihelion (or aphelion, Figure 4.3) of Mercury slowly rotates through an angle of 0.16 degrees per century.

4.3.1 RECONCILIATION OF 43$''$

From our earlier discussion, we understand that planetary orbits should not be exactly closed because they do not move in a pure $1/r^2$ force field with a single center. Besides the Sun, the other planets pull and tug each other all the time, not to mention their moons [72]. What planet would have the largest effect on the precession of Mercury? It is probably not a surprise that Venus does, because it is closest to Mercury, so its force is largest (see Table 4.2). Farther out radially, the effect decreases except for Jupiter, which has the second largest effect because of its mass. Beyond Saturn, the effects are negligible.

The problem is (or was), taking into account all the known perturbations present on Mercury due to these planets (within Newtonian theory of gravity), and after careful number crunching, the precession is expected to be about 531″ for Mercury per century (Table 4.2). That would leave 43″ unaccounted for, or 0.012 degrees/century. This is a very small amount, but well beyond observational uncertainties to be ignored.

As it turns out, the discrepancy can be explained only with Einstein's theory of general relativity. The relativistic correction beautifully accounts for the difference of 43″ exactly [16, 85]. It is one of the major triumphs of Einstein's theory of gravity.

In general relativity, the gravitational field is modified and can be construed as an effective force given by

$$F(r) = -\frac{GMm}{r^2}\left(1+\frac{\lambda}{r^2}\right), \quad (\lambda = 1.096\times 10^{-8}\ \mathrm{AU}^2 \text{ for Mercury}). \tag{4.19}$$

The details of the ingredients going into producing Eq. (4.19) are rather involved and beyond our scope [40]. Compared to Eq. (4.1), though, there is an additional $1/r^4$ term in the relativistically modified force. This term destroys the perfect inverse-square law. It causes the orbit to precess.

4.3.2 THE RUNGE-LENZ VECTOR

At first glance, it seems that we can calculate the precession of Mercury due to the relativistic effect by just tracking the rotation of the aphelion (Figure 4.5). Although this is possible, it is harder than it sounds. Because the parameter λ in Eq. (4.19) is very small, it is like searching for a needle in a haystack. If we were to use the actual value of λ directly in our simulation, we would have found that, for any moderately practical step size h, the change of the aphelion after one revolution would be insignificant compared to the displacement in one time step. Careful fitting and extrapolation would be needed to find the exact position of the aphelion. Is there a better way? In this case, there is: the Runge-Lenz vector [39, 47].

For observational purposes, of course, it might be natural to want to track the change of the aphelion (or perihelion) to discern any precession. This would be unnecessary in a simulation if we have another way to measure the precession. Fortunately, we do. There is a quantity known as the Runge-Lenz vector that serves our purpose well. The Runge-Lenz vector is defined as [40]

$$\vec{A} = \vec{p}\times\vec{L} - mk\hat{r}, \tag{4.20}$$

where $\vec{p} = m\vec{v}$ is the momentum, $\vec{L}$ the angular momentum as usual, m the mass and k the force constant given in Eq. (4.1).

We know energy E and $\vec{L}$ are constants of motion, a fact very useful to test if our programs are performing correctly (see Project P4.1). As it turns out, $\vec{A}$ is also a constant of motion in a pure inverse-square law of force, that is, $d\vec{A}/dt = 0$ (see Exercise E4.4). Using vector product rules, we find that $\vec{A} = 0$ for circular orbits. For ellipses, $\vec{A}$ points in the direction from the origin (center of force) to the perihelion, see Figure 4.6.

When the force is not a pure inverse-square in r, as is the case in Eq. (4.19), the Runge-Lenz vector is no longer a constant of motion. Rather, it will rotate with the perihelion. This helps us greatly: unlike finding the aphelion, calculating the Runge-Lenz vector involves no fitting. We can now just follow $\vec{A}$ directly in our simulation, and the angle it sweeps through gives the amount

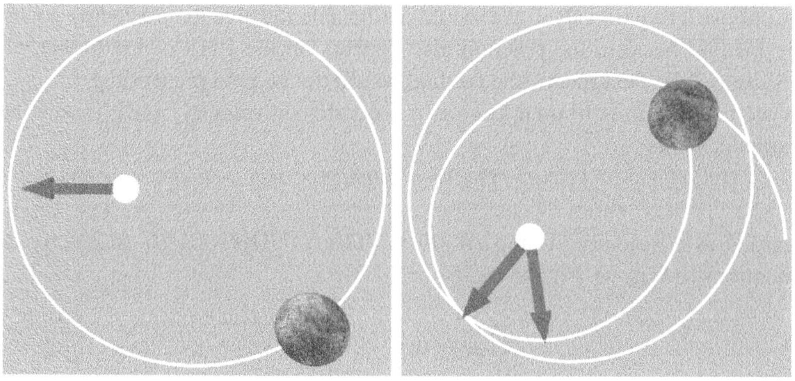

FIGURE 4.6 The Runge-Lenz vector (arrows). It is constant in a closed orbit in a pure $1/r^2$ force field (left). For other forces where the orbit is open, it rotates with the perihelion (right).

of precession. For example, suppose we start our simulation with $\vec{A}$ pointing in the negative x-direction. In time, it will develop a y-component. We can evaluate $\vec{A}$ (both components A_x and A_y) at any time using Eq. (4.20). The angle of precession can be found by $\theta = \tan^{-1}(A_y/A_x)$.

4.3.3 THE LEAPFROG METHOD WITH TIME TRANSFORMATION

If we went ahead and calculated the Runge-Lenz vector with the leapfrog method, we would be disappointed with the results: We would see mostly noise for any meaningfully practical step size h. Again, the problem of the needle-in-a-haystack bites us. Decreasing h is not a preferred solution because it causes more round-off error. But there is something more. Because Mercury's relatively large eccentricity (Table 4.1), its orbit is quite elliptic. Close to the perihelion, the force and speed are much larger than the averages. This large variation is what causes the noise. The physical reason, and the insight we gain, is that the force has a singularity at $r = 0$ and the leapfrog integrator somehow "feels" it. In fact, it is possible to induce artificial precession in a pure $1/r^2$ force law if the orbit is elliptic enough and the step size is not small enough. We need to smooth out, or "regularize," the singularity.

An effective way to deal with the problem is by a time transformation. The idea is that the step size is adjusted so that near the closest approach (perihelion), we take smaller steps. Let us consider the vector equations of motion equivalent to Eqs. (2.39) and (2.40)

$$\frac{d\vec{r}}{dt} = \vec{v}, \tag{4.21}$$

$$\frac{d\vec{v}}{dt} = \vec{a}(\vec{r}). \tag{4.22}$$

We introduce a time transformation from the real time t to a fictitious time s

$$\frac{ds}{dt} = \Omega(r). \tag{4.23}$$

Now rewrite Eqs. (4.21) and (4.22) in terms of s using the chain rule $\frac{d}{dt} = \frac{ds}{dt}\frac{d}{ds} = \Omega\frac{d}{ds}$,

$$\frac{d\vec{r}}{ds} = \vec{v}/\Omega(r), \tag{4.24}$$

$$\frac{d\vec{v}}{ds} = \vec{a}(\vec{r})/\Omega(r). \tag{4.25}$$

Suppose we choose a function $\Omega(r)$. We would not be able to solve the preceding equations using the leapfrog method because the position also appears on the RHS of Eq. (4.24), not velocity alone (see Section 2.4). That means the method would not be area preserving.

What we need to do is to treat $\Omega(r)$ as a generalized "velocity" [63]. Let us introduce an auxiliary variable, W, as

$$W = \Omega(r). \tag{4.26}$$

We will regard W as a "velocity." Using $dW/ds = \nabla\Omega(r) \cdot \vec{v}/\Omega(r)$ and Eq. (4.23), we can add two auxiliary equations to Eqs. (4.24) and (4.25)

$$\frac{d\vec{r}}{ds} = \vec{v}/W, \tag{4.27a}$$

$$\frac{dt}{ds} = 1/W, \tag{4.27b}$$

$$\frac{d\vec{v}}{ds} = \vec{a}(\vec{r})/\Omega(r), \tag{4.27c}$$

$$\frac{dW}{ds} = \nabla\Omega(r) \cdot \vec{v}/\Omega(r). \tag{4.27d}$$

The preceding equations are in the proper form for the leapfrog method if we treat $\vec{r}$ and t as the generalized "coordinates" and $\vec{v}$ and W as the generalized "velocities." Equations (4.27a) and (4.27b) are the exact analogue to Eq. (2.44), and Eqs. (4.27c) and (4.27d) to Eq. (2.45). We are free to choose any form of $\Omega(r)$. A good choice for planetary motion (and other $1/r^n$ forces) is

$$\Omega(r) = \frac{1}{r}. \tag{4.28}$$

It ensures that when r is small, $\Omega(r)$ is large, and dt is small by Eq. (4.23). Furthermore, it is easy to calculate the gradient, $\nabla\Omega(r) = -\vec{r}/r^3$. With Eq. (4.28), we can write the leapfrog algorithm with time transformation analogous to Eqs. (2.41) to (2.43)

$$\vec{r}_{1/2} = \vec{r}_0 + \vec{v}_0 \frac{h}{2W_0}, \tag{4.29a}$$

$$t_{1/2} = t_0 + \frac{h}{2W_0}, \tag{4.29b}$$

$$\vec{v}_1 = \vec{v}_0 + \vec{a}(\vec{r}_{1/2})\, r_{1/2} h, \tag{4.29c}$$

$$W_1 = W_0 - \vec{r}_{1/2} \cdot \frac{\vec{v}_0 + \vec{v}_1}{2} \frac{h}{r_{1/2}^2}, \tag{4.29d}$$

$$\vec{r}_1 = \vec{r}_{1/2} + \vec{v}_1 \frac{h}{2W_1}, \tag{4.29e}$$

$$t_1 = t_{1/2} + \frac{h}{2W_1}. \tag{4.29f}$$

As usual, the subscripts 0 and 1 denote values at the beginning and the end of a time step h, respectively. However, here the time is the fictitious (transformed) time s, not the actual time t. The actual time is returned in t_1. The difference $t_1 - t_0$ is not constant, giving us the desired effect of smoothing out the singularity near the origin.

Here is the leapfrog method with time transformation, Eqs. (4.29a) to (4.29f).

Program listing 4.2 Leapfrog with time transformation (`leapfrog_tt.py`)

```
def leapfrog_tt( lfdiffeq,  r0,  v0,  t0,  w0, h):
    """ vectorized  leapfrog_tt  with time transformation,
        Omega=1/r, that solves general (r,v) ODEs as:
        dr[i]/dt = f[i](v), and dv[i]/dt = g[i](r).
        User supplied  lfdiffeq (id,  r,  v,  t) returns
        f[i](r)  if  id=0, or g[i](v)  if  id=1 """
    # 1st step: calc r at h/2
    hw = h/(2.0*w0)                            # half h/2w0
    t1  = t0 + hw
    r1  = r0 + hw*lfdiffeq(0,  r0,  v0,  t0)   # id=0, get r_1/2
    r2  = np.dot(r1, r1)                       # get r^2=x*x+y*y+z*z
    r12  = np.sqrt(r2)                         # r_1/2

    # 2nd step: calc v1 using r at h/2
    v1 = v0 + h*r12*lfdiffeq(1,  r1,  v0,  t1) # id=1 for g(r) at h/2
    rdotv = np.dot(r1, v0+v1)/2.               # r . v_1/2
    w1 = w0 - rdotv*h/r2                       # w_0 - r . v_1/2 h/r^2

    # 3rd step: calc r by another 1/2 step using v1
    hw = h/(2.0*w1)
    t1  = t1 + hw
    r1  = r1 + hw*lfdiffeq(0,  r1,  v1,  t1)   # get r_1 at t+h
    return r1, v1,  t1,  w1
```

The module `leapfrog_tt()` uses the same `lfdiffeq` format as required by the standard `leapfrog` (Section 2.4). Except for additional elements of time transformation and calculations of dot products with the NumPy function `np.dot`, it works very much the same way also. We will use the transformed leapfrog integrator for sensitive problems such as the precession of Mercury. As discussed in later chapters, it is also a very useful method for N-body systems, including molecular dynamics and atomic reactions (Section S:12.5.1). We include `leapfrog_tt` in the library file `ode.py` with other ODE solvers.

4.3.4 SIMULATION OF DIFFERENTIAL PRECESSION

Precession due to the Relativistic Correction

Finally we have all the necessary parts to calculate the relativistic correction to the precession of Mercury. Putting it all together, we give the complete simulation in Program 4.3 (Section 4.D).

Run the simulation with animation on, and enter a λ large enough, say 0.01, to see exaggerated effect of precession. It should be similar to Figure 4.6. You should also see that the motion of the Runge-Lenz vector is not uniform: it rocks back and forth, with the mean motion forward (counterclockwise), in the same direction as the circulation of Mercury.

Now set `animation=False`, and enter the true value $\lambda = 1.1 \times 10^{-8}$ when prompted. After the simulation ends, we should see a plot showing the angle of precession differential in time over one century in Figure 4.7. It is seen that right at the century mark, the precession is exactly $43''$, a remarkable success of Einstein's theory of gravity. Equally remarkable is the precision of observational astronomy, which allows the detection of this tiny amount. Numerically, the leapfrog method with time transformation is sufficiently accurate to pick up this amount and, at the same time, preserves the phase space properties. It is also quite robust. For example, we could increase the step size by a factor of 10 and still get reasonably accurate results.

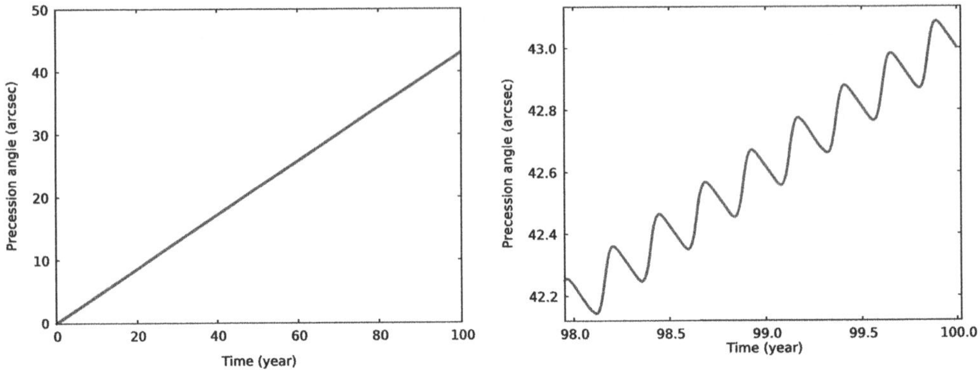

FIGURE 4.7
Differential precession of Mercury, that is, the rotation of the Runge-Lenz vector. A zoom-in of the last two years is shown on the right.

The full curve (Figure 4.7, left) is not as straight as it looks. It has small-scale oscillations that become visible in the enlarged figure. Because these oscillations are so small (a fraction of an arcsecond), we should ask ourselves, are they real? Could they be artifacts of numerics due to such factors as limited accuracy or incorrect algorithm, for example?

In numerical work, we always have to be mindful of potential pitfalls that could produce artificial effects not present in the actual system. In our case, we can check whether the effect remains for a larger perturbation. As seen from the animation with $\lambda \sim 0.01$ discussed earlier, the back-and-forth rocking motion of the Runge-Lenz vector shows that the oscillations are not restricted to small perturbations. Other checks including conservation laws and step sizes can be performed to see if the oscillations are affected. The answer is no (see Project P4.1). In all likelihood, we can exclude numerical error (noise) as the cause.

Once we are certain that the oscillations are real, what is the physical reason for them? We note that there are about four oscillations per year, or the period is about 1/4 years. This is just the orbital period of Mercury (Table 4.1). Therefore, the oscillations might be connected to the orbital period. With animation turned on and with an artificially large $\lambda \sim 0.01$, we can observe the phenomenon visually. Let us assume that Mercury starts at the aphelion going counterclockwise, with the Runge-Lenz vector $\vec{A}$ pointing at the perihelion ($A_y = 0$). The y-component, A_y, has two terms from Eq. (4.86), $A_y/m^2 = -v_x L - GMy/r$.

Roughly,[6] when v_x turns negative, A_y increases, and when y increases, A_y decreases. Near the aphelion, Mercury is farthest away from the Sun, and the attractive relativistic term causes v_x to turn negative faster than the increase in y (more precisely y/r). The net effect is that A_y increases. This causes the clockwise (retrograde) rotation of $\vec{A}$. Near the perihelion (past the $x = 0$ plane), the opposite occurs, and $\vec{A}$ rotates counterclockwise. Because the perturbation is greater near the perihelion, the counterclockwise rotation is larger than the clockwise rotation, resulting in a net precession in the same direction as Mercury's circulation (ccw, prograde). It then follows that if the sign of the perturbation is reversed (repulsive), the precession will be in the opposite direction. This can be verified through an exercise.

Central Field Approximation and Precession due to Other Planets

Mercury interacts with other planets through pairwise forces that are not central relative to the Sun. A general description would need to take into account all the planets simultaneously (a many-body problem). But if the effects are small, and if we are interested in only averaged values like precession over a period of time, we could approximate each pairwise force as a central force and treat it individually. The idea is that Mercury moves around the origin (Sun)

[6] A more quantitative explanation of the direction of precession is given in Ref. [61].

Table 4.2 Precession of Mercury due to the Planets (Arcseconds/Century).

Venus	Earth	Mars	Jupiter	Saturn	Uranus	Neptune	Total
267.4	94.7	2.38	157.2	7.58	0.141	0.0437	529.4

rapidly and the average force points in the radial direction. In such cases, we speak of central field (or mean field) approximations, which are very useful and important in many areas of physics.

If we can construct the force between Mercury and a planet in a central field approximation, we can calculate the precession of Mercury due to that force with the same methodology as the relativistic effect discussed earlier. One way to obtain the force is as follows. Imagine the mass of a planet, M_p, is distributed uniformly over a coplanar ring of radius a (semimajor axis) [72]. By symmetry, the force between Mercury and the ring is central, that is, $\vec{F} = F(r)\hat{r}$. The potential is given by an elliptic integral [49] (see Exercise E4.6)

$$V(r) = -\frac{GM_p m}{2\pi a}\int_0^{2\pi} \frac{d\theta}{\sqrt{1+\rho^2-2\rho\cos\theta}}, \qquad \rho = r/a. \tag{4.30}$$

Because Mercury is the innermost planet and $\rho < 1$, this elliptic integral can be expanded in a power series. The first few terms are

$$V(r) = -\frac{GM_p m}{a}\left[1+\frac{1}{4}\rho^2+\frac{9}{64}\rho^4+\frac{25}{256}\rho^6+\cdots\right] \tag{4.31}$$

The force can be found via the gradient $-\nabla V$ as

$$\vec{F} = \frac{GM_p m}{2a^3}\left[1+\frac{9}{8}\frac{r^2}{a^2}+\frac{75}{64}\frac{r^4}{a^4}\right]\vec{r}. \tag{4.32}$$

This force is repulsive (radially outward), and the leading term behaves like Hooke's law with a negative spring constant. Two factors determine the magnitude of the force, the mass of the planet M_p and its average radius a. Accordingly, with everything being equal, we expect either a larger mass or a smaller semimajor axis will cause a greater precession. The actual amount depends on the interplay between the two parameters.

Using the force in the central field approximation (4.32) in Program 4.3 with the parameters from Table 4.1, we obtain the results of Table 4.2.

The results vary over three orders of magnitude. Venus, the closest one, has the largest effect, followed by Jupiter, the most massive one. In between, the order reflects the balance between the radius and mass. Note that the results for Earth are obtained with the combined mass of Earth and the moon. Neptune, the outermost planet, gives the smallest contribution, 0.0437″, because of its large radius and moderate mass. This is three orders of magnitude smaller than the relativistic contribution (43″). If 43″ was a needle in a haystack, Neptune's contribution might be a drop in a lake. Yet, the leapfrog method with time transformation is robust enough to crunch it out accurately and efficiently. Also, although the central field approximated force (4.32) is positive, the precession is still counterclockwise, the same as the relativistic correction, even though the force there (4.19) is negative. We will investigate these questions and related aspects in Project P4.3.

Scaling Law in the Central Field Approximation

As the numbers listed in Table 4.2 jump up and down with different planets, they are not particularly insightful for discovering trends or revealing more general properties. As Eq. (3.33) from

Chapter 3 showed, scaling laws are very useful in this regard. Whenever possible, we should try to find scaled variables and plot the results in these variables to show the general trends.

What is a good scaled variable to use in this case? The amount of precession depends on the combination of the mass and radius. The prefactor in the potential (4.31) suggests M_p/a, but the force (4.32) points to M_p/a^3. Because the force is directly involved in the simulation, the latter is more appropriate. Therefore, for small perturbations, we expect precession to scale linearly with M_p/a^3, that is, a linear scaling law.[7]

When we go ahead and calculate the scaled variable M_p/a^3 for the planets using Table 4.1, we find that it also varies over three orders of magnitude similar to the precession results in Table 4.2. If we were to display precession versus the scaled variable as linear plots, the data points would be clustered at two locations, the lower-left and upper-right corners. Thus, to more evenly spread out the data points in the whole range and to help visually identify any trends, we should plot the data using log-log scales as we did with the semilog plot in Figure 2.7, for example. The results are shown in Figure 4.8.

Most data points in Figure 4.8 fall on a straight line (trend line), clearly showing a linear scaling law. Small deviations are seen for Venus, Earth, and Mars. This is due to their relative proximity to Mercury, causing the second-order term in Eq. (4.32) to have a nonnegligible effect on the deviation. Nonetheless, Venus and Jupiter, the closest and the most massive planets, are vastly different and yet nearly identical in the scaled variable, 6.47 vs. 6.78, respectively (Figure 4.8). Therefore, they cause a similar amount of precession according to the scaling law. The predicative power and the physical insight of scaling laws make them very useful in practice and in our understanding.

Because each planet produces precession in the same counterclockwise direction, the total precession due to all planets is additive and is equal to 529.4″ (Table 4.2). The result is in very good agreement with the accepted observational data of 531″ [72, 85]. The relative error is about 0.3%, with most of it coming from Venus. What are the sources of error? We have assumed the planets are coplanar with Mercury, which is only approximate. The inclination angles of the

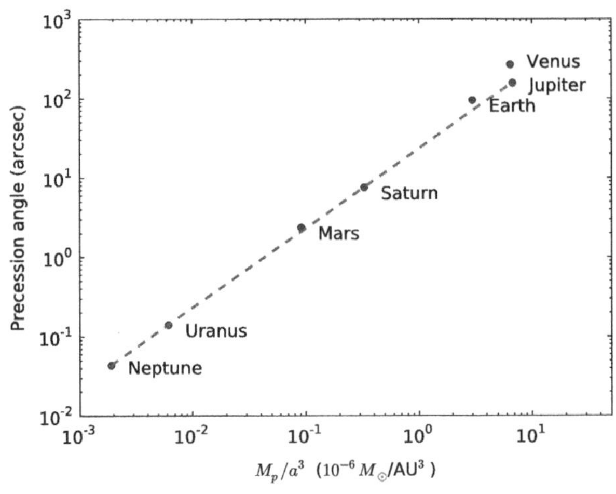

FIGURE 4.8 The precession of Mercury due to other planets as a function of the scaled variable M_p/a^3. The dashed trend line indicates a linear scaling law expected for small perturbations in the central field approximation.

[7]This is a special case of power scaling laws, $y = cx^\alpha$, where x is the scaled variable, α the exponent (or power), and c a constant (usually unimportant). If $x' = \lambda x$, then $y' = \lambda^\alpha y$.

planets differ from each other by a few degrees. Another source is the assumption of circular orbits (rings). From the eccentricities in Table 4.1, except for Pluto, which has negligible effects and Mercury itself, the perturbing planets are nearly, but not perfectly, circular. Last, we have included only a few terms in the series expansion (4.31). This will affect only the close planets though, such as Venus and Earth.

4.4 STAR WOBBLES AND EXOPLANETS

We have seen from Section 4.2 that a two-body problem can be reduced to an effective one-body problem in the center of mass frame. The actual motion, of course, still consists of two bodies pulling on each other (with forces of equal magnitude and opposite direction) and revolving around the center of mass. In the case of a star-planet system, the star barely wobbles due to the large mass ratio.

Nonetheless, the star's wobbling is important to a current topic of considerable interest: the discovery and detection of extrasolar planets, or exoplanets, including Earth-like planets and planets in the habitable zone supporting an atmosphere and liquid water.[8] Unlike planets at the doorstep in our solar system, direct observation of exoplanets is very difficult because they are at astronomical distances, small and faint, and in sharp contrast to the brightness of their host stars. Indirect methods of detection are necessary, including measuring the wobbles of stars (radial velocity), the light blocked off when exoplanets pass in front of the stars (transiting planets), and light bending and brightening due to gravitational fields (microlensing), among others.

4.4.1 RADIAL VELOCITY METHOD

We discuss the method of detecting the wobbles, known as the radial velocity (RV) method. The RV method relies on the fact that light from stars moving about the center of mass will be Doppler-shifted. The Doppler shifts depend on the velocity of the star along the line of sight, that is, the radial velocity from the observer's perspective. Light from the star will be blue-shifted moving toward us, and red-shifted moving away. By measuring the Doppler profile, one can infer the velocity profile of orbital motion. From there, we can determine the orbital parameters of the exoplanets, including the period, eccentricity, and the lower limit of the mass. The RV method is responsible for a majority of exoplanets discovered early on [14].[9]

Referring to the position vectors in Figure 4.2 and Eq. (4.4) (with the results from Exercise E4.1), the wobbling velocity of the star, $\vec{v}_* \equiv \dot{\vec{r}}_1$, is related to the orbital velocity $\vec{v} = \dot{\vec{r}}$ as

$$\vec{v}_* = -\frac{m}{m+M}\vec{v}. \tag{4.33}$$

For a given star-planet system, we can obtain $\vec{v}$ from Program 4.1 and therefore $\vec{v}_*$. The result for the Sun–Jupiter system is shown in Figure 4.9. One component of the Sun's velocity is plotted, $\vec{v}_{*y}$ in this case, which is equivalent to the edge-on view of the orbital motion. The velocity is periodic ($\sim$ 4330 days, Jupiter's period) and nearly sinusoidal because the orbit is almost circular ($e = 0.0484$, see Table 4.1). It would be sinusoidal for circular orbits. The magnitude of the velocity is roughly 10 m/s, typical of the pull by planets of Jupiter's size on a star of about one solar mass. For smaller planets like Earth, the magnitude is about 10 cm/s.

Also shown in Figure 4.9 is an observational RV dataset for a star in the Henry Draper catalog, HD 139357. It shows an almost sinusoidal curve, clearly indicating the presence of an

[8]The latest estimate based on the Kepler mission is that one fifth of Sun-like stars in our galaxy have an Earth-like planet, totaling to tens of billions of habitable exoplanets.

[9]As of this writing, 1022 exoplanets have been confirmed, 548 of which were discovered with the RV method [66].

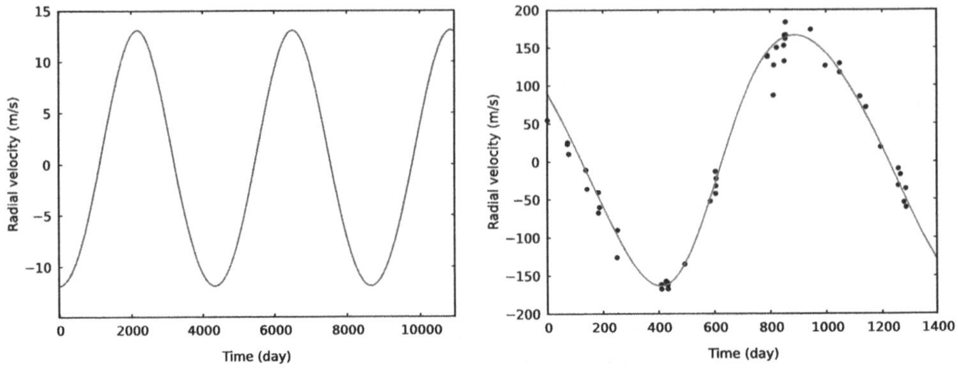

FIGURE 4.9 The wobbling velocity of the Sun due to Jupiter (left) and of HD 139357 due to its exoplanet (right). The solid curve is a fit described in text.

exoplanet moving in a nearly circular orbit around the star, with a period ~ 1100 days. This exoplanet was discovered in 2009. From the RV dataset, the lower limit of the exoplanet's mass is determined to be about 9.8 Jupiter mass. The fact that only a mass limit can be established, and not the mass itself, has to do with observing geometry and information on the inclination angle, or lack thereof.

The geometry is illustrated in Figure 4.10. The space-fixed coordinates (the sky) are denoted by XYZ, where the XY plane defines the space plane. We choose the X-axis along the line where the Kepler orbit of the exoplanet intersects the space plane (ascending node in astronomy nomenclature). The inclination angle, i, is the angle between the orbital plane and the space plane. It is equal to the angle between the Z-axis and the normal of the orbit (direction of angular momentum $\vec{L}$). We use ω to denote the angle between the aphelion and the X-axis. Except for the azimuthal angle around Z (not shown), which is unimportant due to rotational symmetry, i and ω determine the orientation of the Kepler orbit. As usual, θ represents the angle between the radial vector of the exoplanet and the aphelion, the same as in Figure 4.3.

The observer's line of sight to the star runs parallel to the Z-axis. Therefore, only the Z component of the star's velocity, v_{*z}, is relevant and measurable by Doppler shift methods. From the observer's perspective, v_{*z} appears as a radial velocity (away or toward), thus the name of the RV method.[10] We have to determine the relationship between the wobbling velocity v_{*z} and the orbital velocity $\vec{v}$ in terms of orbital parameters and the angles i and ω.

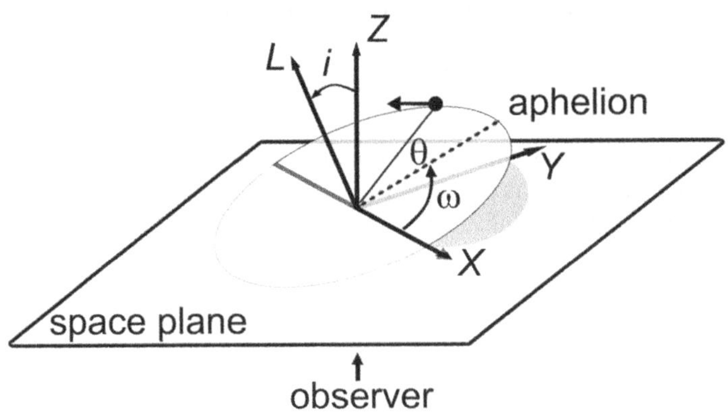

FIGURE 4.10 Observing geometry for radial velocity measurements.

[10]This is not to be confused with the radial velocity $\dot{r}$ in Eq. (4.8), which represents the orbital velocity of the planet directed radially.

The radial velocity of the star is derived in Section 4.C. For a star with a single planet, the result from Eq. (4.85) is

$$v_{*z} = -\mathcal{V}[\cos(\omega + \theta) - e\cos\omega] + C, \quad \mathcal{V} = \frac{m\sin i}{m+M}\frac{2\pi a}{T\sqrt{1-e^2}}, \tag{4.34}$$

where C is a constant representing the velocity of the center of mass, T the period, and a the semimajor axis. The first term indicates that v_{*z} oscillates with a period T and an amplitude $|\mathcal{V}|$.

From fitting a given RV dataset like the one shown in Figure 4.9 according to Eq. (4.34), parameters such as T, e, and $\mathcal{V}$ can be obtained. If the mass of the star M is independently known, the mass of the exoplanet m can be determined from

$$\frac{m\sin i}{m+M} \simeq \frac{m\sin i}{M} = \sqrt{1-e^2}\left(\frac{T}{2\pi GM}\right)^{1/3}\mathcal{V}. \tag{4.35}$$

We have assumed $m/M \ll 1$ and used Kepler's third law (4.18) to eliminate a. Equation (4.35) shows that we can determine the product $m\sin i$ only. Because the inclination angle i is generally unknown, the RV method yields only a lower limit for the exoplanet mass ($\sin i \le 1$).

If the period T is given in years, M in solar masses, and $\mathcal{V}$ in m/s, the numerical value for the exoplanet mass in Earth masses is

$$m\sin i = 11\sqrt{1-e^2}\,(TM^2)^{1/3}\mathcal{V}. \quad \text{(in Earth mass)} \tag{4.36}$$

This formula is useful for a rough estimate of an exoplanet's mass.

Take for instance the exoplanet of the star HD 139357. The star is known to be about 400 light-years away and a mass $M \sim 1.3M_{\odot}$. We can estimate the other parameters by inspecting Figure 4.9 to obtain: $T \sim 1100$ days = 3 years, and $\mathcal{V} \sim 160$ m/s. The RV curve is almost sinusoidal, so the orbit is nearly circular, and $e \sim 0$. Substituting these numbers into Eq. (4.36), we obtain $m\sin i \sim 3000$ Earth mass, or ~ 9.5 Jupiter mass. More careful data fitting gives $m\sin i =$ 9.8 Jupiter mass, which means that the exoplanet is at least 9.8 times more massive than Jupiter.

4.4.2 MODELING RV DATASETS

To accurately model observational RV datasets according to Eq. (4.34), we need to know θ as a function of time t, $\theta(t)$. However, this is not as straightforward as one might expect. Except for circular orbit where θ is linear in t, no simple closed-form solutions exist. We could obtain $\theta(t)$ numerically from simulation codes such as Program 4.3, but that would be the hard way. Fortunately, it is a two-way street here, a hard way and an easy way.

We first write $\dot{r}$ using conservation of energy from Eq. (4.8) and obtain t as

$$t = -\sqrt{\frac{m}{2}}\int_{r_{max}}^{r}\frac{dr}{\sqrt{E - V_{eff}(r)}}. \tag{4.37}$$

This equation can be solved by introducing a new variable ψ through

$$r = a(1 + e\cos\psi), \tag{4.38}$$

where ψ is called the eccentric anomaly.[11] Substituting Eq. (4.38) into (4.37), and after some algebraic details left to Exercise E4.7, we obtain

$$\frac{t}{T} = \frac{1}{2\pi}(\psi + e\sin\psi). \tag{4.39}$$

[11] Physics is not only interesting but also colorful sometimes, judging by some terms used.

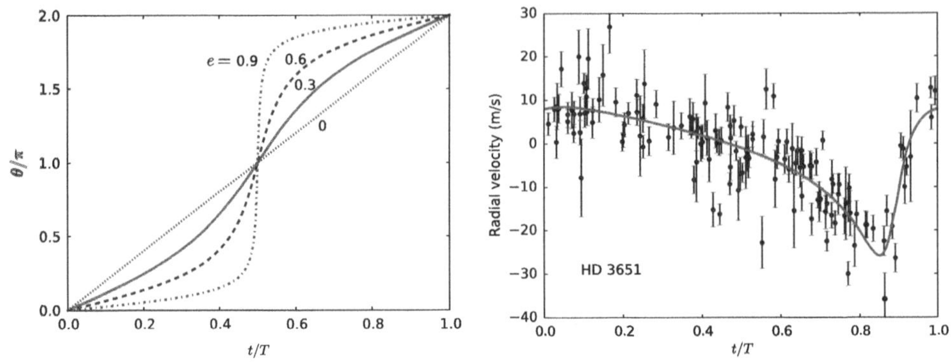

FIGURE 4.11 Left: The relationship between θ and t for four eccentricities. Right: An actual RV dataset of Fischer et al. [30] and a fit.

This is the celebrated Kepler's equation relating t to ψ. We can solve for θ in terms of ψ by combining Eqs. (4.38) and (4.13) to get

$$\cos\theta = \frac{e + \cos\psi}{1 + e\cos\psi}. \tag{4.40}$$

We see from Eq. (4.40) that ψ has the same range as θ, that is, when θ changes from 0 to 2π, so does ψ. Furthermore, the two variables are equal at 0, π, and 2π, no matter the eccentricity e.

The pair of equations (4.39) and (4.40) give the relationship between t and θ. For a given t, Eq. (4.39) can be solved for ψ, which can be substituted into Eq. (4.40) to obtain θ. The relationship for several eccentricities is depicted in Figure 4.11. The dependence is linear for $e = 0$ (circular orbit) as expected. For nonzero e, it becomes increasingly nonlinear. The figure was generated without actually solving Eq. (4.39). The trick is this: first generate a mesh for ψ in $[0, 2\pi]$; second, compute t from Eq. (4.39) for each mesh point and store it in an array; third, compute θ from Eq. (4.40) and store it in another array; finally plot the two arrays. This trick is easy and fast, but the drawback, of course, is that the t values calculated are not predetermined and nonuniform. If we wanted to generate θ for a given t, we would have to solve Eq. (4.39) using a root solver. It's not necessary for our purpose here.

The effect of eccentricity on radial velocities is also shown in Figure 4.11. The dataset is for HD 3651 with an exoplanet reported in 2003 [30]. The curve is not sinusoidal at all. The sharp dip in the radial velocity can be reproduced only with a significant value of eccentricity. Physically, the planet moves much faster near the perihelion than anywhere else in a highly eccentric orbit. This causes a dip (or spike) in the velocity of the recoiling star.

A general best fit of Eq. (4.34) to a given RV dataset is not trivial because of the number of parameters involved. We use least square fitting, `leastsq`, from SciPy as a blackbox. It requires only an error function that returns the difference between our model and the actual data points. The details are given in Program 4.4, yielding the results in Figure 4.11.

The blackbox method is easy to use and gives satisfactory results, but leaves the modeling process a bit opaque. To see various factors at play, we outline an iterative fitting process by simple visual inspection that also works well for our purpose. Starting with an initial guess of parameters $T, \mathcal{V}, e, \omega, C$:

1. Generate a mesh grid for ψ.
2. Adjust the appropriate parameters $T, \mathcal{V}, e, \omega, C$ for better visual fit.
3. Compute t from Eq. (4.39) and θ from Eq. (4.40).
4. Calculate v_{*z} from Eq. (4.34) and plot the results.
5. Inspect the match of the curve and the data, repeat step 2 as necessary.

With code segments from Program 4.4, we can begin to model any datasets and visualize the process. We leave modeling of the dataset shown in Figure 4.11 to Project P4.6. Many other RV datasets can be found at the exoplanet archive [66].

4.5 PLANAR THREE-BODY PROBLEMS

So far we have studied motion of two-body systems reducible to effective one-body motion. If we add just one more body to a two-body system, we have a three-body problem, the simplest of general N-body problems. Even if the pairwise forces obey the pure inverse-square law (4.1), general analytic solutions to the three-body problem are still not possible at present, making simulations indispensable.

From the early development of Newtonian mechanics, efforts have been made to reduce the intractability of the three-body problem. There are nine degrees of freedom in total, three for each body. Going to the center of mass frame reduces that number by three. Further reduction is possible by invoking conservation laws (energy, angular momentum, etc.), but the mathematical difficulties involved are still insurmountable. Special attention, therefore, has been devoted to finding solutions of the planar three-body problem: The motion of all bodies is confined to a two-dimensional plane. This is still an area of ongoing research, especially in the search of periodic orbits. We will discuss several numerical examples to get a glimpse of the planar three-body problem and its intricacies.

4.5.1 EQUATIONS OF MOTION IN THE THREE-BODY PROBLEM

Qualitatively, the motion of three bodies is no more difficult to simulate than one-body motion. Let the three bodies be numbered as $i = 1, 2,$ and 3. Their masses, coordinates, and velocities are m_i, $\vec{r}_i$, and $\vec{v}_i$, respectively. The equations of motion are

$$\frac{d\vec{r}_i}{dt} = \vec{v}_i, \tag{4.41}$$

$$\frac{d\vec{v}_i}{dt} = -Gm_j\frac{\vec{r}_{ij}}{r_{ij}^3} - Gm_k\frac{\vec{r}_{ik}}{r_{ik}^3}. \tag{4.42}$$

It is understood that $i \neq j \neq k$, and $\{ijk\}$ is a permutation of $\{123\}$. The relative coordinates between two bodies are $\vec{r}_{ij} = \vec{r}_i - \vec{r}_j$, and $r_{ij} = |\vec{r}_i - \vec{r}_j|$. Note that $\vec{r}_{ij} = -\vec{r}_{ji}$, a relation useful in programming. Because the gravitational constant G always appears in the product Gm_i, we can set $G = 1$ in our program. It is equivalent to setting the scale of mass.

Equations (4.41) and (4.42) are 12 first-order differential equations of a Hamiltonian system in 2D. Given the initial condition, we can solve them using the leapfrog method. Of course, the more elaborate leapfrog method with time transformation discussed earlier may be extended to the three-body problem as well.[12] However, we are not interested in very small or sensitive processes like precession of Mercury, so its use is unnecessary here. In addition, most of the periodic orbits are physically unstable, and there is little a numerical algorithm can do to rectify that. What remains is to determine the initial condition for a given numerical solution.

4.5.2 EULER'S COLLINEAR MOTION

One of the simplest and most interesting systems is Euler's collinear motion of the three-body problem. It is illustrated in Figure 4.12.

[12]We do this in scattering studies of atomic reactions. See Section S:12.5.

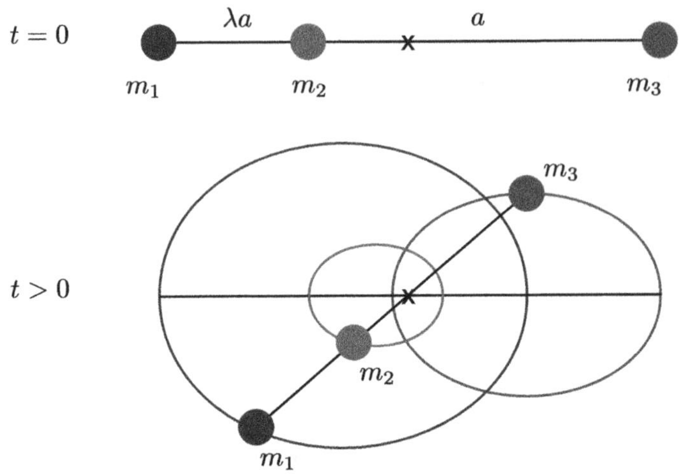

FIGURE 4.12 Euler's collinear motion of the three-body problem.

At any given time, the three bodies are on a straight line, with one body, say m_2, lying in-between the other two bodies, m_1 and m_3. Each moves in an ellipse with the same period. When they reach the aphelion simultaneously on the axis of the ellipses, the distances among them are the largest, and when they pass the perihelion, the distances are the closest. At other times, the distances oscillate between the extrema. The motion is periodic, but unstable. That is to say, if there is a slight deviation from the perfect configuration, the system flies apart from collinear motion.

To determine the initial condition, let us assume the bodies start from the x-axis, thus $y_i = 0$, and only the horizontal positions x_i are needed. To execute Euler's collinear motion, the distances between the bodies must obey certain proportions. Referring to Figure 4.12 (top), let a be the distance between m_2 and m_3, and λa between m_1 and m_2. It follows that

$$x_3 - x_2 = a, \; x_2 - x_1 = \lambda a, \; x_3 - x_1 = (1+\lambda)a. \tag{4.43}$$

The equation of motion for a given body, say m_1, is

$$m_1 \frac{d^2x_1}{dt^2} = -m_1 m_2 \frac{x_1 - x_2}{|x_1 - x_2|^3} - m_1 m_3 \frac{x_1 - x_3}{|x_1 - x_3|^3},$$

where we have set $G = 1$ for convenience. Because of the oscillating nature of the distances, we seek a solution of the form $d^2x_1/dt^2 = -\omega^2 x_1$ (simple harmonic oscillator). Substituting this into the preceding equation and dividing $-m_1\omega^2$ on both sides, we obtain

$$x_1 = \frac{1}{\omega^2}\left[m_2 \frac{x_1 - x_2}{|x_1 - x_2|^3} + m_3 \frac{x_1 - x_3}{|x_1 - x_3|^3}\right]. \tag{4.44}$$

Similarly, we have for the other two bodies

$$x_2 = \frac{1}{\omega^2}\left[m_1 \frac{x_2 - x_1}{|x_2 - x_1|^3} + m_3 \frac{x_2 - x_3}{|x_2 - x_3|^3}\right], \tag{4.45}$$

$$x_3 = \frac{1}{\omega^2}\left[m_1 \frac{x_3 - x_1}{|x_3 - x_1|^3} + m_2 \frac{x_3 - x_2}{|x_3 - x_2|^3}\right]. \tag{4.46}$$

We can combine Eqs. (4.43) to (4.46) to obtain the parameter λ first,

$$\lambda^5(m_2 + m_3) + \lambda^4(2m_2 + 3m_3) + \lambda^3(m_2 + 3m_3)$$
$$+ \lambda^2(-3m_1 - m_2) + \lambda(-3m_1 - 2m_2) + (-m_1 - m_2) = 0. \qquad (4.47)$$

This is known as Euler's quintic equation. We leave as an exercise the verification of Eq. (4.47) and the following results, Eqs. (4.48) to (4.50). Given a set of masses, there is one positive root. The variable a in terms of λ is

$$a^3 = \frac{1}{\omega^2}\left[m_2 + m_3 - \frac{m_1(1+2\lambda)}{\lambda^2(1+\lambda)^2}\right]. \qquad (4.48)$$

With both λ and a found, we can find the initial positions as

$$x_2 = \frac{1}{\omega^2 a^2}\left[\frac{m_1}{\lambda^2} - m_3\right], \; x_1 = x_2 - \lambda a, \; x_3 = -\frac{m_1 x_1 + m_2 x_2}{m_3}. \qquad (4.49)$$

We recognize that the last term means that the center of mass is the origin. The initial velocities (all along y-axis) are

$$v_{1y} = \omega x_1, \; v_{2y} = \omega x_2, \; v_{3y} = \omega x_3. \qquad (4.50)$$

The procedure to simulate Euler's collinear motion is as follows. First, solve Eq. (4.47) for λ using a root finder such as bisection or Newton's methods in Chapter 3. Then compute the initial condition from Eqs. (4.49) and (4.50). Finally, solve the equations of motion (4.41) and (4.42) with the leapfrog method.

The code implementing this strategy is given in Program 4.5 listed in Section 4.D. The program animates the collinear motion of the three-body problem and snapshots are shown in Figure 4.13.

Watching the motion, we should see that at first the bodies move perfectly (to the naked eye) in their own ellipses, retracing over them periodically (Figure 4.13, left). After a while, typically three periods or so depending on the energy, two bodies will be so attracted to each other that they quickly move off the periodic orbits, forming a binary system. The binary partners usually perform choreographically beautiful (albeit mechanical) "dances" (Figure 4.13, right) as they move along. The third body, attracted to the binary system acting roughly as a single body, often turns around. If it does and moves in close, there is a high probability that the dancing partners switch, forming a new binary system. This may continue for a while, but eventually, the final outcome is that a tight binary is formed, and the third body is ejected, moving indefinitely away from the binary system.

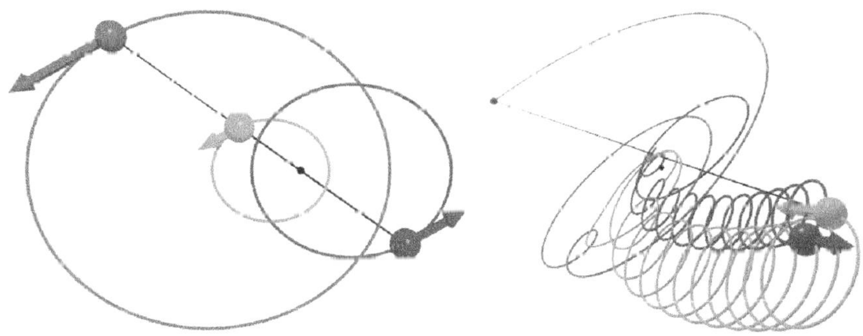

FIGURE 4.13 Euler's collinear motion of the three-body problem at different times. The camera in the right figure is tilted to keep the third body in view.

Gravitationally bound classical three-body systems are mostly unstable and often so sensitive to the initial conditions that a small difference there can lead to totally different outcomes. This behavior is known as classical chaos, a topic we will discuss next in Chapter 5. Our knowledge at present seems to indicate that common many-body systems held by gravity are chaotic, including our solar system [58].

4.5.3 THREE-BODY CHOREOGRAPHY

There are more elaborate periodic orbits that have been found. For example, the following initial condition [84] produces a heart-shaped choreographic orbit shown in Figure 4.14.

$$\begin{aligned}
&m_1 = m_2 = m_3 = 1/3, \\
&x_1 = -0.30805788, y_1 = 0, v_{1x} = 0, v_{1y} = -1.015378093, \\
&x_2 = 0.15402894, y_2 = -0.09324743, v_{2x} = 0.963502817, v_{2y} = 0.507689046, \\
&x_3 = x_2, y_3 = -y_2, v_{3x} = -v_{2x}, v_{3y} = v_{2y}.
\end{aligned} \tag{4.51}$$

The three bodies rotate counterclockwise, so there is a net angular momentum. The starting position of each body is approximately at the position of the body immediately behind it (clockwise) in the left panel of Figure 4.14. The most delicate part occurs near the center of mass (CM, white dot) where the speed is the lowest. Referring to the animation (or Figure 4.14, right), as the red body (closest to the CM) moves in on the upper part of the inner orbit, it slows down, coming to a temporary stop right at the tip (the protrusion) of the inner orbit. At the same time, the other two bodies (light/green and dark/blue) are almost diametrically positioned on the outer orbit, so the net force on the red body is very small. But, there is a slight bend to the diametrical, which provides just enough net force that allows the red body to step back from the tip and continue on to the lower part of the inner orbit (where the body immediately ahead (blue) was previously). However, the motion is unstable. After about two periods, the system breaks up, suffering the same fate of most three-body systems: a tight binary plus a runaway body (not shown in Figure 4.14). But, if we reversed the velocities at any time, the system would run backward to its original configuration precisely like rewinding a videotape (see Exercise E4.9). This is because the leapfrog method preserves time reversibility of the system, which may be unstable but not indeterministic.

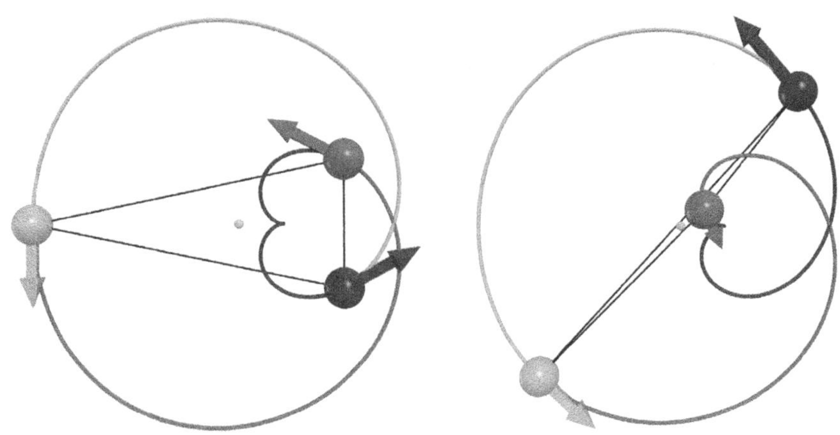

FIGURE 4.14 Choreographic orbit of the three-body problem with the initial condition (4.51). The arrows are proportional to the velocities.

Another interesting orbit is a figure-8 orbit. It is shown in Figure 4.15 produced from the following initial condition [15],

$$\begin{aligned}
&m_1 = m_2 = m_3 = 1,\\
&x_1 = 0.97000436, y_1 = -0.24308753, v_{1x} = -v_{3x}/2, v_{1y} = -v_{3y}/2,\\
&x_2 = -x_1, y_2 = -y_1, v_{2x} = v_{1x}, v_{2y} = v_{1y},\\
&x_3 = 0, y_3 = 0, v_{3x} = -0.93240737, v_{3y} = -0.86473146.
\end{aligned} \tag{4.52}$$

This orbit is symmetric both horizontally and vertically. Unlike the previous configuration, the bodies do not circulate uniformly in one direction. The motion is clockwise on the left half of the orbit (Figure 4.15), and counterclockwise on the right half. The combination is such that there is no net angular momentum. When one body is at the origin, the other two are symmetrically situated on the orbit, all on a straight (slanted) line. In this sense it is like Euler's collinear configuration (Figure 4.13), although the three bodies never line up horizontally. Whenever one body is at the edge of the figure-8 (Figure 4.15, middle), the other two are on the opposite half lined up vertically, forming an isosceles triangle. At this moment, the velocity of the first body is vertical, and the velocities of the other two bodies are such that their x components cancel each other and the sum of their y components cancels the y component of the first body, so that the center of mass velocity remains zero.

Numerical experiments show a rare but remarkable property of this orbit: It is stable. If the orbit is perturbed slightly, the motion remains close to the figure-8 shape. For example, if the position or the velocity of a body (or even its mass) is changed by a small amount, the orbit may be slightly deformed, or it may even precess, but the three-body system does not disintegrate (Exercise E4.9).

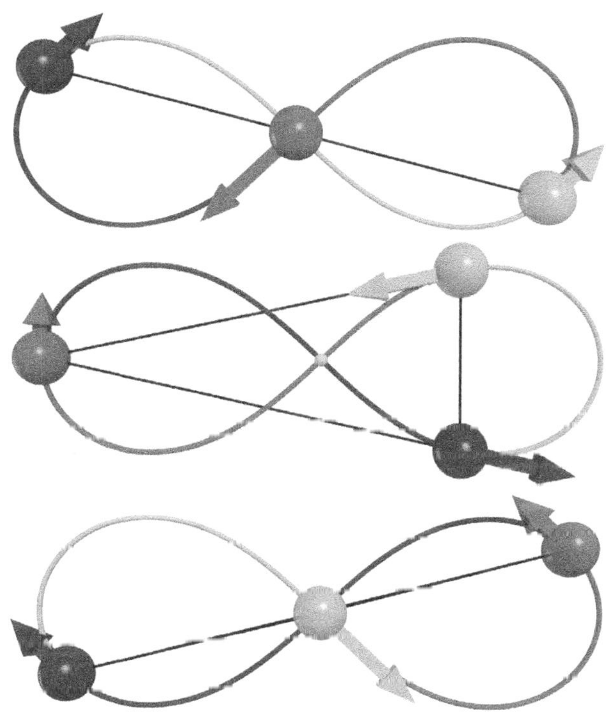

FIGURE 4.15 Choreographic orbit of the three-body problem with the initial condition (4.52). The arrows are proportional to the velocities.

4.6 THE RESTRICTED THREE-BODY PROBLEM

The planar three-body problem does not have any restrictions other than that the motion is confined to a plane. We impose now two additional constraints: the masses of two primary bodies are much larger than the mass of the third body, and the two primaries move in a circular orbit about their center of mass. As a result, the simplified system is called the restricted three-body problem. It is illustrated in Figure 4.16.

Let us call the third body of mass m the test body. Its mass is considered so small compared to the two primaries, that is, $m/M_{1,2} \sim 0$, that it has no effect on the motion of the primaries, which rotate around the center of mass at constant angular velocity ω and constant separation. We are only concerned with the motion of the test body. Though the restricted three-body problem is an idealized situation and no actual systems behave exactly like it, it is still a very good approximation for many systems, including the motion of a satellite (test body) in the Earth–Moon system, or of the asteroids in the Sun–Jupiter system. The simplification allows us to gain much insight from studying such systems.

The motion of the test body may be best understood in the rotating coordinate system of the primaries. For example, if a and b are the respective distances of M_1 and M_2 from the center of mass (the origin, Figure 4.16), then the primaries appear as stationary points at $(-a, 0)$ and $(b, 0)$, respectively, in the rotating coordinates (x-y), though they are rotating in space-fixed coordinates (X-Y). The test body will move around the stationary force centers in the rotating coordinates.

4.6.1 EQUATIONS OF MOTION IN THE ROTATING SYSTEM

The force on the test body is

$$\vec{F} = -GM_1 m\frac{\vec{r}_1}{r_1^3} - GM_2 m\frac{\vec{r}_2}{r_2^3}, \tag{4.53}$$
$$r_1 = \sqrt{(x+a)^2 + y^2}, \; r_2 = \sqrt{(x-b)^2 + y^2},$$

where (x, y) are the coordinates of the test body in the rotating system. The calculation of velocity or acceleration must take the rotating coordinates into account. Details are given in Appendix 4.A. For example, according to Eq. (4.72) the acceleration is

$$\vec{a} = \vec{F}/m - 2\vec{\omega} \times \vec{v} - \vec{\omega} \times (\vec{\omega} \times \vec{r}). \quad \text{(rotating frame)} \tag{4.54}$$

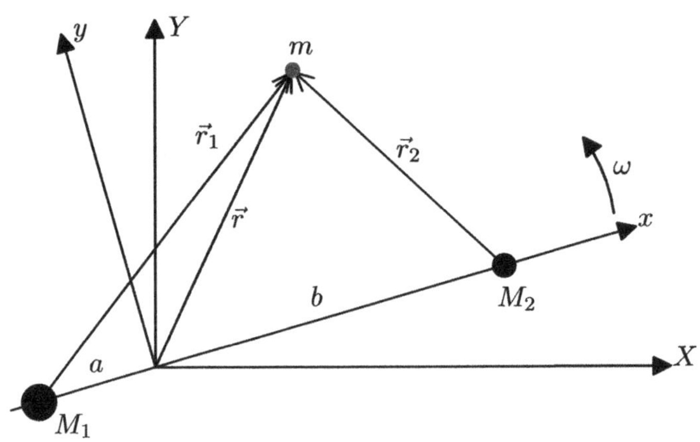

FIGURE 4.16 The restricted three-body coordinate system.

Note that we have dropped the prime (′) according to our coordinate system in Figure 4.16 for clarity. Of course, all variables $\vec{r}$, $\vec{v}$, and $\vec{a}$ refer to the rotating coordinates.

Using the fact $\vec{\omega} = \omega\hat{k}$ and $\vec{A} \times (\vec{B} \times \vec{C}) = (\vec{A} \cdot \vec{C})\vec{B} - (\vec{A} \cdot \vec{B})\vec{C}$, we can simplify the last term in Eq. (4.54) to

$$\vec{\omega} \times (\vec{\omega} \times \vec{r}) = (\vec{\omega} \cdot \vec{r})\vec{\omega} - (\vec{\omega} \cdot \vec{\omega})\vec{r} = -\omega^2\vec{r}.$$

We have used $\vec{\omega} \cdot \vec{r} = 0$ because $\vec{\omega}$ is perpendicular to the plane of motion (x-y plane). Substituting these results and Eq. (4.53) into (4.54), we obtain

$$\vec{a} = -GM_1\frac{\vec{r}_1}{r_1^3} - GM_2\frac{\vec{r}_2}{r_2^3} + \omega^2\vec{r} - 2\vec{\omega} \times \vec{v}. \tag{4.55}$$

The system can be readily integrated with our ODE solvers. However, before doing that, we would like to examine the rich structure contained within the effective potential which will help us better understand the motion.

4.6.2 THE EFFECTIVE POTENTIAL

There are four terms in Eq. (4.55). We recognize that the first two terms are actual forces due to the primaries M_1 and M_2, the third term due to the centrifugal force (or effect), a pseudoforce, and the last term due to the Coriolis effect. Except for the velocity-dependent Coriolis term, the other three terms may be obtained from the gradient of a scalar function of coordinates only, say $V(x, y)$, such that the acceleration can be written as

$$\vec{a} = -\nabla V - 2\vec{\omega} \times \vec{v}. \tag{4.56}$$

We call V the effective potential given by

$$\begin{aligned} V(x, y) &= -\frac{GM_1}{r_1} - \frac{GM_2}{r_2} - \frac{1}{2}\omega^2 r^2 \\ &= -\frac{GM_1}{\sqrt{(x+a)^2 + y^2}} - \frac{GM_2}{\sqrt{(x-b)^2 + y^2}} - \frac{1}{2}\omega^2(x^2 + y^2). \end{aligned} \tag{4.57}$$

The effective potential consists of the actual gravitational potentials and a centrifugal potential that resembles the harmonic oscillator potential but with a negative spring constant (see Eq. (4.32)). The force is therefore repulsive and pushes radially outward. This is because the motion is being described in the rotating system. We could imagine that, to keep the test body stationary in the rotating system, a real force toward the center, that is, a centripetal force, would be needed to balance an extra (unseen) force to keep it in place. That extra force is called the centrifugal force.

Units for Restricted Three-Body Problems

The effective potential (4.57) (and equations of motion) are valid for any unit system. To further analyze its properties quantitatively, we should adopt a unit system most convenient for the restricted three-body (RTB) problem at hand. Analogous to planetary motion, we choose the RTB units as shown in Table 4.3.

Consequently, it follows by definition that $\omega = 2\pi/T$, and from Eqs. (4.16) and (4.17) that

$$GM = G(M_1 + M_2) = 4\pi^2\frac{R^3}{T^2}. \tag{4.58}$$

Table 4.3 RTB Unit System for the Restricted Three-Body Problem.

Mass	$M = M_1 + M_2$, total mass of the system
Length	$R = a + b$, the separation between the primaries
Time	T, rotational period of the primaries

This means that in actual calculations employing the RTB unit system ($M = R = T = 1$), we can set angular velocity $\omega = 2\pi$ and $GM = 4\pi^2$, just as in the planetary unit system.

For the distances, because the center of mass is chosen as the origin (Figure 4.16), we have

$$M_1 a = M_2 b, \ R = a + b.$$

Solving for a and b and introducing the mass parameter α, we have

$$a = \frac{M_2}{M_1 + M_2} R = \alpha R, \ b = \frac{M_1}{M_1 + M_2} R = (1 - \alpha)R, \ \alpha = \frac{M_2}{M_1 + M_2}. \tag{4.59}$$

It is customarily assumed that $M_2 \le M_1$, so $0 < \alpha \le 0.5$. In terms of α, the primaries' masses in our unit system can be expressed as

$$GM_1 = \frac{GM_1}{GM} GM = 4\pi^2(1 - \alpha), \ GM_2 = \frac{GM_2}{GM} GM = 4\pi^2 \alpha. \tag{4.60}$$

Consider the Earth–Moon system as an example. The primaries' masses are $M_1 = 5.974 \times 10^{24}$ kg and $M_2 = 7.348 \times 10^{22}$ kg, respectively. The distance between them is 3.844×10^8 m, and the rotational period is about one month (27.3 days). The absolute units and the RTB units are listed in Table 4.4.

We note that the RTB units are elastic. In other words, when the restricted three-body system changes, for example, to the Sun–Jupiter system, the units of mass, length, and so on, have different numbers on the absolute scale, even though they may all be 1 in the RTB units.

Table 4.4 RTB Units of the Earth–Moon System.

Quantity	Absolute units	RTB units
Unit of mass	$M_1 + M_2 = 6.047 \times 10^{24}$ kg	1
Unit of length	$R = 3.844 \times 10^8$ m	1
Unit of time	$T = 2.36 \times 10^6$ s (27.3 days)	1
Unit of ω	2.66×10^{-6} rad/s	2π
Unit of velocity	163 m/s	1
Unit of acceleration	6.91×10^{-5} m/s^2	1
Mass parameter α	$M_2/(M_1 + M_2) = 0.0121$	0.0121
Earth's mass	$GM_1 = 3.98 \times 10^{14}$ m^3/s^2	$0.988 \times 4\pi^2$
Moon's mass	$GM_2 = 4.90 \times 10^{12}$ m^3/s^2	$0.0121 \times 4\pi^2$
Earth's position	$-a = -4.67 \times 10^6$ m	−0.0121
Moon's position	$b = 3.80 \times 10^8$ m	0.988

The Lagrange Points

We can simplify the effective potential (4.57) by rewriting it in the newly introduced RTB unit system as

$$V(x,y) = -4\pi^2 \left[\frac{1-\alpha}{\sqrt{(x+\alpha)^2+y^2}} + \frac{\alpha}{\sqrt{(x-1+\alpha)^2+y^2}} + \frac{x^2+y^2}{2} \right]. \tag{4.61}$$

One single mass parameter α fully determines the effective potential. The potential surface for $\alpha = 0.121$ is shown in Figure 4.17. This α is about ten times larger than that of the actual Earth–Moon system to show some fine features.

Figure 4.17 is produced with Program 4.6. The program is explained in detail there. However, the block (lines 21 to 34) can be used as a template for making surface and contour plots.

The surface plot (Figure 4.17, top) shows several expected features. The potential has two holes (singularities) at the primaries, M_1 at $(-a, 0)$ and M_2 at $(b, 0)$. It also goes to $-\infty$ at large r due the centrifugal potential $-r^2/2$. In between, there is a broad C-shaped ridge surrounding the bigger body M_1. The formation of the ridge is due to the balance of the gravitational and centrifugal potentials. As the ridge nears the other body M_2, a hole is "burned" into it, and a saddle point is created between M_1 and M_2.

The contour plot (Figure 4.17, bottom) shows that there are finer structures embedded along the ridge. There are five points where the potential is locally maximum. They are known as the Lagrange points, labeled L_1 to L_5.[13] Three of them, L_1, L_2, and L_3, are located on the x-axis, and the other two, L_4 and L_5, are off the axis but symmetrically located about it, one above and one below. On closer look, the locations of the primaries and L_4 (or L_5) form an equilateral triangle.

The gradient of the potential is graphically illustrated by the arrows. Because $\vec{F} = -\nabla V$, the arrows are proportional to the force (hence acceleration) on the test body. We see large gravitational forces around each of the primaries, as well as large centrifugal forces at large distances. Near the Lagrange points, the forces are small. Right at the Lagrange points, the forces are exactly zero. If we were to place a test body at any point L_1 to L_5 exactly and perfectly at rest (in the rotating system), it would remain there indefinitely. However, the Lagrange points are unstable, because forces in the neighborhood point away from them. Any slight deviation (or perturbation) would most likely lead the test body to move away. The existence of Lagrange points is a result of balance between the attractive gravitational and the repulsive centrifugal forces. Without rotation, all Lagrange points but one would vanish. Only L_2, the saddle point, would remain. The effect of the potential surface is wholly dynamic.

For small α, all five Lagrange points lie close to the unit circle $r = 1$. While the locations of L_4 and L_5 can be found analytically, the locations of L_1, L_2, and L_3 need to be solved numerically. But for $\alpha \ll 1$, the following series expansion can be used (Exercise E4.11)

$$\begin{aligned} &L_1 : \left(-1 - \frac{5\alpha}{12}, 0\right), \; L_2, L_3 : \left(1 \mp \sqrt[3]{\frac{\alpha}{3}}, 0\right), \\ &L_4, L_5 : \left(\frac{1}{2} - \alpha, \pm\frac{\sqrt{3}}{2}\right). \quad \text{(exact)} \end{aligned} \tag{4.62}$$

[13]There are variations about the labeling of L_1, L_2, and L_3. We sequentially label them from left to right, agreeing with the convention in Ref. [40].

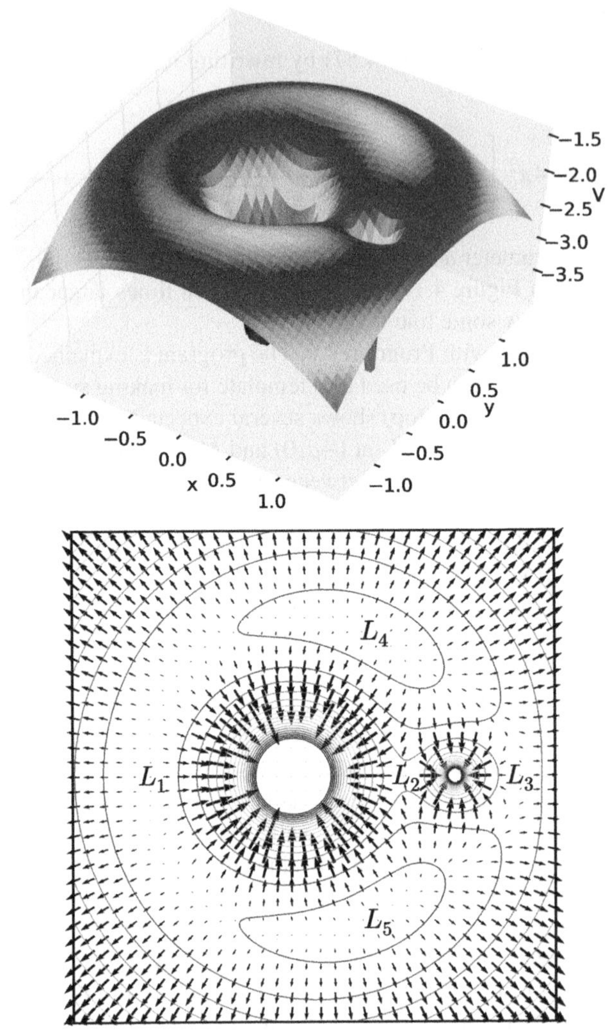

FIGURE 4.17 The effective potential of a restricted three-body system with $\alpha = 0.121$. The arrows on the contours indicate the force from the potential.

Because the existence of the Lagrange points are due to the balance between gravity and rotation, it might appear curious that the locations (4.62) depend on α only and not on other parameters such as the angular velocity ω. The reasons are twofold. First, the mass parameter α as defined by Eq. (4.59) depends on the masses M_1 and M_2. Because ω also depends on the masses via Eq. (4.16), α is related to ω indirectly. Second, as discussed earlier, the RTB unit system (Table 4.3) is elastic. This hides some parameters that would be explicit if we used absolute units.

4.6.3 MOTION AROUND THE LAGRANGE POINTS

As discussed earlier, with the effective potential alone, none of the Lagrange points are stable. However, we must also consider the Coriolis effect (4.56), $-2\vec{\omega} \times \vec{v}$, when the test body is not stationary. The combination of the force from the effective potential and the Coriolis effect makes it possible to have stable periodic orbits in the neighborhood of the Lagrange points.[14]

Let us examine that possibility by focusing on the equipotential island surrounding L_4 in Figure 4.17 (bottom). All around the island, the force from the effective potential (the arrows) points away from L_4. Suppose the test body moves clockwise around the edges. On the outer edge, it will be moving to the right. Recall that $\vec{\omega}$ points out of the page, and the Coriolis effect, $-2\vec{\omega} \times \vec{v}$, points inward. It is thus possible for the Coriolis effect to cancel the outward force, such that the net acceleration is toward L_4. On the inner edge, the velocity is reversed. So does the Coriolis effect. The net result is the same: it counterbalances the inward force from the effective potential. Because no energy is added by the Coriolis effect, $(\vec{\omega} \times \vec{v}) \cdot \vec{v} = 0$, it should be possible for the test body to move about L_4 periodically with the right initial conditions. As it turns out, stable periodic motion does happen if $\alpha < 0.03852$ [79].

The Trojan Asteroids

Both the Earth–Moon and the Sun–Jupiter systems support stable periodic orbits. We start with an actual example for the motion of asteroids in the Sun–Jupiter system. Most asteroids in our solar system are in the so-called asteroid belt. But there are two groups of asteroids that revolve around the Sun at about the same radius and with the same period as Jupiter (Figure 4.18, left). The two groups are symmetrically located with respect to the Sun–Jupiter radial vector: one group 60° ahead and another 60° behind. They are known as the Trojan asteroids (or simply Trojans) [95]. Because their locations form equilateral triangles with the Sun and Jupiter just like the Lagrange points L_4 and L_5 (Figure 4.18, right), we suspect there is a connection.

Though Jupiter is the most massive planet, the mass parameter for the Sun–Jupiter system is $\alpha = 0.0009542$ (see Table 4.1), still well within the limit for stable orbits. If the Trojans move in clockwise orbits around L_4 and L_5 as indicated in Figure 4.18 (right), stable orbits can be formed.

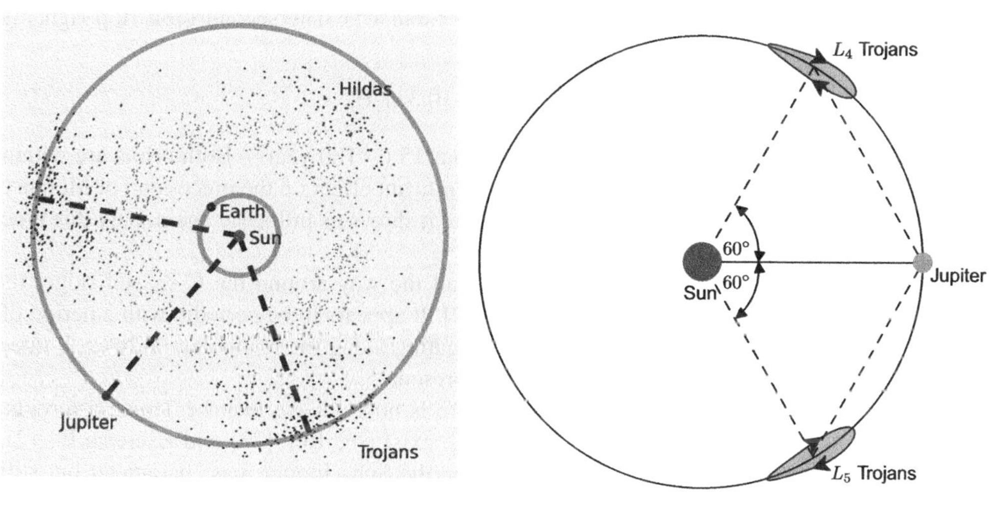

FIGURE 4.18 The Trojan asteroids and their connection to the Lagrange points L_4 and L_5.

[14]This should not be confused with the ability of a central potential to support closed orbits discussed in Section 4.2.5. The forces involved here are velocity-dependent and not central, and the orbits are not closed in the space-fixed coordinates.

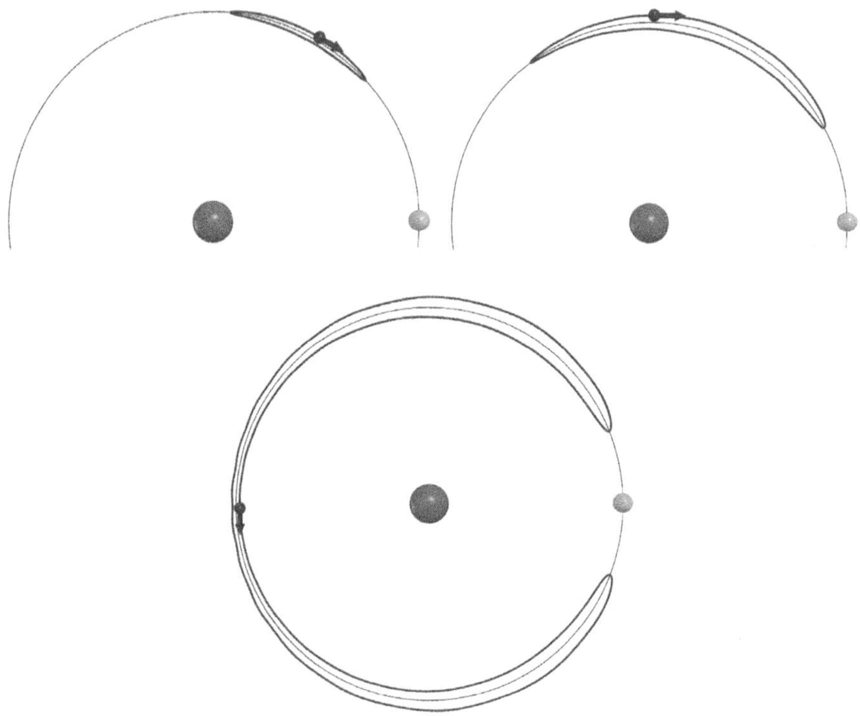

FIGURE 4.19 The orbits of Trojan asteroids near the Lagrange point L_4 (top) and along the C-ridge (bottom).

Program 4.7 (Section 4.D) simulates stable orbits around Lagrange points. Three orbits are shown in Figure 4.19. The first orbit (top left), also the smallest, corresponds to the initial condition as given in the program. It is a stable periodic orbit with a period of about 13 time units (Jupiter orbital periods), or about 150 years (using Table 4.1). The second orbit (top right) is produced with the initial condition

$$\vec{r},\ \vec{v} = [0.0, 1.035],\ [0.331, 0].$$

It is approximately periodic and has a period of about 15 ($\sim$ 180 years). Both orbits are within the Trojan group. We have seen that Jupiter has a significant effect on the precession of Mercury because it is the most massive planet (Table 4.2). Here it shows its influence again, only this time its lasting effect is on the asteroids.

The largest orbit (Figure 4.19, bottom) goes all the way around the C-shaped ridge. Its initial condition is $\vec{r},\ \vec{v} = [0.524, 0.92],\ [0.49, -0.27]$. It appears to be periodic with a period of 33 ($\sim$ 390 years). The motion is very delicate, navigating along the equipotential lines. It does not seem to be connected to any asteroid motion at present.

As for the Earth–Moon system, although there is no evidence like the Trojan asteroids, numerical results show that stable periodic orbits do exist (see Ref. [73] and Exercise E4.12). The mass parameter is $\alpha = 0.0121$, much larger than the Sun–Jupiter mass parameter but still within the limit for the existence of stable periodic orbits. Therefore, it is theoretically possible to put a satellite or space station around L_4 or L_5. However, with the Sun being very close and perturbing the motion, the long-term stability of these orbits is questionable.

All three orbits move clockwise (↻) around the potential ridge. Motion in the restricted three-body system is highly directional. Reversing the velocity of a stable periodic orbit in the restricted three-body problem results in unstable and nonperiodic motion (see Project S:P4.2). In contrast, reversing the velocity would have no effect on a Kepler orbit.

4.6.4 ORBITAL RESONANCE

As viewed in the space-fixed frame where the Sun is at rest, Jupiter and the Trojan asteroids have the same orbital period. We say their periods are commensurate. We can think of Jupiter's large effect on the Trojan asteroids as a result of the commensurability, or orbital resonance. If we consider the ratio of the orbital periods, we will find that orbital resonance exists if that ratio is a low rational number. In the case of Trojans and Jupiter, the ratio is 1/1. When orbits are in resonance, there is a periodic interaction ("kicks") between the orbits, which does not get averaged out and can lead to enhanced effects in the long run.

Trojans are not the only asteroids in resonance with Jupiter, though they are the most prominent. There is another group of asteroids known as the Hilda group, which has a period ratio of roughly 2/3 to Jupiter. The Hildas are in three clusters forming a triangle about the Sun (see Project P4.7).

Here we discuss another interesting case of orbital resonance between Pluto and Neptune.[15] The mass parameter of the primaries, namely the Sun and Neptune, is very small, $\alpha = 0.0000515$. The ratio of Pluto's period to Neptune's is very close to 3/2 (see Table 4.1). For every three revolutions of Neptune around the Sun, Pluto makes two revolutions. Like the Trojan asteroids, the motion of Pluto is mainly determined by the Sun. However, considering the low commensurability ratio, we expect significant orbital resonance effect from the perturbation of Neptune. Also, because of Pluto's relatively high eccentricity (see Table 4.1), its orbit near the perihelion is inside Neptune's orbit (Figure 4.20, top). The question arises as to whether Pluto's orbit is stable and if an "accidental" collision may occur between Pluto and Neptune that might knock Pluto out.

Let us investigate these questions by running our simulation (see details in Project S:P4.3). We can run Program 4.7 as is, with a different α and initial conditions, of course. But if we combine it with Program 4.6, we can observe the orbit in relation to the potential contours and the Lagrange points, which will be helpful in understanding what is going on. All we need to do is to copy parts of the code that do the numerical integration `accel()` and `r3body()`, integrate with RK4 as usual, and record the positions x and y, and plot them on the same figure as the potential contours when the integration ends. The results are displayed in Figure 4.20. The initial condition used is [48]

$$\vec{r},\ \vec{v} = [0.607396, -0.777497],\ [0.680684, 0.531809].$$

This starting position is close to L_5.

The top figure shows the orbit for four Pluto periods. Starting near L_5 and the perihelion, we can see that Pluto's orbit dips inside Neptune's orbit by a small amount. It is about 0.013 length units when Pluto is closest to the Sun. The length unit is the distance between Neptune and the Sun, or 30 AU from Table 4.1. That means Pluto penetrates Neptune's orbit by about 0.39 AU, or nearly 40% the orbital radius of Earth. After one Pluto period, Pluto is again closest to the Sun but at the crossover loop diagonally opposite from the starting position near L_5. It takes two periods for Pluto to return to the starting position. If we look carefully at the crossover loop near L_5, we find that the orbit is not closed. The loop has rotated to the left. During this time, Neptune has revolved around the Sun three times.

Letting the program run for longer times, we obtain the complete orbit shown in Figure 4.20 (bottom). We see the continued pattern of the rotation of the orbit, or precession. The precession occurs over an angle of 76°, and remarkably, it reverses direction after that and returns to the starting position to complete the cycle, which then repeats itself. The oscillatory back-and-forth precession is also called libration [17]. Half a libration cycle takes about 60 time units, or 10,000 years, making a full libration cycle approximately 20,000 years.

[15] Pluto and Neptune have different inclination angles. We ignore that difference here and assume they move in the same plane.

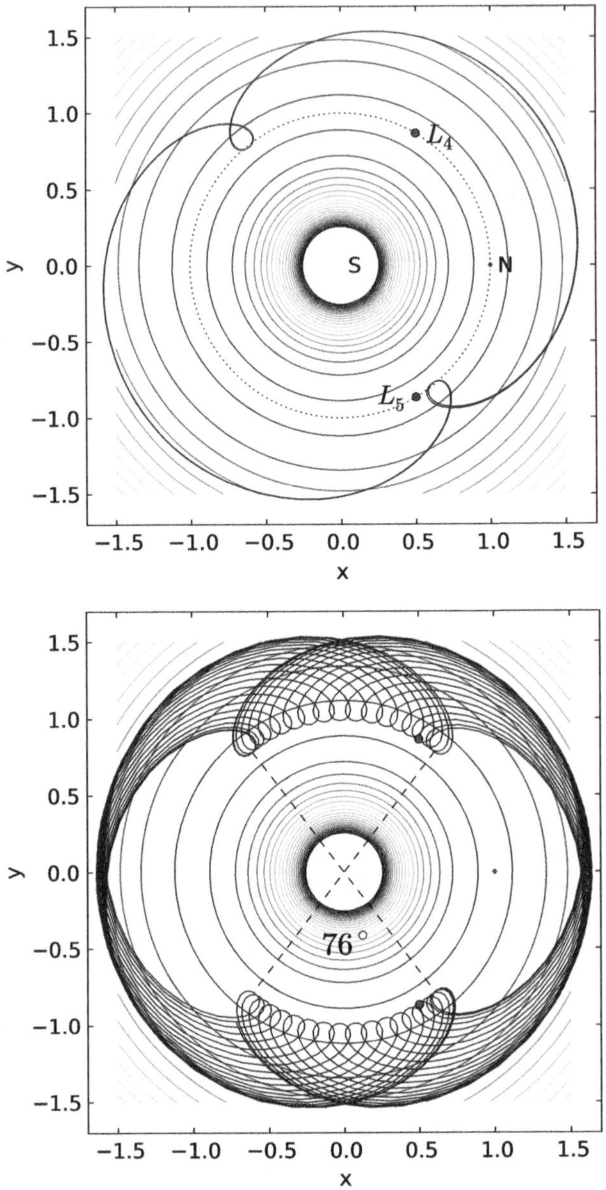

FIGURE 4.20 Orbits of Pluto in the Sun–Neptune system, $\alpha = 0.0000515$, overlaid on the equipotential contours. Top: After several periods. The dotted line is the circular orbit of Neptune (viewed in a nonrotating frame). Bottom: After one half of libration (60 time units, or $\sim$ 10,000 years).

We can understand the reason of the precession as follows. When Pluto is at the crossover loop nearer to Neptune, say at the Lagrange point L_5, it is behind Neptune (taking counterclockwise direction as the forward). The forward pull from Neptune increases Pluto's energy slightly. As a result, its period also increases. The next time Pluto comes back (two revolutions around the Sun), it will be at a later time, and Neptune will be farther ahead, so the distance between Neptune and Pluto's perihelion (tip of the crossover loop) increases. Thus, Pluto appears to precess to the left, or rotate clockwise. In effect, Neptune repels Pluto, or more accurately the nearer loop, away.

Eventually, Pluto gets ahead of Neptune at the perihelion because of the commensurate orbital periods (the other crossover loop near L_4). Then the backward pull from Neptune will decrease Pluto's energy and its period, which means that Neptune will be farther back, again effectively repelling the nearer loop. The cycles continues, giving rise to back and forth oscillations, or libration. The energy also oscillates with the libration period and an amplitude of about 0.4% (see Project S:P4.3).

Because of librational motion, at the closest approach to the Sun, Pluto gets no closer than 52° to Neptune in angular separation. Of course, they do line up radially when passing each other, for example, at (1.6, 0) in Figure 4.20 (bottom). But when that happens, they are radially far apart ($\sim$ 0.6 length units, or 18 AU). As a result, if Pluto and Neptune are close radially, they are far apart in angular separation, and vice versa. Therefore, a collision between them would not occur. In this case, orbital resonance adds dynamic stability to Pluto, and Pluto will stay in its current orbit for the foreseeable future.

We may then ask, how did Pluto end up in its current orbit? It is possible that Pluto, at the time it was formed, happened to be right where it is now, locked in a stable configuration due to Neptune's perturbation. It is also possible, arguably more likely, that Pluto evolved into its current orbit over a long period of time (billions of years). Evidence suggests that Pluto's motion may be chaotic [88]. It may well be that we are seeing a Pluto that has been marching on its chaotic path all along.

CHAPTER SUMMARY

Starting with an animated motion of a planet, we briefly reviewed useful properties of central field motion and Kepler's orbits. We then discussed precession of Mercury due to corrections of the general theory of relativity. Because the effect is small, we introduced the Runge-Lenz vector as a sensitive indicator of precession. An accurate and area-preserving method, the leapfrog method with time-transformation, was implemented to simulate this process. We found an interesting effect that precession is oscillatory, not monotonic. Precession of classical origin due to other planets was also discussed within the framework of central field approximation.

We also described the radial velocity method for the detection of exoplanets. We presented two ways to model RV datasets, least square and visual fit, and applied them to observational datasets.

The planar three-body problem has been discussed, including Euler's collinear motion and several interesting choreographic solutions of gravitational systems. We studied the restricted three-body problem and analyzed in detail the structure of the Lagrange points. Motion around the Lagrange points was simulated. We presented the application to the Trojan asteroids in the Jupiter–Sun system. We also discussed orbital resonance and the stability of Pluto in the Neptune–Sun system.

Throughout this chapter, we have applied several visualization techniques using the 3D elements of VPython and Matplotlib. We have consistently presented the results as dynamic animations or informative graphs with these techniques. They are contained in blocks of sample codes that can be used as basic templates for integrating simple but effective visualization elements into other programs.

4.7 EXERCISES AND PROJECTS

Exercises

E4.1 With the coordinate transformation (4.4), show that (a) the equations of motion for two bodies interacting with the internal force (4.1) are given by Eq. (4.5); and (b)* the two-body Hamiltonian $\mathcal{H} = \frac{1}{2}mv_1^2 + \frac{1}{2}Mv_2^2 + V(|\vec{r}_1 - \vec{r}_2|)$ can be reduced to $\mathcal{H} = H_{CM} + H$, where $H_{CM} = \frac{1}{2}(m+M)v_{CM}^2$ is the center of mass motion, and H is the effective one-body Hamiltonian given by $H = \frac{1}{2}\mu v^2 + V(r)$.

E4.2 Write a program to plot the effective potential of the Earth–Sun system, Eq. (4.8). Use Table 4.1 to determine the angular momentum L. Use planetary motion units. Adjust the scales so the structure near the minimum is clearly displayed.

E4.3 (a) Modify Program 4.1 to simulate the motion of a planet under the force $F = -k/r^{2+\delta}$. Run the program using the same initial condition as in Figure 4.4, and with $\delta = 0.2$ and 0.5. Compare your results with Figure 4.4. Optionally, add an arrow object to the animation of the planet to indicate instantaneous velocity using the `arrow()` function of VPython.

(b) When $\delta \neq 0$, open orbits are real. However, numerical problems can cause unphysical artifacts, including open orbits that should be closed. Let us use the same program, but set $\delta = 0$ and the initial condition to $\vec{r}, \vec{v} = [0.8, 0], [0, 3.5]$ (for exaggerated effect). First run it with $h = 0.001$, then with $h = 0.004$. Discuss your observations. Should we always "trust" the computer results?

E4.4 (a) Prove that the Runge-Lenz vector is a constant, that is, $d\vec{A}/dt = 0$, in the inverse-square law of force (4.1).

(b) Consider an additional perturbation force, $\vec{f}$, such that the total force is $\vec{F}_{tot} = \vec{F} + \vec{f}$, where $\vec{F}$ is given by Eq. (4.1). Show that the Runge-Lenz vector obeys

$$\frac{d\vec{A}}{dt} = \vec{f} \times \vec{L} + \vec{p} \times (\vec{r} \times \vec{f}).$$

As usual, $\vec{p}$ and $\vec{L}$ are the momentum and angular momentum, respectively.

E4.5 Suppose that, instead of $\Omega(r) = 1/r$, we choose $\Omega(r) = 1/r^2$ as the time transformation. Rewrite the corresponding equations of the leapfrog method to Eqs. (4.29a) to (4.29f). Discuss the merits and shortcomings of this time transformation.

E4.6 (a) A total mass M is uniformly distributed over a ring of radius a. A point particle of mass m is placed at a distance r from the center and in the plane of the ring. Show that the potential energy between the particle and the ring is given by Eq. (4.30).

(b) Expand the potential for $r/a < 1$ and $r/a > 1$, keeping four terms in each case. Verify Eq. (4.31), and explain why only even terms appear. You may wish to use SymPy to expand the integrand,

```
In [1]: from sympy import *
In [2]: init_printing(pretty_print=True) # neater output
In [3]: r, x = symbols('rho x')
In [4]: f=series(1/sqrt(1+r*r-2*r*cos(x)),r,0,7) # expand to 7th
```

Inspect `f`, then integrate $x \in [0, 2\pi]$ in SymPy (see Exercise E2.7).

(c) Compute the value M_p/a^3 for each planet in Table 4.2 using the parameters in Table 4.1. Plot precession vs. M_p/a^3 on linear scales. Compare with Figure 4.8 and discuss the differences.

E4.7 (a) Obtain from Eq. (4.35) a relation similar to Eq. (4.36) but with $m \sin i$ given in Jupiter masses, T in days, M in solar masses, and $\mathcal{V}$ in m/s.

(b) Carry out the algebraic details to prove Eq. (4.39) from (4.37).

(c) Verify Eq. (4.79), and reduce it to Eq. (4.80) using conservation of angular momentum (4.6) and Kepler's third law (4.18).

E4.8 (a) Solve Eqs. (4.44) to (4.46) to obtain Euler's quintic equation (4.47).

(b) Write a program to plot Eq. (4.47) for a few sets of masses, say $[m_1, m_2, m_3] = [1, 1, 1]$ and $[1, 2, 3]$. How many positive roots are there in each case? Numerically solve for the positive root(s) in each case using either the bisection or Newton's methods discussed in Chapter 3.

E4.9 (a) The periodic orbit shown in Figure 4.14 is unstable but reversible. First find its period. One way to do so is to add a text box to Program 4.5 using the `label()` function of VPython, see Program 4.3

for an example. Output the time information, and record the period T when the three bodies return to their initial positions (4.51). To make it easier, you can pause the program by intercepting key presses with `scene.kb.keys` and `scene.kb.getkey()`. The system should start to come apart after roughly two periods. Wait until the bodies are well off their orbits, and reverse the velocities of all three bodies. This may be done as a conditional statement based on time or by detecting certain keystrokes. You should observe that the bodies retrace their paths and go back to the initial positions in the same amount of time.

(b) The figure-8 orbit shown in Figure 4.15 is stable. It means that if we start the motion from a slightly different initial condition, the orbit will remain close to the original orbit. Verify that this is true. Run Program 4.5 with the initial condition (4.52). Next change the initial condition by a small amount, say $x_1' \to x_1 + 0.1$, and run the program again. Finally, repeat with $y_3' \to 0.05$. Explain your observations.

E4.10 Make a table similar to Table 4.4 showing the units of various quantities in the restricted three-body units and in the absolute units for the Neptune–Sun system. You may need the data from Table 4.1.

E4.11 (a) Set up the equations for the Lagrange points where $\nabla V = 0$ from Eq. (4.61). Solve these equations to obtain the lowest-order expansions for the collinear ones L_1, L_2, and L_3 and the exact expressions for L_4 and L_5. Show that they agree with Eq. (4.62).

(b) Write a program to solve the equations you obtained in part (a) for the collinear Lagrange points L_1, L_2, and L_3. Use the Earth–Moon system as an example. Compare the numerical results with the series expansions.

E4.12 The Earth–Moon system supports stable periodic orbits around the Lagrange points L_4 and L_5. Explore these orbits with the following initial conditions,

$$\vec{r},\ \vec{v} = [0.49786, 0.88335],\ [0.16698, -0.09218];$$

$$[0.51286, 0.90933],\ [0.42897, -0.20988].$$

(a) For each orbit, find the period, and estimate its length along the arc and its width where it is widest. Convert them to absolute units. From your estimates, find the approximate area of the smaller orbit. If that area were to be colonized, what would be the population density if the entire population on Earth (7 billion) were to be moved? Compare it with the current population density.

(b) Check the stability of the orbits by varying the initial positions slightly, say to $\pm 10\%$. You may also play around with the velocities. Refer to [73] for a general strategy to finding initial conditions of periodic orbits.

Projects

P4.1 (a) The artificial precession as observed in Exercise E4.3 could be eliminated by using a smaller h, which would lead to slower program speed and more round-off error. Though speed is not a problem in this case, it could be crucial in more complex situations. A better choice is to use a leapfrog method with time transformation. Repeat part (b) of Exercise E4.3 with `leapfrog_tt()`. Remember to initialize W_0 before entering the loop (see Program 4.3). Try different step sizes. You should see that, no matter how large h is, there is no precession. Explain why.

(b) Perform checks to make sure that the program works correctly and the oscillations in Figure 4.7 are not numerical artifacts. First, modify Program 4.3 to calculate and record the energy and angular momentum in the main loop. Ignore mass because it is an overall scale factor (see Eq. (4.86)), and use a large $\lambda \sim 10^{-2}$ for exaggerated effects. It is best to define a separate function for energy. Run the program for $t < 10$, and plot the energy and the z-component (L_z) of angular momentum as a function of t. Make sure your potential is correct.

To get a better view of the fluctuations, plot the relative error $|E(t) - E_0|/|E_0|$ and $|L_z(t) - L_0|/L_0$, where E_0 and L_0 are the initial values computed before the loop. The output should be analogous to Figure 2.7. Are they conserved as expected?

(c) On the precession figure, use the zoom-in feature to expand the scale so the oscillations are clearly visible. Measure the amplitudes and periods. Repeat the simulation by doubling and halving h. Check whether the amplitudes or the periods change.

Perform the same checks by changing the sign of λ, that is, enter the number -1.1×10^{-8} when prompted. Does the direction of precession change? Describe your observations and conclusions.

P4.2 Though Mercury has the largest precession due to general relativity, other planets also experience this effect to a lesser degree. For example, the parameter for Venus is $\lambda = 2.1 \times 10^{-8}$ AU^2 (see Eq. (4.19)). Change Program 4.3 accordingly to calculate the precession of Venus per century due to relativistic correction. Refer to Table 4.1 for the initial condition at the aphelion. Plot the results in a similar fashion to Figure 4.7. Note the amount of precession and the amplitude and the period of oscillations. Expand the horizontal scale to observe the oscillations. They should be sinusoidal, so count the period carefully (two crests, for example).

For Earth, the parameter is $\lambda = 2.9 \times 10^{-8}$ AU^2. Based on the preceding results for Venus and for Mercury (Figure 4.7), make a prediction about the amount of precession and the period of oscillations we would expect. Write down your predictions. Repeat the simulation for Earth. How do the results compare with your predictions? Briefly explain.

Compare the results for Venus and Earth with Mercury. The parameters λ for Venus and Earth are larger than for Mercury by a factor of 2 and 3, respectively. Why are the results not proportional to λ? Also discuss the differences in the amplitudes and periods of the oscillations. Explain why the oscillations are sharper (more sinusoidal) for Venus and Earth than for Mercury. Offer any other observations you may have.

P4.3 As we have discussed, the main contribution to the precession of Mercury is from other planets. This amount must be subtracted from the total precession to find the contribution from general relativity.

(a) Calculate the precession due to each planet in the central field approximation using the force (4.32) with the parameters from Table 4.1. Reproduce the results of Table 4.2. Make sure your results are converged with respect to the step size h. One way to check for convergence is to run the simulation with $h = 0.002$, for example, and then reduce it by half to $h = 0.001$. The results should agree to with certain error you set, say 1% for instance.

(b) Observe that the net precession is counterclockwise, which is in the same direction for the relativistic case. However, the force (4.32) in the central field approximation is repulsive, whereas the relativistic effect (4.19) is attractive. Investigate why forces of opposite signs produce precession in the same direction, because only then can we add them up. Do this by switching on the animation, and watch carefully how the Runge-Lenz vector swings in relations to where Mercury is in the orbit. Explain your observations. The results in Exercise E4.4 may help.

P4.4 If the force on a planet does not obey the inverse-square law, even for only a part of the trajectory, the orbit is open and will precess. Let us investigate this effect by modeling the Sun as a uniform sphere of radius R instead of a point source. If the planet is outside the sphere, the force is the usual gravity. But inside the sphere, the force will change to a form different from $1/r^2$.

(a) Show that the gravitational potential and force between a point mass m and a uniform sphere of radius R and mass M are given by

$$V = \begin{cases} \dfrac{GMm}{2R}(r^2/R^2 - 3), & r \le R, \\ -GMm/r, & r > R, \end{cases} \quad ; \quad F = \begin{cases} -GMmr/R^3, & r \le R, \\ -GMm/r^2, & r > R. \end{cases}$$

(b) Modify the force in `mercury()` in Program 4.3 according to the preceding expression. Set $R = 0.3$, leave everything else the same, and run the program. Does the Runge-Lenz vector move? Why?

(c) Compute the perihelion distance of Mercury r_{min} from Eq. (4.14). Set R just above r_{min}, and run the program. Does the Runge-Lenz vector move now? Experiment with R values slightly above r_{min} and determine the value that produces the same precession (43″) as the relativistic effect. Discuss your findings. Does it surprise you?

(d) Let us have fun making a Runge-Lenz clock. With the same initial conditions, find the value of R such that the hour hand (Runge-Lenz vector) ticks exactly 12 times in a full circle. Alternatively, you can achieve the same effect by varying the eccentricity of the orbit while keeping R fixed. Set $R = 0.2$ and vary the initial vertical velocity v_y. Find the value of v_y when the hour hand moves 12 times in one revolution. What would be the eccentricity of the corresponding Kepler orbit?

P4.5 Simulate the wobbling motion of a star-planet system with animation of the two-body problem, and discover a mystery planet.

(a) With Program 4.1 as a basic template, add the Sun as a movable object. Do not include the Sun as a dynamic body in the numerical integration. We still assume an effective one-body problem. We only need to obtain the positions of the Sun and the planet ($\vec{r}_2$ and $\vec{r}_1$ in Figure 4.2) in the center of mass (CM) from the relative planet position stored in `r`, which is already available in line 26 of Program 4.1.

Assume $\vec{R} = 0$, and solve Eq. (4.4) for $\vec{r}_1$ and $\vec{r}_2$ in the CM (see Exercise E4.1). Animate the Sun and the planet by updating their positions in the CM. Make sure the modified code runs correctly for the Sun–Earth system.

(b) Note that due to the large Sun–Earth mass ratio, the Sun barely moves. Now change the mass of the planet, say 0.4 solar mass. Run the code again and observe the visible wobbles. You may also need to change the initial conditions depending on the mass ratio.

(c) Plot the radial velocity of the Sun, v_{*z}, in the actual Sun–Earth system similar to Figure 4.9, assuming edge-on view. Discuss the results, such as the shape and magnitude of the curve.

(d) Make your own mystery planet by assuming certain mass, say anywhere from 1 to 10,000 Earth mass, and initial condition (for added challenge, choose your initial condition such that the orbit has a high eccentricity). Run your program and generate an RV dataset accordingly. Exchange your dataset with another classmate (or team), and try to visually fit the dataset according to Eq. (4.34). Calculate the planet mass from your fit. Compare your value with the value your classmate used to generate the dataset. Do they match? Why or why not?

P4.6 A subset of the radial velocity dataset for HD 3651 plotted in Figure 4.11 is given in Table 4.5.

The star has 0.8 solar masses, and the wobbling period is assumed to be 62 days. The data points were observations from Keck Observatory spread over the years. They are converted (phased) to fall within one period.

Table 4.5 Radial Velocity (m/s) vs. Time (day) for HD 3651.

t	6.3	6.8	11.3	12.5	23.5	25.7	27.6	29.4
v_{*z}	13.9	7.4	9.6	2.2	5.7	1.3	0.3	−5.4
29.4	32.3	34.7	36.2	36.7	38.3	38.6	39.1	39.7
1.2	−3.3	1.5	10.9	−3.1	0.4	−5.9	−1.2	−1.5
40.6	42.2	43.6	50.8	54.1	55.5	56.3	56.8	61.3
−12.1	−5.2	−12.8	−18.8	−15.6	−26.4	−0.5	−1.2	5.9

(a) Follow the procedure outlined in Section 4.4.2 to generate a visual fit similar to the curve shown in Figure 4.11. Use Program 4.4, but turn off least square fit by commenting out line 38 and set time shift to 0 (line 27). Vary $\mathcal{V}$, e, ω, and C but keep T and time shift fixed. As the parameters are being adjusted, watch the effect of each parameter, and try to isolate them to achieve a better match. Once the curve takes the shape of the dataset, adjust the parameters separately and in small steps to visually fine tune the fit, focus on the shape and ignore an apparent horizontal shift for now. Describe the effect of each parameter.

(b) You should see that the fitted curve is shifted from the data, that is, the dip should be at a different location, even though the overall shapes match very well. If you would just shift your curve left or right, it would look quite like the fit in Figure 4.11. Explain the physical reason for the shift. Adjust the time shift to get the best fit. Write down the parameters, and calculate the lower limit of the exoplanet's mass, in Earth and in Saturn masses. Compare with observed values [30].

(c) Run Program 4.4 in full to obtain a least square fit. Compare the best results between visual and automatic fits. Optionally, switch to the reduced dataset of Table 4.5 or the self-generated dataset from Project P4.5, and discuss any difference in the fit.

P4.7 Similar to the Trojans, there is another group of asteroids known as the Hildas whose motion around the Sun is heavily influenced by Jupiter. The Hildas (Figure 4.18) do not move around the Lagrange points L_4 or L_5. Rather, like Pluto, they are influenced by orbital resonance with Jupiter. The Hildas have a period ratio of roughly 2/3 to Jupiter.

(a) Explore the motion of the Hilda asteroids with the following initial condition

$$\vec{r},\ \vec{v} = [0.64772, 0],\ [0, 4.2902].$$

Find the period of motion. First plot the orbit in the rotating frame, then plot it in the nonrotating space frame where the Sun is at rest. You would need to make a coordinate transformation from the rotating frame to the space frame using the rotation matrix (4.74). Compare the orbits between the frames.

(b) Normally the orbit would also have a librational motion. The following initial condition will show substantial libration,

$$\vec{r},\ \vec{v} = [0.78997, 0],\ [-.87208, 1.9879].$$

Run the program for 10 time units. Note the angles of libration. Modify the initial condition slightly, and describe how the librational motion changes.

4.A ROTATING FRAMES AND RATE OF CHANGE OF VECTORS

Consider the coordinate systems of a space-fixed reference frame and a reference frame rotating at constant angular velocity ω about the common z-axis. Let the coordinates be (x, y) and (x', y') in these reference frames, respectively, as shown in Figure 4.21. We assume that at $t = 0$ the axes (e.g., x and x') coincide, so the angle of rotation is $\theta = \omega t$ at later times.

From Figure 4.21, the relationship between the primed unit vectors in the rotating frame and the unprimed ones in the space frame can be written as

$$\hat{\imath}' = \cos\theta\hat{\imath} + \sin\theta\hat{\jmath},\ \hat{\jmath}' = -\sin\theta\hat{\imath} + \cos\theta\hat{\jmath}. \tag{4.63}$$

Because of rotation, the primed unit vectors $\hat{\imath}'$ and $\hat{\jmath}'$ change with time. By differentiating the preceding equation with the help of $d\theta/dt = \omega$, the rate of change is

$$\frac{d\hat{\imath}'}{dt} = \omega\hat{\jmath}',\quad \frac{d\hat{\jmath}'}{dt} = -\omega\hat{\imath}'. \tag{4.64}$$

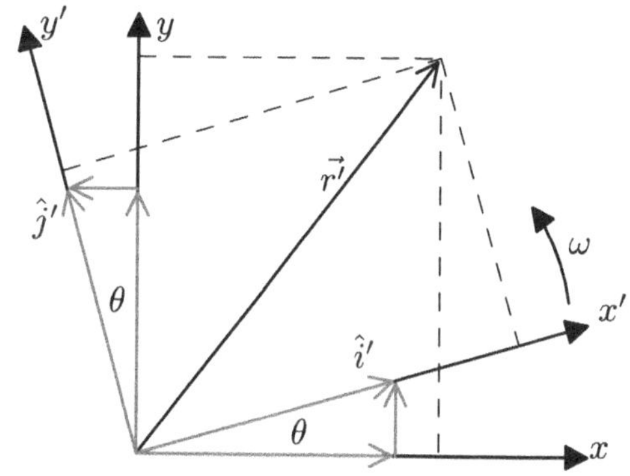

FIGURE 4.21 The space reference frame and the rotating reference frame.

Now, let us consider a position vector $\vec{r'}$ in the rotating frame with components x' and y'. The same vector in the space frame will be denoted as $\vec{r}$. Because we are talking about the same vector, only different notations, we have

$$\vec{r} = \vec{r'} = x'\hat{i}' + y'\hat{j}'.$$

To find the rate of change, we differentiate both sides simultaneously

$$\frac{d\vec{r}}{dt} = \frac{dx'}{dt}\hat{i}' + \frac{dy'}{dt}\hat{j}' + x'\frac{d\hat{i}'}{dt} + y'\frac{d\hat{j}'}{dt}. \tag{4.65}$$

The left side refers to the rate of change in the space frame. The first two terms on the right hand side are the rate of change due to the change of coordinates x' and y' as measured in the rotating frame, and the last two terms are due to the change of the primed unit vectors. Let us denote this more clearly by rewriting Eq. (4.65) as

$$\left(\frac{d\vec{r}}{dt}\right)_{\text{space}} = \left(\frac{d\vec{r'}}{dt}\right)_{\text{rot}} + \omega(x'\hat{j}' - y'\hat{i}'), \tag{4.66}$$

where $(d\vec{r}/dt)_{\text{space}}$ is the velocity in the space frame ($\vec{v}$), and $(d\vec{r'}/dt)_{\text{rot}} = dx'/dt\,\hat{i}' + dy'/dt\,\hat{j}'$ the velocity in the rotating frame ($\vec{v'}$). We have also used Eq. (4.64). We can write the second term in Eq. (4.66) as a vector product by introducing $\vec{\omega} = \omega\hat{k}'$. This leads to a more compact form

$$\left(\frac{d\vec{r}}{dt}\right)_{\text{space}} = \left(\frac{d\vec{r'}}{dt}\right)_{\text{rot}} + \vec{\omega} \times \vec{r'}, \tag{4.67}$$

or equivalently and in a more familiar form,

$$\vec{v} = \vec{v'} + \vec{\omega} \times \vec{r'}. \tag{4.68}$$

In deriving Eq. (4.67), we have used the position vector as an example and restricted the angular velocity in the z direction. It is actually valid for an arbitrary vector and angular velocity. The more general rule is

$$\left(\frac{d}{dt}\right)_{\text{space}} = \left(\frac{d}{dt}\right)_{\text{rot}} + \vec{\omega} \times . \tag{4.69}$$

Equation (4.69) should be read this way: The derivative of a vector in the space frame is equal to the derivative of the (same) vector in the rotating frame plus the cross product of the angular velocity and the vector. This is a very useful rule.

For instance, suppose we wish to calculate acceleration from Eq. (4.68). We have

$$\vec{a} = \frac{d\vec{v}}{dt} = \left(\frac{d\vec{v}'}{dt}\right)_{\text{space}} + \left(\frac{d(\vec{\omega}\times\vec{r}')}{dt}\right)_{\text{space}}. \tag{4.70}$$

Applying Eq. (4.69) to each of the two terms on the right-hand side, we have

$$\left(\frac{d\vec{v}'}{dt}\right)_{\text{space}} = \left(\frac{d\vec{v}'}{dt}\right)_{\text{rot}} + \vec{\omega}\times\vec{v}',$$

$$\left(\frac{d(\vec{\omega}\times\vec{r}')}{dt}\right)_{\text{space}} = \left(\frac{d(\vec{\omega}\times\vec{r}')}{dt}\right)_{\text{rot}} + \vec{\omega}\times(\vec{\omega}\times\vec{r}')$$

$$= \vec{\omega}\times\vec{v}' + \vec{\omega}\times(\vec{\omega}\times\vec{r}').$$

We have used $d\vec{\omega}/dt = 0$ and $(d\vec{r}'/dt)_{\text{rot}} = \vec{v}'$. Substituting them into Eq. (4.70) and using the fact that $(d\vec{v}'/dt)_{\text{rot}} = \vec{a}'$ is the acceleration in the rotating frame, we obtain the relationship for $\vec{a}$ and $\vec{a}'$ as

$$\vec{a} = \vec{a}' + 2\vec{\omega}\times\vec{v}' + \vec{\omega}\times(\vec{\omega}\times\vec{r}'). \tag{4.71}$$

Let $\vec{F}$ be the net force on a particle of mass m. The equation of motion in the space frame (inertial) is $\vec{F} = m\vec{a}$. Expanded in the rotating frame (noninertial), it becomes

$$\vec{F} - 2m\vec{\omega}\times\vec{v}' - m\vec{\omega}\times(\vec{\omega}\times\vec{r}') = m\vec{a}'. \tag{4.72}$$

Besides the actual force, this equation has two additional force-like terms. The velocity-dependent term, $-2m\vec{\omega}\times\vec{v}'$, is the Coriolis effect, and the other term, $-m\vec{\omega}\times(\vec{\omega}\times\vec{r}')$, is the centrifugal force.

4.B ROTATION MATRICES

Sometimes it is necessary to transform the coordinates from the rotating frame to the space frame. This can be done using $(x, y) = (\vec{r}'\cdot\hat{\imath}, \vec{r}'\cdot\hat{\jmath})$ and Eq. (4.63). The results may be expressed conveniently in the form of a rotation matrix as

$$\begin{bmatrix} x \\ y \end{bmatrix} = \begin{bmatrix} \cos\theta & -\sin\theta \\ \sin\theta & \cos\theta \end{bmatrix} \begin{bmatrix} x' \\ y' \end{bmatrix}. \tag{4.73}$$

The inverse transformation is

$$\begin{bmatrix} x' \\ y' \end{bmatrix} = \begin{bmatrix} \cos\theta & \sin\theta \\ -\sin\theta & \cos\theta \end{bmatrix} \begin{bmatrix} x \\ y \end{bmatrix}. \tag{4.74}$$

These rotation matrices work on any vectors, not just position vectors.

Denoting the rotation matrix by A, we can write Eq. (4.74) as

$$(\vec{r})' = A\vec{r}, \quad A = \begin{bmatrix} \cos\theta & \sin\theta \\ -\sin\theta & \cos\theta \end{bmatrix}. \tag{4.75}$$

Note that we are careful with the notation $(\vec{r})'$ rather than $\vec{r}'$. The reason is that the transformation has dual interpretations. Here, we are following the interpretation that the operator A transforms the components of the vector in the space frame (unprimed) into the components of the same vector in the rotating frame (primed). In other words, A is interpreted to act on the coordinate system, the vector itself is unchanged, and only the components are transformed from the unprimed frame to the primed frame, which is rotated counterclockwise by an angle θ.

The second interpretation is that the operator A acts on the vector $\vec{r}$, transforming it into a new vector $\vec{r}'$, both in the same reference frame. The new vector represents a clockwise rotation of the original vector by an angle θ. We would represent this operation as $\vec{r}' = A\vec{r}$. Both interpretations involve identical mathematical operations. In the current context, we follow the first interpretation for convenience, that is, the operator A relates the components of the same vector between rotated coordinates.

In three dimensions, the rotation matrices about the x-, y-, and z-axes follow straightforwardly from Eq. (4.74),

$$A_x = \begin{bmatrix} 1 & 0 & 0 \\ 0 & \cos\theta & \sin\theta \\ 0 & -\sin\theta & \cos\theta \end{bmatrix}, \quad A_y = \begin{bmatrix} \cos\theta & 0 & -\sin\theta \\ 0 & 1 & 0 \\ \sin\theta & 0 & \cos\theta \end{bmatrix},$$

$$A_z = \begin{bmatrix} \cos\theta & \sin\theta & 0 \\ -\sin\theta & \cos\theta & 0 \\ 0 & 0 & 1 \end{bmatrix}. \tag{4.76}$$

The inverse transformation can be obtained by changing the sign of the angle $\theta \to -\theta$ because a rotation by an angle θ is exactly canceled by a reverse rotation by $-\theta$, resulting in no change (the identity transformation). If A_i^{-1} represents the inverse transformation, we can write $A_i^{-1}(\theta) = A_i(-\theta)$, and $A_i^{-1}A_i = A_iA_i^{-1} = \mathbb{1}$, $i = x, y$, or z.

Successive rotations can be represented by a product of rotation matrices. For example, a rotation of α about the x-axis followed by a second rotation of β about the z-axis can be written as

$$A = A_z(\beta)A_x(\alpha). \tag{4.77}$$

The order is important for finite rotations, and the final matrix is built from right to left in the order of rotations. The inverse matrix is $A^{-1} = A_x^{-1}(\alpha)A_z^{-1}(\beta) = A_x(-\alpha)A_z(-\beta)$.

4.C RADIAL VELOCITY TRANSFORMATION

Our goal is to transform the velocity in the orbital plane from the center of mass frame to the space-fixed frame where observation is made. As Figure 4.3 shows, the relative position in the orbital plane is

$$x = r\cos\theta, \quad y = r\sin\theta. \tag{4.78}$$

Using Eq. (4.13), the velocity components can be obtained as,

$$v_x = \dot{x} = -r^2\dot{\theta}\,\frac{\sin\theta}{a(1-e^2)}, \quad v_y = \dot{y} = r^2\dot{\theta}\,\frac{\cos\theta - e}{a(1-e^2)}. \tag{4.79}$$

With the help of Eqs. (4.6) and (4.12), being careful to replace m by reduced mass μ, and using the exact form of Kepler's third law (4.18), we can reduce the orbital (relative) velocity to

$$v_x = -\frac{2\pi a}{T\sqrt{1-e^2}}\sin\theta, \quad v_y = \frac{2\pi a}{T\sqrt{1-e^2}}(\cos\theta - e), \tag{4.80}$$

where, as before, T is the period, a the semimajor axis, and e the eccentricity.

Let $\vec{v}'$ and $\vec{v}'_*$ denote the velocities of the planet and the star in the center of mass, respectively. Then $\vec{v}' - \vec{v}'_* = \vec{v}$, $M\vec{v}'_* + m\vec{v}' = 0$. Solving for $\vec{v}'_*$, we have

$$\vec{v}'_* = -\frac{m}{m+M}\,\vec{v}. \tag{4.81}$$

It is identical to Eq. (4.33) except for the primed notation, which will be convenient for the transformation below. Substituting Eq. (4.80) into (4.81), we obtain the wobbling velocity of the star in the center of mass frame

$$v'_{*x} = \frac{m}{m+M}\frac{2\pi a}{T\sqrt{1-e^2}}\sin\theta,$$

$$v'_{*y} = -\frac{m}{m+M}\frac{2\pi a}{T\sqrt{1-e^2}}(\cos\theta - e). \tag{4.82}$$

Next, we need to transform the wobbling velocity from the center of mass frame (primed) to the space-fixed frame (unprimed). This can be done conveniently by a rotation matrix. As Figure 4.10 illustrates, the orbit orientation is the result of two successive rotations. Imagine the aphelion (x'-axis) and the normal (the $\vec{L}$ vector, or z'-axis) of the orbital plane are initially aligned with X and Z, respectively. The final orientation of the orbit can be obtained as: a first rotation by an angle i about the x'-axis, and a second rotation by an angle ω about the z'-axis. The rotation matrix follows from Eq. (4.77), $A = A_z(\omega)A_x(i)$. Recall that A will transform a vector from unprimed frame (space) to the primed frame (center of mass). What we need is the inverse transformation from the center of mass frame to the space frame, that is, $A^{-1} = A_x^{-1}(i)A_z^{-1}(\omega) = A_x(-i)A_z(-\omega)$. Using the rotation matrices (4.76), we have the inverse transformation

$$A^{-1} = A_x(-i)A_z(-\omega) = \begin{bmatrix} \cos\omega & -\sin\omega & 0 \\ \sin\omega\cos i & \cos\omega\cos i & -\sin i \\ \sin\omega\sin i & \cos\omega\sin i & \cos i \end{bmatrix}. \tag{4.83}$$

Transforming the velocity vector in Eq. (4.82) by A^{-1}, we obtain the velocity of the star in the space-fixed frame

$$\vec{v}_* = A^{-1}\vec{v}'_* = \frac{m}{m+M}\frac{2\pi a}{T\sqrt{1-e^2}}\begin{bmatrix} \sin(\omega+\theta) - e\sin\omega \\ -\cos i\,[\cos(\omega+\theta) - e\cos\omega] \\ -\sin i\,[\cos(\omega+\theta) - e\cos\omega] \end{bmatrix}. \tag{4.84}$$

This is the desired result. We pick up a Z component in the process because the orbit is tilted due to the inclination angle. The inclination angle $i = 90°$ corresponds to the edge-on view, and $i = 0°$ the face-on view.

We can make two extensions to the result. First, if the center of mass is moving with velocity $\vec{V}_{CM}$, we must add it to $\vec{v}_*$. Second, if the star has multiple planets, and if planet–planet interaction is negligible,[16] then the total wobbling velocity is a vector sum of contributions from all the planets. For example, for the Z component v_{*z} of a star with N planets , we have from Eq. (4.84)

$$v_{*z} = -\frac{1}{M_T}\sum_{j=1}^{N}\frac{2\pi a_j m_j \sin i_j}{T_j\sqrt{1-e_j^2}}[\cos(\omega_j+\theta_j) - e_j\cos\omega_j] + C, \tag{4.85}$$

where $M_T = M + \sum_j m_j$ is the total mass, and $C = V_{CM,z}$ is the velocity of the center of mass along Z. Equation (4.85) can be used to fit the Doppler velocity profile to obtain the orbital parameters.

[16]This amounts to the independent particle approximation. It is a good approximation, particularly if we are primarily interested in the wobbling motion of the star.

Program listing 4.3 Precession of Mercury (`mercury.py`)

```
import ode, numpy as np          # get ODE solvers, numpy
import visual as vp              # get VPython modules for animation
import matplotlib.pyplot as plt  # get matplotlib plot functions

def mercury(id, r, v, t):        # eqns of motion for mercury
    if (id == 0): return v       # velocity, dr/dt
    s = vp.mag(r)
    return -GM*r*(1.0 + lamb/(s*s))/(s*s*s)  # acceleration, dv/dt

def set_scene(r):      # r = init position of planet
    # draw scene, mercury, sun, info box, Runge-Lenz vector
    scene = vp.display(title='Precession of Mercury',
                       center=(.1*0,0), background=(.2,.5,1))
    planet= vp.sphere(pos=r, color=(.9,.6,.4), make_trail=True,
                      radius=0.05, material=vp.materials.diffuse)
    sun   = vp.sphere(pos=(0,0), color=vp.color.yellow,
                      radius=0.02, material=vp.materials.emissive)
    sunlight = vp.local_light(pos=(0,0), color=vp.color.yellow)
    info = vp.label(pos=(.3,-.4), text='Angle') # angle info
    RLvec = vp.arrow(pos=(0,0), axis=(-1,0,0), length = 0.25)
    return planet, info, RLvec

def go(animate = True):                          # default: True
    r, v = np.array([0.4667, 0.0]), np.array([0.0, 8.198]) # init r, v
    t, h, ta, angle = 0.0, 0.002, [], []
    w = 1.0/vp.mag(r)                            # W_0 = Ω(r)

    if (animate): planet, info, RLvec = set_scene(r)
    while t<100:                                 # run for 100 years
        L = vp.cross(r, v)                       # L/m = r × v
        A = vp.cross(v, L) - GM*r/vp.mag(r) # scaled RL vec, Eq. (4.86)
        ta.append(t)
        angle.append(np.arctan(A.y/A.x)*180*3600/np.pi) # arcseconds
        if (animate):
            vp.rate(100)
            planet.pos = r                       # move planet
            RLvec.axis, RLvec.length = A, .25    # update RL vec
            info.text='Angle": %8.2f' %(angle[-1]) # angle info
        r, v, t, w = ode.leapfrog_tt(mercury, r, v, t, w, h)

    plt.figure()          # make plot
    plt.plot(ta, angle)
    plt.xlabel('Time (year)'), plt.ylabel('Precession (arcsec)')
    plt.show()

GM = 4*np.pi*np.pi       # G*Msun
# lamb=relativistic correction, global, used in 'mercury()'
lamb = input('Please enter lambda, eg: 0.01, or 1.1E-8 :> ')
go(animate = True)       # set to False to speed up calc. for plots
```

The equations of motion due to the relativistic correction (4.19) are computed by `mercury()`. It uses vector concepts and is nearly identical to `earth()` in Program 4.1, with the only difference in the correction factor $1 + \lambda/r^2$. Also, the animation scene is contained in a standalone function `set_scene()`.

The first block in the main program `go()` initializes the position and velocity for Mercury according to Table 4.1, as well as W_0. If `animate` is true, the simulation will set up the Mercury–Sun system by calling `set_scene()`, which draws the scene, including an arrow representing the Runge-Lenz vector plus a text box (`label`) that will display the angle of precession. In each iteration of the main loop, the scaled angular momentum and Runge-Lenz vectors are computed as

$$\vec{L}/m = \vec{r} \times \vec{v}, \ \vec{A}/m^2 = \vec{v} \times \vec{L}/m - GM\vec{r}/r. \tag{4.86}$$

By this scaling, mass drops out. The angle of precession is found by $\arctan(A_y/A_x)$ and converted into arcseconds from radians. We use full vector operations provided by VPython for $\vec{L}$ and $\vec{A}$. Note that the components of a VPython vector such as $\vec{A}$ can be accessed by index or subscript, for example, `A[0]` is the same as `A.x`.

Next, animation is effected as necessary. This involves changing the Mercury position and updating the direction of the Runge-Lenz vector. The current precession angle (`angle[-1]`) is converted to a float string[17] (line 38) and displayed in the text box. Last, the integrator `leapfrog_tt()` advances the system forward by one step h, in transformed time s. Upon exiting the main loop, the results are displayed in a graph.

Program listing 4.4 Radial velocity data modeling (`rvfit.py`)

```
import numpy as np                      # get numpy
import matplotlib.pyplot as plt         # get plot functions
from scipy.optimize import leastsq      # least square fit

def rv(V, e, omega, C, theta):      # radial velocity
    return -V*(np.cos(omega + theta) - e*np.cos(omega)) + C

def nearest(a, b):      # find indices of nearest differences
    diff = np.subtract.outer(a, b)        # find diff: a=[], b=[] or scalar
    return np.argmin(np.abs(diff), 0)     # nearest index

def time_theta(e, tshift, fit=True):      # calc time and theta
    psi = np.linspace(0., 2*np.pi, 200) # psi grid
    t = (psi+e*np.sin(psi))/(2*np.pi) + tshift        # Kepler eqn
    over, under = t>1.0, t<=1.0           # truth arrays
    t = np.concatenate((t[over]-1, t[under]))         # wrap around
    theta = np.arccos((e+np.cos(psi))/(1+e*np.cos(psi))) # calc theta
    theta[psi>np.pi] = 2*np.pi-theta[psi>np.pi]           # remap to [0,2*pi]
    theta = np.concatenate((theta[over], theta[under]))  # wrap around
    idx = nearest(t, time)                # index of data points
    return (t[idx], theta[idx]) if fit else (t, theta)

def error(p, v, tshift):              # error function for leastsq
    t, theta = time_theta(p[e], tshift)
    return v - rv(p[V], p[e], p[omega], p[C], theta)

```

[17]The float-to-string conversion works as `'format' % (expressions)`. Here, `format` is a string containing the C-style format specifier `%W.PT` where `W` is the field length, `P` the precision (decimal places), `T` the number type. All except `%` and the number type are optional. In the present case, `'%8.2f'` specifies a float of length 8 and 2 decimal places.

```
   T, tshift = 62.23, 0.38          # T=period (day), expt. with time shift
28 time, vel, err = [], [], []
   with open('hd3651.txt', 'r') as file:  # read HD 3651 dataset
30     for line in file:
           if line.strip() and line.strip()[0]!='#':   # not blank/comment
32             t, v, er = eval(line)          # comma separated fields
               time.append(t), vel.append(v), err.append(er)
34 time, vel = np.array(time)/T, np.array(vel)          # scaled time

36 p = [10., 0.5, 1., 1.]        # initial guess: [V, e, omega, C]
   V, e, omega, C = range(4)  # para labels [0-3] for easy ref
38 p, flag = leastsq(error, p, args=(vel, tshift))       # Go fit

40 t, theta = time_theta(p[e], tshift, 0)       # calc fitted results
   vfit = rv(p[V], p[e], p[omega], p[C], theta)
42
   plt.figure()
44 plt.errorbar(time, vel, err, fmt='o')       # data and fit
   plt.plot(t, vfit, lw=2)
46 plt.xlabel('$t/$T'), plt.ylabel('Radial velocity (m/s)')
   plt.text(0.10, -30, 'HD 3651')
48 plt.ylim(-40, 30), plt.show()              # set ylim
```

The purpose of this program is to model radial velocity (RV) datasets. Starting with a set of parameters, the overall strategy is: Compute the radial velocity over a time grid, match it with the nearest observation points, find the error at these points, and input the error to a least square fitting routine, which repeats the process until it hopefully converges to an optimal set of parameters.

The main program starts (line 27) by setting the period T and a time shift `tshift` (change the value to see its effects, see following). The period is not treated as a free parameter, and the time shift is chosen manually. We do so to reduce the parameter space, which is still considerable (four, see Eq. (4.34)).

The next block reads in the RV dataset from a file. The file is opened with `with...as` (line 29), a handy Python construct that automatically cleans up after operations within the block (e.g., closing the file in this case). For each line in the file, after being stripped of white space, if it is not blank (zero length) or a comment ("#" as the first character), `eval()` is called to parse the data fields assumed to contain three comma-separated values, namely time, velocity, and error bar, all added to the respective lists. After processing the file, the time and velocity lists are converted to ndarrays, `time` and `vel`, respectively, for vector operations throughout. Time is also scaled so it is between 0 and 1.

As the initial guess, the parameters $[V, e, \omega, C]$ from Eq. (4.34) are assigned to a list and indexed for easy reference (e.g., `p[e]` instead of `p[1]` to avoid confusion). We are ready now to enter `leastsq`, a least square routine from SciPy, which requires a user-supplied function computing the difference between the model and the data, an initial guess, plus optional parameters (such as velocity and time shift). We name this function `error()`.

Upon entering `error()`, the first function called is `time_theta()`, which accepts as arguments the eccentricity, time shift, and a Boolean variable (default true) indicating fitting or evaluation. The logic of this function is a bit more subtle than usual (and it may take a little digging to understand it), but its purpose is clear: to match a pregenerated time grid with the nearest data points for error computation later. We use the NumPy function, `linspace(start,end,num)`, to generate a 1D grid over ψ. It returns an array of `num` evenly spaced points between `[start,end]` inclusively ($[0, 2\pi]$ in this case). Using Kepler's equation (4.39), the time grid `t` is created from array `psi` in element-wise operations. Here, it is necessary to shift time by a trial amount `tshift` because the starting time in the exoplanet orbit is unknown. To reduce the fitting complexity, we keep this parameter manually adjusted as stated earlier (see Project P4.6).

Once shifted, some points will be over the maximum time 1, so they need to be wrapped around. This is done in line 16 via `np.concatenate` to combine the two halves of array `t`. The first half of the original array is found by `t[over]`. The condition `over = t>1.0` yields a NumPy truth array whose elements are true if the elements of `t` are greater than 1 (hence we subtract 1 before wrapping) and false otherwise. This truth array is then used to take only true elements with advanced indexing (see Section 1.D). Similarly, the second half is found by `t[under]` with `under = t<=1.0`.

Likewise, the `theta` (θ) grid is created from ψ via Eq. (4.40). Because arccos is multivalued, θ thus computed will be in the range $[0, \pi]$ for $0 \le \psi < \pi$, and in $[\pi, 0]$ for $\pi \le \psi < 2\pi$. Because both θ and ψ should vary over the same range $[0, 2\pi]$ (see Eq. (4.40)), we need to remap the second half of θ correctly. We do this in the same way as for `t` above, replacing the respective condition with `psi>np.pi`, and remapping the second half to $[\pi, 2\pi]$ by subtracting itself from 2π. Finally, the θ values are wrapped around like time `t` earlier, so they are synchronized with each other.

Now, we need to match time `t` nearest to data points `time` to compute the difference at approximately the same (orbital) time. For this task, we define `nearest(a,b)`, which accepts `a` as a 1D array and `b` as either a 1D array or a scalar, and returns the indices of the elements of `a` that are nearest to the elements of `b`. For example,

```
In [2]:   nearest([1., 2., 3., 4.], [1.2, 2.7])
Out[2]:   array([0, 2])
```

The pair of numbers in `b` are closest to the first and the third elements of `a`, respectively, so the returned index array is $[0, 2]$.

The function `nearest()` uses the NumPy `outer` method, which turns most functions to work in an outer-product manner (see Program 8.4 for the outer function). Here (line 9), the `np.subtract` function modified by the `outer` method computes the difference between every possible pair of elements in `t` and `time` in such a way that a 2D array is formed such that `diff[m,n]=t[m]-time[n]`. On the next line we locate the indices of the nearest data points using `np.argmin`, which returns the indices of the minimum absolute differences along axis 0 (down a column). This basically amounts to a table lookup. On return from `nearest()`, the variable `idx` now contains the indices of `t` closest to the data points `time`. Last, depending on the fitting flag in the inline `if-else` statement being true or false, only the indexed or all values of time and θ are returned. The false condition should be used after a successful fit to evaluate the full results.

Back to the `error()` function, we can now calculate the difference between the velocity data and the model with the function `rv()` according to Eq. (4.34) using the fit parameters. The difference, an array of length equal to the number of data points, is passed to `leastsq` for further optimization.

Assuming a successful fit, we calculate the results of our model using the fitted parameters. The data and the fit are plotted, with error bars and smooth curves, respectively (see Figure 4.11). The y-axis limit is set manually (line 48).

The combination of the table-lookup strategy and the least square routine `leastsq` seems to be stable and should work well for moderately elliptic orbits. It can become inadequate if the eccentricity is high. In such cases, it may be better to solve Kepler's equation (4.39) to obtain ψ at each data point exactly.

Program listing 4.5 Three-body motion (`3body.py`)

```
import ode, rootfinder as rtf          # ode, root solvers
import visual as vp, numpy as np  # VPython, numpy

def threebody(id, r, v, t):              # Eqns of motion for 3-body
    if (id==0): return v                 # return velocity array
    else:                                # calc acceleration
        r12, r13, r23 = r[0]-r[1], r[0]-r[2], r[1]-r[2]
        s12, s13, s23 = vp.mag(r12), vp.mag(r13), vp.mag(r23)
        a = [-m2*r12/s12**3 - m3*r13/s13**3,       # dv_i/dt, Eq. (4.42)
             m1*r12/s12**3 - m3*r23/s23**3,
             m1*r13/s13**3 + m2*r23/s23**3]
        return np.array(a)               # return accel array

def quintic(x):        # Euler's quintic equation, Eq. (4.47)
    return -m1-m2 + x*(-3*m1-2*m2 + x*(-3*m1-m2
           + x*(m2+3*m3 + x*(2*m2+3*m3 + x*(m2+m3)))))

def dquintic(x):       # derivative
    return -3*m1-2*m2 + x*(2*(-3*m1-m2) + x*(3*(m2+3*m3)
            + x*(4*(2*m2+3*m3) + x*5*(m2+m3))))

def init_cond(scale):              # collinear initial condition
    r, v = np.zeros((3,2)), np.zeros((3,2))  # y = [r1, r2, r3], same for v
    x = rtf.newton(quintic, dquintic, 1.,2.e-16)  # solve for λ
    a = (m2+m3-m1*(1+x+x)/((x*(1+x))**2))**(1./3.)

    r[1,0] = (m1/(x*x)-m3)/(a*a)                  # non-zero x only
    r[0,0] = r[1,0]-x*a
    r[2,0] = -(m1*r[0,0] + m2*r[1,0])/m3          # CoM at 0
    v[0,1], v[1,1] = scale*r[0,0], scale*r[1,0]  # non-zero Vy only
    v[2,1] = -(m1*v[0,1] + m2*v[1,1])/m3          # CoM at rest
    return r, v

def set_scene(R, r):               # create bodies, velocity arrows
    vp.display(title='Three-body motion', background=(1,1,1))
    body, vel = [], []             # bodies, vel arrows
    c = [(1,0,0), (0,1,0), (0,0,1), (0,0,0)]      # RGB colors
    for i in range(3):
        body.append(vp.sphere(pos=r[i],radius=R,color=c[i],make_trail=1))
        vel.append(vp.arrow(pos=body[i].pos,shaftwidth=R/2,color=c[i]))
    line, com = vp.curve(color=c[3]), vp.sphere(pos=(0,0), radius=R/4.)
    return body, vel, line

def run_3body(scale):
    t, h, ic, cycle, R = 0.0, 0.001, 0, 20, 0.1  # anim cycle, R=obj size
    r, v = init_cond(scale)
    body, vel, line = set_scene(R, r)         # create objects
    while True:
        vp.rate(1000)
        r, v = ode.leapfrog(threebody, r, v, t, h)
        ic = ic + 1
        if (ic % cycle == 0):            # animate once per 'cycle'
```

```
        for i in range(3):           # move bodies, draw vel, path, lines
54          body[i].pos = r[i]  # bodies
            vel[i].pos, vel[i].axis = body[i].pos, v[i]
56          vel[i].length = R*(1+2*vp.mag(v[i]))  # scale vel vector
        line.pos = [body[i].pos for i in [0,1,2]]    # lines
58
m1, m2, m3 = 1., 2., 3.                # masses, global
60 run_3body(scale = input('enter scale, eg 0.7 :> '))
```

The heart of the program is `threebody()`, which returns the equations of motion of the coplanar three-body problem, Eqs. (4.41) and (4.42). It uses the same vector concept as in `earth()` of Program 4.1. There are three bodies, so `r` and `v` are 3×2 arrays (line 23) holding two-component position and velocity vectors, respectively, of the three bodies. For instance, the position vectors are stored as $\mathtt{r} = [[x_1, y_1], [x_2, y_2], [x_3, y_3]]$, so $\mathtt{r}[0] = [x_1, y_1]$ gives the position vector of body 1, and $\mathtt{r}[2, 1] = y_3$ gives the y-component of body 3. This simplifies the calculation of acceleration, which again takes the algebraically written form of Eq. (4.42), and at the same time satisfies the requirement of the leapfrog integrator. We have also used Newton's third law with $\vec{r}_{ij} = -\vec{r}_{ji}$.

The next two functions, `quintic()` and `dquintic()`, return the quintic equation (4.47) and its derivative. They are used for finding the root λ by Newton's method, which is particularly suited for our purpose here because it stably converges to the only root faster than the bisection method and does not require an initial bracket.

The initial condition is calculated in `init_cond()`. After initializing the position and velocity vectors as 3×2 NumPy arrays described earlier, it goes on to find λ, as well as a. Because the three bodies are initially on the horizontal axis rotating about the center of mass, all vertical positions and horizontal velocities are zero. The nonzero components are calculated from Eqs. (4.49) and (4.50). The parameter ω is omitted and defaults to 1 without loss of generality. We introduce a scale factor, `scale`, which basically controls the energy and the shape of the ellipse. When `scale=1`, the orbits are circular. Too small, the ellipses are very elongated and too large, the motion becomes unbound. A useful range is between 0.5 and 1.3. Note that the y velocity component of body 3, `v[2,1]`, is set more accurately assuming the center of mass at rest.

The main function `run_3body()` takes the input `scale` and obtains the initial condition. It calls `set_scene()`, which creates the three bodies, velocity vectors as arrows, all appended to the respective lists, and the line connecting the bodies. In the main loop, `leapfrog()` is used to integrate the motion. Rather than animating the motion after each step during which the bodies move only slightly, the scene is updated every `cycle` number of steps, controlled by the counter `ic` whenever `ic` modulus `cycle` is zero. Animation is updated by changing the current positions of the bodies (line 54), as well as the position, direction, and the length (proportional to speed) of the velocity arrows. The lines connecting the bodies are redrawn by setting the points in the curve to the bodies' positions (line 57). Last, the masses, $m_1 : m_2 : m_3 = 1 : 2 : 3$, are set as global variables because they are used in several functions.

Program listing 4.6 Plotting the effective potential (`r3body_veff.py`)

```
import numpy as np                  # get numpy functions
import matplotlib.pyplot as plt     # get matplotlib plot functions
from mpl_toolkits.mplot3d import Axes3D # for surface plots

def potential(x, y):      # returns effective potential, without 4pi^2
    r1 = np.sqrt((x+a)*(x+a)+y*y) # use np.sqrt for array operations
```

```
    r2 = np.sqrt((x-b)*(x-b)+y*y)
    V = -(1.0-a)/r1 - a/r2 - 0.5*(x*x+y*y)
    return np.maximum(V, -4.0)   # low cutoff value

def makeplot():
    n, s = 201, 5                        # grid points and stride
    x = np.linspace(-1.5, 1.5, n)        # same x,y grids
    x, y = np.meshgrid(x, x)             # make meshgrid, ie, 1D -> 2D

    V = potential(x, y)                  # compute V, -grad(V)
    Fy, Fx = np.gradient(-V)             # swap Fx, Fy since V[i,j]=V[x,y]
    big = Fx*Fx + Fy*Fy > 0.003          # truth array for A[i,j] >.003
    Fx[big], Fy[big] = 0, 0              # cut off large values

    fig = plt.figure()    # add surface plot with strides and color map
    axis=fig.add_subplot(111, projection='3d') # add subplot
    axis.plot_surface(x, y, V, rstride=3, cstride=3,
                      linewidth=0, cmap=plt.cm.spectral)
    axis.set_xlabel('x'), axis.set_ylabel('y'), axis.set_zlabel('V')

    plt.figure()                           # draw contours and arrows
    plt.subplot(111, aspect='equal')       # square picture
    plt.contour(x, y, V, 26)               # 26 contour lines
    plt.xticks([], []), plt.yticks([], []) # omit ticks

    # draw arrows on strided grid to space them out, via array slicing
    plt.quiver(x[::s, ::s], y[::s, ::s], Fx[::s, ::s], Fy[::s, ::s],
               width=0.005, minshaft=2, minlength=0)  # arrow shape
    txt = ['$L_1$', '$L_2$', '$L_3$', '$L_4$', '$L_5$'] # Lag. pts label
    loc = [(-1.1,-.07), (.5,-.07), (1.2,-.07), (.3,.8), (.3,-1)] # label pos
    for i in range(len(txt)):
        plt.text(loc[i][0], loc[i][1], txt[i], fontsize=16)

    plt.show()

alpha = 0.121           # globals
a, b = alpha, 1.0-alpha
makeplot()
```

This program computes and plots the scalar potential and vector force fields (Figure 4.17). The function `potential()` evaluates the effective potential (4.57). It might seem to take on only scalar input x and y, but in fact it works equally well on array input. This is because only NumPy universal functions `sqrt` and `maximum` are used (Section 1.D). Therefore, `potential()` is a ufunc, and it is more efficient to generate the potential on a grid on the fly.

The next routine `makeplot()` makes the plots. First we use `np.linspace` to generate a 1D grid as in Program 4.4. The function `meshgrid()` transforms two 1D-arrays into two 2D-arrays (or mesh). If `X,Y=meshgrid(x,y)`, then `X` consists of repeating rows of `x` and `Y` repeating columns of `y`. Therefore, a grid point `(x[i],y[j])` would be given by `(X[j,i], Y[j,i])` after the transformation. For example, let x=[1, 2, 3, 4] and y=[7, 8, 9], then

$$\mathtt{X} = \begin{bmatrix} 1 & 2 & 3 & 4 \\ 1 & 2 & 3 & 4 \\ 1 & 2 & 3 & 4 \end{bmatrix}, \quad \mathtt{Y} = \begin{bmatrix} 7 & 7 & 7 & 7 \\ 8 & 8 & 8 & 8 \\ 9 & 9 & 9 & 9 \end{bmatrix}.$$

If we choose `i=1,j=2`, the grid point is `(x[1],y[2])=(2,9)`, which is the same as `(X[2,1],Y[2,1])`.

The actual statement `x,y=meshgrid(x,x)` (line 14) means that we have a square grid, stored in 2D arrays. With array inputs to `potential(x,y)`, the potential is calculated at every grid point and returned in an array with the same shape as the input arrays automatically. In effect, each statement in `potential()` does implicit looping over two-nested loops. All this magic occurs because `potential()` is a universal function.

Having obtained the potential on the grid, we compute its gradient (line 17) using the NumPy `gradient` function, which returns the x and y components. Note that we have to swap the `Fy` and `Fx` components because x runs across columns and y across rows, which are the second and first dimensions, respectively, in matrix notation.[18]

Just as we set a low cutoff value for the potential (line 9) to avoid singularities when plotting, we also set a high cutoff value for the gradient to suppresses very big arrows in plotting the force vectors. Line 18 builds a truth array (Boolean) `big` whose elements are true if the magnitude exceeds the specified value, and false otherwise. We then invoke advanced indexing, using the Boolean array as the index (see Section 1.D), to set to zero the elements of the gradient corresponding to true elements in `big` (line 19).

With both the potential and force fields properly prepared, we proceed to plot them with Matplotlib. The first figure (line 21) adds a subplot with 3d projection and calls `plot_surface()` from the `Axes3D` package to make the surface plot of the potential. The `rstride` and `cstride` parameters use fewer data points by skipping rows and columns, making the plot easier to manipulate and figure saves smaller. The line width `linewidth` (or `lw`) is set to zero to suppress the wire mesh. In the second figure (line 27), a contour plot of equal aspect ratio is produced with `contour()`. In addition, a force vector field is overlaid on the contours with the `quiver()` function. Again, points are skipped via slicing so the arrows are not too crowded. The paired `'$'` characters in the text labels (line 35) typesets the text in LaTeX math mode. The block (lines 21 to 34) can be used as a basic template for plotting any data.

Program listing 4.7 Restricted three-body (`r3body.py`)

```python
import ode, numpy as np  # get ODE solvers, numpy
import visual as vp      # get VPython modules for animation

def r3body(y, t):    # equations of motion for restricted 3body
    r, v = y[0], y[1]
    r1, r2 = r - [-a,0], r - [b,0]   # rel pos vectors
    acc = -GM*(b*r1/vp.mag(r1)**3 + a*r2/vp.mag(r2)**3) # Eq. (4.55)
    acc += omega**2*r + 2*omega*np.array([v[1], -v[0]]) # Coriolis term
    return np.array([v, acc])

def set_scene(r):    # r = position of test body
    vp.display( title ='Restricted 3body', background=(1,1,1))
    body = vp.sphere(pos=r, color=(0,0,1), radius=0.03, make_trail=1)
    sun = vp.sphere(pos=(-a,0), color=(1,0,0), radius=0.1)
    jupiter = vp.sphere(pos=(b, 0), color=(0,1,0), radius=0.05)
    circle = vp.ring(pos=(0,0), color=(0,0,0), thickness=0.005,
                     axis=(0,0,1), radius=1)     # unit circle
    return body
```

[18]The same problem arises in electrostatic potentials on a grid, see Program 7.2.

```
20 def restricted_3body(y):                # y = [r, v] expected
       testbody = set_scene(y[0])
22     t, h = 0.0, 0.001
       while True:
24         vp.rate(2000)
           y = ode.RK4(r3body, y, t, h)
26         testbody.pos = y[0]

28 GM, omega = 4*np.pi**2, 2*np.pi        # G(M1+M2), omega, RTB units
   alpha = 0.0009542                      # Sun-Jupiter system
30 a, b = alpha, 1.0-alpha
   r, v = [0.509046,0.883346], [0.162719,-0.0937906]     # init pos, vel
32 restricted_3body(np.array([r, v]))
```

The program simulates restricted three-body motion in the Sun–Jupiter system. It is very similar to Program 4.5. But it uses the RK4 method rather than the leapfrog integrator because the motion involves a velocity-dependent force (4.55). Accordingly, `r3body()` has the same structure as `baseball()` in Program 3.8. It takes the input `y` containing the position and velocity vectors in an ndarray and returns the velocity and acceleration vectors. The Coriolis term in Eq. (4.55) is simplified to $-2\vec{\omega} \times \vec{v} = 2\omega(v_y\hat{i} - v_x\hat{j})$ because $\vec{v} = v_x\hat{i} + v_y\hat{j}$ and $\vec{\omega} = \omega\hat{k}$. This is used in line 8.

Most of the VPython objects are the same as in Program 4.5. We use `vp.ring()` to represent the unit circle.

CHAPTER 5 Nonlinear Dynamics and Chaos

We have thus far dealt with systems whose time evolution is described by a set of ordinary differential equations (ODEs). Given initial conditions, we can integrate the ODEs and obtain numerical solutions at any point in the future. When the outcome is uniquely determined by the initial condition, we say the system is deterministic. Newtonian systems such as projectile and planetary motion are deterministic. For the most part (except unstable few-body systems alluded to in Section 4.5), we did not have to worry about any unpredictable results, in the sense that if we propagate the system from two nearby initial conditions, we should expect the final results to be close, at least not catastrophically different.

Starting with Poincaré's work on celestial mechanics, that notion of predictability had been challenged. But the field has rapidly accelerated in the last half century with what Edward Lorenz thought was a computation error when the computer generated results showed a totally unexpected outcome. Since then, numerical computation has led to the discovery of an entirely new field of science, *deterministic chaos*, often considered a glory of computational science. The field of chaos covers everything from physical systems to biological ones, from surprisingly simple models to complex ones [35]. When chaos occurs, the system becomes unpredictable and loses long-term predictability, in the sense that if we start the evolution of the same system (same ODEs) with slightly different initial values, the outcomes are totally different. The so-called butterfly effect, stemming from the seminal work of Lorenz on hydrodynamic flows, refers to extreme sensitivity to initial conditions in weather in which a small disturbance may cause a storm half a world away. It is qualitatively different than a coin toss where random outcomes are a result of not catastrophic difference but the lack of precise initial conditions, assuming negligible air resistance (e.g., in vacuum).

Because these systems require knowledge of ODEs only, we can apply the well-tested techniques covered so far to study them before turning attention to systems described by partial differential equations starting in the next chapter.

A necessary ingredient for chaos to occur is nonlinearity in the system dynamics. Broadly speaking, if a system responds to input in a nonlinear (or disproportionate) fashion, we say it exhibits nonlinear dynamics. In this chapter, we will study nonlinear dynamics and chaos in systems from discrete logistic maps to continuous systems including driven nonlinear oscillators and the Lorenz model. We will examine how chaos develops, how to characterize it, and how to predict the unpredictable.

5.1 A FIRST MODEL: THE LOGISTIC MAP

We begin with the simplest system exhibiting chaotic behavior, the so-called logistic map. We will explore this system in greater detail, for it helps us introduce several important concepts with clarity.

The logistic map is defined as

$$x_{n+1} = 4rx_n(1 - x_n), \quad 0 \le x_n \le 1,\ 0 < r \le 1. \quad \text{(logistic map)} \tag{5.1}$$

This is a discrete, nonlinear system where given the value x_n at stage n, its next value x_{n+1} can be calculated (mapped). Here r is called the control parameter.[1] If we introduce the map function as

$$f(x) = 4rx(1 - x), \tag{5.2}$$

[1]The factor of 4 in Eq. (5.1) is included to make things tidy. Other sources treat $r' = 4r$ as the control parameter (e.g., Ch. 18 of Ref. [4]).

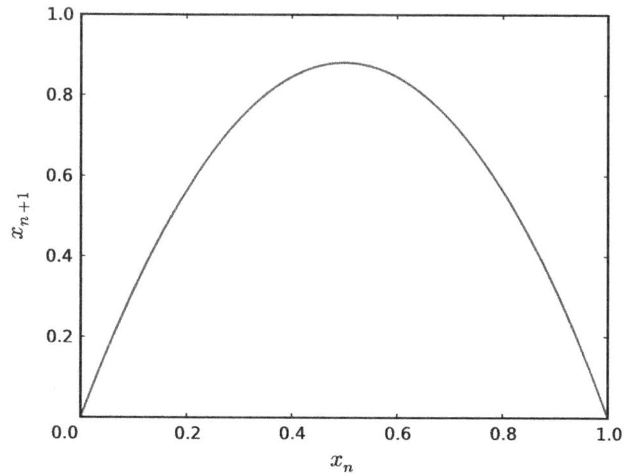

FIGURE 5.1 The map function of the logistic map.

then Eq. (5.1) can be written as

$$x_{n+1} = f(x_n). \tag{5.3}$$

We speak of this equation as a 1D map, which maps an existing point into a new one, deterministically. The map function $f(x)$ is shown in Figure 5.1. Because the maximum of $f(x)$ is at $x = 1/2$ with $f_{max} = r$, the series will remain bounded according to Eq. (5.1).

What could the logistic map represent, you may wonder? One way is to view it as the cycles of an insect population. For example, we could think of x_n as representing the scaled population (in some unit) of a colony of bugs at time n. The population in the next cycle, x_{n+1}, would crudely depend on two things: the number of existing bugs (x_n) for reproduction, and the amount of resources available (e.g., food), $1 - x_n$, where we assume the resource decreases at a rate proportional to the bug population. Then r would be a combination of birthrate and consumption rate. The detail of the model does not matter, but the nonlinear term (x_n^2) in Eq. (5.1) does.

5.1.1 MAP ITERATION

For a given control parameter r and initial value x_0, a series of map iterates may be determined as

$$x_1 = f(x_0), \quad x_2 = f(x_1) = f(f(x_0)) \equiv f^{(2)}(x_0), \quad x_3 = f^{(3)}(x_0), \ldots \tag{5.4}$$

We will use the shorthand $f^{(n)}$ to denote the nth iterate of the map as

$$x_n \equiv f^{(n)}(x_0) = f(x_{n-1}) = \overbrace{f(f(\cdots f(x_0)))}^{n}. \tag{5.5}$$

We can explore the map iterates using Program 5.1. The program first reads the input for x_0 and r from the user, then calculates the map iterates up to a specified number of steps. At each step, the value x_n is also format-printed to the console, four values per line, controlled by the end variable (line 8). Like in Program 4.3, the format string "`'%8f' %x`" prints 8 digits of the float `x`. The inline `if-else` statement produces either a space, or a new line (`'\n'`) every four steps. The results are plotted at the end.

Program listing 5.1 Logistic map (`logisticmap.py`)

```
from __future__ import print_function        # use print() as function
import matplotlib.pyplot as plt      # get matplotlib plot functions

x, r = input('enter x0, r: ')
n, xn = 40, []
for i in range(n):
    xn.append(x)                      # new line every 4 steps
    print('%8f' %x, end='\n' if len(xn)%4==0 else ' ') # inline if
    x = 4*r*x*(1.0-x)                 # next iteration, logistic map

plt.figure()
plt.plot(range(n), xn, '--o')        # plot dashed line with 'o' symbol
plt.xlabel('$n$'), plt.ylabel('$x_n$')
plt.ylim(0,1), plt.text(33, .1, 'r='+ repr(r))   # add text
plt.show()
```

We show in Figure 5.2 representative plots for several values of the control parameter r. For all values of r except the last, x_n (bugs population) undergoes an initial transient phase, then settles into a periodic pattern. For the smallest $r = 0.2$, the population quickly drops to zero and stays there. Too low a birthrate or too little resource cause the colony of bugs to go extinct. Increasing r to 0.7 yields a stable population. In both cases the system is in a period one cycle.

For the next three larger r values, the population repeats after two, four, and eight steps, respectively. We speak of period two, four, and eight cycles. For example, the first 24 points for $r = 0.88$ and $x_0 = 0.8$ as printed from the program are

```
0.800000 0.563200 0.865940 0.408629
0.850613 0.447289 0.870220 0.397539
0.843046 0.465764 0.875874 0.382689
0.831558 0.493043 0.879830 0.372168
0.822479 0.513945 0.879315 0.373542
0.823709 0.511148 0.879563 0.372882
```

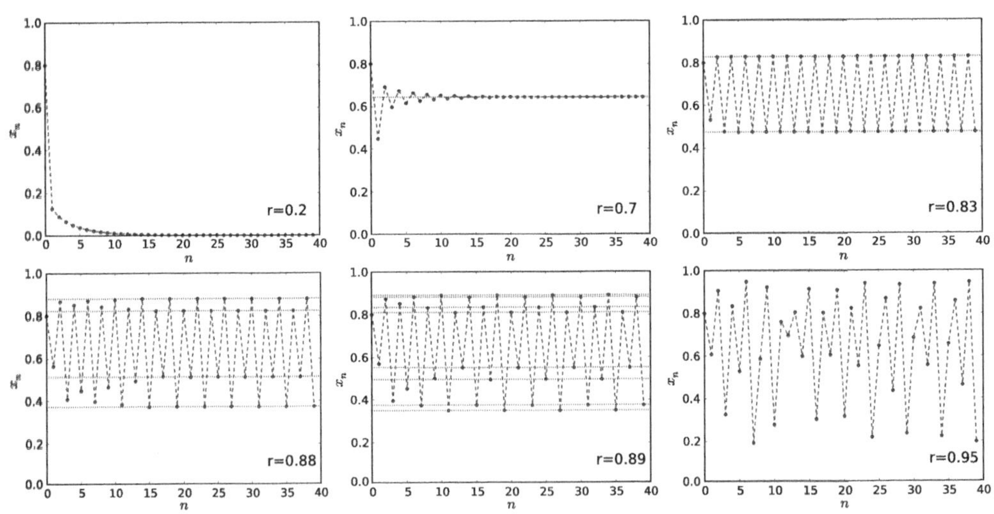

FIGURE 5.2 The values x_n as a function n of the logistic map for different values of the control parameter r. The starting value is the same, $x_0 = 0.8$.

We can see that after initial transients, the series quickly converges to a cycle of period four, that is, the same value appearing after every four steps.

Here interesting dynamics are starting to emerge for increasing r, as cyclic populations start to appear. These periodic values are known as fixed points. Without the nonlinear term (x_n^2) in the logistic map (5.1), the only fixed point would be zero. Increasing r further to 0.95, no apparent periodic structure is recognizable. As we will soon discover, the logistic map has entered the chaotic region.

5.1.2 FIXED POINTS AND ATTRACTORS

As we have just seen, only fixed points appear to remain after transients. Let x^* represent the fixed points. For period-one cycles, the value repeats after one step, so we have

$$x_{n+1} = x_n = x^*. \quad \text{(period-one cycle)} \tag{5.6}$$

We can rewrite the preceding relation using the map function (5.2) and (5.3) to get

$$x^* = f(x^*) = 4rx^*(1 - x^*). \tag{5.7}$$

Solving the preceding equation gives us two fixed points for period-one cycles as

$$x^* = 0, \ 1 - 1/4r. \quad \text{(period-one fixed points)} \tag{5.8}$$

Similarly, period two cycles repeat every two steps,

$$x_{n+2} = x_n, \quad x^* = f^{(2)}(x^*). \quad \text{(period two)} \tag{5.9}$$

Correspondingly, the fixed points are given by two successive applications of the map function like in Eq. (5.4), resulting in a fourth-order (quartic) equation. The four roots are the fixed points of the period-two cycle,[2]

$$x^* = \begin{cases} 0, 1 - 1/4r, \\ \left[1 + 4r \pm \sqrt{16r^2 - 8r - 3}\,\right]/8r, \quad r \geq 3/4. \end{cases} \quad \text{(period two)} \tag{5.10}$$

The first two fixed points are also those of the period-one cycle because if a value repeats after one step, it automatically repeats after two steps.

For higher-period cycles, it gets progressively messy to find the fixed points algebraically, and it would be easier to do so numerically. We can also locate the fixed points graphically, as illustrated in Figure 5.3 for periods one and two (see Exercise E5.2).

The map functions are drawn as curves for $f(x)$ and $f^{(2)}(x)$, respectively, for periods one and two. Note the double hump in the period-two curve. The intersections of the diagonal lines and these curves are the fixed points: 2 for period one, 4 for period two, both including the point 0.

Figure 5.3 also demonstrates how the fixed points are attracted to, starting from $x_0 = 0.1$ in both cases. For period one, the map iterates converge to the larger fixed point ~ 0.64 instead of 0, though the latter is closer to the starting value. For period two, something more interesting happens: The iteration marches toward a fixed point (~ 0.7) quickly at first but does not stay there. Rather, it moves away and ends up at another fixed point (~ 0.47). This explains the appearance of these two points in Figure 5.2 for the corresponding r values.

The fixed points that the iterates eventually converge to are called attractors. All the periodic points observed in Figure 5.2 after the initial transients are attractors. However, not all fixed points

[2] Many of the results can be obtained with the aid of SymPy. See Exercise E5.1.

are attractors. It begs the question: What is it that leads the iterates toward some fixed points but away from others? What makes a fixed point an attractor? The answer lies in the stability of fixed points.

5.1.3 STABILITY OF FIXED POINTS

For convenience, let us formally introduce the (compound) map function of period n as

$$g_n(x) = f^{(n)}(x), \tag{5.11}$$

where $f^{(n)}$ represents n successive operations defined in Eq. (5.5). It follows that for the fixed points of period n, x^*, we have

$$x^* = g_n(x^*) = f(g_{n-1}(x^*)). \tag{5.12}$$

For stability analysis, let us pick a point x from the small neighborhood surrounding a fixed point x^*. As illustrated in Figure 5.4, if x starts at a distance ϵ away from x^*, it will be mapped in the next step to $x_n = g_n(x)$, which is at a distance ϵ_n from the fixed point. When projected on the x-axis via the projection line $y = x$ (similar to Figure 5.3), x_n will be at a horizontal distance ϵ_n from x^*.

The distances are related to the local properties of the map function $g_n(x)$. Graphically from Figure 5.4, we see that $\epsilon_n = h = \epsilon|\tan\theta|$. For small ϵ, $|\tan\theta| \simeq |g_n'(x^*)|$, and we have

$$\epsilon_n = |g_n'(x^*)|\,\epsilon. \tag{5.13}$$

We can get the same result mathematically (Exercise E5.3).

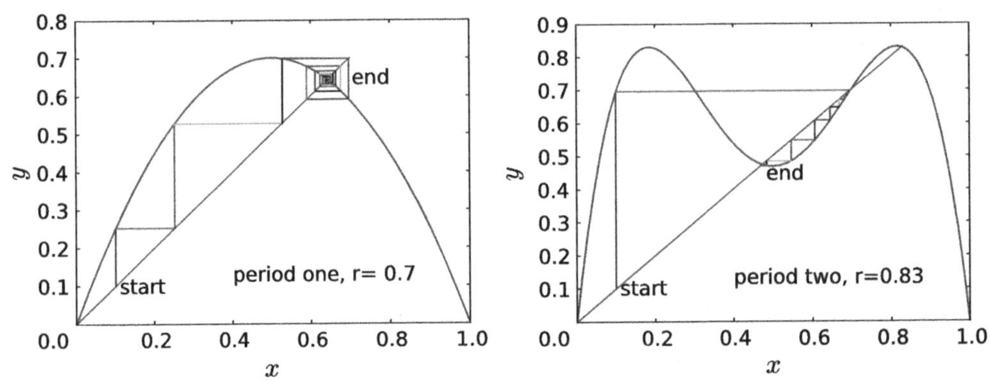

FIGURE 5.3 The fixed points and their convergence for periods one and two.

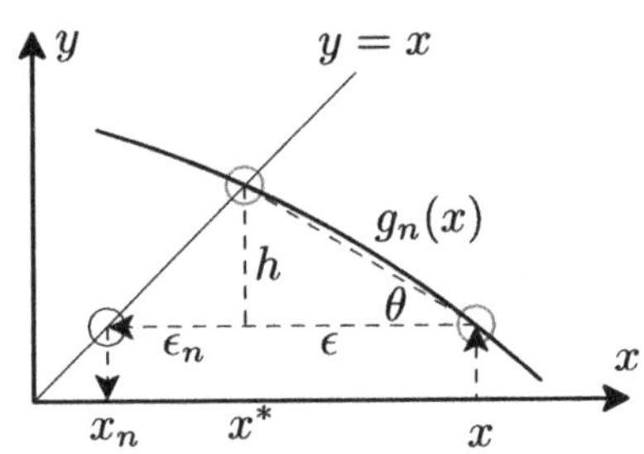

FIGURE 5.4 Deviations around a fixed point. If x is at a small distance ϵ (horizontally) from the fixed point x^*, the value x_n in the next cycle of period n can be found from the mapping function $g_n(x)$ (the arrows).

If a small deviation leads back to the fixed point, that is, if $\epsilon_n < \epsilon$, the fixed point will be stable. Conversely, if $\epsilon_n > \epsilon$, the fixed point is unstable because points in that neighborhood will move away from it. Therefore, the condition for the stability of a fixed point depends on g_n' as,

$$\frac{\epsilon_n}{\epsilon} = |g_n'(x^*)| \begin{cases} < 1, & \text{stable;} \\ > 1, & \text{unstable;} \\ = 1, & \text{neutral.} \end{cases} \tag{5.14}$$

If the derivative is zero, $g_n'(x^*) = 0$, the fixed point is said to be superstable because the convergence is much faster than the exponential rate in normal cases.

The calculation of g_n' can be carried out numerically, or analytically by chain rule with the aid of Eq. (5.12),

$$g_n'(x) = \frac{df(g_{n-1})}{dx} = f'(g_{n-1})g_{n-1}'(x) = f'(g_{n-1})f'(g_{n-2})..f'(x), \tag{5.15}$$
$$g_1'(x) = f'(x) = 4r(1-2x).$$

For example, the derivatives for period one at the fixed points (5.8) are

$$f'(x^*) = \begin{cases} 4r, & \text{at } x^* = 0, \\ 2-4r, & \text{at } x^* = 1 - 1/4r. \end{cases} \tag{5.16}$$

According to Eq. (5.14), the first fixed point ($x^* = 0$) is stable, that is, $|g_1'(0)| < 1$, for $r < 1/4$, and the second one is stable for $1/4 < r < 3/4$. It is for this reason that in Figure 5.2 the population drops to zero for $r = 0.2$ and reaches a finite value for $r = 0.7$ (Figures 5.2 and 5.3).

By the same token, only the last two fixed points of period two in Eq. (5.10) are stable (Exercise E5.4), as the first two are from period one (0 and ~ 0.7 in Figure 5.3, right panel), which as we have seen are unstable at $r > 3/4$.

The regions of stability of the fixed points are summarized in Figure 5.5 over the full range of the control parameter. At a given r, one of the fixed points of period one is stable until

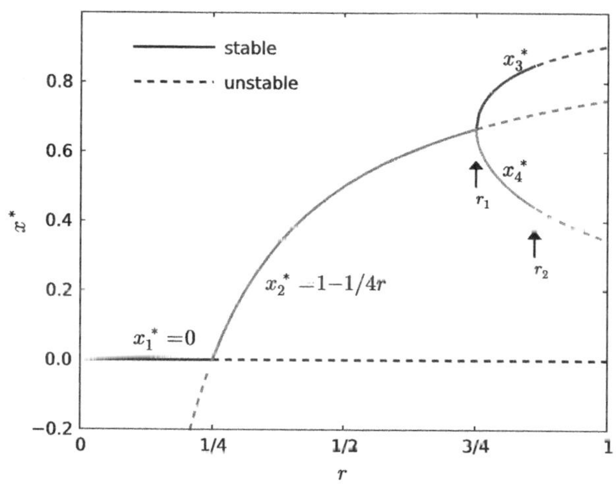

FIGURE 5.5 Regions of stability of the fixed points of periods one and two.

$r = r_1 = 3/4$ when both become unstable. Filling the void are two new fixed points from period two, which double the number of stable fixed points.

However, the new stable fixed points themselves become unstable at $r = r_2 \sim 0.862$ (see Exercise E5.4). What then? Will period doubling continue? Figure 5.2 suggests yes. We will discuss how next.

5.1.4 PERIOD DOUBLING AND ROUTE TO COMPLEXITY

To explore the region beyond r_1 in Figure 5.5, we map the stable fixed points (attractors) as a function of r in small increments with Program 5.2.

Program listing 5.2 Period doubling (`perioddbl.py`)

```
import matplotlib.pyplot as plt     # get matplotlib plot functions
ntrans, nsteps = 1000, 200          # num. of transients and to keep
r, rend, dr, x, xa = 0.7, 1.0, 0.001, 0.5, [0.0]*nsteps
plt.figure()
while r <= rend:
    for i in range(ntrans): x = 4*r*x*(1-x)     # discard transients
    for i in range(nsteps): xa[i], x = x, 4*r*x*(1-x) # keep rest
    plt.plot([r]*nsteps, xa, 'b,')              # blue pixel markers
    r = r + dr
plt.xlabel('r'), plt.ylabel('x'), plt.show()
```

For each r, we iterate the map for `ntrans` steps and discard these points to get rid of transients. The number of steps it takes for the transients to pass would depend on r. For simplicity we use a fixed, large value, as it is adequate for our purpose. The next `nsteps` iterates are presumably the attractors. They are retained and plotted.

The result is a period doubling diagram in Figure 5.6. It shows one fixed point belonging to period one up to $r_1 = 0.75$, where it becomes unstable and is taken over by two stable fixed points of period two, in accordance with earlier observations (Figure 5.5). They remain stable to $r_2 \sim 0.862$, which is the onset of period four. Period doubling, or bifurcation, continues at closer and closer values of r, for example, $r_3 \sim 0.886$ for period eight, and $r_4 \sim 0.891$ for period sixteen (barely visible in Figure 5.6).

Further period doubling (16, 32, etc., not noticeable at the scale shown in Figure 5.6) occurs so close to each other and so fast that, within a narrow window around $r \sim 0.9$, the period becomes infinite. Subsequently, no apparent structure remains other than whole bands being filled. It shows that period doubling is the route to complexity. Although our example is the logistic map, it turns out to be universal of nonlinear systems. As we will see shortly, the logistic map becomes chaotic for $r > 0.9$.

Figure 5.6 shows the static structure after initial transients have been discarded. The dynamics of period doubling may be more fully appreciated in a 3D representation, which also includes temporal changes. Figure 5.7 shows a snapshot from a VPython animation (Exercise E5.5). With time (n) increasing out of the page, the initial transients give way to steady "finger" sheets of bifurcation. At a given r, the series of points bounce alternately between the sheets but never successively between adjacent fingers. For instance, if we label the four branches of the period-4 cycle as 1 to 4 from top to bottom, the sequence is $1 \to 4 \to 2 \to 3...$, without direct jumps between adjacent pairs (1,2) and (3,4). As we will see later (Figure 5.8), branch 3 is closest to $x = \frac{1}{2}$. If we start from there, we will arrive at the adjacent finger (branch 4) two steps later, that is, half a cycle. This is true for higher cycles. This observation is important for our discussion of universal features including the Feigenbaum constants next.

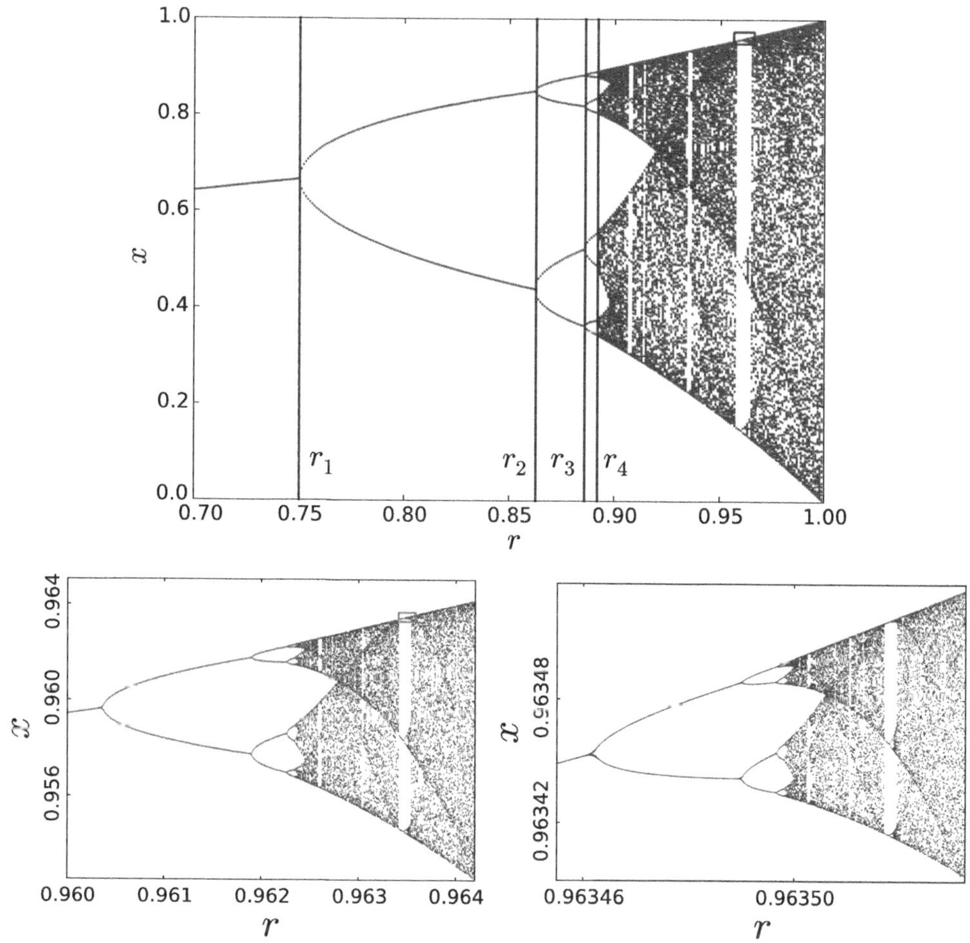

FIGURE 5.6 Period doubling diagram and self similarity. Vertical lines indicate the locations of period doubling. The boxed region in the upper figure is magnified in the lower left figure, whose own box is magnified in the lower right figure.

5.1.5 UNIVERSAL FEATURES

It turns out that several features in the logistic map are universal to a large family of nonlinear maps. One universal constant known as the Feigenbaum δ-number [28] can be estimated from Figure 5.6. Slightly beyond r_4, the logistic map has reached the limit of period doubling series, that is, the order of periodic doubling becomes infinite. Let us denote that limit as $r_\infty \simeq 0.8924864$ (see Project P5.2 and Ch. 18 of Ref. [4]). The r_n value for the onset of period 2^n is related to r_∞ and δ as

$$r_n \simeq r_\infty - c\delta^{-n}, \tag{5.17}$$

where c is some constant. Eliminating c gives

$$\delta = \frac{r_n - r_{n-1}}{r_{n+1} - r_n}, \quad n \to \text{large}. \tag{5.18}$$

This says that the Feigenbaum number is the ratio of adjacent distances in r between successive bifurcations. Using the four approximate r values stated earlier, we find the ratio to be roughly $\delta \sim 4.7$. More accurate evaluation of δ requires more precise determination of the r value. The exact Feigenbaum number is $\delta = 4.6692016...$ (see Ref. [12], Project P5.2).

Self-similar diagrams at different scales can be seen in Figure 5.6 in the magnification of selected regions. The boxed region in the original, upper figure is magnified and shown in the

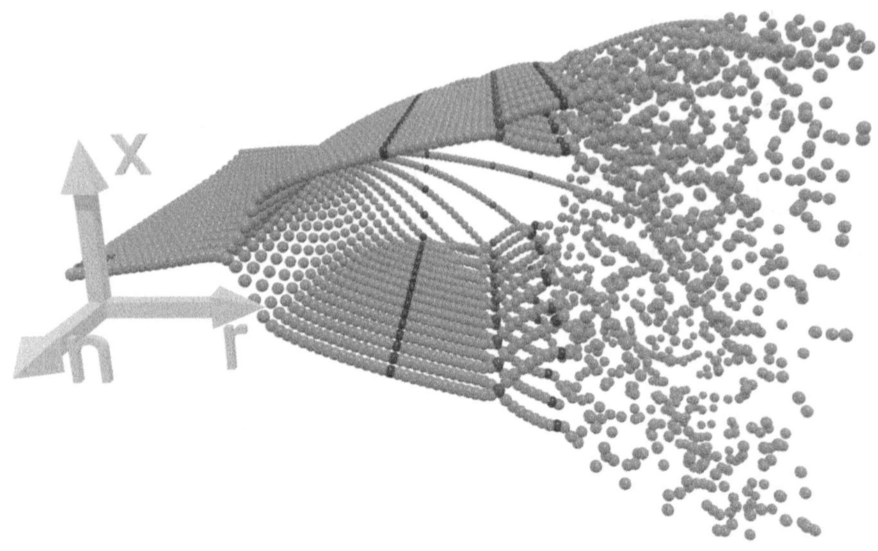

FIGURE 5.7 3D view of period doubling.

lower-left figure. The magnified region is very similar to the original diagram. We can enlarge a subregion in the already-magnified diagram, which again shows strong resemblance to its predecessor. This phenomenon is known as self similarity, that is, the overall structure, including period doubling and onset of chaos, is replicated on the finer and finer scale.

It seems clear from Figure 5.6 that the successive bifurcations are copies spawned from prior ones, only shifted in location and shrunk in scale. Intuitively, we expect to be able to rescale, or renormalize, one bifurcation and reproduce it universally. Indeed, self-similarity is a universal phenomenon in nonlinear dynamics and can be better understood with renormalization theory discussed in Section S:5.A. For example, the approximate r_n values and the δ number can be estimated from renormalization (Project S:P5.1). The beauty of the logistic map is that it exhibits these universal features while keeping things simple and accessible [19].

Feigenbaum's α Number

Another result of renormalization is that we can estimate the second Feigenbaum number, the α-number, which measures the scaling in the vertical direction in the period doubling diagram (Figure 5.6).

The α-number is a constant defined as the limiting ratio

$$\alpha = \lim_{n\to\infty} \frac{d_n}{d_{n+1}} = -2.5029078..., \tag{5.19}$$

where d_n is the vertical displacement from the fixed point $x^* = 1/2$ to the nearest branch of the period-2^n cycle in the bifurcation diagram (see Figure 5.8). The negative sign in Eq. (5.19) refers to the displacement d_n flipping directions between the cycles.

Let r_n^* be the value of the control parameter of the period-2^n cycle, where one of the fixed points is $x^* = 1/2$. The map function is an extremum at this point (see Figure 5.1 or 5.3), and its derivative is zero. Therefore, this fixed point is superstable, hence the r^* designation. For example, the first two r values for the superstables are $r_0^* = \frac{1}{2}$ and $r_1^* = (1 + \sqrt{5})/4$ (Project P5.2).

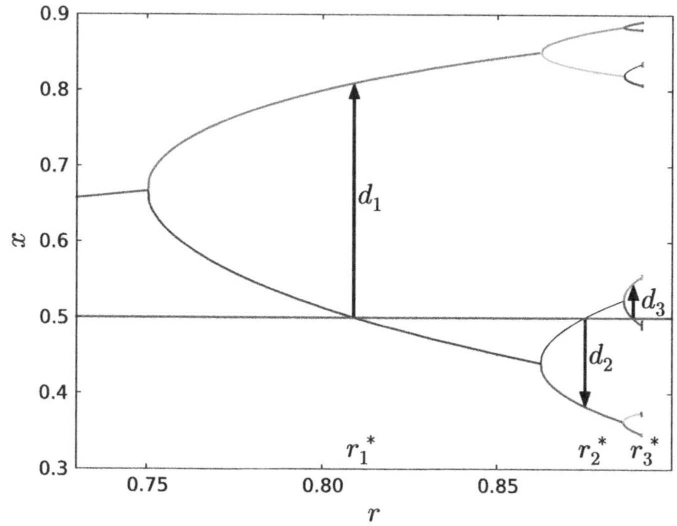

FIGURE 5.8 Vertical displacements and the Feigenbaum's α number.

In general, r_n^* for the period-2^n cycle satisfies

$$f^{(2^n)}\left(\frac{1}{2}\right) - \frac{1}{2} = 0. \tag{5.20}$$

It is understood that $f^{(2^n)}$ depends implicitly on r_n^*.

The vertical displacement d_n starts from x^* and ends at the nearest branch point of the bifurcation tree, which occurs exactly at the halfway point through the cycle per our discussion of Figure 5.7. In other words, the displacement is given by

$$d_n = \frac{1}{2} - f^{(2^{n-1})}\left(\frac{1}{2}\right). \tag{5.21}$$

Knowing d_n, we can calculate α from Eq. (5.19). There is an easier, approximate way, from renormalization, where we obtain $\alpha = -2.24$, compared to the exact value ~ -2.50 in Eq. (5.19) (see Project S:P5.2).

5.2 CHAOS

As we have seen, orderly period doubling ceases to exist for $r \geq r_\infty$, and the bifurcation diagram looks rather "chaotic." How does the abrupt change happen? Can we quantify it? Earlier we gave a general definition of chaos as the sensitive dependence on initial conditions and loss of long-term predictability. With the logistic map detailed so far, we are in a position to demonstrate this and in the process to give an operational definition of chaos.

5.2.1 SENSITIVITY TO INITIAL CONDITIONS

Suppose we start the logistic map with two initial values differing by a small amount and iterate the two series in parallel. How do the series compare? Program 5.3 (Appendix 5.A) lets us explore this question.

The heart of the program iterates the two series separately and simultaneously and records their differences as follows.

```
x2 = x1 + 0.01                          # initial difference
for i in range(n):
    xn1.append(x1), xn2.append(x2), diff.append(abs(x1-x2))
    x1, x2 = 4*r*x1*(1.0-x1), 4*r*x2*(1.0-x2)    # parallel iterates
```

Results generated from the program with different control parameters are shown in Figure 5.9 with a small difference in the initial values. For the control parameter $r = 0.8$, the initial difference is rapidly reduced. The two series overlap each other and are practically identical after a couple of steps, yielding the period-2 cycle as first seen in Figure 5.2.

The situation is quite different for the larger $r = 0.9$ case, where the difference between them grows quickly. By around $n = 5$, the two series are so far apart that they are no longer recognizable as coming from two nearby initial values. In other words, they become unpredictable. We note that the system is still deterministic, but something happens at this r that makes it very sensitive to the initial conditions.

We can get a better idea of what is happening here by studying the difference between the series. It is shown in Figure 5.10 for a small but arbitrary initial difference, produced with the same program.

Although there is some scatter in the data, the overall trend is that the initial difference shrinks exponentially (note the semilog scale) for $r = 0.8$, but grows exponentially for $r = 0.9$. In the former case, the difference becomes insignificant, and we can predict where the system will be even if the initial conditions are not known precisely. However, in the latter case, any difference in

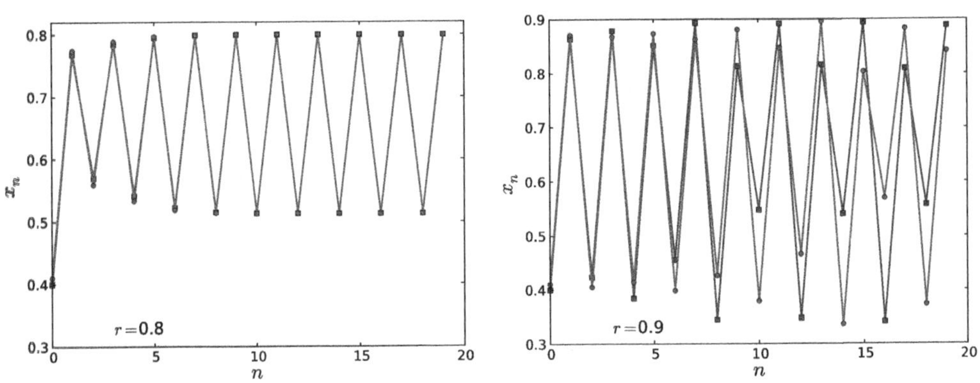

FIGURE 5.9 Logistic map with slightly different initial conditions for two control parameters. The initial values differ by 0.01.

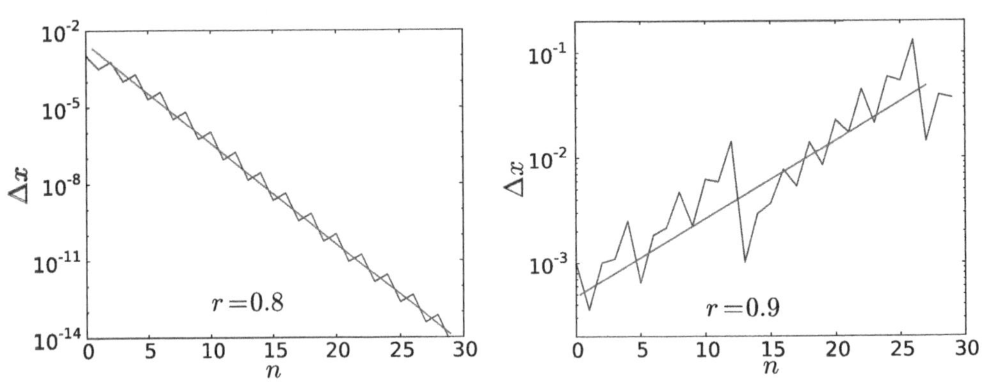

FIGURE 5.10 The difference between series with slightly different initial conditions for two control parameters. The initial values differ by 0.001.

the initial conditions, no matter how small, will lead to such exponentially diverging outcomes that any long-term prediction is meaningless. It is for this reason that we saw the different behaviors in Figure 5.9. We stress that it is not due to inaccuracies in numerics; it is a characteristic of nonlinear systems in the chaotic regime.

5.2.2 A MEASURE OF CHAOS

We are now ready to discuss a more precise definition and a quantitative measure of deterministic chaos. Chaos is characterized by extreme sensitivity to initial conditions. Figure 5.10 suggests that the separation between nearby initial points will shrink or grow exponentially. Let us denote two series differing slightly in the initial value by ϵ,

$$x_0 \to x_1(x_0)... \to x_n(x_0), \quad x_0+\epsilon \to x_1(x_0+\epsilon)... \to x_n(x_0+\epsilon),$$
$$\text{and } \Delta x(n) = |x_n(x_0+\epsilon) - x_n(x_0)|. \tag{5.22}$$

To first order, we expect that the separation will behave as

$$\Delta x \simeq \begin{cases} c\,\epsilon\, e^{\lambda n}, & \text{discrete map,} \\ c\,\epsilon\, e^{\lambda t}, & \text{continuous time,} \end{cases} \tag{5.23}$$

where c is some proportional constant. The factor λ is called the Lyapunov exponent. Accordingly, systems can be classified as

$$\text{Lyapunov exponent } \lambda = \begin{cases} > 0, & \text{chaotic;} \\ < 0, & \text{regular, nonchaotic;} \\ = 0, & \text{indeterminate.} \end{cases} \tag{5.24}$$

The meaning of λ refers to the average rate of exponential convergence or divergence of nearby trajectories. We can use λ as a quantitative measure of chaoticity and give an operational definition of chaos: A system is chaotic if the Lyapunov exponent is positive. The more positive the λ, the more chaotic, or disorder. Conversely, the more negative the λ, the more predictable the system.

Note that the Lyapunov exponent should not depend on the initial separation in the limit $\epsilon \to 0$. In fact, given the values of Δx_1 and Δx_2 at t_1 and t_2, we can eliminate $c\epsilon$ and obtain λ from Eq. (5.23),

$$\lambda \simeq \frac{\ln \Delta x_2 - \ln \Delta x_1}{t_2 - t_1}. \tag{5.25}$$

Accurate determination of λ requires an infinitesimal initial separation and a very long evolution time.

In most cases, λ can be determined only numerically. For example, using the data in Figure 5.10 and Eq. (5.25), we can estimate that $\lambda \sim (\ln 10^{-13} - \ln 10^{-3})/28 = -0.8$ for $r = 0.8$, and $\lambda \sim +0.2$ for $r = 0.9$. These estimates are accurate only to their signs, and their magnitudes are crude because, numerically, we start with a finite initial separation and end in finite evolution time. In practice, the separation is limited by the range of the system, $|\Delta x| < 1$ for the logistic map, and cannot grow indefinitely in Eq. (5.23). As a result, it is important that we measure the slope (such as in Figure 5.10) before the separation levels off and preferably average it over some initial conditions.

5.2.3 LYAPUNOV EXPONENT OF THE LOGISTIC MAP

It turns out that the Lyapunov exponent of the logistic map can also be determined analytically. From Eqs. (5.22), (5.23), and (5.11), we have the separation of two nearby points after n iterations

$$\Delta x(n) = |g_n(x_0 + \epsilon) - g_n(x_0)| = c\epsilon\, e^{\lambda n}. \tag{5.26}$$

Solving for λ yields

$$\lambda = \frac{1}{n} \ln \left| \frac{g_n(x_0 + \epsilon) - g_n(x_0)}{\epsilon} \right| - \frac{1}{n} \ln c. \tag{5.27}$$

In the limit of small $\epsilon \to 0$ and large $n \gg 1$, the argument of the logarithm function becomes a derivative g_n' at x_0, and the second term is negligible,

$$\lambda = \frac{1}{n} \ln \left| \frac{dg_n(x_0)}{dx_0} \right| = \frac{1}{n} \ln \left| g_n'(x_0) \right|. \quad \text{(for large } n\text{)} \tag{5.28}$$

Recalling the chain rule (5.15), we have

$$\lambda = \frac{1}{n} \ln \left| f'(g_{n-1}) f'(g_{n-2}) .. f'(g_0) \right|$$
$$= \frac{1}{n} \sum_{i=0}^{n-1} \ln |f'(g_i)|, \quad g_0 = x_0. \quad \text{(for large } n\text{)} \tag{5.29}$$

To use Eq. (5.29), we choose an initial x_0, wait some steps for transients to pass, generate g_i recursively from Eq. (5.12), and calculate $f'(g_i)$ with Eq. (5.15), for some suitably large n. You may be worried that λ obtained this way may depend on the initial value x_0. The dependence is very weak, if we take care to let the transients pass, and if n is sufficiently large. Essentially, the summation in Eq. (5.29) will wash out any trace of x_0, in both regular and the chaotic regions.

Program 5.4 (Appendix 5.A) calculates and plots λ. Experiment with `ntrans` (as well as x_0) a little, as sometimes it takes a while for the transients to pass. When λ is independent of all the parameters for a given r (say to within 3 or 4 digits), we know it has converged.

The results as a function of r are shown in Figure 5.11. For the control parameter up to $r \sim 0.892$, the Lyapunov exponent is negative, so the dynamics are regular. Bifurcations occur

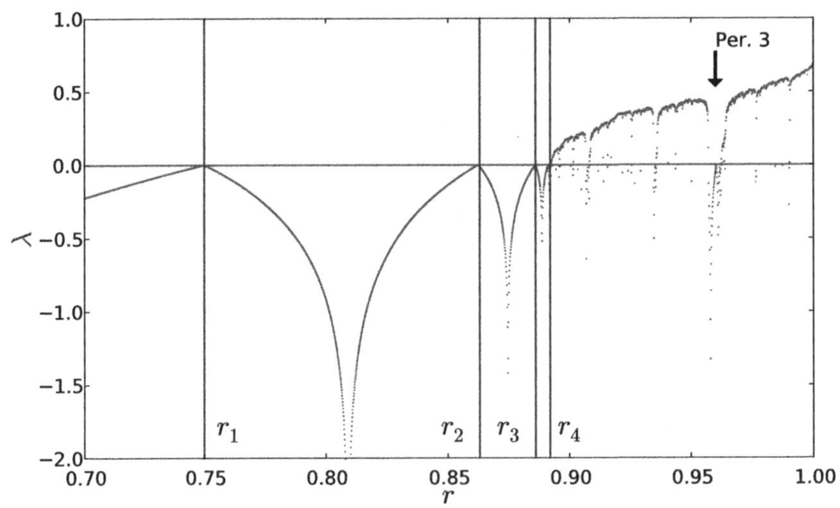

FIGURE 5.11 Lyapunov exponent of the logistic map. Vertical lines indicate onsets of period doubling, at the same locations as those marked in Figure 5.6. Period-3 occurs around $r \sim 0.96$ where λ dips below zero.

at the cusps when λ comes up to zero (but not crossing zero) and then abruptly falls back down. The locations of the cusps are the same as the vertical lines in Figure 5.6. Afterward, λ is largely positive and increasing, signaling increasing disorder or chaoticity. But embedded in the land of chaos are narrow windows of stable islands where λ dips down to negative values. One of these contains the stable fixed points of period three marked by the arrow. There are other regions embedded as well, including periods five, six, seven, and so on. (See Project P5.1.)

5.3 A NONLINEAR DRIVEN OSCILLATOR

The logistic map lets us explore the concept of chaos in a simple model system. We now discuss a real, continuous physical system, the simple pendulum, a classic apparatus (Figure 5.12) made up of a weight connected to a pivot by a thin rod. It oscillates periodically about the equilibrium, the vertical line. This system exhibits chaos that can be experimentally measured.

The equation of motion for a damped pendulum is

$$\ddot{\theta} + b\dot{\theta} + \omega_0^2 \sin\theta = 0, \tag{5.30}$$

where ω_0 is the natural angular frequency dependent on the rod length and gravitational acceleration, b the damping coefficient due to viscous friction or linear drag. The $\sin\theta$ term comes from the tangential gravitational force. Without damping and in the small angle approximation, Eq. (5.30) reduces to the simple harmonic oscillator (2.47).[3] However, the nonlinearity present in this term is crucial for nonlinear dynamics.

With damping, the pendulum would stop moving unless an external driving force is applied. We then speak of a nonlinear, driven oscillator. Including the driving force, and converting to a set of first-order ODEs, we have the equations of motion

$$\frac{d\theta}{dt} = \omega, \quad \frac{d\omega}{dt} = -\omega_0^2 \sin\theta - b\omega + F_d \cos(\omega_d t), \tag{5.31}$$

where F_d is the driving force amplitude, and ω_d the driving angular frequency.

We will assume scaled units for F_d and b according to the units of time or frequency. If time is in seconds, then F_d will be in $1/s^2$, and b in $1/s$. Their units will be omitted unless otherwise noted.

In comparison to the logistic map, F_d plays the role of the control parameter. We expect that for some F_d values, the system should exhibit chaotic behavior, as does the logistic map for certain values of r. To investigate the behavior of the system, we solve Eq. (5.31) numerically with an ODE solver next.

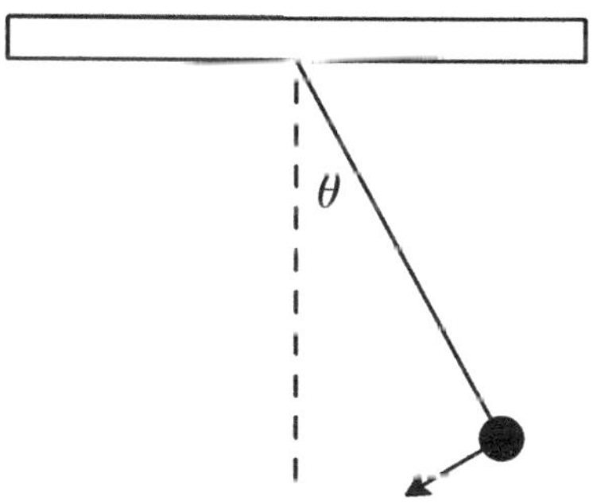

FIGURE 5.12 The nonlinear pendulum.

[3]Strictly speaking, ω_0 is the angular frequency only in this limit. For the nonlinear, driven oscillator, it is best to view ω_0 as parameter, against which the strength of the driving force is to be compared.

5.3.1 REGULAR AND CHAOTIC MOTION

Program 5.5, named "Nonlindro," integrates the equations of motion of the nonlinear driven oscillator. The system dynamics (5.31) is calculated in the function `pendulum()`,

```
def pendulum(y, t):  # y = [theta, omega], omega_0 = 1
    return [y[1], -ma.sin(y[0]) - b*y[1] + fd*ma.cos(omega_d*t)]
```

The two-element input array `y[]` contains the angular displacement and angular velocity as $y = [\theta, \dot{\theta}]$.

The results, $\theta(t)$ and $\omega(t)$, are shown in Figure 5.13 for two driving amplitudes. For $F_d = 0.7$, the angular displacement and angular velocity both undergo an initial transient period. During this period, the oscillator loses track of where it came from due to damping and adjusts the motion to the driving force. After the transients, it oscillates regularly and periodically with the same frequency as the driving frequency. This is akin to pushing a child on a swing at regular intervals. No matter what the initial conditions were, the final oscillations are the same. The system has no memory effect (we will see this again in a linear driven oscillator, Section 6.1).

At a larger $F_d = 1.1$, the motion looks rather irregular. There is no clearly defined transient period, and the oscillator never settles into a regular motion. The angle $\theta(t)$ shows wild swings over the top, exceeding the regular range $|\theta| < \pi$. The same is true for the angular velocity, which shows no apparent structure. We suspect that, although the motion is regular for $F_d = 0.7$, it is chaotic for the larger driving force.

To confirm this, we need to calculate the Lyapunov exponent in each case. We can do this by modifying program Nonlindro slightly to start two oscillators with a small difference in the

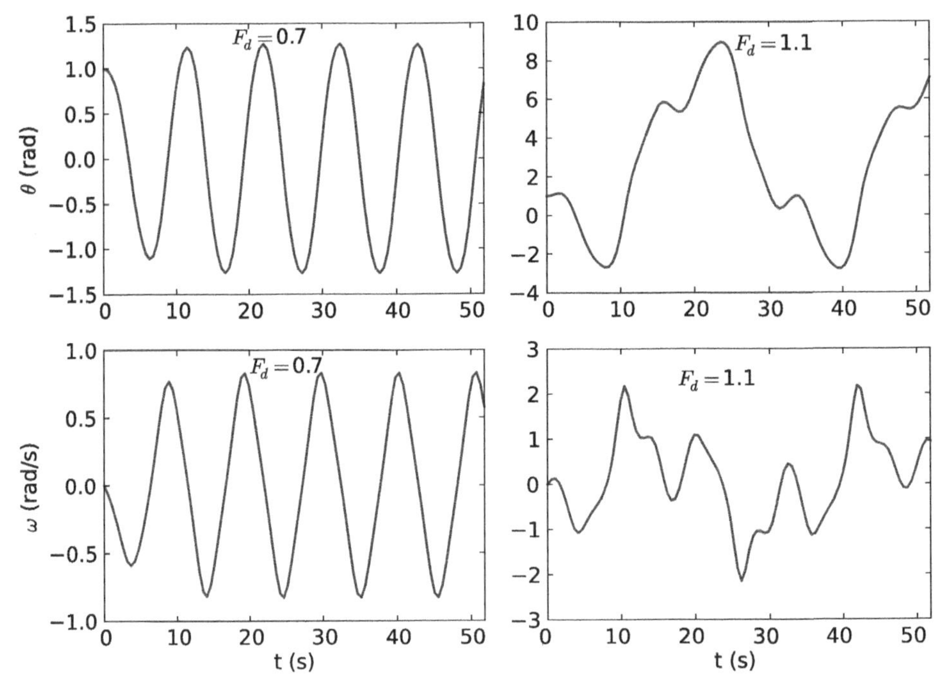

FIGURE 5.13 The angular displacement (θ) and angular velocity (ω) as a function of time for two driving force amplitudes, $F_d = 0.7$ and $1.1\ \mathrm{s}^{-2}$. The other parameters are $\omega_0 = 1\ \mathrm{rad/s}^2$, $b = 0.5\ \mathrm{s}^{-1}$, and $\omega_d = 0.6$ rad/s.

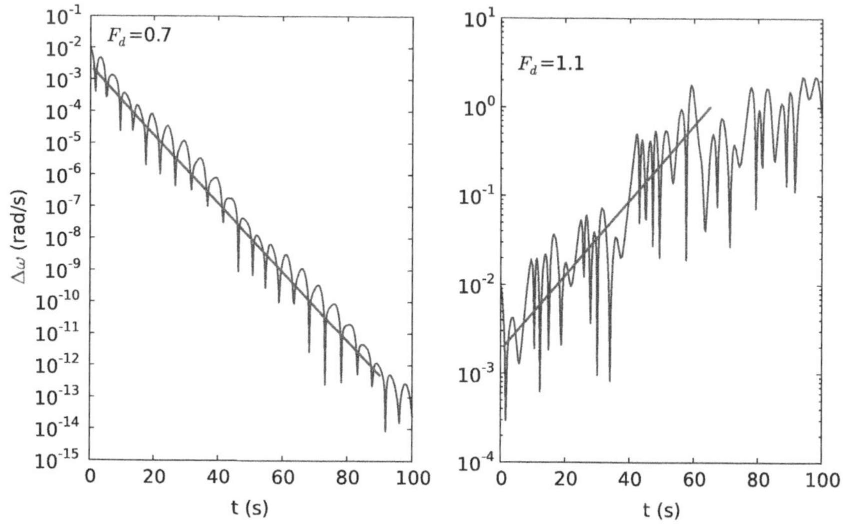

FIGURE 5.14 The difference in ω from slightly different initial conditions for two driving amplitudes. The parameters are the same as in Figure 5.13. The initial conditions are $\theta_1(0) = \theta_2(0) = 0.6$, and $\omega_1(0) = 0$ and $\omega_2(0) = 0.01$.

initial conditions and see how the difference develops in time. We leave this to Project P5.4 and only focus on discussing the results.

Figure 5.14 shows how the difference in angular velocity, $\Delta\omega = |\omega_1(t) - \omega_2(t)|$, changes over time from two nearby initial conditions. For the smaller perturbation of $F_d = 0.7$, the difference becomes smaller with increasing time. There are oscillations in the data due to the oscillating nature of the system, just as there was scatter in the logistic map. But the overall trend is clear: The rate of convergence is exponential. The slope of the trend line, the Lyapunov exponent, is negative. Thus the motion is regular, nonchaotic, for the smaller F_d.

However, for the larger driving force $F_d = 1.1$, the difference grows exponentially with time. Here, the data fluctuates more wildly than before because of the stronger perturbation. The difference also seems to saturate a little, reaching a plateau for $t \sim 70$ to 100 s. This usually happens in any system that is bound physically, like in the logistic map where $|\Delta x| < 1$. In this case, the amount of energy in the system must be finite. Therefore, the difference cannot grow indefinitely. No matter. Here, it is the initial trend of the separation before saturation occurs that is important in measuring the Lyapunov exponent, or the chaoticity. The trend line drawn between 0 to 60 s certainly gives a positive Lyapunov exponent.

After all, if the trajectories diverge exponentially, the system is extremely sensitive to the initial condition, and after a short time they will "forget" where they came from. This, however, is different than the no-memory effect mentioned in the regular motion earlier. There, every point converged to the same motion. Here, no two points lead to the same motion. Put another way, the current path could have come from any point in the chaotic region. We cannot trace backward to reconstruct the original path, even though the system is deterministic.

Quantitatively, estimating from both trend lines in Figure 5.14, we find the Lyapunov exponents to be $\lambda \sim -0.25$ s^{-1} for $F_d = 0.7$ s^{-2}, and $\lambda \sim +0.1$ s^{-1} for $F_d = 1.1$ s^{-2}. This means that, over one period of the driving force $(2\pi/\omega_d)$, the separation between two nearby trajectories is reduced by a factor ~ 0.07 in regular motion and increased by a factor ~ 3 in chaotic motion. Both these numbers confirm our expectations from Figure 5.13.

In the chaotic regime, we can think of $\tau = 1/\lambda$ as the time scale of trajectory divergence. Predictions of motion beyond τ would be meaningless.

5.3.2 THE PHASE SPACE AND THE POINCARÉ MAP

As we will see shortly, another useful way to study nonlinear dynamics is to plot the phase space trajectories showing θ and ω together. As mentioned earlier (Section 2.4), phase space is a concept taken from Hamiltonian mechanics, a multidimensional space encompassing all coordinates and momenta of a dynamical system. In the case of the driven oscillator, it refers to the θ and ω space. Sample trajectories calculated with a slightly modified Nonlindro are shown in Figure 5.15.

At $F_d = 0.7$ the path starts from the initial point and quickly converges to an ellipse in which the motion repeats indefinitely. This is not surprising because we know from earlier discussions that the motion is regular at this driving amplitude. This ellipse is an attractor, though it is not a single point for a continuous system. On the other hand, at $F_d = 1.1$, the path looks very complex. Here the numerical values of θ have no limits (see Figure 5.13), although as an observable it should be within $[-\pi, \pi]$. We thus remap θ whenever it exceeds this range as

$$\theta = \theta + 2\pi \;\text{ if } \theta < -\pi, \quad \theta = \theta - 2\pi \;\text{ if } \theta > \pi. \tag{5.32}$$

The path wanders all over the diagram like spaghetti, and there is no stable attractor for it to evolve around. If we increase F_d to 1.2, the system returns to a simple pattern again, though it is not obvious from the plot because of the continuous nature of the variables.

To make more sense out of continuous changes, a better idea is not to plot every point. Rather, we look at them at discrete times. The result is a distribution of points taken at specific time intervals, also known as a Poincaré map. To produce a Poincaré map, we plot only the points

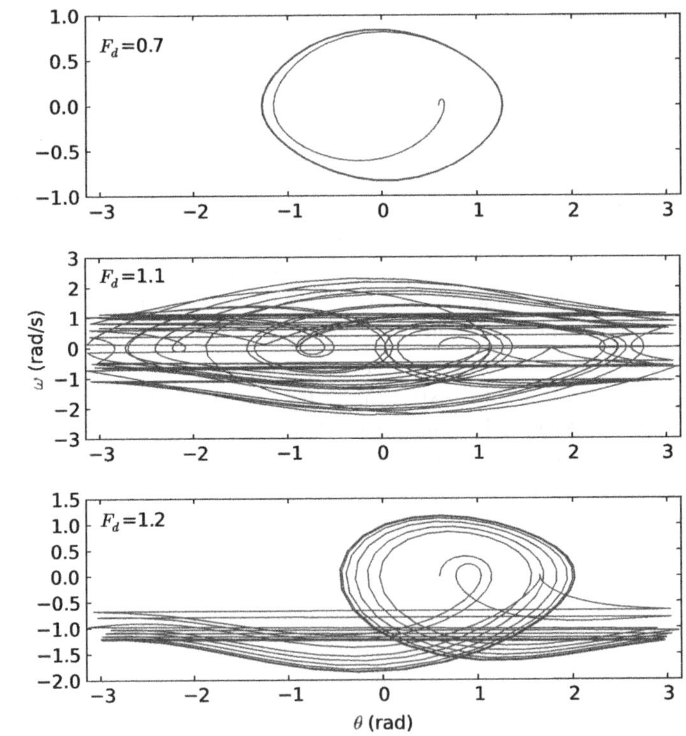

FIGURE 5.15 Phase space plots for different driving force amplitudes $F_d = 0.7, 1.1$, and 1.2. Other parameters are the same as in Figure 5.13.

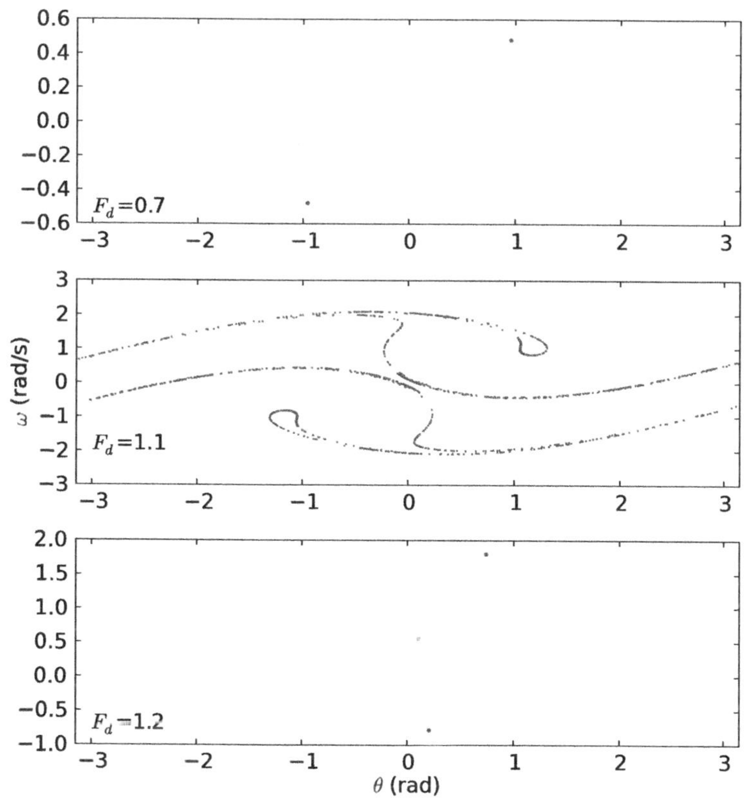

FIGURE 5.16 A Poincaré map of the nonlinear driven oscillator. The plots are generated with the same parameters as in Figure 5.15.

in synchronization with the driving force, for example, when it is maximum or minimum, after the initial transients pass. The technique is used in Program 5.6. The results are shown in Figure 5.16.

For regular motion ($F_d = 0.7$), the plot is simply remarkable: only two points, corresponding to two samplings in each period of the driving force, one at the beginning of the period, and one at the halfway point. Incidentally, at $F_d = 1.2$ we again see two points, even though it looks more complex in Figure 5.15. We can guess that the motion must be regular too. In these Poincaré maps, regular motion repeats periodically as discrete points. It is similar to stroboscopic imaging. For example, a spinning disc may rotate too fast to see the marking on it. But, if we were to shine light on it at periodic intervals, then instead of seeing a blur, some patterns could be recognized. Poincaré maps are indispensable for understanding complex dynamics, especially in higher dimensions.

For $F_d = 1.1$, there are many more points in the Poincaré map, but they trace out relatively simple structures. The points do not fill up the graph; rather, they follow a disjointed line, like a shoreline. When a point is recorded, we know it will be on the shoreline, though not exactly where. This shows that even chaotic motion obeys certain rules. We can predict the unpredictable, so to speak.

The appearance of shoreline structures is a result of the trajectory meandering through available phase space. It is a signature of chaos. To understand how it comes about, it is helpful to think of the Poincaré map as a cut of a topological surface, like a twisted donut, in a multidimensional space. We can imagine the phase space trajectory moving along and wrapping around the surface.

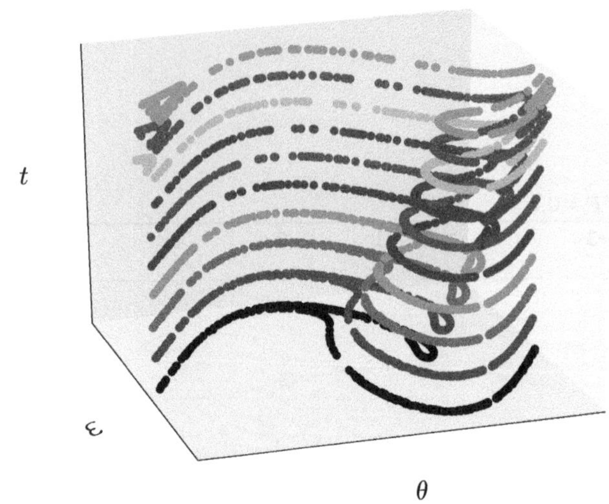

FIGURE 5.17 The Poincaré surface of section of the nonlinear oscillator.

When a cut is made, the cross section of the surface, the shoreline structure, is exposed, and that is what we see. It is called the Poincaré surface of section.

Figure 5.17 shows the Poincaré surface of section of the nonlinear driven oscillator for $F_d = 1.1$. It represents several cuts, or layers, of the Poincaré map. The first layer (bottom) is recorded at the start of the period, the second layer one step after the start, the third one two steps after, and so on. The bottom layer is the same as the Poincaré map shown Figure 5.16 (half of it to be precise, because only one point per period is taken).

The shape of this surface of section is relatively simple. In other chaotic systems it can be quite complex. Generally, the surface over which the phase space trajectory travels is referred to as a torus. The Poincaré maps represent cross sections of these tori. In more than three dimensions of phase space, the Poincaré surface of section is an abstract multidimensional surface. For regular motion, the surface of section reduces to a curve.

5.3.3 BIRFURCATION DIAGRAM

Figure 5.16 tells us that the motion is regular at $F_d = 0.7$, chaotic at 1.1, and regular again at 1.2. The fact that there are regions of regular motion below and above this particular value of $F_d = 1.1$ should not surprise us. We have seen similar islands of regular motion sandwiched between chaotic regions in the logistic map (see Figure 5.6).

As in the logistic map, a bifurcation diagram can give us an overview of regions of regular and chaotic motion. We use the same discrete sampling technique to calculate the bifurcation diagram for the driven oscillator. For each F_d, we wait for the transients to pass and record data points in synchronization with the driving force. Program 5.6 can be used for this purpose; only the graphing part needs to be modified.

The results are shown in Figure 5.18. We see period doubling structures similar to Figure 5.6. There are period-2 cycles in the beginning, which bifurcate into period-4, period-8, and so on, and quickly becomes period infinite, that is, chaotic. The chaotic region is followed by windows of regular motion again ($F_d \sim 1.05$ and 1.2). This all looks very much like what we have in the logistic map. We again see that the route to complexity is through period doubling.

The enlarged figure is so strikingly similar to the logistic map (see Figure 5.6) that we could have mistaken for it without looking carefully at the graph labels. Two totally different systems, yet quite similar structures. It shows that deterministic chaos is not randomness, has universal properties, and plays by a set of predictable rules.

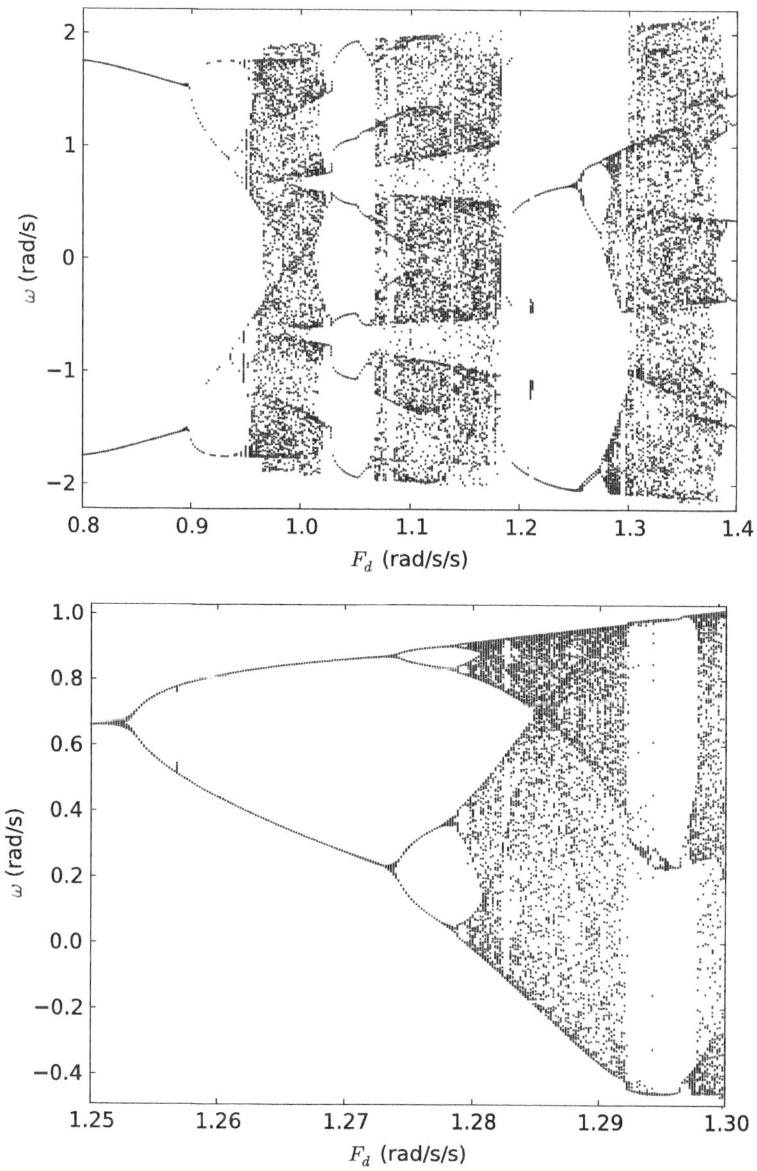

FIGURE 5.18 Period doubling diagram of the nonlinear driven oscillator with the same parameters as in Figure 5.13. The bottom figure is an enlargement of the top figure in the window $1.25 \le F_d \le 1.3$ and $-0.5 \le \omega \le 1$.

5.4 THE LORENZ FLOW

Simple as they appear, the two examples we have discussed so far would be far less understood or illuminating as to their chaotic nature without digital computation. Since Lorenz's work, computational studies have firmly established chaos as a new field of science.

In the seminal paper published in 1963 on hydrodynamic flow [59], Lorenz wrote: "it is found that nonperiodic solutions are ordinarily unstable with respect to small modifications, so that slightly differing initial states can evolve into considerably different states." The term *chaos* was not a common scientific vocabulary then, and the science of chaos was yet to be advanced. Today, we recognize, unmistakenly, that Lorenz was speaking about chaos. He is widely credited with the discovery of chaos.

Lorenz was studying convection problems of hydrodynamic flow in climatic and atmospheric physics. Given the very limited computing power at that time by today's standard,[4] Lorenz modeled the process using a much simplified version of the Navier-Stokes equations, which are essentially Newton's laws expressed in continuous variables like density, temperature, pressure, and so on, to describe fluid dynamics. The result is the Lorenz model, which consists of three ODEs as

$$\frac{dx}{dt} = -\sigma(x - y), \tag{5.33a}$$

$$\frac{dy}{dt} = x(r - z) - y, \tag{5.33b}$$

$$\frac{dz}{dt} = xy - bz. \tag{5.33c}$$

Equations (5.33a) to (5.33c) approximately describe the convective flow of fluid between two plates kept at a temperature difference. The variable x represents the flow intensity, y the temperature difference between the rising and falling currents, and z the temperature gradient. The parameters σ, r, and b are positive constants.[5] Given the idealization of the Lorenz model, we do not expect it to accurately describe realistic flow, which surely requires a great deal more variables and parameters than present in the model. However, the model does capture one of the most important features in the actual system: the instability and sensitivity to initial conditions. For this reason, we will treat the model as a nonlinear physical system and not expect it to predict the exact behavior of actual systems.

Though simplified, the Lorenz model exhibits complex behaviors typical of nonlinear systems. Note the nonlinear terms $-xz$ in Eq. (5.33b) and xy in Eq. (5.33c). In σ, r, and b, we have three control parameters. Relative to the previous examples, this is a rather large parameter space to explore. We will follow Lorenz and use fixed values $\sigma = 10$, $b = 8/3$, and vary r as our sole control parameter. We will see that as r is varied, the motion will transition between regular and chaotic behaviors.

5.4.1 SOLUTIONS OF THE LORENZ MODEL

To follow the dynamics, we can use the RK4 method to integrate the Lorenz equations as we did for the nonlinear oscillator because the model is dissipative (Eq. (5.33c), for example). We can follow any of the variables, as they behave similarly for a given control parameter. We choose to show the results of y for three values of r in Figure 5.19 (see Project P5.5).

For the two smaller r values, y increases rapidly in the beginning, reaches a peak value, then drops sharply. Because it represents the temperature difference of rising and falling currents, convective flow is too far out of equilibrium when the difference becomes large, causing subsequent instabilities in the form of rapid rises and falls. After the initial phase, y oscillates around an equilibrium point with reduced amplitude (fixed point) and eventually settles to that point like a damped oscillator.

For $r = 28$, however, the behavior of y after the initial phase is very different. It oscillates with increasing amplitude, reaching a peak at $t \sim 17$. Thereafter, it falls and rises sharply, sometimes having two or three peaks of the same sign before flipping to the other side. It never reaches a steady state. The motion becomes irregular and chaotic.

---

[4]Lorenz carried out his calculation on a Royal McBee LGP-30 computer, the size of a chest freezer weighing 700 pounds and classified as a "desk" computer. It could perform one iteration of calculations per second. Its speed was probably around 10 FLOPS. Typical speed of today's PCs is around 100 GigaFLOPS.

[5]The Lorenz model approximates the Rayleigh-Bénard flow. The parameter σ is the Prandtl number, r the scaled Rayleigh number, and b the scaled dimension of the system.

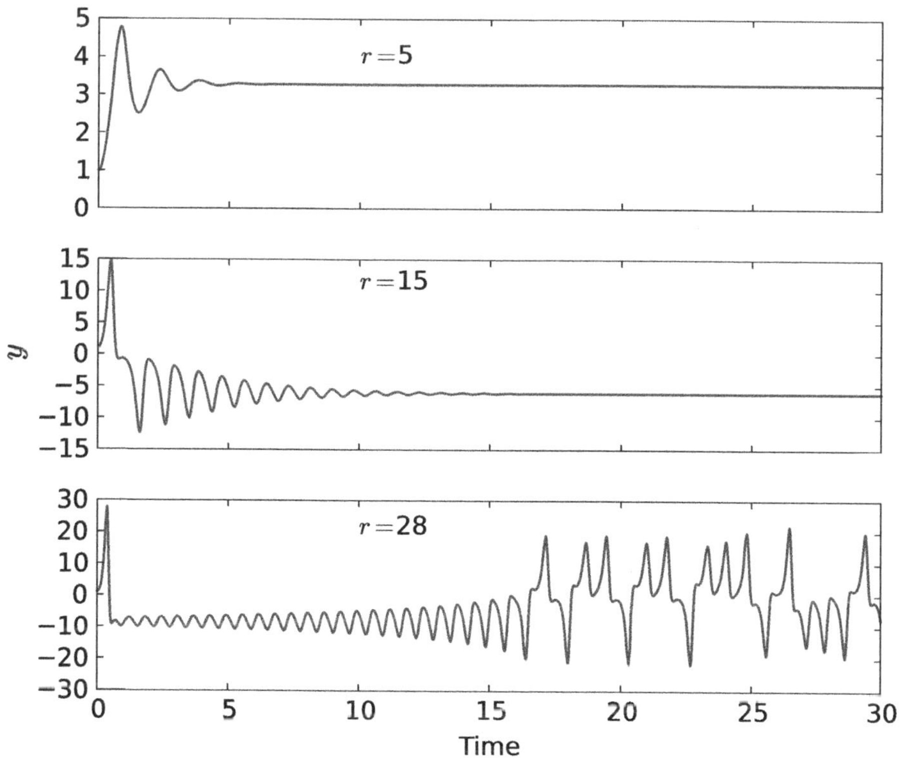

FIGURE 5.19 The variable y of the Lorenz model as a function of time for three values of r. The initial condition is $(0, 1, 0)$.

At this stage, the system becomes very sensitive to initial conditions or small perturbations. In fact it is so sensitive that one must take care to make sure the results are converged. For the $r = 28$ results shown in Figure 5.19, calculations with the RK4 method converge only for step size of ~ 0.005 or smaller.

Incidentally, in his original calculation, Lorenz used a quasi second-order method with a step size of 0.01. Compared to our converged results, Lorenz's calculation was accurate only up to about $t \sim 17$. There is considerable disagreement between the two results after that point. However, this does not alter the conclusions Lorenz drew: long-term predictions in this system are not feasible. The key factor is that when the system is chaotic, a small numerical error will be amplified exponentially just like a small separation between two trajectories. In this sense, numerical accuracy is less critical in assessing long-term trends, as long as the results are statistically representative of the chaotic nature of the problem (numerical accuracy still needs to be treated with care, of course). Weather systems, though much more complicated than the idealized Lorenz model, behave similarly in the essential point, that is: the fundamentally chaotic nature of weather prevents reliable long-term forecasts.[6]

5.4.2 THE ANATOMY OF A STRANGE ATTRACTOR

The wild oscillations observed earlier seem to be random by the variable y alone. But, when viewed in the three-dimensional phase space, the trajectories are in fact meandering between two regions known as a strange attractor, tracing out a butterfly-like figure, illustrated in Figure 5.20.

[6]We should take multiday forecasts such as 10-day planners with a grain of salt, especially in weather-prone regions like New England.

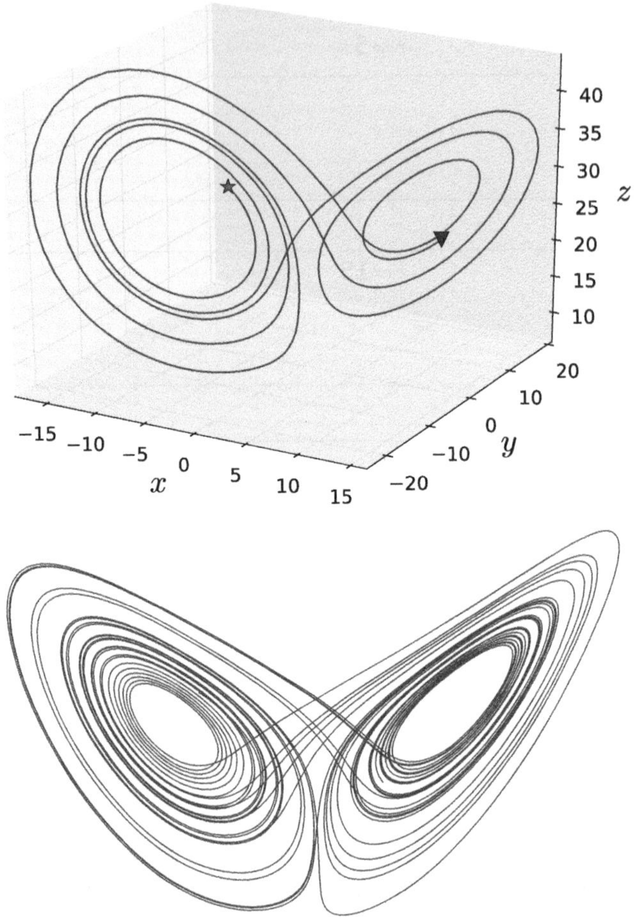

FIGURE 5.20 Strange attractor of the Lorenz model. The plot with fewer windings starts at (★) and ends at (▷). The initial condition is $(-7, -2, 30)$ and the parameters are $\sigma = 10$, $\beta = 8/3$, and $r = 28$.

Two "eyes" of the attractor (or steady state solutions, Exercise E5.6) are located roughly at the centers of the butterfly's wings, on the opposite side of the x-z plane. From the starting point, the trajectory spirals outward from one eye for a few rounds, then crosses the x-z plane to the neighborhood of the other eye, where it swirls around about two times before crossing the x-z plane again toward the first eye. Thereafter, the cycle repeats. Because the attractor is in the chaotic regime, they show sensitive dependence on initial conditions. This behavior seems abnormal and is one reason the attractor has been called a strange attractor.

After many crossings, a beautiful figure emerges like a butterfly. We can better visualize the structure of the butterfly with VPython. Figure 5.21 shows one 3D view where we are peering down in the negative z direction, with the x-axis to the right and y-axis up. We can clearly see the spirals about the two eyes and the crossings through the x-z and y-z planes. The structure formed by the points is approximately the shape of the Poincaré surface of section.

Note, however, the trajectory never crosses itself. This is a general property of phase space for any system. Also note from Figure 5.21 that there is a "thickness" to the spirals, that is, adjacent spirals are out of plane with each other. So the space traversed by the trajectory has a "volume" structure. However, as we will discuss shortly, It has a fractal dimension greater than 2 but less than 3, meaning it is a volume-like object with zero volume. This is the other technical reason the attractor is labeled strange: It has a fractal dimension (Section 5.6).

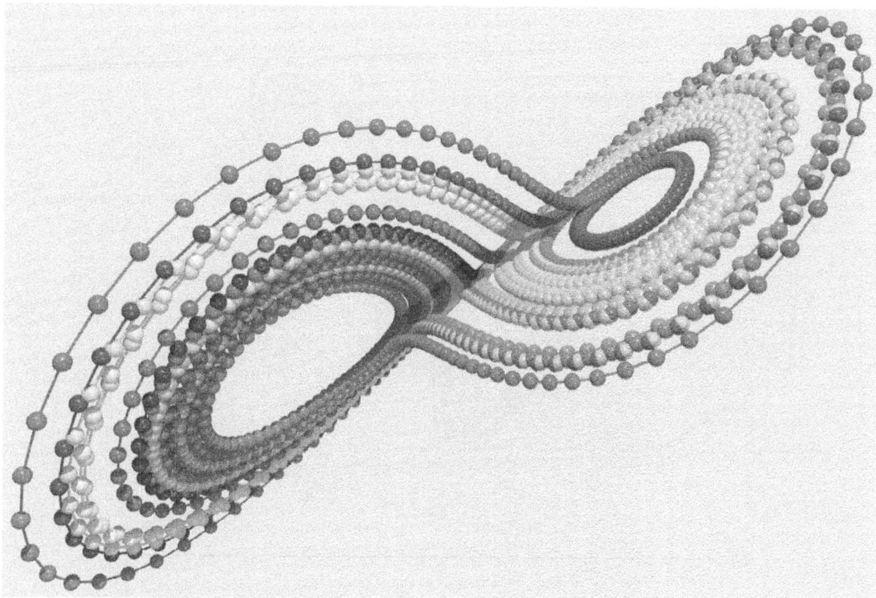

FIGURE 5.21 The butterfly trajectory of the Lorenz model, with the same initial condition and parameters as in Figure 5.20. The camera points to the negative z-axis (into the page), with the x-axis to the right and y-axis up.

Because the weather system is nonlinear and chaotic under certain conditions, a small difference in its initial conditions may lead to totally different outcomes. This is sometimes referred to as the butterfly effect. The flapping of a butterfly's wings somewhere in the world may cause either nothing, or a storm somewhere else. The unpredictability is not just due to the limitation of computer modeling. More powerful computers and more refined initial conditions can improve weather forecasts, but will not alter the inherit unpredictability and limited long-range forecasts of weather systems.

As we learned from the nonlinear driven oscillator, even chaotic motion is predictable in the Poincaré map (see Figure 5.16). How do we make such a map for the Lorenz model, which does not have a periodic driving force? We do this by making a cut along a given plane in phase space, say y-z plane, such that every time the trajectory crosses this plane, we plot a point at (y, z). When plane-crossing happens, $x = 0$. This means that when we integrate the equations of motion (5.33a) to (5.33c), we keep track of the value of x, and if it changes signs, we plot the pair of values (y, z).

A Poincaré map for the strange attractor is shown in Figure 5.22. As the trajectory moves between the wings of the strange attractor, it crosses the y-z plane numerous times (Figure 5.22, top). We mark the position on the plane at each crossing. In the end, the marked points form a "V" shaped distribution (Figure 5.22, bottom).

Suppose the trajectory is spiraling around the eye on the negative side of the x-axis at a given moment. By following the trajectory, we observe that it will move to the positive side of the x-axis by crossing on the right branch of the "V" and returning to the other side on the left branch. The order is never violated. Over a long time, there are equal number of crossings on each branch. However, there is no telling how many spirals around an eye it will execute before crossing, or where exactly on the "V" it will cross (other than the branch).

A Poincaré map is just one slice of the Poincaré surface of section (Figure 5.17). The one shown in Figure 5.22 is a slice of the Poincaré surface of section for the Lorenz model at $x = 0$.

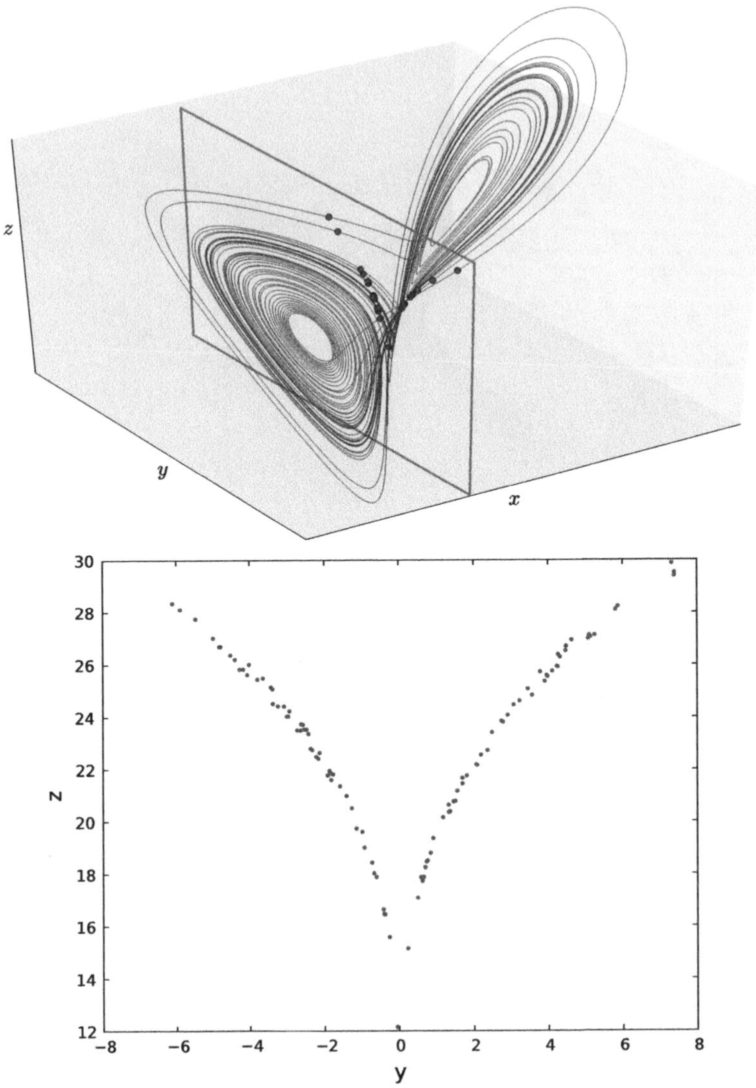

FIGURE 5.22 The making of (top) and the finished Poincaré map of the Lorenz model. The parameters are the same as in Figure 5.21.

We can build up a complete surface of section by making a series of Poincaré maps on planes parallel to the y-z plane at regular intervals of x. The finished surface of section should resemble Figure 5.21 (see Project P5.5).

5.5 POWER SPECTRUM AND FOURIER TRANSFORM

We have discussed ways to characterize chaotic dynamics in several nonlinear systems. They include the Lyapunov exponent, the Poincaré map, and the nonperiodicity of the trajectories. There is another powerful method, the power spectrum.

The power spectrum method can reveal frequencies in the system. For regular, periodic motion, we expect that the spectrum will be dominated by discrete frequencies. If the system becomes chaotic, what do we expect to see in the spectrum?

Power spectra can be obtained via the Fourier transform defined by

$$g(\omega) = \frac{1}{\sqrt{2\pi}} \int_{-\infty}^{\infty} dt f(t) \exp(-i\omega t). \tag{5.34}$$

The inverse Fourier transform can be written as

$$f(t) = \frac{1}{\sqrt{2\pi}} \int_{-\infty}^{\infty} d\omega\, g(\omega) \exp(i\omega t). \tag{5.35}$$

The functions $g(\omega)$ and $f(t)$ are uniquely related to each other by Eqs. (5.34) and (5.35). Furthermore, they satisfy the following integral relation,[7]

$$\int_{-\infty}^{\infty} |f(t)|^2\, dt = \int_{-\infty}^{\infty} |g(\omega)|^2\, d\omega. \tag{5.36}$$

If the function $f(t)$ represents the amplitude of a wave, we expect the energy carried by the wave to be proportional to the square of the amplitude, $|f(t)|^2$. We can therefore interpret $|g(\omega)|^2$ (or simply $|g(\omega)|$) as the energy density in frequency domain, namely, the power spectrum.

As an example, we show the power spectra of the logistic map at several control parameters in Figure 5.23. They are obtained by Fourier transform explained later. At the two smaller values of r, we see discrete peaks. The number of peaks corresponds to period-4 and period-16 cycles, respectively. This agrees with the period of stable fixed points in regular motion. By far the largest contribution comes from $\omega = 0$ component. It is a result of nonsinusoidal data.

Increasing r by just a small amount $r = 0.892 \rightarrow 0.893$, we see the spectra change qualitatively. There is a "noisy" background beneath the peaks. The logistic map is weakly

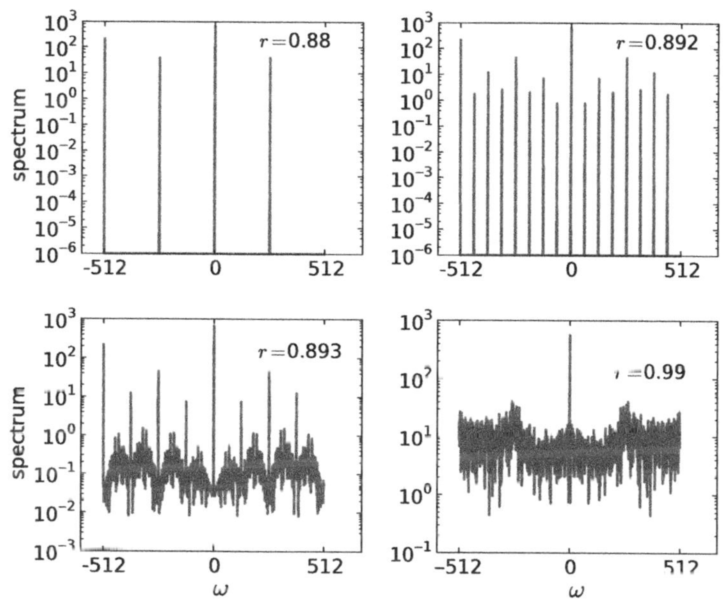

FIGURE 5.23 The power spectra of the logistic map at different values of r.

[7] It is a result of the convolution theorem and is also known as the Parseval's relation.

chaotic because it is at the boundary between regular and chaotic regions (Figure 5.6). Some quasi-periodic trajectories still persist. In the case of fully chaotic motion ($r = 0.99$), the spectrum is nearly continuous, except the $\omega = 0$ peak, which is always present. There are no periodic trajectories, so the power spectrum resembles the spectrum of white noise. In general, when a system becomes chaotic, the power spectrum will be dominated by white noise whose frequencies extend the full range.

The FFT Method

The spectra are generated by fast Fourier transform (FFT) of the map iterates. The data is the same as that used in making Figure 5.6. For numerical work, a discrete version of the continuous Fourier transform (5.34) is required. The discrete Fourier transform (DFT) takes the form

$$g_m \equiv g(\omega_m) = \sum_{k=0}^{N-1} f(t_k)\exp(-i\omega_m t_k), \tag{5.37}$$

where $f(t_k)$ are the N sampled data points at equidistant intervals t_k over a period of T. The discrete angular frequencies are

$$\omega_m = 2\pi m/T, \quad m = 0, 1, ..., N-1. \tag{5.38}$$

The direct approach to DFT as given by Eq. (5.37) is straightforward to implement. For each component g_m, the summation over k requires N operations (multiplications plus exponential function evaluations). To compute all N components, N^2 operations are needed. For large N, it is increasingly inefficient and slow.

The FFT algorithm can compute the Fourier transform in $N \ln N$ operations [18]. With the FFT method, one can achieve large gains in speed. It is one of the few truly significant algorithms discovered in digital computing that has had a tremendous impact on computational physics and nearly every field of scientific computing. Though FFT is widely available in nearly every programming language, it is well worth the effort to understand how it works to not only appreciate its power and elegance, but also to have a portable and flexible FFT function and to be able to modify it if necessary.

The FFT method is described separately in Section S:5.B, and the Python codes are shown in Program S:5.1. From now on we will include these functions in our FFT library `fft.py`, and `import fft` before use.

Figure 5.23 is produced with the FFT library. Because the Fourier transforms are generally complex, the power spectrum plots the absolute magnitude of the transformed data. The use of FFT is not limited to calculating the power spectrum. It can be used to find the onset of period doubling, for example. Exploration of power spectra of other systems are left to projects.

5.6 FRACTALS

Earlier we mentioned that the strange attractor of the Lorenz model apparently has a fractional dimension, that is, noninteger dimensions like one- or two-dimensional space that we are used to. For example, the attractor has a dimension ~ 2.05, so it is strange. How does this happen? A rough explanation is that the region around the attractor is chaotic in some directions but regular in others, that is, some Lyapunov exponents (λ) are positive and some negative. So the area is stretched (positive λ) in some directions but compressed in others (negative λ), making the area irregular and nonmeasurable in normal integer dimensions.

Table 5.1 The Number of Measuring Units as a Function of Unit Size for the Koch Curve.

Unit size l	1	1/3	1/9	...	$1/3^n$
Number of units $N(l)$	1	4	16	...	4^n

To understand fractional dimensions, suppose we wish to measure the length of a line, L. Typically, we would choose a measuring unit (scale), say l, and count how many units are needed to cover the line, say $N(l)$, then we would have the answer, $L = N(l) \times l$. So normally, we expect $N(l) \propto 1/l^1$. We say the dimension of a line is 1. For an area, we could use a unit square of area $\sigma = l^2$ as the basic measuring unit and would expect the number of unit squares to cover an area to vary as $N(\sigma) \propto 1/\sigma = 1/l^2$. This is then a two-dimensional space (surface).

In general we define the dimension of an object as

$$N(l) \propto l^{-d_f}, \tag{5.39}$$

where l is the scale, and $N(l)$ the number of basic units required to cover the object. The dimension is d_f, also known as the Hausdorff dimension. For regular lines, $d_f = 1$ and for surfaces, $d_f = 2$ as discussed earlier.

Note that Eq. (5.39) is another case of scaling laws, which we have seen to be very useful in prior cases (see Eq. (3.33) and Figure 4.8).

To obtain d_f, we just need to measure an object with two different scales l and l', count the numbers of units needed in each case, $N(l)$ and $N(l')$, and determine d_f as

$$d_f = -\frac{\ln N(l)}{\ln l}, \quad \text{or } d_f = -\frac{\ln(N(l)/N(l'))}{\ln(l/l')}, \quad l, l' \to 0. \tag{5.40}$$

Let us see what happens if we apply Eq. (5.40) to several self-similar objects. First, let us start with a line of unit length, cut out the middle third, and form a triangle, as shown in Figure 5.24. As we repeat the process indefinitely, a self-similar curve is formed. It is called a Koch curve. To determine the dimension of the curve, we count the number of units needed to cover the curve at various unit sizes, and we get Table 5.1.

According to Eq. (5.40), the dimension of this Koch curve is

$$d_f = -\ln 4^n / \ln(1/3^n) = \ln 4 / \ln 3 \sim 1.2618.$$

So it is not an integer dimension, but a fractional one. It is between 1 and 2, between a regular line and a regular area. We can think of it as an object of infinite length but zero area. Indeed, the length of the Koch curve does go to infinity as we add more and more triangles. We see that the process produces a self-similar object that has a fractional dimension, or a fractal. We could construct other types of fractals such as the Sierpinski carpet, which consists of taking out the middle third of a square repeatedly, or the Cantor set, and so on, all having fractional dimensions (Exercise E5.7 and Exercise E5.8).

Fractal Dimensions of Attractors

How do we define the fractal dimension of a strange attractor like that shown in Figure 5.21, which does not have a well-defined shape?

It turns out a rather elegant yet simple measure is more suitable for chaotic trajectories. Consider the trajectory spiraling around the strange attractor in the Lorenz model (see Figure 5.21).

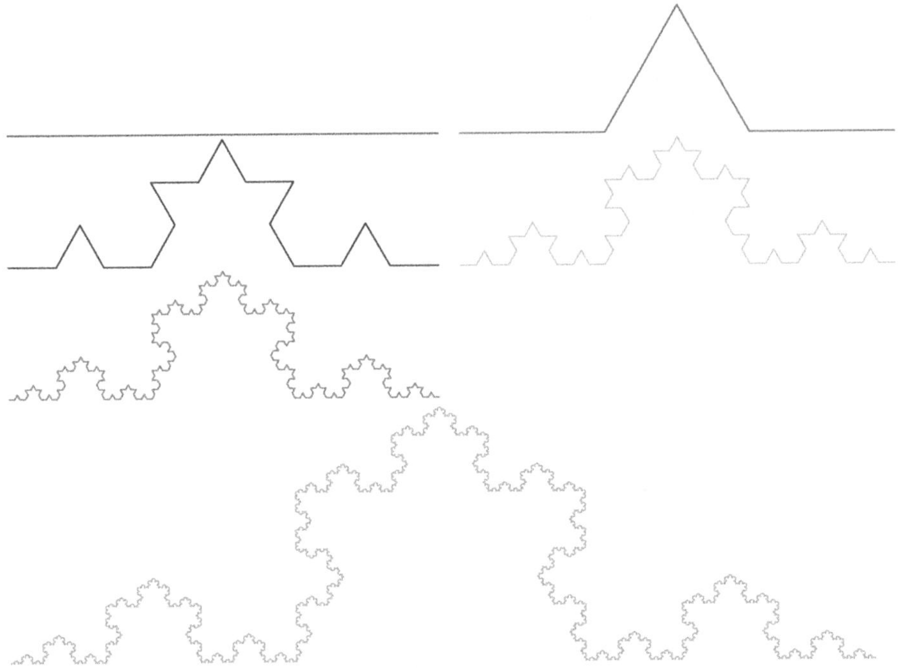

FIGURE 5.24 The Koch curve. The last level is enlarged by a factor of two.

Let us pick a point P on the trajectory and place a sphere of radius R at that point, like one of those depicted in the figure. Because the trajectory is chaotic, it will travel over all the allowed volume in phase space. If we wait long enough, the trajectory is bound to return to the neighborhood of the sphere.

We can count the number of times the trajectory enters the sphere over a given amount of time. Let us denote it by $N_P(R)$. We expect that $N_P(R)$ will increase with increasing R. Let us assume a power-law relation,

$$N_P(R) = CR^{d_c}, \quad \text{(correlation dimension)} \tag{5.41}$$

where C is a proportional constant, and d_c is called the correlation dimension [86].

Roughly, correlation dimension measures the frequency (or density) of the same trajectory visiting the neighborhood around point P. To find d_c, vary the radius and record $N_P(R)$ for different values of R. We can then plot $N_P(R)$ as a function of R on a log-log scale. Experiments show that over an intermediate range of R, the slope is a straight line. This slope is d_c, the correlation dimension.

The value of d_c thus obtained will have a weak dependence on the location of the point P. To get a more accurate value, this procedure is repeated at different points, and we take the average over them. By this method, we can obtain the correlation dimension of the strange attractor, ~ 2.05, see Project P5.6. It is thus a fractal with infinite area but zero volume, like an infinitely crumbled piece of (elastic) paper.

Mandelbrot Fractals

There are many kinds of fractals. A common feature among them is self-similarity. We have seen self-similarity in chaotic systems as well. Perhaps the most well-known ones are the Mandelbrot fractals, shown in Figure 5.25. The Mandelbrot fractals are generated by the equation, which

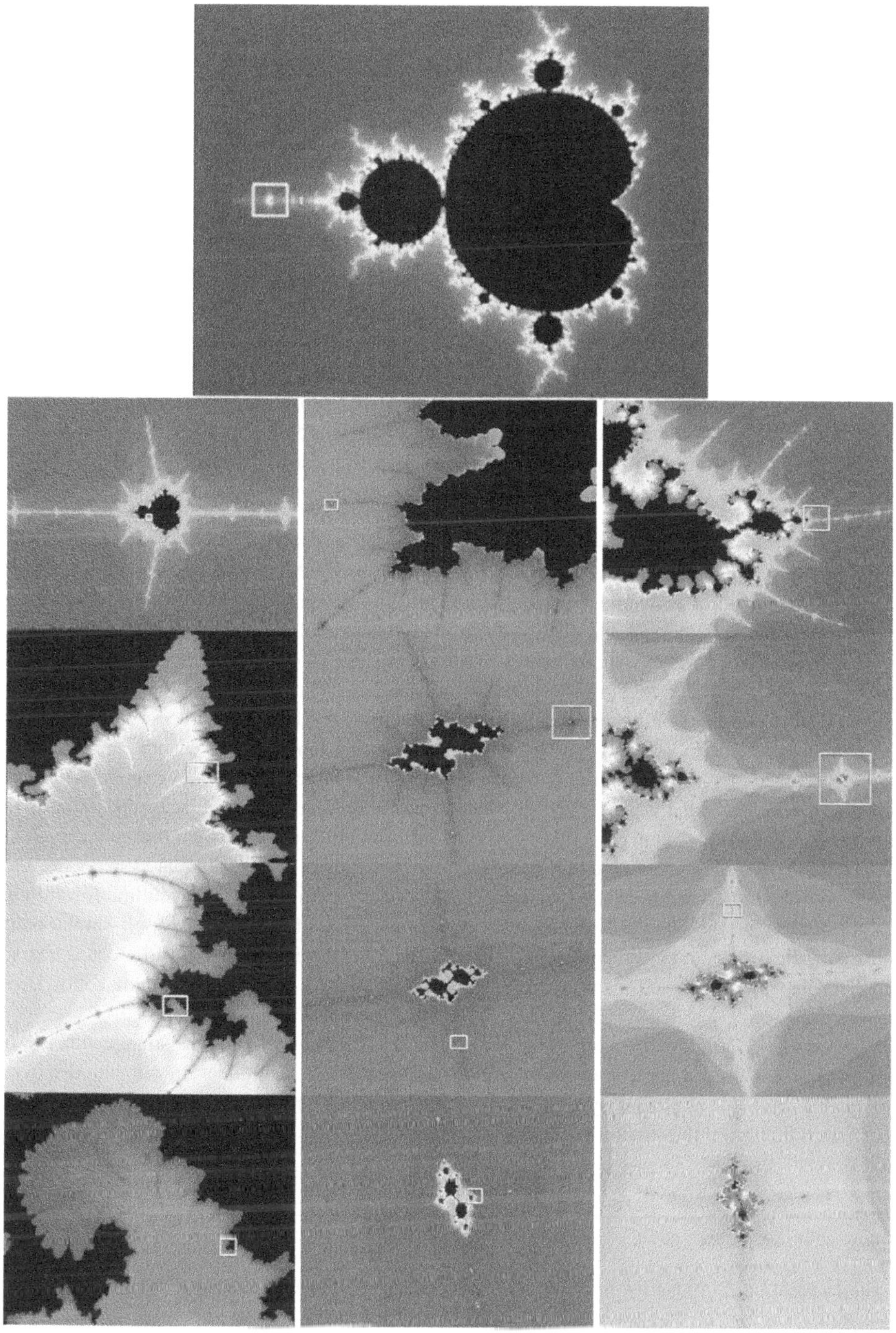

FIGURE 5.25 Successive levels (column major) of the Mandelbrot fractals.

defines the Mandelbrot set,

$$z_0 = 0, \quad z_{n+1} = z_n^2 + c, \quad |z| < \infty \text{ as } n \to \infty. \tag{5.42}$$

Here c is a complex number. The number c is in the Mandelbrot set if for large iterations z_n remains bounded.

To produce the Mandelbrot fractals, we pick a pair of numbers (x, y) and set $c = x + iy$. We then iterate Eq. (5.42) and see if $|z_n|$ remains finite. It can be shown that if $|z_n| > 2$, then it will grow with each iteration and tend to infinity, so this c is not in the set. We record the number of steps it takes until $|z_n|$ reaches the value 2 and color-code that point accordingly (Program 5.7).

Figure 5.25 shows successive magnifications of the Mandelbrot fractals starting with the top figure. The smaller figures are displayed in column-major order. The white box in each picture is blown up in the next level. We see at each scale, self-similar structures repeat themselves. By the last level shown, we have reached the limit of double precision. Renormalization would be required to go deeper. The level of complexity is infinitely deep.

Modify Program 5.7 to explore different parts of the domain. It is not possible to see the complete structure of the set. No one has, or ever will. Have fun making fractals, your own, beautiful, unique fractals. Who knew doing science would be so much fun?

CHAPTER SUMMARY

Through the logistic map, this chapter opened our discussion on nonlinear dynamics and a window to the rich and bewildering world of chaotic dynamics in deterministic systems. We carefully examined the logistic map with a mixture of numerical and analytical results, which enabled us to introduce several important concepts in the description of chaotic dynamics. We described periodic orbits revolving around stable fixed points. The fixed points become unstable as the control parameter increases, destroying the old periodic orbits but creating new ones of higher order in a series of period doubling. This process continues until at a critical point when all periodic orbits have been destroyed. The system becomes chaotic.

We discussed ways to characterize and quantify chaos, including the Lyapunov exponent, phase space trajectories, the Poincaré map and surface of section, and the power spectrum with the introduction of FFT. All these concepts are universal and demonstrated in nonlinear systems we presented, such as the driven pendulum and the Lorenz model of flow. Through the study of these exemplar systems, we see that chaotic motion is unpredictable in the long term, but it is not randomness. The evolution toward chaotic motion mostly follows the same pattern, the destruction of periodic orbits, the birth of self-similar but more complex periodic orbits, and eventually, sometimes abruptly, the onset of chaos. Besides chaos characterization, there are ways for limited chaos control (e.g., see [91]) because even chaotic systems follow deterministic laws.

Self-similarity at infinitely subdivisible levels leads to fractals, including the strange attractor of the Lorenz model. We used a variety of visualization and numerical techniques to aid our discovery of the interesting world of nonlinear dynamics and chaos, including unique perspectives in visualizing the Poincaré maps in 3D and the structure of the butterfly attractor.

5.7 EXERCISES AND PROJECTS

Exercises

E5.1 (a) Expand Eq. (5.9) for the fixed points of period 2, and show explicitly that the quartic equation is

$$x^* = 4r[4rx^*(1 - x^*)][1 - 4rx^*(1 - x^*)]. \tag{5.43}$$

(b) Solve the preceding equation by using explicit formulas for the roots of quartic equations, or, via a more insightful approach, by successively reducing it to a cubic and then a quadratic equation by re-using the period-one fixed points (5.8). Do it by hand, or better yet, let SymPy work for us (for free),

```
In [1]:  from sympy import *
In [2]:  x, r = symbols('x r')
In [3]:  f = 4*r*x*(1-x)
In [4]:  f2= 4*r*f*(1-f)
In [5]:  factor(f2-x)            # factor expression
```

After factoring out these terms, the resulting equation should look like

$$(x^* - x_1)(x^* - x_2)(x^{*2} + bx^* + c) = 0, \tag{5.44}$$

where $x_1 = 0$ and $x_2 = 1 - 1/4r$ are the fixed points of period one (5.8), $b = -1 - 1/4r$, and $c = (1 + 1/4r)/4r$. Solving the remaining quadratic equation yields the results, Eq. (5.10).

E5.2 (a) Write a program that generates the cobweb diagram showing the convergence toward attractors as illustrated in Figure 5.3. The procedure is as follows: plot the map function for a given cycle, for example, $y = f(x)$ for period one, and $f^{(2)}(x)$ for period two; place the initial point on the projection (diagonal) line, $y = x$; draw a vertical line to find its next value on the $y(x)$ curve; draw a horizontal line to the projection line $y = x$ for the seed value for the next cycle; and repeat the last two steps.

(b) Apply your program to other cases such as $f^{(3)}(x)$ and $f^{(4)}(x)$. Optionally, find the fixed points with SymPy, and compare with the graphical results.

E5.3 Prove the relation (5.13) mathematically. Start with

$$\epsilon_n = |x_n - x^*| = |g_n(x) - x^*|.$$

Assume x is in the small neighborhood of x^*, that is, $x = x^* + \epsilon$, and expand $g_n(x)$ in powers of ϵ.

E5.4 Show that the last two fixed points of period-2 (5.10) are stable for $3/4 < r < (\sqrt{6} + 1)/4$.

E5.5 Create a program to visualize the dynamics of period doubling using VPython. It should produce output like the sample Figure 5.7. Generate the time series for a given r, and vary it from ~ 0.7 to 1. Describe how the points move between the branches. Annotate the path for a representative value of r if possible.

E5.6 Find the steady state solutions of the Lorenz model (5.33a) to (5.33c), that is, solutions when $dx/dt = dy/dt = dz/dt = 0$. Analyze their stabilities.

E5.7 A Cantor set can be formed this way (Figure 5.26): start with a line, cut out the middle third, remove the middle thirds of the remaining two segments, and repeat this process indefinitely. Calculate the fractal dimension of the set. Discuss its meaning.

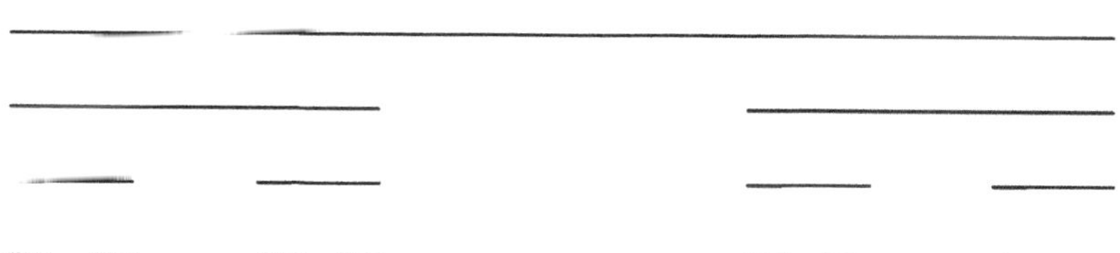

FIGURE 5.26 The first three levels of the Cantor set.

E5.8 Figure 5.27 shows the Sierpinski carpet. Describe how it is made, and calculate its dimension. Discuss your result.

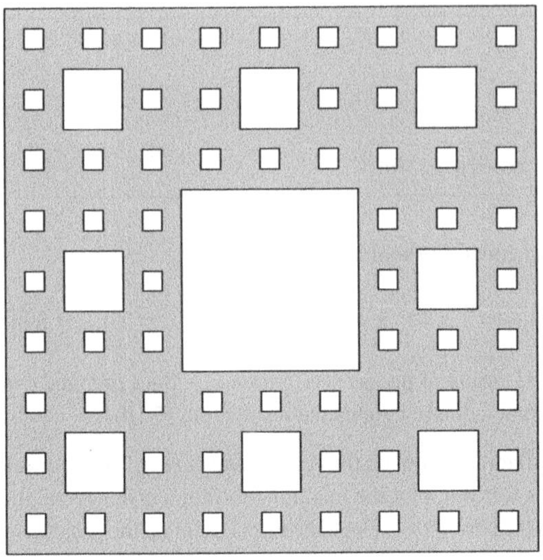

FIGURE 5.27 The first three iterations of the Sierpinski carpet.

E5.9 Generate a Julia set by solving $f(z) = z^3 + 1$ with Newton's method as follows. Pick a point (x, y) in the complex plane as our initial guess and set $z_1 = x + iy$. Iterate Newton's method (3.18) explicitly as

$$z_{n+1} = z_n - \frac{z_n^3 + 1}{3z_n^2}.$$

The iterations will converge to one of three roots (attractors), namely $[-1, \exp(\pm i\pi/3)]$. Check which one after each step within some tolerance. Record the number of iterations, and put a point at (x, y) using a color scheme. For instance, color the point in hues of red, green, or blue, depending on which root it converged to. Explore the basins of attraction and their boundaries. Discuss your observations regarding self-similarity and connectedness of the points.

Projects

P5.1 We have seen from Figure 5.6 that there is a period-3 stable window embedded in the chaotic region of the logistic map around ~ 0.96. Why does it not appear at smaller values of r? Explore this and related questions of the period-3 orbits.

(a) Plot the map function $f^3(x)$ at $r = 0.959$, and the straight line $y = x$. Find the intersection of $x = f^3(x)$. How many fixed points do you see? Which ones are stable?

(b) Repeat the above for $r = 0.95$. Are there any stable fixed points?

(c) Decrease r from 0.959. At each r generate a power spectrum of the map iterates after discarding transients. What is the lower limit of r when no period-3 cycles exist? Compare your results with $(1 + \sqrt{8})/4 \sim 0.9571$.

(d)* Investigate the other visible stable windows around $r \sim 0.935, 0.926$, and 0.907, corresponding to periods-5, 7, and 6, respectively.

P5.2* We take up the task of calculating the r values for the superstable fixed points in the logistic map. Direct calculations are useful in understanding the universal properties, including the Feigenbaum constants and the rate of convergence.

(a) Show analytically that the first two values for period-1 and period-2 cycles are $r_0^* = \frac{1}{2}$ and $r_1^* = (1 + \sqrt{5})/4$, respectively.

(b) Write a program to calculate the higher r_n^* series by solving Eq. (5.20) with Newton's root finder. The following suggested strategy should be useful. For a given $n \geq 2$:

- Define $g(r) = f^{(2^n)}(\frac{1}{2}) - \frac{1}{2}$, with $f^{(2^n)}(\frac{1}{2})$ iterating according to Eq. (5.5).
- Define $h(r) = dg(r)/dr$, and evaluate the derivative in a way analogous to the chain rule (5.15), but with respect to r (not x). Again, iterate (recursion-like) the intermediate $g_i \equiv x_i$ as earlier.
- Set the initial guess as

$$r = r_{n-1}^* + \frac{r_{n-1}^* - r_{n-2}^*}{\delta_{n-1}},$$

  where δ_{n-1} is the latest δ-number computed according to (5.18).
- Call Newton's root finder with g, h, and the initial r. Cross fingers and record the returned value. That is r_n^*.
- Repeat preceding steps for the next n, and so on. See how far you can push to higher n, or how close to r_∞.

(c) Compute the α numbers at each n. Make a table listing both the δ and the α numbers. Plot the results. Is the convergence uniform? Where does it break down? Why? Discuss these and other observations you have made.

P5.3 Investigate the properties of the sine map,

$$x_{n+1} = r \sin \pi x_n, \quad 0 < r \leq 1, \quad 0 \leq x \leq 1.$$

(a) Plot the map function and compare it to the logistic map.

(b) Find the fixed points of period-1 and period-2 cycles. Do this with a root finder. Discuss their stability.

(c) Generate a bifurcation diagram. Note where the bifurcations occur. Locate the r values for the onset of period-2^n cycles for $n = 1$ to 5. Do the first two values agree with your analysis above? Discuss and compare your results with the logistic map.

(d) Use the FFT technique to find the higher r_n values. What is the onset of chaos? Compute the Feigenbaum δ number with these values.

(e) Calculate the Lyapunov exponent of the map. Discuss similarities and differences with the logistic map.

(f) Pick two representative r values corresponding to a positive and a negative Lyapunov exponent, respectively. Iterate the sine map with a small difference in the initial conditions, and obtain the Lyapunov exponent graphically. Compare your direct, graphical results with the numerical results shown earlier.

P5.4 We will carry out further investigation of the nonlinear driven pendulum in this project. It is assumed that you will program each part as necessary.

(a) The Lyapunov exponents shown in Figure 5.14 are from only one initial condition. Compute these values more accurately by averaging results from several different initial conditions. Space them out over the ranges of θ and ω. How do the results change? How sensitive are they?

(b) For chaotic motion at $F_d = 1.1$, generate Poincaré maps from a number of different initial conditions, say 5 or more. Graph the data on the same plot. Are the shorelines different from Figure 5.16? Do they fill up the phase space? Explain.

(c) Produce the bifurcation diagram in Figure 5.18.

(d) Calculate the power spectra at $F_d = 0.7$ and $F_d = 1.1$. Use at least 1024 points. What is the difference between the two? Pick a few interesting regions from the bifurcation diagram, calculate and compare the spectra.

(e)* Visualize chaotic motion of two pendula through animation. Write a program to interactively display their positions. Introduce a small difference in their initial conditions, and see them either "attract" each other in regular motion or diverge from each other in chaotic motion. Add key detection to change F_d interactively (e.g., "up" to increase and "down" to decrease).

P5.5 Write a program to explore the Lorenz model and the strange attractor.

(a) Compare your results with Figure 5.19 for y. What is the characteristic time scale of the problem? What is your step size? How do you check convergence?

(b) Once your results are converged, calculate and plot the results for x and z for the same r values. Also compute the power spectra and analyze them.

(c)* Make a butterfly diagram and a Poincaré map (Figure 5.22). Construct the Poincaré surface of section by producing a series of Poincaré maps at several surfaces of constant x. Stack them next to each other using '3d' projection in Matplotlib as

```
ax = plt.subplot(1, 1, 1, projection='3d')
    ......
    ax.plot ([x]*len(y), y, z, 'o')
```

where y and z are the points for the Poincaré map at x.

(d)* Repeat parts (a) and (b) using RK45n (or if you used it already, switch to RK4n). Compare their performance, and the values of step size when the results are first converged.

P5.6 We quoted 2.05 as the fractal dimension of the strange attractor in the Lorenz model. Let us confirm it by computing the correlation dimension from Eq. (5.41) in this project. We will first try a simple approach to make things work. Once it is working, we can try a more complex method.

(a) Start a trajectory in the vicinity of the strange attractor, say at $(-7, -2, 30)$. You can use the program developed in Project P5.5 if applicable. Integrate the Lorenz equations with $r = 28$. Wait for transients to pass; 1,000 steps would suffice if the step size is 0.01.

(b) Take the first point after the transients as our center point P, with position (X, Y, Z). Let us imagine we make three spheres of different radii R_i ($i = 1, 2,$ and 3) centered at P. For example, radii of 1, 2, and 3 work well. Initialize three counters that will keep track of the numbers the trajectory has entered the spheres.

(c) After every step, check if the current position of the trajectory is inside any of the three spheres. A point is inside if

$$(x - X)^2 + (y - Y)^2 + (z - Z)^2 < R_i^2.$$

If so, increment the corresponding counter. The total number of integration steps should be such that the counter for the smallest sphere should be 100 or so. Roughly 20,000 to 30,000 steps are enough.

(d) When the integration stops, plot the three counter values versus the radii on a log-log scale. Calculate the three slopes between the three points (1 to 2, 2 to 3, and 1 to 3). Take the average, and this is an approximate correlation dimension at P. It should be 5% to 10% within the value 2.

(e) Once the preceding steps are working, try the same procedure for different center points. Average all the values, and it should be approaching 2.05.

(f)* Now let us try a dozen or so different-sized spheres. The radii should differ by some constant factor so that when plotted on the log-log scale, they spread out evenly. The factor should be such that the largest radius should be no more than half the size of the attractor. Note that if the trajectory is inside a sphere of given size, then it is also inside the spheres of larger sizes (outer spheres). Discuss the shape of your curve. What happens at the two extremes of the radii? Compare your results with the three-sphere results.

(g)* Repeat part (f), but instead of spheres, use boxes of different sizes. Are the results comparable? Comment on your findings.

P5.7 The double-well oscillator is a particle moving in a potential with two minima on either side of a local maxima, $V(x) = \frac{1}{4}x^4 - \frac{1}{2}x^2$. The equation of motion is

$$\ddot{x} + b\dot{x} - x + x^3 = F_d \cos(\omega_d t),$$

where b, F_d, and ω_d have the same meaning as those in the driven nonlinear pendulum (5.31). It is a useful model for two-state systems.

(a) Sketch the potential, then graph it. Find the natural frequency for small oscillations without the driving force.

(b) Predict how the particle will move if F_d is zero, small, and large. Write down your predictions.

(c) Simulate the double-well oscillator, assuming fixed values $b = 0.25$ and $\omega_d = 1$ (in some scaled units). Plot the position and velocity curves (x-t and v-t), for $F_d = 0.1$, 0.2, and 0.4. Describe the results, and how they compare with your predictions.

(d) Compute the power spectrum for each F_d above. Relate your results above to the power spectra.

(e)* Construct Poincaré maps, one for each F_d. What do they represent? Are any of them strange attractors? Quantify your answers.

5.A PROGRAM LISTINGS AND DESCRIPTIONS

Program listing 5.3 Initial value dependence of logistic map (`logisdiff.py`)

```
import matplotlib.pyplot as plt     # get matplotlib plot functions

x1, r = input('enter x1, r; eg .4, .7 : ')
x2 = x1 + 0.01                      # initial difference
n, xn1, xn2, diff = 20, [], [], []                    # buffers
for i in range(n):
    xn1.append(x1), xn2.append(x2), diff.append(abs(x1-x2))
    x1, x2 = 4*r*x1*(1.0-x1), 4*r*x2*(1.0-x2)      # parallel iterates

plt.figure()                        # plot the series
plt.plot(range(n), xn1, 's-', range(n), xn2,'o-')  # squares & circles
plt.xlabel('$n$'), plt.ylabel('$x_n$')
plt.text(3, .01+min(xn1),'$r=$' + repr(r))

plt.figure()                        # plot the difference
plt.plot(range(n), diff)
plt.semilogy()                      # semilog scale
plt.xlabel('$n$'), plt.ylabel('$\Delta x$')
plt.show()
```

This program calculates and plots the difference of the logistic map from two initial values as shown in Figures 5.9 and 5.10. Note the latter is on a semilog scale.

Program listing 5.4 Lyapunov exponent of logistic map (`lyapunov.py`)

```
import matplotlib.pyplot as plt      # get matplotlib plot functions
import math as ma                    # get math functions

def lyapunov(r):                     # compute Lyapunov exponent
    sum, x, ntrans, nsteps = 0.0, 0.5, 2000, 2000     # try diff num.
    for i in range(ntrans):          # let transients pass
        x = 4*r*x*(1-x)
    for i in range(nsteps):          # sum up next n steps
        x = 4*r*x*(1-x)
        dfdx = 4.0*r*(1.0-x-x)       # Eq. (5.15)
        sum += ma.log(abs(dfdx))
    return sum/float(nsteps)         # lambda

ra, lyap, r, dr = [], [], 0.7, 0.0001
while r < 1.0:
    r = r + dr
    ra.append(r), lyap.append(lyapunov(r))

plt.figure()
plt.plot(ra, lyap, ',')              # ','=pixels
plt.axhline(0.), plt.ylim(-2, 1)     # draw horiz. line, set y limits
plt.xlabel('r'), plt.ylabel('λ')
plt.show()
```

The Lyapunov exponent is computed from the analytic expression (5.29) in Program 5.4.

Program listing 5.5 Nonlindro: Nonlinear driven oscillator (`nonlindro.py`)

```
import matplotlib.pyplot as plt      # get matplotlib plot functions
import ode, math as ma               # get ODE, math functions

def pendulum(y, t):  # y = [theta, omega], omega_0 = 1
    return [y[1], -ma.sin(y[0]) - b*y[1] + fd*ma.cos(omega_d*t)]

def solution(n_periods):        # find solutions for n_periods
    bins = 40                   # number of points per period
    t, y, h = 0.0, [1.0, 0.0], 2*pi/(omega_d*bins) # init values
    ta, theta, omega = [], [], []
    for i in range(n_periods*bins):
        ta.append(t), theta.append(y[0]), omega.append(y[1])
        t, y = t+h, ode.RK4n(pendulum, y, t, h)
    return ta, theta, omega

b, omega_d = 0.5, 0.6           # damping coeff., driving frequency
subnum, pi = 1, ma.pi           # subplot number, pi
plt.figure()
for fd in [0.7, 1.1]:
    ax1 = plt.subplot(2, 2, subnum)         # 2x2 subplots
    ax2 = plt.subplot(2, 2, subnum+2)
    ta, theta, omega = solution(n_periods = 5)
    ax1.plot(ta, theta), ax2.plot(ta, omega)
    if (subnum == 1):                       # subplot specific label
        ax1.set_ylabel('$\\theta$ (rad)')
```

```
        ax2.set_ylabel('$\\omega$ (rad/s)')
    subnum, fdtxt = subnum + 1, '$F_d=$'+repr(fd)
    ax1.text(17, max(theta), fdtxt), ax2.text(17, max(omega), fdtxt)
    plt.xlabel('t (s)')
plt.show()
```

Program "Nonlindro" integrates the equations of motion of the nonlinear driven oscillator (5.31), which are calculated in `pendulum()`. The two-element array `y` contains the angular displacement and angular velocity as $y = [\theta, \dot{\theta}]$. It returns the derivatives in a list to be used with nonvectorized ODE solvers because the two ODEs are simple, and speed is important in such calculations like the bifurcation diagram (see Project P5.4). The integration uses the nonvectorized Runge-Kutta `RK4n`, which is faster than vectorized `RK4` for a handful of ODEs. Because energy is not conserved, there is no need to use symplectic integrators such as the leapfrog method. The parameters `b`, `fd`, and `omega_d` are global variables defined in the main program.

To maintain accuracy, the step size must be small compared to the characteristic time. In this case, it should be the smaller of the two time scales: the natural period $2\pi/\omega_0$, and the period of the driving force $2\pi/\omega_d$. In practice, we aim to achieve consistent accuracy by specifying a fixed number of points per period (line 8) and calculate the step size accordingly (line 9).

The results are plotted in four subplots (Figure 5.13) made up of two rows by two columns (line 20) arranged as

1	2
3	4

Each subplot is positioned by a subplot number, `subnum`, that ranges from 1 to 4, starting from the top left, increasing across and then down.

Program listing 5.6 Poincaré map (`poincare.py`)

```
import matplotlib.pyplot as plt      # get matplotlib plot functions
import ode, math as ma               # get ODE, math functions

def remap(x):            # remap theta to [-pi,pi]
    if (abs(x) > pi): x = (x - 2*pi if x>0 else x + 2*pi)
    return x

def pendulum(y, t):  # y = [theta, omega], omega_0 = 1
    return [y[1], -ma.sin(y[0]) - b*y[1] + fd*ma.cos(omega_d*t)]

def poincare(transient, n_periods):     # transient periods, n_periods
    bins = 40                        # number of points per period
    t, y, h = 0.0, [0.6, 0.0], 2*pi/(omega_d*bins) # init values
    theta, omega = [], []
    for i in range(bins*transient):     # discard transients
        t, y = t+h, ode.RK4n(pendulum, y, t, h)
        y[0] = remap(y[0])
    for i in range(bins*n_periods):
        if (i%(bins//2) == 0):          # record every half a period
            theta.append(y[0]), omega.append(y[1])
```

```
21          t, y = t+h, ode.RK4n(pendulum, y, t, h)
            y[0] = remap(y[0])
23
        return theta, omega
25
    b, omega_d = 0.5, 0.6           # damping coeff., driving frequency
27  subnum, pi = 1, ma.pi           # subplot number, pi
    plt.figure()
29  for fd in [0.7, 1.1, 1.2]:
        theta, omega = poincare(transient = 20, n_periods = 400)
31      ax = plt.subplot(3, 1, subnum)    # 3x1 subplots
        ax.plot(theta, omega, '.'), ax.set_xlim(-pi, pi)
33      if (subnum == 2): ax.set_ylabel('$\\omega$ (rad/s)')
        subnum, fdtxt = subnum + 1, '$F_d=$'+repr(fd)
35      ax.text(-3, min(omega), fdtxt)

37  plt.xlabel('$\\theta$ (rad)')
    plt.show()
```

The `poincare()` function takes as input two parameters: the first specifies the number of periods to wait for transients, and the second the number of periods to record data. Data is recorded only at discrete points in a given period, at the beginning and in the middle in this case (line 19). Results are displayed in multiple plots with manual control of x-axis limits.

Program listing 5.7 Mandelbrot fractal (`fractal.py`)

```
import numpy as np, matplotlib.pyplot as plt
from numba import jit                   # comment out if jit not available
@jit                                    # just-in-time compiling
def mandelbrot(c, maxi):
    z = c
    for m in range(maxi):               # maxi=max iterations
        z = z*z + c
        if (z.real*z.real + z.imag*z.imag >=4.0): break
    return m

xl, xr, yb, yt = -2.2, 0.8, -1.2, 1.2          # box size
nx, ny, maxi = 600, 481, 100
x, y = np.linspace(xl,xr,nx), np.linspace(yb,yt,ny)
fractal = np.zeros((ny, nx, 3))                 # fractal RGB image
for i in range(nx):
    for k in range(ny):
        m = mandelbrot(x[i] + 1j*y[k], maxi)    # point in complex plane
        fractal[k, i] = [m, 2*m, 3*m]           # RGB color mix
plt.imshow(fractal/maxi, extent=[xl,xr,yb,yt])  # plot as image
plt.show()
```

The program computes the Mandelbrot fractal using complex arithmetic. In Python, complex numbers are formed by combining real values with the pure imaginary number $i = \sqrt{-1}$ represented by `1j`. If `z` is a complex number, `z.real` and `z.imag` return the real and imaginary parts, respectively.

The box size can be changed to look at different parts of the domain. At finer scales, the maximum iteration needs to be increased. The fractal is stored as an RGB image with a color mixing scheme (line 18, experiment a bit to your liking). The image is displayed by `plt.imshow()` on line 19. It is similar to a contour plot but may be visually more effective in cases like this.

Mapping the Mandelbrot set is computationally intensive. The bottleneck is in determining if a point is in the set. It is an ideal candidate for a speed boost. Here we use Numba for simplicity. The `@jit` statement instructs Numba to compile `mandelbrot` at runtime. The speedup factor is about six in this case, quite good considering how effortless it is on our part. If Numba is not installed, comment out the two lines. There are other remedies for speed boost (see Section 1.3.2). Another example is given in Program S:11.4.

CHAPTER 6 Oscillations and Waves

Periodic motion is ubiquitous in nature, from the obvious like heartbeats, planetary motion, vibrations of a cello string, and the fluttering of leaves, to the less noticeable such as the vibrations of a gong or shaking of buildings in the wind. Oscillations and periodicity are the unifying features in these problems, including nonlinear systems in Chapter 5. The motion is back and forth repeating itself over and over and can cause waves in a continuous medium.

We will study periodic motion including oscillations and waves in this chapter. There are two motivating factors for studying these phenomena at this juncture. Physically, waves involve independent motion in both space and time. We can see waves moving along a string (space) and also feel the vibrations in time at a fixed point. We can better understand wave motion after understanding particle motion. Mathematically, waves are described by partial differential equations (PDEs) with at least two independent variables such as space and time. We are equipped to tackle PDEs only after studying ordinary differential equations (ODEs) first in the previous chapters.

We begin with the discussion of a single, damped harmonic oscillator and work our way to oscillations of small systems such as triatomic molecules. We then discuss the displacement of a string under static forces. Finally, simulations of waves on a string and on a membrane are discussed.

To solve these problems, we introduce matrix algebra and eigenvalue formulation of linear systems. We also discuss finite difference and finite element methods for boundary value problems involving ordinary and partial differential (wave) equations. Various visualization techniques including animation of waves on strings and membranes using objects in our VPython modules are also presented.

6.1 A DAMPED HARMONIC OSCILLATOR

The best-known oscillator in the physical world is also the simplest: the simple harmonic oscillator (2.46), with its ubiquitous acronym—SHO. Figure 6.1 captures several frames of the animated motion of an SHO over half a period. It is produced with Program 6.1, a slightly modified version of Program 2.1.

Program listing 6.1 Simple harmonic oscillator (`sho.py`)

```
import visual as vp

ball  = vp.sphere(pos=(-2,-2,0), radius=1, color=vp.color.green) # ball
wall  = vp.box(pos=(-4,-2,0), length=.2, height=4, width=4)     # wall
floor  = vp.box(pos=(0,-3.1,0), length=8, height=0.2, width=4) # floor
spring  = vp.helix(pos=wall.pos, thickness=0.1, radius=0.5)       # spring

h, v = 0.1, 0.0                              # step size,  initial  velocity
while True:
    vp.rate(50)
    ball.pos.x = ball.pos.x + v*h
    v = v - ball.pos.x*h                     # Euler-Cromer method
    spring.length = ball.pos.x + 3           # stretch spring (offset by 3)
```

FIGURE 6.1 Animated motion of a simple harmonic oscillator.

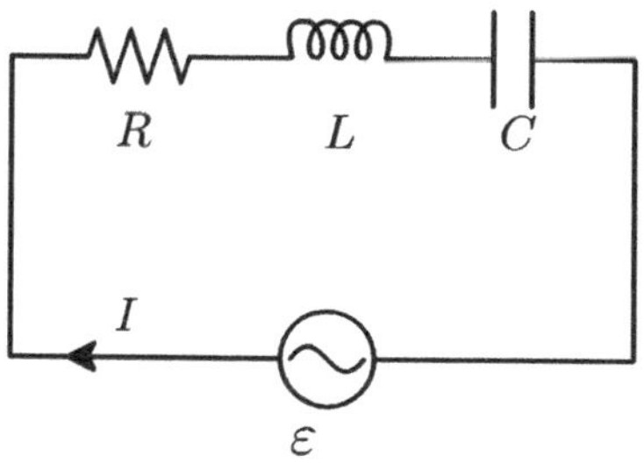

FIGURE 6.2 *RLC* circuit.

The spring is drawn as a `helix` object in VPython (line 6), which has its position at the tail end of the coil (at the wall in this case) and extends in the direction of its axis vector (default to the x-axis, (1,0,0)). The length of the spring is changed as it is being compressed or stretched by the ball (line 13, plus an offset from the wall to the equilibrium position). The integration uses the Euler-Cromer method (line 12), a first-order symplectic integrator for conservative systems (Section 2.4, and Exercise E2.6). The numerical method is stable, and the SHO will oscillate periodically and perpetually according to Eq. (2.47) because there is no dissipation.

Realistic oscillators are dissipative, or damped, due to forces like friction or air resistance. A damped harmonic oscillator of mass m and spring constant k may be described by the linearized version of Eq. (5.30) (small oscillations and a change of variable $\theta \to x$) as

$$\frac{d^2x}{dt^2} + 2\gamma\omega_0\frac{dx}{dt} + \omega_0^2 x = 0. \tag{6.1}$$

Two generic, system-independent constants are introduced

$$\omega_0 = \sqrt{\frac{k}{m}} = \text{natural frequency}, \quad \gamma = \frac{b}{2m\omega_0} = \text{damping coefficient}, \tag{6.2}$$

where b is the linear coefficient in the damping force, $-bv$ (see Eq. (3.3)).

Given ω_0 and the dimensionless γ, Eq. (6.1) is the universal equation for a damped harmonic oscillator. It occurs frequently in physics and engineering problems and has wide ranging applications. For example, an *RLC* circuit (Figure 6.2) behaves exactly like the mechanical oscillator we assumed in Eq. (6.1). The only difference is that the displacement x would be replaced with the charge q (Exercise E6.2).

6.1.1 CRITICAL, OVER- AND UNDERDAMPING

Let us first examine the solutions of Eq. (6.1) numerically using RK4 for several cases of damping as shown in Figure 6.3.

With no damping, the oscillation is sinusoidal as expected. When damping is small, the oscillation persists but the amplitude is reduced. When damping is further increased to $\gamma \geq 1$,

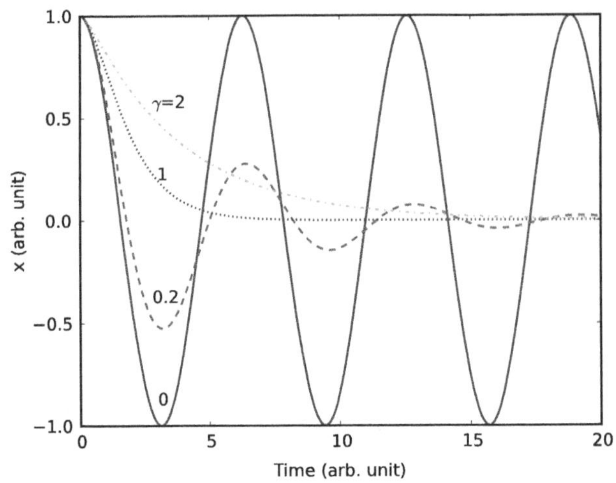

FIGURE 6.3 The damped harmonic oscillator for different damping coefficients γ: no damping, $\gamma = 0$; underdamping, $\gamma = 0.2$; critical damping, $\gamma = 1$; and overdamping, $\gamma = 2$.

the oscillation is totally wiped out, as the amplitude does not cross the zero axis even once. For the smaller $\gamma = 1$, damping happens quicker and more effective than the larger $\gamma = 2$. This seems counterintuitive at first, but it is all about the balance of forces. If damping is large, the spring force needs a longer time to restore the system to equilibrium, losing all energy and getting stuck at the equilibrium. The smaller the damping, the faster the process. However, there must be a threshold where if the damping is further reduced, the oscillator will be unable to lose all its energy by the time it returns to equilibrium. If it still has a positive amount of energy left when it reaches the equilibrium position, it will keep going, crossing zero to the other side of the equilibrium. Indeed, this threshold exists ($\gamma = 1$, Figure 6.3) and is called critical damping. Below the threshold $\gamma < 1$, the oscillation continues, but with reduced peaks and valleys in each cycle, and eventually stops.

The preceding numerical results can be confirmed by analytical analysis [10, 31]. The general solution $x(t)$ to Eq. (6.1) is (Exercise E6.3)

$$x(t) = \exp(-\gamma\omega_0 t)\left[A\exp\left(\sqrt{\gamma^2-1}\,\omega_0 t\right) + B\exp\left(-\sqrt{\gamma^2-1}\,\omega_0\right)\right], \qquad (6.3)$$

where A and B are constants, determined by the initial conditions $x(0)$ and $\dot{x}(0)$. Without damping ($\gamma = 0$), Eq. (6.3) reduces to a perfect harmonic (2.47).

Because of the overall damping factor $\exp(-\gamma\omega_0 t)$, the solution will eventually reach zero. We expect this because damping will dissipate all the energy in the system so it stays put at equilibrium. How it gets there, however, is not trivial and is quite interesting, as discussed earlier following Figure 6.3.

Qualitatively, the exponential damping factor is determined by the smaller exponent (A term) in Eq. (6.3), $\gamma - \sqrt{\gamma^2-1}$ for $\gamma \geq 1$. Its maximum is 1 when $\gamma = 1$, the critical damping – the fastest damping possible. If $\gamma \gg 1$, then $\gamma - \sqrt{\gamma^2-1} \sim 1/2\gamma$. The larger the γ, the less effective the damping. This is overdamping.

If $\gamma < 1$, the factor $\sqrt{\gamma^2-1}$ becomes imaginary, so the terms in the brackets of Eq. (6.3) are oscillatory (sinusoidal) functions of time. This leads to underdamping of the kind

$$x(t) = A'\exp\left(-\gamma\omega_0 t\right)\cos(\sqrt{1-\gamma^2}\,\omega_0 t + \varphi), \quad (\gamma < 1), \qquad (6.4)$$

where A' and φ are new constants related to A and B. Both overdamping and underdamping increase the time it takes to reach equilibrium.

6.1.2 DRIVEN OSCILLATOR AND RESONANCE

In our discussion so far, the system loses energy continuously and stops at equilibrium on its own. What happens if energy is added to the system via an external driving force? We then speak of a driven, linear oscillator. How will it be different from a driven nonlinear oscillator of Section 5.3?

Let the driving force be $F(t)$. We simply add it to Eq. (6.1) to obtain the equation of a driven oscillator

$$\frac{d^2x}{dt^2} + 2\gamma\omega_0\frac{dx}{dt} + \omega_0^2 x = F(t). \tag{6.5}$$

An interesting case involves a periodic driving force $F(t) = F_d\cos(\omega_d t)$. Given ω_0, there are essentially three parameters we can play with: the damping coefficient γ, the driving frequency ω_d, and amplitude F_d. The amplitude F_d has dimensions of force/mass, but its units are arbitrary depending on the units of x and t.

The results of the driven linear oscillator obtained with RK4 are shown in Figure 6.4, where we vary the driving frequency ω_d but keep γ and F_d fixed.

The oscillations start out in irregular fashion, but eventually settle into regular sinusoidal oscillations. This happens quicker for critical damping and overdamping, somewhat slower for underdamping. The irregular part of the oscillation is due to transient solutions and the regular part due to steady state solutions. If we look at the steady state solutions carefully, we find that the oscillation frequency is the same as the driving frequency. This is because damping will cause any initial perturbations to subside, and in the end, the system is responding to the driving force only. Mathematically, the transient solution is the general solution (6.3) to the homogeneous equation (6.1), and the steady state solution is a particular solution to the inhomogeneous equation (6.5) [10].

Unlike the chaotic nonlinear oscillator (Section 5.3), the response of the linear oscillator is proportional to the driving amplitude (see Exercise E6.4). Incidentally, there is no nonlinear dynamics, hence no chaos, in a driven linear oscillator.

Though the driving amplitude is the same in each case, we see a big difference in the steady state amplitudes. For the case shown with relatively small damping, the amplitude is significantly enhanced when the driving frequency is close to the natural frequency, $\omega_d \simeq \omega_0$. This is resonance. The driving force is in near lockstep with the natural oscillation that energy is most easily absorbed, producing an enhanced amplitude.

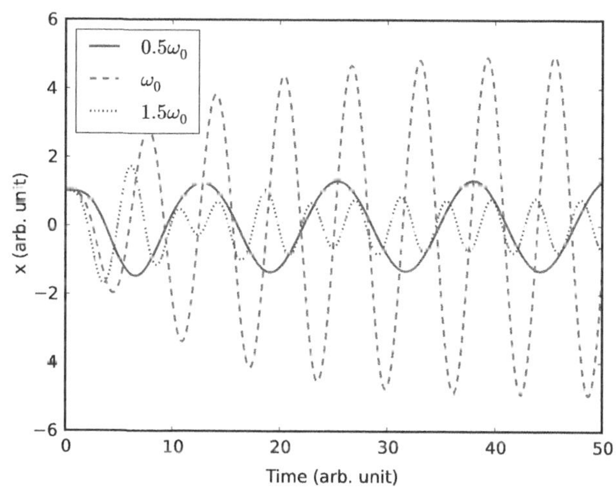

FIGURE 6.4 The driven harmonic oscillator for different driving frequencies: $\omega_d = 0.5\omega_0$, ω_0, and $1.5\omega_0$, with $\gamma = 0.1$ and $F_d = \omega_0^2$ in each case.

The general resonance frequency where the amplitude is maximum occurs at (Exercise E6.4)

$$\omega_r = \sqrt{1 - 2\gamma^2}\,\omega_0. \tag{6.6}$$

Resonance is an important phenomenon in physics and in nature.[1] We will see it in quantum transitions (Sections 8.4.2 and S:8.2) and scattering (Sections 12.4.1 and S:8.1.2).

6.2 VIBRATIONS OF TRIATOMIC MOLECULES

Instead of external driving forces, coupled oscillators can drive each other with internal forces, conserving the energy of the system. Like the simple two- and three-body problems seen before (Section 4.2), Figure 6.5 shows two systems consisting of two and three particles, respectively. We assume the particles are connected by identical springs with spring constant k and unstretched length l. The spring forces could represent bonds in diatomic or triatomic molecules such as CO (C–O) or CO_2 (O–C–O), for instance.

Displacement from Equilibrium

For the two-particle system, the equations of motion are

$$\ddot{x}_1 = -k(x_1 - x_2 + l)/m_1, \quad \ddot{x}_2 = -k(x_2 - x_1 - l)/m_2. \tag{6.7}$$

The addition (or subtraction) of l takes into account that, in Hooke's law $F = -kx$, the variable x refers to the stretching (or compression) of the spring from the unstretched length (l). We can make Eq. (6.7) tidier by measuring the displacement from equilibrium positions. Let x_1^* and x_2^* represent the equilibrium positions of particle 1 and 2, respectively. We then have $x_2^* - x_1^* = l$. Introducing new displacement variables as

$$u_1 = x_1 - x_1^*, \quad u_2 = x_2 - x_2^*, \tag{6.8}$$

we can rewrite Eq. (6.7) as

$$\ddot{u}_1 = -k(u_1 - u_2)/m_1, \quad \ddot{u}_2 = -k(u_2 - u_1)/m_2. \tag{6.9}$$

Using the center of mass (CM) coordinates similar to Eq. (4.4),

$$U = \frac{m_1 u_1 + m_2 u_2}{m_1 + m_2}, \quad u = u_1 - u_2, \tag{6.10}$$

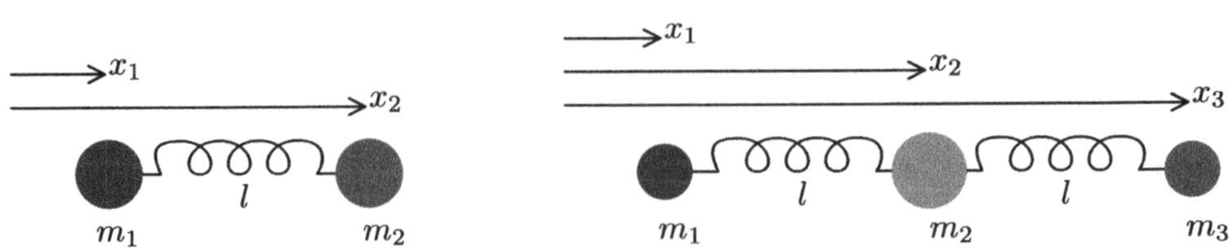

FIGURE 6.5 The diatomic and triatomic molecules.

[1]The Millennium Bridge in London had to be closed just two days after opening in June 2000 due to resonant swaying. The bridge swayed laterally when crowds walked over it, causing them to sway in step to keep balance and leading to resonant vibrations. Fixed with more damping, it reopened 20 months later, nicknamed the "wobbly bridge."

we obtain

$$\ddot{U} = 0, \quad \ddot{u} + \omega^2 u = 0, \quad \text{with } \omega = \sqrt{k/\mu}, \quad \mu = \frac{m_1 m_2}{m_1 + m_2}. \tag{6.11}$$

The oscillation (u) of two particles is effectively one SHO with angular frequency ω and reduced mass μ. It is a perfect harmonic (2.47).

6.2.1 NUMERICAL SOLUTIONS AND POWER SPECTRUM

A pair of true coupled oscillators are contained in the three-particle system in Figure 6.5, referred to as the triatomic molecule from now on. We can write down the equations of motion directly in displacement variables u_i from Eq. (6.8) as

$$\begin{aligned} m_1\ddot{u}_1 &= -k(u_1 - u_2), \\ m_2\ddot{u}_2 &= -k(u_2 - u_1) - k(u_2 - u_3), \\ m_3\ddot{u}_3 &= -k(u_3 - u_2). \end{aligned} \tag{6.12}$$

Let us first examine the numerical solutions of Eq. (6.12), which may be solved with a code like Program 6.5. The results are displayed in Figure 6.6. The parameters used are $k = 1$, $m_1 = m_3 = 1/4$, $m_2 = 2/5$, in arbitrary units.[2]

The initial condition was set to $(u_1, u_2, u_3) = (-0.2, 0.2, 0)$ and zero velocities. Only the displacement u_1 is shown (the other two are qualitatively similar). We see regular, periodic oscillations, but they are not sinusoidal. The positive and negative displacements are not symmetric about the equilibrium. The period appears to be 2π for this set of initial conditions.

The nice periodicity in the data tells us that it can be decomposed into a Fourier series to reveal the fundamental frequencies in it. We calculate the power spectrum of the displacement u_1 using the FFT method and show the results in Figure 6.7.

A total of 1,024 data points are used in the Fourier transform. We interpret the spectrum as ranging from negative to positive frequencies, or ω_m in $[-512, 511]$ (see Sections 5.5 and S:5.B).

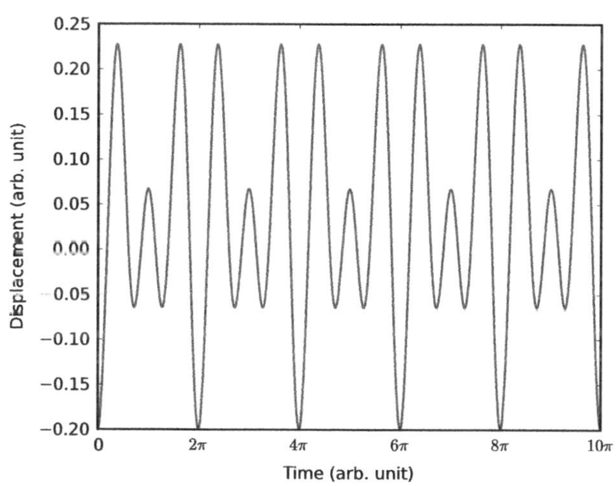

FIGURE 6.6 The displacement of particle 1 in the triatomic molecule.

[2]The time and length scales are determined by the choice of units for k and m_i. We can set them to arbitrary units without loss of generality. If they are in SI units, for example, then time and length would be in seconds and meters, respectively.

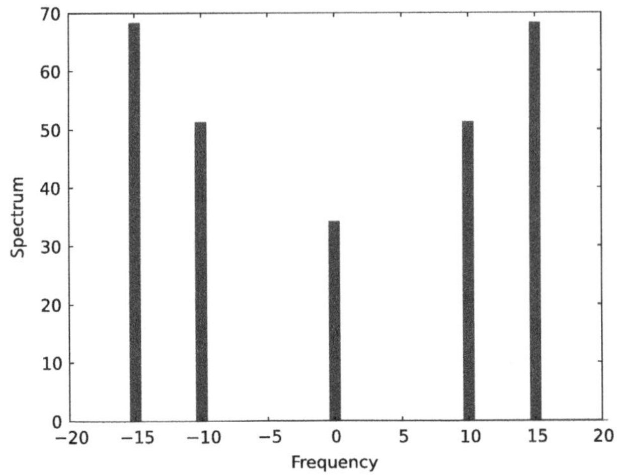

FIGURE 6.7 The spectrum of the displacement of particle 1 in the triatomic molecule.

There are only five peaks in the spectrum, symmetrically distributed about the origin at $\omega_m = 0$, ± 10, and ± 15. Using the time interval of the data points, we can figure out that they correspond to actual frequencies 0, ± 2, and ± 3. Because a positive–negative frequency pair $\pm\omega$ are physically indistinguishable, we conclude that there are three frequencies in the oscillation: $\omega = 0$, 2, and 3. Had we done the same spectrum analysis of the other two particles, we would have found the same frequencies. As we will see next, these are the exact frequencies of the triatomic molecule.

6.2.2 NORMAL MODES

The simulation result in Figure 6.7 shows that the triatomic molecule have discrete frequencies. We can also find these frequencies without numerical integration by a totally different approach: the linear algebraic method. It represents a class of powerful methods involving systems of linear equations and matrix algebra. We introduce this method here and will rely on it heavily in subsequent problems.

We expect that the solutions to Eq. (6.12) are a superposition of fundamental harmonics (2.47), $u_i = a_i \cos(\omega t + \varphi)$ (see Sec. 3.12 of Ref. [10]). Substituting this form of solution into Eq. (6.12) and rearranging the terms, we obtain

$$\begin{aligned} ka_1 - ka_2 \qquad\quad &= m_1\omega^2 a_1, \\ -ka_1 + 2ka_2 - ka_3 &= m_2\omega^2 a_2, \\ -ka_2 + ka_3 &= m_3\omega^2 a_3. \end{aligned} \tag{6.13}$$

These linear equations may be expressed in matrix form,

$$\begin{bmatrix} k & -k & 0 \\ -k & 2k & -k \\ 0 & -k & k \end{bmatrix}\begin{bmatrix} a_1 \\ a_2 \\ a_3 \end{bmatrix} = \omega^2 \begin{bmatrix} m_1 & 0 & 0 \\ 0 & m_2 & 0 \\ 0 & 0 & m_3 \end{bmatrix}\begin{bmatrix} a_1 \\ a_2 \\ a_3 \end{bmatrix}. \tag{6.14}$$

Let $\lambda = \omega^2$ and

$$A = \begin{bmatrix} k & -k & 0 \\ -k & 2k & -k \\ 0 & -k & k \end{bmatrix}, \quad B = \begin{bmatrix} m_1 & 0 & 0 \\ 0 & m_2 & 0 \\ 0 & 0 & m_3 \end{bmatrix}, \quad \mathbf{u} = \begin{bmatrix} a_1 \\ a_2 \\ a_3 \end{bmatrix}. \tag{6.15}$$

Equation (6.14) can be more conveniently written as

$$A\mathbf{u} = \lambda B\mathbf{u}, \text{ or } [A - \lambda B]\mathbf{u} = 0. \qquad (6.16)$$

The matrix A comes from the coupling forces and is so appropriately named the stiffness matrix because of the spring constants. Note that A is symmetric about the diagonal, that is, $A_{ij} = A_{ji}$. The matrix B is due to kinetic energy. For general small oscillations, A and B can be determined systematically from kinetic and potential energies (see Project P6.2).

Equation (6.16) is a so-called (generalized) eigenvalue equation. As we shall see shortly, only discrete, characteristic values of λ are allowed by Eq. (6.16). Moreover, once they are determined, the solutions $\mathbf{u}$ can also be found, one for each λ value. A characteristic value of λ is called an eigenvalue, and the corresponding solution $\mathbf{u}$ is called an eigenvector, a column vector.

We wish to determine the eigenvalues λ and the eigenvectors $\mathbf{u}$. Linear algebra dictates that (e.g., see Sec. 3.8 of Ref. [10]), aside from the trivial solution $\mathbf{u} = 0$, Eq. (6.16) has nonzero solutions if and only if the determinant is zero

$$\det|A - \lambda B| = 0, \quad \text{or} \quad \begin{vmatrix} k - \lambda m_1 & -k & 0 \\ -k & 2k - \lambda m_2 & -k \\ 0 & -k & k - \lambda m_3 \end{vmatrix} = 0. \qquad (6.17)$$

The problem now is to solve for λ from Eq. (6.17), then substitute it back into Eq. (6.16) to obtain $\mathbf{u}$, one for each λ. So an eigenvalue equation like (6.16) determines both the eigenvalues and the eigenvectors.

A direct way to solve Eq. (6.17) is to expand the determinant, which is a cubic polynomial of λ,

$$(k - \lambda m_1)(2k - \lambda m_2)(k - \lambda m_3) - k^2(k - \lambda m_1) - k^2(k - \lambda m_3) = 0. \qquad (6.18)$$

Thus we will have three eigenvalues, or eigenfrequencies. For the parameters used in Figure 6.7, we obtain three eigenfrequencies, $\omega_1 = 0$, $\omega_2 = 2$, and $\omega_3 = 3$ (see Exercise E6.5), in agreement with the Fourier analysis.

Substituting the eigenfrequencies back into Eq. (6.13), we obtain the corresponding unnormalized eigenvectors as

$$\mathbf{u} = \begin{bmatrix} 1 \\ 1 \\ 1 \end{bmatrix}, \quad \begin{bmatrix} -1 \\ 0 \\ 1 \end{bmatrix}, \quad \begin{bmatrix} -4/5 \\ 1 \\ -4/5 \end{bmatrix}, \qquad (6.19)$$

and the final solutions are (up to some phase factors)

$$\begin{bmatrix} u_1 \\ u_2 \\ u_3 \end{bmatrix} = \begin{bmatrix} 1 \\ 1 \\ 1 \end{bmatrix} \cos(0 \cdot t), \quad \begin{bmatrix} 1 \\ 0 \\ -1 \end{bmatrix} \cos(2t), \quad \begin{bmatrix} 4/5 \\ -1 \\ 4/5 \end{bmatrix} \cos(3t). \qquad (6.20)$$

Equations (6.19) and (6.20) are the normal modes. They are the fundamental harmonic oscillations of the triatomic system.

Figure 6.8 shows graphically the motion associated with the normal modes (run Program 6.5 to see animated motion). In mode 1, $\omega_1 = 0$, all three particles have the same, constant displacement. There is no oscillation at all. This is the motion of the CM, a pure translation.

In mode 2 ($\omega_2 = 2$), the two particles on the sides oscillate about the center particle, which stays at rest. The outer particles move in opposite directions of each other with the

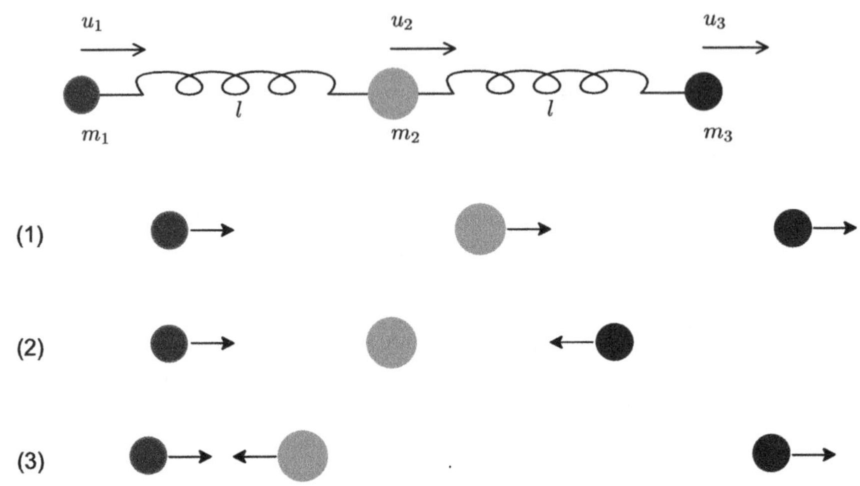

FIGURE 6.8 The normal modes of the triatomic molecule.

same amplitude. This is the symmetric stretch. Because the center particle does not participate in the vibration, the frequency is the same as an SHO, namely, $\sqrt{k/m_1} = 2$ (recall $k = 1, m_1 = m_3 = 1/4$ in our example).

All three particles vibrate in mode 3 ($\omega_3 = 3$), with the outer particles moving in unison and the center particle vibrating in the opposite direction. This is the asymmetric stretch. The amplitudes of the outer particles are the same, but are smaller than the amplitude of the center particle. Their ratio is such that the CM is at rest in the middle, the same as in mode 2. Any oscillation of the triatomic molecule is a linear superposition of the three normal modes. This is what we have seen in Figure 6.6.

6.2.3 EIGENFREQUENCIES AND MATRIX DIAGONALIZATION

Direct calculation of eigenfrequencies via determinant expansion as used earlier works for a small system, but quickly becomes unmanageable for larger systems. For an N-particle system, the matrices A, B, and the determinant matrix will have dimensions $N \times N$. There would be $N!$ terms in the determinant alone (merely $\sim 3.6 \times 10^6$ for $N = 10$), not to mention a highly oscillatory polynomial. In such cases, matrix diagonalization methods explained next should be used.

To illustrate this, let us assume that in Eq. (6.15) matrix A is already diagonal, and matrix B is the identity matrix.[3] Then the determinant in Eq. (6.17) is

$$\begin{vmatrix} k-\lambda & 0 & 0 \\ 0 & 2k-\lambda & 0 \\ 0 & 0 & k-\lambda \end{vmatrix} = 0. \tag{6.21}$$

Finding eigenvalues of the preceding equation are simple: They are just the diagonal elements of A.

So, the goal of matrix diagonalization is to nudge a nondiagonal A toward diagonal form while preserving the eigenvalues. It is a complicated business. One of the simpler methods is the Jacobi transformation algorithm for symmetric matrices. It uses a series of rotation transformations to successively eliminate the off-diagonal elements. When the latter are all zero (or close to it), the diagonal elements are the eigenvalues.

[3]This would be the case if the particles have the same unity mass, and the couplings between them are neglected.

Let us illustrate this approach with a very simple example. Suppose $N = 2$, and B is the identity matrix. Then, matrix A and the eigenvalue equation are

$$A = \begin{bmatrix} a_{11} & a_{12} \\ a_{21} & a_{22} \end{bmatrix}, \quad \det|A - \lambda \mathbb{1}| = \begin{vmatrix} a_{11} - \lambda & a_{12} \\ a_{21} & a_2 - \lambda \end{vmatrix} = 0. \tag{6.22}$$

The Jacobi diagonalization method uses a pair of rotation matrices to transform A to A' as

$$A' = R(-\theta)AR(\theta), \tag{6.23}$$

where $R(\theta)$ is the rotation matrix (4.75). The rotation $R(\theta)$ and its inverse $R(-\theta)$ pair are necessary so the transformation is orthogonal and preserves the eigenvalues [10].

Denoting $c = \cos\theta$, $s = \sin\theta$, and carrying out the matrix multiplication in Eq. (6.23), we arrive at[4]

$$A' = \begin{bmatrix} c^2 a_{11} - 2sc\, a_{12} + s^2 a_{22} & sc(a_{11} - a_{22}) + (c^2 - s^2)a_{12} \\ sc(a_{11} - a_{22}) + (c^2 - s^2)a_{12} & s^2 a_{11} + 2sc\, a_{12} + c^2 a_{22} \end{bmatrix}, \tag{6.24}$$

where we have used the symmetry $a_{21} = a_{12}$.

We can eliminate the off-diagonal elements by choosing the rotation angle to satisfy

$$\sin\theta\cos\theta(a_{11} - a_{22}) + (\cos^2\theta - \sin^2\theta)a_{12} = 0,$$

which yields

$$\theta = \frac{1}{2}\tan^{-1}\frac{2a_{12}}{a_{22} - a_{11}}, \quad \theta = \frac{\pi}{4} \text{ if } a_{11} = a_{22}. \tag{6.25}$$

Substituting this θ back into the diagonal elements gives us the eigenvalues.

Suppose we have for A

$$A = \begin{bmatrix} 2 & -1 \\ -1 & 2 \end{bmatrix}, \quad \det|A - \lambda \mathbb{1}| = \begin{vmatrix} 2 - \lambda & -1 \\ -1 & 2 - \lambda \end{vmatrix} = 0,$$

which has exact eigenvalues $\lambda_1 = 1$ and $\lambda_2 = 3$. According to Eq. (6.25), $\theta = \pi/4$, which, when substituted into Eq. (6.24), gives the transformed matrix A'

$$A' = \begin{bmatrix} 3 & 0 \\ 0 & 1 \end{bmatrix}.$$

We see that the diagonal elements are the correct eigenvalues when the off-diagonals are zero.

We can essentially apply the Jacobi transformation to any symmetric $N \times N$ matrix, eliminating one pair of off-diagonal elements each time.[5] But there is more to it than meets the eye. When $N > 2$, eliminating one off-diagonal pair introduces nonzero contributions to other off-diagonal elements, though their magnitudes decrease successively [71]. Complications such as stability and convergence must also be addressed when N is moderately large. There are other complicating factors including condition of the matrix, and so on. Clearly, the art of solving eigenvalue problems is very involved.

---

[4] You may also use SymPy's symbolic matrix manipulation to verify the results.

[5] The rotation matrix would be an $N \times N$ identity matrix except for rows i and j: $a_{ii} = a_{jj} = \cos\theta$, $a_{ij} = -a_{ji} = \sin\theta$.

For practical reasons, discussing other algorithms and designing codes that can tackle general eigenvalue problems reliably and efficiently would give us a very limited return within the scope of the text. This is one of the few areas where we defer to "blackbox" solvers for eigenvalue problems. Fortunately, there are several well-tested and widely available open-source packages for such purposes, including BLAS (Basic Linear Algebra Subprograms) and LAPACK (Linear Algebra Package), both available from the Netlib Repository (netlib.org).

We can access BLAS and LAPACK routines in Python via the SciPy libraries (Section 1.3.2), which make available linear algebra functions, among other things, at the speed of compiled Fortran codes. For our purpose, the relevant function is `eigh()`, which solves for eigenvalues and eigenvectors of symmetric (or Hermitian) matrices. The following code solves the eigenvalues of our triatomic molecule. It requires the SciPy library (Section 1.B).

Program listing 6.2 Diagonalization of triatomic systems (`eigh.py`)

```
   import numpy as np
2  from scipy.linalg import eigh        # Hermitian eigenvalue solver

4  k = 1.0
   m1, m2, m3 = 1./4., 2./5., 1./4.
6  A = np.array([[k, -k, 0], [-k, 2*k, -k], [0, -k, k]])
   B = np.array([[m1, 0, 0], [0, m2, 0], [0, 0, m3]])
8
   lamb, u = eigh(A, B)          # eigenvalues and eigenvectors
10 print (np.sqrt(lamb))         # print omega
```

The program prepares the matrices A and B according to Eq. (6.15) as NumPy arrays, the expected input format of SciPy routines. The symmetric eigenvalue solver (line 9) takes A and B as input and returns a pair of arrays: `lamb` is a 1D array containing the eigenvalues, and `u` a 2D array filled with the eigenvectors. The i-th column `u[:,i]` is the eigenvector corresponding to the eigenvalue `lamb[i]` (see Program 6.6). Executing the program gives the following output:

```
[1.47451053e-08 2.00000000e+00 3.00000000e+00]
```

We recognize the last two elements are the exact eigenvalues. The first one, however, is not exactly zero. The reason is that $\omega = \sqrt{\lambda}$, and λ was "zero," as far as the computer is concerned, on the order of machine accuracy 10^{-16} for double precision (Section 1.2).

For larger matrices, it is more efficient to use sparse matrix functions from SciPy.[6] We will encounter this and other linear algebra routines in the sections and chapters ahead.

6.3 DISPLACEMENT OF A STRING UNDER A LOAD

Suppose we attach an extra atom to the triatomic system of Figure 6.5 and wish to know how this will affect its motion. Well, this does not pose a serious problem. We could follow the same procedure and integrate a four-particle system in place of a three-particle one (6.12). What about five-, six-, or N-particle systems, where N may be very large (think $N \sim 10^{23}$)? We now have a string of atoms forming a continuous system that can support wave motion. There is no hope of treating

[6] Aside from efficiency derived from special algorithms for sparse matrices, we can also reduce memory bandwidth of the program, making it more efficient by minimizing the amount of data being moved, which may be as important as computational speed.

this many atoms individually even with all the computing resources in the world combined. Nor would we want to if we want to make sense of collective vibrational and wave motion. Here, we must treat the string as a continuous mass and use concepts such as mass density found in continuum mechanics.

6.3.1 THE DISPLACEMENT EQUATION

Before studying vibrations on a string, we discuss a simpler problem: the shape of an elastic string in equilibrium under a static, external force. Consider a piece of string between x and $x + \Delta x$ illustrated in Figure 6.9.

We denote the transverse displacement by u. The piece is held by three forces, the tension to the left $T_1 = T(x)$ and to the right $T_2 = T(x + \Delta x)$, and the external transverse load f, which is defined as force per unit length. The equilibrium condition is that the longitudinal (horizontal) and transverse (vertical) forces cancel each other, namely,

$$T(x)\cos\theta_1 = T(x + \Delta x)\cos\theta_2 \equiv T, \qquad \text{(longitudinal)} \qquad (6.26a)$$

$$T(x)\sin\theta_1 = T(x + \Delta x)\sin\theta_2 + f\Delta x. \qquad \text{(transverse)} \qquad (6.26b)$$

Equation (6.26a) tells us that the longitudinal tension, $T = T(x)\cos(\theta)$, is the same throughout the string. It is a parameter measuring how stretched (tautness) the string is.

Rewriting Eq. (6.26b) in terms of T gives us

$$T\tan\theta_1 = T\tan\theta_2 + f\Delta x. \qquad (6.27)$$

The net transverse tension force is $T\tan\theta_2 - T\tan\theta_1$. Because $\tan\theta_1 = u'(x)$, and $\tan\theta_2 = u'(x + \Delta x)$, we can rearrange Eq. (6.27) to read

$$T\frac{u'(x + \Delta x) - u'(x)}{\Delta x} + f = 0. \qquad (6.28)$$

In the limit $\Delta x \to 0$, the ratio in the first term becomes a second derivative of u, and we have

$$Tu'' + f = 0, \quad \text{or} \quad T\frac{d^2u(x)}{dx^2} = -f(x). \qquad (6.29)$$

This equation describes the displacement of an elastic string under an external load $f(x)$. It is an example of a problem reducing to a differential equation under infinitesimal changes mentioned in Chapter 2.

Given $f(x)$, Eq. (6.29) can be solved if boundary conditions are specified. For instance, if $f(x) = f_0$ is a constant, then

$$u(x) = a + bx - \frac{f_0}{2T}x^2, \quad \text{(constant force)} \qquad (6.30)$$

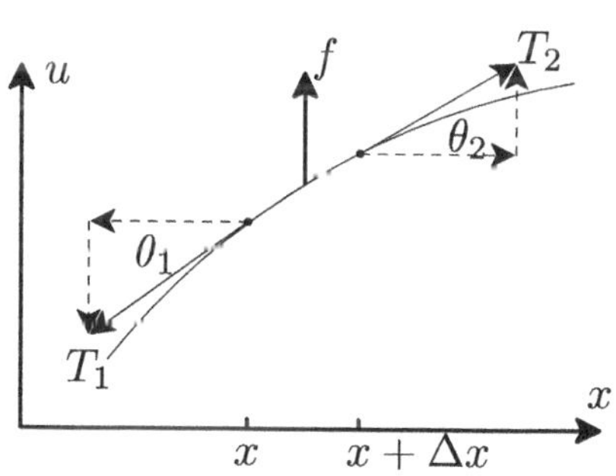

FIGURE 6.9 The displacement of an elastic string under tension and external forces.

where a and b are constants to be determined by the boundary conditions $u(0)$ and $u(L)$ at the ends 0 and L, respectively. We see that it is a parabola.

6.3.2 FINITE DIFFERENCE METHOD

For a general $f(x)$, we can solve Eq. (6.29) using the finite difference method (FDM). Without loss of generality, we assume the two ends of the string are separated by a distance (L) of unity at $x = 0$ and $x = 1$. We divide the distance into N equal intervals, each size h (Figure 6.10). There are $N + 1$ grid points, including $N - 1$ interior points and two boundary points. Over the grid, let $x_i = ih$, $u(x_i) = u_i$, and $f(x_i) = f_i$, $i = 0, 1, 2, ..., N$. The boundary points are at u_0 and u_N.

Once space is discretized, we can approximate the derivatives by numerical differentiation using a three-point central difference scheme. By adding or subtracting the Taylor series such as in Eqs. (2.9) and (2.10), we obtain the first and second derivatives as

$$\frac{du(x)}{dx} = \frac{u(x+h) - u(x-h)}{2h} + O(h^2), \tag{6.31a}$$

$$\frac{d^2u(x)}{dx^2} = \frac{u(x+h) - 2u(x) + u(x-h)}{h^2} + O(h^2). \tag{6.31b}$$

Equation (6.31a) is the same as the midpoint method (2.13), and the error in both cases is second order in h.

Substituting the discretized second derivative (6.31b) into (6.29) at grid point i, we obtain the FDM displacement equation

$$u_{i+1} - 2u_i + u_{i-1} = -h^2 f_i / T, \quad i = 1, 2, ..., N - 1. \tag{6.32}$$

We have simplified the left-hand side (LHS) by moving h^2 and T to the right-hand side (RHS). We also restrict the grid points to the interior points only. This is a three-term recursion relation, a result of three-point approximation of the second derivative. Even so, all the $N - 1$ equations must be solved simultaneously because a single grid point can propagate its influence through its neighbors to its neighbors' neighbors, and so on.

Defining $g_i = -h^2 f_i / T$, we can write out Eq. (6.32) explicitly, ending up with

$$\begin{aligned}
-2u_1 + u_2 \qquad\qquad &= g_1 - u_0, \\
u_1 - 2u_2 + u_3 \qquad &= g_2, \\
\vdots \qquad\qquad\quad &\quad \vdots \\
u_{N-3} - 2u_{N-2} + u_{N-1} &= g_{N-2}, \\
u_{N-2} - 2u_{N-1} \qquad &= g_{N-1} - u_N.
\end{aligned} \tag{6.33}$$

The first and the last rows on the LHS have two terms because we have moved to the RHS the boundary values that are given and need not to be solved. That is how the boundary conditions determine the solutions. The other rows have three terms each.

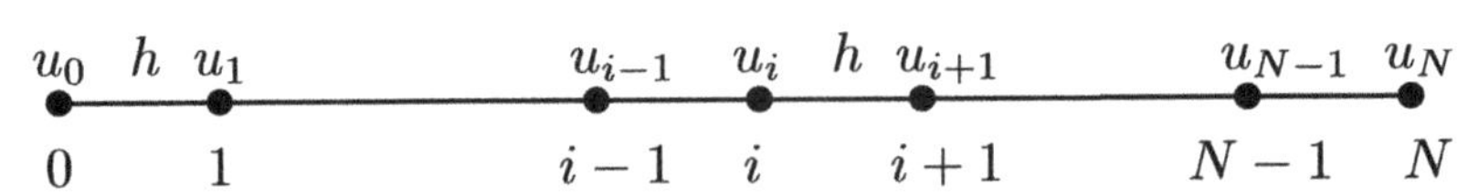

FIGURE 6.10 The finite difference grid.

In matrix form, we have a tridiagonal system,

$$\begin{bmatrix} -2 & 1 & & & \\ 1 & -2 & 1 & & \\ & \ddots & \ddots & \ddots & \\ & & 1 & -2 & 1 \\ & & & 1 & -2 \end{bmatrix} \begin{bmatrix} u_1 \\ u_2 \\ \vdots \\ u_{N-2} \\ u_{N-1} \end{bmatrix} = \begin{bmatrix} g_1 - u_0 \\ g_2 \\ \vdots \\ g_{N-2} \\ g_{N-1} - u_N \end{bmatrix}. \quad (6.34)$$

We can express it in compact matrix form

$$A\mathbf{u} = B, \quad (6.35)$$

where A is the $(N-1, N-1)$ tridiagonal matrix (stiffness like), B the column matrix on the RHS $(N-1, 1)$, and $\mathbf{u}$ as usual is a column vector containing the $N-1$ unknowns $u_1, u_2, \ldots u_{N-1}$.

Equation (6.34) (or (6.35)) is a linear system of equations. It can be solved by linear equation solvers once the load $f(x)$ is given and thus B known. For example, assuming zero boundary values $u_0 = u_N = 0$, matrix B for a constant load $f(x) = f_0$ is simply

$$B = -\frac{h^2}{T} f_0 \begin{bmatrix} 1 \\ 1 \\ \vdots \\ 1 \\ 1 \end{bmatrix}. \quad \text{(FDM, constant load)} \quad (6.36)$$

6.3.3 SOLUTIONS OF LINEAR SYSTEM OF EQUATIONS

There are well-known methods for solving a linear system of equations, including Gauss elimination and a closely related variant, the Gauss-Jordan elimination. These methods plus the iterative Gauss-Seidel method are discussed in Appendix S:6.A. Programs implementing these methods are included in Program S:6.1.

However, in keeping with our approach to eigenvalue problems, we will use LAPACK routines via SciPy, which provides a general-purpose solver `solve(A, B)`. Next we illustrate its use in solving Eq. (6.34).

First let us consider the external force to be constant, $f(x) = f_0$. The following program solves Eq. (6.34) for this case.

Program listing 6.3 String under external forces (`stringfdm.py`)

```
from scipy.linalg import solve        # SciPy linear eqn solver
import numpy as np, matplotlib.pyplot as plt

N, u0, uN= 20, 0., 0.              # number of intervals, boundary values
x = np.linspace(0., 1., N+1)       # grid
h, T, f = x[1]-x[0], 1.0, -1.0     # bin size, tension, load

A = np.diag([-2.]*(N-1))           # diagonal
A += np.diag([1.]*(N-2),1) + np.diag([1.]*(N-2),-1) # off diagonals

B = np.array([-h*h*f/T]*(N-1))         # B matrix
B[0], B[-1] = B[0]-u0, B[-1]-uN        # boundary values, just in case

```

```
   u = solve(A, B)                          # solve
15 u = np.insert(u, [0, N-1], [u0, uN])     # insert BV at 1st and last pos

17 plt.plot(x, u, label='$f=-1$'), plt.legend(loc='center') # legend
   plt.xlabel('$x$'), plt.ylabel('$u$'), plt.show()
```

After setting the number of intervals, the boundary values, and related parameters, the program generates matrix A according to Eq. (6.34). First, the diagonal of A is created using `np.diag(a,k)` (line 8), which returns a 2D array with the k-th (default 0) diagonal given by the 1D input array a. It can also extract the diagonal from a 2D input array (not used here). The next line fills the upper and lower diagonals ($k = \pm 1$, and one fewer element than the diagonal), respectively, completing the tridiagonal matrix A. Next, matrix B is created (recall $g_i = -h^2 * f/T$, Eq. (6.36)), and the boundary values are accounted for. Once A and B are prepared, `solve()` from SciPy is called to obtain the solution u (line 14). It contains the displacements at the interior grid points.

Before plotting the results, the two boundary values are inserted into array u, one at the beginning and one at the end. The pair of lists `[0,N-1]` and `[u0,uN]` in the `np.insert` function (line 15) will insert u0 before the first element (index 0) and uN before (the yet nonexistent) index $N-1$, which is one position behind the last element, effectively appending to the end (recall that the length of u was $N-1$, with the last element at index $N-2$). Finally, the results are plotted with legends from the labels given to the curves (line 17).

Figure 6.11 shows the results for two external forces: a constant force $f = -1$, and a linear force $f = -2x$ (per unit length, of course). The tension parameter is $T = 1$ in both cases, and the space between the end points were divided into twenty intervals, $N = 20$. Because the forces are negative, the string droops downward as it is being pulled down by the forces. For the constant force, the shape is a perfect parabola as expected from the analytic solution (6.30) and is symmetric about the center point.

For the linear force, the magnitude of the load increases linearly with x, so the symmetry is broken. The string is sheared toward the right end. The bottom of the valley is shifted to $x \sim 0.6$, instead of at the center in the case of constant load. The shape can be described by a cubic polynomial. Even though the FDM method is second order and should be exact only for a second-order polynomial, it turns out that the solutions are actually exact in this case due to a quirk.

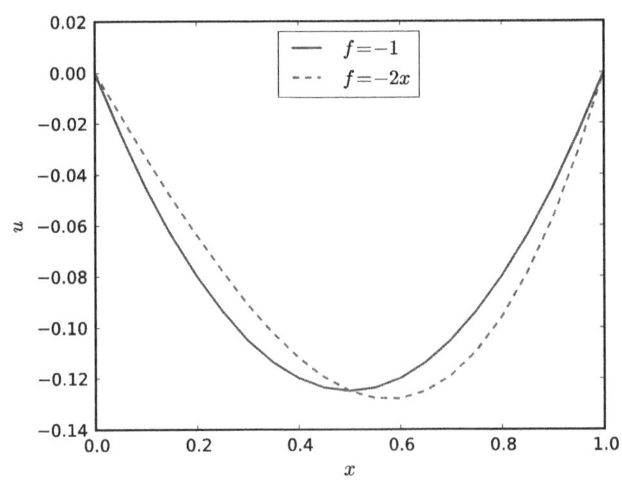

FIGURE 6.11 The displacement of an elastic string under external forces.

The FDM method can be used for any continuous load distribution $f(x)$. If the load is a point source acting on a single point, say a Dirac function $f(x) = f_0\delta(x - x_0)$ (see Section S:5.1, Figure S:5.1), then the point x_0 must be treated separately. A different method to be described next, however, can deal with point sources gracefully.

6.4 POINT SOURCE AND FINITE ELEMENT METHOD

The finite element method (FEM) is complementary to FDM, but is more versatile and advantageous for certain problems such as point sources just mentioned and irregular domains in higher dimensions. The FEM is commonly used in science and engineering.

Next we introduce FEM in 1D space, which will also be extended to 2D in Chapters 7 and 9. The development of FEM consists of four steps: choosing the elements and basis functions, converting a differential equation to integral form, applying boundary conditions, and finally assembling system matrices and solving the resultant linear system of equations.

6.4.1 FINITE ELEMENTS AND BASIS FUNCTIONS

In FEM, the space is divided into discrete elements. Figure 6.12 shows an example with $N = 4$ elements labeled e_1 to e_4. Each element e_i has a length h_i, which need not be the same. To keep things tidy, we will assume the elements are identical, $h_1 = h_2 = \cdots = h_N = h$. There are $N + 1$ grid points, called nodes in FEM. The positions of the nodes are $x_i = ih$, same as in FDM.

The difference between FEM and FDM is that, rather than discretizing the differential operator, the FEM tries to interpolate the solution over the domain. To that end, we expand the solution (displacement) u in a basis set as

$$u(x) = \sum_i u_i \varphi_i(x), \tag{6.37}$$

where u_i are the expansion coefficients to be determined, and φ_i the basis functions.

In general, the basis functions are arbitrary. For instance, orthogonal basis functions are often used in physics (Sections 9.4 and S:8.2). However, it is crucial to FEM that the basis functions be chosen such that only neighboring basis functions overlap (the reason will be clear in a moment).

The simplest function between two points is a straight line. We construct a basis function φ_i centered at node i, consisting of two lines and making a tent-shaped function (Figure 6.12).

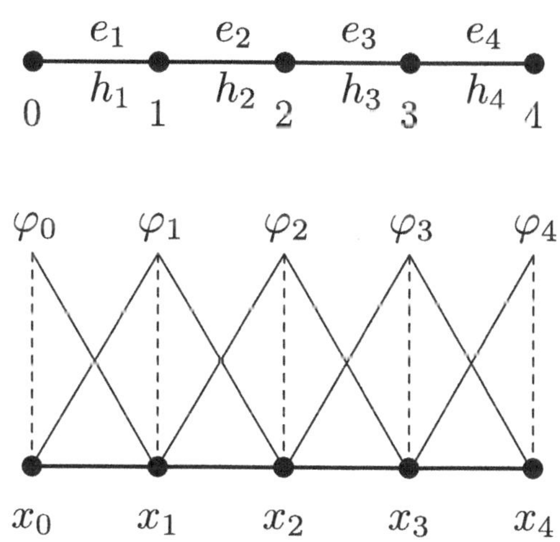

FIGURE 6.12 The finite elements and tent basis functions.

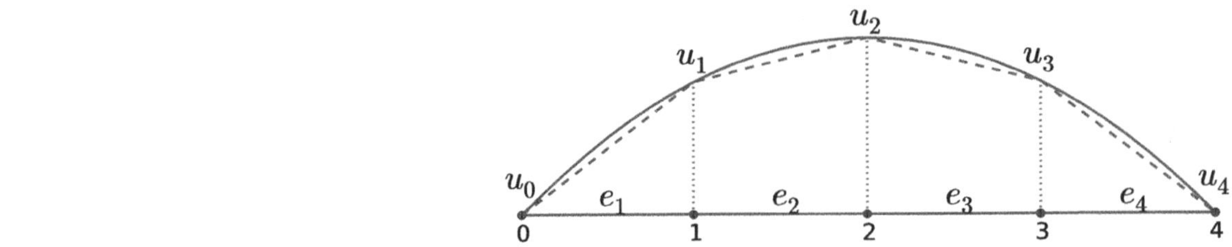

FIGURE 6.13 Interpolation (dashed line) of $4x(1-x)$ (solid line) using FEM basis functions.

In terms of node number i and width h, we can write it as

$$\varphi_i(x) = \begin{cases} \frac{x-x_{i-1}}{h}, & x_{i-1} \le x < x_i, \\ \frac{x_{i+1}-x}{h}, & x_i \le x < x_{i+1}, \\ 0, & x < x_{i-1} \text{ or } x \ge x_{i+1}. \end{cases} \tag{6.38}$$

The basis function φ_i is nonzero only in the two elements sharing the common node i. Functions of this type are said to be compact support. Furthermore, φ_i is unity at node i and zero at other nodes,

$$\varphi_i(x_j) = \delta_{ij}, \tag{6.39}$$

where δ_{ij} is the Kronecker delta function, $\delta_{ij} = 1$ if $i = j$, and 0 if $i \neq j$.

Therefore, the solution u (6.37) at node i is

$$u(x_i) = \sum_j u_j \varphi_j(x_i) = \sum_j u_j \delta_{ji} = u_i. \tag{6.40}$$

In other words, the value of the solution at x_i is just the expansion coefficient, u_i.

Figure 6.13 illustrates the interpolation of the function $y = 4x(1-x)$ (logistic map) over the unity interval with four finite elements, spanned by five basis functions. Setting $u_i = y(x_i)$, the values u_0 to u_4 are $\left[0, \frac{3}{4}, 1, \frac{3}{4}, 0\right]$, so the interpolated function is $u(x) = \frac{3}{4}\varphi_1(x) + \varphi_2(x) + \frac{3}{4}\varphi_3(x)$. Over each element, the function is approximated by a straight line. It shows that the more elements are used, the more accurate the interpolation.

We will need the derivatives of the basis functions later. They are

$$\varphi_i'(x) = \begin{cases} \frac{1}{h}, & x_{i-1} \le x < x_i, \\ -\frac{1}{h}, & x_i \le x < x_{i+1}, \\ 0, & x < x_{i-1} \text{ or } x \ge x_{i+1}. \end{cases} \tag{6.41}$$

The derivative is discontinuous at node i. The discontinuity at the nodes is carried over to interpolation (Figure 6.13).

6.4.2 INTEGRAL FORM OF SOLUTIONS

Having chosen the basis functions, we next turn our attention to finding the coefficients u_i. Instead of solving Eq. (6.29) directly, FEM solves the integral form (also called the weak form) of the equation [7].

Multiplying both sides of Eq. (6.29) by $\varphi_j(x)$ and integrating over the interval $[a, b]$ (which are the end points of the domain) leads to

$$\int_a^b \frac{d^2u(x)}{dx^2}\varphi_j(x)dx = -T^{-1}\int_a^b f(x)\varphi_j(x)dx. \tag{6.42}$$

Integrating the LHS by parts, we have

$$\text{LHS} = q_j - \int_a^b u'(x)\varphi_j'(x)dx, \quad q_j = u'(b)\varphi_j(b) - u'(a)\varphi_j(a). \tag{6.43}$$

The quantity q_j is zero unless $j = 0$ or $j = N$ because of the basis function evaluated at the boundaries, that is, $q_j = u'(b)\delta_{jN} - u'(a)\delta_{j0}$.

We differentiate Eq. (6.37) as $u'(x) = \sum u_i\varphi_i'(x)$ and substitute it into Eq. (6.43), obtaining

$$\text{LHS} = q_j - \sum_i u_i \int_a^b \varphi_i'(x)\varphi_j'(x)dx. \tag{6.44}$$

Now let

$$A_{ij} = \int_a^b \varphi_i'(x)\varphi_j'(x)dx, \quad p_j = \int_a^b f(x)\varphi_j(x)dx. \tag{6.45}$$

Because the basis functions and their derivatives have no overlap unless they share a common node, we see that $A_{ij} = 0$ unless $i = j$ or $j \pm 1$. This is the reason the basis functions are chosen to be compact support, so the matrix A_{ij} is sparse. It is also symmetric, $A_{ij} = A_{ji}$.

With Eqs. (6.44) and (6.45), Eq. (6.42) can be expressed as

$$\sum_i A_{ij}u_i = T^{-1}p_j + q_j, \quad j = 0, 1, 2, ..., N. \tag{6.46}$$

This is a system of $N + 1$ linear equations, one for each node. Explicitly, Eq. (6.46) can be expanded as

$$\begin{aligned}
&A_{00}u_0 + A_{01}u_1 && = T^{-1}p_0 + q_0, \\
&A_{10}u_0 + A_{11}u_1 + A_{12}u_2 && = T^{-1}p_1, \\
&\vdots && \vdots \\
&A_{N-1,N-2}u_{N-2} + A_{N-1,N-1}u_{N-1} + A_{N-1,N}u_N && = T^{-1}p_{N-1}, \\
&A_{N,N-1}u_{N-1} + A_{N,N}u_N && = T^{-1}p_1 + q_N.
\end{aligned} \tag{6.47}$$

Equation (6.47) is the final form of the FEM approach. Except for the first and last rows with two terms each, all other rows have three terms, reflecting the fact only neighboring elements contribute to matrix A_{ij}.

6.4.3 BOUNDARY CONDITIONS

The unknowns u_i in Eq. (6.47) depend on the boundary conditions, q_0 and q_N. There are two types of boundary conditions.

The Neumann boundary condition specifies the derivatives, $u'(a)$ and $u'(b)$. According to Eqs. (6.39) and (6.43), we replace $q_0 = -u'(a)$ and $q_N = u'(b)$ in Eq. (6.47). All $N+1$ node values, u_0 to u_N, can be determined in the Neumann boundary problem as

$$\begin{bmatrix} A_{00} & A_{01} & & & \\ A_{10} & A_{11} & A_{12} & & \\ & \ddots & \ddots & \ddots & \\ & & A_{N-1,N-2} & A_{N-1,N-1} & A_{N-1,N} \\ & & & A_{N,N-1} & A_{N,N} \end{bmatrix} \begin{bmatrix} u_0 \\ u_1 \\ \vdots \\ u_{N-1} \\ u_N \end{bmatrix} = T^{-1} \begin{bmatrix} p_0 - u'(a) \\ p_1 \\ \vdots \\ p_{N-1} \\ p_N + u'(b) \end{bmatrix}. \tag{6.48}$$

The other type is the Dirichlet boundary condition, in which the boundary values $u(a) = u_0$ and $u(b) = u_N$ are specified. We are mainly interested in this type of boundary conditions here. Because u_0 and u_N are already given, we need not to solve for them. We therefore cross off the first and the last rows in Eq. (6.47). Furthermore, we move the first term $A_{10}u_0$ in the second row, and the last term $A_{N-1,N}u_N$ in the second to last row, to the RHS. The remaining rows form a linear system of $N-1$ equations, which determine the $N-1$ interior node values, u_1 to u_{N-1},

$$\begin{aligned} A_{11}u_1 + A_{12}u_2 &= \frac{p_1}{T} - A_{10}u_0, \\ A_{21}u_1 + A_{22}u_2 + A_{23}u_3 &= \frac{p_2}{T}, \\ \vdots \qquad & \quad \vdots \\ A_{N-2,N-3}u_{N-3} + A_{N-2,N-2}u_{N-2} + A_{N-2,N-1}u_{N-1} &= \frac{p_{N-2}}{T}, \\ A_{N-1,N-2}u_{N-2} + A_{N-1,N-1}u_{N-1} &= \frac{p_{N-1}}{T} - A_{N-1,N}u_N. \end{aligned} \tag{6.49}$$

In matrix form, let

$$A = \begin{bmatrix} A_{11} & A_{12} & & & \\ A_{21} & A_{22} & A_{23} & & \\ & \ddots & \ddots & \ddots & \\ & & A_{N-2,N-3} & A_{N-2,N-2} & A_{N-2,N-1} \\ & & & A_{N-1,N-2} & A_{N-1,N-1} \end{bmatrix}, \tag{6.50}$$

$$B = \begin{bmatrix} T^{-1}p_1 - A_{10}u_0 \\ T^{-1}p_2 \\ \vdots \\ T^{-1}p_{N-2} \\ T^{-1}p_{N-1} - A_{N-1,N}u_N \end{bmatrix}, \qquad \mathbf{u} = \begin{bmatrix} u_1 \\ u_2 \\ \vdots \\ u_{N-2} \\ u_{N-1} \end{bmatrix}. \tag{6.51}$$

The solutions in the Dirichlet boundary problems are given by

$$A\mathbf{u} = B. \quad \text{(Dirichlet type)} \tag{6.52}$$

In both FDM and FEM, we end up with a linear system like Eq. (6.35) or (6.52). The difference lies in how the boundary conditions are treated and how the system matrices are set up.

6.4.4 ASSEMBLING SYSTEM MATRICES

Matrix A in FEM is independent of load. From Eqs. (6.41) and (6.45), we find that the integral over the domain is effectively over the elements e_i and e_j only,

$$A_{ij} = \int_a^b \varphi_i'(x)\varphi_j'(x)dx = \int_{x_{i-1}}^{x_{i+1}} \varphi_i'(x)\varphi_j'(x)dx$$
$$= \begin{cases} \frac{2}{h}, & i=j, \\ -\frac{1}{h}, & i=j\pm 1, \\ 0, & \text{otherwise.} \end{cases} \tag{6.53}$$

The full matrix A reads

$$A = \frac{1}{h}\begin{bmatrix} 2 & -1 & & & \\ -1 & 2 & -1 & & \\ & \ddots & \ddots & \ddots & \\ & & -1 & 2 & -1 \\ & & & -1 & 2 \end{bmatrix}. \tag{6.54}$$

This is the stiffness matrix in FEM (Dirichlet type), with a dimension $(N-1)\times(N-1)$. It is the same as the FDM stiffness matrix (6.34) (for identical elements of same size h).

For a given load, matrix B will be evaluated according to Eqs. (6.45) and (6.51), where p_i is given by

$$p_i = \int_a^b f(x)\varphi_i(x)dx = \int_{x_{i-1}}^{x_{i+1}} f(x)\varphi_i(x)dx. \tag{6.55}$$

Here we have again used the fact that $\varphi_i(x)$ is nonzero only if $x_{i-1} \le x < x_{i+1}$. For constant load $f(x) = f_0$, $p_i = hf_0$. Assuming $u_0 = u_N = 0$, matrix B is

$$B = \frac{h}{T}f_0\begin{bmatrix} 1 \\ 1 \\ \vdots \\ 1 \\ 1 \end{bmatrix}. \quad \text{(FEM, constant load)} \tag{6.56}$$

Once A and B have been prepared, we can obtain the solutions by solving either Eq. (6.48) for the Neumann boundary type or (6.52) for the Dirichlet boundary type. Comparing the system matrices between FDM and FEM in this case, we find that they are the same (besides the placement of h), so the solutions are identical. It turns out they are also identical for linear load, and differences start to develop only for quadratic and higher order load.

6.4.5 POINT SOURCE SOLUTIONS

For a point source, the FEM is easier to apply. For example, consider the following load,

$$f(x) = f_0 + \alpha\delta(x - x_n). \tag{6.57}$$

The Dirac δ function (see Section S:5.1 and Figure S:5.1) makes direct evaluation of $f(x_n)$ impossible in FDM. In FEM, however, it poses no special difficulty for this integrable singularity, as

$$p_n = \int_{x_{n-1}}^{x_{n+1}} (f_0 + \alpha\delta(x - x_n))\varphi_n(x)dx = f_0h + \alpha\varphi_n(x_n) = f_0h + \alpha, \tag{6.58}$$

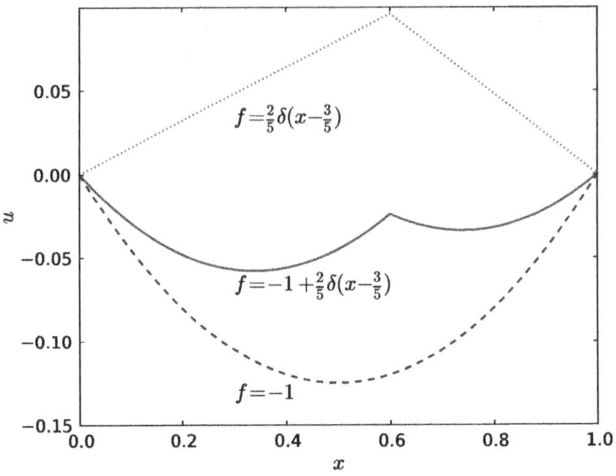

FIGURE 6.14 FEM solutions for combinations of a point source load and a constant load.

where we have used the property $\varphi_n(x_n) = 1$ from Eq. (6.39). All elements of B are the same as Eq. (6.56) except at node n, where it is $(hf_0 + \alpha)/T$. Solving for $\mathbf{u}$ using this modified B, we obtain the results shown in Figure 6.14.

With only a single point-source force, the solution consists of two segments of straight lines. It is what we expect if we pull a string and hold it in position. The "kink" at the point of the applied force indicates a discontinuity of the first derivative, that is,[7]

$$u'(x_n^+) - u'(x_n^-) = -\alpha. \tag{6.59}$$

If the force is suddenly withdrawn, as if plucking the string, we expect waves to be formed on the string. This is discussed shortly.

When there is an additional constant load, the same discontinuity exists. On each side of x_n, the string is a shallower curve relative to the constant force alone, akin to pulling up a suspended cable.

6.5 WAVES ON A STRING

When a string, held by a point-like force as depicted in Figure 6.14, is released, wave motion will be generated on it. If the string is coupled to a sound chamber such as that of a cello, vibration energy of the string will be transferred to the chamber, and musical notes can be heard. Modeling the energy loss would be similar to but more complicated than damping in a harmonic oscillator (Section 6.1). We are interested instead in describing an ideal vibrating string without dissipation.

6.5.1 THE WAVE EQUATION

The starting point for the wave equation is Figure 6.9 and Eq. (6.28). The net (transverse) force per unit length on the piece of string from x to $x + dx$ is given by Eq. (6.28),

$$f_{net} = Tu'' + f, \tag{6.60}$$

where T is the tension, f the external load. For a static string, the net force is zero, giving us the displacement equation (6.29).

[7]We use the shorthand for $x^{\pm} = x \pm \epsilon$, where ϵ is a positive infinitesimal.

For a vibrating string, this is no longer the case. The displacement depends on both position x and time t, $u = u(x, t)$. From Newton's second law, the net force (per unit length) is equal to the linear mass density (ρ) times acceleration. The acceleration at x is given by $\partial^2 u/\partial t^2$, the second derivative of displacement holding x constant. Replacing u'' by $\partial^2 u/\partial x^2$ in Eq. (6.60), we obtain

$$T\frac{\partial^2 u}{\partial x^2} + f = \rho\frac{\partial^2 u}{\partial t^2}. \tag{6.61}$$

The preceding equation describes the general motion of a string under an external load f.

For a free vibrating string, $f = 0$, and we have the following wave equation,

$$\frac{\partial^2 u(x,t)}{\partial x^2} = \frac{1}{v^2}\frac{\partial^2 u(x,t)}{\partial t^2}, \quad v = \sqrt{\frac{T}{\rho}}. \quad \text{(wave on string)} \tag{6.62}$$

Here, v is the speed of the wave. It increases with increasing tension or decreasing mass density.[8]

So far we have mostly dealt with ODEs. The wave equation (6.62) is a PDE. In general, it is much harder to solve PDEs than ODEs because a PDE has at least two independent variables, whereas an ODE only one. We have systematic methods such as the Runge-Kutta scheme that are applicable to most initial value problems involving ODEs. For PDEs, however, few such widely applicable approaches exist. The correct method depends on the kind of problems, and the type of boundary conditions.

6.5.2 ANALYTICAL PROPERTIES

Before solving Eq. (6.62) numerically, we describe analytic properties as a guide to our subsequent discussions. If $u(x)$ is a solution at $t = 0$, then $u(x \pm vt)$ is also a solution. Because Eq. (6.62) is linear, we can form a superposition of counterpropagating waves as

$$u(x,t) = Af(x - vt) + Bg(x + vt), \tag{6.63}$$

with A and B being constants. The function $f(x - vt)$ describes a traveling waveform to the right, for if $\phi = x - vt$ is kept constant, then a fixed value $f(\phi)$ moves to the right (positive x direction) with speed v. Conversely, $g(x + vt)$ is a traveling wave to the left. Whether f or g (or both) is present depends the initial condition.

A useful technique for PDEs is the separation of variables that can help reduce the dimensionality. Let us try a solution that is a product of space and time variables separately,

$$u(x,t) = X(x)Y(t). \tag{6.64}$$

Substitute it into Eq. (6.62) and rearrange the resulting equation so that x and t are on the opposite sides as

$$\frac{1}{X}\frac{d^2X}{dx^2} = \frac{1}{v^2}\frac{1}{Y}\frac{d^2Y}{dt^2}. \tag{6.65}$$

For Eq. (6.65) to hold for all x and t that can vary independently, each side must be a constant, call it $-k^2$,

$$\frac{1}{X}\frac{d^2X}{dx^2} = -k^2 = \frac{1}{v^2}\frac{1}{Y}\frac{d^2Y}{dt^2}. \tag{6.66}$$

[8] For instance, the speed on the thickest string of a cello (C string) is $v \sim 100$ m/s. Assuming a mass density of $\rho \sim 1.5 \times 10^{-2}$ kg/m for a metal string, the tension is $T \sim 150$ N, a strongly strung string!

The solutions are

$$X(x) = A\sin(kx + \phi), \quad Y(t) = B\sin(vkt + \theta), \tag{6.67}$$

from which we can construct the general solution

$$u(x,t) = C\sin(kx + \phi)\sin(vkt + \theta). \tag{6.68}$$

We have four constants (C, k, ϕ, θ) to be determined by the initial condition $[u(x,0), \partial u(x,0)/\partial t]$ and by the boundary condition $[u(0,t), u(L,t)]$.

Let us consider the case where the two ends of a string are held fixed at zero, $u(0,t) = u(L,t) = 0$. Equation (6.68) dictates that

$$\sin(\phi) = 0, \quad \sin(kL + \phi) = 0. \tag{6.69}$$

The first term can be satisfied by choosing $\phi = 0$. The second term implies

$$k_n = \frac{n\pi}{L}, \quad n = 1, 2, 3, ... \tag{6.70}$$

meaning that only discrete values of wavevector k are allowed.

Let

$$\lambda_n = \frac{2\pi}{k_n}, \quad L = n\frac{\lambda_n}{2}, \quad f_n = \frac{v}{\lambda_n} = \frac{nv}{2L}. \tag{6.71}$$

The corresponding discrete solution follows from Eq. (6.68),

$$u_n(x,t) = C\sin(\frac{2\pi x}{\lambda_n})\sin\left(2\pi f_n t + \theta\right). \tag{6.72}$$

This is a standing wave. The λ_n and f_n defined in Eq. (6.71) are the fundamental wavelengths and frequencies (eigenfrequencies). The lowest fundamental frequency is $v/2L$, and higher harmonics are integer multiples of that. Equation (6.72) is called an eigenmode. The general solution is a linear superposition of all eigenmodes like the normal modes (Section 6.2.2).

6.5.3 FDM FORMULATION

Let us try to numerically solve the wave equation using FDM. We need to discretize both space and time. As before, let the space interval be h, and $x_m = mh$, $m = 0, 1, 2, ..., N$. For time, let Δt be the time interval, and $t_n = n\Delta t$, $n = 0, 1, 2, ...$ The displacement at x_m and t_n is written as $u(x_m, t_n) = u_m^n$, so the subscript refers to space, and the superscript to time.

We use the second-order approximation for the derivatives (6.31b),

$$\frac{\partial^2 u(x,t)}{\partial x^2} = \frac{u(x+h,t) - 2u(x,t) + u(x-h,t)}{h^2}, \tag{6.73a}$$

$$\frac{\partial^2 u(x,t)}{\partial t^2} = \frac{u(x,t+\Delta t) - 2u(x,t) + u(x,t-\Delta t)}{(\Delta t)^2}. \tag{6.73b}$$

Evaluating the preceding at x_m and t_n, and substituting into Eq. (6.62), we have

$$\frac{u_{m+1}^n - 2u_m^n + u_{m-1}^n}{h^2} = \frac{1}{v^2}\frac{u_m^{n+1} - 2u_m^n + u_m^{n-1}}{(\Delta t)^2}. \tag{6.74}$$

The terms on the LHS refer to different space grids at the same time grid, whereas the reverse is true on the RHS.

For stepping forward, we solve for u_m^{n+1} from Eq. (6.74), obtaining

$$u_m^{n+1} = 2(1-\beta^2)u_m^n - u_m^{n-1} + \beta^2\left(u_{m-1}^n + u_{m+1}^n\right), \quad \beta = \frac{v\Delta t}{h}. \tag{6.75}$$

This is the FDM form of the wave equation. It is a recursion relating the solution at t_{n+1} to the solutions at previous two steps t_n and t_{n-1} (for all space grid points). The wave equation is a second-order differential equation, so we expect to have two initial values to determine the solution.

6.5.4 VON NEUMANN STABILITY ANALYSIS

As discussed earlier (Section 1.2.2), recursion relations notoriously tend to be bimodal; they are either very stable or very unstable. If it is unstable, the recursion relation is useless numerically, as a small error will be amplified exponentially, totally overwhelming the true solution. Therefore, before we plunge ahead solving Eq. (6.75) iteratively (or using any recursions), we need to be sure of its stability. We can check this from the von Neumann stability analysis.

Essentially, we represent the solution u_m^n as

$$u_m^n = \rho^n \exp(ikx_m) = \rho^n \exp(ikmh), \tag{6.76}$$

where ρ and k are constants. The recursion is stable if $|\rho| \le 1$.

To find ρ, we substitute Eq. (6.76) into (6.75). After cancellations and simplification, we obtain

$$\rho^2 - 2(1-2\beta^2 s^2)\rho + 1 = 0, \quad s = \sin(kh/2). \tag{6.77}$$

This is the von Neumann characteristic equation. It has two roots,

$$\rho_\pm = 1 - 2\beta^2 s^2 \pm \sqrt{(1-2\beta^2 s^2)^2 - 1}. \tag{6.78}$$

The dividing line is $|1-2\beta^2 s^2| = 1$, corresponding to $\beta s = 1$, or $v\Delta t s = h$. If $|1-2\beta^2 s^2| > 1$, or $v\Delta t s > h$, we have two real roots. However, one of them is always greater than unity, so it is unstable.

If $|1-2\beta^2 s^2| \le 1$, or $v\Delta t s \le h$, the two roots are complex conjugates of each other,

$$\rho_\pm = 1 - 2\beta^2 s^2 \pm i\sqrt{1-(1-2\beta^2 s^2)^2}, \tag{6.79}$$

such that the magnitude is unity

$$|\rho_\pm|^2 = (1-2\beta^2 s^2)^2 + 1 - (1-2\beta^2 s^2)^2 = 1, \quad \text{if } v\Delta t\, s \le h. \tag{6.80}$$

Because $s = \sin(kh/2) \le 1$, the sufficient condition for stability is

$$v\Delta t \le h, \quad \text{or } \beta \le 1, \quad \text{(stability condition)} \tag{6.81}$$

Physically, this means that the time step must be smaller than the time it takes for the wave to travel one space grid.

It is instructive to note that if $v\Delta t s$ is slightly greater than h, then Eq. (6.81) will not be satisfied near the maximum $s = \sin(k_{max}h/2) = 1$, that is, at $k_{max}h/2 = \pi/2$, or,

$$k_{max} = \pi/h. \quad \text{(maximum wavevector)} \tag{6.82}$$

This is just the Nyquist sampling theorem (S:5.44) (Section S:5.B). If we take $\Delta k = 2k_{max}$ and $\Delta x = h$ as the uncertainties in wavevector and in position, respectively, we can restate Eq. (6.82) as $\Delta p \Delta x = \hbar \Delta k \Delta x = 2\pi\hbar$, which satisfies Heisenberg's uncertainty principle. Moreover, using $k = 2\pi/\lambda$, we have from Eq. (6.82)

$$\lambda_{min} = 2h, \tag{6.83}$$

which is the minimum wavelength representable over space intervals of h.

6.5.5 SIMULATION OF WAVE MOTION

From the preceding analysis, we can now solve Eq. (6.75) iteratively and maintain stability by choosing $\beta \le 1$, the single parameter in the wave equation. We use two arrays to hold the solutions at times t_{n-1} and t_n, advance the solution to t_{n+1} and store it in a third array, then swap the appropriate arrays to advance to t_{n+2}, and so on. This is done in Program 6.4.

Program listing 6.4 Waves on a string (`waves.py`)

```
import numpy as np, visual as vp, vpm

def wavemotion(u0, u1):
    u2 = 2*(1-b)*u1 - u0                    # unshifted terms
    u2[1:-1] += b*( u1[0:-2] + u1[2:] )     # left, right
    return u2

def gaussian(x):
    return np.exp(-(x-5)**2)

L, N = 10.0, 100              # length of string, number of intervals
b = 1.0                       # beta^2

x, z = np.linspace(0., L, N+1), np.zeros(N+1)
u0, u1 = gaussian(x), gaussian(x)

scene = vp.display( title='Waves',background=(.2,.5,1),center=(L/2,0,0))
string = vpm.line(x, u0, z, (1,1,1), 0.05)    # string

while(True):
    vp.rate(100), vpm.wait(scene)             # pause if key press
    u2 = wavemotion(u0, u1)
    u0, u1 = u1, u2                           # swap u's
    string.move(x, 2*u0, z)                   # redraw string
```

To start the iterative process, we need the initial solutions at two successive steps. There are several ways to do this. A practical way is to choose u^0, then shift it by an amount proportional to $v\Delta t$ and set it to u^1. We use the simplest choice in Program 6.4: We set both to the same Gaussian.

A more important issue is how to handle boundary conditions. When $m = 0$, for instance, we need u^n_{-1} from Eq. (6.75), which is not defined on our grid. The simplest choice is to keep

I_{j-1} [0 1 2 3 4]

I_j [0 1 2 3 4]

I_{j+1} [0 1 2 3 4]

FIGURE 6.15 Conceptual alignment of arrays by shifting and slicing.

both ends fixed, $u_0^n = u_N^n = 0$, so we do not have to worry about them. This is the approach taken in Program 6.4. Or we could use periodic conditions, $u_{-1}^n = u_N^n$ and $u_{N+1}^n = u_0^n$. It is equivalent to identical strings placed to both the left and right sides of the grid. A string with one end fixed and one end open could also be considered (Project P6.6).

Rather than looping through the array index explicitly, `wavemotion()` iterates Eq. (6.75) implicitly via element-wise operations of NumPy arrays. Whereas the former may seem more direct, the latter is more suitable in terms of speed and simplicity. At a given grid m, the last term in Eq. (6.75) involves the sum of the left and right neighbors, $u_{m\mp1}^n$. It suggests a sum of the form $H_j = aI_{j-1} + bI_j + cI_{j+1}$ can be done via shifting and slicing, illustrated in Figure 6.15 for a five-element array.

We can see from the shaded elements that, except for the end points, the interior elements (middle row) are lined up with their neighbors with the shifted arrays. We could accomplish the conceptual shifting via slicing of NumPy arrays (Section 1.D). Guided by Figure 6.15, we can find H_j by

```
H[1:-1] = a*I[:3] + b*I[1:-1] + c*I[2:]
```

Note that I[1:−1] is the same as I[1:4] for a five-element array.

Using this strategy, `wavemotion()` has the uncanny resemblance to the written form of Eq. (6.75), reproducing it almost verbatim: line 4 corresponds to the first two terms in Eq. (6.75) and the next line to the third term via slicing. We gain not only speed, but also simplicity. Because discretized operators like Eq. (6.31b) always involve coupling of close neighbors, we will often encounter such situations. As a result, element-wise operations with slicing will be the preferred method from now on.

The main program animates waves on a string with both ends fixed. The pair of statements (lines 18 and 24) work together to create and redraw the string using the VPM (VPython modules, Program S:6.4) library functions `line()` and `line.move()`, respectively. The animation can be paused by a key press (line 21).

Figure 6.16 displays the results from Program 6.4. It shows a single Gaussian peak in the beginning. As time increases, it splits into two smaller peaks, traveling in opposite directions. When they reach the boundaries, there is no possibility of moving forward. The tension is downward, pulling the string in that direction. As a result, the displacement is inverted, and the waves move toward each other until they merge into one. Thereafter, they split again, repeating the cycle.

The fact that there are two counterpropagating waves is a result of our choice of initial conditions and the principle of linear superposition (6.63). By choosing the two initial waves to be the same, we are in effect forcing two waves into one. It shows we can choose the kind of wave by setting the appropriate initial conditions, from traveling waves to standing waves.

We have used a combination of h and Δt such that $\beta = 1$ in Program 6.4. From the von Neumann stability analysis (6.81), this is the maximum β for stable solutions. Could we obtain more accurate solutions if we had used a smaller Δt? Our experience and expectation would tell us yes. As it turns out, we would be wrong in this case: We would get worse, less accurate solutions.

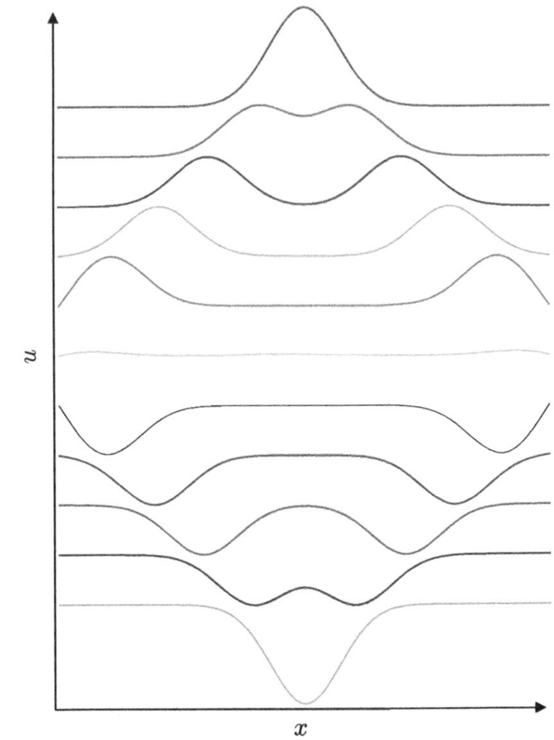

FIGURE 6.16 Traveling waves on a string. Time increases from the top down.

Due to a quirk in the 1D wave equation (6.62), we actually obtain the exact solution if $\beta = 1$. The quirk has to do with the fact that $\partial^4 u/\partial x^4 = v^{-4}\partial^4 u/\partial t^4$ [24]. By a happy accident, this leads to total cancellation of the third- and higher-order terms in the discretized wave equation (6.62) if $\beta = 1$. For 1D waves, the optimal value is $\beta = 1$, which simplifies Eq. (6.75) to

$$u_m^{n+1} = u_{m+1}^n + u_{m-1}^n - u_m^{n-1}. \tag{6.84}$$

This result is almost too simple to be true.

However, the exact algorithm is striking in the symmetry about space and time. Rewriting Eq. (6.84) as

$$u_m^{n+1} + u_m^{n-1} = u_{m+1}^n + u_{m-1}^n, \tag{6.85}$$

it tells us that the sum of the amplitudes at a grid point m immediately before and after the current time is equal to the sum of the amplitudes at the current time n immediately ahead and behind the current grid point.

6.6 STANDING WAVES

Like the normal modes for a system of particles, standing waves are the fundamental building blocks of waves on a string. Any wave can be represented as a linear superposition of these fundamental harmonics (6.72).

Standing waves can be formed by proper choice of initial conditions, and their time-dependent motion can be simulated with Program 6.4. Even though analytically solvable (Eq. (6.71)), we are interested in computing the fundamental frequencies using the same FDM scheme as the simulations.

The standing waves satisfy Eq. (6.66)

$$\frac{d^2X}{dx^2} = -k^2X, \tag{6.86}$$

subject to the boundary condition $X(0) = X(L) = 0$. Equation (6.86) is analogous to the static string equation (6.29) if $T = 1$ and $f(x) = k^2X$. Thus, we can solve it by following the same steps as in Section 6.3.2. In particular, the FDM equation (6.34) as applied to Eq. (6.86) becomes

$$\begin{bmatrix} -2 & 1 & & & \\ 1 & -2 & 1 & & \\ & \ddots & \ddots & \ddots & \\ & & 1 & -2 & 1 \\ & & & 1 & -2 \end{bmatrix} \begin{bmatrix} X_1 \\ X_2 \\ \vdots \\ X_{N-2} \\ X_{N-1} \end{bmatrix} = -h^2k^2 \begin{bmatrix} X_1 \\ X_2 \\ \vdots \\ X_{N-2} \\ X_{N-1} \end{bmatrix}. \tag{6.87}$$

This is an eigenvalue problem, $A\mathbf{X} = \lambda\mathbf{X}$, analogous to Eq. (6.16) but simpler because $B = \mathbb{1}$. It can be solved using the same method as in Program 6.2 with the eigenvalue solver `eigh()`. This is done in Program 6.6. The results for the first few standing waves are shown in Figure 6.17.

The FDM results are plotted as symbols and the analytic solutions as curves (sine waves, Eq. (6.72)) for a string of unit length. The FDM results and the analytic solutions are normalized at the first point, X_1. A total of twenty grid segments ($N = 20$) are used in the FDM calculation. We recognize the fundamental wavelengths $\lambda_n = 2/n$ at harmonic frequencies f_1 to $4f_1$ where $f_1 = v/2$ (recall $L = 1$). The FDM waveforms agree exactly with the analytic solutions.

The FDM eigenvalues from Program 6.6 are listed in Table 6.1. They are slightly off in comparison to the exact results. The relative error increases with n. The most accurate result is

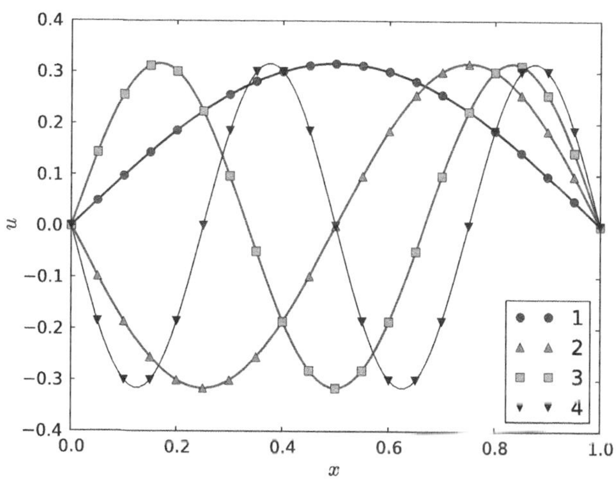

FIGURE 6.17 The first four harmonics on a string. The symbols are FDM results, and the solid curves are exact solutions.

Table 6.1 The FDM Eigenwavelengths and Relative Error.

Harmonic n	1	2	3	4	...	19
Exact λ_n	2	1	2/3	1/2	...	2/19
FDM λ_n	2.00206	1.00412	0.67288	0.50832	...	0.15757
Relative error	0.00103	0.00412	0.00931	0.0166	...	0.497

for the lowest harmonic $n = 1$. This is expected because we are approximating the full spectrum by a finite number of spectral lines $(N - 1)$. Because higher harmonics would require more grid points, we expect larger errors for higher n. Increasing N will improve the accuracy of the lower harmonics.[9]

It is remarkable that, even though the eigenvalues are only approximate, the FDM waveforms are exact as seen earlier (Figure 6.17). The exact waveform holds even for the highest $n = N - 1$, while the relative error in λ_n is fairly large, $\sim 50\%$. The reason has to do with the special form of the eigenequation (6.86), a coincidence that led to the FDM producing exact solutions for the 1D wave equation (6.62) at optimal value of $\beta = 1$.

Finally, let us consider why only discrete frequencies are allowed. Imagine we pluck the string close to the left end point. The displacement will travel along the string at a fixed speed v in (6.62). By the time it reaches the right end point, the displacement must be zero, for otherwise it does not satisfy the boundary condition and is not allowed. This means that in the time interval $\Delta t = L/v$, an allowed wave must complete an integer n number of half sinusoidal periods, that is, $\Delta t = nT/2$, or $T = 2L/nv$. In terms of frequency, it is $f = nv/2L$, same as Eq. (6.71). We see that discrete frequencies arise naturally from the wave equation. We will see essentially the same idea leading to discrete energies in quantum mechanics (Section 9.1.1).

6.7 WAVES ON A MEMBRANE

When the vibrating medium is a surface rather than a string, two-dimensional waves are created. With a slight modification, we can generalize the one-dimensional wave equation (6.62) to two dimensions as

$$\frac{\partial^2 u(x,y,t)}{\partial x^2} + \frac{\partial^2 u(x,y,t)}{\partial y^2} = \frac{1}{v^2}\frac{\partial^2 u(x,y,t)}{\partial t^2}. \quad \text{(wave in 2D)} \tag{6.88}$$

In effect, the differential operator $\partial^2/\partial x^2$ is replaced by the Laplacian in 2D. Equation (6.88) describes two-dimensional waves on a membrane such as a drum. Holding either x or y constant, we recover the 1D wave equation in the other direction.

Discretizing $\partial^2/\partial y^2$ in the same way as Eq. (6.73a), the Laplacian is

$$\frac{\partial^2 u}{\partial x^2} + \frac{\partial^2 u}{\partial y^2} \simeq \frac{u(x+h_x,y,t) - 2u(x,y,t) + u(x-h_x,y,t)}{h_x^2} + \frac{u(x,y+h_y,t) - 2u(x,y,t) + u(x,y-h_y,t)}{h_y^2}, \tag{6.89}$$

where h_x and h_y are the grid sizes in x and y directions, respectively.

Assuming $h_x = h_y = h$, and letting $u^n_{i,j} = u(ih, jh, n\Delta t)$, we have the 2D FDM wave equation $(\beta = v\Delta t/h)$

$$u^{n+1}_{i,j} = 2\left(1 - 2\beta^2\right) u^n_{i,j} - u^{n-1}_{i,j} + \beta^2\left(u^n_{i,j-1} + u^n_{i,j+1} + u^n_{i-1,j} + u^n_{i+1,j}\right). \tag{6.90}$$

Equation (6.90) may be solved in a similar way to Eq. (6.75), except the displacement arrays are two dimensional. The code is given in Program 6.7. The core function `wave2d()` is shown here.

```
u2 = 2*(1-b-b)*u1 - u0
u2[1:-1,1:-1] += b*( u1[1:-1,0:-2] + u1[1:-1,2:]      # left, right
                  + u1[0:-2,1:-1] + u1[2:,1:-1] )  # top, bottom
```

[9]The same phenomenon exists in quantum mechanics. See Section 9.6.2.

The waves in the previous two iterations are stored on a 2D grid as $u^0_{i,j}$ and $u^1_{i,j}$ (`u0` and `u1`), respectively. Like the 1D case (Section 6.5.5), the solution for the next step $u^2_{i,j}$ is obtained via element-wise operations. The first line accounts for the first two terms in Eq. (6.90). The third term, which is proportional to the sum of the four nearest neighbors, is computed in the next two lines via slicing, again bearing striking simplicity and resemblance to the written form. This is a natural extension of the 1D case, as illustrated by the following 4×4 array.

`u[1:-1,1:-1]`   `u[1:-1,0:-2]`

$$\begin{bmatrix} u_{00} & u_{01} & u_{02} & u_{03} \\ u_{10} & \boxed{u_{11} \quad u_{12}} & & u_{13} \\ u_{20} & u_{21} & u_{22} & u_{23} \\ u_{30} & u_{31} & u_{32} & u_{33} \end{bmatrix}, \quad \begin{bmatrix} u_{00} & u_{01} & u_{02} & u_{03} \\ \boxed{u_{10} \quad u_{11}} & & u_{12} & u_{13} \\ u_{20} & u_{21} & u_{22} & u_{23} \\ u_{30} & u_{31} & u_{32} & u_{33} \end{bmatrix}. \tag{6.91}$$

Slicing on the left side of line 2 in the preceding code selects the center subarray (boxed) from the left array in Eq. (6.91). It contains the interior points to be updated (compare with Figure 6.15). The boundary points where $u_{i,j} = 0$ are excluded. The first term in the brackets on the right side of line 2 is depicted in the right array of Eq. (6.91). Here, slicing refers to the boxed subarray, which is shifted to the left by one column relative to the center box. Therefore, entries in `u[1:-1,0:-2]` are the left neighbors to `u[1:-1,1:-1]`. Similarly, the other three terms on

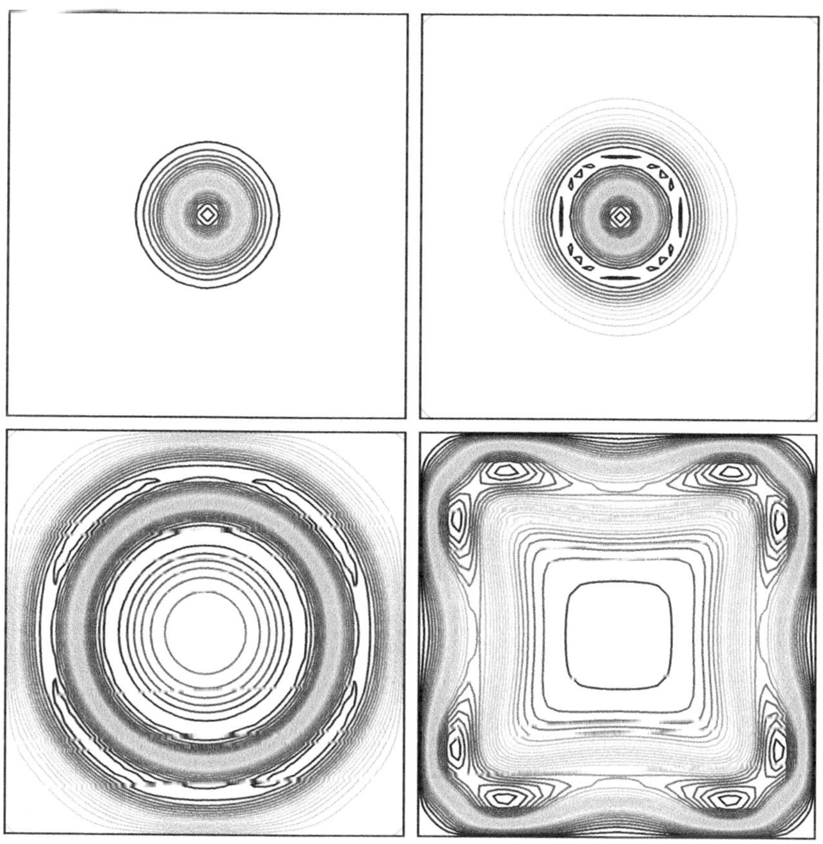

FIGURE 6.18 Waves on a membrane at different times.

the right side of the statement (lines 2 and 3) correspond to the right, top, and bottom neighbors, respectively. Here, left-to-right refers to moving across columns and top-to-bottom to running down the rows.

Through slicing and element-wise operations, the algorithm (6.90) is thus expressed succinctly in two compact statements. Because no explicit looping is used, computation is fast and highly efficient. We will encounter similar situations later and use the same technique (e.g., Program 6.8 and Section 7.2.1).

The results from Program 6.7 are shown in Figure 6.18 at several instants as contour plots and snapshots of the animated wave are shown in Figure 6.19. The initial wave is a product of symmetric Gaussians in both directions, localized at the center of the membrane (Figure 6.18, upper left). As time increases, it spreads outward in the form of circular waves (Figure 6.18, upper right). It is rather like the ripple effect caused by a pebble hitting a water surface.

When the waves hit the boundary wall where the membrane is fixed, they are no longer circular. Instead, an inversion occurs, similar to that observed in waves on a string at the end points. But it is more complicated for 2D waves because the wavefront reaches the boundary at different times. The wavefront first hits the central portion of the walls (Figure 6.18, lower left). By the time it spreads toward the corner of the walls, the central portion has already inverted and started going toward the center (Figure 6.18, lower right). The wavefronts are distorted by the boundary walls.

After many encounters with the boundary, the wave develops numerous nodes and crests. It is thoroughly mixed, and the original Gaussian shape is unrecognizable. This is shown in Figure 6.19. Depending on the initial and boundary conditions, many interesting

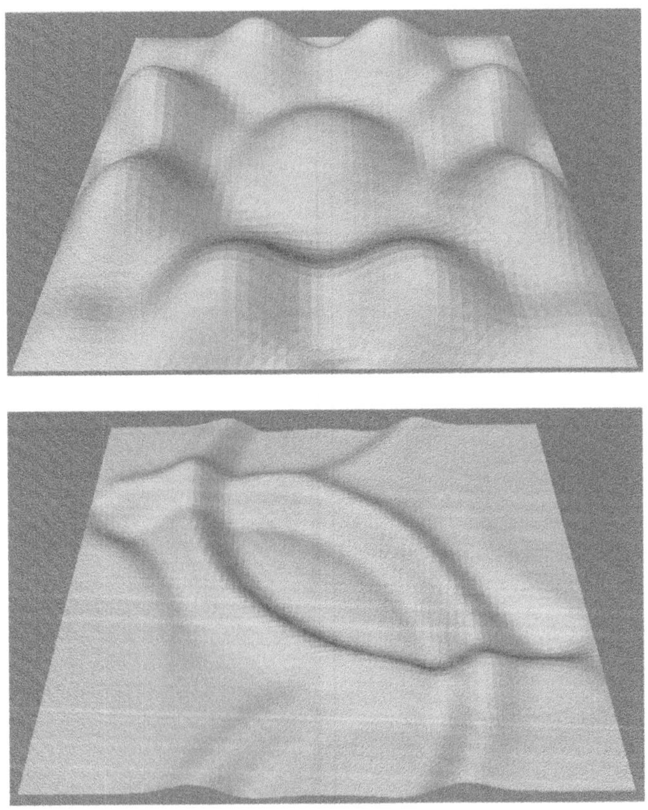

FIGURE 6.19 Snapshots of animated waves on a membrane.

landscapes can occur, none repeating themselves, even after $t \gg L/v$. For more explorations, see Project P6.8.

Of course, to maintain stability of evolution, we must be sure the algorithm is stable. A value of $\beta = 0.5$ was used in generating the preceding results. For 2D waves, the range of stability of β is slightly different than for 1D waves. A von Neumann stability analysis can be carried out as discussed in Section 6.5.4. It is found that the FDM solutions are stable for $\beta \leq 1/\sqrt{2}$. The number 2 under the square root comes as a result of two-dimensional space. Details are left to Exercise E6.8.

6.8 A FALLING TABLECLOTH TOWARD EQUILIBRIUM

The ideal waves on a membrane will oscillate forever. A realistic wave would suffer dissipation. We are interested in modeling such a realistic wave on a surface such as a tablecloth.

We construct a tablecloth model as a two-dimensional array of particles arranged on a grid shown in Figure 6.20. There are N rows, and each row has M particles, for a total of $M \times N$ particles. Nearest neighbors are connected by springs. The particle at column i (x) and row j (y) interacts with its four nearest neighbors: left, right, top, and bottom. We assume initially the grid is laid out in the horizontal x-y plane. The force on the particle at (i, j) includes four spring forces plus gravity and damping as

$$\vec{F}_{i,j} = \vec{f}_{left} + \vec{f}_{right} + \vec{f}_{bottom} + \vec{f}_{top} - b\vec{v}_{i,j} - mg\hat{k}. \tag{6.92}$$

The equations of motion for this system of particles are qualitatively the same as Eq. (S:6.5) (or (4.41) and (4.42) with Hooke's law) and can be solved accordingly. The included Program 6.8 closely mirrors Program S:6.2. It combines the numerical integration technique of Program S:6.2 and the visualization technique of Program 6.7. The visual representation of motion is based on dividing the particle grid (Figure 6.20) into triangular meshes, approximating a vibrating surface.

Program 6.8 can be used to simulate 2D waves with or without damping, for small or large amplitudes. It can also accommodate various types of boundary conditions, as well as constant or variable mass density with a slight modification. Figure 6.21 shows the results of a uniform surface (tablecloth) held at the corners and falling under its own weight. The tablecloth was initially flat and at rest. As it falls, it gains both kinetic energy and elastic potential energy. Toward

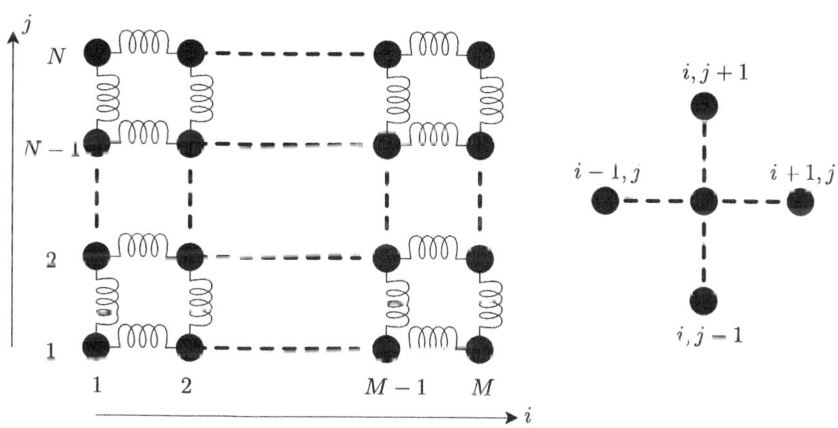

FIGURE 6.20 An array of particles on a grid. The right figure shows the nearest neighbor interactions.

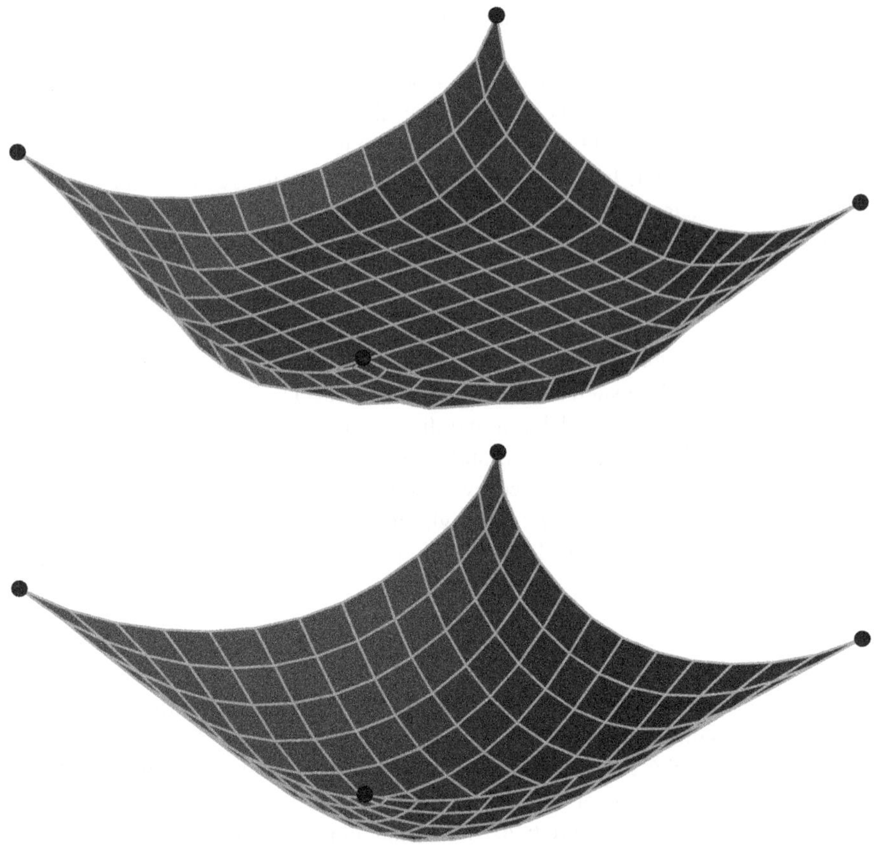

FIGURE 6.21 A falling tablecloth held at the corners.

the bottom, it rebounds, and oscillations start to propagate throughout the grid. Except the corner points, all the boundary points are free. For zero or small damping, very complex waves can be formed. After a while, the tablecloth settles into equilibrium. The overall shape is like the catenary (Figure S:6.1) spun around the central axis.

Our model can also be used to study relaxation time, rate of energy dissipation, boundary effects, and so on. We will revisit the model for energy minimization in thermal physics (see Section 11.3).

Computationally, modeling a vibrating surface is much more expensive for a 2D particle array than for a 1D string. The core of Program 6.8 is `cloth()`, which computes the forces in the equations of motion. If N is the linear dimension, the number of operations scales like N^2 in 2D. For medium to large N, it would be a speed bottleneck for an interpretative language like Python if explicit looping were used (Section 1.3.2). But, like Program 6.7, `cloth()` uses implicit looping of NumPy arrays with efficient vectorized operations, enabling us to simulate 2D waves without the need of language extensions.

Programs 6.7 and 6.8 are prime examples where Python with NumPy provides the best of both worlds: the versatility and elegance of Python programming and the computational speed of compiled codes. Because this combination results in a flexible yet efficient workflow, use of NumPy (and SciPy) is indispensable in the chapters ahead. In most cases, this is sufficient. If extra speed boost is needed, we can turn to language extensions (see Program 5.7 for Mandelbrot fractals and Program S:11.4 for molecular dynamics).

CHAPTER SUMMARY

We began discussion of oscillations and waves with a damped, driven harmonic oscillator, including critical, over- and underdamping, and resonance. We discussed the oscillations of small systems such as triatomic molecules, its power spectrum, and normal modes motion. Reduction to linear systems of equations, eigenvalue problems, and their solutions are described using the SciPy linear algebra library.

The displacement of a string, representing the continuum limit of an array of particles under static forces, was discussed. The finite difference method (FDM) was applied to solving the displacement equation. We also introduced the basic idea of the finite element method (FEM) for 1D problems. In Chapters 7 and 9, we will extend the FEM to 2D problems in classical and quantum physics.

Simulations of one- and two-dimensional waves on a string and on a membrane as described by second-order PDEs were presented last. We discussed stability of solutions in terms of von Neumann stability conditions for the FDM scheme. We also presented a model using a grid of particles to approximate an oscillating membrane.

We have illustrated the use of several composite objects from the VPython modules (VPM) library to visualize oscillations and wave motion. The VPM library defines several classes of objects, including strings, slinkies, nets (grids), and meshes (triangular faces). These objects will be used in subsequent chapters.

6.9 EXERCISES AND PROJECTS

Exercises

E6.1 (a) Modify Program 6.1 so the oscillation is in the vertical direction. Do not forget gravity. How is the motion different from the horizontal case? Where is the equilibrium position? Explain.

(b) Show analytically that the only effect of gravity is to shift the equilibrium position of the oscillator.

E6.2 Show that the *RLC* circuit in Figure 6.2 can be described by

$$\frac{d^2q}{dt^2} + 2\gamma\omega_0\frac{dq}{dt} + \omega_0^2 q = \varepsilon(t)/L. \qquad (6.93)$$

Here q is the charge, $\varepsilon(t)$ the driving voltage. Find γ and ω_0 in terms of resistance R, inductance L, and capacitance C.

E6.3 The solutions to Eq. (6.1) may be obtained by assuming a trial solution of the form $x(t) = \exp(\lambda t)$. Substitute it into Eq. (6.1), and solve the resulting characteristic equation for λ. Obtain Eq. (6.3) by a linear superposition of $\exp(\lambda_\pm t)$ (see Sec. 8.5 of Ref. [10]).

E6.4 (a) Assume a driving force of the form $F = F_d \exp(i\omega t)$, and a trial solution $x = A(\omega)\exp(i(\omega t + \phi))$ for Eq. (6.5). Show that $A(\omega)$ is proportional to F_d.

(b) Verify the resonance condition (6.6) by finding the ω_r for which A is maximum.

E6.5 (a) Solve Eq. (6.18) for λ in the special case $m_1 = m_3 = m$ and $m_2 = M$. (b) Verify that if $k = 1$, $m = 1/4$, and $M = 2/5$, there are three eigenfrequencies $\omega_i = 0$, 2, and 3, $i = 1$ to 3. Remember $\lambda = \omega^2$. (c) Solve Eq. (6.13) for each ω_i to obtain the eigenvector $[a_{i1}, a_{i2}, a_{i3}]$. You will only be able to determine uniquely two of the three components. Normalize it such that the largest entry is unity. (d)* All the preceding can be studied with SymPy as well. For example, factor Eq. (6.18) to show that zero is always a root (see Exercise E5.1). Next solve it symbolically for arbitrary masses.

E6.6 (a) Program 6.5 shows the animated motion of the triatomic molecule in the symmetric vibration mode. Change the initial condition such that it represents the asymmetric vibrational motion.

(b) Start with an arbitrary initial condition, analyze the motion using the FFT method, and confirm the results in Figure 6.7.

(c) Replace the Euler-Cromer method with the standard Euler method, and run the simulation again. What is the long-term consequence?

E6.7 (a) Verify the FEM matrix A_{ij} in Eq. (6.53).

(b) Evaluate p_i given in Eq. (6.55) for linear and quadratic loads, $f_1 = a + bx$ and $f_2 = cx^2$, respectively. Express your results in terms of the node index i and grid spacing h.

E6.8 Carry out a von Neumann stability analysis of the FDM equation (6.90) for waves in 2D to determine the stability condition for $\beta = v\Delta t/h$.

E6.9 Starting from Eq. (6.68), derive the wavevectors for standing waves subject to one fixed end and one open end, Eq. (6.94). Do the same for two open ends. Sketch the first few standing waves in both cases.

E6.10 Suppose a membrane is open on one side, say $x = L$. Derive the relation describing the open boundary condition in 2D.

E6.11 We usually choose higher-order methods over lower-order ones, hoping to gain better accuracy and efficiency. Most of the time, it works out this way. However, higher-order methods are not always better. Strictly speaking, higher orders translate into better accuracy only if the problem is sufficiently smooth. For highly oscillatory problems like those we are dealing with in this chapter, a higher-order method can be worse than a lower-order one.

As an example, run Program 6.8 with the fourth-order Runge-Kutta `RK4` and the fifth-order Runge-Kutta-Fehlberg `RK45n` integrators. Observe the system's behavior with the latter. Is it physical? Vary the step size, watch where the problem starts, and analyze what caused the behavior (by plotting the energy of the problematic part, for example). Find ways to resolve the problem.

Projects

P6.1 Consider the longitudinal oscillations of an N-particle array, Figure 6.22. Assume the interaction between neighboring particles obeys Hooke's law, and all the masses are the same.

(a) Set up the equations of motion for the system, and the A and B matrices as in Eq. (6.14).

(b) Let $N = 3$, what kind of motion do you predict? Briefly write down your predictions. Integrate the equations with an ODE solver. Choose a suitable k and m, for example, unity, and use an arbitrary initial condition, say $u_1 = 0.2$, all others including velocities are zero. Optionally animate the motion of the particles with VPython.

Plot the displacement of a particle as a function of time, $u_3(t)$, for instance. Discuss your results vs. predictions.

(c) Obtain the power spectrum of the data plotted earlier. Identify the frequencies in the spectrum, remembering a $\pm$ frequency pair counts as one. Make sure you have enough data points that should be a power of two. Are the frequencies as you expected? Hint: To obtain a "clean" spectrum, adjust your step size so that the peaks are well separated and not "contaminated."

(d) Calculate the normal modes of the system. Compare the eigenfrequencies with your power spectrum results. Explain the motion associated with the normal modes.

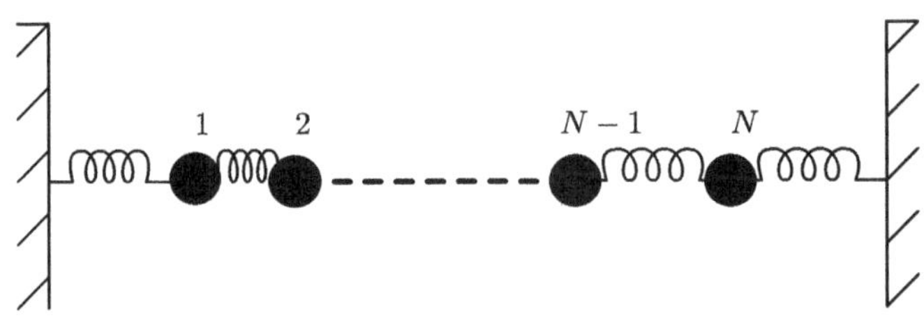

FIGURE 6.22 A particle array.

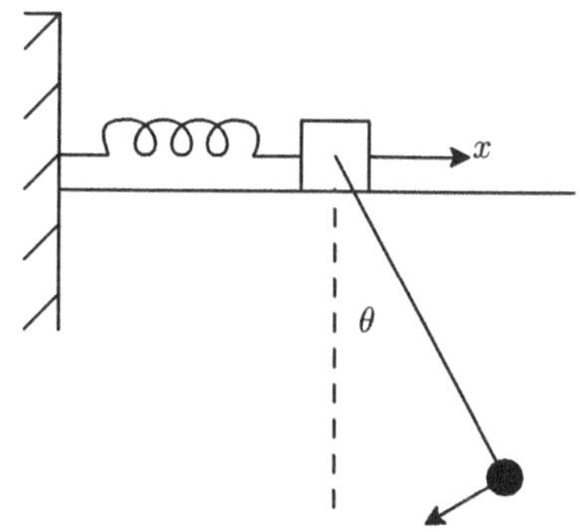

FIGURE 6.23 Two coupled oscillators.

(e) Repeat the preceding calculations for $N = 5$. Identify all harmonics. Compare and discuss your results with the exact value, $\omega_i = 2\omega_0 \sin(i\pi/(2N+2))$, $i = 1, 2, ..., N$, where $\omega_0 = \sqrt{k/m}$.

P6.2 Two coupled oscillators are shown in Figure 6.23. The horizontal oscillator (mass m_1 and spring constant k) moves on a frictionless surface. A pendulum (mass m_2 and cord length l) is connected to m_1 via a cord.

(a) Show that the kinetic and potential energies are

$$T = \frac{1}{2}m_1\dot{x}^2 + \frac{1}{2}m_2(\dot{x}^2 + l^2\dot{\theta}^2 + 2l\dot{x}\dot{\theta}\cos\theta),$$
$$V = \frac{1}{2}kx^2 + m_2 gl(1 - \cos\theta).$$

(b) Take the small oscillation limit, expand T and V in powers of θ and retain terms only to second order in any combination of x, θ, $\dot{x}$, and $\dot{\theta}$.

(c) Let $q_1 = x$ and $q_2 = \theta$, and define

$$V_{ij} = \frac{\partial^2 V}{\partial q_i \partial q_j}, \quad T_{ij} = \frac{\partial^2 T}{\partial \dot{q}_i \partial \dot{q}_j}.$$

Compute the A and B matrices as

$$A = \begin{bmatrix} V_{11} & V_{12} \\ V_{21} & V_{22} \end{bmatrix}, \quad B = \begin{bmatrix} T_{11} & T_{12} \\ T_{21} & T_{22} \end{bmatrix}.$$

This is how A and B in Eq. (6.16) may be found for general small oscillations [40]. Calculate the eigenfrequencies and eigenvectors. Describe the normal modes.

P6.3 This project and the next one (Project P6.4) should ideally be treated as team projects. Each individual team member can focus on one primary project, but work as a team exchanging ideas and results.

(a) Run Program 6.3 for the constant load $f(x) = -2$. Compare your results with Figure 6.11 ($f = -1$), and discuss the shapes of the strings.

(b) Assume a linear load $f(x) = -2x$. Work out the exact solution from Eq. (6.29).

(c) Modify your program to calculate the displacement numerically for the linear load. You need to change matrix B. Plot both numerical and analytical results. Compare each other and with Figure 6.11 for the linear case.

(d) Predict what the shape will look like for a quadratic load, $f(x) = -3x^2$. Sketch it. Carry out the FDM calculation. How did your prediction turn out? Explain.

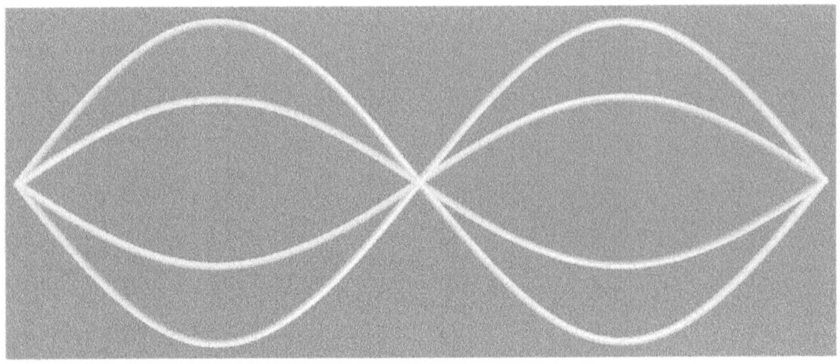

FIGURE 6.24 Standing waves on a string.

P6.4 Investigate FEM solutions for the displacement of a string under Dirichlet boundary conditions, $u_0 = u_N = 0$. It is helpful if you had gone through Project P6.3 individually or as a team.

(a) Modify Program 6.3 for the constant load, and make sure you can reproduce the FDM results shown in Figure 6.11. You can decide whether or not to move the factor h in Eq. (6.54) to the RHS of Eq. (6.52). Verify that doing so does not alter the solutions.

(b) Consider the combination of constant and Dirac point-source loads shown in Figure 6.14. Reproduce the graphs. Your FEM program should be considered validated.

(c) Calculate the displacement for the same linear and quadratic loads in Project P6.3. Results from Exercise E6.7 would be helpful. Compare the results. Plot the error between FDM and FEM solutions. Explain the discrepancy.

P6.5 (a) Set up a traveling Gaussian wavepacket on a string. Do this by changing the initial condition in Program 6.4 to the following:

`u1 = gaussian(x+h)` or `u1 = gaussian(x-h)`

Which would result in a right-traveling wave? Observe the actual wave. Is it a pure traveling wave?

(b) Let us change `u1 = gaussian(x+2*h)`. Predict the kind of wave that will be formed. Discuss the actual waves you see.

(c) Figure 6.24 shows the second harmonic on the string. Implement initial conditions that will produce the standing wave, as well as the first, third, and fourth harmonics.

P6.6 Study waves on a string with an open end. Suppose one end of the string is tied to a small, frictionless, massless ring that can move freely along a transverse thin rod at the boundary (say at $x = L$).

(a) By considering the balance of forces similar to Eqs. (6.26a) to (6.26b), show that an open end (or free) boundary is equivalent to

$$\left.\frac{\partial u(x,t)}{\partial x}\right|_L = 0. \tag{6.94}$$

(b) Simulate waves on a string with a closed end and an open end. Modify the FDM scheme in Program 6.4 to take into account the boundary condition above and $u_0 = 0$. Plot the waves as presented in Figure 6.16. What are the differences when one end is open?

P6.7 Calculate the fundamental wavelengths of standing waves on a string with either one or two open ends.

(a) Assume only one open end, so the boundary condition is $X(0) = 0$, and $\partial X/\partial x|_{x=1} = 0$. Implement the boundary condition to obtain the appropriate eigenequation analogous to Eq. (6.87). Plot the first four harmonics, discuss and compare your results to the exact values.

(b) Repeat above for two open ends.

P6.8 There are many more modes of oscillation with two-dimensional waves on a membrane. Let us sample a few of them.

(a) The waves shown in Figure 6.18 propagate outward from the center. Suppose we start the wave off center at $(L/2, L/4)$. What do you expect the wave motion to be? Modify Program 6.7 and simulate this case. Describe your results.

(b) What if the initial wave is asymmetric? Simulate this effect by introducing different Gaussian widths in the x and y directions. Which direction does the wave broaden faster? Why?

(c) Is it possible to set up a pure traveling wave on a membrane equivalent to such a wave on a string (Project P6.5)? Make an educated guess. Test your hypothesis by changing `u1` to a Gaussian (symmetric) displaced in the diagonal direction by a distance of $v \times dt$ relative to `u0`. Describe the resulting wave you see. Explain the wake.

(d) What would standing waves look like in 2D? Set up a standing wave that is the first harmonic in the x direction and the second harmonic in the y direction.

(e) So far we have assumed fixed boundary conditions. Now suppose we use periodic boundary conditions in the x direction, that is, $u(-h, y, t) = u(L, 0, t)$ and $u(L + h, y, t) = u(0, y, t)$. Implement this boundary condition in your program. Discuss the differences with fixed boundary conditions.

(f)* Simulate the waves on a membrane with an open boundary condition on one side, say at $x = L$.

P6.9 Let us explore the falling tablecloth model in some detail.

(a) Study Program 6.8 to understand the organization of the code and the workings of the function `cloth()`.

(b) Add a function that calculates the energy, E, of the system at a given time. The energy should include the kinetic and potential (elastic and gravitational) energies. Calculate the energy in the main loop, and plot it as a function of time for a given damping factor, say $b = 0.2$. Test that your code works correctly by checking some limits, for example, without damping ($b = 0$ in Eq. (6.92)), the energy should be conserved.

Find the final energy, E_f, when the system reaches equilibrium. Here equilibrium is defined as when the energy has reached a certain limit relative to the final energy, say 99%. Record the net energy loss as $\Delta E = E_i - E_f$, where E_i is the initial energy.

(c) Vary b from 0.1 to 1 in steps of 0.1. For each b, simulate the system until it reaches equilibrium. Record the time it takes for the system to lose half of the amount ΔE. Denote this time as the relaxation time. Plot the relaxation time as a function of b. Discuss your results. Can you classify the system in terms of under- and overdamping?

For each b, also graph the change in energy (dE) as a function of energy difference $E - E_f$ at each step. Characterize the trend between the rate of energy loss dE and $E - E_f$. Analyze the relationship between relaxation times and rate of energy loss.

(d) Suppose we want to change the boundary condition. Instead of fixing only the corners, let us set the particles on opposite sides of the square fixed, for example, at $x = 0$ and $x = L$. Predict how this will affect the relaxation time and the rate of energy loss.

Change the boundary condition in `cloth()`, and run the simulation again. Compare the results with your predictions, and briefly explain.

(e)* Currently, `cloth()` calculates the forces from the right and top neighbors directly, then uses the third law for the left and bottom neighbors (Figure 6.20). Recode this function so the process is reversed, that is, compute the forces from the latter directly, then apply the third law.

6.A PROGRAM LISTINGS AND DESCRIPTIONS

Program listing 6.5 Triatomic vibrations (`triatom.py`)

```
1  import visual as vp, numpy as np
2
3  def accel(u):
4      return -k*np.array([u[0]-u[1], 2*u[1]-u[0]-u[2], u[2]-u[1]])/m
5
6  col = [(1,0,0), (0,1,0), (0,0,1)]                     # RGB colors
7  x0, atom, r = np.array([-3., 0., 3.]), [0]*3, 1.0
8  for i in range(3):
9      atom[i] = vp.sphere(pos=(x0[i],0,0), radius=r, color=col[i]) # atoms
10 floor = vp.box(pos=(0,-1.1,0), length=8, height=0.2, width=4)  # floor
11 s1 = vp.helix(pos=(x0[0],0,0), thickness=0.1, radius=0.5)       # spring
12 s2 = vp.helix(pos=(x0[1],0,0), thickness=0.1, radius=0.5)
13
14 h, k, m = 0.1, 1.0, np.array([1./4, 2./5, 1./4])
15 u, v = np.array([-1.,0.,1.]), np.zeros(3)    # symmetric init cond
16 while True:
17     vp.rate(50)
18     u = u + v*h                              # Euler-Cromer method
19     v = v + accel(u)*h
20     x = x0 + u
21     for i in range(3):                       # animation
22         atom[i].pos = (x[i],0,0)
23     s1.axis, s1.pos, s1.length= (1,0,0), x[0]+r, x[1]-x[0]-2*r   # move
24     s2.axis, s2.pos, s2.length= (1,0,0), x[1]+r, x[2]-x[1]-2*r
```

Program 6.5 simulates the motion a triatomic system by solving Eq. (6.12) using the Euler-Cromer method. Other than dealing with three atoms, it works just like Program 6.1. The springs are attached to and moving with atoms 1 and 2 (line 23). They compress or stretch to fill the distances to their right neighbors, atoms 2 and 3, respectively.

Program listing 6.6 Standing waves on a string (`stringfun.py`)

```
from scipy.linalg import eigh       # Hermitian eigenvalue solver
import numpy as np, matplotlib.pyplot as plt

N = 20                        # number of intervals
h, s = 1./N, 5                # bin size, skip

A = np.diag([-2.]*(N-1))                                  # diagonal
A += np.diag([1.]*(N-2),1) + np.diag([1.]*(N-2),-1) # off diagonals

A = -A/(h*h)                  # solve -h^-2 AX = k^2 X
lamb, u = eigh(A)             # so eigenvalues lamb = k^2

x, sty = np.linspace(0.,1., s*N + 1), ['o','^','s','v']
fig, f, pi = plt.figure(), np.sin, np.pi             # some aliases
print (2*pi/np.sqrt(lamb[:len(sty)]))                # wavelength Eq. (6.71)
```

```
for i in range(1, len(sty)+1):
    ui = np.insert(u[:,i-1], [0, N-1], [0., 0.])  # insert BV
    plt.plot(x, f(i*pi*x)*ui[1]/f(i*pi*x[s]))     # normalize soln
    plt.plot(x[::s], ui, sty[i-1], label=repr(i))

plt.xlabel('x'), plt.ylabel('u')
plt.legend(loc='lower right'), plt.show()
```

This program computes the eigenvalues of the wavevector k^2 on a string of unit length. After preparing matrix A the same way as in Program 6.3 using `np.diag`, it is scaled by $-1/h^2$, such that the eigenvalues are actually for k^2 instead of $-h^2k^2$. This is convenient because they will be positive and ordered in ascending order. Because this is a normal eigenvalue problem, matrix B is omitted from `eigh()`.

The first four eigenvectors, stored in column i of `u[:,i]` for eigenvector i, are plotted with legends (mode number). As in Program 6.3, boundary values are inserted before plotting. To compare with analytical solutions directly, they are normalized at the first point from the boundary. A skip factor s is used to ensure that FDM symbols do not obscure the smooth analytical curves.

Program listing 6.7 Waves on a membrane (`waves2d.py`)

```
import numpy as np, visual as vp, vpm

def wave2d(u0, u1):
    u2 = 2*(1-b-b)*u1 - u0
    u2[1:-1,1:-1] += b*( u1[1:-1,0:-2] + u1[1:-1,2:]      # left, right
                       + u1[0:-2,1:-1] + u1[2:,1:-1] )    # top, bottom
    return u2

def gaussian(x):
    return np.exp(-(x-5)**2)

L, N = 10.0, 40                # length of square, number of intervals
b = 0.25                       # beta^2

x = np.linspace(0., L, N+1)                      # mesh grid
x, y = np.meshgrid(x, x)
u0, u1 = gaussian(x)*gaussian(y), gaussian(x)*gaussian(y)

scene = vp.display(title='2D waves', background=(.2,.5,1),
                   center=(L/2,L/2,0), up=(0,0,1), forward=(1,2,-1))
mesh = vpm.mesh(x, y, 2*u0, (.9,.8,0), (1,0,0))  # mesh

while(True):
    vp.rate(40), vpm.wait(scene)                 # pause if key press
    u2 = wave2d(u0, u1)
    u0, u1 = u1, u2                              # swap u's
    mesh.move(x, y, 2*u0)                        # redraw mesh
```

The preceding code works in nearly the same way as Program 6.4. The only differences are that the arrays u0, u1, and u2 are two dimensional, and the animation uses a mesh redrawn at each time step. The core function `wave2d()` relies on 2D slicing as described in Section 6.7. The

VPM mesh object (line 21) is created with `mesh(x,y,z, topcolor, bottomcolor)` over the meshgrid `(x,y)` and displacement `z`. It is animated with the `mesh.move()` method (line 27).

Program listing 6.8 A falling tablecloth (`tablecloth.py`)

```
1  import ode, vpm, numpy as np, visual as vp
2
3  def force(r):          # force of particle pair, with relative pos r
4      s = np.sqrt(np.sum(r*r, axis=-1))              # distance
5      s3 = np.dstack((s, s, s))                      # make (m,n,3) array
6      return -spring_k*(1.0 - spring_l/s3)*r        # Hooke's law
7
8  def cloth(Y, t):       # tablecloth
9      r, v, f = Y[0], Y[1], np.zeros((N,M,3))
10
11     rtop = r[0:-1, :] - r[1:, :]                   # rel pos to top neighbor
12     rright = r[:, 0:-1] - r[:, 1:]                 # rel pos to right neighbor
13     ftop, fright = force(rtop), force(rright)      # forces from top, right
14     f[0:-1, :] = ftop                    # force from top
15     f[:, 0:-1] += fright                 # force from right
16     f[1:, :] -= ftop                     # below, left: use 3rd law
17     f[:, 1:] -= fright
18     a = (f - damp*v)/mass + gvec
19     v[0,0], v[0,-1], v[-1,0], v[-1,-1]=0, 0, 0, 0    # fixed coners
20     return np.array([v,a])
21
22 L, M, N = 2.0, 15, 15                    # size, (M,N) particle array
23 h, mass, damp = 0.01, 0.004, 0.01        # keep damp between [.01,.1]
24 x, y = np.linspace(0,L,M), np.linspace(0,L,N)       # particle grid
25 r, v = np.zeros((N,M,3)), np.zeros((N,M,3))
26 spring_k, spring_l = 50.0, x[1]-x[0]     # spring const., relaxed length
27 r[:,:, 0], r[:,:, 1] = np.meshgrid(x,y)              # initialize pos
28 Y, gvec = np.array([r, v]), np.array([0,0,-9.8])     # [v,a], g vector
29
30 scene = vp.display(title='Tablecloth', background=(.2,.5,1),
31                    up=(0,0,1), center=(L/2,L/2,-L/4), forward=(1,2,-1))
32 vp.points(pos=[(0,0,0),(0,L,0),(L,L,0),(L,0,0)], size=50)   # corners
33 x, y, z = r[:,:,0], r[:,:,1], r[:,:,2]                      # mesh points
34 net = vpm.net(x, y, z, vp.color.yellow, 0.005)              # mesh net
35 mesh = vpm.mesh(x, y, z, vp.color.red, vp.color.yellow)
36
37 while (1):
38     vp.rate(100), vpm.wait(scene)        # pause if key pressed
39     Y = ode.RK4(cloth, Y, 0, h)
40     x, y, z = Y[0,:,:,0], Y[0,:,:,1], Y[0,:,:,2]
41     net.move(x, y, z), mesh.move(x, y, z)
```

We can use this program to simulate a falling tablecloth made of a 2D array of particles. The particles are initially arranged on a $(M \times N)$ grid (see Figure 6.20) but can move in 3D space. Like Program S:6.2 for 1D, neighboring particles interact through Hooke's force. The function `force()` computes this force given the relative position vector `r` between the particles. It first calculates the distance squared (line 4) by summing over the three components via `np.sum` along the last axis (i.e., $x^2 + y^2 + z^2$), followed by the square-root. Because `r` is actually a 3D array (see following), the distance is a 2D array after the sum. Next the distance grid is stacked depth-wise with `np.dstack` (laying over each other) to form a 3D array, so that it has the correct shape for element-wise operations when multiplied by `r` to obtain the force (line 6). That is to say, every component of $\vec{r}$ is divided by r^3, akin to $\vec{r}/r^3 = (x/r^3, y/r^3, z/r^3)$, for every particle at once. We need to do this because automatic broadcasting is ambiguous here.

The core function is `cloth()`, which returns the equations of motion. The input `Y=[r,v]` contains the position `r` and velocity `v` vectors, respectively, each a $(N, M, 3)$ array corresponding to the particles on the grid. Like Program 6.7, implicit looping via slicing is heavily used. For instance, we find the relative positions of the particles to the neighbors on top (one row above, Figure 6.20) by subtracting the rows shifted up by one from the current rows (line 11). The relative positions to the neighbors on the right can be obtained in a similar fashion. The forces can then be calculated from these relative positions and recorded accordingly (lines 14 and 15). We use Newton's third law (line 16) for the forces from the neighbors below and to the left, cutting the computational time by half. Note that slicing keeps the proper boundary conditions throughout. For example, the last column does not have neighbors to the right, so it is excluded correctly in `f[:,0:-1]` (line 15).

Once the forces are known, acceleration can be obtained for all particles at once. Finally, the velocities of the corner particles are set to zero (line 19), so they never move. The result is returned in a 4D array `[v,a]`. This again reaffirms the vector nature of our ODE solvers (Section 2.3.2) that can handle scalar (one ODE) and multidimensional vector equations automatically and transparently.

Visual representation of the tablecloth consists of a net overlaid onto a mesh. Like Program 6.7, we create a net with the VPM function `net()` which takes as input x, y, and z grid points, color, and thickness of the links. It is moved with `net.move()` as usual.

CHAPTER 7 Electromagnetic Fields

In the simulation of few-body problems in Chapter 4, gravity was the sole interaction at large distances. If the particles are charged, they will interact via electromagnetic forces, which are much stronger than gravity at the same distance. For example, the ratio of the Coulomb force to gravity between a proton and an electron is about 10^{39} at a given separation, making gravity utterly negligible. In fact, macroscopic properties of all ordinary matter are determined by electromagnetic interactions. Forces such as tension in a string (Section 6.3), spring forces, friction, and so on, are all manifestation of electromagnetic forces between atoms.

Like gravity, electromagnetic interactions occur through action at a distance via electromagnetic fields described by classical field equations, the Maxwell's equations. In this chapter, we are interested in the simulation of electric potentials and fields in electrostatics and propagation of electromagnetic waves. Both scalar and vector fields in electromagnetism will be studied by solving the appropriate partial differential equations (PDEs) first encountered in Chapter 6. We discuss two mesh-based methods for solving the 2D Poisson and Laplace equations and related boundary value problems: an iterative self-consistent (relaxation) method, and a non-iterative finite element method first introduced for 1D problems in Chapter 6. Complementing mesh-based methods, we introduce a meshfree method using radial basis functions for solving PDEs. Finally, we model the vector fields and electromagnetic radiation due to electric and magnetic dipoles.

7.1 THE GAME OF ELECTRIC FIELD HOCKEY

Let us illustrate electric fields via a game of electric field hockey. The puck is electrically charged and moves in a rectangular playing area, the rink. Instead of contact forces by sticks involving a great number of atoms, Coulomb forces from stationary charges placed around the rink and at the center control the motion of the puck. A particular configuration of charges is illustrated in Figure 7.1.

The goal of the game is to score goals, of course. Two such goals are placed at the opposite ends of the rink. The rule of the game is simple: shoot the puck from one end with a given velocity; a point is awarded if it goes into the goal at the opposite end. If the puck goes out of bounds, your turn ends.

Figure 7.1 shows three shots, two misses and one score. The success rate depends on the initial condition (skill) and the underlying physics for a given charge configuration. The configuration in Figure 7.1 consists of seven charges: five are positive, one at each corner and one at the center; two negative ones are placed in the middle of each side. In this example, the puck is negatively charged, so the forces on the puck are repulsive from the two negative charges on the sides and attractive from the corners and the center.

In the first instance, the puck was aimed at the lower side. When it got close to the negative charge, the repulsive force slowed it down and turned it back up, resulting in a miss. The other instances had the same speed but different launching angles. In each case, the puck moved toward the center charge, orbiting around it once under its attractive force. The puck with the smaller angle suffered a greater pull, overrotated, and went out of bounds from below, another miss. The third one had just the right pull and was catapulted toward the goal and a score.

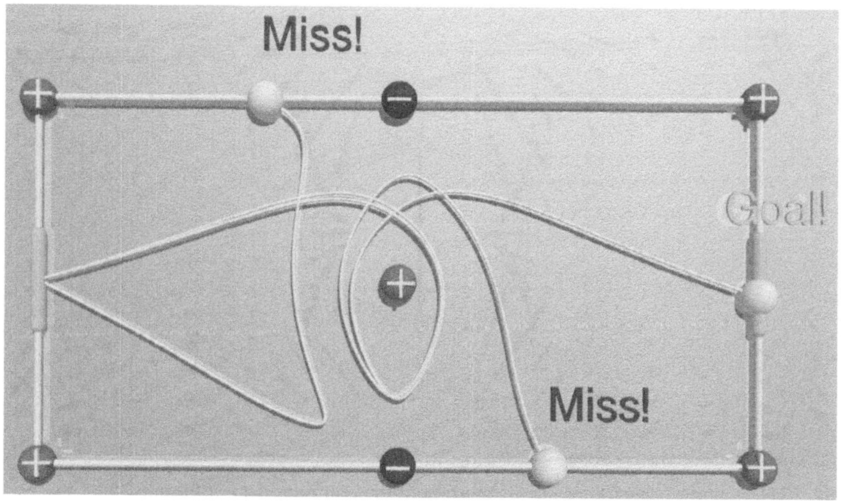

FIGURE 7.1 Three shots in a game of electric field hockey.

The hockey game simulator is given in Program 7.1. Most of the code is to set up visual animation. The actual game dynamics is modeled in `hockey()`, which returns the equations of motion as

$$\frac{d\vec{r}}{dt} = \vec{v}, \quad \frac{d\vec{v}}{dt} = \frac{kq}{m} \sum_i Q_i \frac{\vec{r} - \vec{r}_i}{|\vec{r} - \vec{r}_i|^3}, \tag{7.1}$$

where q is the charge of the puck, Q_i and $\vec{r}_i$ the charge and position, respectively, of the stationary charges. The factor $k = 1/4\pi\varepsilon_0 = 9 \times 10^9$ Nm^2/C^2 is the electrostatic constant for the Coulomb force analogous to gravity (4.1). In the simulation, we set $k/m = 1$ for simplicity. This can always be done with a proper choice of units for charge and mass (Exercise E7.1).

The main program integrates the equations of motion (7.1) and checks whether the puck is in bounds after each step. If it is out of bounds, the code decides whether it is a miss or a goal.

As seen in Chapter 4 in the restricted three-body problem (Figure 4.17), the motion of the particle is more readily understood in terms of the potential surface it navigates on. Figure 7.2 shows the equipotential lines and forces (electric field) on the puck. The figure is plotted with Program 4.6, where the effective potential is replaced by the Coulomb potentials between the puck and stationary charges, $V = kq \sum_i Q_i/|\vec{r} - \vec{r}_i|$ (Project P7.1).

The five positive charges at the corners and the center act like sinkholes the puck is attracted to. There are two saddle points, one on each side of the playing field, situated approximately in the middle of each half. Therefore, there is a minimum energy threshold for the puck to overcome the saddle point. Once over the threshold, the puck must navigate through the repulsive and attractive fields, and the trajectory can become very complex, as some in Figure 7.1 illustrate. This is true especially at lower energies when the puck must orbit around the center charge to escape to the other side. Occasionally, the particle can even become trapped to the center of the field, orbiting indefinitely. At higher energies, it is possible to deflect directly off the negative charge in the middle of the side and score a goal.

For the configuration of point charges considered here, it is straight forward to compute the potential surface and electric fields (forces) from Coulomb's law. For other charge distributions (including continuous distributions) and arbitrary boundary conditions, we need to approach the problem differently. We discuss this next.

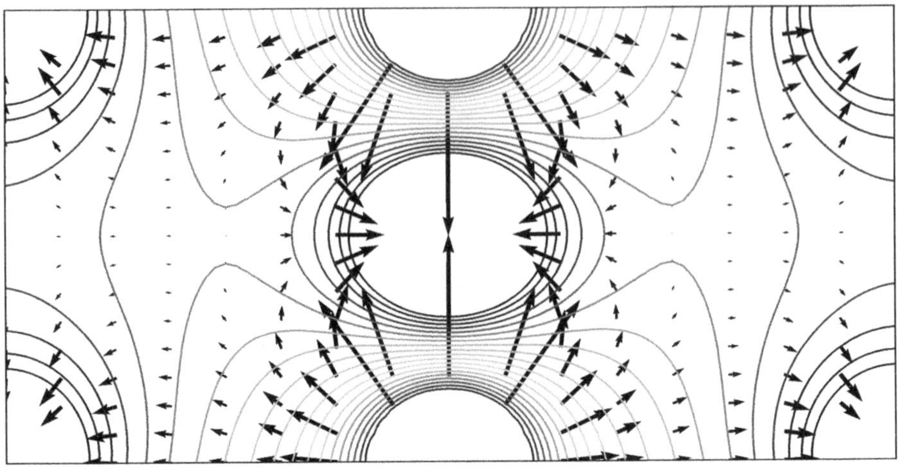

FIGURE 7.2 The potential and forces corresponding to the configuration in Figure 7.1.

7.2 ELECTRIC POTENTIALS AND FIELDS

In electrostatics, Gauss's law tells us that the electric flux through a closed surface is equal to the enclosed charge divided by a constant ε_0 (permittivity of free space). We can also express this law in a differential form using vector calculus. This differential form of Gauss's law is called the Poisson equation [43],

$$\frac{\partial^2 V}{\partial x^2} + \frac{\partial^2 V}{\partial y^2} + \frac{\partial^2 V}{\partial z^2} = -\frac{\rho}{\varepsilon_0}, \quad \text{(Poisson equation)} \tag{7.2}$$

where V is the electric potential, and ρ the charge density. The electric field can be obtained from V by

$$\vec{E} = -\nabla V. \tag{7.3}$$

If the region within the boundary is charge free, that is, $\rho = 0$ everywhere inside, Eq. (7.2) reduces to the Laplace equation. We are mainly interested in the two-dimensional Laplace equation,

$$\frac{\partial^2 V}{\partial x^2} + \frac{\partial^2 V}{\partial y^2} = 0. \quad \text{(Laplace equation in 2D)} \tag{7.4}$$

Solutions to Eq. (7.4) are uniquely determined by the boundary conditions.

Like the displacement and wave equations discussed in Chapter 6, there are two types of boundary conditions: the Neumann and Dirichlet types. Under the Neumann boundary condition, the derivative, or more accurately the gradient, of V is specified on the boundary. This is equivalent to specifying the electric field on the boundary. On the other hand, the Dirichlet boundary condition deals with the values of V itself on the boundary. There can also be a mixture of both types of boundary conditions. Next, we will discuss solutions of Eq. (7.4) assuming Dirichlet boundary conditions where the values of V are given on the boundary points unless noted otherwise.

7.2.1 SELF-CONSISTENT METHOD

The general idea of self-consistent methods is to solve a given problem iteratively, starting from an initial guess. In each iteration, we use the previous iteration to obtain an improved solution. The process continues until the solution becomes self-consistent or relaxed toward the "equilibrium" value (see Section S:6.1.2 for the catenary problem solved with this method).

We will follow this approach here to solve the Laplace equation by relaxation. We shall first discretize Eq. (7.4), which is the same as the left-hand side (LHS) of the wave equation (6.88). Accordingly, we can discretize the Laplacian over a 2D grid just like Eq. (6.89). Dropping the time variable from Eq. (6.90), the Laplace equation in finite difference method (FDM) is

$$\frac{V_{i+1,j} + V_{i-1,j} + V_{i,j+1} + V_{i,j-1} - 4V_{i,j}}{h^2} = 0, \tag{7.5}$$

or equivalently,

$$V_{i,j} = \frac{1}{4}\left[V_{i-1,j} + V_{i+1,j} + V_{i,j-1} + V_{i,j+1}\right]. \tag{7.6}$$

As before, h is the grid size (we assume the same h for x and y), and we have used the following convention

$$x_i = ih, \quad y_j = jh, \quad V(x_i, y_j) \equiv V_{i,j},$$
$$i = 1, ..., M, \quad j = 1, ..., N. \tag{7.7}$$

Equation (7.6) states that the correct, self-consistent solution at any point is just the average of its four nearest neighbors. It is independent of the grid size h (though the accuracy is). Its simplicity is strikingly similar to Eq. (6.85) for an exact wave on a string.

Accordingly, if we find a solution satisfying Eq. (7.6) and the boundary condition, we have obtained the correct solution. So here is how the self-consistent method works: We start with an initial guess, say zero everywhere, and set the boundary values; then we go through every grid point not on the boundary, assigning the average of its four neighbors as its new value. In the beginning, we expect large changes in the grid values. But as the iterations progress, they should become more self-consistent, and only small changes will occur. We repeat the process until the solution has converged.

Program 7.2 implements this strategy (Section 7.A) to calculate the potential of two finite parallel plates kept at opposite potentials and placed inside a grounded box. The details of the program are described following the listing, but the central function is `relax()`, which finds the self-consistent solution by relaxation.

```
1 def relax(V, imax=200):          # find self-consistent soln
2     for i in range(imax):
3         V[1:-1,1:-1] = ( V[1:-1,:-2] + V[1:-1,2:]      # left, right
4                       + V[:-2,1:-1] + V[2:,1:-1] )/4    # top, bottom
5         V = set_boundary(V)    # enforce boundary condition
6         draw_pot(V), vp.rate(1000)
7     return V
```

The potential V is stored on a 2D grid as $V[i,j]$ where, by convention of Eq. (7.7), indices i and j, respectively, refer to the x and y directions. The key statement is line 3. It uses the same slicing and element-wise operations as in 2D waves (see explanation in Section 6.7 and Eq. (6.91)) to compute the average of the four nearest neighbors, in remarkable resemblance to Eq. (7.6). Besides elegantly expressing the algorithm in one concise statement, we gain speed and efficiency because otherwise two nested loops would be required without implicit looping. Furthermore, the averaged values are assigned back to the same array without the need for a second array.

The statement is repeated many times, and the difference between iterations becomes progressively smaller. After each iteration, the boundary condition on the plates is applied, and the solution is visually displayed as colored points (Figure 7.3). Toward the end, very little change occurs, and we say the solution has become self-consistent, or it has converged. Finally, the potential and electric fields are plotted in 7.4. See Program 7.2 for the overall flow and explanation.

Figure 7.3 shows the progress of the solution as a function of iterations. At the start, the trial solution is set to zero everywhere except on the plates. After 20 iterations (top right), the potential has visibly spread to the immediate areas surrounding the plates. The outward spreading continues, although at a slower rate, as seen at iteration 60 (bottom left) and 200 (bottom right). After the last picture, it is hard to spot changes to the naked eye between iterations. But it is in fact surely getting closer and closer to the correct self-consistent solution, only too slowly (see following). This standard approach is known as the Jacobi scheme.

The potential contour lines plus electric field vectors and surface plots are shown in Figure 7.4. Constant contour lines of positive potential wrap around the top plate and those of negative potential around the bottom plate, as expected. In between the plates, we have approximately equidistant potential lines, consistent with the surface plot, which shows a roughly linear slope inside as well.

Consequently, the electric field vector inside the plates is nearly constant, pointing from top to bottom. Near the edges, there are large fringe fields. Outside the plates, the "leaked" electric field is considerably weaker. In contrast, in the ideal case of infinite parallel plates, the electric field would be constant inside the plates and zero outside. The large fringe fields and nonzero fields outside are due to the plates being finite. They should become increasingly less significant if the plate size increases.

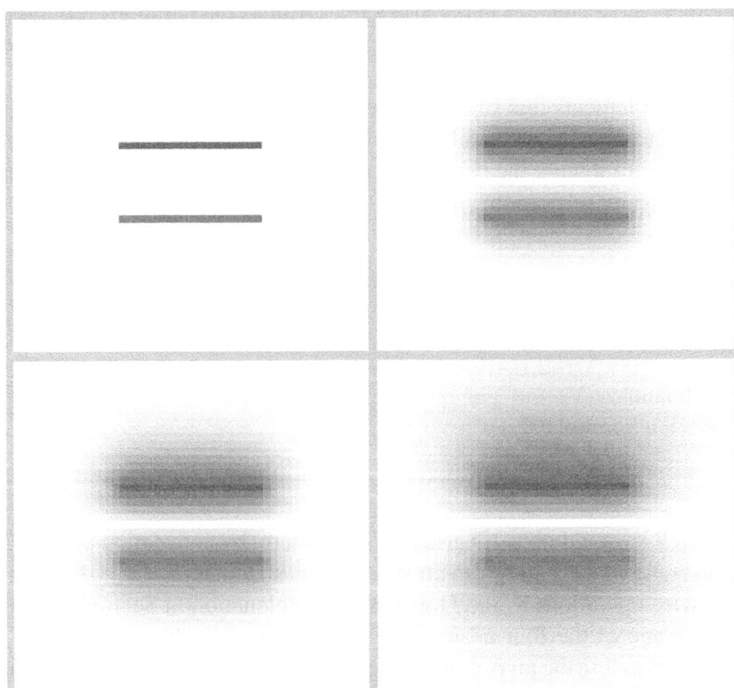

FIGURE 7.3 Relaxation of the self-consistent method.

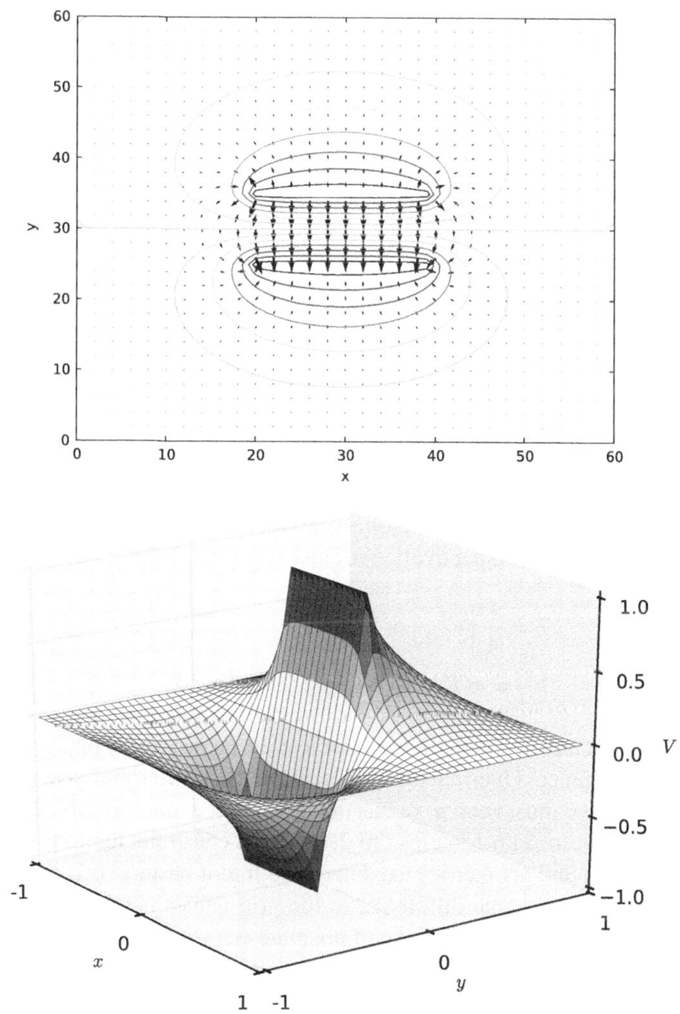

FIGURE 7.4 The potential and electric field of two finite parallel plates.

7.2.2 ACCELERATED RELAXATION

At first, we can see visual changes in the standard relaxation shown in Figure 7.3 because our initial guess, zero everywhere except on the two plates, is not close to the self-consistent (correct) solution. After a while, smaller changes occur with each iteration, and the rate of convergence is rather slow. For instance, for the grid size used above (60×60), the solution is within a fraction of a percent of the correct solution in about the first 50 iterations. Thereafter, accuracy improves at a low power law (Figure 7.5). To gain one more significant digit, approximately 25 times more iterations are required. For larger grid sizes, the standard relaxation scheme is time consuming and becomes impractical for calculations requiring higher accuracy. The root cause of the problem is that change at a particular grid point propagates slowly to neighbors far away, like in a diffusion process.

Another scheme, called the Gauss-Seidel (GS) method, makes the system relax faster. Instead of using only old values before the current iteration, the GS scheme uses the current values of the nearest neighbors as they become available in the current iteration. For example, if we start from the first row and go across the column in each row, the modified loop in `relax()` in the GS scheme for one iteration would look like

```
for i in range(1, M-1):
    for j in range(1, N-1):
        if (not_on_plates( i,j )):
            V1[i,j]=(V1[i-1,j] + V0[i+1,j] + V1[i,j-1] + V0[i,j+1])/4
```

Here, `V0` is the old solution of the last iteration, and `V1` the new solution. The function `not_on_plates()` checks that the point (i, j) is not on the plates (boundary) before updating. Because `V1[i-1,j]` and `V1[i,j-1]` (previous row and column of the current iteration) are available by the time the grid point (i, j) is being calculated, we make use of them. It mixes old and new values and should converge faster. Indeed, it is faster than the Jacobi scheme by about a factor of two. However, its advantage does not stop there.

It turns out that the GS scheme is suitable for accelerated relaxation (successive overrelaxation) through the introduction of an acceleration parameter, α. Let V^0 and V^1 be the values of the last (old) and current iterations, respectively. We can express the accelerated relaxation method as

$$T = \frac{1}{4}\left[V^1_{i-1,j} + V^0_{i+1,j} + V^1_{i,j-1} + V^0_{i,j+1}\right],$$
$$V^1_{i,j} = \alpha(T - V^0_{i,j}) + V^0_{i,j}, \quad 0 < \alpha < 2. \tag{7.8}$$

If $\alpha = 1$, we recover the Gauss-Seidel scheme. For $\alpha < 1$, the process is underrelaxed and usually results in slower convergence. Overrelaxation is achieved for $\alpha > 1$ and generally speeds up the convergence. However, we must keep $\alpha < 2$ as the method becomes unstable beyond 2. Even so, we need to be careful not to set α too close to 2, as it can cause fluctuations (noise) that do not vanish. For the grid sizes and geometries used here, we find it best to keep $\alpha \sim 1.85$.

Figure 7.5 shows the average difference of the grid values between successive iterations, $\sum_{i,j} |V^1_{i,j} - V^0_{i,j}|$, which we take as a measure of absolute error in the solution. The error analysis was performed on the parallel plates system. The initial guess was the same as in the standard (Jacobi) and the accelerated GS methods.

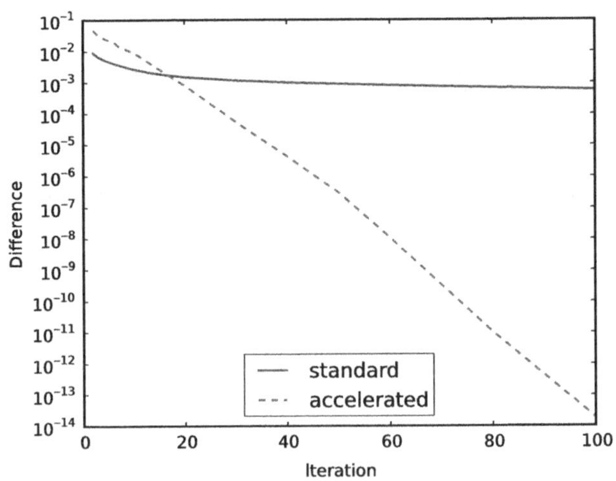

FIGURE 7.5 The error of the standard and accelerated Gauss-Seidel methods.

At the beginning, the Jacobi method produced smaller differences than the accelerated GS method, presumably due to the mixing of old and new values in the latter. The slope of the error, however, is much steeper in the accelerated GS method. On the semilog scale, it is approximately a straight line, indicating a fast, exponential rate of convergence. Fully self-consistent solutions within machine accuracy are obtained in about 100 iterations. The standard relaxation method converges very slowly as a power scaling. The advantage of accelerated GS method producing accurate solutions efficiently is evident.[1]

It is curious why the accelerated method works so efficiently. Physically, we imagine a situation analogous to the falling tablecloth (Section 6.8). The standard method would correspond to uninterferred relaxation into equilibrium, and the accelerated method to forced relaxation. The role of the parameter α can be thought of as an applied force, in addition to gravity, that pulls the tablecloth into equilibrium faster than gravity alone.

7.3 LAPLACE EQUATION AND FINITE ELEMENT METHOD

The self-consistent method solves the Laplace equation iteratively by successive relaxation. We can also solve it noniteratively with other methods. One such method is the finite element method (FEM) introduced in Chapter 6 for 1D problems. In this section we will extend the FEM to 2D to treat the Laplace equation. The FEM offers several advantages including easy adaptation to irregularly shaped domains and accommodating different types of boundary conditions. More broadly, the FEM is widely used in science and engineering problems.

However, it takes some preparation to develop and understand the FEM framework. Hence, it is helpful to be familiar with 1D FEM first (Section 6.4), as our discussions follow a similar approach here in 2D.

The FEM development is organized in five steps:

- Choose the elements and basis functions.
- Set up meshes and tent functions for interpolation.
- Reduce the Laplace equation to integral form.
- Impose the boundary conditions.
- Build the linear system of equations (system matrix).

To see FEM in action first, you can skip ahead to Section 7.4 for a worked example and sample Program 7.3. You may also switch back and forth between FEM development and application.

7.3.1 FINITE ELEMENTS AND BASIS FUNCTIONS IN 2D

We wish to obtain the solution in a two-dimensional domain, a region surrounded by an arbitrary boundary. As in 1D, we expand the solution u in a basis set as

$$u(x, y) = \sum_i u_i \phi_i(x, y). \tag{7.9}$$

Analogous to Eq. (6.37), u_i is the expansion coefficient and ϕ_i the tent function to be formed with basis functions. Note that in 2D, we make a distinction between tent functions and basis functions for clarity.

---

[1] Accurate to within the discretization scheme, of course, which in the present case is of second-order accuracy (6.89).

In 1D, the basic elements were line segments, and the basis functions were linear functions between neighboring elements and zero elsewhere (Figure 6.12). In 2D, we can form a surface with a minimum of three points, so the simplest element is a triangle. We will divide the domain into triangular meshes, or elements. The vertices of the triangles are the nodes (we will use the terms *nodes* and *vertices* interchangeably in this section). Each node will belong to at least one element, but can belong to many more.

What basis function do we want to use over each element? Like the 1D case, the simplest function is a linear function in both x and y, which is an equation describing a plane. Because there are three nodes for each element, there will be three basis functions covering an element, one for each node.[2]

Sample basis functions are shown in Figure 7.6. Each basis function is a triangular plane defined to be unity at its own node and zero at the other two. The basis function is nonzero only over the element containing the node. It is defined to be zero outside the element (compact support).

Let the vertices of the triangle be (x_i, y_i), $i = 1$, 2, and 3. The basis function for node 1 can be written as [75]

$$\varphi_1(x, y) = \frac{x_2y_3 - x_3y_2 + x(y_2 - y_3) + y(x_3 - x_2)}{2A}, \tag{7.10}$$
$$2A = x_1(y_2 - y_3) + x_2(y_3 - y_1) + x_3(y_1 - y_2).$$

The constant $2A$ is the magnitude of the cross product of the two vectors from vertex 1 to vertices 2 and 3, respectively, and is equal to twice the area of the triangle. We can verify

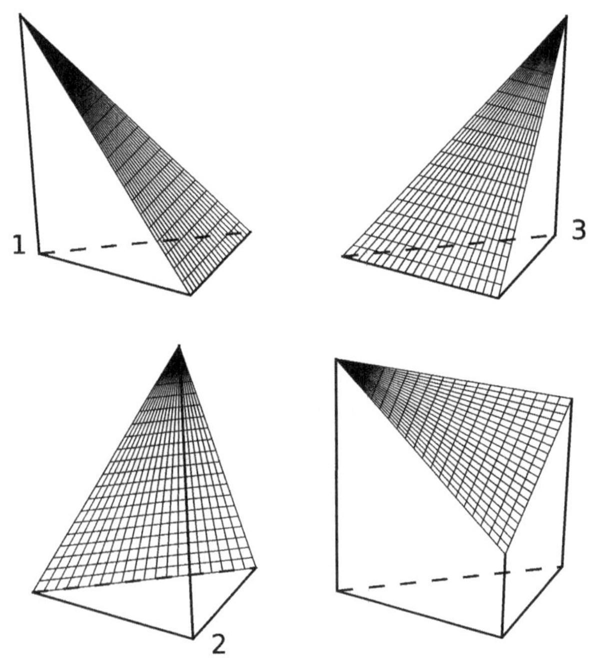

FIGURE 7.6 An element and its basis functions.

[2] We assumed one basis function in 1D for each node (tent function, Figure 6.12). Technically, we could split the tent into two, so we would also have two basis functions for each element.

that $\varphi_1(x_1, y_1) = 1$ and $\varphi_1(x_2, y_2) = \varphi_1(x_3, y_3) = 0$ at the nodes, and $\varphi_1(x, y) = 0$ along the base opposite node 1 (Figure 7.6, top left). The basis function $\varphi_1(x, y)$ is the shaded triangular plane, which can be viewed as being formed by pulling one vertex of an elastic triangle up vertically, while holding the base down. Similarly, two other surfaces, $\varphi_2(x, y)$ and $\varphi_3(x, y)$, can be formed about nodes 2 and 3, respectively.

Let us denote the element by e. We label the nodes on element e as $[1, 2, 3]$ in counter-clockwise order. Also, let $[i, j, k]$ represent a cyclic permutation of $[1, 2, 3]$, including $[2, 3, 1]$ and $[3, 1, 2]$. With these notations, we can generalize Eq. (7.10) to be

$$\alpha_i = x_j y_k - x_k y_j,\ \beta_i = y_j - y_k,\ \gamma_i = -(x_j - x_k),$$
$$\varphi_i^e(x, y) = \frac{\alpha_i + \beta_i x + \gamma_i y}{2A_e}, \tag{7.11}$$

where A_e is the area of the element, the same as A in Eq. (7.10).

Suppose we want to represent an arbitrary flat surface over element e. We can do so with a linear combination of the basis functions

$$f(x, y) = c_1\varphi_1^e(x, y) + c_2\varphi_2^e(x, y) + c_3\varphi_3^e(x, y).$$

This is illustrated in Figure 7.6 (bottom right). By varying the coefficients c_i, the patch can be stretched and oriented in any direction. Because our solution $u(x, y)$ will ultimately be represented by the basis functions (through tent functions to be discussed next), $u(x, y)$ will be composed of flat, triangular patches over the domain.

7.3.2 MESHES, TENT FUNCTIONS AND INTERPOLATION

We illustrate with an example how the meshes, basis functions, and the tent functions fit together. Figure 7.7 shows a ¬-shaped domain, a rectangle with a missing corner. We divide it into four elements (meshes), numbered from e_1 to e_4. There are six nodes in total, all on the boundary. Node 1 is a member of all four elements, nodes 2 and 6 of a single element, and nodes 3 to 5 of two elements.

We associate a tent function with each node for interpolation as given by Eq. (7.9). A tent function is just a collection of basis functions from all elements containing the node. Two tent functions are drawn in Figure 7.8.

The first tent function at node 1 consists of four basis functions because node 1 belongs to all four elements. This tent function would be closed except for the fact that the node is on the

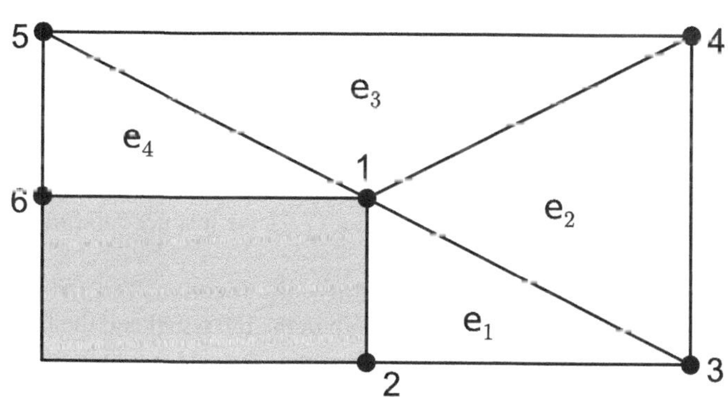

FIGURE 7.7 The triangular meshes in a domain (unshaded).

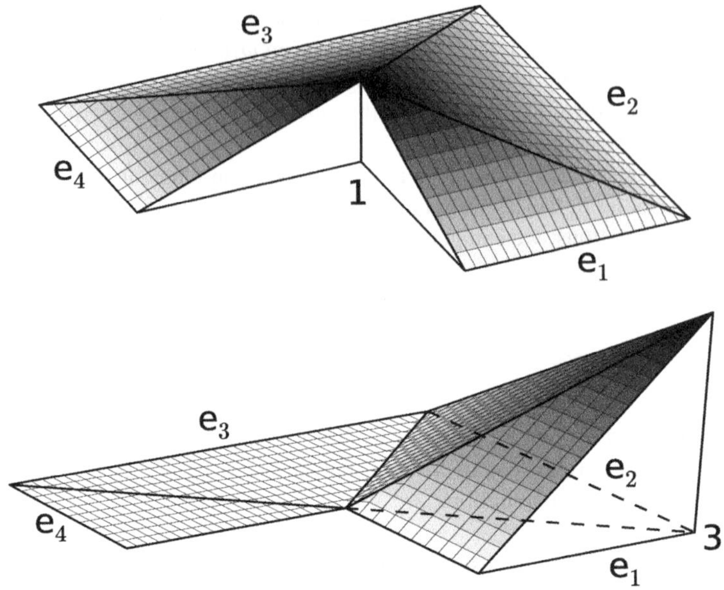

FIGURE 7.8 The tent functions at nodes 1 and 3.

boundary, and the lower-left corner is therefore open. We can express this tent function in terms of the basis functions as

$$\phi_1(x, y) = \sum_{e=e_1}^{e_4} \varphi_1^e(x, y). \tag{7.12}$$

The tent function is continuous, though not smooth, across the element boundaries (the ridges). Because the basis functions are zero outside their own element boundaries, the value of ϕ_1 at a given point (x, y) can be found on one, and one only, basis function of the element containing that point.

This is true for all points except those on the ridges and the exact node point for which the tent function is defined. To avoid double counting, we require that only one term in the sum (7.12) is actually selected: the basis function of the first element determined to contain the point. The value of the tent function at the node is unity by definition.

Likewise, tent functions at other nodes can be built in the same way. For example, the tent function at node 3 (Figure 7.8, bottom) has two basis functions, one each from elements e_1 and e_2. There are six such tent functions, one for each node. We note that two tent functions overlap each other only if the two nodes are directly connected by an edge. For instance, the two tent function shown in Figure 7.8 overlap because nodes 1 and 3 are connected by the common edge between e_1 and e_2. In contrast, the tent functions at node 3 and node 5 would not overlap.

Suppose we are given the values of a function at the nodes u_i, and we wish to interpolate the function over the whole domain. Then from Eq. (7.9), we just need to run the sum over all the tent functions thus defined.

To see it in action, we show the interpolation of the function $f(x, y) = \sin(\pi x)\sin(\pi y)$ over the unit square in Figure 7.9. The meshes are shown in the left panel and the function (wireframe) and its interpolations (solid pyramids) in the right panel.

Two different sets of meshes are used. In the first instance, the domain is divided into eight elements, with one internal node and eight boundary nodes. The values of $u_i = f(x_i, y_i)$ are zero

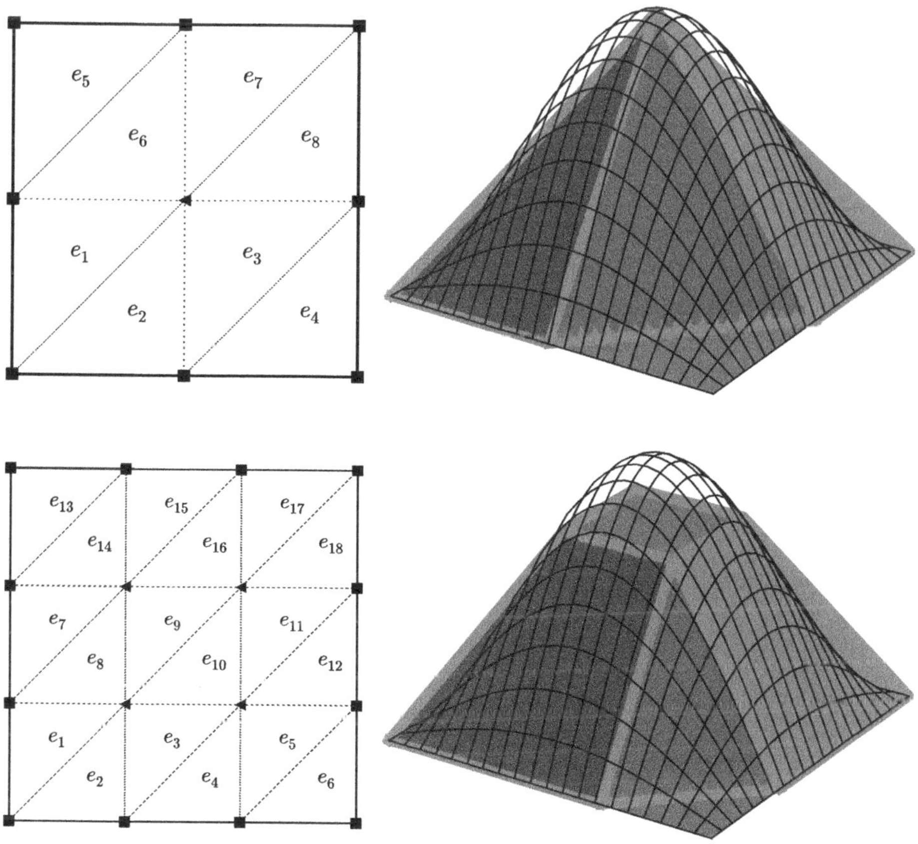

FIGURE 7.9 Interpolation with triangular elements.

at the boundary nodes and 1 at the lone internal node (the center). Of the nine tent functions, only the one at the center contributes. The interpolation is therefore given by this tent function, which is a six-sided pyramid. The front and back facets have been sliced off because all node values of elements e_4 and e_5 are zero. The interpolation is below the actual function over most of the domain, except at the other two edges along the diagonal (between e_1/e_2 and e_7/e_8).

The second set (Figure 7.9, second row) increases the number of meshes to 18. Now there are four internal nodes with the same value $\frac{3}{4}$. Summing up the four tent functions gives us a flat-top pyramid. Again the front and back edges corresponding to elements e_6 and e_{13} are sliced off, but by a smaller amount than before. There are no improvements over the other two edges. However, overall fill volume has improved, mainly due to the facets becoming closer to the function.

We mention two observations. First, if we increase the number of meshes to make them finer, the interpolation would become better. However, one thing remains: The interpolation is made up of flat, finite patches that are continuous across element boundaries, but their derivatives are discontinuous.[3] Second, the way we have divided the domain up along one diagonal of the unit square breaks some symmetries of the original function. This impacts the quality of the

[3]This is the reason we must reduce second-order PDEs to an integral form involving only first derivatives.

interpolation and of the solution when symmetries are important in a problem. Mesh generation should take these factors into account.

7.3.3 REDUCTION TO A LINEAR SYSTEM OF EQUATIONS

Because the gradient of the FEM interpolation is discontinuous across element boundaries, we cannot work directly with second-order equations like the Laplace equation in differential form. Rather, like in 1D, we convert the problem into one involving only first derivatives.

Let us consider the Poisson equation in 2D (change V to u in Eq. (7.2)),

$$\nabla^2 u = \frac{\partial^2 u}{\partial x^2} + \frac{\partial^2 u}{\partial y^2} = f(x,y), \quad f(x,y) = -\frac{\sigma}{\varepsilon_0}, \tag{7.13}$$

where σ is the surface charge density. Similar to Eq. (6.42), we multiply both sides of Eq. (7.13) by $\phi_j(x,y)$ and integrate over the domain S as

$$\int_S \phi_j(x,y)\nabla^2 u(x,y)\, dxdy = \int_S f(x,y)\phi_j(x,y)\, dxdy. \tag{7.14}$$

Using the identity $v\nabla^2 u = \nabla \cdot (v\nabla u) - \nabla v \cdot \nabla u$, we can reduce Eq. (7.14) to

$$\int_S (\nabla \cdot (\phi_j \nabla u) - \nabla u \cdot \nabla \phi_j)\, dxdy = \int_S f\phi_j\, dxdy. \tag{7.15}$$

The first term can be simplified with Green's theorem, which converts a surface integral of divergence over S into a line integral along the boundary C

$$\int_S \nabla \cdot \vec{F}\, dxdy = \oint_C \vec{F} \cdot \hat{n}\, dl. \tag{7.16}$$

The contour of integration C encloses the domain S in the counterclockwise direction, and the normal vector $\hat{n}$ points outward and perpendicular to the tangent of the contour [10].

With the help of Eq. (7.16), we can rewrite Eq. (7.15) as

$$q_j = \oint_C \phi_j \nabla u \cdot \hat{n}\, dl, \quad p_j = \int_S f\phi_j\, dxdy,$$
$$\int_S \nabla u \cdot \nabla \phi_j\, dxdy = q_j - p_j. \tag{7.17}$$

Note that for the Laplace equation, $f(x,y) = 0$, so $p_j = 0$, but we keep it for generality. We substitute the interpolation (7.9) into (7.17) to arrive at

$$\sum_i u_i \int_S \nabla \phi_i \cdot \nabla \phi_j\, dxdy = q_j - p_j. \tag{7.18}$$

Let

$$A_{ij} = \int_S \nabla \phi_i \cdot \nabla \phi_j\, dxdy, \tag{7.19}$$

we can express Eq. (7.18) for each node j as

$$\sum_i A_{ij} u_i = q_j - p_j, \quad j = \text{all nodes}. \tag{7.20}$$

This is a linear system of equations, to be solved subject to boundary conditions.

7.3.4 IMPOSING BOUNDARY CONDITIONS

We can specify two types of boundary conditions in Poisson (or Laplace) problems: the values of the electric field, ∇u (Neumann type), or the values of the potential, u (Dirichlet type). Using two parallel plates as an example, the Neumann type would correspond to specifying the electric field on the plates (analogous to setting up surface charge densities), and the Dirichlet type to keeping the plates at certain voltages. We are mostly interested in the latter here and restrict our following discussion to Dirichlet boundary values.

Suppose node k is on the boundary with a given value u_k. Because this node no longer needs to be solved for, we remove the kth equation from Eq. (7.20) and move the term $A_{jk}u_k (j \neq k)$ in other equations to the right-hand side (RHS) so the new system is

$$\sum_{i \neq k} A_{ij}u_i = q_j - p_j - A_{jk}u_k, \quad j \neq k. \tag{7.21}$$

We have one fewer equation in (7.21) than (7.20).

We can eliminate all other boundary nodes the same way, crossing off the equations corresponding to these nodes and moving them to the RHS in the rest, so we end up with

$$\sum_{\text{intrnl}} A_{ij}u_i = q_j - p_j - \sum_{\text{bndry}} A_{jk}u_k, \quad j = \text{internal nodes only}. \tag{7.22}$$

The sums on the LHS and the RHS involve internal and boundary nodes only, respectively.

There is one more simplification we can make. Because the index j in Eq. (7.22) runs over internal nodes only, and the tent function centered on an internal node is zero on the boundary (see Figure 7.8 and discussion following it), q_j as given in Eq. (7.17) is zero because the line integral is on the boundary where $\phi_j = 0$. Therefore, we can drop q_j from Eq. (7.22). This is an expected result because we are concerned with the Dirichlet boundary condition here, and q_j involves the Neumann type (∇u on the boundary).

Expressing Eq. (7.22) in matrix form, we have

$$\begin{bmatrix} A_{11} & A_{12} & \cdots \\ A_{21} & A_{22} & \cdots \\ \vdots & \vdots & \ddots \end{bmatrix} \begin{bmatrix} u_1 \\ u_2 \\ \vdots \end{bmatrix} = \begin{bmatrix} -p_1 - b_1 \\ -p_2 - b_2 \\ \vdots \end{bmatrix}, \quad b_j = \sum_{\text{bndry}} A_{jk}u_k. \tag{7.23}$$

It is understood that indices of the matrix entries refer to the internal nodes. This is the final FEM result. Solving Eq. (7.23) will give us the FEM solutions uniquely determined by the boundary condition. For the Laplace equation, we can also drop p_j's from Eq. (7.23), leaving only b_j's.

7.3.5 BUILDING THE SYSTEM MATRIX

The matrix entries A_{ij} in Eq. (7.19) involve the gradient of the tent functions ϕ_i and ϕ_j which are composed of basis functions. The evaluation of A_{ij} will necessarily operate on the basis functions of the elements containing the nodes i and j.

To make this explicit, let us denote the list of elements connected to node i by $e(i)$. For example, referring to Figure 7.7, four elements contain node 1, so the list for node 1 is $e(1) = \{e_1, e_2, e_3, e_4\}$; and for node 2, $e(2) = \{e_1\}$, and so on.

Similar to Eq. (7.12), the tent functions ϕ_i and ϕ_j can be written as

$$\phi_i = \sum_{e(i)} \varphi_i^{e(i)}, \quad \phi_j = \sum_{e(j)} \varphi_j^{e(j)}, \tag{7.24}$$

where the basis functions are given in Eq. (7.11). Substitution of ϕ_i and ϕ_j into Eq. (7.19) yields

$$A_{ij} = \sum_{e(i),e(j)} \int_S \nabla\varphi_i^{e(i)} \cdot \nabla\varphi_j^{e(j)}\, dxdy. \tag{7.25}$$

This result tells us that only basis functions in the elements containing nodes i and j contribute to A_{ij}. Furthermore, because a basis function is zero outside its own element, there will be no contribution to A_{ij} unless $e(i)$ and $e(j)$ refer to the same element. In other words, A_{ij} will be zero unless nodes i and j are directly connected.

It seems that a straightforward way to calculate A_{ij} is to proceed as follows. For each node, we build a list containing the elements the node belongs to. When evaluating the sum in Eq. (7.25), we just go through the lists associated with the pair of nodes. But this node-oriented approach would be cumbersome and inelegant. For instance, additional bookkeeping would be necessary to decide whether a pair of nodes are directly connected, as well as what the common edges are.

A more natural way is the element-oriented approach. We specify each element by listing the three nodes at its vertices (in counterclockwise order). Then, we iterate through the elements, computing the contributions from each element ($e = [i, j, k]$) to the six entries of matrix A: three diagonals $A^e_{ii}, A^e_{jj}, A^e_{kk}$ and three off-diagonals $A^e_{ij}, A^e_{ik}, A^e_{jk}$ plus the symmetric pairs A^e_{ji}, and so on. After iterating over all the elements, all contributions will have been accounted for, and the matrix A will be completely built.

Let us illustrate the element-oriented approach using the mesh shown in Figure 7.7 and reproduced in Figure 7.10. The six nodes are numbered 1 to 6 and have coordinates (x_i, y_i), $i = 1$ to 6. The four finite elements are specified as $e_1 = [1, 2, 3]$, $e_2 = [1, 3, 4]$, and so on. Let

$$A^e_{ij} = \int_{S_e} \nabla\varphi^e_i \cdot \nabla\varphi^e_j\, dxdy. \tag{7.26}$$

This is the contribution to A_{ij} from element e. It is symmetric about the diagonal, $A^e_{ij} = A^e_{ji}$. Note that we have changed the integration domain S to S_e, the domain over element e.

Figure 7.10 shows the contributions to off-diagonal entries of matrix A. We start from e_1 with nodes $[1, 2, 3]$. It contributes to matrix A along the three edges as $A^{e_1}_{12}$, $A^{e_1}_{13}$, and $A^{e_1}_{23}$. There are also symmetric pairs ($A^{e_1}_{21} = A^{e_1}_{12}$, etc.) and diagonal entries (e.g., $A^{e_1}_{11}$, not shown in Figure 7.10).

After e_1 has been processed, entries A_{12} and A_{23} and their symmetric pairs are complete. No contributions will come from other elements (besides e_1) because node 2 is connected only

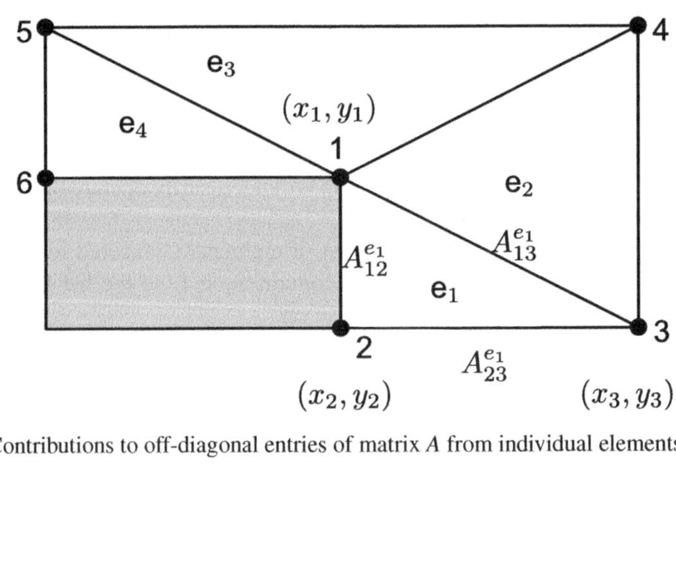

FIGURE 7.10 Contributions to off-diagonal entries of matrix A from individual elements.

to nodes 1 and 3. By the same token, after processing e_2, all entries associated with e_1 (plus A_{34} with e_2) are complete with the exception of A_{11}, which awaits further contribution from elements e_3 and e_4. For example, the entry A_{13} will be complete to become $A_{13} = A_{13}^{e1} + A_{13}^{e2}$. We see that after going through every element, all entries of matrix A will be complete.

We can accomplish the matrix-building process with the following pseudo code:

```
for e in [e1, e2, e3, e4]:
    for m = 0,1,2:
        for n = m to 2:
            i = e[m], j = e[n]
            A_ij = A_ij + A^e_ij from (7.27)
            if i ≠ j:
                A_ij = A_ji
```

The two inner loops cycle through the nodes of a given element, assigning the contribution to the appropriate entry of A, and taking advantage of symmetry about the diagonal.

It is now just a matter of evaluating Eq. (7.26) to complete the construction of matrix A. The gradient may be obtained from Eq. (7.11) as $\nabla\varphi_i^e = (\beta_i\hat{x} + \gamma_i\hat{y})/2A_e$, which is a constant vector over element e. Therefore, Eq. (7.26) becomes

$$A_{ij}^e = \int_{S_e} \frac{\beta_i\hat{x} + \gamma_i\hat{y}}{2A_e} \cdot \frac{\beta_j\hat{x} + \gamma_j\hat{y}}{2A_e}\, dxdy$$

$$= \frac{\beta_i\beta_j + \gamma_i\gamma_j}{4A_e^2} \underbrace{\int_{S_e} dxdy}_{A_e} = \frac{\beta_i\beta_j + \gamma_i\gamma_j}{4A_e}. \qquad (7.27)$$

Once matrix A is prepared, the linear system (7.23) is readily solved.

7.3.6 MESH GENERATION

Mesh generation is a major effort in FEM. The quality of the mesh can affect the quality of solutions and computational efficiency. There are special Python packages dedicated to just mesh generation. We will discuss the basic idea and the data structure in the generation of triangular meshes.

Let us consider a rectangular domain shown in Figure 7.11, over which the Laplace equation is to be solved. We divide the domain into N intervals along the x-axis and M intervals along the y-axis. The grid will have $N \times M$ subrectangles and $(N + 1) \times (M + 1)$ grid points, that is, nodes. Each subrectangle is subdivided into two triangles along the diagonal. Each triangle is a finite element, so there are a total of $2N \times M$ elements. Figure 7.11 shows a mesh grid where the unit square is divided with $N = M = 3$. There are 16 nodes (consecutively numbered), including 12 boundary nodes, 4 internal nodes, and 18 elements.

We store the nodes as a master list of coordinates ($n \times 2$ array). The elements are also stored in a list, each entry having three numbers: the indices of the nodes (vertices) in counterclockwise order. The boundary and internal nodes are indexed to the master list as well. Table 7.1 illustrates these variables and sample values.

A node may belong to several elements, for example, node 5 is a member of six elements, $e_{0-2,7-9}$. The coordinates of an element's vertices may be determined by finding the indices associated with the nodes. For example, the second vertex of the first element e_0 is

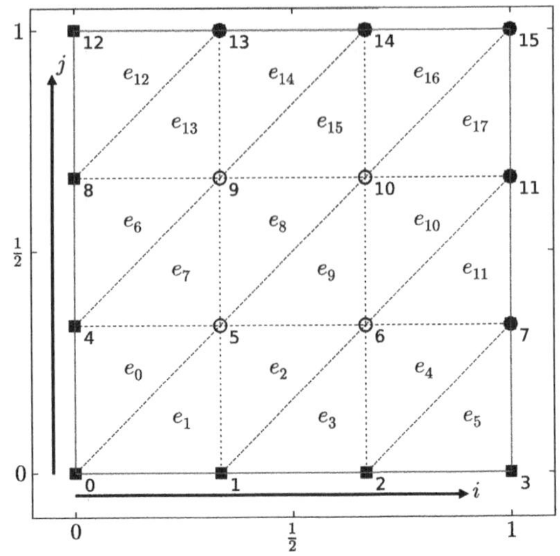

FIGURE 7.11 The mesh over a rectangular domain. Solid symbols are boundary nodes, and open symbols are internal nodes.

Table 7.1 FEM Variables and Data Structure of the Mesh in Figure 7.11.

Name	Variable	Example
nodes	node	$[[0,0], \left[\frac{1}{3},0\right], \left[\frac{2}{3},0\right],\ldots,\left[\frac{1}{3},1\right], \left[\frac{2}{3},1\right], [1,1]]$
elements	elm	[[0, 5, 4], [0, 1, 5],...,[10, 15, 14], [10, 11, 15]]
boundary points	bp	[0, 1, 2, 3, 4, 7, 8, 11, 12, 13, 14, 15]
internal points	ip	[5, 6, 9, 10]

node `elm[0][1]=5`, and the coordinates are `node[elm[0][1]]`, or $\left[\frac{1}{3}, \frac{1}{3}\right]$ from Table 7.1 and Figure 7.11. See Figure 9.24 for another sample mesh.

With the sample mesh of Figure 7.11, the dimension of the full matrix is 16×16. After removing the boundary nodes, the actual linear system (7.23) to be solved would be 4×4. As we will see next, the matrix is banded (Figure 7.12), and most entries are zero (see Project P7.5).

7.4 BOUNDARY VALUE PROBLEMS WITH FEM

7.4.1 A WORKED EXAMPLE

We present a worked example of boundary value problems with FEM as applied to the Laplace equation over a rectangular domain. We assume a unit square such as the one shown in Figure 7.11. The boundary conditions are that the potential is zero on the left and bottom sides, and a quadratic curve on the right and top sides. They are (we make a notation change from V to u to be consistent with the FEM development)

$$u(0, y) = u(x, 0) = 0, \quad u(1, y) = y^2, \quad u(x, 1) = x^2. \tag{7.28}$$

The FEM program designed to solve the Laplace problem with these boundary conditions is given in full in Section 7.A (Program 7.3). The program has two key parts: mesh generation and system-matrix construction.

Mesh generation is discussed in detail following the program. The construction of matrix A is done as follows.

```
def fem_mat(node, elm):      # fills matrix A_ij = ⟨∇φ_i · ∇φ_j⟩
    A = np.zeros((len(node),len(node)))
    for e in elm:
        (x1,y1), (x2,y2), (x3,y3) = node[e[0]], node[e[1]], node[e[2]]
        beta, gama = [y2-y3, y3-y1, y1-y2], [x3-x2, x1-x3, x2-x1]
        ar = 2*(x1*(y2-y3) + x2*(y3-y1) + x3*(y1-y2))
        for i in range(3):
            for j in range(i,3):
                A[e[i],e[j]] += (beta[i]*beta[j] + gama[i]*gama[j])/ar
                if (i != j): A[e[j],e[i]] = A[e[i],e[j]]    # symmetry
    return A
```

The routine goes through the element list `elm`. For each element, the coordinates of the vertices are extracted from the `node` list, and the parameters β_{0-2}, γ_{0-2} and A_e are calculated according to Eq. (7.11). The contributions to matrix A, A^e_{mn}, where m and n are the nodes of the current element, are computed from Eq. (7.27). Three of the six off-diagonal entries are computed, the other three are obtained via symmetry.

We show in Figure 7.12 a sample system matrix with $N = 10$. The boundary points have been removed, leaving the matrix dimension at 81×81. The matrix is sparse, consisting of three bands. We expect a sparse matrix because A_{ij} is nonzero only if the nodes i and j share a common edge (directly connected). From Figure 7.11, we see a node has a maximum of seven connections (six neighboring nodes plus self). However, Figure 7.12 shows that a row (or column) has at most five nonzero entries, two less than the number of connections.

The reason turns out to be that our finite elements are all right triangles. As such, the entries A_{ij} on the hypotenuse are zero. Take node 5 in Figure 7.11 as an example, $A_{5,0}$ and $A_{5,10}$ are zero.

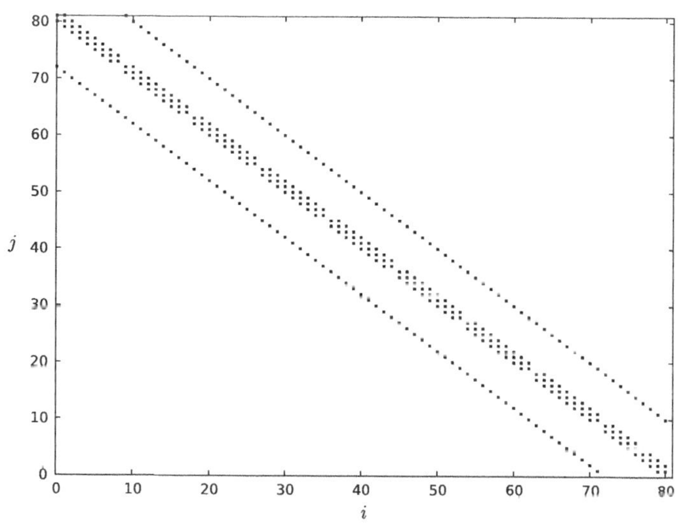

FIGURE 7.12 The system matrix for a mesh like Figure 7.11 with $N = 10$.

We can visualize that the basis function at node 5, $\varphi_5^{e_0}$, changes along the x direction only, whereas the basis function at node 0, $\varphi_0^{e_0}$, changes along the y direction only, such that their gradients are perpendicular to each other, resulting in zero contribution to $A_{5,0}$. The same reason applies to contributions to $A_{5,0}$ from e_1 and contributions to $A_{5,10}$ from e_8 and e_9.

The sparseness of the matrix means that the number of nonzero matrix entries scales linearly with N^2, rather than N^4. With special storage schemes, we could deal with a large number of finite elements before running out of memory (see Section 8.3.1 for band matrix representation).

We have the necessary ingredients to solve for the potential values at the internal nodes. The rest of Program 7.3 is described following its listing. We show the results in Figure 7.13. The potential slopes down monotonically from its highest point at (1, 1) and is symmetric about the ridge running down the diagonal line to the origin. The slope is greatest near the top, producing the greatest electric fields there (arrows on the contour plot).

The electric field near the top and right sides initially points away from the sides near the highest point. Further left and down, it becomes weaker, and its directions turn toward each side. By the midpoint, they are parallel to the respective sides, but eventually point into the sides past the midpoint. On the left and bottom sides, however, the electric fields are perpendicular to the boundaries because the potentials are constant on these sides. As the contour lines extend closer to the bottom-left corner, they start to mimic the shape of the walls, running nearly parallel to them and bending more sharply near the origin.

To approximate the solution near the corner more accurately, finer meshes would be necessary. How fine? The answer lies in the accuracy of FEM solutions considered next.

7.4.2 ACCURACY AND ERROR OF FEM SOLUTIONS

To analyze the accuracy and error of FEM solutions, we consider an example that admits compact analytic solutions. We assume the same unit-square domain as in the preceding worked example but with sinusoidal boundary conditions as

$$u(0, y) = u(x, 0) = 0, \quad u(1, y) = \sin(\pi y), \quad u(x, 1) = \sin(\pi x). \tag{7.29}$$

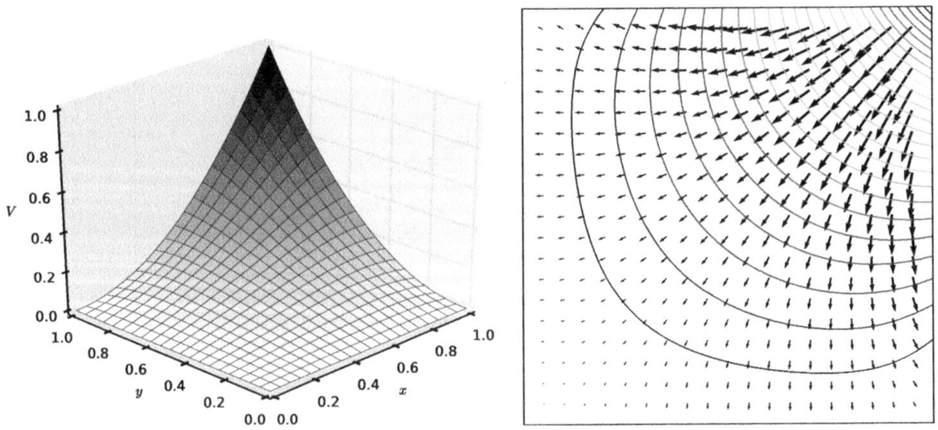

FIGURE 7.13 The potential and fields by FEM of a unit square with boundary conditions (7.28).

The analytic solution can be shown to be

$$u(x, y) = \frac{1}{\sinh \pi} \left[\sin(\pi x) \sinh(\pi y) + \sin(\pi y) \sinh(\pi x) \right]. \tag{7.30}$$

We only need to change the `mesh()` routine slightly to take into account the new boundary conditions to obtain the FEM solutions from Program 7.3. The results are displayed in Figure 7.14.

The boundary conditions dictate that the solutions be zero at all four corners. Because of this, there is a saddle point in the potential surface near $\left(\frac{3}{4}, \frac{3}{4}\right)$ where the potential shows considerable variation, in contrast to the monotonic solutions of Figure 7.13. The contour lines as well as electric fields (not shown in Figure 7.14) show symmetry about the diagonal line $y = x$. Away from the saddle point and toward the left and bottom sides, the potential distribution approaches the boundary values, a situation similar to that in Figure 7.13.

To assess the accuracy of FEM, we show in Figure 7.15 the absolute difference between the analytic and FEM solutions (Project P7.6). At each node, we compute the difference as

$$Z(x, y) = u_{exact}(x, y) - u_{FEM}(x, y),$$

where u_{exact} is the analytic solution (7.30) and u_{FEM} the FEM solution (Figure 7.14). The parameter $N = 40$ is used in the FEM calculation, corresponding to grid spacing of 0.025 and 3,200 finite elements. The error is displayed as a color-coded image using `plt.imshow()` (see Program 7.4).

The difference ranges between 0 and $\sim 3 \times 10^{-4}$ and is always positive. The FEM results are consistently below the analytic ones, even though the solution is not monotonic as discussed earlier. The smallest error occurs near the boundary as expected. The largest error is slightly off the saddle point where the potential values are largest away from the boundaries.

In Figure 7.16, we show the average error of the FEM solutions as a function of N. The error decreases with increasing N, but the slope after the second point ($N = 4$) is a straight line. Because the plot is on a log-log scale, it indicates that the error decreases as a power law.

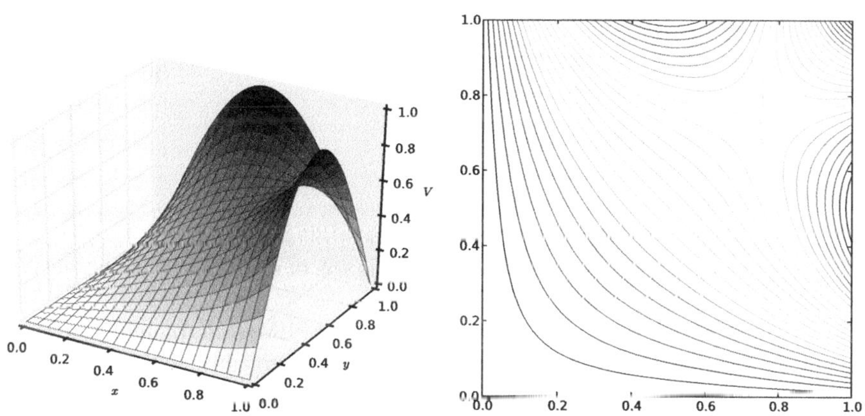

FIGURE 7.14 FEM solutions with sinusoidal boundary conditions (7.29).

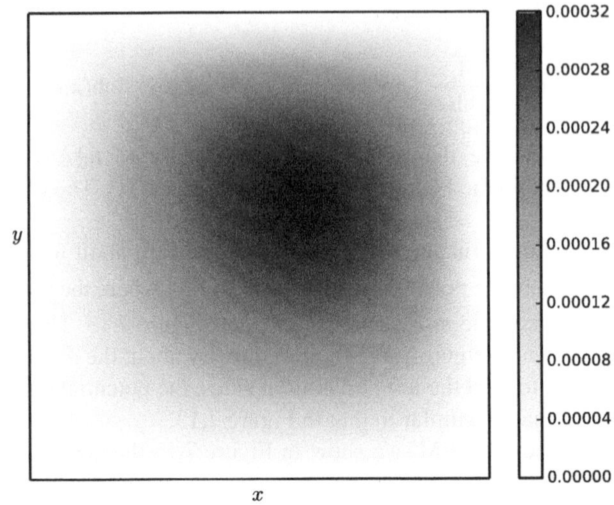

FIGURE 7.15 The difference between the analytic and FEM solutions for boundary conditions (7.29).

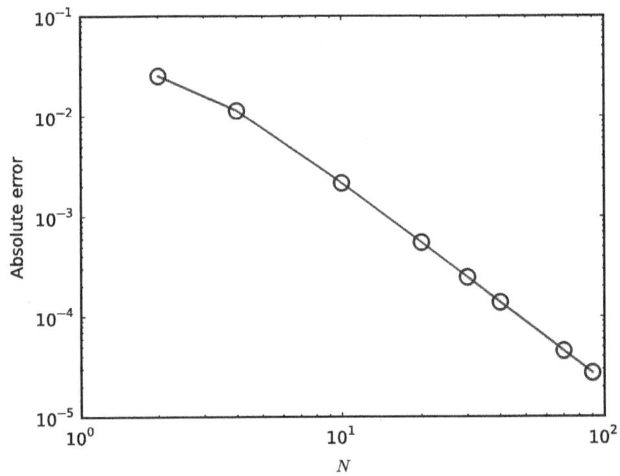

FIGURE 7.16 Accuracy of FEM solutions.

Careful examination of the trend shows that the slope is -2, or error $\sim N^{-2}$. Doubling N reduces the error by a factor of 4.

Because the grid spacing $h \sim 1/N$, we conclude that FEM with triangular meshes is second order in accuracy, the same as FDM. In hindsight, we probably could have guessed this because the linear interpolation in FEM is like the trapezoid rule, which is also second order (Section 8.A.1). Higher-order polynomial basis functions could also be used, but the resulting system matrix would be more complex [38].

7.5 MESHFREE METHODS FOR POTENTIALS AND FIELDS

The methods we have discussed so far all involve certain ways of discretizing space into grids or meshes, over which the differential operators are approximated. The accuracy of mesh-based methods can generally be improved by reducing the mesh size h.

There is another class of meshfree methods taking a different approach. Rather than discretizing space, the solutions are interpolated over a chosen set of basis functions. We discuss one meshfree method here, namely, the radial basis function method.[4] The advantage of this method is that it is fast, is straightforward to implement, and can easily handle irregular domains. It may also be readily scaled up to higher dimensions. The drawback is that its accuracy depends on the quality and size of the basis set, and it is trickier to systematically improve accuracy.

7.5.1 RADIAL BASIS FUNCTION INTERPOLATION

The basic idea of meshfree methods is scattered data interpolation [96]. We assume that, like in FEM, the solution u can be approximated by expansion in a distance-based radial basis function (RBF) set $\phi_i(\vec{r})$ as

$$u(\vec{r}) = \sum_i a_i \phi_i(\vec{r}). \tag{7.31}$$

where a_i are the expansion coefficients.

There are many possible functional forms for ϕ_i. We cite two commonly used RBFs as example, the Gaussian (GA) and the multiquadric (MQ),

$$\text{GA.}\quad \phi_i(\vec{r}) = e^{-\epsilon^2|\vec{r}-\vec{r}_i|^2}, \qquad \text{MQ:}\quad \phi_i(\vec{r}) = \sqrt{1+\epsilon^2|\vec{r}-\vec{r}_i|^2}, \tag{7.32}$$

where ϵ is the shape (control) parameter and $\vec{r}_i$ are the nodes (centers) scattered over the domain of interest.

We can use Eq. (7.31) for scattered data interpolation (Figure 7.17). For example, suppose we have four known data points, u_i at $\vec{r}_i$, $i = 1$ to 4. The scattered data points are not restricted to any particular geometry. The coefficients a_i can be determined by requiring

$$u(\vec{r}_i) = u_i, \tag{7.33}$$

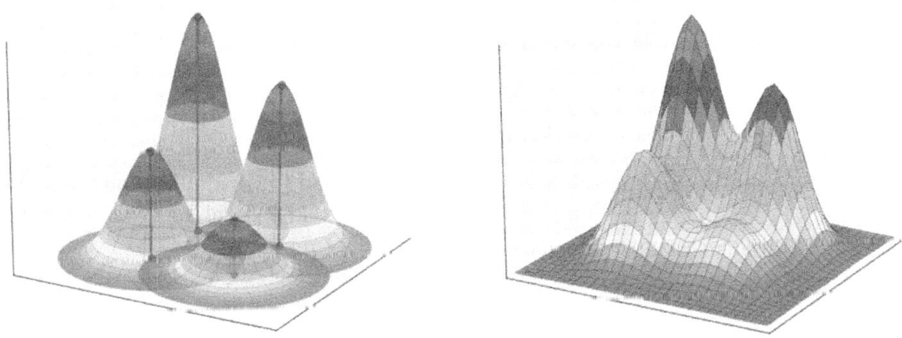

FIGURE 7.17 Scattered RBFs (left) and the resultant interpolation (right).

[4]Other meshfree methods are discussed in solutions of quantum systems in Chapters 8 and 9.

where we have taken the nodes to coincide with the scattered data points. The result can be expressed as a matrix equation

$$\begin{bmatrix} \phi_1(\vec{r}_1) & \phi_2(\vec{r}_1) & \phi_3(\vec{r}_1) & \phi_4(\vec{r}_1) \\ \phi_1(\vec{r}_2) & \phi_2(\vec{r}_2) & \phi_3(\vec{r}_2) & \phi_4(\vec{r}_2) \\ \phi_1(\vec{r}_3) & \phi_2(\vec{r}_3) & \phi_3(\vec{r}_3) & \phi_4(\vec{r}_3) \\ \phi_1(\vec{r}_4) & \phi_2(\vec{r}_4) & \phi_3(\vec{r}_4) & \phi_4(\vec{r}_4) \end{bmatrix} \begin{bmatrix} a_1 \\ a_2 \\ a_3 \\ a_4 \end{bmatrix} = \begin{bmatrix} u_1 \\ u_2 \\ u_3 \\ u_4 \end{bmatrix}. \tag{7.34}$$

Equation (7.34) can be solved for the RBF expansion coefficients a_{1-4}. The matrix entries are defined at the RBF collocation points as

$$\phi_i(\vec{r}_j) = \phi(|\vec{r}_i - \vec{r}_j|) = \phi_j(\vec{r}_i), \tag{7.35}$$

where ϕ is either a GA or MQ RBF in Eq. (7.32). Because the RBFs are based on the radial distance $r_{ij} = |\vec{r}_i - \vec{r}_j|$ from the nodes, the system matrix is positive and definite. Furthermore, its determinant is invariant under the exchange of any pair of nodes because the scalar distances r_{ij} remain the same. Therefore, the system matrix is well conditioned, and the solution is assured to be nonsingular (Exercise E7.6). In fact, the tent basis functions in the 1D FEM (Figure 6.12, Eq. (6.38)) are linear RBFs and share some of the same traits, except they are not infinitely differentiable like the GA or MQ. In general, however, if the basis functions were position (rather than distance) based, the aforementioned properties would no longer hold. In this respect, scattered data interpolation by RBFs has unique advantages.

Figure 7.17 illustrates an example consisting of four Gaussian RBFs centered at the scattered data points. After obtaining the expansion coefficients a_{1-4} (vertical lines at the nodes in Figure 7.17, left), the resultant interpolant (7.31) can be used to approximate the solution over the whole domain.

7.5.2 RBF COLLOCATION METHOD

The RBF collocation method is a way to solve PDEs. It is based on differentiating a RBF interpolant, from which the differentiation operator matrices can be built [27]. As an example, we use the method to solve the 2D Poisson equation with the Gaussian RBF (equally applicable to 3D).

We consider a region surrounded by an arbitrary boundary (the domain). Let N be the total number of scattered nodes. Of these, we assume the number of internal nodes is n, and the number of boundary nodes is $N - n$.

Substituting the interpolant (7.31) into the Poisson equation (7.13), we obtain

$$\nabla^2 u = \sum_i a_i \nabla^2 \phi_i(\vec{r}) = f(\vec{r}). \tag{7.36}$$

The collocation method, also known as Kansa's method [52], stipulates that the Poisson equation (7.36) be satisfied at the internal nodes $\vec{r}_j$, $j \in [1, n]$, and simultaneously, the boundary conditions be also satisfied at the boundary nodes $\vec{r}_j$, $j \in [n+1, N]$.

For simplicity, we assume Dirichlet boundary conditions. We have for the collocation method

$$\sum_i a_i \nabla^2 \phi_i(\vec{r})|_{\vec{r}=\vec{r}_j} = f_j, \quad j = 1, 2, \ldots, n, \tag{7.37a}$$

$$\sum_i a_i \phi_i(\vec{r}_j) = b_j, \quad j = n+1, n+2, \ldots, N, \tag{7.37b}$$

where $f_j = f(\vec{r}_j)$, and b_j are the potential values on the boundary. Casting Eqs. (7.37a) and (7.37b) into matrix form, we have

$$A\mathbf{a} = \mathbf{b}, \tag{7.38}$$

where $\mathbf{a} = [a_1, a_2, ..., a_N]^T$ is a column vector to be determined, and

$$A = \begin{bmatrix} D_{11} & D_{12} & \cdots & D_{1N} \\ \vdots & \vdots & \vdots & \vdots \\ D_{n1} & D_{n2} & \cdots & D_{nN} \\ \phi_{n+1,1} & \phi_{n+1,2} & \cdots & \phi_{n+1,N} \\ \vdots & \vdots & \vdots & \vdots \\ \phi_{N1} & \phi_{N2} & \cdots & \phi_{NN} \end{bmatrix}, \quad \mathbf{b} = \begin{bmatrix} f_1 \\ \vdots \\ f_n \\ b_{n+1} \\ \vdots \\ b_N \end{bmatrix}. \tag{7.39}$$

The differentiation operators and RBF values are defined as

$$D_{ij} = \nabla^2 \phi_i(\vec{r})|_{\vec{r}=\vec{r}_j} \equiv \nabla_j^2 \phi_i(\vec{r}_j), \quad \phi_{ij} \equiv \phi_i(\vec{r}_j). \tag{7.40}$$

The system matrix A is square but asymmetric. It can be evaluated given the specific RBF (GA or MQ) and the nodes. If the operator is different than the Laplacian, it can be modified in a straightforward way. The vector $\mathbf{b}$ consists of the charge densities f_j and the boundary values b_j. Once the necessary parameters are known, we can obtain the solutions $\mathbf{a}$ by solving Eq. (7.38) in the same manner as Eq. (7.23).

Let us apply the RBF collocation method to solve the Laplace equation with Gaussian RBFs. The system geometry is a disk (roughly an octagon at the current node spacing) in the center of a square box, Figure 7.18. We assume the box is kept at zero potential and the disk at a constant potential (say 1). We simply distribute the nodes uniformly over the domain (23 × 23), though other choices are also possible (e.g., the Halton sequence [27]). The number of internal nodes is $n = 344$, and the number of boundary nodes is $N - n = 185$. The $\mathbf{b}$ vector in (7.39) consists of zeros in the top n rows and ones in the bottom $N - n$ rows.

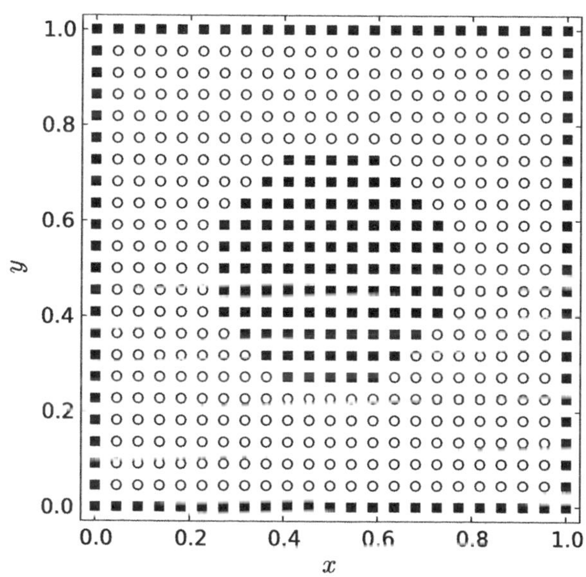

FIGURE 7.18 The disk-in-a-box configuration. There are $n = 344$ internal nodes (o) and 185 boundary nodes (■) for a total of $N = 529$ nodes. The potential is constant on the octagon and zero along the walls.

The required differentiation operator is (Exercise E7.6)

$$\nabla^2 \phi_i(\vec{r}) = \nabla^2 e^{-\epsilon^2|\vec{r}-\vec{r}_i|^2} = 4\epsilon^2 e^{-\epsilon^2|\vec{r}-\vec{r}_i|^2}\left(\epsilon^2|\vec{r}-\vec{r}_i|^2 - 1\right). \tag{7.41}$$

It can be readily evaluated to form the system matrix A given the nodes and the shape parameter ϵ. The value of ϵ turns out to be critical to the accuracy and success of the RBF method. It is known that the accuracy depends on ϵ, sometimes very sensitively, given the choice of the RBF type and the placement of the nodes [55]. Typically, the extreme values, either very small or very large ϵ, lead to very bad approximations. It is unknown, however, where the sweet spot is in general. Sometimes we must find it by trial and error. From the applied physicists' perspective, we resort to physics to guide us, for example, Eq. (6.82). Suppose h is the typical distance between the nodes. We expect that the resolution should be such that $\epsilon h \sim 1$, or $\epsilon \sim 1/h$.

Figure 7.19 shows the results of the RBF collocation method from Program 7.4. It is structurally similar to the FEM code and is fully explained in Section 7.A. The value $\epsilon = 20$ is used (the actual spacing is $1/22$). However, test shows that values from $\sim$ 10 to 25 work as well. The potential is highest, and approximately constant, within the disk (or the octagon), and decreases to zero toward the walls as expected. It also shows the correct symmetry for the configuration.

Compared to FDM, the accuracy of the RBF method is good (Project P7.9). Considering that sharp corners usually require higher density of nodes for the scattered data interpolation to be accurate, the RBF results with a modest number of uniform nodes in the present case are satisfactory. The RBF method is fast and easier to use than the FEM. The disadvantage is that it is more subtle to control accuracy systematically.

With FDM or FEM, we can progressively reduce the error with smaller meshes (within the limits of machine accuracy or memory). In the RBF method, we expect that increasing the number of nodes will lead to more accurate results. However, if the number of nodes changes, we need to change the shape parameter accordingly, whose optimal value is not known a priori. And the situation may be different for different RBFs (Project P7.9).

Finally, the location of the nodes may represent a potential degree of freedom that can be exploited to our advantage, similar to the location of the abscissa in Gaussian integration

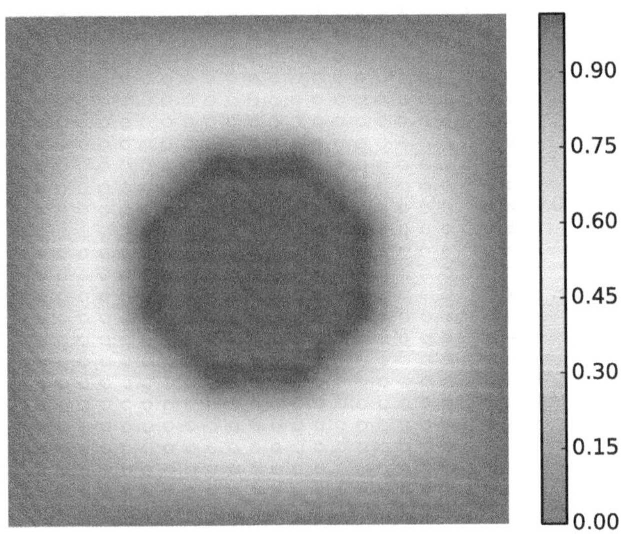

FIGURE 7.19 The potential for the disk-in-a-box configuration.

(Section 8.A.3). For instance, it is thought that some other independent way (equilibrium points, standing wave nodes, etc.) may foretell a set of good nodes from bad ones, proof again that sometimes numerical computing is indeed a form of art.

7.6 VISUALIZATION OF ELECTROMAGNETIC FIELDS

We have discussed electrostatic problems where the electric charges are stationary. In this section we are interested in visualization of electromagnetic vector fields produced by either steadily moving charges or accelerating charges. The vector fields can be effectively visualized using the tools and techniques we have employed so far.

7.6.1 MAGNETIC FIELDS DUE TO CURRENTS

In magnetostatics, the source of magnetic fields is moving charges, that is, electric currents. Like Coulomb's law for electric fields, the Biot-Savart law describes the magnetic fields due to a current segment as

$$d\vec{B} = \frac{\mu_0}{4\pi} \frac{I d\vec{s} \times \vec{r}_{rel}}{r_{rel}^3}, \tag{7.42}$$

where I is the current, $d\vec{s}$ the current-carrying line segment (Figure 7.20), $\vec{r}_{rel} = \vec{r} - \vec{r}_s$ the position vector relative to $d\vec{s}$, and $\mu_0/4\pi = 10^{-7}$ T·m/A. The total magnetic field $\vec{B}$ is the sum over all segments, and for a continuous current distribution such as shown in Figure 7.20, it approaches a line integral.

As a first example, we simulate the magnetic field of a closed loop. We consider the loop as a unit circle placed in the x-y plane, forming a magnetic dipole.[5]

Figure 7.21 displays the magnetic fields calculated from Eq. (7.42) in Program 7.5 using 36 circular segments. Because it is rotationally symmetric about the z-axis, we show a cut of the field in the x-z plane, perpendicular to the loop. The current comes out of the page in the left eye in Figure 7.21 and goes into the page in the right eye. The magnetic field is strongest inside the loop (between the eyes) and weakens significantly outside it. This is because the magnetic fields from all line segments are additive inside, but are canceling outside due to the vector

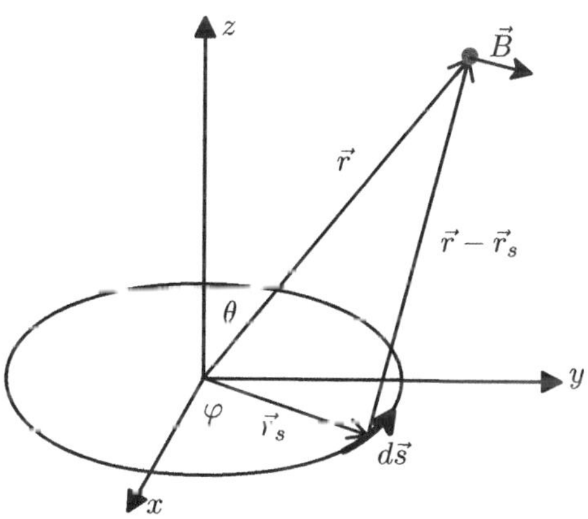

FIGURE 7.20 Magnetic fields due to a current loop.

[5]Unlike electric fields, the basic units producing magnetic fields are magnetic dipoles. There is at present no evidence of magnetic monopoles equivalent to electric charges.

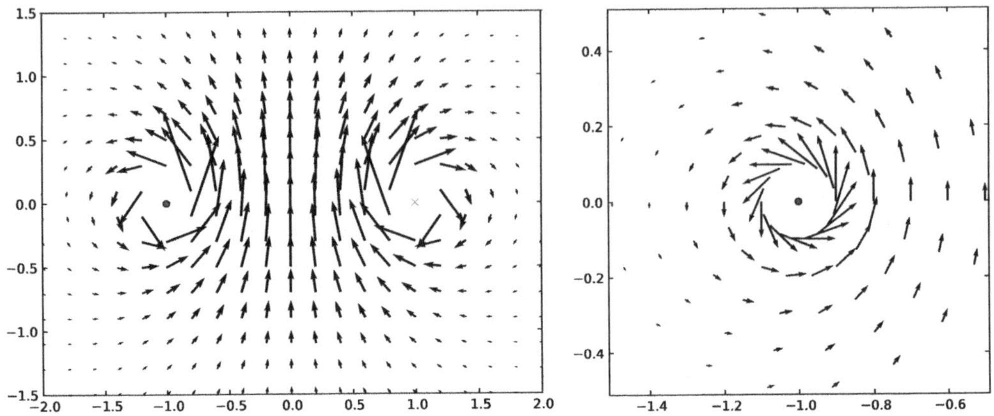

FIGURE 7.21 Left: magnetic dipole fields; right: zoom-in around the wire.

product nature of Eq. (7.42). Along the central z-axis, the magnetic field is perpendicular to the loop.

3D Vector Field Visualization

By swinging around the central z-axis in Figure 7.21, we could imagine how the magnetic fields look like in 3D. But they can also be directly visualized using VPython. This can give a new perspective to viewing vector fields. Figure 7.22 shows two 3D examples of the magnetic fields of a dipole and of a long, current-carrying wire.

For the dipole, the calculation of the magnetic fields remains the same, and only the representation is different. To preserve symmetry, the spatial points are chosen in an axially symmetric manner (see Program 7.6).

For the long, current-carrying wire, we use the analytic results for the magnetic fields,

$$B = \frac{\mu_0 I}{2\pi r}. \quad \text{(long wire)} \tag{7.43}$$

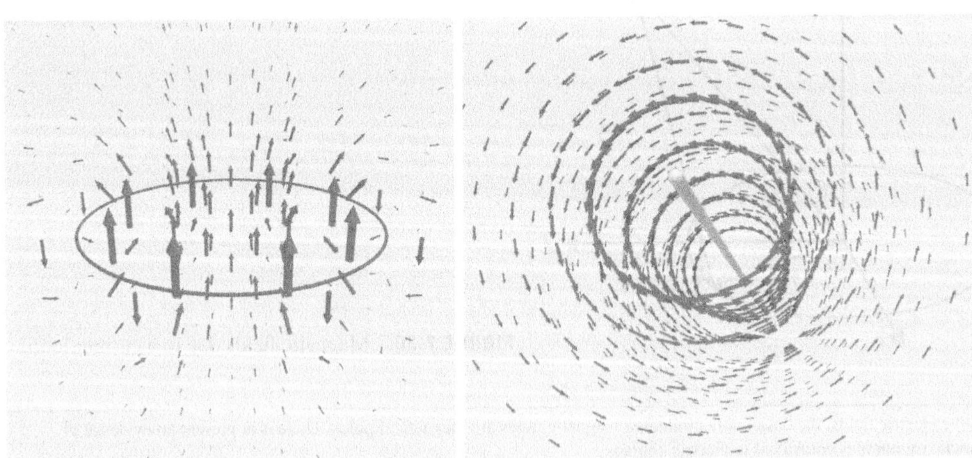

FIGURE 7.22 Magnetic fields of a dipole and of a long, current-carrying wire.

The direction of the current is out of the page, so the magnetic fields are in counterclockwise tangential directions as given by the right-hand rule. In both examples, we can rotate the scene to view the fields from arbitrary viewpoints.

7.6.2 ELECTROMAGNETIC RADIATION OF AN OSCILLATING DIPOLE

In electrodynamics, if moving charges are also accelerating, for example, moving along a curved path or oscillating back and forth, they emit electromagnetic radiation. One of the most important sources of radiation is an oscillating electric dipole. This can be thought of as a pair of opposite charges oscillating sinusoidally about their center. Let us assume the center at the origin and the oscillation in the z direction. The electric dipole may be written as

$$\vec{p} = p_0 \cos(\omega t)\hat{z}, \tag{7.44}$$

where p_0 is the magnitude of the electric dipole, and ω the angular frequency. The resulting electromagnetic fields in 3D visualization obtained from Program 7.7 are shown in Figure 7.23.

Away from the dipole in the radiation zone ($kr \gg 1$), the electromagnetic fields have particularly simple expressions [50],

$$\begin{aligned}\vec{B} &= \frac{p_0 k^2}{4\pi\varepsilon_0 c}\frac{\cos(kr - \omega t)}{r}\,\hat{r}\times\hat{z},\\ \vec{E} &= c\,\vec{B}\times\hat{r}.\end{aligned} \tag{7.45}$$

Here c is the speed of light, $k = \omega/c = 2\pi/\lambda$ the wavevector (λ the wavelength).

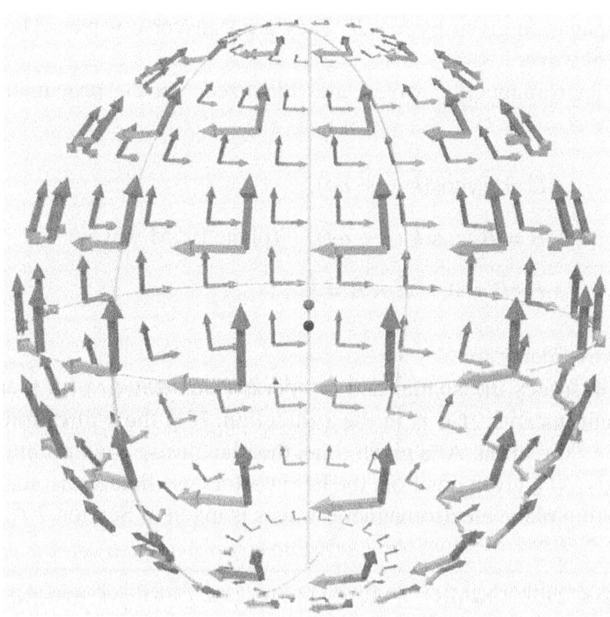

FIGURE 7.23 Electromagnetic fields of an oscillating dipole, with $\vec{E}$ and $\vec{B}$ in the longitudinal and latitudinal directions, respectively.

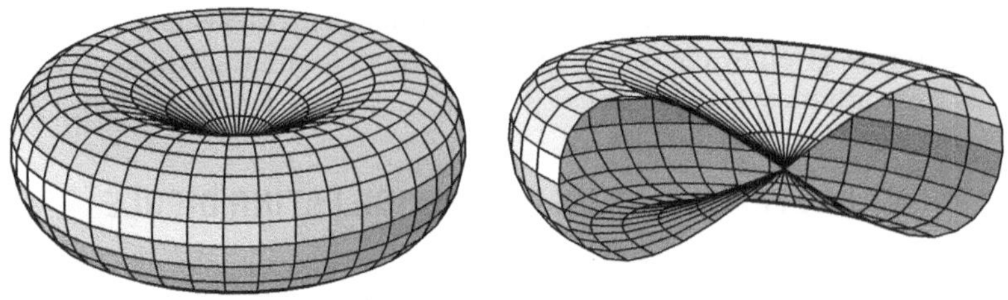

FIGURE 7.24 Angular distribution of electric dipole radiation and a cross-sectional view (right).

The magnitudes of the electromagnetic fields decrease inversely proportional to r. This ensures that the radiative power, proportional to $\vec{E} \times \vec{B}$, is constant through a spherical surface on average.[6] This is in contrast to electrostatic dipole electric fields which fall off much faster at $1/r^3$ at large distances.

On the spherical surface of a given radius r, the magnitudes vary as $B, E \propto \sin\theta$, with θ being the polar angle (see Figure 7.20). Therefore, the radiation is anisotropic and highly directional. Figure 7.24 shows the angular distribution of field magnitudes from Program 7.8.

The pattern displayed in Figure 7.24 is characteristic of a dipole distribution. The fields are zero at the north and south poles along the dipole axis ($\theta = 0$ and π) and maximum at the equator. Furthermore, as Figure 7.23 shows, the electric field $\vec{E}$ is longitudinal and magnetic field $\vec{B}$ latitudinal, such that they are perpendicular to each other at a given point. Their cross product is along the radial vector, the direction of field propagation. As a result, the radiative power is largest near the equator in the direction perpendicular to the electric dipole.

Plane Electromagnetic Waves

As seen earlier, the electric and magnetic fields in free space are perpendicular to each other, with the direction of propagation given by $\vec{E} \times \vec{B}$. They are transverse waves. The dipole radiation is an outgoing spherical wave.

The plane electromagnetic waves are idealized waves propagating in free space described by

$$\begin{aligned} &\vec{E} = \vec{E}_0 \cos(\vec{k} \cdot \vec{r} - \omega t), \\ &\vec{B} = \vec{B}_0 \cos(\vec{k} \cdot \vec{r} - \omega t), \quad \text{(plane E\&M waves)} \\ &\vec{E} \cdot \vec{B} = 0, \quad \hat{E} \times \hat{B} = \hat{k}, \end{aligned} \tag{7.46}$$

where $\vec{k}$ is the 3D wavevector.

Figure 7.25 illustrates the spatial and temporal visualization of a plane electromagnetic wave at four different instants. If $\vec{E}$ is in the x direction, $\vec{B}$ in the y direction, then the direction of propagation is the z direction. At a given time, the plane wave is an infinite train of sinusoidal wave along the z-axis. At a given position, the field vectors oscillate sinusoidally and indefinitely. Program 7.9 simulating plane electromagnetic waves is given in Section 7.A.

[6]The power in an electromagnetic field is given by the Poynting vector, $\vec{S} = \mu_0^{-1} \vec{E} \times \vec{B}$, with units of power/area.

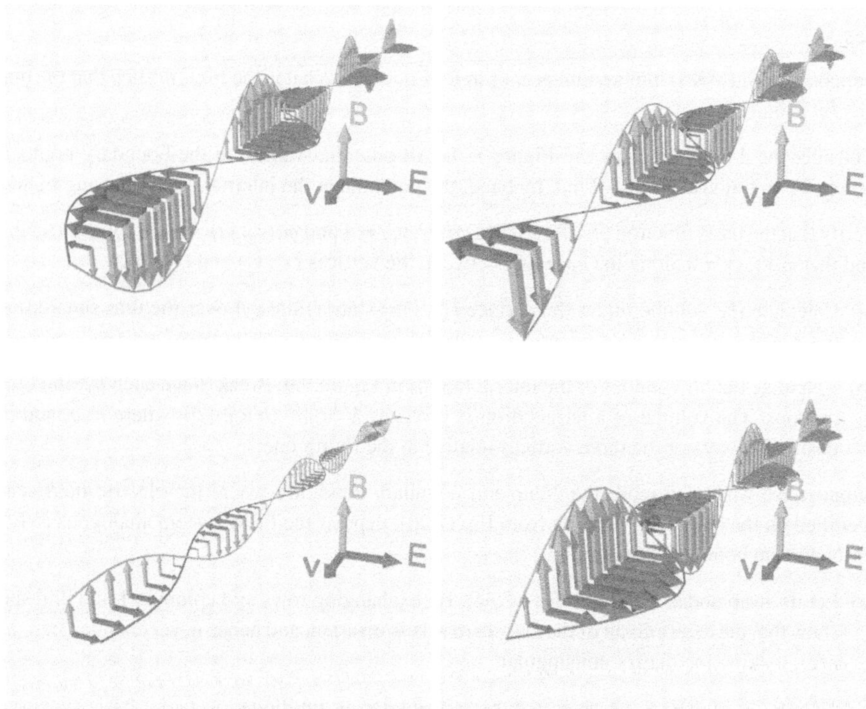

FIGURE 7.25 Plane electromagnetic waves.

CHAPTER SUMMARY

We introduced electric fields of point charges through the simulation of the electric hockey game. For the general boundary value problems in electrostatics, we studied numerical solutions of Poisson and Laplace equations. We presented two methods, the iterative, self-consistent FDM and the noniterative FEM.

The self-consistent FDM is simple to implement and yields accurate solutions. The simplicity of the algorithm, namely, the potential at a given point is equal to the average of its neighbors, offers a clear physical insight to the meaning of the Laplace equation and enables us to visualize the relaxation process. The standard self-consistent iteration suffers from slow convergence, but that can be overcome to a certain extent via accelerated schemes such as the Gauss-Seidel or the overrelaxation methods. It is less straightforward, however, to deal with Neumann boundary conditions or point charge distributions.

We described the 2D FEM for boundary value problems as applied to the Poisson equation from a physicist's point of view for practical use, rather than mathematical rigor. The FEM is more adapt to handling different types of boundary conditions and charge distributions. Its versatility is not limited to electrostatics, and we will encounter it again in quantum eigenvalue problems.

We introduced the meshfree RBF collocation method for boundary value problems. The method is fast and can easily handle irregular domains and mixed boundary conditions. It can be scaled up to higher dimensions without drastically altering the basic framework.

We have also modeled the vector fields and electromagnetic radiation due to electric and magnetic dipoles, using both 2D vector graphs and 3D representations.

7.7 EXERCISES AND PROJECTS

Exercises

E7.1 Suppose the SI (MKS) unit system is chosen for Eq. (7.1). What is the mass m such that the magnitude of k/m is unity?

E7.2 Consider the 2×2 grid shown in Figure 7.28. All potential values on the boundary points are zero except $V_5 = 1$ at grid point 5. Find, by hand, the potential at the internal point V_4 using FDM.

E7.3 Verify that the basis function (7.10) satisfies $\varphi_1(x_1, y_1) = 1$ and $\varphi_1(x_2, y_2) = \varphi_1(x_3, y_3) = 0$ at the nodes, and that $\varphi_1(x, y) = 0$ along the base line between the vertices (x_2, y_2) and (x_3, y_3).

E7.4 (a) Calculate the volume under the surface $f(x, y) = \sin(\pi x)\sin(\pi y)$ over the unit square $0 \le x \le 1$, $0 \le y \le 1$.

(b) Calculate the fill volumes of the interpolations in Figure 7.9. Break down each pyramid into tetrahedrons. The volume of a tetrahedron is given by $V = |\vec{a} \cdot (\vec{b} \times \vec{c})|/6$, where $\vec{a}$, $\vec{b}$, and $\vec{c}$ are the position vectors of the three vertices relative to the fourth one.

E7.5 Suppose we wish to impose the Neumann boundary condition where the electric field vector $\vec{E}$ is specified on the boundary. Starting from Eq. (7.20), express the FEM system analogous to Eq. (7.23) for Neumann boundary conditions.

E7.6 (a) Let us swap nodes 1 and 2 in Eq. (7.34). By exchanging rows and columns 1 and 2, respectively, show that the determinant of the system matrix is invariant and hence never changes sign or crosses zero, making the matrix nonsingular.

(b) Derive $\nabla\phi$ and $\nabla^2\phi$ where ϕ is either a Gaussian or multiquadric radial basis function (7.32). Verify Eq. (7.41).

(c) Assume Neumann boundary conditions, that is, $\nabla_n u$ is specified along the normal at the boundary nodes. Obtain the corresponding form of the system matrix (7.39).

E7.7 Plot the magnetic filed patterns of the following currents, two of which are shown in Figure 7.26. We are interested only in the relative magnitudes and assume all currents have unity magnitude. Predict the field patterns in each case, and use VPython to generate a 3D vector field. Then choose an appropriate cross section that best represents the symmetry, and use `quiver()` to produce a cross-sectional (2D) visual representation.

(a) Two parallel, long wires separated by a distance $d = 1$ and carrying parallel and opposite currents, respectively.

(b) Two semicircular loops of radii 1 and 2 (Figure 7.26(b)).

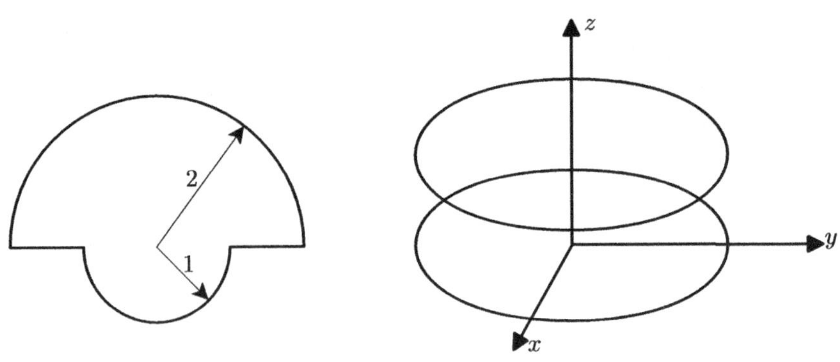

(b) (c)

FIGURE 7.26 Current loops, Exercise E7.7.

(c) Two coaxial circular loops, Figure 7.26(c), stacked a distance $d = 1$ apart.

(d) A solenoid of radius 1 and length 2 with N equispaced turns, $N \geq 10$.

E7.8 Modify the hockey game (Program 7.1) and add a magnetic field perpendicular to the playing surface. Simulate the game to include the Lorentz force $\vec{F} = q\vec{v} \times \vec{B}$. Vary B from ~ 0.2 to 1. Compare the trajectories with and without the magnetic field.

E7.9 Simulate the temporal and spatial variation of electric dipole radiation.

(a) First, modify Program 7.8 so the sinusoidal time oscillation at a given r is simulated.

(b)* Program both the r and t dependence (7.45) in select directions, say along polar angles 30°, 60°, and 90°. How are they different from the plane electromagnetic waves (Figure 7.25)?

Projects

P7.1 Generate the potential surface (Figure 7.2) of the hockey field. Use Program 4.6, and replace the effective potential by the Coulomb potentials between the puck and stationary charges. What would the potential surface look like if the charge of the puck is positive? Is it possible to score a goal?

P7.2 (a) Implement the Gauss-Seidel acceleration and overrelaxation methods discussed in Section 7.2.2, using Program 7.2 as the basis. Verify that your program produces the same results as the standard self-consistent method for the parallel plates.

(b) For the overrelaxation method, vary the acceleration parameter α between $1 \leq \alpha < 2$. Record the number of iterations, $n(\alpha)$, required to reach a given accuracy, say 10^{-10}. Plot $n(\alpha)$ as a function of α. Discuss the results and trends.

P7.3 Consider the geometries shown in Figure 7.27. Assume in each case the central object is kept at a constant potential (1) and the potential is zero on the rectangular boundary. Take the size of the object to be about one-third of the domain size.

(a) Predict the potential lines and electric field patterns in each geometry, and sketch them.

(b) For each geometry, find the numerical solutions with the standard FDM relaxation method. Use a grid dimension of 30 or more (for larger grid dimensions, use the overrelaxation method). Display the solutions graphically.

(c)* No doubt you will have noticed that the square solution seems to be made up of two halves: One is the wedge solution and the other its reflection about the diagonal. In fact, the Laplace equation is linear, and solutions corresponding to piece-wise boundary conditions are additive.

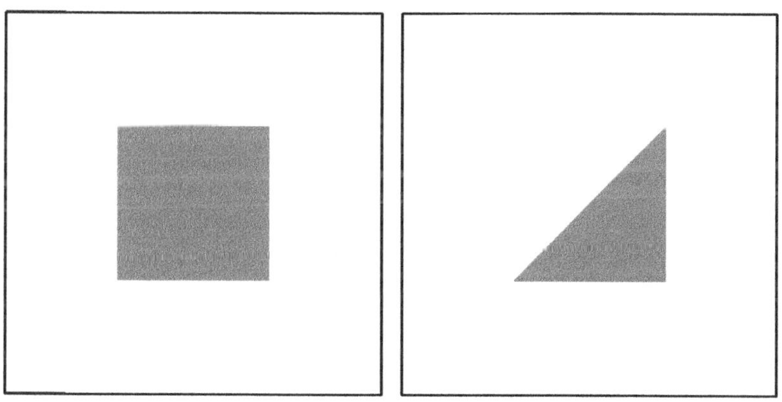

FIGURE 7.27 The square and wedge geometries for Project P7.3.

Let u_1 and u_2 be the solutions for the boundary conditions b_1 and b_2, respectively. Then, $u = u_1 + u_2$ is the full solution for the combined boundary condition $b = b_1 + b_2$. We call this the additivity rule.

In the geometries of Figure 7.27, change the boundary condition of the wedge and of the wedge reflected about the diagonal so the combined effect is the same as the square. Find the two solutions, and verify numerically the preceding additivity rule.

(d) Using $E_\perp = \sigma/\varepsilon_0$, compute the surface charge densities on the square and the wedge, respectively. Find a way to visualize σ.

P7.4 The Laplace operator may be approximated by a higher-order, nine-point scheme as

$$\nabla^2 V_{i,j} = -\frac{1}{12h^2}[60V_{i,j} - 16(V_{i-1,j} + V_{i+1,j} + V_{i,j-1} + V_{i,j+1}) + (V_{i-2,j} + V_{i+2,j} + V_{i,j-2} + V_{i,j+2})] + O(h^4).$$

(a) With the preceding Laplacian, obtain the FDM equation equivalent to Eq. (7.6). Let us call this approximation FDM9.

(b) Devise a standard relaxation scheme for FDM9, program and test it. Use Program 7.2 as the base program for the same parallel plates system. In particular, decide how to implement the boundary condition. One choice is to treat the points beyond the boundary the same as their nearest boundary point. Compare the FDM9 results with FDM for the same grid size.

(c) Once your program is working, apply FDM9 to other geometries, such as a square with Eq. (7.29). Study convergence and accuracy of FDM9 by varying the grid sizes and comparing with the exact solutions. Plot the scaling of error vs. grid size analogous to Figure 7.16.

(d)* Apply the standard Gauss-Seidel acceleration and overrelaxation methods to FDM9. Investigate the rate of convergence in each case.

P7.5 Consider a domain of eight finite elements shown in Figure 7.28.

(a) Make a table listing the nodes, elements, and internal and boundary nodes according to the data structure outlined in Table 7.1.

(b) Sketch the tent functions at nodes 0, 1, 2, and 4. Optionally draw them as surface plots with Matplotlib, or as `faces` objects with VPython (see VPM mesh objects, Program S:6.4).

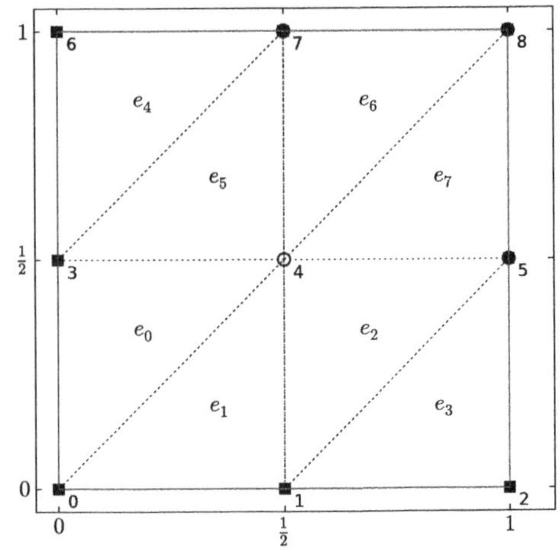

FIGURE 7.28 A sample FEM mesh (Project P7.5).

(c) Verify that the system matrix (7.25) is given by

$$A = \frac{1}{2}\begin{bmatrix} 2 & -1 & 0 & -1 & & & & & \\ -1 & 4 & -1 & 0 & -2 & & & \text{\huge 0} & \\ 0 & -1 & 2 & 0 & 0 & -1 & & & \\ -1 & 0 & 0 & 4 & -2 & 0 & -1 & & \\ & -2 & 0 & -2 & 8 & -2 & 0 & -2 & \\ & & -1 & 0 & -2 & 4 & 0 & 0 & -1 \\ & & & -1 & 0 & 0 & 2 & -1 & 0 \\ & \text{\huge 0} & & & -2 & 0 & -1 & 4 & -1 \\ & & & & & -1 & 0 & -1 & 2 \end{bmatrix}.$$

Check the first and second row entries by calculating them explicitly.

(d) Explain why the entries A_{13} and A_{15} are zero. Indicate what nodes are involved. The entry $A_{55} = 8/2 = 4$ is the largest. What is the reason?

(e) Suppose we wish to solve the Laplace equation, and all boundary values are zero except $u_5 = 1$ at node 5. Obtain the FEM equation (7.23) and solve it. Discuss your results and compare with the FDM result from Exercise E7.2.

P7.6 Investigate the FEM solution for the boundary condition (7.29).

(a) Impose the boundary condition by modifying Program 7.3. Only minimal change to `mesh()` is required. Reproduce the solutions shown in Figure 7.14.

(b) Based on the contour plot, sketch the electric fields. Calculate the actual electric field vectors, and plot them over the contour plot. Discuss your results.

(c) Compare the numerical and analytic results (7.30), and plot the absolute error as shown in Figure 7.15 for a given grid dimension $N \sim 40$. In a separate graph, plot the relative error in the interior region. Compared to the absolute error, is the relative error localized? Can you tell if the FEM solution is symmetric about the diagonal line?

(d) Vary N from 2 to ~ 100 (push to the highest limit possible), doubling N each time. For each N, compute the average error as

$$\Delta = N^{-2} \sum_{i,j} \left| u_{exact}[i,j] - u_{FEM}[i,j] \right|,$$

and plot Δ vs. N as in Figure 7.16. Describe the trend.

(e)* Study the accuracy of the self-consistent method with FDM by repeating the last two parts. Use the code from Project P7.2 if available.

P7.7 Solve the boundary value problems in Figure 7.27 with FEM.

(a) For the square problem, use the same mesh as presented in Program 7.3, set the inner boundary values to 1 but exclude the area within the inner square. For the wedge, generate a mesh as shown in Figure 7.29, a sample mesh with $N = 10$. The outer boundary values are zero on the box (■) and the inner boundary values are 1 (•) on the wedge. Plot the solutions using the Matplotlib function `tripcolor()` (see Program 9.8). Compare the FEM and FDM solutions (Project P7.3 if available) for a given $N \geq 30$.

(b) The wedge is symmetric about the upper-left to lower-right diagonal. Does the FEM preserve this symmetry? Devise a way to check it.

(c)* Generate a mesh so that the dividing line of each subsquare is parallel to the upper-left to lower-right diagonal line. Repeat the preceding calculation, and discuss your results.

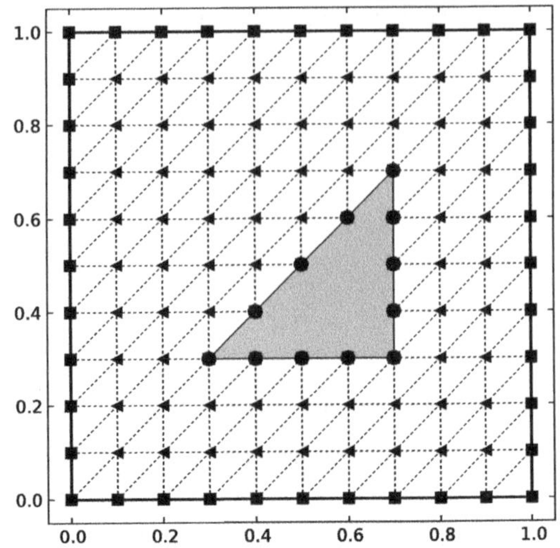

FIGURE 7.29 The FEM mesh for the wedge ($N = 10$).

P7.8 Solve the Poisson equation with FEM for a point charge inside a unit square that is kept at zero potential. Treat the point charge at (x_0, y_0) as a Dirac charge distribution $f(x, y) = \delta(x - x_0)\delta(y - y_0)$.

(a) Assume the charge is just off the center of the square, $(x_0, y_0) = (\frac{1}{2} + \epsilon, \frac{1}{2} - \epsilon)$. Choose a mesh (Figure 7.11) with an even N, say $N = 10$. The charge would be in only one element just to the southeast of the center (in Figure 7.28, it would be e_2).

Prepare the FEM matrices in Eq. (7.23). When evaluating the column matrix on the RHS, the only nonzero p_j's are on the three nodes of the element containing the charge. Furthermore, because $\epsilon \sim 0$, only one is nonzero (using the 2×2 mesh above, it would be p_4). Solve Eq. (7.23), and display the solutions graphically. Are the results as expected?

(b) How would the results change if the charge is exactly at the center, $(x_0, y_0) = (\frac{1}{2}, \frac{1}{2})$? Explain your answer both physically and numerically.

(c)* Write a program that accepts the charge at an arbitrary location. Optionally, generalize it so it is able to handle n charges at different locations. Test it with $n = 4$.

P7.9 Study the RBF collocation method pertaining to the solutions of the Laplace and Poisson equations.

(a) Perform an error analysis of the RBF method. For the same grid size, carry out calculations using FDM and RBF methods, Programs 7.2 and 7.4, respectively, for the disk-in-the-box configuration (Figure 7.18). Either combine the programs, or write the results to files for separate processing (for file handling, see Program 4.4 on text I/O or Program 9.7 using `pickle`). Plot the absolute error as shown in Figure 7.15. Repeat the calculations for several shape parameters $\epsilon = 10$ and 15. Discuss your finding. Optionally, perform the same analysis by comparing with the FEM results.

(b) Instead of constant potential, suppose the disk is uniformly charged with a surface charge density of 1 (i.e., $\sigma/\epsilon_0 = 1$). Change the initialization function in Program 7.4 to solve the Poisson equation. Calculate the potential and explain the results.

(c) Modify Program 7.4 to use the multiquadric RBF (7.32) instead of the Gaussian RBF (use result of Exercise E7.6 if available). Repeat the above investigations.

(d)* Apply the RBF method to the configuration in Figure S:7.3. Compare results with Project S:P7.1. Optionally time the program execution speed for the FDM and RBF methods to compare their performance (exclude plotting, and use `time` library). Program profiling is discussed in Section S:8.B.

7.A PROGRAM LISTINGS AND DESCRIPTIONS

Program listing 7.1 Electric field hockey (`hockey.py`)

```
   import visual as vp, numpy as np, ode
 2
   def hockey(Y, t):                  # return eqn of motion
 4     accel = 0.
       for i in range(len(loc)):
 6         accel += Q[i]*(Y[0]-loc[i])/(vp.mag(Y[0]-loc[i]))**3
       return [Y[1], q*accel]         # list for non-vectorized solver
 8
   a, b, w = 1., 0.5, 0.125                      # rink size, goal width
10 q, qcor, qmid, qcen = -1.0, 1.0, -2., 2.    # Qs: puck, cor., mid, cen.
   Q = [qcor, qmid, qcor, qcor, qmid, qcor, qcen]   # charges, locations
12 loc = [[-a, b], [0, b], [a, b], [a,-b], [0,-b], [-a,-b], [0,0]]

14 scene = vp.display(title='Electric hockey', background=(.2,.5,1))
   puck  = vp.sphere(pos=(-a,0,0), radius = 0.05, make_trail=True) # trail
16 rink  = vp.curve(pos=loc[:-1]+[loc[0]], radius=0.01)      # closed curve
   goalL = vp.curve(pos=[(-a,w,0),(-a,-w,0)], color=(0,1,0), radius=.02)
18 goalR = vp.curve(pos=[( a,w,0),( a,-w,0)], color=(0,1,0), radius=.02)
   for i in range(len(loc)):          # charges, red if Q>0, blue if Q<0
20     color = (1,0,0) if Q[i]>0 else (0,0,1)
       vp.sphere(pos=loc[i], radius = 0.05, color=color)
22
   v, theta = input('enter speed, theta; eg, 2.2, 19:') # try 2.2, 18.5
24 v, theta = min(4, v), max(1,theta)*np.pi/180.       # check valid input
   Y = np.array([[-a,0], [v*np.cos(theta), v*np.sin(theta)]])
26 while True:
       vp.rate(200)
28     Y = ode.RK45n(hockey, Y, t=0., h=0.002)
       x, y = Y[0][0], Y[0][1]
30     if (abs(x) > a or abs(y) >b):
           txt = 'Goal!' if (x > 0 and abs(y) < w) else 'Miss!'
32         vp.label(pos=(x, y+.2), text=txt, box=False)
           break
34     puck.pos = Y[0]
```

Most of the program is to set up visual animation. The dynamics of motion (7.1) is modeled in `hockey()`. The input `Y` holds the position and velocity vectors of the puck. The motion is in the x-y plane, so all vectors are two-dimensional. Calculation of acceleration utilizes vector operations with NumPy arrays. The velocity and acceleration vectors are returned as a list of NumPy arrays for nonvectorized ODE solvers.

The playing area is a rectangle bounded by $|x| \le a$ and $|y| \le b$ with the origin at the center of the field. After the charges and their locations are initialized, the scene is drawn. Each charge is represented as a sphere, including the puck, which is set up to leave a trail as it moves (line 15).

The speed and launching angle are read from the user, with an upper speed limit and a lower cutoff for the angle. The latter is to prevent the puck from falling into the sink hole at the center of the field. The main loop integrates the motion with the `RK45n` solver for improved accuracy. Because `hockey()` returns a list of vectors, `RK45n` can be readily used, because vector objects represented by NumPy arrays can be manipulated like scalars in arithmetics (addition, subtraction, and multiplication). The rest of the loop checks collisions with the wall, decides if a goal has been scored, and the outcome is labeled at the end.

Program listing 7.2 Self-consistent method for parallel plates (`laplace.py`)

```
import matplotlib.pyplot as plt
from mpl_toolkits.mplot3d import Axes3D
import visual as vp, numpy as np, copy

def relax(V, imax=200):          # find self-consistent soln
    for i in range(imax):
        V[1:-1,1:-1] = ( V[1:-1,:-2] + V[1:-1,2:]         # left, right
                       + V[:-2,1:-1] + V[2:,1:-1] )/4    # top, bottom
        V = set_boundary(V)    # enforce boundary condition
        draw_pot(V), vp.rate(1000)
    return V

def set_boundary(V):                 # set interior BC values
    V[w:2*w, top], V[w:2*w, bot] = 1., -1.  # plates voltage
    return V

def draw_pot(V):                     # refresh potential, slow version
    for i in range(M):
        for j in range(N):
            q = abs(V[i,j])       # graduated color mix
            if (V[i,j] >0): grid.color[N*i + j] = (q, .5*q, .2*q)
            else:              grid.color[N*i + j] = (.2*q, .5*q, q)

def draw_efield(V, scale):           # draw electric field
    Ex, Ey = np.gradient(-V)
    Emag = np.sqrt(Ex*Ex + Ey*Ey)
    for i in range(2, M-1, 2):
        for j in range(2, N-1, 2):
            vp.arrow(pos=(i,j), axis=(Ex[i,j], Ey[i,j]),
                     length=Emag[i,j]*scale)
        vp.rate(100)
    return Ex, Ey

M, N, s = 61, 61, 10                 # M x N = grid dim, s = point size
w, d, h = M//3, N//6, N//2          # plates width, separation, half N
bot, top = h - d//2, h + d//2       # bottom and top plates

scene = vp.display(width=M*s, height=N*s, center=(M//2,N//2))
grid = vp.points(pos=[(i,j) for i in range(M) for j in range(N)], # grid
                 color=[(0,0,0)]*(M*N), size=s, shape='square')

V = np.zeros((M,N))                  # initialze V on grid, apply BC
V = set_boundary(V)
V = relax(V)                         # solve by relaxation

Ex, Ey = draw_efield(V, scale = 16)
V, Ex, Ey = np.transpose(V), np.transpose(Ex), np.transpose(Ey)
X, Y = np.meshgrid(range(M), range(N))
```

```
50 plt.figure()                    # Fig.1, contour plot
   plt.contour(V, 14)
52 plt.quiver(X[::2,], Y[::2,], Ex[::2,], Ey[::2,],      # stride 2 in y dir
              width=0.004, minshaft=1.5, minlength=0, scale=10.)
54 plt.xlabel('x'), plt.ylabel('y')

56 plt.figure()                    # Fig.2, surface plot
   ax = plt.subplot(111, projection='3d')
58 ax.plot_surface(X, Y, V, rstride=1, cstride=1, cmap=plt.cm.jet, lw=0)
   ax.set_xlabel('x'), ax.set_ylabel('y'), ax.set_zlabel('V')
60 plt.show()
```

The code implements the self-consistent method to solve the Laplace equation for two parallel plates held at a potential difference. It solves the problem by iterative FDM over a grid of $M \times N$.

The main program sets the grid size and the parallel plates dimensions and visually represents the grid by square points (line 39). It then initializes the potential array and applies the boundary values. The boundary conditions include the two parallel plates held at values of ± 1 (in arbitrary units) and the four sides of the box held at zero. The plates run horizontally from $M/3$ to $2M/3$. The vertical positions of the bottom and top plates are $(N \mp -d)/2$, respectively, with separation d. The unit of distance is also arbitrary because it is scalable.

The heavy work is done in `relax()`, which attempts to find the self-consistent solution via relaxation. It works the same way as `waves2d()` in Program 6.7 with slicing as discussed in Section 6.7. It iterates for a maximum of `imax` times (default value can be changed by the caller), applying the boundary conditions and redrawing the potential grid after each iteration. The function `draw_pot()` displays the potential on the grid as hues of orange or blue colors for positive and negative potentials, respectively, similar to graduated colors used in Program 5.7.

Once the solution is returned, `draw_efield()` is called to calculate and draw the electric field using the NumPy gradient function. VPython arrows representing the electric field are scaled to their magnitudes. Contour and surface plots are also generated. Similarly, vector electric fields are also overlaid on the contour plot.

One note about an unfortunate source of confusion caused by inconsistent mathematical conventions. We are used to the normal convention of labeling a point by its coordinates (x, y), or (i, j) on a grid, with i running horizontally and j vertically. However, after conceptually setting up the problem this way, when it is time to represent the potential grid as a 2D array, we face a dilemma: `V[i,j]` means a vertical position `i` (row) and horizontal position `j` (column) in the interpretation of a matrix. This is opposite to our conceptual setup. We had this problem in Program 4.6 for the effective potential of the restricted three-body motion.

We believe it is important to keep conceptual clarity, so we will set up the problem in the normal (coordinate) convention and do calculations accordingly. But when it comes to graphing the results, we have to transpose the arrays first. Fortunately, the NumPy function `np.transpose` eases the pain considerably.

Speaking of clarity, we chose code readability over speed in `draw_pot()`, which updates the colors of the grid one point at a time inside a double loop. This slows down the simulation because it is called repeatedly, and explicit looping is relatively slow in Python (Section 1.3.2). For the grid size used in this program, `draw_pot()` actually uses most of the computing time because the core computation in `relax()` is already highly efficient. This will only get worse for larger grid sizes.

Fortunately, we can eliminate the bottleneck if we wish to retain visualization. We do so by entirely eliminating the double loop as follows:

```
def draw_pot(V):          # refresh potential, fast version
    hue, neg = np.dstack((V, V*.5, V*.2)), V<0. # stack a 3D array
    hue[neg] = - hue[neg][:, [2,1,0]]          # swap R,B for neg pot
    grid.color = hue.reshape(M*N,3)            # reshape to (M*N,3)
```

To associate each point with an RGB color code, we use a base code $(1, 0.5, 0.2)$ and scale it by the potential to obtain shades of orange. First, we stack the potential grid depth-wise (on top of each other) to form a 3D array ($M \times N \times 3$), with appropriate weighing by the base color components at each point. Next, color codes for negative potentials, indicated by the truth array `neg`, are turned positive. Their red and blue components are also swapped at the same time to make shades of blue. All this is done with advanced indexing (Section 1.D). Finally, the array is reshaped to a 2D array ($MN \times 3$) before assignment to the grid.

Now the visualization code is as fast and efficient as the core computation, and we can take on larger grids, at the expense of code readability. Use the new `draw_pot()` in Program 7.2, quadruple the grid size (say 128×128, also remember to halve the point size), and run!

Program listing 7.3 Laplace equation by FEM (`laplacefem.py`)

```
from scipy.linalg import solve
from mpl_toolkits.mplot3d import Axes3D
import matplotlib.pyplot as plt, numpy as np

def mesh(L, N):                 # generate mesh
    elm, bv = [], []            # elements: [[n1,n2,n3],..], bndry value
    x, y = np.linspace(0,L,N+1), np.linspace(0,L,N+1)  # same x,y grids
    ndn = np.arange((N+1)*(N+1)).reshape(N+1,N+1)  # node number
    node = [[xi, yj] for xi in x for yj in y]         # nodes
    for j in range(N):
        for i in range(N):
            elm.append([ndn[i,j], ndn[i+1,j+1], ndn[i,j+1]])  # upper
            elm.append([ndn[i,j], ndn[i+1,j], ndn[i+1,j+1]])  # lower

    ip = ndn[1:-1,1:-1].flatten()       # internal nodes
    bp = np.delete(ndn, ip)             # boundary nodes=all-internal
    for p in bp:
        bv.append((node[p][0]*node[p][1])**2)    # boundary values
    return node, elm, bp, ip, bv, x, y

def fem_mat(node, elm):       # fills matrix A_ij = <∇φ_i · ∇φ_j>
    A = np.zeros((len(node),len(node)))
    for e in elm:
        (x1,y1), (x2,y2), (x3,y3) = node[e[0]], node[e[1]], node[e[2]]
        beta, gama = [y2-y3, y3-y1, y1-y2], [x3-x2, x1-x3, x2-x1]
        ar = 2*(x1*(y2-y3) + x2*(y3-y1) + x3*(y1-y2))
        for i in range(3):
            for j in range(i,3):
                A[e[i],e[j]] += (beta[i]*beta[j] + gama[i]*gama[j])/ar
                if (i != j): A[e[j],e[i]] = A[e[i],e[j]]    # symmetry
    return A
```

```
32
   L, N = 1.0, 40                        # length of square, number of intervals
34 node, elm, bp, ip, bv, x, y = mesh(L, N)   # generate mesh
   ip.sort()                                  # sort ip, just in case
36 A, b = fem_mat(node,elm), np.zeros(len(ip)) # build matrices
   for j in range(len(ip)):
38     b[j] = np.dot(A[ip[j], bp], bv)          # boundary condition Eq. (7.23)

40 A = np.delete(A, bp, axis=0)   # delete rows specified by bp
   A = np.delete(A, bp, axis=1)   # delete cols specified by bp
42 u = solve(A, -b)               # solve

44 u = np.concatenate((u, bv))      # combine internal+boundary values
   all = np.concatenate((ip, bp))   # internal+boundary nodes
46 idx = np.argsort(all)            # index sort nodes
   u = np.take(u, idx)              # now u[n] is the value at node n
48 u = np.reshape(u, (N+1, N+1))    # reshape grid for graphing
   x, y = np.meshgrid(x, y)
50
   plt.figure()
52 ax = plt.subplot(111, projection='3d')
   ax.plot_surface(x, y, u, rstride=1, cstride=1,
54                 linewidth=0, cmap=plt.cm.jet)
   ax.set_xlabel('x'), ax.set_ylabel('y'), ax.set_zlabel('V')
56 plt.figure()
   plt.subplot(111, aspect='equal')
58 plt.contour(x, y, u, 26)
   plt.show()
```

Mesh generation is done through `mesh()` in accordance to the data structure outlined in Table 7.1. We divide the rectangular domain into N intervals in both x and y directions (Figure 7.11). A node $n_{i,j}$ is a grid point at (i,j), which has the coordinates (x_i, y_j), with $x_i = ih$ and $y_j = jh$, $i, j = 0, 1, \ldots, N$, and $h = 1/N$ being the grid spacing. The nodes, numbered from 0 to $N(N+2)$, are reshaped to a $(N+1, N+1)$ array, `ndn`. As illustrated in Figure 7.11, the node numbers increase from left to right horizontally then up the rows. The node number at grid (i,j) is $n_{i,j} = j \times (N+1) + i$. The (x, y) coordinates of the nodes are stored in a nested list `node` (line 9) in the same order as the node numbers.

We next build the triangular elements. Each element is a list of three node numbers defining the triangle. The process traverses the subsquares (Figure 7.11) from the bottom row moving left to right (i-loop) then going up the rows (j-loop). For a subsquare whose lower-left corner is at (i,j), two elements are appended to the `elm` list (lines 12 and 13): the upper one (above diagonal) with nodes $[n_{i,j}, n_{i+1,j+1}, n_{i,j+1}]$, and the lower one (below diagonal) with $[n_{i,j}, n_{i+1,j}, n_{i+1,j+1}]$.

We can obtain the internal points from the nodes `ndn` by taking only the interior points via slicing, excluding points on the four sides (line 15). The resultant nodes are flattened to a 1D array and assigned to `ip`. Now, the boundary points `bp` can be found by deleting the internal nodes `ip` from all nodes `ndn` (line 16). This is actually done by advanced indexing because `ip` is an integer array. The array `ndn` is flattened first, then elements indexed by `ip` are removed.

Having found `bp`, we can evaluate the boundary values `bv` according to Eq. (7.28). Let (x, y) be the coordinates of a given boundary point. Because it is on the boundary, the expression x^2y^2 yield a zero value at the bottom and left sides, x^2 at the top ($y = 1$), and y^2 on the right side ($x = 1$), so they are the correct boundary conditions.

For other domain or boundary configurations, the details of mesh generation may be different, but the general flow is similar. In fact, meshes generated by other means including Python meshing packages such as MeshPy can be used equally well (see Programs 9.7 and 9.9).

The function `fem_mat()` for building the system matrix has been explained in the text. The main program obtains the nodes, elements, and boundary conditions by calling `mesh()`. As a precaution, the internal points `ip` are sorted (line 35) upon return. Once we have the mesh structure, the system matrix A is built. The boundary terms b_j are computed from Eq. (7.23) using the `np.dot` function for the inner product $b_j = \sum_k A_{jk}u_k$ over the boundary points k (recall the p_j's are zero for the Laplace equation, no charges).

Next, we delete the rows and columns corresponding to the boundary points from matrix A. This is done via the `np.delete` function (line 40), which removes entries contained in `bp` from the dimension specified by the `axis` parameter: 0 for rows and 1 for columns. Finally, the linear system is solved with `solve()`.

The returned solution is stored in the array `u` in the order of the internal nodes after the removal of boundary nodes from matrix A. We must be certain what internal node corresponds to a given index in `u`. This is why we sort the internal nodes (`ip`) before deletion as a precaution. If they are already in sorted order, this step would be unnecessary.

This is important as we combine the internal and boundary values next (lines 44 to 47) for graphing (or other purposes such as data analysis). First, both the node values and the nodes themselves are concatenated. Because the internal (`ip`) and boundary (`bp`) nodes already correspond to the right order in `u` and `bv` arrays, respectively, the concatenated nodes in `all` correlate one-to-one with the node values in the concatenated array `u`. Next, we index-sort the nodes in array `all` using `np.argsort`, which returns the indices `idx` pointing to the order that would sort the array. The same `idx` will also sort the concatenated array `u` such that the sorted node values correspond to the sorted nodes in `all`. Thus, we sort `u` by taking the elements in the order in `idx` (line 47). Now, `u` contains the node values in the correct order, that is, the value at node `n` is given by `u[n]`.

Last, we reshape the 1D array `u` into the shape of our 2D grid (line 48) for surface and contour plotting at the end. If the grid is not square, use the Matplotlib function `tripcolor()`, which works well for plotting data over triangular meshes. See Program 9.8.

Program listing 7.4 Laplace equation by RBF (`laplacerbf.py`)

```python
from scipy.linalg import solve
import matplotlib.pyplot as plt, numpy as np

def initialize (L, R, N=22):            # calc nodes and boundary values
    inode, bnode, b = [], [], []        # intnl, bndry nodes, bndry values
    x, y = np.linspace(0,L,N+1), np.linspace(0,L,N+1)  # same x,y grids
    for j in range(N+1):
        for i in range(N+1):
            r2 = (x[i]-L/2.)**2 + (y[j]-L/2.)**2
            if (i==0 or i==N or j==0 or j==N or r2<=R*R):
                bnode.append([x[i], y[j]])          # bndry node
                b.append(1. if r2<=R*R else 0.)
            else:
                inode.append([x[i], y[j]])
    ni, node = len(inode), inode+bnode              # combine nodes
    return np.array(node), [0.]*ni+b, ni, len(node)

def phi(i, r):                  # Gaussian φ_i and ∇²φ_i
    r2, e2 = (node[i,0]-r[0])**2 + (node[i,1]-r[1])**2, eps*eps
```

```
20      f = np.exp(-e2*r2)
21      return f, 4*e2*f*(e2*r2-1)
22
23  def rbf_mat(ni, nt):          # fills matrix A_ij = ∇_j^2 φ_i or φ_ij
24      A = np.zeros((nt, nt))
25      for j in range(nt):
26          f, df = phi(np.arange(nt), node[j])  # vector operation
27          A[j, :] = df if j < ni else f
28      return A
29
30  def usum(a, x, y):            # calc u at (x,y)
31      u = 0.
32      for i in range(len(a)):
33          u += a[i]*phi(i, [x, y])[0]      # [0]= first return value
34      return u
35
36  L, R, eps = 1.0, 0.25, 20.    # sizes of square, disk; shape parameter
37
38  node, b, ni, nt = initialize(L, R)      # ni, nt = num. intrnl/tot nodes
39  A = rbf_mat(ni, nt)
40  ai = solve(A, b)              # solve
41
42  x = np.linspace(0, L, 41)     # same x,y plotting grids
43  x, y = np.meshgrid(x, x)
44  u = usum(ai, x, y)
45
46  plt.figure()
47  img = plt.imshow(u, cmap=plt.cm.jet)  # plot as image
48  plt.colorbar(img), plt.axis('off')
49  plt.show()
```

We solve the Laplace problem with the RBF collocation method in Program 7.4. The program is structurally similar to Program 7.3. The first function `initialize()` sets up the nodes and boundary values, playing the role of mesh generation in FEM. It creates a uniform node distribution along x and y directions. The two loops iterate through the rows j and columns i. The function keeps two lists of nodes, internal or boundary nodes. At each node, it checks whether it is internal or on the boundary. If a node is within a radius R of a disk at the center of the box, or if it is on the walls of the box, it is added to the boundary nodes list (line 11). The next line sets the boundary value to 1 or 0, depending on its being on the disk or the walls. Otherwise, the node is added to the internal nodes list. For finite number of nodes, the disk is actually a polygon. In this case, it is approximately an octagon (Figure 7.18).

At the end, the internal and boundary nodes are added to create the total nodes list on line 15. When lists are added, the elements are combined in the same order to form a larger list, essentially like concatenating them. The node is returned as a NumPy array for vector operations (line 26, see following), together with the column vector **b**, which is formed by adding to the boundary values list a number of zeros (zero charge density) equal to the number of internal nodes. The other two returned values are the numbers of internal and total nodes. Note that the internal nodes come first and the boundary nodes second.

The next function `phi()` computes the Gaussian RBF value $\phi_i(\vec{r})$ and its derivative (Laplacian) value $\nabla^2\phi_i(\vec{r})$. It is called by `rbf_mat()` to form the system matrix A. It iterates over the rows, and for each row, it calculates all the entries in the entire row (all columns) via vector operation (line 26). To do so, it passes a vector list of indices in place of a single index, and because the node is stored as a NumPy array, `phi()` interprets this as an implicit element-wise vector operation. This is much faster than going through columns one by one. The next line fills the A matrix with the derivatives for the internal nodes and RBF values the for boundary nodes. The last function `usum()` just calculates the interpolant (7.31) given the expansion coefficients **a**. Because `phi()` returns a list of values, it picks the first value in line 33, which is the RBF value.

The main code sets the parameters, including the box size, radius of the disk, and the shape parameter ϵ. Then it generates the nodes and boundary values, the system matrix, and solves the linear system to obtain the expansion coefficients **a**. The potential is calculated over a grid for plotting. The result is shown as a color-coded image via `imshow()` on line 47, with a color bar and the axis (frame) turned off.

Program listing 7.5 Magnetic dipole fields (`bdipole.py`)

```
   import visual as vp, numpy as np
 2 import matplotlib.pyplot as plt

 4 def bdipole(r):                       # calc magnetic fields at r⃗
       b, rel = vp.vector(0,0,0), r-rs
 6     for i in range(len(ds)):          # sum over segments
           b += vp.cross(ds[i], rel[i])/vp.mag(rel[i])**3    # ∑ I ds⃗ × r⃗/r³
 8     return b

10 divs, m, n = 36, 21, 16              # circle divisions, space grid
   phi = np.linspace(0., 2*np.pi, divs+1)[:-1]               # omit last pt
12 rs = np.column_stack((np.cos(phi), np.sin(phi), 0*phi))   # segments
   ds = np.column_stack((-np.sin(phi), np.cos(phi), 0*phi))  # tangents ds⃗
14 x, z = np.linspace(-2, 2, m), np.linspace(-1.5, 1.5, n)
   x, z = np.meshgrid(x, z)
16
   bx, bz, big = np.zeros((n,m)), np.zeros((n,m)), 100.
18 for i in range(n):
       for j in range(m):
20         b = bdipole(vp.vector(x[i,j], 0., z[i,j]))
           if (vp.mag(b) < big):                 # exclude large fields
22             bx[i,j], bz[i,j] = b.x, b.z

24 plt.subplot(111, aspect='equal')
   plt.quiver(x, z, bx, bz, width=0.004, minshaft=1.5, minlength=0)
26 plt.plot([-1], [0], 'bo', [1], [0], 'bx')       # eyes
   plt.show()
```

The code calculates and displays the magnetic fields due to a circular loop in the x-y plane. The function `bdipole()` assumes the loop divided into a number of segments $d\vec{s}$ at positions $\vec{r}_s$. We approximate the magnetic field at a given location $\vec{r}$ by summing over all the segments, each contributing an amount $d\vec{s} \times \vec{r}_{rel}/r_{rel}^3$ ($\vec{r}_{rel} = \vec{r} - \vec{r}_s$, Figure 7.20). This is conveniently accomplished via the vector product operations `vp.cross` in VPython.

The main program divides the current loop into `divs` segments and hence `divs+1` points for the variable `phi` between 0 and 2π. The position vector $\vec{r}_s$ of each segment on the unit circle and the associated tangent vector ($d\vec{s}$) are computed and stored in `rs` and `ds`, respectively, as 2D arrays of vectors. These vectors are formed by column stacking the (x, y, z) components using the NumPy function `np.column_stack` (line 12), which puts 1D arrays as columns side-by-side to make a 2D or 3D array. Note that the tangent vector is just a rotation of the unit vector $\hat{j}'$ in Eq. (4.63).

The magnetic fields are plotted using the `quiver()` function. The length of the arrow indicates the relative magnitude of the magnetic field, so that we need not worry about units (such as the units of current or the length of the segments). The approach works well for grid points not too close to the loop itself. Because of the $1/r^2$ divergence, the small neighborhood around the wire is treated separately. There, the magnetic fields diverge and would dominate the graph. For this reason, a number `big` is set to exclude these regions. To accurately represent these regions, we need to plot the fields along concentric circles around the wire as shown in Figure 7.21 because they then have comparable magnitudes.

Program listing 7.6 Magnetic fields of a current-carrying wire (`longwire.py`)

```
1  import visual as vp, numpy as np
2  
3  m, n = 6, 30                                # r, phi intervals
4  r = np.linspace(0.15, 0.4, m)
5  phi = np.linspace(0., 2*np.pi, n+1)
6  r, phi = np.meshgrid(r, phi)                # n x m grid
7  
8  x, y = r*np.cos(phi), r*np.sin(phi)         # pos and dirs of B-fields
9  ax, ay = -np.sin(phi), np.cos(phi)          # tangential vectors
10 
11 scene = vp.display( title='Magnetic fields', background=(.2,.5,1))
12 wire = vp.curve(pos=[(0,0,.6), (0,0,-.6)], radius = 0.01)
13 
14 scale, q = 0.008, 1.0
15 for z in np.linspace(.5, -.5, 6):
16     for i in range(n):
17         for j in range(m):
18             vp.arrow(pos=(x[i,j], y[i,j], z), axis=(ax[i,j], ay[i,j]),
19                      length=scale/r[i,j], color=(q, q, 0))
20     q = q - .05
```

The magnitudes of the magnetic fields of a long current-carrying wire vary as $1/r$ (7.43), and their directions (line 9) are tangential to concentric cylinders around the wire. The latter are calculated similarly to Program 7.5 with the aid of Eq. (4.63). The color intensity variable q dims the VPython arrows further into the page. This helps with depth perception because as they move further back (and out), the arrows are too small to judge from light reflection.

Program listing 7.7 Electric dipole fields (`edipole.py`)

```
import visual as vp, numpy as np

r, scale, m, n = 0.5, 0.05, 11, 19                # parameters
scene = vp.display( title='Electric dipole', background=(.2,.5,1),
                    forward=(0,-1,-.5), up=(0,0,1))
zaxis = vp.curve(pos=[(0,0,-r),(0,0,r )])
qpos = vp.sphere(pos=(0,0,.02), radius=0.01, color=(1,0,0))
qneg = vp.sphere(pos=(0,0,-.02), radius=0.01, color=(0,0,1))
c1 = vp.ring(pos=(0,0,0), radius=r, axis=(0,0,1), thickness=0.002)
c2 = vp.ring(pos=(0,0,0), radius=r, axis=(0,1,0), thickness=0.002)
```

```
12 theta, phi = np.linspace(0, np.pi, m), np.linspace(0, 2*np.pi, n) # grid
13 phi, theta = vp.meshgrid(phi, theta)
14 rs = r*np.sin(theta)
15 x, y, z = rs*np.cos(phi), rs*np.sin(phi), r*np.cos(theta)   # coord.
16 for i in range(m):
17     for j in range(n):
18         rvec = vp.vector(x[i,j], y[i,j], z[i,j])
19         B = scale*vp.cross(rvec, vp.vector(0,0,1))/(r*r)     # r⃗×ẑ/r^2
20         E = vp.cross(B, rvec)/r                               # B⃗×r⃗/r
21         vp.arrow(pos=rvec, axis=E, length=vp.mag(E), color=(1,1,0))
22         vp.arrow(pos=rvec, axis=B, length=vp.mag(B), color=(0,1,1))
```

The program first draws the z-axis, a pair of charges representing the dipole and two great circles in the x-y and x-z planes. Real work begins at line 12, which divides the sphere into grids over the polar (θ) and azimuthal (φ) angles. The grid is converted into Cartesian coordinates (line 15), where the field vectors will be drawn.

The double loop calculates the electromagnetic fields from Eq. (7.45) at a fixed time, so the factor $\cos(kr - \omega t)$ is omitted and absorbed into the scaling factor `scale`, which also controls the proportion of the arrows. We use VPython's vector product operations to simplify the calculations.

Program listing 7.8 Angular distribution of dipole radiation (`dipole.py`)

```
1  import numpy as np
2  import matplotlib.pyplot as plt
3  from mpl_toolkits.mplot3d import Axes3D
4
5  theta = np.linspace(0, np.pi, 21)
6  phi   = np.linspace(0, 2*np.pi, 41)
7  theta, phi = np.meshgrid(phi, theta)
8  x = np.sin(theta)*np.cos(phi)
9  y = np.sin(theta)*np.sin(phi)
10 z = np.sin(theta)*np.cos(theta)       # z=radial times cos(theta)
11
12 plt.figure()
13 ax = plt.subplot(111, projection='3d', aspect=.5)
14 ax.plot_surface(x, y, z,  rstride=1, cstride=1, color='w')
15 plt.axis('off')
16 plt.show()
```

We first generate a grid over polar and azimuthal angles, from which we calculate the Cartesian grids, $x = \sin\theta\cos\varphi$ and $y = \sin\theta\sin\varphi$, over the unit circle. The dipole radiation pattern is a polar plot revolved around the z-axis. At a given θ, the magnitude (radial distance) is proportional to $r \sim \sin\theta$, and as a result, $z = r\cos\theta \sim \sin\theta\cos\theta$ (line 10). The actual surface is made with `plot_surface` with an aspect ratio less than one to flatten the donut-shaped surface.

Program listing 7.9 Plane electromagnetic waves (`plane.py`)

```
import visual as vp, numpy as np, vpm

t, dt, v, L = 0.0, 0.01, 0.2, 1.0          # time, speed, wavelength
n, scale, E, B = 61, 0.2, [], []           # grid, scale
```

```
6  scene=vp.display(title='Electromagnetic waves', background=(.2,.5,1),
                    center=(0,0,L), forward=(-.4,-.3,-1))
8  ax, ay, az = (1,0,0), (0,1,0), (0,0,1)             # axes and labels
   vp.arrow(pos=(.2, -.2, L), axis=ax, length=0.2, color=(1,1,0))
10 vp.arrow(pos=(.2, -.2, L), axis=ay, length=0.2, color=(0,1,1))
   vp.arrow(pos=(.2, -.2, L), axis=az, length=0.2, color=(1,1,1))
12 vp.label(pos=(.45, -.2, L), text='E', box=False, height=30)
   vp.label(pos=(.25, -.0, L), text='B', box=False, height=30)
14 vp.label(pos=(.2, -.15, L+.3), text='v', box=False, height=30)

16 idx, z = np.arange(n), np.linspace(-L, 2*L, n)  # order of vectors
   mag = scale*np.sin(2*np.pi*z/L)                   # sine envelope
18 ewave = vp.curve(color=(1,1,0), pos=np.column_stack((mag,0*z,z)))
   bwave = vp.curve(color=(0,1,1), pos=np.column_stack((0*z,mag,z)))
20 for i in idx:
       E.append( vp.arrow(pos=(0, 0, z[i]), axis=ax, length=mag[i],
22                        color=(1,1,0)) )
       B.append( vp.arrow(pos=(0, 0, z[i]), axis=ay, length=mag[i],
24                        color=(0,1,1)) )
   while True:
26     vp.rate(100), vpm.wait(scene)                  # hit a key to pause
       t, mg = t + dt, mag*np.cos(t)                  # sinusoidal wave
28     for i in range(n):
           E[i].pos.z += v*dt                         # traveling wave
30         B[i].pos.z += v*dt
           if (E[i].pos.z > 2*L):                     # wrap around
32             E[i].pos.z, B[i].pos.z = -L, -L
               idx = np.insert(idx, 0, i)[:-1]        # move to end
34
           E[i].axis, B[i].axis = ax, ay              # reset axis to
36         E[i].length, B[i].length = mg[i], mg[i]   #   draw correctly
           id = idx[i]
38         ewave.pos[i] = (mg[id], 0, E[id].pos.z)   # envelope curves
           bwave.pos[i] = (0, mg[id], B[id].pos.z)
```

After setting up the axes and labels without boxes, sinusoidal electromagnetic field vectors are created along the z-axis. The E vectors are in the x direction, and B vectors in the y direction. We also place envelope curves over the tips of the vectors. The points on the curves are created by column stacking as in Program 7.5. In this case, the points in the E-curve are in a $(n \times 3)$ array formed by stacking E magnitudes for x, n zeros for y, and n grid points for z. It is similar for the B curve.

The propagation of the electromagnetic wave is simulated by moving the vectors in the z direction and varying their magnitudes sinusoidally. When the vectors at the wavefront move beyond the set z limit, we assume they fall off the front and append them to the end, effectively enforcing periodic condition. To effect this strategy, an index variable `idx` is created to maintain the order of the vectors. When they move out of the front range, they are inserted back to the other end of the array (line 33).

Finally, a caution on working with VPython arrows. As of VPthon version 5.7x, when specifying arrows with a negative length, the axis is flipped. This is a useful behavior, and we have used it to our advantage. On the other hand, if we are working with the assumption that they are relative to a fixed direction, we must reset the axis to ensure this consistent default behavior.

CHAPTER 8 Time-Dependent Quantum Mechanics

Can a particle be at two places simultaneously? Could a cat be dead and alive at the same time? If we asked these questions in previous chapters on classical physics, the answer would seem to be obviously silly. However, in the context of quantum physics, the answer is neither obvious nor as silly as it may seem. Enter the fascinating world of quantum physics at the atomic scale where classical physics breaks down, and where few classical beings like us really have any intuitive experience. Even now, quantum mechanics is still somewhat mysterious, though its validity has been confirmed by available experiments for almost a century since its firm establishment.

The student of classical physics begins by studying motion. To gain a feel for quantum physics, it is important that we also begin with motion in the quantum world. What is free fall motion like in quantum mechanics? How about a simple harmonic oscillator? By studying, simulating, and observing motion in quantum physics, we begin to build up intuition and to understand the conceptual underpinning of quantum mechanics. This may be done even without prior knowledge of quantum mechanics, at least for Sections 8.2 and 8.3, if we treat the Schrödinger equation that governs the quantum world as just a wave equation analogous to wave motion discussed in Section 6.5.

However, quantum motion is not as direct or as easy to study as classical motion. The Schrödinger equation is a partial differential wave equation, and its solutions are not local like Newton's second law. Unlike deterministic Newtonian mechanics, quantum mechanics is both deterministic and probabilistic. As a result, quantum mechanics suffers from what may be called the "unsolvability" problem, that is, meaningful time-dependent quantum systems that are analytically solvable are few and far in between.

This is where numerical simulations are particularly valuable to reveal aspects of quantum mechanics that cannot be easily perceived. We will describe several methods suitable for simulating time-dependent quantum systems. We begin with a direct method that converts the Schrödinger equation into a system of ordinary differential equations (ODEs), and use it to study the quantum motion of a simple harmonic oscillator. This is the most direct and straightforward approach. Next, we discuss a split-operator method to deal with motion in open space such as free fall. We also introduce a two-state model to study transitions and to illustrate the concept of coherent states. Finally, we briefly discuss quantum waves in 2D and quantum revival.

8.1 TIME-DEPENDENT SCHRÖDINGER EQUATION

In quantum mechanics the notion that a particle's position and velocity can be simultaneously determined is abandoned. Instead, due to particle-wave duality, a particle is described by the wave function, $\psi(x, t)$ [44]. The interpretation of the wave function is such that $|\psi|^2 dx$ is the probability of finding the particle in space from x to $x + dx$ at time t. For a normalized wave function, the probability over all space is unity,

$$\int_{-\infty}^{\infty} |\psi(x, t)|^2 \, dx = 1. \tag{8.1}$$

The wave function evolves in time according to the time-dependent Schrödinger equation (TDSE)

$$-\frac{\hbar^2}{2m}\frac{\partial^2\psi(x,t)}{\partial x^2} + V(x,t)\psi(x,t) = i\hbar\frac{\partial\psi(x,t)}{\partial t}. \tag{8.2}$$

Here, $V(x, t)$ is the potential, $\hbar$ the rationalized Planck constant, and m the mass of the particle. The TDSE is fundamental to quantum mechanics as Newton's second law is to classical mechanics.

Compared to waves on a string (6.62), the TDSE for matter waves is first order in time and second order (same) in space.[1] Furthermore, it contains the unit imaginary number $i = \sqrt{-1}$, making the time-dependent wave function necessarily complex.

Except for a few trivial interaction potentials, no analytic solutions exist for the wave function $\psi(x, t)$. The simplest case is the motion of a free particle, $V(x, t) = 0$, whose solution is called a plane wave, $\psi(x, t) = A\exp(i(kx - \omega t))$. It is similar to plane electromagnetic waves (Figure 7.25), where k is the wave vector and $\omega = \hbar k^2/2m$. The wave function at different times is shown in Figure 8.1 (animated in Program 8.1).

The real and imaginary parts of the wave function are sinusoidal waves, phase shifted by $\frac{\pi}{2}$. However, the probability density is constant (a bit of oddity already).[2] A plane wave has a well-defined momentum ($\hbar k$) and no uncertainty, $\Delta p = 0$. According to Heisenberg's uncertainty principle $\Delta p \Delta x \geq \hbar/2$, it has an infinite uncertainty in position, that is, no beginning or ending. This gives rise to the constant probability density over all space. It is a useful idealization of a free particle that does not exist in reality.

In most cases when the potential is nonzero, Eq. (8.2) must be solved numerically using appropriate computational methods. In such cases, a sensical unit system should be used as discussed next.

Atomic Units

Given the small constants such as $\hbar$ in the TDSE, it would be cumbersome to use SI units in actual calculations. For quantum systems in atomic, molecular, or solid state physics, a natural unit system is the atomic units (a.u.). It is based on the Bohr model of the hydrogen atom.

In the a.u. system, the units for mass and charge are the electron mass m_e and basic charge e, respectively. The length is measured in Bohr radius a_0, and energy is chosen equal to the

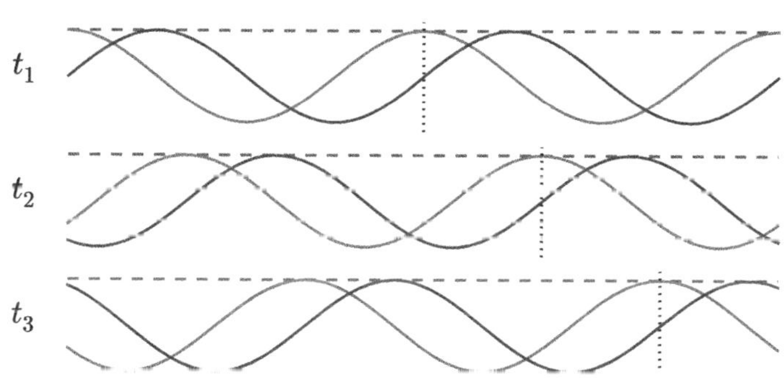

FIGURE 8.1 The wave function (solid curves: real and imaginary parts) and probability density (dashed) of a free particle. Dotted line: a wavefront.

[1] If you are unfamiliar with Eq. (8.2) but familiar with wave motion, it may be best to think of it as a wave equation for now and explore its behavior purely as a partial differential equation (PDE).

[2] Many aspects of quantum mechanics do not seem intuitive to us classical beings, making computer simulations of quantum systems all the more expedient and interesting.

Table 8.1 The Atomic Units (a.u.) System.

Quantity	Expression	Conversion to SI units
charge	e	1.60×10^{-19} C
mass	m_e	9.11×10^{-31} kg
length	a_0	5.29×10^{-11} m
energy	E_0	4.36×10^{-18} J
angular momentum	$\hbar$	1.05×10^{-34} J·s
time	$\frac{\hbar}{E_0}$	2.42×10^{-17} s
velocity	αc	2.19×10^{6} m/s
frequency	$\frac{E_0}{\hbar}$	4.14×10^{16} 1/s
electric field	$\frac{E_0}{ea_0}$	5.14×10^{11} V/m

magnitude of the potential energy of the electron at the radius a_0 in the Bohr atom (Hartree). They can be most conveniently expressed in terms of the universal fine structure constant α and speed of light c as

$$\alpha = \frac{1}{4\pi\varepsilon_0}\frac{e^2}{\hbar c} \simeq \frac{1}{137}, \quad \text{(fine structure constant)} \tag{8.3}$$

$$a_0 = \frac{\hbar}{\alpha m_e c} = 0.529\ \text{Å}, \quad E_0 = \alpha^2 m_e c^2 = 27.2\ \text{eV}.$$

Given the chosen units for mass, charge, length, and energy, all other units can be derived from these four. Their relationship and numerical values (in SI) are summarized in Table 8.1.

We can express the TDSE in a.u. for an electron as (Exercise E8.1)

$$-\frac{1}{2}\frac{\partial^2\psi(x,t)}{\partial x^2} + V(x,t)\psi(x,t) = i\frac{\partial\psi(x,t)}{\partial t}. \quad \text{(TDSE in a.u.)} \tag{8.4}$$

Comparing Eqs. (8.2) and (8.4), we see that the net effect in switching to atomic units is dropping all constants, equivalent to setting $\hbar = m_e = e = 1$ in the equations.[3] In the rest of this chapter, we will use atomic units unless otherwise noted. We also assume the particles to be electrons.

8.2 DIRECT SIMULATION

We first consider a direct method for solving the TDSE (and other time-dependent PDEs alike) by converting it to a set of ODEs.[4] This will enable us to start simulating quantum systems with minimum delay using standard ODE solvers.

8.2.1 SPACE DISCRETIZED LEAPFROG (SDLF) METHOD

We discretize space into grid points $x_j = jh$, where h is the grid size, and $j = 0, 1, ..., N$. The second-order differential operator $\partial^2/\partial x^2$ is approximated the same way as before, Eq. (6.73a), such that

$$\frac{\partial^2\psi(x,t)}{\partial x^2} \simeq \frac{\psi(x+h,t) - 2\psi(x,t) + \psi(x-h,t)}{h^2}. \tag{8.5}$$

[3] For Coulomb potentials, we can set $e^2/4\pi\varepsilon_0 = 1$ in atomic units.

[4] This technique is also known as the method of lines.

Let $\psi_j(t) = \psi(jh, t)$ and $V_j(t) = V(jh, t)$ as usual, we obtain the space discretized TDSE by substituting Eq. (8.5) into (8.4),

$$\frac{d\psi_j(t)}{dt} = \frac{i}{2h^2}\left(\psi_{j+1}(t) - 2\psi_j(t) + \psi_{j-1}(t)\right) - iV_j(t)\psi_j(t). \tag{8.6}$$

This is a set of $N + 1$ ODEs, to be solved given the initial wave function over space, $\psi_j(0)$. This way, the solution to the TDSE is reduced an initial value problem.

Before we can integrate Eq. (8.6), we need to consider the fact the Schrödinger equation preserves the normalization (8.1), so an ODE solver that preserves the norm (or flux) should be used. We have discussed in Chapter 2 that the leapfrog method has this special property. It is our method of choice here.

To rearrange Eq. (8.6) suitable for leapfrog integration, we separate the wave function into its real and imaginary parts, R_j and I_j, respectively. Let

$$\psi_j(t) = R_j(t) + iI_j(t). \tag{8.7}$$

Substitution of Eq. (8.7) into (8.6) yields (Exercise E8.2)

$$\frac{dR_j}{dt} = -\frac{1}{2h^2}I_{j-1} + \left(\frac{1}{h^2} + V_j\right)I_j - \frac{1}{2h^2}I_{j+1}, \tag{8.8a}$$

$$\frac{dI_j}{dt} = \frac{1}{2h^2}R_{j-1} - \left(\frac{1}{h^2} + V_j\right)R_j + \frac{1}{2h^2}R_{j+1}. \tag{8.8b}$$

We have assumed that the potential is real, as physical potentials are.

Besides being real, Eqs. (8.8a) and (8.8b) are in the proper form for leapfrog integration. The derivative of the real part depends only on the imaginary part, and vice versa. If we regard R_j as the general "coordinates," and I_j as the general "velocities," these equations mirror exactly the pair of equations (2.44) and (2.45), with R_j corresponding to r_j and I_j to v_j.

Let us express Eqs. (8.8a) and (8.8b) in a more compact matrix form. Let

$$a_j = \frac{1}{h^2} + V_j, \quad b = -\frac{1}{2h^2}, \text{ and}$$

$$\mathbf{R} = \begin{bmatrix} R_0 \\ R_1 \\ \vdots \\ R_N \end{bmatrix}, \quad \mathbf{I} = \begin{bmatrix} I_0 \\ I_1 \\ \vdots \\ I_N \end{bmatrix}, \quad A = \begin{bmatrix} a_0 & b & & & \\ b & a_1 & b & & \\ & \ddots & \ddots & \ddots & \\ & & & b & a_N \end{bmatrix}. \tag{8.9}$$

The matrix A is tridiagonal, so it is a band, sparse matrix.

In terms of these matrices, Eqs. (8.8a) and (8.8b) become

$$\frac{d\mathbf{R}}{dt} = A\mathbf{I}, \quad \frac{d\mathbf{I}}{dt} = -A\mathbf{R}. \tag{8.10}$$

Note the opposite signs in front of the matrix A. This result is formally equivalent to Eqs. (2.39) and (2.40) in that the change of the "velocity" vector $\mathbf{I}$ depends on the "acceleration," $-A\mathbf{R}$, which is a function of "position" vector $\mathbf{R}$ only, like in a Hamiltonian system. In fact, the analogy goes further: Eq. (8.10) describes a general "harmonic oscillator" in the same way Eqs. (2.39) and (2.40) do a simple harmonic oscillator, recalling $a(x) = -\omega^2 x$ in the latter. Consequently, we can formally view $\mathbf{R}^T\mathbf{R} + \mathbf{I}^T\mathbf{I}$ as a general energy, and conservation of energy in a Hamiltonian system requires that

$$\frac{d}{dt}\left(\mathbf{R}^T\mathbf{R} + \mathbf{I}^T\mathbf{I}\right) = 0, \tag{8.11}$$

which is a statement on the conservation of probability (8.1). We can prove this result analytically (Exercise E8.2).

Accordingly, we can formally port over the leapfrog method from Eqs. (2.41) and (2.42) to

$$\begin{aligned} \mathbf{R}_{\frac{1}{2}} &= \mathbf{R}_0 + A\mathbf{I}_0\frac{\Delta t}{2}, \\ \mathbf{I}_1 &= \mathbf{I}_0 - A\mathbf{R}_{\frac{1}{2}}\Delta t, \\ \mathbf{R}_1 &= \mathbf{R}_{\frac{1}{2}} + A\mathbf{I}_1\frac{\Delta t}{2}, \end{aligned} \tag{8.12}$$

where Δt is the step size. Equation (8.12) is the space discretized leapfrog (SDLF) method. It advances the solution one step forward, from $(\mathbf{R}_0, \mathbf{I}_0)$ to $(\mathbf{R}_1, \mathbf{I}_1)$, and more important, preserves the norm.

When the initial wave function is given at $t = 0$, we can march forward to obtain the wave function at later times. But, with what step size? As we saw earlier (Section 6.5.4), we also need to check the stability of an algorithm whenever finite differencing is involved in time stepping. A similar von Neumann stability analysis can be carried out for the SDLF (Exercise E8.3), and the stability criterion is

$$\Delta t < \lambda h^2, \quad \text{(stability condition, SDLF)} \tag{8.13}$$

where $0 < \lambda < 1$ is a coefficient dependent on the potential V. For free particle motion ($V = 0$), $\lambda = 1$. In general, the greater the potential, the smaller the λ. From experimentation, $\lambda = \frac{1}{2}$ should work for most cases.

Recall that for waves on a string, the stability condition (6.81) is $\Delta t \propto h$. Why is $\Delta t \propto h^2$ in the present case? As mentioned earlier, the Schrödinger equation is first order in time, whereas the wave equation is second order in time. This accounts for the h^2 dependence mathematically. Physically, the Schrödinger equation is similar to the diffusion equation. In diffusion, typified by Brownian motion such as a drop of ink in water, the particles spread out at a rate $\langle x^2 \rangle \propto t$ (see Section 10.3 and Eq. (10.40)). Equation (8.13) requires that the time step must be such that a particle does not exceed the physically allowed diffusion distance.

8.2.2 IMPLEMENTATION

We have everything in place to start implementing the direct simulation method. We need to write a derivative function, similar to Program 4.3, appropriate to the leapfrog integrator. First we declare two arrays, R and I, to hold the real and imaginary parts of the wave function over the spatial grid. They will be initialized to the wave function at $t = 0$ and stepped forward according to Eqs. (8.8a) to (8.8b).

It seems straightforward enough to program this. Depending on whether dR_j/dt or dI_j/dt is needed, it requires only one loop to sum up the three terms in each equation. But (there is always a but, isn't there? this but is different this time), consider this: each loop needs about N operations, and because the grid size is $h \sim N^{-1}$, the time step Δt required for stable integration scales as $\Delta t \propto h^2 \sim N^{-2}$. It means to advance the solution over some time interval (say 1 unit), the number of operations scales like $N_{op} \sim N^3$. Suppose $N = 1000$, then $N_{op} \sim 10^9$. This is large and taxing for an interpretive language like Python.

Thus, we need to make the aforementioned loop as efficient as possible. Fortunately, as in previous situations (see Sections 6.5.5, 6.7, or 7.2.1), the NumPy library makes it easy.[5] There are

[5] An alternative is to use Numba, or call compiled codes in Fortran or C/C++ using F2Py, Weave, or Cython (see Section 1.3.2, and Programs 5.7 and S:11.3).

two approaches. The first is to use fast matrix multiplication, realizing that A in Eq. (8.12) is tridiagonal and banded. The second is to use element-wise operations with array slicing to avoid explicit looping as in Program 6.4. We choose the second approach here for it is simpler and just as fast.

Observing from Eq. (8.8a), for instance, dR_j/dt depends on I_j and its left and right neighbors, I_{j-1} and I_{j+1}, respectively. We can thus obtain $d\mathbf{R}/dt$ at once by adding three arrays (up to some scaling constants) as: **I**, **I** right-shifted by one, and **I** left-shifted by one. This is the same method as in 1D waves discussed in Section 6.5.5 and illustrated in Figure 6.15.

The end points can be resolved assuming a given boundary condition. We will use periodic boundary conditions. Consider the actual system consisting of $N+1$ grid points with ψ_j, $j = 0, 1, ..., N$. Imagine identical copies of the system are placed next to each other along the x-axis. If we step past the right end point of the (actual) system, we will be at the first (left-most) point of the (virtual) system to the right. Likewise, if we go the other way and move past the left end point, we will be at the right-most point of the system to the left. We can express the periodic boundary condition as

$$\psi_{N+1} = \psi_0, \quad \psi_{-1} = \psi_N. \quad \text{(periodic BC)} \tag{8.14}$$

With Eq. (8.14), the actual derivative function, `sch_eqn()`, is given here.

```
1  def sch_eqn(id, R, I, t):                # return dR/dt, dI/dt of Sch. eqn.
2      b = 0.5/(h*h)
3      if (id == 0):                        # gen. velocity
4          tmp = -b*I                       # 1st/3rd term in Eq. (8.8a)
5          dydt = (2*b + V)*I               # 2nd term
6      else:                                # gen. acceleration
7          tmp = b*R                        # 1st/3rd term in Eq. (8.8b)
8          dydt = (-2*b - V)*R              # 2nd term
9
10     dydt[1:-1] += tmp[:-2] + tmp[2:]     # add bψj-1, bψj+1 to grid j
11     dydt[0]  += tmp[-1] + tmp[1]         # 1st point, periodic BC
12     dydt[-1] += tmp[-2] + tmp[0]         # last point
13     return dydt
```

It returns $d\mathbf{R}/dt$ if `id` is zero, else $d\mathbf{I}/dt$. In each case, a temporary array `tmp` is used to hold the first and the third terms of Eq. (8.8a) or (8.8b) before shifting, and the middle term is stored in `dydt`. The potential array `V` is built in the main program (Program 8.2). Then, the interior elements are added via slicing discussed earlier. Note `tmp[:-2]` implies all elements from the beginning up to (but not including) the second last, and `tmp[2:]` from the third to the end. The end points are treated separately according to the periodic boundary condition (8.14), which means that the left neighbor of the first point is the same as the last point, `tmp[-1]` (line 11), and the right neighbor of the last point is the same as the first point, `tmp[0]`.

Similarly, we can use the vectorized leapfrog code (Program 2.6), which succinctly expresses the algorithm (8.12). In addition, it has the advantage of speed gain. Each of the three steps in the leapfrog method is an implicit loop and can be computed much faster for large arrays, as is the case here. Profiling (Section S:8.B) shows that we gain about 50x speedup compared to explicit looping for $N = 500$.

8.2.3 QUANTUM HARMONIC MOTION

The complete code is given in Program 8.2, which yields the following results (slightly modified for plotting, Project P8.1). Figure 8.2 displays the motion of an electron in the simple harmonic

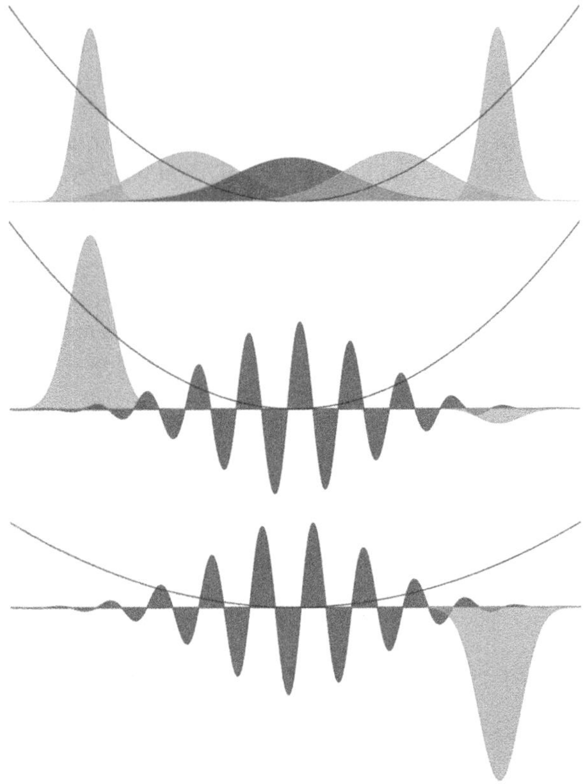

FIGURE 8.2 The probability density (top), the real and imaginary parts of the wave function (middle and bottom, respectively), in the SHO potential.

oscillator (SHO) potential (2.46), $V(x) = \frac{1}{2}x^2$. The initial wave function is a normalized Gaussian wavepacket,

$$\psi(x, 0) = \left(\sigma\sqrt{\pi}\right)^{-1/2} \exp\left[-\frac{(x - x_0)^2}{2\sigma^2} + ik_0x\right]. \tag{8.15}$$

The parameters x_0 and σ are the center and width of the Gaussian, respectively. The variable k_0 is the average momentum (at $t = 0$) of the wavepacket.

The results shown in Figure 8.2 are for $\sigma = \frac{1}{2}$, $x_0 = -3$ and $k_0 = 0$. The spatial range is set to $-10 \le x \le 10$, divided into $N = 500$ intervals (for a total of 501 grid points), so $h = 0.04$ and $\Delta t = \frac{1}{2}h^2 = 0.008$. This value of N gives us good balance between accuracy and speed. The top graph shows the probability distribution at five equidistant locations. The potential is also drawn for easy reference. The wavepacket moves to the right, accelerates, and broadens as it approaches the center. Past the center, it starts to slow down and becomes narrower until the right end point, where the wavepacket looks nearly identical to the original one (refocusing). Afterward, it reverses direction and moves toward the starting position.

The middle and bottom graphs in Figure 8.2 show the real and imaginary parts, R and I, respectively, of the wave function at three positions (left end, center, and right end). Initially, there is only the real part. With increasing t, both parts develop and become more oscillatory toward the center. The peaks of R and the nodes of I coincide with each other (vice versa), making the probability density smooth throughout. Being slightly offside, the real part does not vanish at the right end, where the imaginary part dominates. We would not have suspected this behavior based on the appearance of the probability density graph alone.

The evolution of the probability density is shown in Figure 8.3 as a function of space and time simultaneously. We can see more clearly from the surface plot the periodic motion of the wavepacket as it zigzags back and forth in time. The width of the wavepacket also oscillates periodically. We can observe in the contour plot the periodicities of both the probability density and the width more clearly. The period is close to 2π, which matches very well with the classical period of the SHO, $2\pi/\omega$ from Eq. (6.11), when the mass and spring constant are unity. We can also understand these oscillations as a result of quantum revival, discussed in Section 8.5.

Based on the results so far, we should expect that the average position, that is, the expectation value of position $\langle x\rangle$, to give a quantitative depiction of the periodic motion. Given the wave function $\psi(x, t)$, we can calculate the expectation value for position as

$$\langle x\rangle = \int_{-\infty}^{\infty} \psi^*(x, t)\, x\psi(x, t)\, dx. \tag{8.16}$$

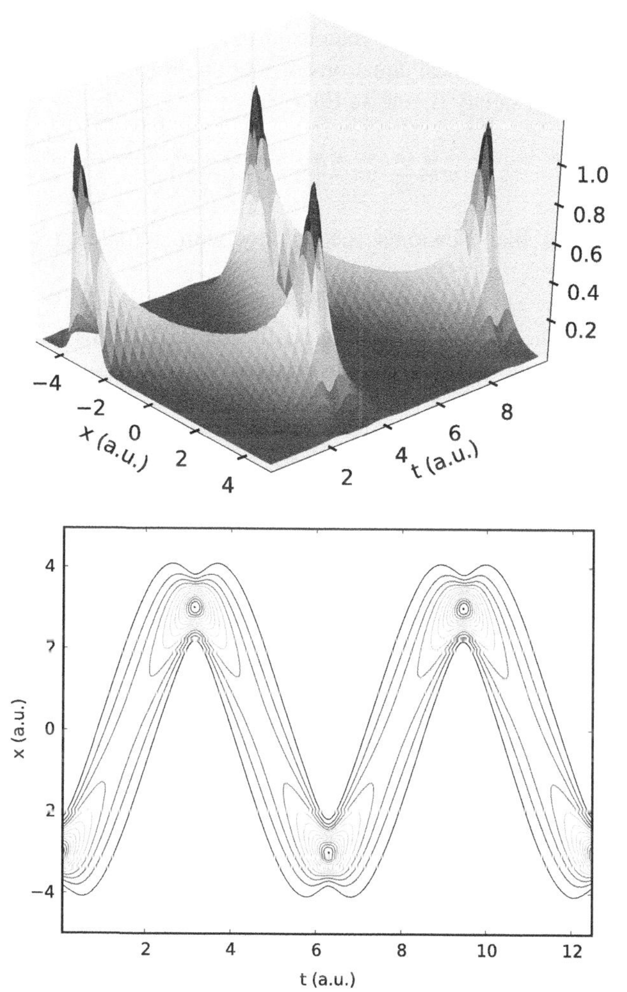

FIGURE 8.3 The surface and contour plots of probability density.

For momentum p and kinetic energy T, respectively, we have

$$\langle p\rangle = m\frac{d\langle x\rangle}{dt} = -i\hbar\int_{-\infty}^{\infty}\psi^*\frac{\partial\psi}{\partial x}\,dx = \hbar\,\mathrm{Im}\int_{-\infty}^{\infty}\psi^*\frac{\partial\psi}{\partial x}\,dx, \tag{8.17}$$

$$\langle T\rangle = \langle\frac{p^2}{2m}\rangle = -\frac{\hbar^2}{2m}\int_{-\infty}^{\infty}\psi^*\frac{\partial^2\psi}{\partial x^2}\,dx = \frac{\hbar^2}{2m}\int_{-\infty}^{\infty}\left|\frac{\partial\psi}{\partial x}\right|^2\,dx. \tag{8.18}$$

The last expressions in Eqs. (8.17) and (8.18) are useful for numerical calculations (see Exercise E8.5 and Project P8.2). For time-dependent systems, the preceding expectation values are generally a function of t.

We need to evaluate the integral (8.16) numerically. Like solving ODEs, numerical integration is a common procedure in scientific computing. We discuss several integration techniques in Section 8.A.

Because the wave function is discretized over a regular grid, it is simplest to use Simpson's rule (Project P8.1) to evaluate Eq. (8.16). The expectation value $\langle x\rangle$ as a function of time is graphed in Figure 8.4. As expected, it indeed follows a sinusoidal pattern like the classical SHO (2.47). From visual inspection, we infer $\langle x\rangle \simeq -A\cos t$.

The quantity $\langle x\rangle$ provides a bridge between classical and quantum mechanics through the Ehrenfest theorem, which describes the dynamics of physical observables governed by a set of relationships very similar to classical equations of motion. For instance, the expectation values of position and momentum satisfy (Exercise E8.4)

$$m\frac{d^2\langle x\rangle}{dt^2} = \frac{d\langle p\rangle}{dt} = -\left\langle\frac{\partial V}{\partial x}\right\rangle. \tag{8.19}$$

We see that this is just like Newton's second law with a force given by $F = -\langle\partial V/\partial x\rangle$ (see Eq. (2.46)).

In our case, $V = \frac{1}{2}x^2$, thus $F = -\langle x\rangle$. Substituting it into Eq. (8.19) yields $\langle x\rangle \sim -A\cos t$, just as observed earlier. Therefore, we can numerically confirm Eq. (8.19), $m\langle\ddot{x}\rangle = F$. So classical mechanics is recovered for space-averaged observables.[6]

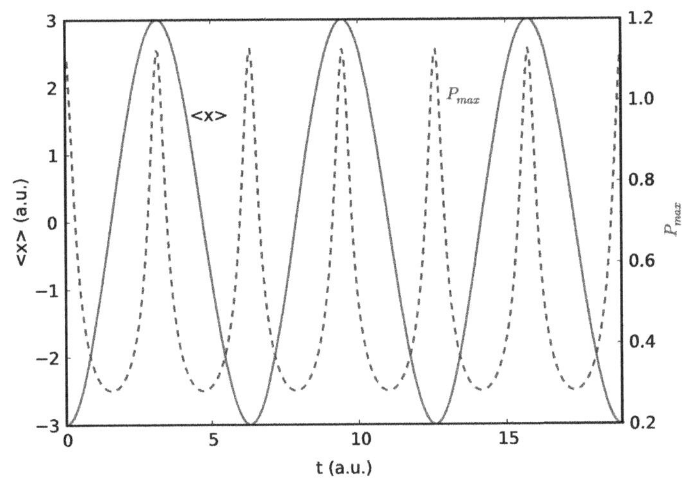

FIGURE 8.4 The expectation value of position (solid line) and the maximum of the probability density (dashed line).

[6]"It is weird that quantum mechanics is not weird," Nevin (a student of mine) thus exclaimed after seeing quantum mechanics in action and working like classical mechanics.

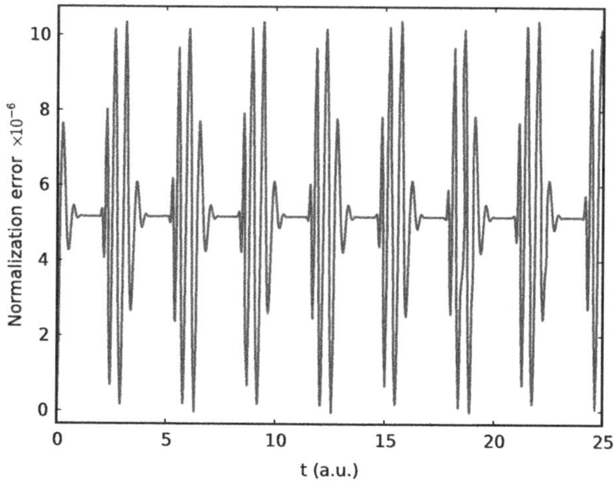

FIGURE 8.5 The normalization error, $1 - \int |\psi|^2 dx$.

Also shown in Figure 8.4 is the value of maximum probability density (right y-axis) as a function of time. The maximum value, P_{max}, refers to the peak of the distribution, which is approximately at the center of the wavepacket. We can think of P_{max} as a measure of how quickly the wavepacket broadens because it is inversely proportional to the width (roughly). P_{max} starts from the peak value at $t = 0$ when the Gaussian is at its narrowest. It reaches the lowest value at the moment the center of the Gaussian is at $x = 0$, where the wavepacket is broadest, before recovering its peak value at the right end. The cycle repeats afterward, so the period of P_{max} is half the period of the SHO. Because the SDLF method preserves the norm, the peak values of P_{max} stays constant.

We can keep track of normalization continuously by computing Eq. (8.1) at regular time intervals. Figure 8.5 shows the deviation of normalization from unity, that is, $1 - \int |\psi| dx$. The absolute error is on the order of 10^{-6} given the same grid size h and step size Δt as in Figure 8.2. The absolute error is less significant than its oscillatory behavior: It shows that the error is bounded, much like what we saw for the energy of the SHO using the leapfrog method (Figure 2.7). Here we see again that the leapfrog method, being finite order and symplectic, exactly preserves a numerically inexact result. Overall, the SDLF is a direct, robust method for solving the TDSE.

It is interesting to observe the absolute error exhibiting a period equal to half of the SHO's period, just like P_{max}. Correlating with Figure 8.4, we see that the error is largest when the wavepacket is at the end points when it is narrowest. A sharper wave function requires a denser grid to maintain accuracy. Making h or Δt smaller can help reduce the magnitude of the error. A more effective method is to make our numerical scheme unitary (norm preserving) to begin with, an approach we take next.

8.3 FREE FALL, THE QUANTUM WAY

The SDLF is an explicit method that marches forward given the initial wave function. It is direct and robust, provided that the stability condition (8.13) is fulfilled, which is also its major limitation. In this section, we discuss a first-order split-operator method, simply referred to as the split-operator method, which removes this limitation. The method is particularly useful for motion in open space such as scattering problems or unidirectional motion like free fall. We apply it to simulate the free fall of a quantum particle.

8.3.1 THE SPLIT-OPERATOR METHOD

To develop the split-operator method, we rewrite Eq. (8.4) more concisely utilizing the Hamiltonian H,

$$i\frac{\partial\psi(x,t)}{\partial t} = H\psi(x,t), \quad H = -\frac{1}{2}\frac{\partial^2}{\partial x^2} + V(x), \tag{8.20}$$

where we assume for the moment that the potential is time-independent.

Formally, we can write the solution to Eq. (8.20) as [80]

$$\psi(x,t) = e^{-iH(t-t_0)}\psi(x,t_0). \tag{8.21}$$

It is understood that the function of an operator $\hat{q}$ is defined through its Taylor series

$$e^{\hat{q}} = 1 + \hat{q} + \frac{\hat{q}\hat{q}}{2!} + \ldots \tag{8.22}$$

If the effect of $\hat{q}$ is small, we can retain the first few terms for accurate representation of the operator function.

The operator $e^{-iH(t-t_0)}$ is called the evolution operator for it advances the wave function from t_0 to t. The evolution operator is unitary, that is, it preserves the normalization at all time, $\int |\psi(x,t)|^2\, dx = \int |\psi(x,t_0)|^2\, dx$ (see Exercise E8.6). Because its exact solution is unknown, we must approximate it.

Let $\Delta t = t - t_0$, $\psi^0 = \psi(x,t_0)$, and $\psi^1 = \psi(x, t_0 + \Delta t)$. We have from Eq. (8.21)

$$\psi^1 = e^{-iH\Delta t}\psi^0. \tag{8.23}$$

Assuming Δt is small, we expand $e^{-iH\Delta t}$ according to Eq. (8.22), and to first order we have

$$\psi^1 \simeq (1 - iH\Delta t)\psi^0. \tag{8.24}$$

This is Euler's method for the TDSE. But, this form turns out to be numerically unusable because it is nonunitary and unstable no matter how small Δt is (Exercise E8.3).

The split-operator method, also known as the Crank-Nicolson method originally applied to heat conduction problems [21], takes a symmetric approach by splitting the exponential operator in halves and applying the first-order truncation to each half as

$$e^{-iH\Delta t} = e^{-i\frac{1}{2}H\Delta t}e^{-i\frac{1}{2}H\Delta t} = \frac{1}{e^{i\frac{1}{2}H\Delta t}}e^{-i\frac{1}{2}H\Delta t} \simeq \frac{1 - i\frac{1}{2}H\Delta t}{1 + i\frac{1}{2}H\Delta t}. \tag{8.25}$$

It can be shown that this result is accurate to first order in Δt and agrees with the Euler method (8.24) (Exercise E8.6). However, though an approximation, Eq. (8.25) is unitary,

$$\left[\frac{1 - i\frac{1}{2}H\Delta t}{1 + i\frac{1}{2}H\Delta t}\right]^{\dagger}\frac{1 - i\frac{1}{2}H\Delta t}{1 + i\frac{1}{2}H\Delta t} = \frac{1 + i\frac{1}{2}H\Delta t}{1 - i\frac{1}{2}H\Delta t}\,\frac{1 - i\frac{1}{2}H\Delta t}{1 + i\frac{1}{2}H\Delta t} = 1. \tag{8.26}$$

Physically, it means that the split operator preserves the normalization of the wave function regardless of the size of Δt.

Substituting Eq. (8.25) into (8.23) and acting on both sides by the operator in the denominator, we obtain

$$\left(1 + i\frac{1}{2}H\Delta t\right)\psi^1 = \left(1 - i\frac{1}{2}H\Delta t\right)\psi^0. \tag{8.27}$$

To look at this relationship another way, we have

$$\psi^1 = \psi^0 - iH\Delta t\frac{\psi^1 + \psi^0}{2}. \tag{8.28}$$

If we were to replace ψ^1 on the right-hand side (RHS) by ψ^0, we would recover Euler's method (8.24). The effect of the split-operator method is to replace ψ^0 in Euler's method by the average $(\psi^1 + \psi^0)/2$. This little trick makes a big difference.

To construct a numerical scheme, we discretize the Hamiltonian operator in Eq. (8.27) the same way as in SDLF over the spatial grid $x_j = jh$. Let $\psi_j^0 = \psi(x_j, t_0)$, $\psi_j^1 = \psi(x_j, t_0 + \Delta t)$. After some algebra (Exercise E8.7), the final result is

$$2\left[i\frac{2h^2}{\Delta t} - 1 - h^2V_j\right]\psi_j^1 + \psi_{j-1}^1 + \psi_{j+1}^1 =$$
$$2\left[i\frac{2h^2}{\Delta t} + 1 + h^2V_j\right]\psi_j^0 - \psi_{j-1}^0 - \psi_{j+1}^0. \tag{8.29}$$

Unlike the explicit SDLF method (8.12), the split-operator method is an implicit method: The value ψ_j^1 depends not only on previous values ψ_j^0 at t_0, but on the current values $\psi_{j\pm1}^1$ at t as well. Rather than simply marching forward, we have to invert Eq. (8.29) for all values of ψ_j^1 simultaneously at each step. This is the price we pay, in exchange for absolute stability guaranteed in Eq. (8.29) (Exercise E8.8).

Equation (8.29) had first been solved using clever recursion techniques [37]. We use a much simpler matrix method. Let us assume that, for the time interval of interest, the wave function at the boundaries are so small as to be negligible, $\psi_0 = \psi_N = 0$ (fixed ends). This can be interpreted in two ways. One is that the spatial range is so large that the particle (wavepacket) is far away from either end of the box. The other is that the boundaries are hard walls (infinite potentials) so the particle does not penetrate them. We will revisit this point shortly in discussing the results (Section 8.3.2).

Let

$$A = \begin{bmatrix} a_0 & 1 & & \\ 1 & a_1 & 1 & \\ & \ddots & \ddots & \ddots \\ & & 1 & a_N \end{bmatrix}, \quad \mathbf{u}^{0,1} = \begin{bmatrix} \psi_0^{0,1} \\ \psi_1^{0,1} \\ \vdots \\ \psi_N^{0,1} \end{bmatrix} \tag{8.30}$$

$$a_j = 2\left[i\frac{2h^2}{\Delta t} - 1 - h^2V_j\right], \quad B = -A^*,$$

and assuming fixed-end boundary conditions, we can express Eq. (8.29) in matrix form as

$$A\mathbf{u}^1 - B\mathbf{u}^0. \tag{8.31}$$

Equation (8.31) is the final result of the split-operator (Crank-Nicolson) method in matrix formulation. Given the wave function $\mathbf{u}^0$ at t_0, we obtain the new wave function $\mathbf{u}^1$ at $t + \Delta t$ by solving a linear system of equations. Because we must do this at each step, it would be very costly in general. However, we are helped by two facts: The matrix A is a tridiagonal, band matrix, and it is time-independent (if $V = V(x)$). The tridiagonal nature of A is due to approximating $\partial^2/\partial x^2$ by a three-point scheme (8.5).

We can use special matrix solvers designed for band matrices in the SciPy library, which reduce the number of operations linearly in N rather than N^2 in general. To use band matrix solvers, we first need to express the matrix A in the required format.

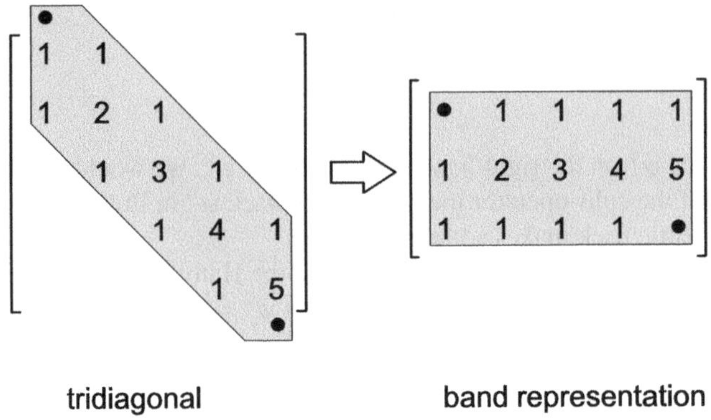

FIGURE 8.6 A tridiagonal matrix (5 × 5) and its band representation (3 × 5).

Figure 8.6 illustrates an example of representing a tridiagonal matrix as a band matrix. Our original matrix (A, source) has a maximum of three nonzero elements in each column. The band representation (B, target) is chosen to be a matrix with three rows and the same number of columns. The resultant matrix is just the band of the original matrix being flattened. The first and the last elements of the target, B_{11} and B_{35}, respectively (•), are dummy fillers not used in actual computation, so they can be set arbitrarily. For general band matrices, we just need to increase the number of rows to represent it properly. If u is the number of elements above the diagonal, then $A_{ij} = B_{i-j+u+1,j}$, $i \le j$. For instance, we have $u = 1$ for the tridiagonal matrix, such that $A_{34} = B_{14} = 1$, and $A_{44} = B_{24} = 4$, and so on.

The actual implementation of the split-operator method is straightforward. It consists of six lines shown next, taken from the full Program 8.3.

```
1 A = np.ones((3, N+1), dtype=complex) # prepare band matrix A, B
2 A[1,:] = 2*(2j*h*h/dt - 1 - h*h*V)
3 dB = - np.conjugate(A[1,:])             # diagonal of B
4 ......
5    C = dB*psi                           # prepare RHS Eq. (8.31)
6    C[1:-1] -= (psi[:-2] + psi [2:])
7    psi = solve_banded((1,1), A, C)      # band matrix solver
```

The first line creates a $3 \times (N + 1)$ matrix, data-typed complex and filled with ones initially. The second line replaces the second row with the complex diagonal elements from Eq. (8.30) using vector operation, completing the band representation of matrix A. The third line creates a 1D array containing the diagonal of matrix B in Eq. (8.30) only, which is the negative complex conjugate of the diagonal of A.

Next, we prepare the RHS of Eq. (8.31). Rather than direct matrix multiplication, we obtain it from the RHS of Eq. (8.29) because SciPy does not yet have a banded matrix multiplication routine. Line 5 computes the first term on the RHS of Eq. (8.29), that is, the diagonal elements of $B\mathbf{u}^0$ in Eq. (8.31). This is why we did not need the full matrix B earlier. The next line subtracts the last two terms, the left and right neighbors, respectively, using the same shift-slicing technique as before (Section 8.2.2). For fixed ends, we neglect the boundary points because $\psi_0^0 = \psi_N^0 \simeq 0$. Now, we have completed the RHS of Eq. (8.31).

The last line (line 7) calls the SciPy band matrix solver, `solve_banded()`, to obtain the solution ψ^1 at $t + \Delta t$. The function requires a minimum of three parameters as

```
solve_banded((l,u), A, b)
```

It solves a linear system, $A\mathbf{x} = \mathbf{b}$, assuming A is in band matrix format. The `(l,u)` pair specifies the lower and upper diagonals, that is, the numbers of elements below and above the diagonal in the band, respectively, such that the band width is `l+u+1`. For our tridiagonal matrix, we have `(l,u)=(1,1)` in line 7.

8.3.2 QUANTUM FREE FALL

As a test case, we apply the split-operator method to the free fall of a quantum particle. We choose the linear potential to be

$$V(x) = ax. \tag{8.32}$$

The parameter a plays the same role as the gravitational acceleration in free fall, mgy. Such a scenario exists for electrons in a constant electric field.

Figure 8.7 shows the results from Program 8.3. The initial wave function is again a Gaussian (8.15), and the acceleration is $a = 2$. The center of the wavepacket falls toward the negative x direction as expected. It also broadens as before, but unlike the SHO potential, there is no refocusing effect in the linear potential, and the wavepacket just keeps broadening.

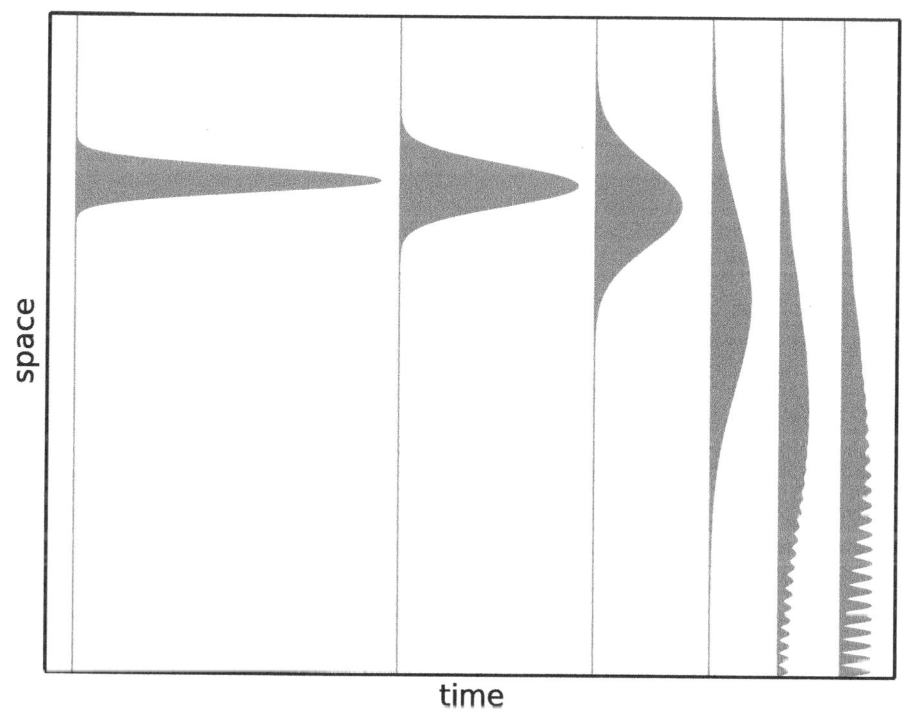

FIGURE 8.7 A quantum particle in free fall: probability density $|\psi(x,t)|^2$.

Because a wavepacket is made up of different momentum components that spread out at different speeds, the broadening is irreversible if there is no refocusing force to restore the motion. The SHO potential provided a restoring force causing the wavepacket to narrow and broaden periodically (Section 8.2.3). The linear potential, however, yields a constant force that accelerates different parts of the wavepacket equally, and the spreading continues unabated.

Another difference between the linear and SHO potentials is that free fall motion is unidirectional and has no lower bound. As a result, when waves inevitably reach the finite boundaries in our simulation, it will be reflected due to our assumption that the boundary values are small, $\psi_0 = \psi_N \simeq 0$. We have seen similar reflections on a vibrating string (Section 6.5.5). This causes the leading edge of the waves to interfere with its self-reflection. The growing ripples we see in the last three frames (Figure 8.7) are self-interference patterns. These patterns are clearly an artifact of our domain being finite, and not real physical effects. In infinite space, the wavepacket would just keep spreading, so the amplitude would become ever smaller while its width ever larger. This cautions us that, in interpreting simulation results or comparing with observations such as experimental data, we should consider boundary effects carefully and take steps to minimize them if possible. Conversely, as stated earlier, we can interpret the results as free fall between hard walls, in which case the interference patterns are real and due to actual reflections.

We note that the split-operator method as implemented initially preserves the normalization (8.1) to a high degree of accuracy. As time increases, error starts to grow due to waves reaching the boundaries. The leading waves have higher momentum so they are highly oscillatory as seen in Figure 8.7. To maintain accuracy, smaller grids are needed (see Project S:P8.1). A more accurate, second-order split evolution operator (SEO) method is developed in Section S:8.1.

Figure 8.8 shows a contour plot of the probability density as a function of time using the same parameters as in Figure 8.7. In addition to broadening and self-interference discussed earlier, we see the center of the wavepacket following a parabolic curve (before reaching the boundary). The classical free fall motion is given by Eq. (2.2), which in our case is $x = 5 - t^2$ ($a = 2$). This is drawn as the dashed line in Figure 8.8. We see good agreement and an interesting connection between quantum waves and classical trajectory motion, which is a another clear demonstration

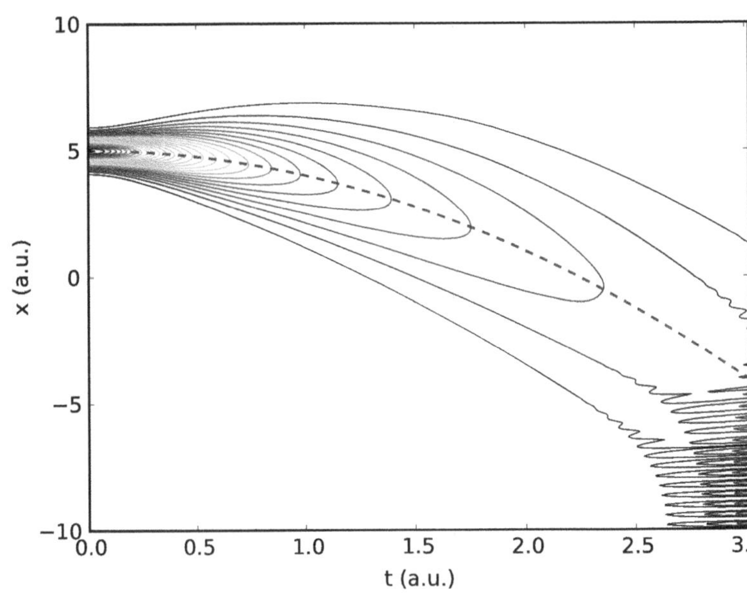

FIGURE 8.8 The probability density of a quantum particle in free fall. The dashed line shows the classical path.

of the Ehrenfest theorem (8.19). A quantum wave moves along the classical path of a particle in terms of average position and force (expectation values).

8.3.3 MOMENTUM SPACE REPRESENTATION

From the motion of the wavepacket in coordinate space, we have learned a good deal about the momentum (velocity) of the particle. Better yet, quantum mechanics allows us to get an entirely different, but equivalent, view from the wave function in momentum space. In a.u., the momentum is equal to the wavevector $\hbar k = k$.

The wave function in momentum space, $\phi(k, t)$, and in coordinate space, $\psi(x, t)$, are related by a pair of reciprocal Fourier transforms (5.34) and (5.35),

$$\phi(k,t) = \frac{1}{\sqrt{2\pi}} \int_{-\infty}^{\infty} \psi(x,t)e^{-ikx}\, dx, \tag{8.33a}$$

$$\psi(x,t) = \frac{1}{\sqrt{2\pi}} \int_{-\infty}^{\infty} \phi(k,t)e^{ikx}\, dk. \tag{8.33b}$$

It is sometimes convenient to use a shorthand notation for these Fourier transforms, $\phi(k) = \mathcal{F}[\psi(x)]$, and $\psi(x) = \mathcal{F}^{-1}[\phi(k)]$ (we have omitted t in the shorthand to denote the actual variables x and k).

Just like $|\psi|^2$ giving the probability in position space, $|\phi(k)|^2$ gives the probability of finding the particle with momentum k. And $\phi(k)$ is normalized if $\psi(x)$ is as well, that is, $\int |\phi(k,t)|^2 dk = \int |\psi(x,t)|^2 dx = 1$.

Because our wave function $\psi(x, t)$ is discrete over the space grid, it is efficient and straightforward to evaluate Eq. (8.33a) with the fast Fourier transform (FFT) method (if you have not done so already, now would be a good time to review Section S:5.B).

Suppose we sample M points at equal intervals $\delta x = h$ over the spatial range (i.e., period) $d = Mh$. From Eqs. (S:5.18) and (S:5.20), and according to Section S:5.B.3, the momentum grid points are determined by

$$k_j = \delta k \times j, \quad j = -\frac{M}{2}, -\left(\frac{M}{2} - 1\right), \ldots, -1, 0, 1, \ldots, \frac{M}{2} - 2, \frac{M}{2} - 1, \tag{8.34}$$

where $\delta k = 2\pi/d$ is the interval in momentum. The range of momentum is $q = M\delta k = 2\pi/h$, again in accordance with Eq. (S:5.21). Like δk, the spatial interval is related to q reciprocally by $\delta x = 2\pi/q$. To use FFT, we make the number of grid points M a power of 2, that is, $M = 2^L$, $L =$ integer. By our convention (Program 8.3), if N is the number of intervals, then $N = M - 1$.

In Eq. (8.34), we are interpreting the Fourier components as made up of positive and negative momentum values. Because the transform has a period M, we have $\phi(k_j) = \phi(k_{j+M})$. Even though all momentum values can be treated as being positive mathematically, we require half of them to be negative from physical point of view. The FFT storage scheme for the components is outlined in Table S:5.1.

The momentum distribution is presented as a profile on the k-axis in Figure 8.9 and as a density contour plot in Figure 8.10 in terms of k and t (Project P8.4). The results are obtained from a slight modification of Program 8.3, where line 37 is replaced by the following,

```
phi = np.array(fft.fft_rec(psi, L))              # FFT
phi = np.concatenate((phi[M//2:], phi[:M//2]))   # re-order neg to pos k
pb = phi.real**2 + phi.imag**2
```

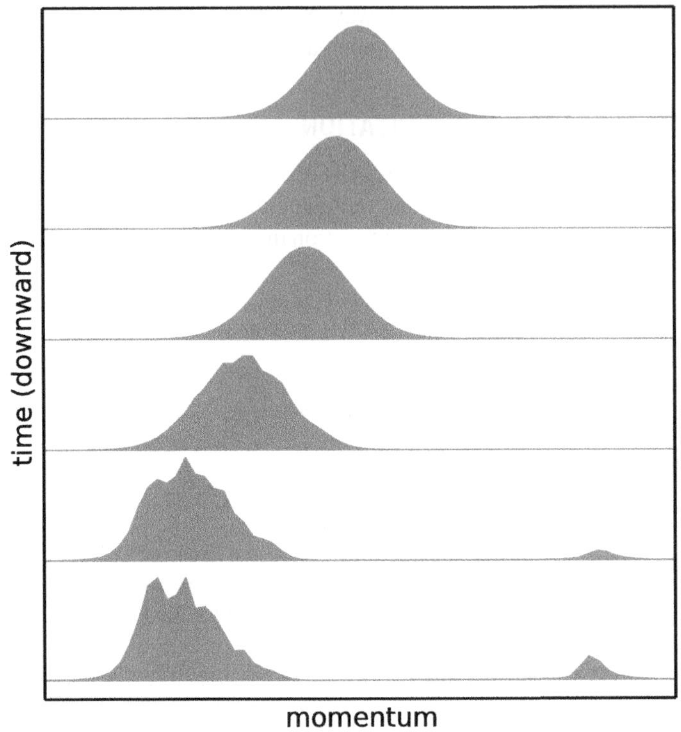

FIGURE 8.9 A quantum particle in free fall: momentum profile $|\phi(k, t)|^2$.

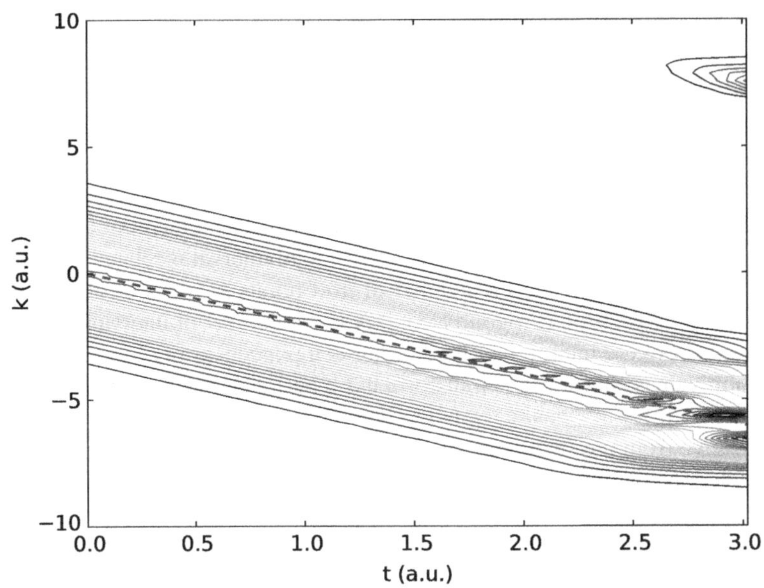

FIGURE 8.10 The momentum density of a quantum particle in free fall. The dashed line shows the classical relation $v = v_0 - at$ ($a = 2$).

The parameter L is equal to 10, so $M = 1024$ and $N = 1023$. After taking the FFT using the recursive function `fft_rec()` from our library, the array `phi[]` contains the resultant momentum wave function. The second line pulls the second half of the array in front of the first half so it is in the order of increasing k per Eq. (8.34) after concatenation (see Eqs. (S:5.41) and (S:5.42)).

Initially the momentum profile is a Gaussian (in k space) because the Fourier transform of a Gaussian is another Gaussian. Furthermore, the widths in coordinate and momentum spaces are related by the uncertainty principle (do not confuse Δx and Δk with sampling intervals δx and δk)

$$\Delta x \Delta k \sim 1. \tag{8.35}$$

In the present case, we had $\Delta x \sim 2\sigma = 1$, so Δk should be on the order of one. This may be assessed from Figure 8.10. If we increase σ, Δk will decrease proportionally, and vice versa.

Both figures show that, at the beginning, the profile keeps its initial shape with no apparent broadening, in stark contrast to position space where broadening happens rather quickly. Imagine the wavepacket was in free space. Because there would be no force, the momentum as well as the kinetic energy of the particle must be conserved, and we would expect no spreading at all.

In free fall, there is a constant acceleration in the negative direction. As a result, we see the entire momentum profile moving in that direction, following the classical linear relation, $v = v_0 - at$ (Figure 8.10), just like its classical path (Figure 8.8). As time increases, the leading edge reaches the boundary, equivalent to hitting a hard wall. Part of the wave is being reflected, and self-interference occurs as discussed earlier. The momentum distribution begins to deform from a perfect Gaussian, and smaller structures develop on top of the Gaussian. Toward the end at $t \sim 3$, we see a spike at a positive momentum $k \sim 7$. In large part the spike is due to the acceleration of the center-left portion of the wavepacket reaching that speed and being reflected by the wall.

The center of the momentum is zero initially, but because $\Delta k \sim 1$, there is a significant portion of the distribution near $k \sim \pm 1$. Classically, the leading edge ($k = v_0 \sim -1$) at $t \sim 3$ would have a displacement of $v_0 t - \frac{1}{2}at^2 \sim -12$ (Figure 8.8), and a velocity of $v = v_0 - at \sim -7$. Its position, assuming an initial position $x_0 = 5$ and a spread of ± 1, would be roughly $x \sim -7$ to -9, just reaching the wall located at -10. Upon colliding with the wall, the velocity would be reversed, giving rise to the spike at $k = -v \sim 7$.

As seen earlier, this effect is due to our domain being finite. If we interpret the physical boundaries as actual hard walls, the particle would bounce up and down, with self-interference making it look less and less like a localized classical particle with time. Interestingly, if we evolve the system long enough until the so-called revival time (8.55), nearly all the distribution will be contained in the spike, and the cycle repeats.[7]

8.4 TWO-STATE SYSTEMS AND RABI FLOPPING

In the last two sections we discussed quantum motion under time-independent potentials. There are situations where a quantum system is subject to a time-dependent potential, causing it to change states, a process called quantum transition. For instance, if a hydrogen atom is exposed to an oscillating laser field, the electron can jump to an excited state. When a system interacts with an external field, it is an open system.

We are interested in quantum transitions in this section. For simplicity and clarity, we will consider only two-state systems here and will keep the model brief. A full, unlimited model with a worked example is developed in Section S:8.2, which should be reviewed for more details.

[7] It would behave like a quantum bouncer, which has been observed experimentally (see J. Gea-Banacloche, *Am. J. Phys.* **67**:776-782 (1999), and G. Della Valle et al., *Phys. Rev. Lett.* **102**:180402 (2009)).

8.4.1 TWO-STATE MODEL FOR TRANSITION

Let H_0 be the unperturbed Hamiltonian of the system in the absence of the laser field, and u_1 and u_2 be its two states, called eigenstates (like eigenmodes in Chapter 6; see Chapter 9 for eigenstate problems). The solutions of the unperturbed system satisfy

$$H_0u_1(x) = E_1u_1(x), \quad H_0u_2(x) = E_2u_2(x), \tag{8.36}$$

where E_1 and E_2 are the eigenenergies of states 1 and 2, respectively. We assume u_1, u_2 and E_1, E_2 are known, for example, they may be the up or down states of a spin, or the lowest two states of a particle in a box (S:8.39). The eigenstates are also orthonormal, that is, (integration limits omitted)

$$\int u_1^* u_1 dx = \int u_2^* u_2 dx = 1, \quad \int u_1^* u_2 dx = \int u_2^* u_1 dx = 0. \tag{8.37}$$

We prepare the system so that the particle is in state 1 at $t = 0$ and then turn on the laser. The particle will respond to the laser force and start mixing with state 2. We expect the wave function of the particle at t, $\psi(x, t)$, to be a superposition of the two states as

$$\psi(x, t) = a_1(t)u_1(x)e^{-iE_1t/\hbar} + a_2(t)u_2(x)e^{-iE_2t/\hbar}, \tag{8.38}$$

where $a_1(t)$ and $a_2(t)$ are the expansion coefficients (generally complex). Note that we have included the exponential time factors of stationary states in Eq. (8.38).

The interpretation of the coefficients is that the probabilities of finding the particle in states 1 or 2 are given by $|a_1(t)|^2$ and $|a_2(t)|^2$, respectively. They are also called occupation (or transition) probabilities. Because the total probability is conserved, we have

$$|a_1(t)|^2 + |a_2(t)|^2 = 1. \tag{8.39}$$

To solve for $a_1(t)$ and $a_2(t)$, we substitute $\psi(x, t)$ of Eq. (8.38) into the Schrödinger equation $i\hbar\partial\psi/\partial t = (H_0 + V(x, t))\psi$, where $V(x, t)$ represents the interaction with the laser field (see Eq. (S:8.40)). After some algebra and using Eq. (8.36) for cancellations (see Section S:8.2), we obtain

$$i\hbar\left(\dot{a}_1u_1e^{-iE_1t/\hbar} + \dot{a}_2u_2e^{-iE_2t/\hbar}\right) = V(x, t)\left(a_1u_1e^{-iE_1t/\hbar} + a_2u_2e^{-iE_2t/\hbar}\right). \tag{8.40}$$

The notation $\dot{a}_{1,2}$ represents derivative with respect to time.

Equation (8.40) is valid for all space and time. To extract the coefficients, we project it to the eigenstates u_1 and u_2. For instance, we can multiply both sides of Eq. (8.40) by u_1^*, and integrate over x. Using the orthogonality relation (8.37), the $\dot{a}_2$ term drops out. The same can be done with u_2^*. The final result after projection is

$$\begin{aligned} i\hbar\dot{a}_1 &= a_1V_{11}(t) + a_2V_{12}(t)e^{-i\omega t}, \\ i\hbar\dot{a}_2 &= a_1V_{21}(t)e^{i\omega t} + a_2V_{22}(t), \end{aligned} \tag{8.41}$$

where $\omega = (E_2 - E_1)/\hbar$ is the transition frequency, and the matrix elements V_{mn} are defined as

$$V_{mn}(t) = \int u_m^*(x)V(x, t)u_n(x)dx, \quad m \text{ or } n = 1, 2. \tag{8.42}$$

Equation (8.41) is the central result of the two-state model (a subset of Eq. (S:8.33)). It can be integrated if the laser field $V(x, t)$ is given.

8.4.2 RABI FLOPPING

As an example, we carry out a calculation for a particle in the box (infinite potential well) with Program S:8.1, which integrates Eq. (8.41) with the RK4 method. The ground and the first excited states are included ($N = 2$), and the laser field interaction is given by Eqs. (S:8.40) and (S:8.41). The laser has a center frequency equal to the transition frequency between the two states and a pulse duration of $\tau = 600$ a.u. The rest of the parameters are kept the same as in Figure S:8.8. The results are shown in Figure 8.11.

The system starts out in state 1, its population P_1 decreases continuously to very nearly zero. At the same time, the population P_2 of state 2 increases in such a way that the net loss in P_1 becomes the net gain in P_2. The two states oscillate back and forth between practically 0 and 1. This behavior may be understood in a simplified picture known as Rabi flopping.

Rabi flopping is an important phenomenon in externally driven two-state systems. To avoid unnecessary clutter, we assume a monochromatic laser field tuned to (resonant with) the transition frequency,

$$F(t) = F_0 \cos(\omega_L t), \quad \omega_L = \omega = (E_2 - E_1)/\hbar. \tag{8.43}$$

This form of the field will help us see more clearly the core idea of Rabi flopping. In numerical calculations such as Figure 8.11, we still use realistically enveloped pulses like Eq. (S:8.41).

For the particle in a box, the transition matrix elements can be written explicitly from Eq. (S:8.44) as

$$V_{11} = V_{22} = 0, \; V_{12} = V_{21} = V_0 \cos(\omega_L t), \tag{8.44}$$

where $V_0 = -16eaF_0/9\pi^2$ (e is electron charge, a box width, F_0 field amplitude). The diagonal matrix elements being zero (or close to zero) is a common property in dipole transitions of many systems.

Substituting Eq. (8.44) into (8.41) gives us

$$\begin{aligned} \dot{a}_1 &= -\frac{iV_0}{2\hbar} a_2(1 + \exp(-2i\omega_L t)), \\ \dot{a}_2 &= -\frac{iV_0}{2\hbar} a_1(1 + \exp(+2i\omega_L t)). \end{aligned} \tag{8.45}$$

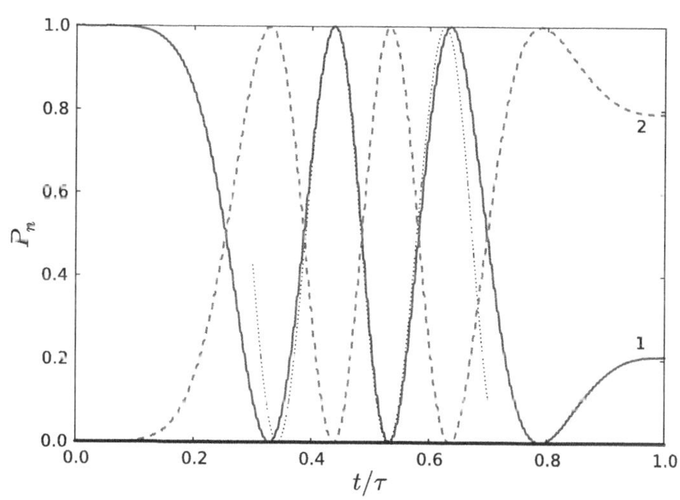

FIGURE 8.11 State population for a laser duration $\tau = 600$. The dotted line shows a shifted Rabi flopping function $\cos^2(\Omega_R t + \varphi)$, Eq. (8.49a).

Equation (8.45) is analytically solvable, but the solutions may be simplified with a useful approximation. Over a time period long compared to $1/\omega_L$, the factors $\exp(\pm 2i\omega_L t)$ in Eq. (8.45) will undergo many oscillations. Their net average values will be close to zero, that is, $\langle \exp(\pm 2i\omega t)\rangle \simeq 0$. We can then drop them from Eq. (8.45). This is called the "rotating wave approximation" (RWA) [64]. Within RWA, we arrive at two simplified equations

$$\dot{a}_1 = -\frac{iV_0}{2\hbar}a_2, \quad \dot{a}_2 = -\frac{iV_0}{2\hbar}a_1. \tag{8.46}$$

Differentiating both sides of Eq. (8.46), we obtain the following equivalent second-order differential equations,

$$\ddot{a}_1 + \Omega_R^2 a_1 = 0, \quad \ddot{a}_2 + \Omega_R^2 a_2 = 0, \quad \Omega_R = \frac{|V_0|}{2\hbar}. \tag{8.47}$$

The quantity Ω_R is the Rabi frequency. It depends on the interaction strength only, not on the laser frequency.

Equation (8.47) is a pair of well-known SHO equations whose solutions, subject to the initial condition $a_1(0) = 1$ and $a_2(0) = 0$, are

$$a_1(t) = \cos(\Omega_R t), \quad a_2(t) = -i\sin(\Omega_R t). \tag{8.48}$$

The occupation probabilities of the two states at time t are

$$P_1(t) = |a_1(t)|^2 = \cos^2(\Omega_R t), \tag{8.49a}$$

$$P_2(t) = |a_2(t)|^2 = \sin^2(\Omega_R t). \tag{8.49b}$$

Clearly, $P_1(t) + P_2(t) = 1$, as required by Eq. (8.39). The two states flip-flop back and forth with the Rabi frequency Ω_R. This is Rabi flopping [74]. The frequency can be controlled through the laser field strength V_0. Rabi flopping occurs in many situations, such as nuclear magnetic resonance, quantum computing, and optics. It is one of the most important ideas in quantum optics.

Going back to Figure 8.11, we can understand the oscillations between 0 and 1 as a result of Rabi flopping. Recall that the results are for a realistic pulse, Rabi flopping will be significant only after the pulse has ramped up. In a complete Rabi cycle the two states are populated and depopulated completely and complementarily. We have superimposed $P_1(t)$ from Eq. (8.49a) in Figure 8.11 (dotted line) to simulate the oscillation for an idealized, infinitely long, monochromatic laser. It is phase shifted to account for the ramping up stage. We can see that during the middle part of the laser there is good agreement between numerical results and the ideal solutions.

Although no simple analytic solutions for Rabi flopping are known with an enveloped laser pulse described by Eq. (S:8.41), our numerical simulations are straightforward and the results demonstrate the universal nature of the important phenomenon (see Project S:P8.9 for Rabi flopping in hydrogen atoms). The oscillation frequency is dependent on the laser strength, and independent of laser frequency. The longer or stronger the pulse, the more cycles of Rabi oscillations. The laser frequency is responsible for the small variations (step structures) as discussed in Section S:8.2.2.

8.4.3 SUPERPOSITION AND THE MEASUREMENT PROBLEM

In Rabi flopping (8.48), we have a system whose wave function ψ is a superposition of two states ϕ_1 and ϕ_2,

$$\psi = \cos(\Omega_R t)\phi_1 - i\sin(\Omega_R t)\phi_2, \tag{8.50}$$

where $\phi_{1,2} = u_{1,2} \exp(-iE_{1,2}t/\hbar)$ (see Eq. (8.38)). At any given time, the wave function is a coherent sum of the two states, that is, both states exist simultaneously within ψ. The probabilities of finding the states are $\cos^2(\Omega_R t)$ and $\sin^2(\Omega_R t)$, respectively. Quantum mechanics prescribes exactly, and deterministically, how the probabilities change with time.

How do we realize the probabilities? Measurement, of course. When we make a measurement, we will find the system in either state 1 or 2, not both. But quantum mechanics never tells us which state we will get, only the probabilities. The system would evolve just fine without any measurement. A measurement seems to force the system to reveal itself. How does it happen? What state was the system in just before we made the measurement? If we regard the two states as a cat being dead or alive, we have the well-known Schrödinger's cat problem.

Because the action of measurement is not modeled in quantum mechanics, the answers to the preceding questions are unknowable within the confines of quantum mechanics. Essentially, this is the measurement problem and is open to interpretation. A commonly accepted view is the Copenhagen interpretation, which holds that the system ceases to be a coherent superposition of states and collapses to one state or the other when a measurement is made. To put it another way, measurement destroys the system. Even though this does not quite answer how the system stops being a superposition of states, it is an answer. In fact, this is one of the intriguing aspects of quantum mechanics, and searching for an answer can get quite philosophical, and metaphysical sometimes. Fortunately, it does not hinder our ability to do quantum mechanics and, least of all, to simulate quantum systems.

8.5 QUANTUM WAVES IN 2D

As seen from earlier, we can create a superposition of states, or a coherent state, with external perturbations like laser excitation. After the external interaction is turned off, the coherent state will continue to evolve but with different characteristics than single stationary states. The numerical methods we discussed earlier also enable us to study coherent states in higher dimensions. In the following we will explore an example in 2D and examine the evolution in both coordinate and momentum spaces.

Let us consider a coherent state prepared at $t = 0$ in a 2D box (infinite potential well) as

$$\psi(x, y, 0) = \sqrt{\frac{2}{\sigma_x}} \cos\left(\frac{\pi}{\sigma_x}(x - x_0)\right) \sqrt{\frac{2}{\sigma_y}} \cos\left(\frac{\pi}{\sigma_y}(y - y_0)\right), \tag{8.51}$$

$$|x - x_0| \le \frac{\sigma_x}{2}, \quad |y - y_0| \le \frac{\sigma_y}{2}.$$

This is a normalized compact wave function that is nonzero only within a rectangular region $\sigma_x \times \sigma_y$ centered at (x_0, y_0).

The state will evolve similarly to the expansion (8.38), which in 2D reads

$$\psi(x, y, t) = \sum_{m,n} a_{mn} \exp(-iE_{mn}t/\hbar) u_m(x) u_n(y). \tag{8.52}$$

The basis functions u_m and u_n are the same as Eq. (S:8.39), and the eigenenergies are $E_{mn} = E_m + E_n$, the sum of individual eigenenergies E_m and E_n from Eq. (S:8.39). We can determine the amplitudes a_{mn} from the wave function at $t = 0$ as (see Exercise E8.9)

$$a_{mn} = \int u_m^*(x) u_n^*(y) \psi(x, y, 0) dx dy = C_m(\sigma_x, x_0) C_n(\sigma_y, y_0),$$

$$C_n(\sigma, z) = \frac{4}{\pi}\sqrt{\frac{\sigma}{a}}\left[1 - \frac{n^2\sigma^2}{a^2}\right]^{-1} \cos\left(\frac{n\pi\sigma}{2a}\right) \sin\left(\frac{n\pi z}{a}\right). \tag{8.53}$$

A time sequence of probability densities $|\psi|^2$ is shown in Figure 8.12. The calculations are done with Program 8.4. We choose a square well of size $a = 10$, and a symmetric coherent state with $\sigma_x = \sigma_y = 2$ placed at the center of the well. This is a relatively narrow wavepacket, and to adequately represent the state, we include $N = 20$ basis states in the expansion in each dimension, so the total number of states is $N^2 = 400$.

Initially the wavepacket broadens as expected (Figure 8.12, top row). Starting $t = 1.4$, we see interference effects on the outer edges of the distribution from reflected waves from the wall. Increasingly intricate interference patterns (quantum quilt) are developed at later times ($t = 2$ and 3), showing the level of complexity a quantum wave is capable of exhibiting. You should play with Program 8.4, for example, setting the initial wavepacket off center, looking at the imaginary part. We had seen 2D classical waves on a membrane (Section 6.7) also showing similarly interesting interference patterns (Figures 6.18 and 6.19). But quantum waves can display more entangled wave pockets (Figure 8.12, middle row) at a comparable level of energy because the time-dependent wave functions are complex.

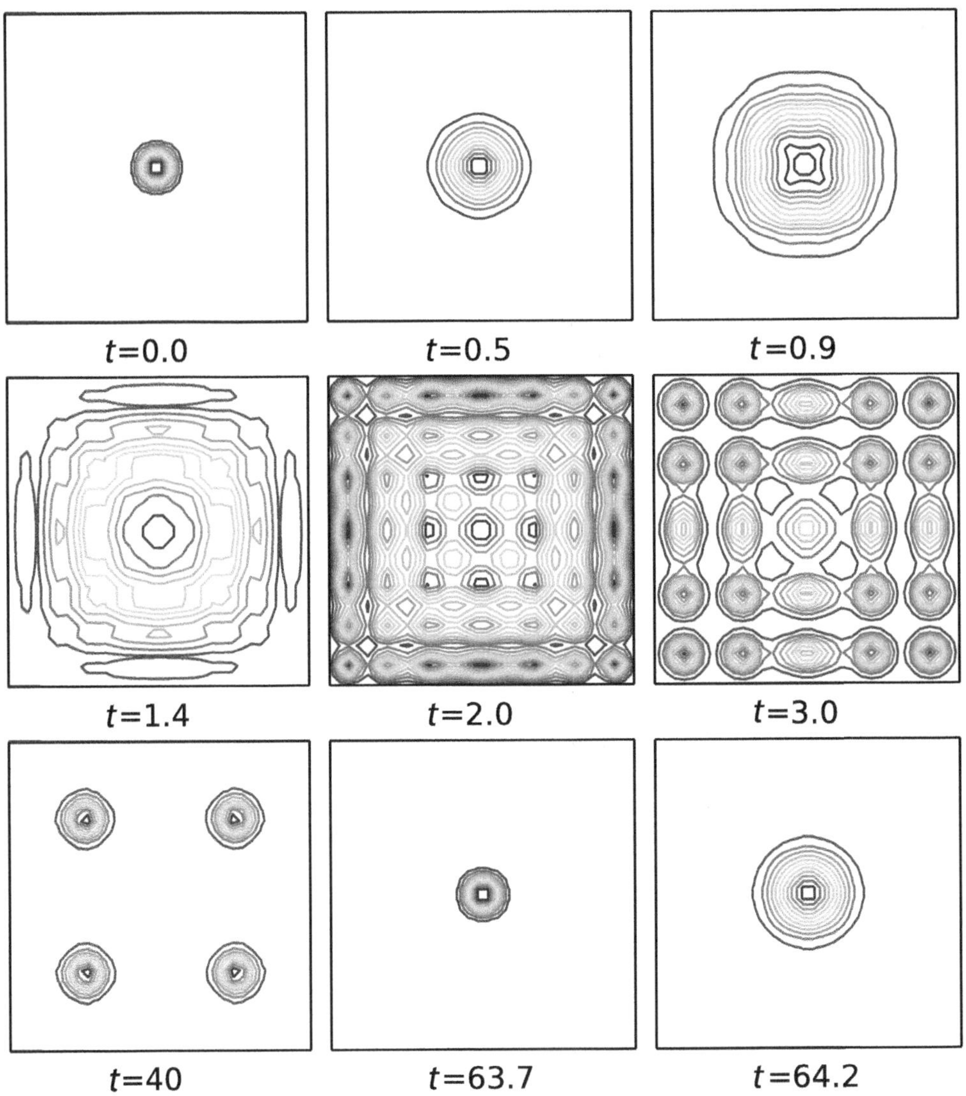

FIGURE 8.12 Probability density in a 2D well. The side length is $a = 5.3$ Å.

As time increases further, the number of pockets decreases and the wave function starts to reassemble itself into the original state at $t = t_R \sim 63.7$. Thereafter, it repeats as if time had been reset periodically at t_R, for example, $\psi(t = 0.5) = \psi(t = 64.2)$. This phenomenon is known as quantum revival, and t_R the revival time (see following).

The corresponding momentum probability density $|\phi(k_x, k_y, t)|^2$ is shown in Figure 8.13. To obtain the momentum wave function, we calculate the 2D FFT of $\psi(x, y, t)$ by inserting the following line in Program 8.4

```
phi = np.fft.fft2(wf)
```

The results show that the momentum distribution changes little in the beginning, $t = 0$ and 0.5. When the leading edge of the wave comes in contact with the wall, the momentum distribution begins to change substantially due to the "external" forces from the wall. The large

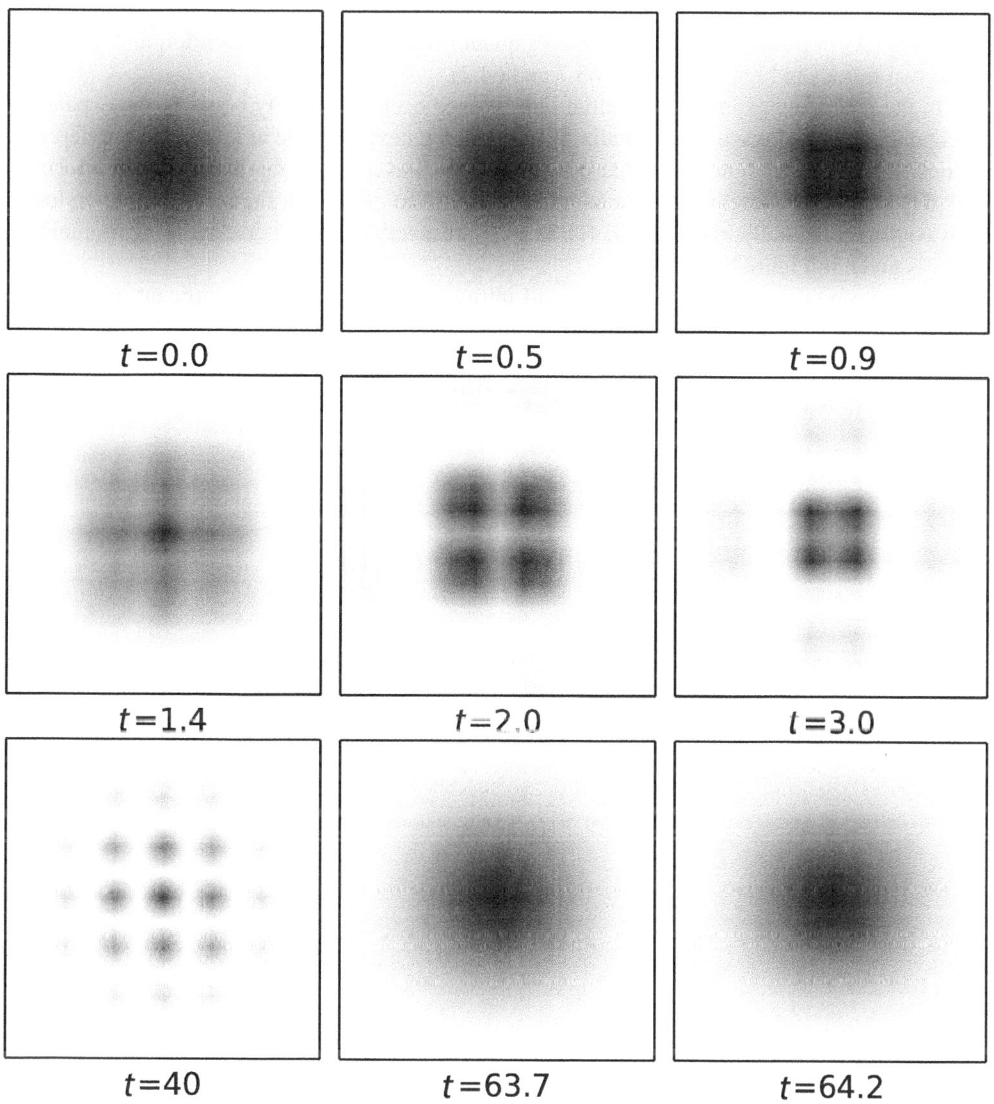

FIGURE 8.13 Momentum density in a 2D well. The k_x or k_y scale is $[-4, 4]$.

momentum components on the perimeter (t =1.4 to 40) are significantly well separated from the initial distribution. All these are in accord with the 1D case (Figure 8.9).

What is new, however, is the clear reciprocal relationship between the position and momentum spaces depicted in Figures 8.12 and 8.13. They are essentially optical diffraction or interference patterns of each other [8]. For instance, we can think of the initial wave function at $t = 0$ in position space as a single aperture. The diffraction pattern through a single aperture would correspond to the momentum distribution. The smaller the aperture, the wider the diffraction spot. At $t = 40$, we can think of the four peaks as small "holes" spaced-out in position space (Figure 8.12, bottom left). The resulting interference pattern in momentum space has rows of small dots stacked over each other, forming a larger square. The size of the square is related to the size of the holes in position space, and the size of the dots to the spacing between the holes. For example, if the size of the holes is reduced, the square would grow, and if the spacing between the holes is reduced, the dots would get bigger. Conversely, we can run the argument backward from momentum space to position space and arrive at similar conclusions.

Of course, interference of waves can be mathematically represented as Fourier transforms. Evolution of 2D quantum waves in Figures 8.12 and 8.13 graphically describes these relationships very clearly. As in position space, we observe quantum revival in the momentum distribution in the last two frames in Figure 8.13 ($t = 63.7$ and 64.2).

In the preceding example, the initial interference occurs primarily between the expanding wave and the reflected wave off the walls. We look in the next example at intrawave interference. We construct the initial state by a superposition of two wavepackets that have the same form as Eq. (8.51) but centered at different locations (Project P8.6). Shown in Figure 8.14 are two localized wavepackets initially placed at (4, 4) and (6, 6), respectively. The corresponding momentum density is given in Figure 8.15.

In position space, the waves merge and intrawave interference is significant up to $t = 1$. Because no external forces are involved at this stage, the momentum distribution remains

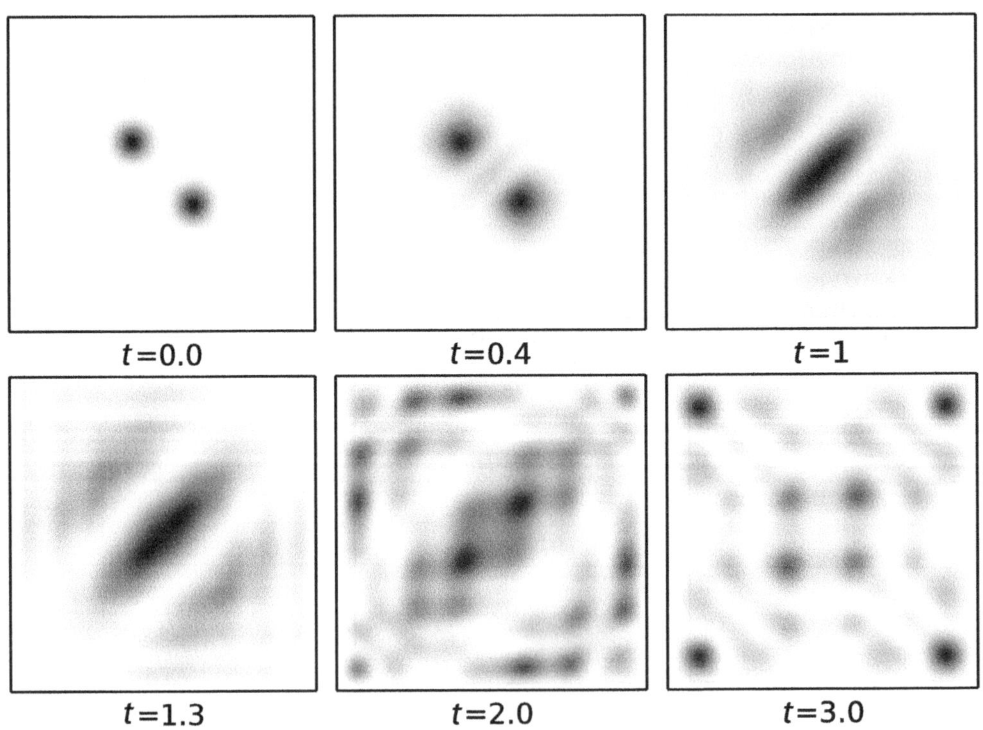

FIGURE 8.14 Probability density in 2D well of two localized wavepackets.

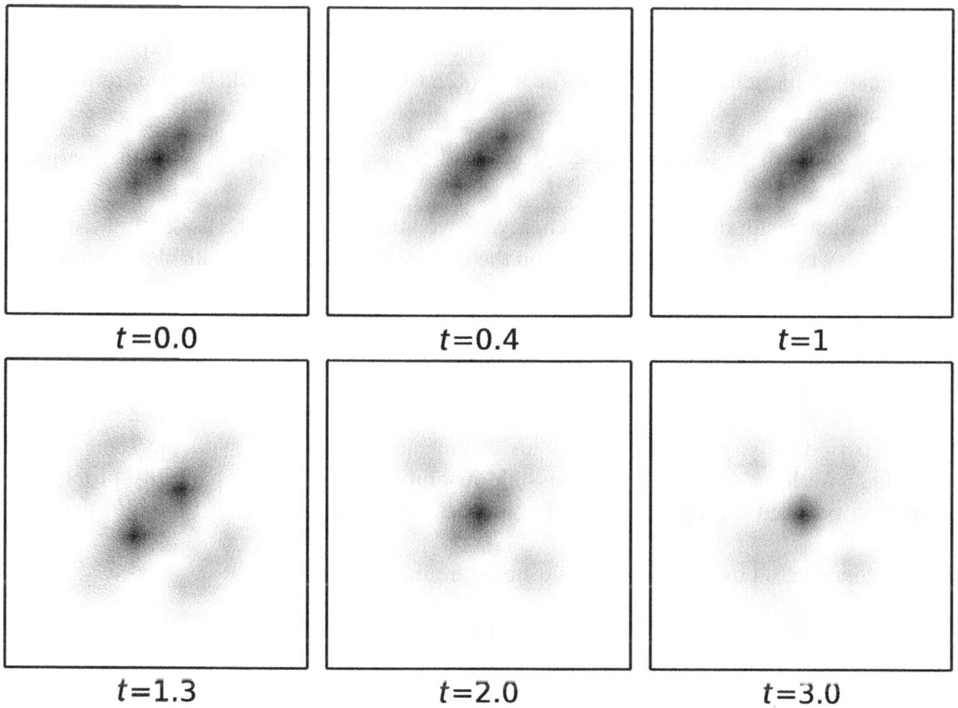

FIGURE 8.15 Momentum density corresponding to Figure 8.14.

substantially the same, and little visible changes are seen between $t = 0$ and 1. Starting from $t = 1.3$, collision and reflection from the walls cause the wave function in both spaces to form highly entangled pockets, as was seen earlier.

Note that in the beginning, the wave function in position space is along the upper-left to lower-right diagonal, but in momentum space the distribution is in bands oriented from the lower-left to upper-right direction, such that the two distributions are approximately perpendicular to each other. The reason can be understood in analogy to optical diffraction alluded to earlier. The lengths and widths of the bands, as well as the fine structures within each band, are related to the size and separation of wavepackets in position space.

Quantum Revival

A coherent state of an isolated, finite quantum system has an interesting property: periodic revival of its wave function. We have seen an example of this from Figures 8.12 and 8.13. Let us consider a coherent state expanded in the basis functions for a particle in a box (S:8.39)

$$\psi(x,t) = \sum_n a_n u_n(x) \exp(-in^2 E_1 t/\hbar), \tag{8.54}$$

where $E_1 = \pi^2\hbar^2/2m_e a^2$ is the ground state energy.

We see that the exponential factors are periodic if $E_1 t/\hbar = 2\pi j$, $j = 1, 2, 3$, and so on. Taking the smallest multiple, we define the revival time as

$$t_R = \frac{2\pi\hbar}{E_1} = \frac{4m_e a^2}{\pi\hbar}. \tag{8.55}$$

The wave function is periodic in t_R, $\psi(t + t_R) = \psi(t)$. For an electron in a box with width $a = 10$, the period is $t_R = 127.3$ a.u.

The period observed from the 2D box (Figure 8.12) is half this value, 63.7. The ground state energy in 2D is twice that in 1D, so the revival time is halved according to Eq. (8.55).

Classically, a particle with energy E (or $v = \sqrt{2E/m_e}$) in a box would bounce from wall to wall periodically at

$$t_{cl} = \frac{2a}{v} = \sqrt{\frac{2m_e}{E}}\, a. \tag{8.56}$$

This classical result depends on the energy, as we intuitively expect. However, the quantal result (8.55) does not show any dependence on the energy of the coherent state at all. It is determined solely by the system parameters. The two results seem inconsistent. Worse, they also seem incompatible with the Bohr-Sommerfeld correspondence principle, which states that the behavior of a quantal system should approach its classical description in the limit of large quantum numbers (semiclassical limit).

This suggests a deeper connection than the wave function alone. Though the wave function revives periodically, physical observables do not necessarily vary with the same revival period. In fact, we have seen hints to a possible resolution of this incompatibility problem in the connection between the expectation value of position and the classical motion in the SHO (Figure 8.4).

To be concrete, let us assume only two states are included,

$$\psi = a_m u_m \exp(-iE_m t/\hbar) + a_n u_n \exp(-iE_n t/\hbar). \tag{8.57}$$

The expectation value of position is given by

$$\langle x \rangle = \frac{a}{2} + 2x_{mn} \cos \frac{(E_m - E_n)t}{\hbar}, \qquad x_{mn} = \int u_m^* x u_n \, dx. \tag{8.58}$$

We see that $\langle x \rangle$ oscillates with a period $2\pi\hbar/|E_m - E_n|$, which is different than the revival period (8.55). What is more, the energies of the states enter the equation. It shows that the oscillation period is inversely proportional to the difference of eigenenergies, not the eigenenergies themselves.[8]

Let us take the semiclassical limit, $m, n \gg 1$ but keep $m = n + 1$, the oscillation period reduces to

$$t_{sc} = \frac{4m_e a^2}{(2n+1)\pi\hbar} \sim \sqrt{\frac{2m_e}{E_n}}\, a, \tag{8.59}$$

where we have used $E_n = n^2\pi^2\hbar^2/2m_e a^2$. This semiclassical result agrees with the classical prediction, Eq. (8.56). In this way, we have resolved the apparent incompatibility in the correspondence principle. This oscillation period t_{sc}, rather than the revival time t_R, should be compared between classical and quantal predictions.

In general, we can construct a coherent state in the semiclassical limit by including a number of eigenstates clustered around $n_0 \gg 1$ [87]. Then, Eq. (8.59) can be generalized to

$$t_{sc} = 2\pi\hbar \left[\frac{dE_{n_0}}{dn_0}\right]^{-1}. \tag{8.60}$$

The period t_{sc} can be much smaller than t_R. We leave further exploration of quantum revival to Project P8.7.

---

[8] Another example is found in the hydrogen transition frequency, which also depends on the difference of eigenenergies.

CHAPTER SUMMARY

We studied numerical simulations of time-dependent quantum systems including dynamic evolution of closed systems and quantum transitions in open systems. We presented several efficient and easy-to-use methods appropriate for solving the TDSE, breaking down the traditional unsolvability barrier in quantum mechanics where very few meaningful time-dependent problems are analytically solvable. This enabled us to learn by doing quantum mechanics rather than by just studying it.

The easiest way to get started is the explicit SDLF method, Program 8.2. This method converts the TDSE into a familiar system of ODEs, and we only need to change the initialization for different energies, potentials, and so on. It is norm preserving and robust, provided the time step satisfies the stability criterion. The drawback is that this limits its speed for very accurate calculations. It is the recommended method because it is simple to understand and program, easy to implement boundary conditions, and requires only knowledge of ODE solvers.

Alternatively, we can use the first-order split-operator method (Crank-Nicolson), which is fast and stable. But it is implicit and requires a banded matrix solver for speed. It is suitable for short-time evolution or weak potentials.

We discussed quantum transitions in external fields within a two-state model. We can obtain state occupation probabilities directly, which oscillate between the two states in a superposition. We also studied coherent states in position and momentum spaces, including quantum revival. In addition, we discussed general numerical integration techniques and banded matrix handling.

8.6 EXERCISES AND PROJECTS

Exercises

E8.1 (a) Check the atomic unit system in Table 8.1, evaluate the numerical values for the fine structure constant α, the Bohr radius a_0, and the Hartree E_0, in Eq. (8.3). Also, prove the identity $m_e a_0^2 E_0/\hbar^2 = 1$.

(b) Let us choose a unit system: x_0, m_0, E_0, and t_0 for length, mass, energy, and time, respectively. Express the variables as $x = x'x_0$, $m = m'm_0$, $t = t't_0$, and $V = E_0V'$. Show that, if the units are chosen to be the atomic units in Table 8.1, the TDSE (8.2) becomes in a.u.

$$-\frac{1}{2m'}\frac{\partial^2\psi(x',t')}{\partial x'^2} + V'(x',t')\psi(x',t') = i\frac{\partial\psi(x',t')}{\partial t'}. \tag{8.61}$$

For electrons, $m' = 1$, and the preceding equation reduces to Eq. (8.4).

E8.2 (a) Fill in the steps from Eq. (8.6) to Eqs. (8.8a) and (8.8b). (b) Prove Eq. (8.11) by explicit differentiation.

E8.3 Analyze the stability condition of the space discretized TDSE. Assume the potential is zero in this exercise.

(a) Apply Euler's method to Eq. (8.6), and carry out a von Neumann stability analysis. Show that there is no stable step size in general.

(b) Repeat the preceding with the leapfrog method to find the stability criterion for the SDLF.

E8.4 The dynamical evolution of an observable O is given by the Ehrenfest theorem [44]

$$\frac{d\langle O\rangle}{dt} = \frac{1}{i\hbar}\langle[O,H]\rangle + \left\langle\frac{\partial O}{\partial t}\right\rangle, \tag{8.62}$$

where $[O,H]$ is the commutator between the operator O and the Hamiltonian H. Using Eq. (8.62), verify Eqs. (8.17) and (8.19).

E8.5 Decompose the wave function into real and imaginary parts as $\psi(x,t) = R(x,t) + iI(x,t)$. Show that:

(a) the expectation value of momentum from Eq. (8.17) is

$$\langle p \rangle = \hbar \int_{-\infty}^{\infty} \left(R\frac{\partial I}{\partial x} - I\frac{\partial R}{\partial x} \right) dx;$$

(b) the expectation value of kinetic energy (8.18) is similarly

$$\langle T \rangle = \frac{\hbar^2}{2m} \int_{-\infty}^{\infty} \left(\left[\frac{\partial R}{\partial x}\right]^2 + \left[\frac{\partial I}{\partial x}\right]^2 \right) dx;$$

(c) the probability current defined as $J(x,t) = -i\frac{\hbar}{2m}\left(\psi^*\frac{\partial \psi}{\partial x} - \psi\frac{\partial \psi^*}{\partial x}\right)$ is

$$J = \frac{\hbar}{m}\left(R\frac{\partial I}{\partial x} - I\frac{\partial R}{\partial x} \right).$$

The results show that if the wave function is purely real or imaginary aside from an overall multiplication constant, then $\langle p \rangle = 0$ and $J = 0$. Stationary states are in this category. For nontrivial time-dependent processes we are dealing with in this chapter, they are nonzero.

E8.6 (a) Let $U = e^{-iH(t-t_0)}$ be the evolution operator. Show that U is unitary, that is, the product with its adjoint is unity, $U^\dagger U = 1$.
Given U and Eq. (8.21), show that $\int |\psi(x,t)|^2\, dx = \int |\psi(x,t_0)|^2\, dx$.

(b) Expand the denominator in Eq. (8.25) in a Taylor series, and show that the split operator agrees with $e^{-iH\Delta t}$ to first order.

E8.7 Using Eq. (8.5), discretize the split operators in Eq. (8.27) to prove Eq. (8.29).

E8.8 Conduct a von Neumann stability analysis of the split-operator result (8.29). Show that it is stable regardless of the time step.

E8.9 Given the initial wave function (8.51), prove the expansion coefficients (8.53).

Projects

P8.1 Explore the evolution of a wavepacket in the SHO potential, and use Program 8.2 as the base program. In this and other projects that follow, assume the wavepacket describes an electron.

(a) Generate the spatial and temporal probability density shown in Figure 8.3. To prepare data, declare two empty lists before the main loop, say `ta` and `Z`. Inside the loop, periodically append the probability density array (over the spatial grid) to `Z` and time to `ta`. This is one slice of the probability density at a given time. The preceding procedure should be the same as in Program 8.3. Terminate the loop after certain amount of time, for example, $t = 8$. Adjust the frequency of sampling so that roughly 50 slices of data are recorded. After exiting the loop, the data `Z` is a list of arrays, that is, a two-dimensional array.

Prepare the space-time grid using `X, Y = np.meshgrid(x, ta)` where `x` is the spatial grid. Make a surface plot as usual (see Programs 8.3 or 7.2, for example). You can exclude the end regions $|x| \gg 1$ where the probability density is small via index slicing (you can safely cut off a quarter from each end without affecting the appearance of the plot). This reduces the amount of unnecessary data to be plotted. Also use `rstride` or `cstride` parameters to control the number of meshes in the surface plot for efficiency and easy manipulation.

With the same data, reproduce the contour plot in Figure 8.3. Also try `plt.imshow()` to change the appearance of the contour plot to an image (see Program 8.4).

(b) Calculate and graph the expectation value of position $\langle x \rangle$ as a function of time. Let pb be the probability density array at t. An easy way to approximate $\langle x \rangle$ is to convert the integral (8.16) into a sum with the NumPy `sum()` function,

```
xexp = h*np.sum(x*pb)
```

Here, h is the grid size. The expression `np.sum(x*pb)` is equivalent to $\sum x_i |\psi_i|^2$. Note that `xexp` is a number (scalar). Put this line inside the main loop, append `xexp` to a list, and plot it at the end. If everything works correctly, you should see a graph like Figure 8.4.

Do a more accurate calculation of $\langle x \rangle$ using Simpson's rule with

```
import integral as itg
......
xexp = itg.simpson(x*pd, h)
```

Compare your results with the above approximation.

P8.2 We will calculate the average momentum in this project. We have seen the oscillatory motion of a wavepacket in the SHO potential. Classically, both position and velocity oscillate complementarily. Even though we cannot specify position and velocity simultaneously in quantum mechanics, we examine the expectation values to see if they exhibit similar behavior.

(a) Predict the momentum as a function of time, assuming the same initial condition as in Figure 8.3. Sketch your $\langle p \rangle$ vs. t curve, and briefly describe your reasoning.

(b) Calculate the average momentum $\langle p \rangle$ from either the modified code from Project P8.1 or Program 8.2. Use the results from Exercise E8.5, and carry out the integral by Simpson's rule. Evaluate the derivatives in the integrand using the central difference formula (6.31a).

Plot the results for time up to 20 a.u. Discuss them and compare with your prediction. Estimate the period from the graph. What function best describes the oscillation? Are they consistent with the average position $\langle x \rangle$ shown in Figure 8.4?

(c)* Study the uncertainty relations between position and momentum. Let $\Delta x = \sqrt{\langle x^2 \rangle - \langle x \rangle^2}$ and $\Delta p = \sqrt{\langle p^2 \rangle - \langle p \rangle^2}$. Calculate $\Delta x \Delta p$ and plot the results to $t = 20$ a.u. What are the maximum and minimum values of the uncertainty? Does it ever violate the uncertainty principle?

P8.3 Simulate the motion of a wavepacket in a rigid box, $V = 0$ if $0 < x < a$ and $V = \infty$ outside the box. Use the SDLF method (Program 8.2), and set zero values $\psi(0, t) = \psi(a, t) = 0$ for the boundary conditions. This may be done either by setting the derivatives at the end points zero in `sch_eqn()`, or by requiring $R_0 = R_N = I_0 = I_N = 0$ after each iteration. Choose a reasonable set of parameters for the width of the box and the Gaussian wavepacket. Suggested values are $a = 5$ and $\sigma = \frac{1}{2}$. Place the center of the initial wavepacket away from the walls so that the wave function is negligible at the boundaries, say $x_0 = 1$ to 2 for the suggested a and σ.

(a) Give the initial wavepacket zero average momentum ($k_0 = 0$). Run the simulation with animation on, and observe the wavepacket spreading and colliding with the walls. Summarize your observations, and briefly explain. Wait long enough to see the wavepacket approximately coming back together, that is, revival of the wave function. You can speed up the process if you set N proportional to a, keeping the grid size $h \sim 0.02$ to 0.04. Add a `vp.label()` so the time t is displayed on the screen. Note the revival time.

(b) Repeat the simulation, but give the wavepacket a nonzero initial momentum, $k_0 = n\pi/a$, with $n =$ 1 to 3. Turn off animation, and modify the program to plot the probability density as a contour plot like Figure 8.3. In addition, calculate and plot the expectation value of position as a function of time (refer to Project P8.1 for a procedure on doing so). Generate a pair of plots for each k_0. The time should be long enough so you can clearly identify a couple of periods in the probability density plot.

Discuss your findings in terms of quantum revival. How do your results depend on energy ($\frac{1}{2}k_0^2$)? What do you expect from classical motion of a particle trapped between two walls? How are the quantal and classical results different?

P8.4 (a) For free fall, compute the expectation value of momentum as a function of time with the split-operator method from Program 8.3. Use the results of Exercise E8.5, and follow the procedure outlined in Project P8.2. Plot the results.

(b) Generate the momentum density plot of Figure 8.10. Modify Program 8.3 to use the FFT algorithm discussed in Section 8.3.2 with the NumPy version, `np.fft()`, for speed. Compare the center of the momentum distribution with results above.

(c)* Obtain the expectation value of momentum as a function of time directly from the momentum wave function of part (b). How do your results compare with those from part (a)? Plot the difference between them, and briefly explain.

P8.5 Consider the absolute linear potential, $V = a|x|$, shown in Figure 8.16.

(a) Solve the motion classically for a given energy E. Find the position as a function of time and determine the period.

(b) Study the motion of a wavepacket with the split-operator method. Choose a reasonable $a \sim 2$. Place the wavepacket like in Program 8.3 at $x_0 \sim 5$ from origin, plot the probability density distribution (contour plots) and the expectation value of position as a function of time for a few periods. Discuss your results, and compare with classical motion at the same energy $E \sim a|x_0|$.

(c)* Repeat the preceding with the SDLF method. Use a conservative λ in Eq. (8.13) for stability. Follow a similar approach to the SHO potential. Discuss and compare with the preceding results.

P8.6 In this project, let us further explore our quantum quilt, the ever changing patterns of quantum waves in 2D.

(a) Modify Program 8.4 to calculate and plot the wave function in momentum space. Use 2D FFT to transform the wave function from coordinate to momentum space (Section 8.5). Make sure to

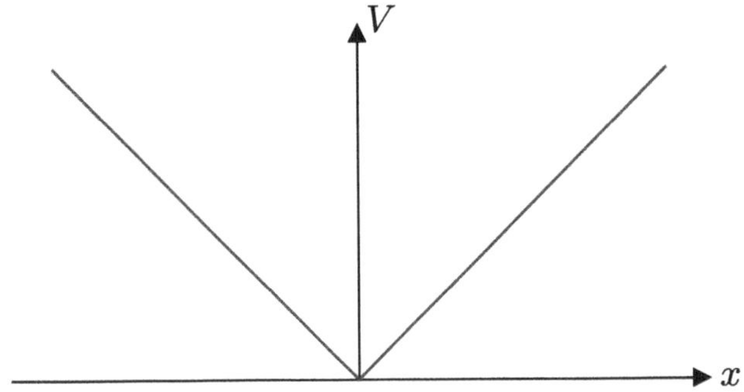

FIGURE 8.16 The wedged linear potential.

reorder momentum as negative and positive components. Also note that most of the momentum distribution is concentrated near $(k_x, k_y) = (0, 0)$. Cut off large $|k_{x,y}|$ values via slicing. You may wish to test the 2D FFT on some simple functions to increase familiarity with reordering. You should see pictures as shown in Figure 8.13.

(b) Explore intrawave interference by constructing the initial state from two wavepackets (8.51) centered at different locations. First assume they are identical. Note that the expansion coefficients are additive from the two wavepackets. Simulate the motion; you should see patterns shown in Figures 8.14 and 8.15.

Next, suppose the wavepacket is narrower in one direction, say $\sigma_x = 0.9$ and $\sigma_y = 1.8$. Predict what the evolution would look like. In particular, predict the momentum distribution before strong interference sets in.

Run the simulation, first with animation on to observe the evolution process. Then, turn off animation, and increase the number of grid points to produce accurate snapshots of the wave function in momentum space. Comment on your results.

P8.7 Construct a wavepacket in the infinite potential well as follows,

$$\psi(x, 0) = \sum_{n=N-m}^{N+m} a_n u_n(x),$$

where $u_n(x)$ are the eigenstates (S:8.39). Set $N = 20$, $m = 5$, and choose equal weights, $a_n = (2m + 1)^{-1/2}$. Evolve the state according to Eq. (8.38) or (S:8.23). Plot the average position, and optionally the probability density, as a function of time. What is the average energy? What is the revival time?

Next change the mixing coefficients, for example, $a_n = (-1)^n(2m + 1)^{-1/2}$ and repeat the calculation. How did the results change?

P8.8 The symmetric anharmonic potential is given by $V = V_0(x^4 - ax^2)$. The potential, graphed in Figure 8.17, is double-welled.

Simulate the motion of a Gaussian wavepacket in the potential. Use your method of choice (including the SEO method, Section S:8.1). Assume $V_0 = 1$ and $a = 4$ for the potential. Place the wavepacket in one of the wells, for example, the left well at $x_0 = -\sqrt{2}$, and set $\sigma = \frac{1}{2}$. Describe what motion is expected before running the simulation. Because of the steep rise of the potential for large $|x|$, the wavepacket is limited to a compact spatial range. For the preceding suggested parameters, it is sufficient to set the range to $-5 \le x \le 5$ in the actual simulation.

Discuss the actual motion. Does the wavepacket stay in one well, or vacillate between the wells? Produce probability density or average position plots, explain their features. Do they show quantum revival?

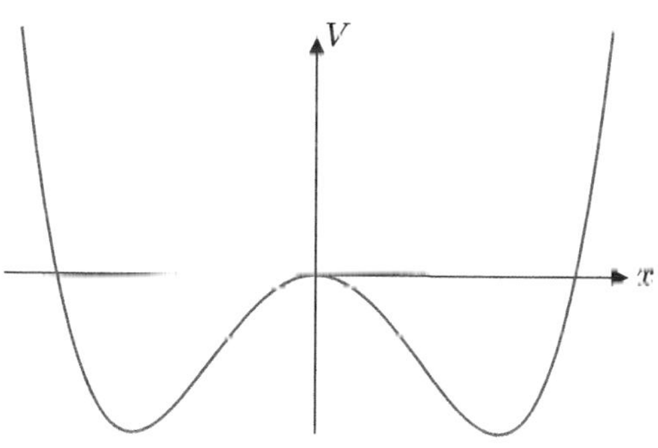

FIGURE 8.17 The anharmonic potential.

8.A NUMERICAL INTEGRATION

The need for numerical evaluation of integrals occurs frequently in scientific computing. In certain cases, computing integrals accurately, especially oscillatory or multiple integrals, is the core operation. It is important to understand the basic technique of numerical integration so we can choose or tailor a technique to balance speed and accuracy.

The basic problem of numerical integration is, given a known integrand $f(x)$, find I as

$$I = \int_a^b f(x)\,dx. \tag{8.63}$$

Most numerical techniques are geared toward finding the approximate area under the curve, shown in Figure 8.18, as efficiently as possible. In other words, we want to use the fewest number of function evaluations to achieve the maximum possible accuracy.

8.A.1 TRAPEZOID RULE

To begin, let us divide the range $[a, b]$ into N equidistant intervals of width h. As usual, let

$$h = \frac{b-a}{N}, \quad x_i = a + ih, \quad f_i \equiv f(x_i), \qquad i = 0, 1, 2, \ldots, N. \tag{8.64}$$

If we connect two neighboring points, say (x_0, f_0) and (x_1, f_1), by a straight line, we obtain a trapezoid. Within this interval, we can use linear interpolation (3.51) for $f(x)$,

$$x = a + \eta h, \quad f(x) \simeq (1-\eta) f_0 + \eta f_1, \quad 0 \le \eta \le 1. \tag{8.65}$$

Then the integral in the first interval of Figure 8.18 is

$$I_1 = \int_{x_0}^{x_1} f(x)dx \simeq h \int_0^1 ((1-\eta)f_0 + \eta f_1)d\eta = \frac{h}{2}(f_0 + f_1) + O(h^3). \tag{8.66}$$

Summing up all N intervals, we have the total integral

$$I = h\left(\frac{f_0}{2} + f_1 + f_2 + \cdots + f_{N-1} + \frac{f_N}{2}\right) + O(h^2). \tag{8.67}$$

This is the trapezoid rule, exact for a linear function. The local error in Eq. (8.66) is $O(h^3)$, and the overall error is $O(h^2)$, or $O(1/N^2)$.

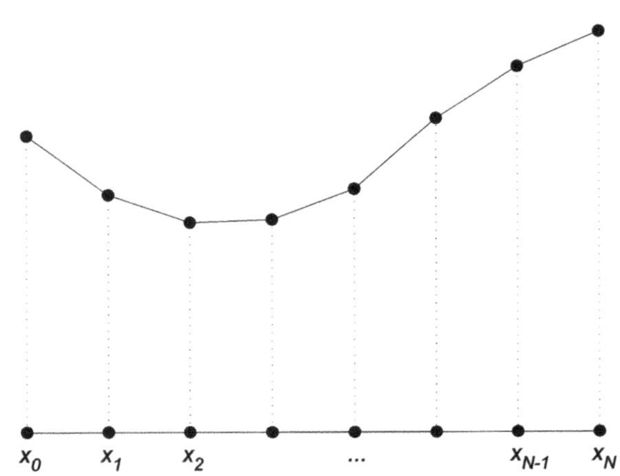

FIGURE 8.18 Numerical integration of $f(x)$ by finding the area under the curve at discrete intervals.

Except for the end points f_0 and f_N that have a coefficient $\frac{1}{2}$, the interior points appear twice in the sum (8.67), so their coefficients are 1. We can rewrite Eq. (8.67) a little differently as (dropping the error)

$$I = \sum_i w_i f_i, \tag{8.68}$$

$$w_0 = \frac{h}{2}, w_1 = h, \cdots, w_{N-1} = h, w_N = \frac{h}{2}. \quad \text{(trapzoid)} \tag{8.69}$$

This way, the integral is reduced to a sum of function values at abscissa x_i with weight w_i.

8.A.2 SIMPSON'S RULE

We can improve the accuracy by using a three-point quadratic interpolation, for example, between x_0 and x_2,

$$x = x_1 + \eta h, \quad f(x) \simeq \frac{1}{2}\eta(\eta - 1)f_0 - (\eta^2 - 1)f_1 + \frac{1}{2}\eta(\eta + 1)f_2, \quad -1 \le \eta \le 1. \tag{8.70}$$

The integral between x_0 and x_2 becomes

$$I_{0-2} = \int_{x_0}^{x_2} f(x)dx = \frac{h}{3}(f_0 + 4f_1 + f_2) + O(h^5). \tag{8.71}$$

Assuming even N and summing up all pairs of intervals, we obtain the total integral

$$I = \frac{h}{3}(f_0 + 4f_1 + 2f_2 + 4f_3 + \cdots + 2f_{N-2} + 4f_{N-1} + f_N) + O(h^4). \tag{8.72}$$

This is Simpson's rule, with error scaling $O(1/N^4)$. It is exact up to cubic polynomials.

Similar to Eq. (8.69), we present Simpson's rule in the abscissa-weight form (8.68) with weights

$$w_0 = w_N = \frac{h}{3}, \quad w_1 = w_3 = \cdots = w_{N-1} = \frac{4h}{3},$$
$$w_2 = w_4 = \cdots = W_{N-2} = \frac{2h}{3}. \quad \text{(Simpson, } N \text{ even)} \tag{8.73}$$

Simpson's rule performs adequately for sufficiently smooth integrands. It is useful for piece-wise continuous functions. If we are integrating over discrete data points, Simpson's rule should be used. We give an implementation in Program 8.5 as part of our numerical integration library.

Let our test integral be $I = \int_0^\pi \sin x \, dx = 2$. If $N = 2$, we have $h = \frac{\pi}{2}$, $x = [0, \frac{\pi}{2}, \pi]$, and $f = [0, 1, 0]$. Calling Simpson's rule with

```
I = integral.simpson(f, h)
```

returns a value 2.0944, an error of $\sim 5\%$. Doubling N gives a result 2.0046, an error of $\sim 0.2\%$.

To further improve the accuracy, we could increase N, or use a higher order interpolation. For instance, the Boole's rule uses five points interpolation,

$$I_{0-4} = \int_{x_0}^{x_4} f(x)dx = \frac{2h}{45}(7f_0 + 32f_1 + 12f_2 + 32f_3 + 7f_4) + O(h^7), \tag{8.74}$$

which can be readily cast into the abscissa-weight form like Eq. (8.68). However, a better alternative is discussed next.

8.A.3 GAUSSIAN INTEGRATION

If we view numerical integration as a process of finding optimal weights, then it effectively is an optimization problem. For both the trapezoid and Simpson's rules, the weights w_i are obtained at equidistant abscissas x_i.

Equidistant abscissas are reasonable, but their locations represent unused degrees of freedom in the optimization process. What if the abscissas are allowed to float? As it turns out, doing so can double the order of accuracy with the same number of nonequidistant abscissas compared to equidistant ones.

The Gaussian integration scheme uses nonequidistant abscissas and associated weights. The well-known Gaussian-Legendre method takes the zeros of the Legendre polynomials as the abscissas. The corresponding weights can be found by table lookup or calculation for a given order n (see Program 8.6). The integral can be obtained from a shifted abscissa-weight sum (see Eq. (S:8.55), Section S:8.A).

For instance, the abscissas and weights for $n = 3$ are $x = [-\sqrt{\frac{3}{5}}, 0, \sqrt{\frac{3}{5}}]$, $w = [\frac{5}{9}, \frac{8}{9}, \frac{5}{9}]$. Evaluating the sum (S:8.55) with these x and w values, we obtain for our test integral $\int_0^\pi \sin(x)\, dx \simeq 2.0014$, an error $\sim 0.07\%$. To obtain a similar level of accuracy with Simpson's rule, we need to roughly double the number of abscissas.

In Program 8.5, we give a Gaussian integration routine for $n = 36$. The routine `gauss(f,a,b)` may be used as a general-purpose integrator. It is sufficiently accurate, provided the integrand f is smooth enough and the range $[a, b]$ is not too great. Furthermore, it can handle integrable singularities at the end points. The routine needs a user-supplied integrand function. The following illustrates its usage for our test integral.

```
import integral as itg, numpy as np

def f(x):
    return np.sin(x)

I = itg.gauss(f, 0, np.pi)
print (I)
```

The returned integral is 2, practically exact within machine precision. The test integral is smooth. In general, we will not be so lucky. The theory of Gaussian integration is discussed in Section S:8.A.

8.A.4 INTEGRATION BY DIFFERENTIATION AND MULTIPLE INTEGRALS

Let us write the integral (8.63) as

$$y(t) = \int_a^t f(x)dx. \tag{8.75}$$

It means that

$$\frac{dy}{dt} = f(t), \quad y(a) = 0, \quad y(b) = I \text{ (the actual integral).} \tag{8.76}$$

We know how to integrate a single ODE, Eq. (8.76), by RK4 for instance. Thus, we start with the initial condition $y(a) = 0$ at $t = a$ (not 0), and integrate to $t = b$, at which point we have found our integral, $I = y(b)$. This method is of low efficiency. But it may be useful in some special situations.

Often we need to do double, triple, or higher multiple integrals. For instance, a double integral looks like

$$I = \int_a^b \int_c^d f(x, y)dxdy, \quad (f(x, y) \neq u(x) \times w(y)). \tag{8.77}$$

We can treat the integral as two nested, single integrals,

$$I = \int_a^b g(x)dx, \quad g(x) = \int_c^d f(x, y)dy. \tag{8.78}$$

Each single integral can be done using Gaussian integration, where x is a parameter in the second integral.

We can do up to 4D or perhaps 5D multiple integrals this way, depending on the integrand. Higher dimensional integrals would require N^d number of points, where N is the abscissa and d the dimension, making it impractical for higher d. In such cases, the integral should be done by Monte Carlo methods (see Section 10.4).

8.B PROGRAM LISTINGS AND DESCRIPTIONS

Program listing 8.1 A quantum plane wave (`qmplane.py`)

```
1  import matplotlib.pyplot as plt, numpy as np
2  import matplotlib.animation as am
3  
4  def updatefig(*args):                    # args[0] = frame
5      wf = np.exp(1j*(x-0.1*args[0]))      # traveling plane wave
6      w = [prob, wf.real, wf.imag]
7      for i in range(len(w)):
8          plot[i].set_data(x, w[i])        # update data
9  
10 x = np.linspace(-2*np.pi, 2*np.pi, 101)
11 wf, prob = np.exp(1j*x), [1]*len(x)
12 fig = plt.figure()
13 plot = plt.plot(x,prob, x,wf.real, x,wf.imag)             # plot object
14 ani = am.FuncAnimation(fig, updatefig, interval=10)       # animate
15 plt.ylim(-1.2, 1.2), plt.show()
```

The program animates a traveling plane wave given by $\psi = e^{i(kx-\omega t)}$, with $k = \omega = 1$ in this case. It uses complex numbers like in Program 5.7, and the identity $|\psi|^2 = 1$ for the probability density.

Three line objects are created and stored in the list `plot` (line 13). The Matplotlib animation function (line 14) calls `updatefig()` at regular intervals (in ms) to update the figure, passing the frame number in the arguments by default. In `updatefig()`, we scale the frame number as time, and update the (x, y) data set for each of the lines in `plot`.

Program listing 8.2 Space discretized leapfrog method (`sdlf.py`)

```
import numpy as np, visual as vp, ode, vpm

def sch_eqn(id, R, I, t):              # return dR/dt, dI/dt of Sch. eqn.
    b = 0.5/(h*h)
    if (id == 0):                      # gen. velocity
```

```
        tmp = -b*I                           # 1st/3rd term in Eq. (8.8a)
        dydt = (2*b + V)*I                   # 2nd term
    else:                                    # gen. acceleration
        tmp = b*R                            # 1st/3rd term in Eq. (8.8b)
        dydt = (-2*b - V)*R                  # 2nd term

    dydt[1:-1] += tmp[:-2] + tmp[2:]         # add bψ_{j-1}, bψ_{j+1} to grid j
    dydt[0]  += tmp[-1] + tmp[1]             # 1st point, periodic BC
    dydt[-1] += tmp[-2] + tmp[0]             # last point
    return dydt

def gaussian(s, x0, x):          # normalized Gaussian, s=width, x0=center
    c = 1.0/np.sqrt(s*np.sqrt(np.pi))
    return c*np.exp(-((x-x0)/s)**2/2)

def initialize (a, b, N):            # set parameters
    x = np.linspace(a, b, N+1)       # grid points
    V = 0.5*x*x                      # SHO potential
    s, x0  = 0.5, -5.0               # width (sigma), center of gaussian
    R, I = gaussian(s, x0, x), np.zeros(N+1)    # real, imag w.f.
    return x[1]-x[0], x,V,R,I        # grid size, h=x[1]-x[0]

a, b, N = -10., 10., 500             # space range [a,b], num. intervals
z, mag = np.zeros(N+1), 4            # zeros, magnifying factor

h, x, V, R, I = initialize (a, b, N)                    # initialization
scene = vp.display(background=(1,1,1), ambient=1)       # set scene
bars = vpm.bars(x, z, z, z, h, 0.05, (1,0,1))           # wave function
line = vpm.line(x, z, z, (1,0,1), 0.02)
pot  = vpm.line(x, V*0.1, z, (0,0,0), 0.02)             # potential line

t, ic, cycle = 0.0, 0, 20            # t, animation cycle
dt = h*h*0.5                         # time step, dt < h*h
while True:
    R, I = ode.leapfrog(sch_eqn, R, I, t, dt)    # main work
    ic += 1
    if (ic % cycle == 0):            # animate
        pb = R*R + I*I               # probability
        line.move(x, mag*pb, z)
        bars.move(x, z, z, mag*pb)
    vp.rate(3000), vpm.wait(scene)
```

The program starts by calling `initialize()`, which initializes the space grid, the potential, and the Gaussian wavepacket returned by `gaussian()`. The probability $|\psi|^2$ is represented by a line on top of a bar graph, which creates a distribution of solid colors. Both the line and bar objects are created from VPM modules of the same name and can be animated with the `move` method. The main loop integrates the Schrödinger equation, `sch_eqn()`, with the vectorized leapfrog method as explained in the text. The time t is not updated in the loop. However, if the potential was time-dependent, it would be necessary to update t and V after each step. A magnification factor is necessary to scale the distribution so it is visible over the large spatial range.

Program listing 8.3 Quantum free fall (`splitop.py`)

```
import numpy as np, visual as vp, vpm
from scipy.linalg import solve_banded
import matplotlib.pyplot as plt

def gaussian(s, x0, x):        # normalized Gaussian, s=width, x0=center
    c = 1.0/np.sqrt(s*np.sqrt(np.pi))
    return c*np.exp(-((x-x0)/s)**2/2)

def initialize (a, b, N):          # set parameters
    x = np.linspace(a, b, N+1) # grid points
    V = 2*x                        # linear potential
    s, x0 = 0.5, 5.0               # width (sigma), center of gaussian
    R, I = gaussian(s, x0, x), np.zeros(N+1)   # real, imag w.f.
    return x[1]-x[0], x,V,R,I   # grid size, h=x[1]-x[0]

a, b, N = -10., 10., 500          # space range [a,b], num. intervals
z, mag = np.zeros(N+1), 6         # zeros, magnifying factor

h, x, V, R, I = initialize (a, b, N)                  # initialization
scene = vp.display(background=(1,1,1), ambient=1) # set scene
bars = vpm.bars(x, z, z, z, h, 0.05, (1,0,1))         # wave function
line = vpm.line(x, z, z, (1,0,1), 0.02)
pot  = vpm.line(x, V*0.05, z, (0,0,0), 0.02)          # potential line

t, ic, cycle = 0.0, 0, 10                  # t, animation cycle
dt, psi = 0.001, R + 1j*I                  # initialize dt, complex psi
ta, pba = [], []                           # time, prob. arrays
A = np.ones((3, N+1), dtype=complex) # prepare band matrix A, B
A[1, :] = 2*(2j*h*h/dt - 1 - h*h*V)
dB = - np.conjugate(A[1,:])                # diagonal of B
while(t<=3):
    C = dB*psi                             # prepare RHS Eq. (8.31)
    C[1:-1] -= (psi[:-2] + psi [2:])
    psi = solve_banded((1,1), A, C)        # band matrix solver
    t, ic = t+dt, ic+1
    if (ic % cycle == 0):
        pb = psi.real**2 + psi.imag**2 # probability
        ta.append(t), pba.append(pb)   # store data
        line.move(x, mag*pb, z)
        bars.move(x, z, z, mag*pb)
    vp.rate(1200), vpm.wait(scene)

(X, Y), ta = np.meshgrid(x, ta), np.array(ta)
plt.figure()
plt.contour(Y, X, pba, 36, linewidths=1)
plt.plot(ta, 5 - ta*ta, 'r--')             # classical free fall
plt.xlabel('t (a.u.)'), plt.ylabel('x (a.u.)')
plt.show()
```

The overall program flow is similar to Program 8.2. Lines 28 to 34 are the core of the first-order split-operator method and have been explained in the main text. The variable pba[] stores the probability distribution as a list of arrays, appended to periodically. It contains data for contour plotting at the end.

Program listing 8.4 Quantum waves in 2D (qmwaves2d.py)

```
import matplotlib.pyplot as plt
import numpy as np, visual as vp, vpm

def initialize (a=10., N=12, nx=40, ny=40): # size, max QM num, grid
    s, x0, y0 = 1.9, a/2, a/2                # width, center of wavepacket
    na, pi = np.arange(1, N+1), np.pi        # QM numbers
    En = pi*pi*na**2/(2*a*a)                 # all eigenenergies
    x, y = np.linspace(0,a,nx), np.linspace(0,a,ny)      # grid
    X, Y = np.meshgrid(x, y)
    umx, uny = np.zeros((N,ny,nx)), np.zeros((N,ny,nx))
    for n in range(N):                       # compute basis functions
        umx[n] = np.sqrt(2/a)*np.sin((n+1)*pi*X/a)
        uny[n] = np.sqrt(2/a)*np.sin((n+1)*pi*Y/a)
    amn0 = expansion(a, na, s, x0, y0)       # expansion coeff.
    return a, N, En, amn0, X, Y, umx, uny

def expansion(a, n, s, x0, y0):        # calc expansion coefficients
    r, q, pi = s/a, n*s/a, np.pi       # warning: 'a/s' must not be integer
    cm = np.cos(q*pi/2)*np.sin(n*pi*x0/a)/((1-q)*(1+q))
    cn = np.cos(q*pi/2)*np.sin(n*pi*y0/a)/((1-q)*(1+q))
    return 16*r*np.outer(cm, cn)/(pi*pi)    # amn(0)

def psi(amn0, t):                      # calc wavefunction at time t
    wf, phase = 0.0, np.exp(-1j*En*t)
    amnt = amn0*np.outer(phase, phase)    # a_mn(t) = a_mn(0)e^(-iE_mn t)
    for m in range(N):
        s = 0.0
        for n in range(N): s += amnt[m,n]*uny[n]  # vector operation
        wf += s*umx[m]                         # Σ a_mn u_m(x)u_n(y)
    return wf

t, num, plot = 0., 1, False              # set plot=False to animate forever
time = [0.0, 0.5, 0.9, 1.4, 2.0, 3.0, 40, 63.7, 64.2]    # snapshots
a, N, En, amn0, X, Y, umx, uny = initialize()

scene = vp.display(background=(.2,.5,1), center=(a/2,a/2,-0.5),
                   up=(0,0,1), forward=(1,2,-1))
mesh = vpm.mesh(X, Y, 0*X, vp.color.yellow, vp.color.red)
info = vp.label(pos=(a/4, .8*a,1), height=20)
while True:
    wf = psi(amn0, t)
    mesh.move(X, Y, wf.real*3)          # show real part, times 3
    info.text='%5.2f' %(t)
    vpm.wait(scene), vp.rate(40)
    if (plot):
        plt.subplot(3, 3, num)
        plt.imshow(np.abs(wf)**2, cmap=plt.cm.jet) # try contourf()
```

```
48        plt.xticks([], []), plt.yticks([], [])          # omit ticks
49        plt.xlabel('t=%s' %(t))
50        if (num > len(time)-1): break
51        t, num = time[num], num+1
52    else:
53        t = t + 0.05
54 if (plot): plt.show()
```

The function `initialize()` initializes some parameters and sets up the space grid. The basis functions $u_m(x)$ and $u_n(y)$ for each state $n = 1$ to N are calculated over the grid. It also calls `expansion()`, which computes the expansion coefficients of the compact wavepacket at $t = 0$ from Eq. (8.53). Both C_m and C_n are calculated first, each a 1D array of length N, then an $N \times N$ matrix a_{mn} is formed with `np.outer` (line 21), which produces an outer product such that $a_{mn} = C_m C_n$, similar to Program S:6.1.

The same outer product is used in `psi()` to calculate the time-dependent amplitudes $a_{mn}(t) = a_{mn} \exp(-i(E_m + E_n)t)$, from which the wave function (8.52) is computed. Because $u_m(x)$ and $u_n(y)$ have been precomputed, the sum in Eq. (8.52) is performed efficiently using element-wise (vector) operations over the space grid.

The main program sets up a VPM mesh to animate the real part of the wave function and displays the time with `vp.label`. If `plot` is set to `True`, nine subplots (3×3) at preselected times are created using `imshow()`, which produces a smooth, contour-like image. The tick marks are suppressed by supplying empty lists (line 48). If `plot` is `False`, the program just steps through time indefinitely.

Program listing 8.5 Numerical integration (`integral.py`)

```
def simpson(f, h):                  # Simpson's rule over discrete points
    n, extra = len(f), 0.0
    if (n-1)%2 != 0:                        # numbers of bins not even
        n = n - 1
        extra = f[-2]+f[-1]                 # trapzoid for last bin

    sum, w = f[0]+f[n-1], 4
    for i in range(1, n-1):
        sum = sum + w*f[i]
        w = 6 - w                           # alternate between 4, 2
    return h*(sum/3. + extra/2.)

def gauss(f, a, b):                 # Order 36: abscissas and weights
    x=[0.9978304624840857, 0.9885864789022123, 0.9720276910496981,
       0.9482729843995075, 0.9174977745156592, 0.8799298008903973,
       0.8358471669924752, 0.7855762301322065, 0.7294891715935568,
       0.6680012365855210, 0.6015676581359806, 0.5306802859262452,
       0.4558639444334203, 0.3776725471196892, 0.2966849953440283,
       0.2135008923168655, 0.1287361038093848, 0.0430181984737086]

    w=[.005565719664245068, 0.01291594728406522, 0.02018151529773513,
       0.02729862149856854, 0.03421381077030707, 0.04087575092364476,
       0.04723508349026585, 0.05324471397775972, 0.05886014424532480,
       0.06403979735501547, 0.06874532383573639, 0.07294188500565307,
       0.07659841064587077, 0.07968782891207167, 0.08218726670433972,
       0.08407821897966186, 0.08534668573933869, 0.08598327567039468]
```

```
    p = 0.5*(b+a)           # map from [a,b] to [-1,1]
    q = 0.5*(b-a)
    n = len(x)
    sum = 0.0
    for i in range(n):
        sum += w[i]*(f(p + q*x[i]) + f(p - q*x[i]))
    return q*sum
```

The function `simpson()` assumes discrete data points. The number of intervals should be even. If it is odd, the first $n-2$ intervals (n = number of data points) are computed by Simpson's rule, and the last one by trapezoid rule. If the latter contribution is not small, the accuracy may be affected.

The Gaussian-Legendre function `gauss()` can be used with any abscissa-weight data. SciPy provides an easy way to generate these data sets from its special functions library. For orthogonal polynomials, the library can return objects called `orthopoly1d` containing the abscissa-weight data. The following code illustrates its use.

Program listing 8.6 Gauss abscissa and weight (`gaussxw.py`)

```
from scipy.special import legendre

n = 36                  # order
p = legendre(n)
x = p.weights[:,0]      # abscissa
w = p.weights[:,1]      # weight

for i in range(n):
    print ('%20.16f %20.16f' %(x[i], w[i]))
```

The data can be used directly in `gauss()`. If the order is even, no modification is necessary (other than dropping the negative abscissas). If it is odd, make sure to include the $x = 0$ point only once in the loop. If the default order here (36) is not enough to give accurate results, our integrand is probably ill-behaved (highly oscillatory, stiff, or singular), or the integration limit is too great. Increasing the order alone rarely cures the problem. A better strategy is to break the range into several subintervals and add up their contributions, or to make a proper change of variable. The latter is particularly important if one end of the integral is infinity, for example,

$$\int_0^a \frac{f(x)}{\sqrt{x}}\,dx \overset{x=t^2}{\longrightarrow} 2\int_0^{\sqrt{a}} f(t^2)\,dt, \qquad \int_a^\infty f(x)dx \overset{x=1/t}{\longrightarrow} \int_0^{1/a} \frac{f(1/t)}{t^2}\,dt. \tag{8.79}$$

Program 8.5 serves as our numerical integration library. If more sophisticated integration technique is required, such as the need for error control, look to SciPy library `scipy.integrate`. It has several routines, including a general-purpose routine quad for single integrals, and `dblquad` and `tplquad` for double and triple integrals, respectively.

Time-Independent Quantum Mechanics

CHAPTER 9

We saw in Chapter 8 that time-dependent quantum motion could be studied in its own right, but some features such as quantum transitions are connected to the stationary states of time-independent quantum systems, where the potential has no explicit dependence on time. Transitions between these states give rise to *discrete* emission lines, which could not be explained by classical physics when they were discovered around the turn of the twentieth century. In fact, one of the most spectacular successes of quantum mechanics was its full explanation of the hydrogen spectra, helping to establish quantum physics as the correct theory at the atomic scale.

In this chapter, we discuss the structure of time-independent quantum systems. We are interested in seeking solutions to the bound states of the time-independent Schrödinger equation (see Eq. (9.1)). We will discuss several methods, beginning with shooting methods for 1D and central-field problems. They are the most direct methods with a high degree of flexibility in choosing the core algorithm, requiring no more than an ordinary differential equation (ODE) solver and a root finder at the basic level. We use it with the efficient Numerov algorithm to study band structure of quasi-periodic potentials, atomic structure, and internuclear vibrations.

We also apply the 1D finite difference and finite element methods introduced previously for boundary value problems to eigenvalue problems, including point potentials in the Dirac δ-atom and δ-molecules. The basis expansion method is introduced as a general meshfree method that depends on the basis functions rather than the space grid.

To simulate quantum dots, we develop a full finite element method (FEM) program to solve the Schrödinger equation in arbitrary 2D domains. The program is robust and simple to use, the core having less than 10 lines of code. We will explore the rich physics in these structures, including the triangle and hexagon quantum dots.

Unless explicitly noted, we use atomic units throughout.

9.1 BOUND STATES BY SHOOTING METHODS

Our study begins with the time-independent Schrödinger equation, which can be obtained from the time-dependent case (8.2) by the separation of space and time variables such as in Eq. (6.64),

$$\psi(x,t) = u(x)\exp(-iEt/\hbar),$$
$$-\frac{\hbar^2}{2m}\frac{d^2u(x)}{dx^2} + V(x)u(x) = Eu(x). \tag{9.1}$$

This is a second-order, homogeneous ODE. For most potentials, the stationary solutions fall into two categories, localized bound states that vanish at infinity (i.e., $u(\pm\infty) = 0$), and continuum states that oscillate forever. As stated earlier, we focus on the bound states in this chapter and will consider continuum states in Chapter 12.

Subject to the boundary condition $u(\pm\infty) = 0$, Eq. (9.1) determines simultaneously the correct solutions and the allowed energies E, in the same way that Eq. (6.86) determines both the shapes and frequencies of standing waves. The shooting method (Section 3.7) helps us see how this works and why only discrete values of E are allowed.

Because Eq. (9.1) is an ODE, it suggests that we could try solving it like a two-point boundary value problem, $u(-\infty) = u(\infty) = 0$, similar to shooting a projectile illustrated in Figure 3.22. Suppose we start with a trial E, integrate from $x = -R$ to $+R$ (R large enough to approximate ∞) with the initial value $u(-R) = 0$, and check the final value $u(R)$. If it is not zero, we adjust E and repeat. When $u(R)$ is zero, we have found the right energy and solution because they satisfy the correct boundary conditions. This in essence is the shooting method.

9.1.1 VISUALIZING EIGENSTATES

The idea of the shooting method is animated in Program 9.1 for the square well potential and illustrated in Figure 9.1 with snapshots taken from the program. We assume the potential is symmetric about the origin for now, that is, $V = -V_0$ if $|x| \le \frac{a}{2}$, and zero elsewhere. The width and depth of the well are $a = 4$ a.u. (2.1 Å) and $V_0 = 4$ a.u. (109 eV), respectively.

Program 9.1 steps through the energy range $-V_0 < E < 0$ at fixed increments. One can show analytically and graphically that no bound states exist outside this range (Project P9.1). At a given energy E, the program integrates the Schrödinger equation from $x = -R$ upward to $x = 0$, instead of $x = +R$. The reason is that as $x \to \pm\infty$, there are two mathematical solutions, one exponentially decaying and another exponentially growing. The former is physically acceptable, but the latter is numerically dominant. A small error will be exponentially magnified and overwhelm the correct solution. We saw the same kind of divergence in the calculation of the golden mean (Section 1.2.2). This means that if we were to integrate to $x = \infty$, the wave function will never be zero at the boundaries and will fail the test $u(\infty) = 0$, no matter how exact the energy E is.

The correct technique to avoid divergence is to integrate inward from both ends: upward from $x = -R$ and downward from $x = R$, and meet somewhere in the middle at a matching

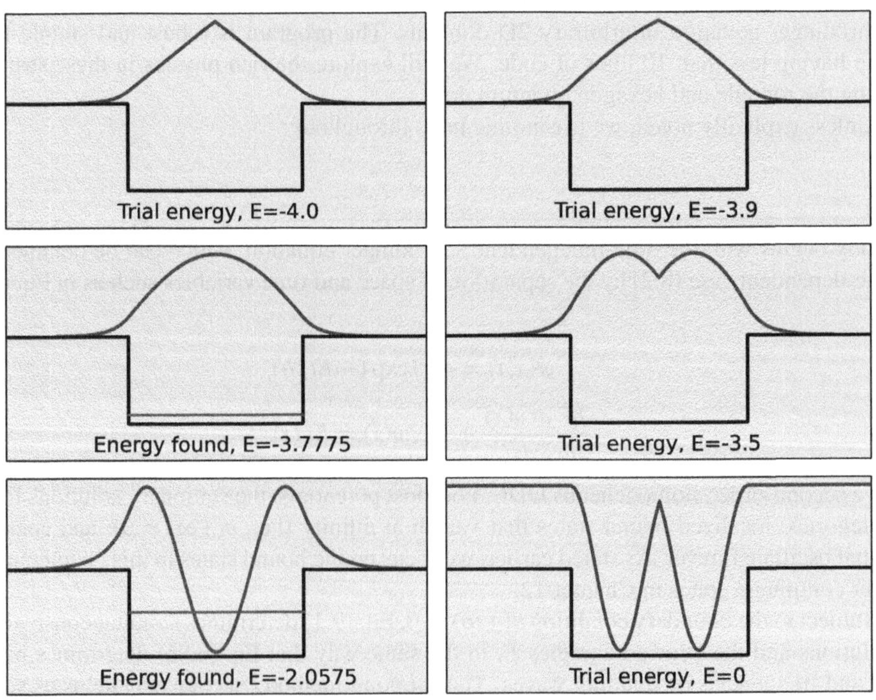

FIGURE 9.1 Visual eigenstates of a square well potential by the shooting method.

point, x_m. We choose $x_m = 0$ in this case. For a symmetric potential, $V(-x) = V(x)$, the solutions are either even or odd, that is, the wave function has definite parities. Because of this, we integrate from $x = -R$ to 0, and obtain the wave function in the other half by symmetry. We will discuss general, asymmetric cases shortly.

If the wave functions match each other smoothly at x_m, we have a solution that satisfies Eq. (9.1), vanishes at $\pm\infty$, is differentiable, and therefore must be a correct solution. For even states shown in Figure 9.1, the derivatives of correct solutions will be zero at the origin. Initially at $E = -V_0 = -4$, the wave functions inside the well are straight lines meeting at $x = 0$, creating a cusp. The energy is too low, and the wave function has not bent enough. The situation improves a little at a higher energy $E = -3.9$, and the wave functions bend downward and are no longer straight lines. But it is still not smooth across the origin. The next frame at $E = -3.78$, the wave functions connect smoothly to each other, and the kink has disappeared. We have found the first allowed energy—an eigenenergy, and with it, the correct solution—an eigenfunction.

Increasing the trial energy further, the wave function overbends. By the time it again becomes smooth at $E = -2.06$, we have found the second, and last, even eigenstate with two nodes. There are no more even states to be found after this. The last frame ($E = 0$) shows a would-be solution with 4 nodes, but it is not smooth at the matching point. There are also a finite number of odd states that can be found similarly (Project P9.1).

This example shows graphically how discrete energies naturally emerge in quantum mechanics. There had been considerable attention to visual representation of eigenstates even before the age of personal computers. In Ref. [23],[1] for instance, a second-order recursion method similar to RK2 had been used to generate a film like Figure 9.1. If we accept normal modes of oscillations (Section 6.2.2) or standing waves (Section 6.6) in classical mechanics as fact of life in the land of waves, these discrete energies in quantum mechanics may not seem so strange anymore. They are a result of finding well-behaved solutions to a wave equation, the Schrödinger equation.

9.1.2 GENERAL SHOOTING METHOD

We just illustrated the idea of the shooting method with the symmetric square well as a simple example. For it to be applicable to arbitrary potentials, we need to generalize the method so that it can deal with any symmetry and can narrow down the precise value of the eigenenergies.

To begin, we introduce an auxiliary variable $w = u'$ to represent the first derivative of the wave function and convert the second-order Schrödinger equation into two first-order ODEs (in a.u.)

$$u' = w,$$
$$w' = 2(V - E)u. \tag{9.2}$$

We solve Eq. (9.2) in the range $x_L \le x \le x_R$ over the grid points $x_j = x_0 + jh$, $j = -N, \ldots, -1, 0, 1, \ldots, N$, $h = (x_R - x_L)/2N$, so $x_0 = \frac{1}{2}(x_L + x_R)$ is the midpoint. For open space, the value of $x_R - x_L$ should be large compared to the characteristic scale of the system. We assume the wave function is zero outside the range. As stated earlier, we must integrate inward from $x = x_L$ and x_R to maintain stability. Usually, we can set the boundary values $u(x_{\pm N}) \equiv u_{\pm N}$ to zero, $u_{-N} = u_N = 0$, and w_{-N} and w_N to a small but arbitrary nonzero value (e.g., 0.1).

[1]The study concluded: "Students ... are weaned on continuous, well-behaved functions and variables. It is therefore often a shock to them when they study quantum mechanics and find, suddenly, discrete variables. The purpose of the computer-generated film described in this paper is to illustrate that these discrete variables are not so peculiar after all—that, in fact, they arise in a natural way from requiring the solution of the Schrödinger equation to be continuous, differentiable, and finite."

The latter affects only the overall scale (normalization) of the wave function, which cancels out, as we will see shortly.

Let $u^{\uparrow}$ and $u^{\downarrow}$ represent the solutions of upward and downward integration, respectively. If the energy is a correct eigenenergy, the matched wave functions will be smooth at the matching point x_m. This means that the wave functions and their derivatives at x_m can differ at most by an overall constant C,

$$u^{\uparrow}(x_m) = Cu^{\downarrow}(x_m), \quad u'^{\uparrow}(x_m) = Cu'^{\downarrow}(x_m). \tag{9.3}$$

Eliminating C, we have an equivalent matching condition

$$u^{\uparrow}(x_m)u'^{\downarrow}(x_m) - u^{\downarrow}(x_m)u'^{\uparrow}(x_m) = 0. \tag{9.4}$$

Equation (9.4) is satisfied only if the trial energy E is an exact eigenenergy. In other words, allowed energies are roots of Eq. (9.4). We can use this fact to locate the exact energy: Pick a trial E, increase E until Eq. (9.4) changes sign, then we know a root has been bracketed. We can narrow the bracket down with a root solver like the bisection method.

Let us define a matching (shooting) function in terms of u and w on the grid,

$$f(E) = u_m^{\uparrow} w_m^{\downarrow} - u_m^{\downarrow} w_m^{\uparrow}, \tag{9.5}$$

where we made E explicit, and $u_m^{\downarrow\uparrow} = u^{\downarrow\uparrow}(x_m)$ as usual. We also substituted $w_m^{\downarrow\uparrow} = u_m'^{\downarrow\uparrow}$. This is the function we will evaluate numerically,[2] in much the same way as Eq. (3.49b) for projectile motion.

Program 9.2 implements the shooting algorithm. Key parts of the program are given here.

```
def sch(psi, x):                    # Schrodinger eqn
    return [psi[1], 2*(V(x)-E)*psi[0]]

def intsch(psi, n, x, h):    # integrate Sch eqn for n steps
    psix = [psi[0]]
    for i in range(n):
        psi = ode.RK45n(sch, psi, x, h)
        x += h
        psix.append(psi[0])
    return psix, psi                # return WF and last point

def shoot(En):                      # calc diff of derivatives at given E
    global E                        # global E, needed in sch()
    E = En
    wfup, psiup = intsch([0., .1], N, xL, h)      # march upward
    wfdn, psidn = intsch([0., .1], N, xR, -h)     # march downward
    return psidn[0]*psiup[1] - psiup[0]*psidn[1]

......
while (E1 < 0):       # find E, calc and plot wave function
    if (shoot(E1) * shoot(E1 + dE) < 0):          # bracket E
        E = rtf.bisect(shoot, E1, E1 + dE, 1.e-8)
    ......
```

[2] A mathematically equivalent form to Eq. (9.5) is $f(E) = u_m^{\uparrow} - u_m^{\downarrow} w_m^{\uparrow} / w_m^{\downarrow}$, but this is not stable for numerical work because the wave function or its derivative can become close to zero, and $f(E)$ could fluctuate wildly.

The function sch() returns the Schrödinger equation (9.2). The wave function and its derivative are contained in a two-element array, psi=$[u, w]$. The main workhorse is the function intsch(). It integrates sch() for n steps from the initial value using the nonvectorized RK45 (Runge-Kutta-Fehlberg) method for faster speed because the ODEs in sch() are simple to evaluate and called repeatedly. A vectorized version would be inefficient for only two ODEs in this case.

The shooting function (9.5) is calculated in shoot(). First, it sets energy E, which is declared to be global because it is required in a different function sch(). Then it marches upward from $x = x_L$ for N steps and downward from $x = x_R$ for an equal number of steps, so the matching point x_m is at the midpoint (origin in this case). Note that in the downward march the step size is $-h$. The main loop scans through the energy. When a root is bracketed, the bisection root finder is called to accurately find the eigenenergy.

As our first test case, we apply the shooting method to the square well potential in Figure 9.1. The results from Program 9.2 are given in Table 9.1 (along with other results to be discussed later). They agree very well with the exact values obtained by numerically solving the analytic eigenequations (Exercise E9.1). The first and third states are even, and the second and fourth odd. We had seen the even states in Figure 9.1.

As well as the shooting method works for the square well, it performs even better for continuous potentials such as the simple harmonic oscillator (SHO) and yields practically exact results (see Project P9.2).

As the second test case, we use the shooting method to calculate the eigenstates of a double square well potential defined here and graphed in Figure 9.2,

$$V = \begin{cases} -V_0, & \frac{b}{2} \le |x| \le \frac{a}{2}, \\ 0, & |x| < \frac{b}{2}, \text{ or } |x| > \frac{a}{2}. \end{cases} \tag{9.6}$$

Table 9.1 Eigenenergies of the Square Well Potential ($a = 4$, $V_0 = 4$).

Method	1	2	3	4	Notes
Exact	−3.77791	−3.11995	−2.05806	−0.69467	
Shooting	−3.77791	−3.11993	−2.05803	−0.69463	$h = 0.032$
FDM (9.3.1)	−3.77769	−3.11948	−2.05849	−0.69833	$h = 0.063$
FEM (9.3.2)	−3.77786	−3.11952	−2.05642	−0.69087	$h = 0.053$
BEM (9.4)	−3.77791	−3.11993	−2.05802	−0.69459	$N = 300$

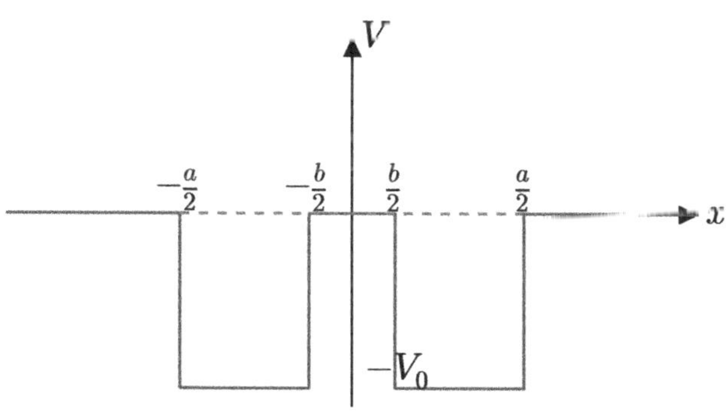

FIGURE 9.2 The double square well potential.

The results from Program 9.2 are shown in Figure 9.3. The parameter values are $a = 4$, $b = 1$, and $V_0 = 6$. We again see four bound states, two even and two odd, the same as the single-well potential. But there are several differences. The wave functions are very different. For the ground state (−4.907), we see a double hump structure rather than a single peak. This is caused by the barrier in the middle. Inside the barrier, the wave function does not oscillate and must either decay or grow exponentially. We can understand this from Eq. (9.2), $u'' = 2(V - E)u$. Below the potential, $V - E$ is positive, so u'' and u have the same sign. If u, and hence its second derivative u'', is positive, the wave function must be concave up, so it will stay positive (remember, no nodes allowed for the ground state). Similarly for negative u, the opposite is true. Therefore, the wave function in each well has to bend and passes a maximum to decrease to zero at $x = \pm\infty$. In the next even state (−2.005), a local peak is formed within the barrier for the same reason, the only way to have two nodes.

In contrast, the odd states look similar to their counterparts for a single-well potential (see Project P9.1). The difference is a change of curvature in the middle of the barrier, with the second derivative changing sign due to u crossing zero (visible in the first excited state at −4.858, but hardly noticeable at −1.69).

The ground and the first excited states are separated in energy by 0.049, which is only 0.8% of the depth of the wells. The wave function of the first excited state looks like that of the ground state if one of its humps is flipped. In fact, we can think of the wave functions as a linear superposition of one basic state shifted to different centers $\pm x_c$,

$$u_{\pm}(x) = \varphi(x + x_c) \pm \varphi(x - x_c). \tag{9.7}$$

The plus and minus signs correspond to the even and odd states, respectively, that is, $u_{\pm}(-x) = \pm u_{\pm}(x)$, assuming $\varphi(-x) = \varphi(x)$. Equation (9.7) represents a linear combination of

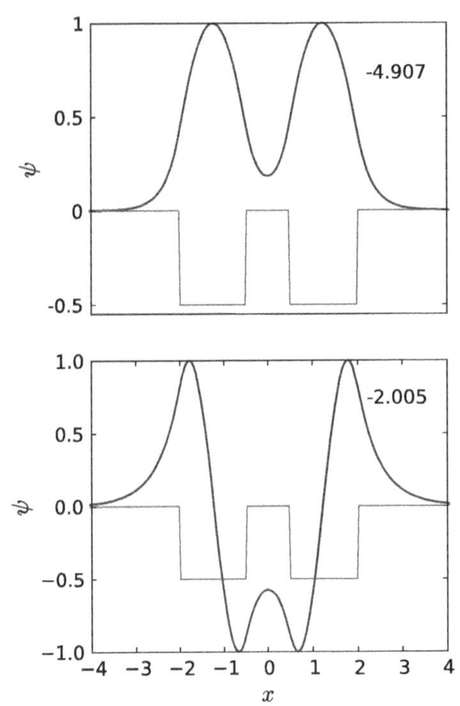

FIGURE 9.3 The eigenstates in the double square well potential by the shooting method.

atomic orbitals (LCAO). The function φ would describe the ground state in one of the wells (single-particle state). As the barrier width becomes large, the two wells are increasingly independent, so the two states become approximately degenerate with the same energy.

9.2 PERIODIC POTENTIALS AND ENERGY BANDS

Suppose we add many potential wells to Figure 9.2 to form an extended periodic system. The shooting method would require a large number of grid points, slowing down the speed. The bottleneck is the calculation of the matching function $f(E)$, Eq. (9.5), which relies on a general ODE solver (RK45 in this case) to integrate the Schrödinger equation. Because $f(E)$ is called repeatedly, the efficiency of the ODE solver is crucial.

As it turns out, we can solve the Schrödinger equation much more efficiently using Numerov's method rather than using a general ODE solver. This method is specialized to second-order ODEs, which do not contain first derivatives explicitly. It is presented in Section 9.A.

To use Numerov's method, we write the Schrödinger equation (9.1) as

$$u'' + \frac{2m}{\hbar^2}(E - V)u = 0. \tag{9.8}$$

The method gives the solution as a three-term recursion (9.59),

$$u_{i+1} = \left(1 + \frac{h^2}{12}f_{i+1}\right)^{-1}\left[2\left(1 - \frac{5h^2}{12}f_i\right)u_i - \left(1 + \frac{h^2}{12}f_{i-1}\right)u_{i-1}\right], \tag{9.9}$$

where $f_i = 2m(E - V(x_i))/\hbar^2$. Equation (9.9) is symmetric whether marching up or down, that is, the subscripts $i \pm 1$ can be swapped with $i \mp 1$ without any change.

The recursion is stable and is accurate to fifth order in h, the same order as the RK45 method. Although it requires three calls to $f(x)$, two of the values can be reused in subsequent iterations, so it needs only one new call per iteration. In contrast, RK45 requires six function calls per step.

However, there are two issues to consider. First, Numerov's method is not self-starting because two seed values are necessary to get started as expected from a second-order ODE. This poses no difficulty for us. For instance, marching downward from the boundary point N, we can set $u_N = 0$ and u_{N-1} to a small value (say 0.1) as before, affecting only the normalization. Now, we can obtain u_{N-2} from Eq. (9.9), then u_{N-3}, and so on. Upward recursion is similar.

Second, Numerov's method does not provide first derivatives on its own. But we need them when matching the wave functions. We can calculate the first derivative using the central difference formula (6.31b). Better yet, we can do it more accurately with a similar formula (9.60), which takes advantage of the special form of the ODE the same way Numerov's method does. Either way, we must march one step past the matching point in both directions to compute the first derivatives. If the matching point is m, we must march upward from $-N$ to $m + 1$, and downward from N to $m - 1$, calculate the derivatives at x_m, and match the solutions with Eq. (9.5). We leave the implementation to Project P9.3.

The speed gain for shooting with Numerov's method is significant. We can use it to extend our study to large systems requiring large number of grid points. As an example, we show the results obtained this way for multiple identical potential wells (1, 2, 4, and 8, Figure 9.4). In each case, the width of the well is $a = 1$, depth $V_0 = 6$, and the thickness of the walls (barriers) between the wells is 0.5.

There are only two states for a single well. For the double well, there are three states in total. The extra state comes from the ground state of the single well, which splits into two adjacent

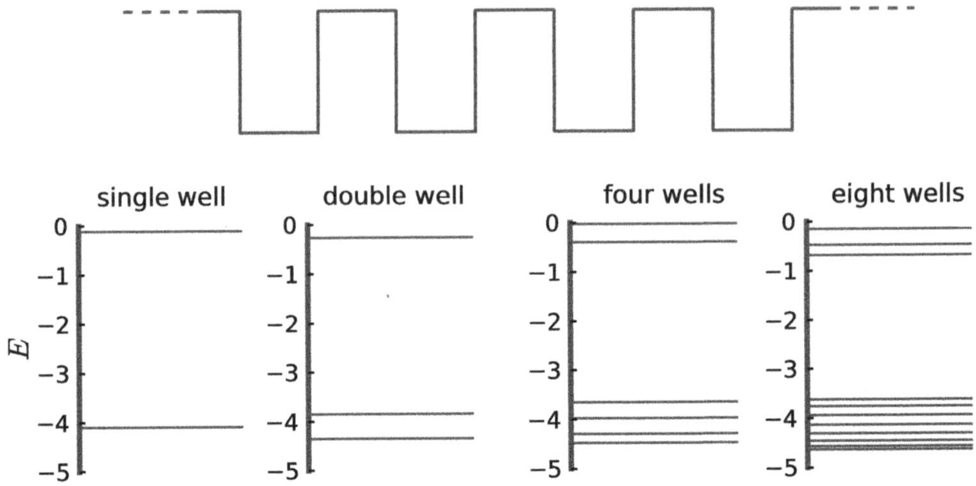

FIGURE 9.4 Multiple potential wells (top) and their eigenenergies forming energy bands.

levels, one slightly below and the other slightly above the original level. The two levels come from the near degeneracy we discussed regarding the double-well potential in Figure 9.3.

The excited state of the single well is too close to zero to split, as one would end up above zero, so it remains one bound state below the original energy. For four wells, we see more splitting, with the states clustering into two groups at the top and bottom. We are beginning to see energy bands. At eight wells, the bands become more well defined and more densely filled, particularly the bottom band. The gap between the bands remains empty.

We can imagine as the number of wells increases, we have a quasi-periodic system with filled bands and empty gaps. This is how energy bands form in solids [6]. Atoms pack together in periodic lattice structures in solids. Multiple states cluster around single atomic states, forming energy bands and gaps. Details may differ, but the qualitative features are the same. Our example shows not only that energy bands form, but how they form, namely via state splitting. Each pair of new states are approximately degenerate. When many nearly degenerate states cluster together, we have completely filled bands.

9.3 EIGENENERGIES BY FDM AND FEM METHODS

The shooting method discussed so far scans the energy range and finds one eigenstate at a time. It is suitable primarily for effectively 1D problems. We describe two other grid-based methods, finite difference method (FDM) and FEM, that solve for all states simultaneously and can be extended to higher dimensions.

9.3.1 EIGENENERGIES BY FDM

We described the FDM method in connection with the displacement of an elastic string under external loads (Section 6.3.2). We just need to make a slight change to convert it from a linear system to an eigenvalue system.

The displacement equation (6.29) and the Schrödinger equation (9.1) are structurally similar. Comparing the two, we identify "tension" $-\hbar^2/2m$ ($-1/2$ in a.u.) and "load" $f(x) = (V(x) - E)u(x)$ for the latter. The function f now depends on the solution u itself, due to the linearity of the Schrödinger equation. This leads to an eigenvalue problem.

We can discretize the kinetic energy operator $\hat{T} = -\frac{1}{2}d^2/dx^2$ and follow the discussion of the FDM method in Section 6.3.2 up to Eq. (6.33). There, the f values on the grid are replaced by

$f_i = V_i u_i - E u_i$, even though E and u_i are unknown yet. In matrix form, let

$$\hat{T} = \frac{1}{2h^2}\begin{bmatrix} 2 & -1 & & & \\ -1 & 2 & -1 & & \\ & \ddots & \ddots & \ddots & \\ & & -1 & 2 & -1 \\ & & & -1 & 2 \end{bmatrix}, \quad \hat{V} = \begin{bmatrix} V_1 & & & & \\ & V_2 & & & \\ & & \ddots & & \\ & & & V_{N-2} & \\ & & & & V_{N-1} \end{bmatrix}. \tag{9.10}$$

Note the negative sign in $\hat{T} = -\frac{1}{2}d^2/dx^2$ is absorbed inside the matrix (compare with Eq. (6.34)), and $\hat{V}$ is the potential energy operator in matrix form. The eigenvalue equation can be expressed as

$$(\hat{T} + \hat{V})\mathbf{u} = E\mathbf{u}. \tag{9.11}$$

This is the same eigenvalue problem as in Section 6.6. In Eqs. (9.10) and (9.11), we assume the usual notations: $V_i = V(x_i)$ on the grid (Figure 6.10), and $\mathbf{u}$ is a column vector containing the $N-1$ unknown wave function values, $u_1, u_2, \ldots, u_{N-1}$. We have excluded the boundary values $u_0 = u_N = 0$ in Eq. (9.11) because the wave function vanishes on the boundary for bound states (Dirichlet boundary condition).

The kinetic energy matrix $\hat{T}$ is tridiagonal (same as Eq. (6.34) except for an overall negative sign), and the potential energy matrix $\hat{V}$ is diagonal. Equation (9.11) may be solved to obtain the eigenenergies and associated eigenvectors using SciPy functions like `eigh()` in Program 6.2, or the more efficient sparse eigenvalue solver, `eigsh()` (see Program 9.3, line 40).

The FDM method is direct and easy to use. The $\hat{T}$ matrix is independent of potential, and the $\hat{V}$ matrix has only diagonal elements $\hat{V}_{ij} = V_i\delta_{ij}$. As a test, we give the results of the FDM method for the square well in Table 9.1. The step size is the same as in the shooting method. The eigenequation (9.11) has $N-1$ eigenvalues ($N = 500$ in the present case). The four lowest eigenvalues are negative and are listed in Table 9.1, so the FDM gives the correct number of true bound states. The rest are positive, and they represent pseudocontinuum states.[3]

The first three bound states agree with the exact results to four digits, and the fourth to two digits. The accuracy (second order) is good except for the fourth bound state. The reason is that the highest bound state is close to the continuum, and coupling to (i.e., mixing with) the pseudocontinuum makes its accuracy worse. The FDM results are less accurate than the shooting method, which is of a higher order with either RK45 or Numerov's method and does not suffer from pseudocontinuum coupling. If we need only a few discrete states or high accuracy, the shooting method is better suited than the FDM. But the FDM is much simpler to use and can produce bound states en masse.

9.3.2 EIGENENERGIES BY FEM

Similar to the FDM, we developed the FEM method and first applied it to the displacement solution of an elastic string. It is equally applicable to quantum eigenvalue problems as we show next. The development of 1D FEM will also help us better understand 2D FEM in the study of quantum dots (Section 9.6).

We start from the integral formulation of FEM in Section 6.4. In complete analogy to the earlier FDM description, we identify from Eq. (6.42) $T = -\frac{1}{2}$ and $f(x) = (V(x) - E)u(x)$ by comparing the displacement equation (6.29) and the Schrödinger equation (9.1). Analogous to

[3]Due to the boundary conditions, these positive energy states are bound states in the (numerically finite) box, mimicking the true continuum states of the actual, infinite system.

Eq. (6.42), we have for the Schrödinger equation in FEM form

$$-\frac{1}{2}\int_a^b \frac{d^2u(x)}{dx^2}\varphi_j(x)dx + \int_a^b V(x)u(x)\varphi_j(x)dx = E\int_a^b u(x)\varphi_j(x)dx, \tag{9.12}$$

where a and b are the boundary points.

Following exactly the same steps from Eq. (6.42), namely, integrating the kinetic energy term by parts, and substituting u' and u from Eq. (6.37), we obtain (Exercise E9.4)

$$\frac{1}{2}\sum_i A_{ij}u_i - \frac{q_j}{2} + \sum_i V_{ij}u_i = E\sum_i B_{ij}u_i, \quad j = 0, 1, 2, ..., N, \tag{9.13}$$

where q_j is given by Eq. (6.43), A_{ij} by Eq. (6.45), and N the number of finite elements (intervals). The first two terms belong to the kinetic energy matrix. The potential matrix V_{ij} and basis overlap matrix B_{ij} are defined as

$$V_{ij} = \int_a^b \varphi_i V \varphi_j dx, \quad B_{ij} = \int_a^b \varphi_i\varphi_j dx. \tag{9.14}$$

As discussed in Section 6.4 (see Eq. (6.53)), even though the integrals are over the whole domain $[a, b]$, they are effectively over the elements e_i or e_j only (see Figure 6.12). Furthermore, because the basis functions do not overlap unless they share a common node, V_{ij} and B_{ij} are nonzero only if $|i-j| = 0, \pm 1$.

Like A_{ij} in Eq. (6.53), the overlap matrix can be evaluated analytically,

$$B_{ij} = \begin{cases} \frac{2}{3}h, & i = j, \\ \frac{1}{6}h, & i = j \pm 1, \\ 0, & \text{otherwise.} \end{cases} \tag{9.15}$$

From Eq. (6.43) and the discussion following it, the value q_j is zero unless $j = 0$ or $j = N$. Because the boundary values $u_0 = u_N = 0$ for bound states are already known, we can remove the indices $j = 0$ and $j = N$ from Eq. (9.13). As a result, we can eliminate q_j entirely from Eq. (9.13).

Collecting A_{ij} from (6.54) and B_{ij} earlier, we can express the kinetic, potential, and overlap matrices as

$$\hat{T} = \frac{1}{2h}\begin{bmatrix} 2 & -1 & & & \\ -1 & 2 & -1 & & \\ & \ddots & \ddots & \ddots & \\ & & -1 & 2 & -1 \\ & & & -1 & 2 \end{bmatrix}, \tag{9.16a}$$

$$\hat{V} = \begin{bmatrix} V_{11} & V_{12} & & \\ V_{21} & V_{22} & V_{23} & \\ & \ddots & \ddots & \ddots \\ & & V_{N-1,N-2} & V_{N-1,N-1} \end{bmatrix}, \tag{9.16b}$$

$$B = \frac{h}{6}\begin{bmatrix} 4 & 1 & & & \\ 1 & 4 & 1 & & \\ & \ddots & \ddots & \ddots & \\ & & 1 & 4 & 1 \\ & & & 1 & 4 \end{bmatrix}. \tag{9.16c}$$

All three matrices are tridiagonal with dimensions $(N-1) \times (N-1)$. Except for a multiplication factor, the $\hat{T}$ matrix in FEM basis is identical to the FDM version, Eq. (9.10).

Finally, we can write the FEM eigenvalue equation as

$$(\hat{T} + \hat{V})\mathbf{u} = EB\mathbf{u}. \tag{9.17}$$

Comparing with the FDM equation (9.11), we have an extra matrix B on the right-hand side (RHS) of Eq. (9.17). Equations of this type are known as the generalized eigenvalue equations (see also Eq. (6.16)). It is common in quantum problems when the basis functions are not orthogonal.

Given a potential $V(x)$, we can evaluate V_{ij} between nodes i and j from Eq. (9.14) analytically if possible or numerically if necessary. In the latter case, it is more convenient to use the element-oriented rather than node-oriented approach.[4] We can rewrite Eq. (9.14) as a sum over all elements e

$$V_{ij} = \int_a^b \varphi_i V \varphi_j dx = \sum_e V_{ij}^e, \quad V_{ij}^e = \int_{x_e} \varphi_i V \varphi_j dx. \tag{9.18}$$

Here, x_e denotes the domain over finite element e (compare with Eq. (7.26)). Except on the boundary, each domain contributes three terms to the $\hat{V}$ matrix: two diagonal entries to the left and right nodes, and one off-diagonal cross-node entry. After going through all finite elements, we will have built the full $\hat{V}$ matrix.

Upon solving Eq. (9.17), we will have both the eigenenergies E and the wave functions $\mathbf{u}$. We can determine any property of the system, such as the expectation values of the kinetic and potential energies (Exercise E9.5),

$$\langle \hat{T} \rangle = \mathbf{u}^T \hat{T} \mathbf{u}, \quad \langle \hat{V} \rangle = \mathbf{u}^T \hat{V} \mathbf{u}, \tag{9.19}$$

where the normalization of the wave function is assumed to be

$$\mathbf{u}^T B \mathbf{u} = 1. \tag{9.20}$$

9.3.3 THE FEM PROGRAM

A complete FEM program is given in Program 9.3. It implements the preceding strategy for an arbitrary potential. The code segment for building the $\hat{V}$ matrix is given next.

```
def Vij(i, j):        # pot. matrix Vij over [x_i,x_i+1] by order-4 Gaussian
    x=np.array([0.3399810435848564, 0.8611363115940525])  # abscissa
    w=np.array([0.6521451548625463, 0.3478548451374545])  # weight
    phi = lambda i, x: 1.0 - abs(x-xa[i])/h          # tent function
    vV, hh = np.vectorize(V), h/2.0                  # vectorize V
    x1, x2 = xa[i] + hh - hh*x, xa[i] + hh + hh*x
    return hh*np.sum(w * (vV(x1) * phi(i, x1) * phi(j, x1) +
                          vV(x2) * phi(i, x2) * phi(j, x2)) )

def V_mat():          # fill potential matrix
    Vm = np.zeros((N-1,N-1))
    for i in range(N):        # for each element     # contribution to:
        if (i>0):  Vm[i-1,i-1] += Vij(i, i)          #    left node
        if (i < N-1): Vm[i,i] += Vij(i+1, i+1)       #    right node
        if (i>0 and i< N-1):
            Vm[i-1,i] += Vij(i, i+1)                 #    off-diagonals
            Vm[i,i-1] = Vm[i-1,i]                    #    symmetry
    return Vm
```

[4] This was discussed in Chapter 7 and illustrated in Figure 7.10 for FEM in 2D. The reason is still true in 1D.

The function `Vij()` calculates V^e_{ij} in Eq. (9.18) over a finite element between x_i and x_{i+1}. We use a fourth-order Gaussian abscissa and weight for numerical integration (see Section 8.A.3). Because our FEM is a second-order method, the relatively low-order Gaussian integration is sufficient. It defines an inline `lambda` function for the tent function centered at node i as

$$\varphi_i(x) = 1 - |x - x_i|/h, \quad x_{i-1} \le x \le x_{i+1}, \tag{9.21}$$

which is equivalent to Eq. (6.38). Then, the Gaussian integration formula (S:8.55) is applied where the sum $\sum w_k f(x_k)$ with $f(x) = V\varphi_i\varphi_j$ is performed vectorially using `np.sum`. For this reason, the potential $V(x)$ is vectorized to ensure it can operate on an array of abscissas (a ufunc).

The next routine `V_mat()` builds the potential matrix by iterating over all finite elements. For each element, it calls `Vij()` to compute its contribution to the diagonal entries V^e_{ii} and $V^e_{i+1,i+1}$ of the left and right nodes, and the off-diagonal entries $V^e_{i,i+1} = V^e_{i+1,i}$. Because the dimension of the $\hat{V}$ matrix is $(N-1, N-1)$, its indices are offset by 1 relative to the element number.

Once the necessary input is prepared, we are ready to solve Eq. (9.17). This is done with the sparse matrix solver from SciPy, `eigsh()`, in Program 9.3. The FEM results for the square well potential are shown in Table 9.1. Its accuracy is comparable to FDM, as expected because both are second order. Usually, the smaller the element size, the better the results. However, per the sharp walls of the well, actual accuracy could suffer because of the discontinuity. For best result, choose N such that the walls bisect the elements covering the walls, or linearly interpolate the potential (Project P9.5). This poses no problem for continuous potentials, for example, FEM results for the SHO are rather accurate. Go ahead and try it: Replace the potential, and start SHOwing!

```
def V(x):   return 0.5*x*x    # potential
```

9.3.4 THE DIRAC δ-ATOM AND δ-MOLECULE

One advantage of the FEM over the FDM is that it is applicable to integrable singular potentials such as the Dirac delta potentials. For instance, for $V = -\alpha\delta(x)$, referred to as the Dirac δ atom, there is only one bound state at $E = -\frac{1}{2}\alpha^2$, and $u = A\exp(-\alpha|x|)$ [44]. With FEM, we get one bound state as well. At $\alpha = 1$, the FEM result is -0.499882 using the same element size as in Program 9.3, $h \sim 0.05$ (Project P9.6).

The wave function of the bound state is shown in Figure 9.5. It is obtained along with eigenenergies from Program 9.3. The discontinuity in the first derivative is accurately reproduced, agreeing with the analytic "cusp" condition

$$u'(0^\pm) = \mp\alpha u(0). \quad \text{(cusp, Dirac } \delta \text{ potential)} \tag{9.22}$$

Equation (9.22) is similar to (6.59) for a point source on a string. The piece-wise basis functions in FEM are flexible enough to handle this kind of discontinuities efficiently.

We can also determine the average kinetic and potential energies from Eq. (9.19) in a straightforward manner (Project P9.6). For $\alpha = 1$, the results are $\langle\hat{T}\rangle = 0.499644$ and $\langle\hat{V}\rangle = -0.999526$, which compare well with the exact values 0.5 and -1, respectively. As expected, $\langle\hat{T}\rangle + \langle\hat{V}\rangle$ is equal to the total energy quoted earlier.

When there are two δ potentials next to each other (Figure 9.6), we have a Dirac δ molecule,

$$V = -\alpha\delta(x + \frac{a}{2}) - \beta\delta(x - \frac{a}{2}). \tag{9.23}$$

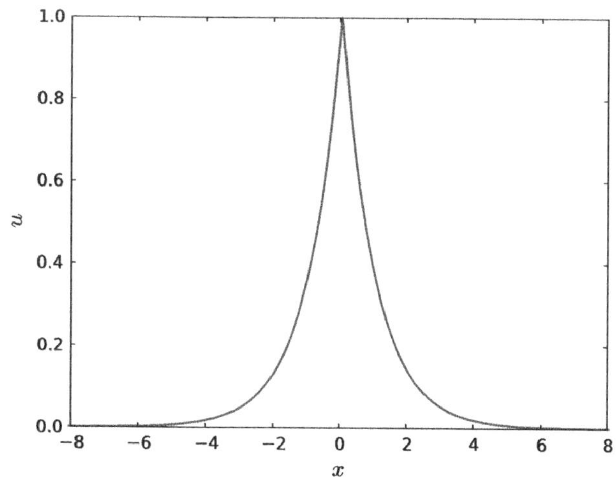

FIGURE 9.5 The bound state wave function of the Dirac δ atom.

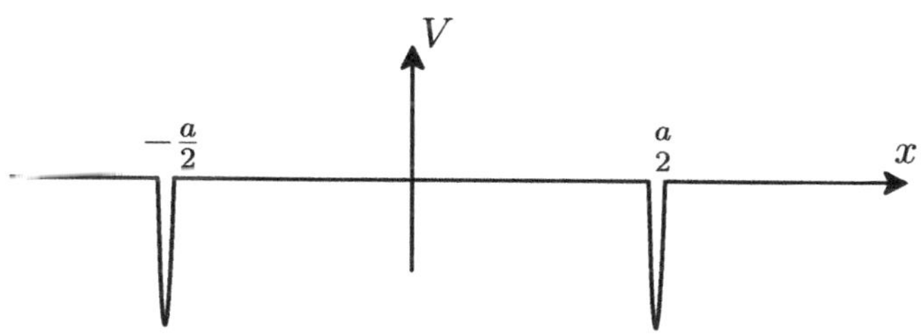

FIGURE 9.6 The Dirac δ molecule.

The analytic eigenequation of the δ molecule is (Exercise E9.6)

$$(k-\alpha)(k-\beta)\exp(2ka)=\alpha\beta, \quad E=-\frac{1}{2}k^2. \quad (\delta \text{ molecule}) \tag{9.24}$$

For arbitrary potential strengths α and β, Eq. (9.24) needs to be solved numerically. There are two limits where the solutions are readily obtained. The first is the separate atom limit, $a \to \infty$. We have two isolated δ atoms, so $k = \alpha$ or β. The second is the united atom limit, $a = 0$, and $k = \alpha + \beta$. This is a single δ atom with an effective potential strength $\alpha + \beta$ as expected.

If the potential strengths are equal, $\alpha = \beta$, we can solve Eq. (9.24) analytically in terms of the Lambert W function discussed in Chapter 3 (Section 3.4). The solutions are [90] (Exercise E9.6)

$$k_{\pm}=\alpha\left[1+\frac{1}{\alpha a}W(\pm\alpha a e^{-\alpha a})\right]. \tag{9.25}$$

We see from Eqs. (9.24) and (9.25) that if we scale the energy E by $E_0 = \frac{1}{2}\alpha^2$, the magnitude of the isolated δ-atom energy, then E becomes a function of the scaled variable αa only. We can make a universal energy-level diagram as a function of the scaled variables, shown in Figure 9.7. For a given α, the diagram shows the energy as a function of internuclear distance a, also known as the molecular potential curve. The δ-molecule has been a useful model in atomic ionization studies [93].

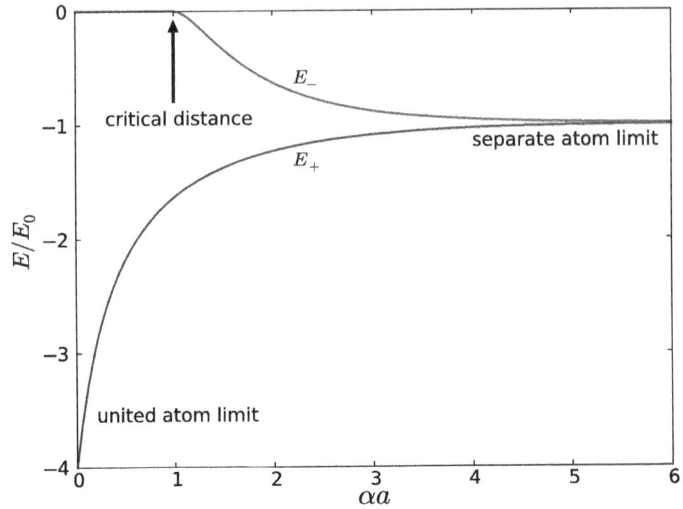

FIGURE 9.7 The universal energy-level diagram of the symmetric Dirac δ molecule, where $E_0 = \frac{1}{2}\alpha^2$.

For large a, the energy approaches $E = -E_0$, the separate atom limit. As discussed earlier, this limit corresponds to two degenerate atomic levels. As a decreases, the degeneracy is lifted and the two levels begin to split: The ground state bends downward and the excited state upward. This is fully analogous to state-splitting in multiple-well potentials (Figure 9.4). Their wave functions would resemble the first two shown in Figure 9.3 for the double-well potential, with the former being symmetric and the latter antisymmetric.[5] We expect that they may be approximately described by the LCAO according to Eq. (9.7) [6].

The molecule supports two bound states until the critical internuclear distance

$$\alpha a_c = 1, \quad \text{(critical distance)} \tag{9.26}$$

below which the excited state ceases to exist. We can see that the critical value a_c comes from Eq. (9.25) by requiring $k_- = 0$ and noting the special value $W(-e^{-1}) = -1$ in Eq. (3.22). The energy E_-, ever increasing with decreasing a, has reached zero at a_c and entered the continuum for $a < a_c$. On the other hand, the energy of the ground state continues to decrease below the critical distance, finally approaching $E_+ = -4E_0$, the united atom limit at $a = 0$.

The case for asymmetric potentials $\alpha \neq \beta$ may not be as tidy analytically, but is certainly as approachable numerically with the FEM. We leave its investigation to Project S:P9.1.

9.4 BASIS EXPANSION METHOD

We had encountered the general idea of the basis expansion method (BEM) previously in electrostatic and time-dependent quantum systems (Sections 7.5 and 8.4.1). The key was that, rather than discretizing space, the solution was expanded in a (finite) basis set. We discuss the same approach for eigenvalue problems. This method is easy to set up and can be very accurate and efficient. Usually, we have the freedom to choose from many basis sets the one that best suits the problem. When a good basis set is used, a relatively small number of basis states can give very accurate results, saving both computing time and memory (compact matrices).

[5]The Dirac δ molecule is a good model for understanding the qualitative features of diatomic molecules such as H_2^+, where the terms "gerade" and "ungerade" are used to describe the symmetric ground and the antisymmetric excited states, respectively.

Let the basis set be $u_n(x)$ satisfying the boundary conditions $u_n(a) = u_n(b) = 0$. Like Eq. (7.31) or (S:8.23), we expand the wave function as

$$u(x) = \sum_n a_n u_n(x). \tag{9.27}$$

Unlike the time-dependent case, the expansion coefficients a_n are constants independent of time.

The Schrödinger equation to be solved is

$$Hu(x) = Eu(x), \quad H = \hat{T} + \hat{V}. \tag{9.28}$$

As before, H is the Hamiltonian representing the sum of kinetic and potential energies defined over the same range $[a, b]$.

We can substitute u into Eq. (9.28) and project onto u_m^* on both sides, following analogous steps from Eq. (S:8.24). Alternatively, we can start from Eq. (9.12) but keep the kinetic energy operator as is without integration by parts. Either way, after projection, we obtain the familiar eigenequation, $(\hat{T} + \hat{V})\mathbf{u} = EB\mathbf{u}$ (Exercise E9.8). Though this is identical in form to Eq. (9.17), the operator representations are different in BEM than in FEM. Here, the kinetic, potential, and overlap matrices are respectively defined as

$$T_{mn} = \int_a^b u_m^* \hat{T} u_n \, dx, \quad V_{mn} = \int_a^b u_m^* V u_n \, dx, \quad B_{mn} = \int_a^b u_m^* u_n \, dx. \tag{9.29}$$

Depending on the basis set, the BEM matrices are not necessarily sparse like in FDM or FEM, but they often are for the problems we are interested in.

The eigenvector $\mathbf{u}$ is a column vector similar to Eq. (6.15), that is, $\mathbf{u} = [a_1, a_2, ...]^T$. The $\hat{T}$ and B matrices depend on the basis set only, but the $\hat{V}$ matrix depends on both the basis set and the potential. If the same basis set is used for different problems, only the potential matrix needs to be evaluated for each problem, while $\hat{T}$ and B remain the same.

If the basis set is orthogonal, then $B_{mn} = \delta_{mn}$, so $B = \mathbb{1}$ is the identity matrix. We assume orthonormal basis sets in the rest of this section. In this case, we need to solve the standard eigenvalue problem (9.11) rather than a generalized eigenvalue problem.

9.4.1 THE BOX BASIS SET

In our first example, let us illustrate the BEM using the familiar basis set of a particle in the box, Eq. (S:8.39). Because they are eigenfunctions of the kinetic energy operator, the $\hat{T}$ matrix is diagonal. The diagonal elements are the eigenvalues associated with u_n, that is, $T_{mn} = (\pi^2\hbar^2 n^2/2m_e L^2)\delta_{mn}$, where L is the box width. The full matrix is

$$\hat{T} = \frac{\pi^2\hbar^2}{2m_e L^2} \begin{bmatrix} 1 & & & & \\ & 4 & & 0 & \\ & & 9 & & \\ & 0 & & 16 & \\ & & & & \ddots \end{bmatrix}. \quad \text{(box basis)} \tag{9.30}$$

As discussed earlier, this is the same for any problem as long as we choose the same box basis set. The potential matrix is specific to a given system. Let us apply the BEM to the common test case, the square well potential. The matrix element is from Eq. (9.29)

$$V_{mn} = -\frac{2V_0}{L} \int_{-a/2}^{a/2} \sin\left(\frac{m\pi}{L}(x - \frac{1}{2}L)\right) \sin\left(\frac{n\pi}{L}(x - \frac{1}{2}L)\right) dx, \tag{9.31}$$

where a and V_0 are the width and depth of the square well as usual. It can be evaluated either numerically or analytically (Exercise E9.9). We have assumed the box is centered at the origin, $-L/2 \le x \le L/2$, so V_{mn} is zero unless $m+n$ is even.

We leave the calculations to Project P9.7 and show the results in Table 9.1. The accuracy depends on the box width L and the number of included states N. We chose the box width $L = 8a$, and used $N = 300$ states. The results are in good agreement with the exact values.

The advantage of using the box basis set is that the basis functions are simple, elementary functions. The required matrices are easy to generate. In fact, V_{mn} as given by Eq. (9.31) can be efficiently evaluated using fast Fourier transform (FFT). But convergence can be slow with increasing N for the upper excited states close to the continuum, mainly caused by coupling to pseudocontinuum states discussed earlier.

9.4.2 THE SHO BASIS SET

The simple harmonic oscillator offers another useful basis set for the BEM. The SHO potential is $V = \frac{1}{2}kx^2 = \frac{1}{2}m_e\omega^2x^2$, where $\omega = \sqrt{k/m_e}$ is the natural frequency.

The SHO is investigated numerically in Project P9.2. Analytically, its eigenenergies are

$$E_n = \left(n + \frac{1}{2}\right)\hbar\omega, \tag{9.32}$$

and the associated eigenfunctions can be written as

$$u_n(x) = A_n H_n(y)e^{-y^2/2}, \quad A_n = (\sqrt{\pi}a_0\, n!\, 2^n)^{-1/2}, \tag{9.33}$$
$$y = \frac{x}{a_0}, \quad a_0 = \sqrt{\frac{\hbar}{m_e\omega}}.$$

In Eq. (9.33), A_n is the normalization constant, and a_0 the characteristic length. Either ω or a_0 serves as a free controlling parameter of the SHO basis set.[6]

The function $H_n(y)$ are the Hermite polynomials [10], and the first few series are

$$H_0 = 1, \quad H_1 = 2y, \quad H_2 = 4y^2 - 2, \quad H_3 = 8y^3 - 12y. \tag{9.34}$$

Higher orders can be obtained from the recurrence formula

$$H_{n+1}(y) = 2xH_n(y) - 2nH_{n-1}(y). \tag{9.35}$$

This formula is stable for upward recursion, in the direction of increasing n. The SHO basis is defined for all space $-\infty < x < \infty$, so the BEM using this basis is applicable in principle to all systems whose boundary conditions are $u(\pm\infty) = 0$.

The kinetic energy operator in the SHO basis set is given by (Exercise E9.10)

$$\hat{T} = \frac{\hbar\omega}{4}\begin{bmatrix} 1 & 0 & -\sqrt{2} & 0 & 0 \\ 0 & 3 & 0 & -\sqrt{6} & 0 \\ -\sqrt{2} & 0 & 5 & 0 & \ddots \\ 0 & -\sqrt{6} & 0 & 7 & 0 \\ 0 & 0 & \ddots & 0 & \ddots \end{bmatrix}. \quad \text{(SHO basis)} \tag{9.36}$$

[6]The wave functions u_n are available from SymPy: sympy.physics.qho_1d.psi_n.

This is a sparse matrix consisting of three single-element bands, the diagonal and two symmetric off-diagonals. Each row or column, other than the first and last two, has three nonzero elements, T_{mm} and $T_{m,m\pm2}$. The diagonal elements are equal to one half of the eigenenergies.

We choose the wedged linear potential, $V(x) = \beta|x|$ (Figure 8.16), as the test case because, like the SHO, it supports only bound states and is also analytically solvable in terms of Airy functions [60] (see Exercise E9.11, Section 9.B). Program 9.4 solves the problem using BEM with the SHO basis set. The results are given in Table 9.2. A total of $N = 20$ basis states are used, and the free parameter is $\omega = 1$.

The analytic (exact) values are also computed in the program and given in the same table. The numerical results agree very well with the analytic results, mostly to six digits. This level of accuracy was obtained with only $N = 20$, a very small and efficient basis set. Of course, increasing N improves the agreement. For a given N, the adjustable parameter ω affects the accuracy as well. It controls the shape of the basis functions (like radial basis functions, Section 7.5.2), which in turn affects the quality of the basis expansion.

It is curious, though, that the accuracy for the ground state is only four digits, worse than for the excited states, when we expect the BEM (and other methods) to give the most accurate results for low-lying states. The wedged potential is an exception to the rule. The reason has to do with the wedge at $x = 0$, which causes a discontinuity in the third derivative of the wave function. For smoother wedges, the agreement becomes better.

Scanning through Table 9.2 reveals that the gap between adjacent energy levels decreases with increasing n. For instance, the gap is $\Delta E_{21} = 1.05$ between $n = 1$ and 2, and $\Delta E_{65} = 0.556$ between 5 and 6. This behavior is compared in Figure 9.8 to two other well-known potentials, the infinite potential well (rigid box) and the SHO. The level spacing increases in the rigid box, remains equidistant in the SHO, and decreases in the linear potential. It suggests that the asymptotic break-even point that determines the gap's behavior is quadratic. If the potential increases faster than x^2 as $x \to \pm\infty$, the gap will increase with increasing n. Right at x^2, the gap is constant. If the potential increases slower than quadratic, the gap will decrease. Realistic potentials are in the last category.

Table 9.2 The First Several Eigenenergies of the Potential $V = |x|$.

n	1	2	3	4	5	6
BEM	0.808655	1.85576	2.57811	3.24461	3.82579	4.38193
Exact	0.808617	1.85576	2.57810	3.24461	3.82572	4.38167

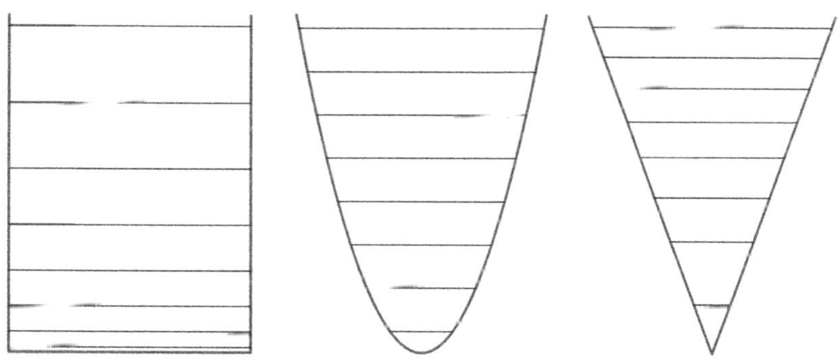

FIGURE 9.8 Energy-level diagrams for a rigid box (left), the SHO (center), and the wedged linear potential (right).

Half-Open Space

The full SHO basis set spans the whole space. For bound systems that exist in the half-open space, $0 \le x < \infty$, the boundary conditions are $u(0) = u(\infty) = 0$. For example, one such system is a particle bouncing between a hard wall and a linear potential similar to quantum free fall in Section 8.3.2. The potential is $V(0) = \infty$, and $V(x) = \beta x$ for $x > 0$ (Figure 9.9).

We can still use the SHO basis set, but only half of it. The even-n basis states of Eq. (9.33) are nonzero at the origin, and must be discarded because they do not satisfy the boundary conditions. The odd-n basis states, however, do satisfy the boundary conditions. They are what we need. In fact, these states are eigenstates of the SHO in the half-open space: $V(0) = \infty$, $V(x) = \frac{1}{2}kx^2$ for $x > 0$. Therefore, they are a complete basis set for the half-open space.

We can use the odd states as is in the BEM, provided we renormalize them by multiplying the normalization constants A_n by a factor $\sqrt{2}$. This accounts for the loss of probability in the other half space. With only odd states, we are effectively removing from the kinetic and potential matrices the rows and columns due to even states (Project S:P9.2).

Similarly, for the potential shown in Figure 9.9, a valid wave function must be zero at $x = 0$. The odd solutions given by Eq. (9.61) satisfy the Schrödinger equation everywhere in the half-open space, including the correct boundary conditions, so they are the eigenfunctions of the system. Accordingly, the valid eigenenergies are those corresponding to odd states. For instance, we can obtain the first several values from Table 9.2 by omitting entries for $n = 1, 3, 5$ and keeping $n = 2, 4, 6$. We can confirm this numerically (Project S:P9.2) with the values determined by the zeros of the Airy function (9.63).

The Morse potential is another interesting case defined in the half space $x \ge 0$ as

$$V(x) = V_0 \left(e^{-2\alpha(x-r_0)/r_0} - 2e^{-\alpha(x-r_0)/r_0} \right). \tag{9.37}$$

This potential approximately describes the interaction between two atoms in a diatomic molecule at internuclear distance x. The parameter r_0 represents the equilibrium distance where the potential is a minimum, $-V_0$ (Figure 9.10). The potential rises rapidly as $x \to 0$, simulating the hard wall (e.g., the Coulomb repulsion), and decreases exponentially at large $x \to \infty$.

Around the minimum, the potential is approximately like an SHO potential, so the Morse potential is a useful model to describe vibrational states of a molecule. The energy levels computed by the BEM (Project S:P9.3) show roughly equidistant level spacing for low-lying states. As energy increases, the potential becomes anharmonic, and the spacing decreases, in accord with our expectation (Figure 9.8) for realistic potentials.

Quantitatively, the results shown in Figure 9.10 were obtained with realistic parameters for molecular hydrogen, H_2 (Project S:P9.3). The BEM code found 17 bound states. The largest gap is $\Delta E_{10} = 0.51$ eV (wavenumber 4100 cm^{-1}) between the ground state and the first excited state,

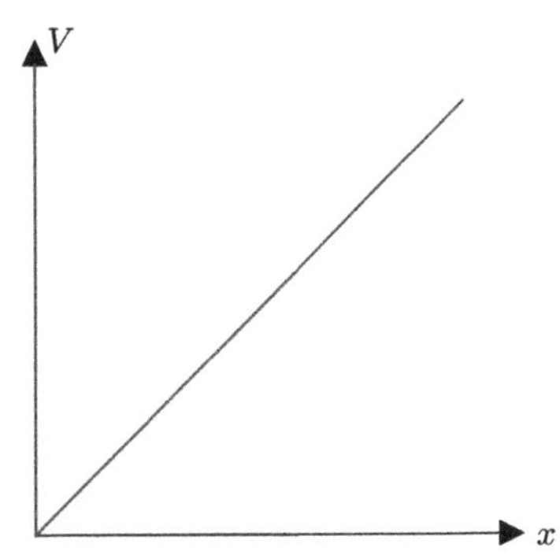

FIGURE 9.9 A hard wall at $x = 0$ plus a linear potential for $x > 0$.

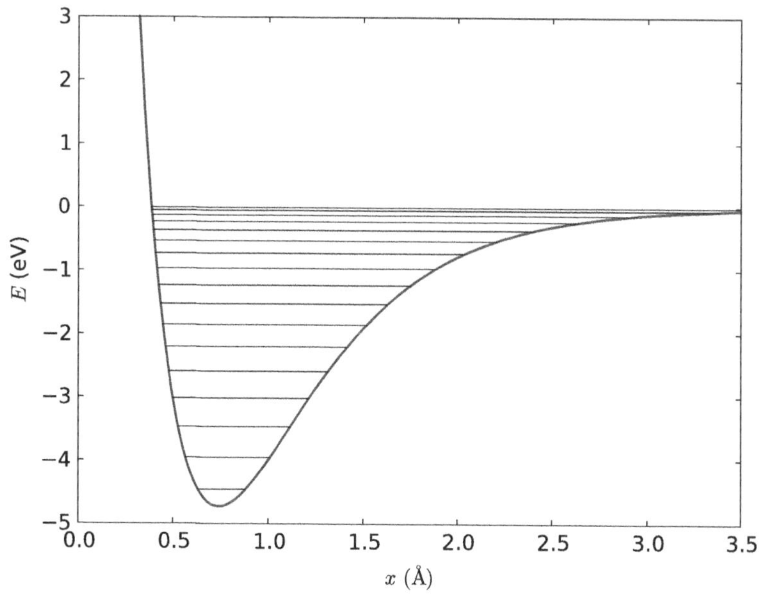

FIGURE 9.10 Energy-level diagram for the Morse potential.

and the smallest gap at the top is 0.04 eV (300 cm^{-1}). These are typical of vibrational excitation energies in molecules. They are larger than thermal energies at room temperature ($\frac{1}{40}$ eV), so vibrational degrees of freedom are usually frozen at these temperatures.

9.5 CENTRAL FIELD POTENTIALS

An important class of systems consists of central field potentials such as the Coulomb potentials and central (mean) field approximations, for example, Section 4.3.4. In these systems, the three-dimensional Schrödinger equation (S:12.1) can be separated into radial and angular solutions, $\psi = R_{nl}(r)Y_{lm}(\theta, \varphi)$. The latter, Y_{lm}, are the universal spherical harmonics [4]. They are eigenfunctions of angular momentum, which is conserved in central fields.

The radial wave function, R_{nl}, satisfies the radial Schrödinger equation

$$\frac{1}{r^2}\frac{d}{dr}\left(r^2\frac{dR_{nl}}{dr}\right) + \frac{2m_e}{\hbar^2}\left(E - V(r) - \frac{l(l+1)\hbar^2}{2m_e r^2}\right)R_{nl} = 0. \tag{9.38}$$

The first several hydrogen radial and angular wave functions are given in Project S:P8.9.

For a general potential, the energy E depends on both n and l, the principal and angular momentum numbers, respectively. For a pure $1/r$ potential like in the hydrogen atom, the energy becomes independent of l, that is, it is degenerate with respect to l (Figure 9.11, left). The Coulomb potential is the only potential having this degeneracy. When modified slightly (Figure 9.11, right), the degeneracy disappears (see Project P9.8).

Equation (9.38) can be simplified further by making a substitution $u = rR$ to arrive at

$$\frac{d^2u}{dr^2} + \frac{2m_e}{\hbar^2}\left(E - V_{eff}(r)\right)u = 0, \tag{9.39}$$

$$V_{eff}(r) = V(r) + \frac{l(l+1)\hbar^2}{2m_e r^2}.$$

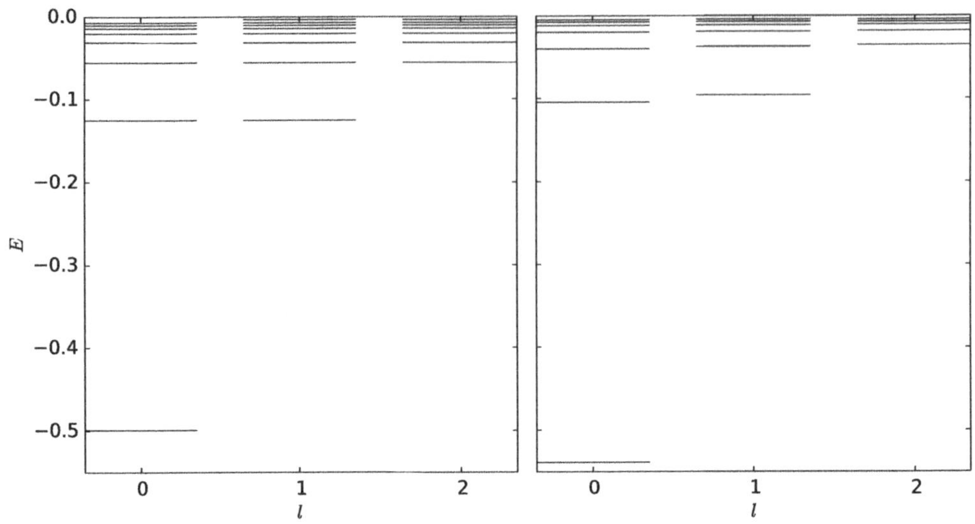

FIGURE 9.11 Energy-level diagrams of hydrogen ($-1/r$, left), and modified potential ($-1/r^{1.1}$, right).

The effective potential V_{eff} is the same as Eq. (4.8) for planetary motion, with the replacement of $L^2 \to l(l+1)\hbar^2$. The limiting behavior of u is $u \overset{r\sim 0}{\to} r^{l+1}$, and $u \overset{r\to\infty}{\to} 0$.

Equation (9.39) is identical to the one-dimensional Schrödinger equation (9.8) with the introduction of V_{eff}. We can apply all the methods discussed so far to (9.39), subject to the boundary conditions $u(0) = u(\infty) = 0$. In particular, Numerov's method is very efficient for solving central potential problems. Program 9.5 combines the Numerov and shooting methods to solve for the eigenenergies and wave functions in central fields such as the hydrogen atom, shown in Figure 9.11 and Figure 9.12, respectively.

The effective potential for each angular momentum number l supports an infinite number of states in hydrogen, all converging toward the continuum $E = 0$. Each state can be labeled by a pair of quantum numbers, n and l. The lowest state in each l-series starts from the second-lowest level of the previous series. For a given n, the maximum value of l is $l_{max} = n - 1$. For a given l (fixed angular momentum), the smallest permitted n is $l + 1$. The results for hydrogen are in good agreement with the exact eigenenergies $E_n = -\frac{1}{2n^2}$.

Even though the Coulomb potential vanishes at $r \to \infty$, the number of bound states is still infinite, unlike most potentials of similar asymptotic behavior. This is attributed to the particularly long-range nature of the Coulomb potential. The rate of decrease as $1/r$ is slow enough that a particle "feels" the effect of the potential even as $r \to \infty$.[7]

The radial wave functions of the first several states of hydrogen are shown in Figure 9.12. They are qualitatively the same for other central fields. The states are labeled by the principal quantum number n followed by a letter according to spectroscopic notation where the values $l = 0, 1, 2, 3,...$ are represented by $s, p, d, f...$, and so on. For the ground state 1s, the wave function is tightly localized around $r = 1$. With increasing n and l, it becomes more spread out. Higher energy pushes the electron further outward, resulting in a higher potential energy and a lower kinetic energy. We can show that on average, the two energies in hydrogen are related by

$$\langle T \rangle = -\frac{1}{2}\langle V \rangle. \quad \text{(virial theorem)} \tag{9.40}$$

[7]Technically, a particle never becomes free in the Coulomb potential, even though it can escape to infinity if its energy is positive. In this case, the continuum wave function is distorted compared to the plane wave representing a truly free particle. See Section 12.B and Exercise S:E12.7.

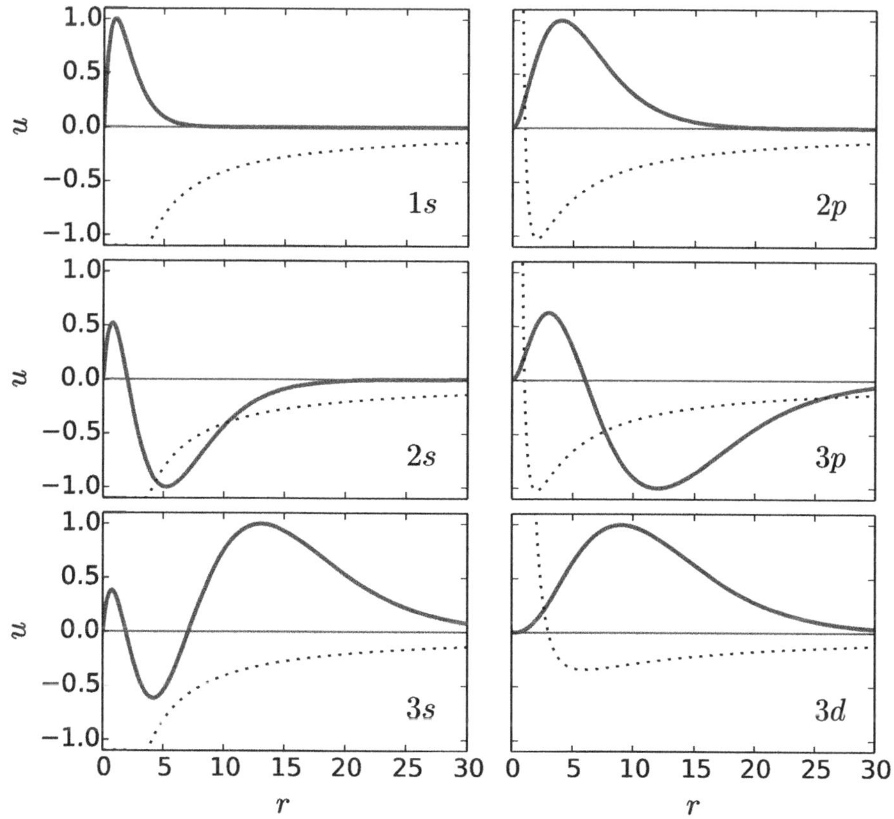

FIGURE 9.12 The radial wave functions of the first six states of hydrogen. The dotted lines represent the effective potential.

The preceding relationship is called the virial theorem. The increase in potential energy is twice the decrease in kinetic energy, so the total energy increases.

We see from Figure 9.12 that for a given n, the wave function becomes more localized toward larger r with increasing l. The centrifugal potential wall keeps the wave function small at small r (e.g., compare $2s$ with $2p$, or $3s$ and $3p$ with $3d$). When $l = l_{max} = n - 1$, there is a single peak centered at $r = n^2$. This result can also be proven analytically (Exercise S:E9.1). It shows two things: the size of the atom scales as n^2, and the radial probability distribution is increasingly confined to a spherical shell.[8] If we make a cut through the shell structure, we would obtain a circular distribution, or a ring. For this reason, states of maximum angular momentum $l = l_{max}$ are called circular Rydberg states [11]. Figure 9.13 displays the radial probability densities of $4s$ to $4f$ states, showing the approach to circular states with increasing l. In the semiclassical limit $n \gg 1$, these circular Rydberg states look increasingly like the circular orbits of classical motion. This is the limit of the Bohr model, a fact used later (see Section S:12.5.2).

We also observe that the number of nodes for a given state is equal to $n - l - 1$, which is zero for circular states such as $3d$. This may seem paradoxical to the notion we have developed so far, which is that the k-th excited state should have k nodes. For three-dimensional systems, the wave function can have nodes in radial as well as angular variables, so the number of nodes

[8]Because $\langle r \rangle$ scales like n^2, numerical integration must extend to large distances for moderate to large n. To maintain efficiency, numerical integration is often done on a logarithmic grid. See Exercise S:E9.2.

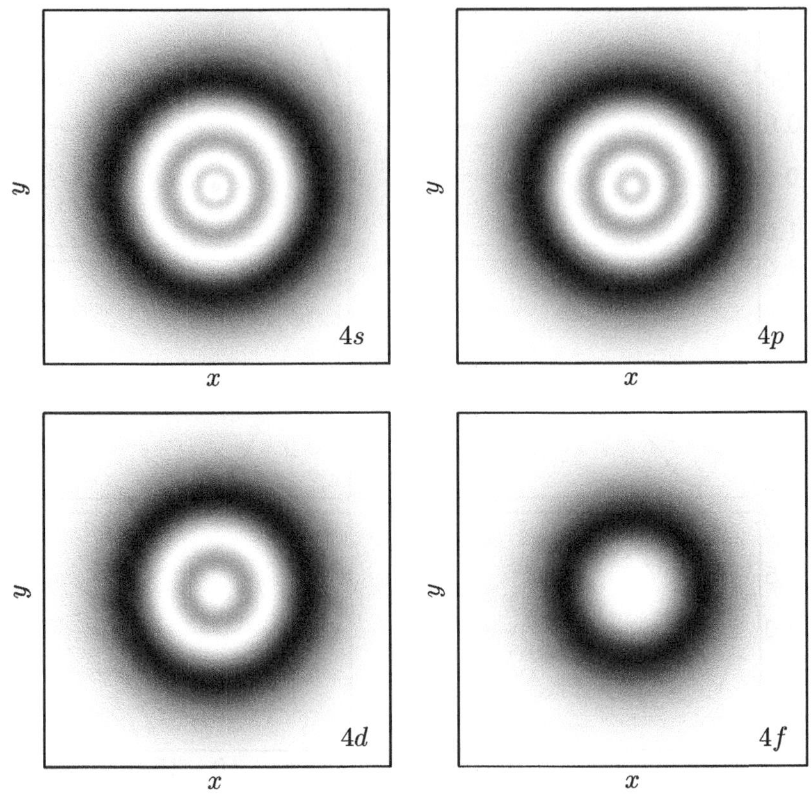

FIGURE 9.13 The probability density for $4s$ to $4f$ states in the x-y plane.

should be the sum in all directions. The fact that the $3d$ state has fewer nodes than the $2s$ state in the radial wave function means that there must be more nodes in the angular wave function for $3d$ than for $2s$. This is in fact true.

In Figure 9.14 we plot the radial and angular probability density distribution, $|u_{nl}Y_{lm}|^2$. Due to azimuthal symmetry, we show a cut in the x-z plane. We choose the circular state $(n, l) = (4, 3)$ and vary the quantum number $m = 0$ to 3. Because m measures the z component of the angular momentum (L_z), we can view different m values as representing the orientation of the classical circular orbit. The maximum L_z occurs when $|m| = l$, and we expect the orbit to be in the x-y plane so the central axis (normal) of the orbit is aligned with the z-axis. Figure 9.14 shows that the probability density in this case ($m = 3$) is mostly confined to the space near the x-y plane where $\theta \sim \frac{\pi}{2}$, or $|z| \sim 0$. Conversely, if $m = 0$, the central axis should be perpendicular to the z-axis. We expect the probability density to be concentrated near the poles, $\theta \sim 0$ and π. This case is depicted in Figure 9.14 (top-left, $m = 0$). For $0 < |m| < l$, the density evolves between the two limits.

Furthermore, we can count the number of nodes in the angular wave function from Figure 9.14. For example, we see three valleys (zero density) between $\theta = 0$ and π in the $m = 0$ case, corresponding to three angular nodes. Because the radial wave function $u_{4,3}$ has zero nodes, the total number of nodes is three in the state $nlm = (4, 3, 0)$. This is equal to the number of radial nodes in the $4s$ state, which has zero angular nodes.

The number of angular nodes has no direct impact when we are only interested in computing the radial wave function. But counting radial nodes in a given l-series is helpful numerically because it lets us know if we have missed any state when searching for eigenenergies.

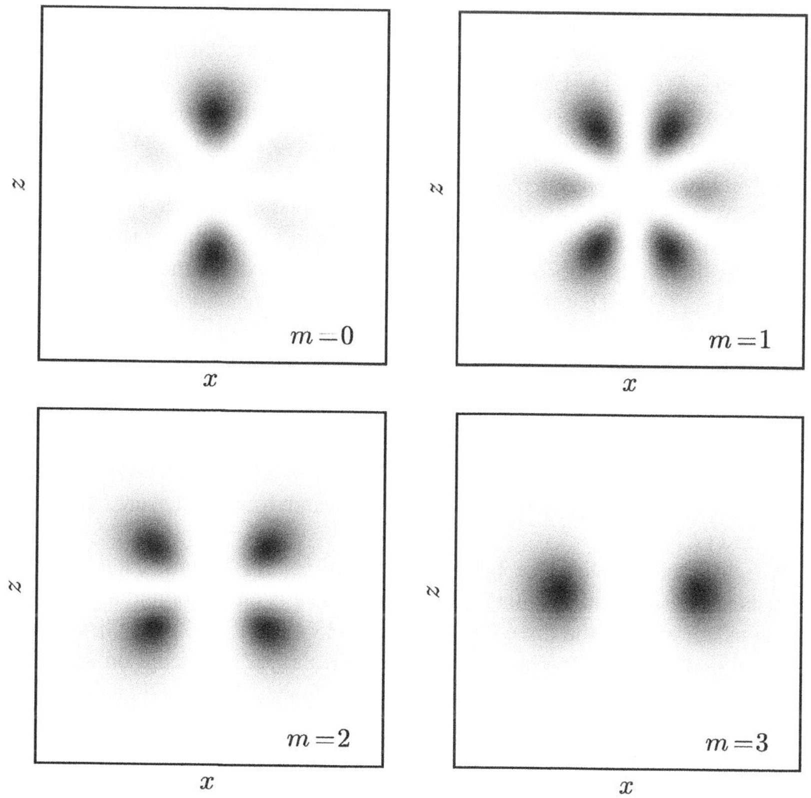

FIGURE 9.14 The probability density of the circular state $(n, l) = (4, 3)$ for $m = 0$ to 3 (different orientations) in the x-z plane.

9.6 QUANTUM DOT

We have discussed solutions to 1D quantum systems and 3D central field problems that are effectively one-dimensional in the radial direction. In this section, we will study 2D quantum systems.

Perhaps the most interesting among them are quantum dots [45], systems confined to sizes in the nanometer range. Some occur in lattice structure, but they are often fabricated in semiconductors using techniques such as precision lithography and are sometimes called designer atoms.[9] Their shapes can be simple or highly irregular. In the latter case, enforcing boundary conditions can be problematic. One of the advantages of the FEM is its ability to adapt to flexible boundaries. We will extend the FEM to solve the Schrödinger equation in 2D and use it to study quantum dots.

9.6.1 FEM SOLUTIONS OF SCHRÖDINGER EQUATION IN 2D

The Schrödinger equation in two-dimensional space is

$$-\frac{1}{2}\nabla^2 u + Vu = Eu, \tag{9.41}$$

where $\nabla^2 = \frac{\partial^2}{\partial x^2} + \frac{\partial^2}{\partial y^2}$ is the Laplacian in 2D.

[9]Quantum dots are literally shining a new light on quantum mechanics. Because the energy levels, hence light emission, can be manually controlled via shape and size, they are beginning to usher in new technologies such as the quantum dot display.

We can follow two paths to develop the FEM for the Schrödinger equation: starting with explicit expansion analogous to Eq. (9.12) in 1D, or using the readily available results from Chapter 7 (Section 7.3) for 2D FEM. We choose the latter.

For the purpose of FEM development, the Schrödinger equation (9.41) can be cast into an equivalent form to the Poisson equation (7.13). If we multiply both sides of Eq. (9.41) by −2 and compare with Eq. (7.13), we can identify the equivalent function in the latter as $f = 2(V - E)u$.

The framework of FEM in 2D had been fully developed in Chapter 7 (Section 7.3) with the Poisson equation as the prime example. Therefore, we can follow the exact development down to a tee for application to the Schrödinger equation. The only difference is in the imposition of boundary conditions (7.22). That is where we start.

Equation (7.22) reads,

$$\sum_{\text{intrnl}} A_{ij}u_i = q_j - p_j - \sum_{\text{bndry}} A_{jk}u_k, \quad i,j = \text{internal nodes only}, \tag{9.42}$$

where the sums on the left-hand side (LHS) and RHS refer to internal and boundary nodes, respectively. As before, u_i is the solution in the expansion (7.9), and A_{ij}, q_j, and p_j are respectively defined in Eqs. (7.19) and (7.17).

We are interested in Dirichlet boundary conditions for bound states, where the wave function vanishes on the boundary, that is, $u_k = 0$ if node k is on the boundary.

Let us consider each of the three terms on the RHS of Eq. (9.42). The first term, q_j, is zero if j is an internal node because the tent function ϕ_j in Eq. (7.17) centered on an internal node is zero on the boundary (see Figure 7.8 and discussion following it, as well as the discussion right after Eq. (7.22)). Hence, q_j drops out. The third term, $\sum_{\text{bndry}} A_{jk}u_k$, is also zero because the sum involves only boundary nodes where $u_k = 0$.

That leaves the second term, p_j, as the only nonzero term. The expression for p_j follows from Eq. (7.17) with f identified earlier for the Schrödinger equation

$$p_j = \int_S f\phi_j u \, dxdy = 2\int_S (V - E)\phi_j u \, dxdy. \tag{9.43}$$

Substituting the basis expansion $u = \sum u_i\phi_i$ from Eq. (7.9) into (9.43), we obtain

$$p_j = 2\sum_i V_{ij}u_i - 2E\sum_i B_{ij}u_i, \tag{9.44}$$

$$V_{ij} = \int_S \phi_i V \phi_j \, dxdy, \quad B_{ij} = \int_S \phi_i\phi_j \, dxdy. \tag{9.45}$$

Like Eq. (9.14) in 1D FEM, V_{ij} is the potential matrix and B_{ij} the overlap matrix between the tent functions ϕ_i and ϕ_j.

Substituting p_j into Eq. (9.42), dividing both sides by 2, and moving the potential matrix to the LHS, we obtain

$$\frac{1}{2}\sum_i A_{ij}u_i + \sum_i V_{ij}u_i = E\sum_i B_{ij}u_i. \quad (i,j = \text{internal nodes only}) \tag{9.46}$$

In matrix form, Eq. (9.46) can be written as

$$(\hat{T} + \hat{V})\mathbf{u} = EB\mathbf{u}, \tag{9.47}$$

where $\hat{T} = \frac{1}{2}A$ is the kinetic energy matrix. This final result is again a generalized eigenvalue problem that is formally the same as Eq. (9.17) for the 1D Schrödinger equation. In retrospect, we may have guessed Eq. (9.47).

We have discussed triangular mesh generation in Section 7.3. On a given mesh, the kinetic and overlap matrices can be calculated from

$$A_{ij} = \sum_e A^e_{ij}, \quad B_{ij} = \sum_e B^e_{ij}. \tag{9.48}$$

It is assumed that the summation runs over all elements e that contain nodes i and j. We have already encountered A^e_{ij} which is given by Eq. (7.27). For B^e_{ij}, we can evaluate it similarly as [75]

$$X = \sum_{k=1}^{3} x_k, \quad Y = \sum_{k=1}^{3} y_k, \quad X_Y = \sum_{k=1}^{3} x_k y_k, \quad X_X = \sum_{k=1}^{3} x_k^2, \quad Y_Y = \sum_{k=1}^{3} y_k^2,$$

$$B^e_{ij} = \frac{1}{12A_e}\left\{ 3\alpha_i\alpha_j + (\alpha_i\beta_j + \alpha_j\beta_i)X + (\alpha_i\gamma_j + \alpha_j\gamma_i)Y \right. \tag{9.49}$$
$$\left. + \frac{1}{4}\left[\beta_i\beta_j(X^2 + X_X) + (\beta_i\gamma_j + \beta_j\gamma_i)(XY + X_Y) + \gamma_i\gamma_j(Y^2 + Y_Y)\right] \right\}.$$

The quantity A_e is the area of the triangle whose vertices are given by (x_k, y_k) in Eq. (7.10). The parameters α_i, β_i, and γ_i define the basis functions over the triangular element e given in Eq. (7.11).

For a given potential, we can evaluate the potential matrix on the mesh by numerical integration. We will assume simple potentials for the quantum dot: either zero or constant in the domain (of course, $V = \infty$ outside where $u = 0$). As a result, the potential matrix either vanishes or is proportional to the overlap matrix B^e_{ij}.[10] Once we have all three matrices, we can solve the generalized eigenvalue problem (9.47).

9.6.2 FEM IMPLEMENTATION

We break down the FEM approach into three separate tasks: mesh generation, matrix preparation, and actual computation.

We treat mesh generation as an independent process. As discussed in Chapter 7 (Section 7.3), proper meshing is important to the quality of the solution (see Table S:9.1) and computational efficiency. We generate all the meshes for our systems in simple domains (see Figure 7.11, Table 7.1, and examples in Programs 7.3 and 9.9). They can also be generated independently by other methods including MeshPy and can be read from a file (see Figure 9.24 and associated sample file format).

On a given mesh, we calculate the A and B matrices defined in Eqs. (9.48), (7.27), and (9.49). The necessary routines are included in the FEM library (Program 9.6). It includes a pair of functions `A_mat()` and `B_mat()` that accept the nodes and elements defined according to the data structure given in Table 7.1.

Actual computation and results presentation are performed with Program 9.7. We consider it a universal FEM program for quantum dots where the potential is zero in the domain. It requires only mesh data stored in a file. Additionally, it calculates eigenenergies and eigenfunctions only once and writes the data to a file. Subsequently, the data can be loaded, analyzed, and graphed without recomputation.

The Triangular Box

As a test, we apply Program 9.7 to the isosceles right triangular box (see Exercise E9.12). This system is exactly solvable so we can directly compare numerical and analytical results.

[10]For other nontrivial potentials, numerical integration is more conveniently carried out by mapping the triangles to a simpler shape such as isosceles right triangles through coordinate transformations.

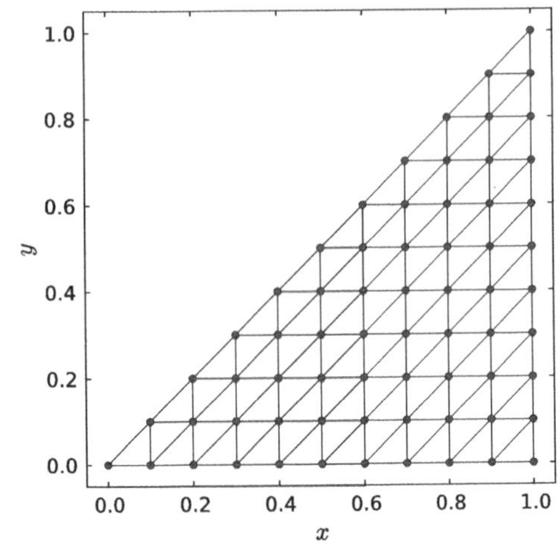

FIGURE 9.15 The FEM mesh for the right triangular box.

A sample mesh is shown in Figure 9.15. The isosceles sides are divided into $N = 10$ intervals each. There are 100 elements. The mesh setup properly preserves the symmetry of the geometry, including reflection symmetry about the bisector of the right angle.

The following actual mesh used is for $N = 20$, or a total of 400 elements. Of 231 total nodes, 171 are internal, so the dimensions of A and B matrices are 171×171, and there are a total of 171 states. The mesh data is stored in a file and is read in by Program 9.7. Figures 9.16 and 9.17 display the eigenenergies and wave functions, respectively, of low-lying states. The latter is obtained from a separate code, Program 9.8, which plots the wave function over a triangle mesh.

Comparing numerical and analytic eigenenergies, we observe good agreement for the lower states, within $\sim 1\%$ for the first five states. Analytically, the solutions can be obtained from the 2D box as

$$u_{mn}(x, y) = \frac{\sqrt{2}}{a}\left[\sin\left(\frac{m\pi}{a}x\right)\sin\left(\frac{n\pi}{a}y\right) - \sin\left(\frac{n\pi}{a}x\right)\sin\left(\frac{m\pi}{a}y\right)\right], \tag{9.50}$$

where a is the side length. In Eq. (9.50), m and n are nonzero integers, and $m < n$. The wave function is a combination of 1D wave functions in the x and y directions (S:8.39), satisfying

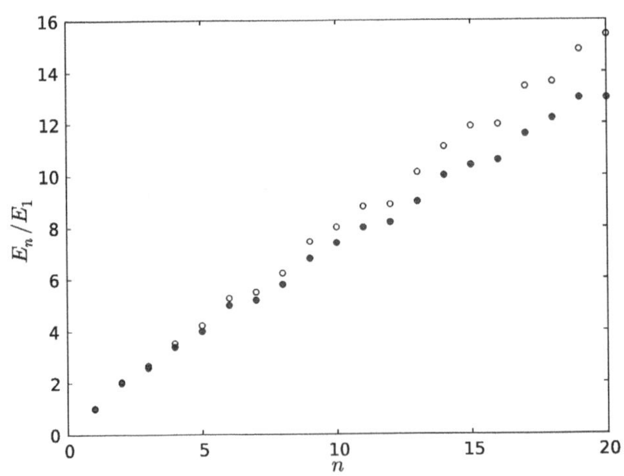

FIGURE 9.16 The first 20 eigenenergies of the right triangular box. The open circles (○) are FEM results, and filled circles (•) analytic results.

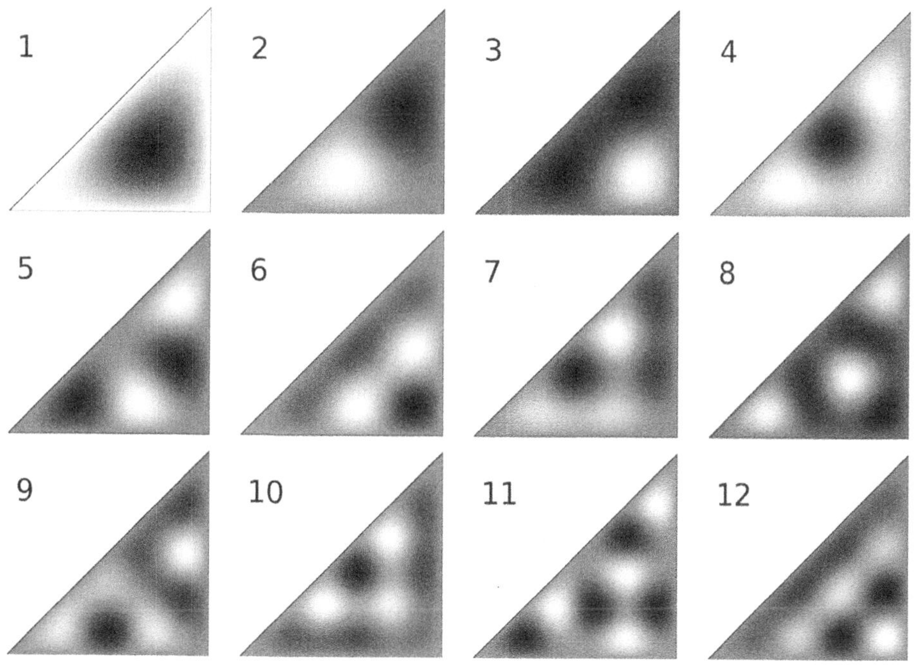

FIGURE 9.17 Wave functions in a triangular box.

the boundary condition that the wave function is zero on the three sides, $u_{mn}(x, 0) = u_{mn}(a, y) = u_{mn}(x, y = x) = 0$. The eigenenergies are given by

$$E_{mn} = \frac{\pi^2\hbar^2}{2m_e a^2}(m^2 + n^2), \quad m < n. \tag{9.51}$$

In contrast to the rectangular box, which has degeneracies, the triangular box removes degeneracy all together, a result of boundary conditions requiring that the wave function vanish on the hypotenuse. This restricts the way the wave function can oscillate. As Figure 9.17 shows, the ground state has one peak in the center of the triangle. With increasing state number, more pockets are formed, which are generally correlated with the state number, but not always. Sometimes, the pockets are broadly connected local maxima or minima (e.g., states 3 and 4). They develop into clear peaks at later stages (e.g., states 4 → 5).

The wave function (9.50) in the triangular box is not separable into a product of x and y variables. Therefore, the concept of nodes is not readily applicable. Instead, we must identify the wave function by nodal lines or the pattern of the extrema. A higher number of extrema corresponds to a higher state. If two states have the same number of extrema, the wave function with sharper and better developed peaks belongs to the higher state.

Going back to the eigenenergies shown in Figure 9.16, the discrepancy between the numeric and analytic results increases toward higher states, and numeric values are always above the analytic values for a given state. At the 20th state, the error is about ∼ 15%. The difference comes from two sources, the accuracy of the FEM and the related number of representable states. The difference can be reduced with a finer mesh, which in turn increases the number of states in the FEM representation. This will delay the onset of large errors, but they will happen eventually.

In other words, the fundamental problem remains, which is to represent a system having an infinite number of states with a finite number of representable states in a numerical method. We had seen the same phenomenon for standing waves on a string (Section 6.6). Essentially, we are trying to mimic an infinite Hilbert space with a finite number of states in our basis set. So it is really not the fault of the numerical method. What we must keep in mind is that higher states have larger errors. In the present case, we can trust the energies of states up to 8 at 1% accuracy, or about 5% of the total states (171). A general guide requires a statistical analysis of energy levels (Section S:9.1).

9.6.3 THE HEXAGON QUANTUM DOT

The hexagon quantum dot is very interesting. Its shape is highly symmetric and appears naturally in semiconductors such as GaAs or in a honeycomb as the basic cell.

Figure 9.18 shows a sample mesh of the hexagon quantum dot. The side length, assumed to be 1, is divided into $N = 2$ intervals for illustration. It is generated with Program 9.9. The elements are equilateral triangles, preserving all the rotation and reflection symmetries. Note that the node number starts from the bottom-left corner and increases consecutively from left to right and bottom to top, until the end of the center row, where it jumps to the top-left corner. This has to do with the way the mesh was made, in which the lower half was generated first and the upper half obtained by reflection (see explanation of Program 9.9).

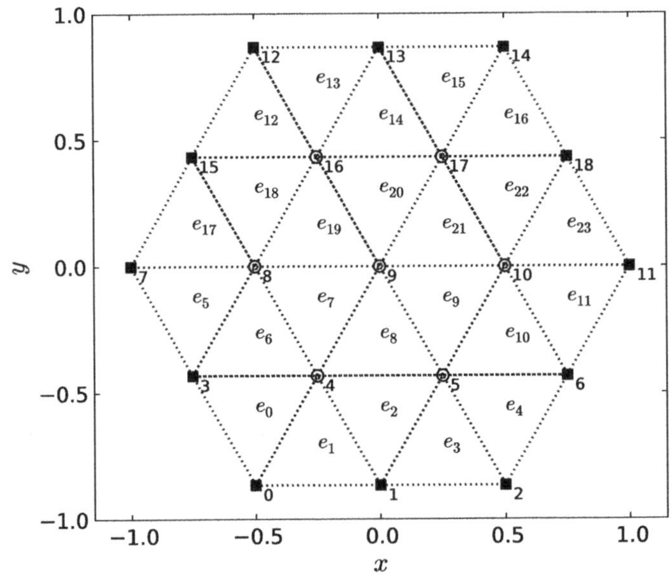

FIGURE 9.18 Sample mesh for the hexagon quantum dot.

Table 9.3 The Lowest 12 Eigenenergies of the Hexagon Quantum Dot.

1 to 6	3.57866	9.07223	9.07223	16.24645	16.24645	18.77309
7 to 12	23.85887	26.37310	30.12305	30.12305	35.16621	35.16621

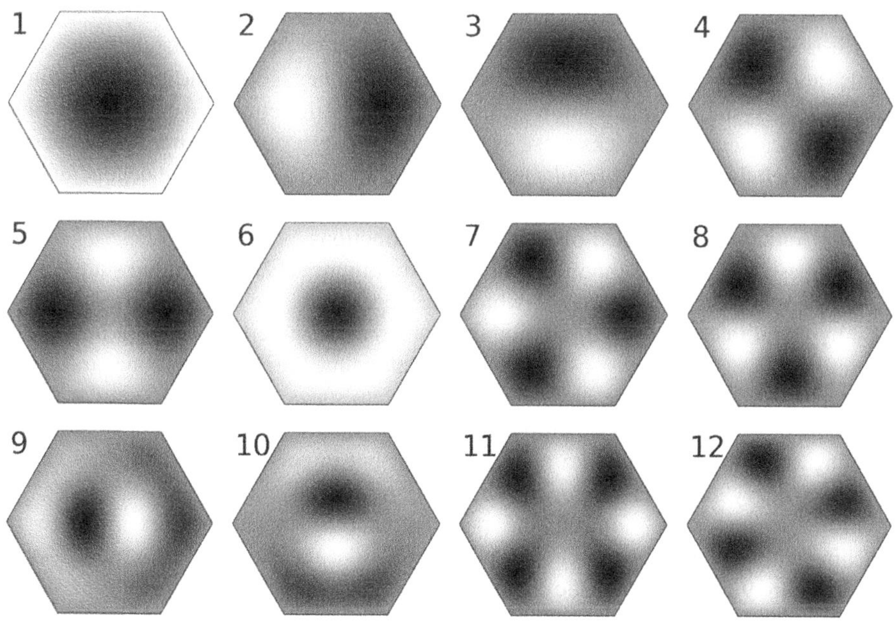

FIGURE 9.19 Wave functions in the hexagon quantum dot.

The number of elements is equal to $6N^2$. We use more elements than shown in Figure 9.18 in actual computation. We have performed a series of calculations with $N = 10$ to 40. This is done in part to gauge accuracy and to understand the rate of convergence. In Table 9.3 and Figure 9.19, we show eigenenergies and wave functions calculated with $N = 40$ and 9600 elements. The hexagon quantum dot does not yield analytic solutions in general. Analytic solutions can be obtained only for a limited subset of highly symmetric states. Even then, they are not in closed form [57]. Judging from the convergence sequence, the relative error in the eigenenergies of Table 9.3 is about $\sim 10^{-3}$.

The energy spectrum is very interesting. The ground state, which must be nondegenerate, has an energy 3.57866. With regard to geometry, we can compare this value to three wells of comparable size, two circular wells—an inner circular well of diameter $\sqrt{3}$ that just fits inside the hexagon and an outer one of diameter 2 that just encloses the hexagon, and a rectangular well of dimensions $2 \times \sqrt{3}$. The eigenenergies of the circular wells are related to the zeros of the Bessel function, and that of the rectangular well is well known. In order, their values are 3.85546, 2.89159, and 2.87863, respectively (Exercise S:E9.5). The inner circular well gives the best fit to the hexagon, and its energy is the closest. It is higher than the actual energy because its area is smaller than the area of the hexagon. This is due to the uncertainty principle, which generally dictates a larger momentum for a smaller well, thus a higher energy. We see the same trend for the outer circular well with an area π and the rectangular well with an area $2\sqrt{3} = 3.4641$.

Immediately above the ground state, the first and second excited states are degenerate, as are the next pair, the third and fourth excited states. After states 6 to 8, which are nondegenerate, degeneracy resumes for the next two pair of states, (9, 10) and (11, 12). The pattern continues for later states.

The high frequency of degeneracy is a reflection of the high symmetry of the hexagon structure. It is symmetric under rotations of $\frac{\pi}{3}$ about the center and under reflections about the "diagonals" and bisectors.[11] In addition, we can imagine our hexagon is one unit in an infinite honeycomb. Then the system has translation symmetry in several directions. In contrast, the triangular quantum dot discussed earlier lacks these symmetries and, as a result, shows no degeneracy.

Even though the numerical results may have only limited accuracy, the degenerate energies are exactly equal. This is because our mesh preserves the exact symmetry of the system. We expect this symmetry to be carried over to the wave functions.

The corresponding wave functions are shown in Figure 9.19. The ground state has a single maximum as expected. States 2 and 3, which are degenerate, have a peak and a valley each, separated by nodal lines. In state 2, the nodal line runs along a vertical bisector and divides the hexagon into two irregular pentagons. In state 3, it is along the horizontal diagonal and divides the hexagon into two trapezoids. The next pair of degenerate states 4 and 5 have similar saddle-like patterns slightly rotated off each other, but no simple nodal lines. State 4 has two straight nodal lines running along a diagonal and a bisector (like $\times$ but in perpendicular directions), whereas state 5 has two curvy nodal lines running roughly along two bisectors (like $\mathcal{X}$).

With increasing state number, the patterns become more complex. State 6 has no side-touching nodal lines, but the central peak looks like the peak of the ground state stuck in a bowl. Figure 9.20 compares the two more clearly. The very similar patterns of states 7

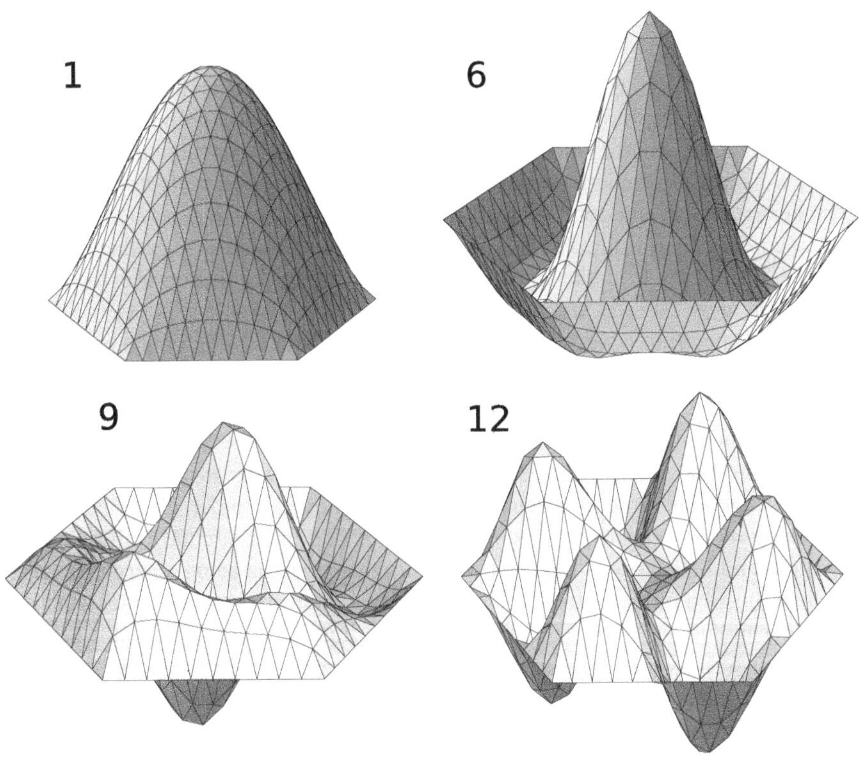

FIGURE 9.20 Surface plot of select wave functions from Figure 9.19.

[11]Here we ascribe the term *diagonals* to lines connecting the opposite vertices, and *bisectors* to lines connecting the centers of opposite edges.

and 8 suggest they might be degenerate, but are in fact not. A closer examination of state 8 shows that each of the six peaks or valleys is completely enclosed by an equilateral triangle, so they are the ground state of the equilateral triangle quantum dot. This hexagon state consists of 6 triangular states stitched together. The last two degenerate pairs of states, (9, 10) and (11, 12), show the rapid transition from underdeveloped extrema to sharp peaks and valleys (see Figure 9.20). Electron densities of quantum dots have been imaged experimentally [26], and the single-electron states form the basis for understanding many-electron quantum dots [77].

CHAPTER SUMMARY

The focal area of this chapter is the simulation of time-independent quantum systems. We studied systems such as simple wells, linear potentials, central field potentials, and quantum dots. We discussed several methods for solving the Schrödinger equation to obtain the energy spectrum and the wave function. Solutions to any given potential including singular potentials can be found using at least one of the methods presented, making the study of any quantum system within our grasp. Furthermore, each of the methods can be used with minimum modification, at most requiring the user to supply a custom potential function.

With little preparation, we can use the standard shooting method, first introduced for projectile motion, with an ODE solver and a root finder (e.g., RK4 and bisection) to obtain solutions directly. A more efficient approach is to replace the ODE solver with Numerov's method for increased speed and accuracy, while maintaining robustness and versatility.

We also discussed the application of the familiar finite difference and finite element methods previously used for boundary value problems to eigenvalue problems. For 1D eigenvalue problems, the FDM is slightly simpler than the FEM, but has no performance advantage over the FEM. Except for pedagogical reasons, the FEM is preferred because it is able to deal with point potentials.

The basis expansion method can be used as a general meshfree method that depends on the basis functions rather than a space grid because space is not discretized. With an appropriate basis set to a given problem, this method can lead to very accurate solutions even for large systems.

We have devoted considerable effort to the study of quantum dots in this chapter and developed a full FEM program to solve the Schrödinger equation in 2D for arbitrary boundaries. The program is robust and simple to use, including a small FEM library `fem.py` that can be used for other general purposes. One may start using the code first and going through its development after you are familiar with its flow and operation.

Throughout, we have made use of several functions (Airy, Bessel, Hermite, etc.) from the SciPy special function library. We have also discussed file I/O for storing data and avoiding recalculation. Effective graphing techniques have been demonstrated using Matplotlib's triangle plotting capability, which proved to be an excellent fit for working with data over triangular meshes such as in FEM.

9.7 EXERCISES AND PROJECTS

Exercises

E9.1 Solve for the eigenenergies of the square well potential shown in Figure 9.1. The eigenequations for even and odd states are given by (in a.u.)

$$\sqrt{-E} - \sqrt{V_0 + E}\tan(\frac{a}{2}\sqrt{V_0 + E}) = 0, \quad \text{(even)}$$

$$\sqrt{-E} + \sqrt{V_0 + E}\cot(\frac{a}{2}\sqrt{V_0 + E}) = 0. \quad \text{(odd)}$$

Assume $a = 4$ and $V_0 = 4$. Also, convert your results to eV and J, respectively. Does it make sense to use the latter?

You can find the roots one by one using the bisection or Newton's method, or SciPy's equation solver `fsolve` (see Exercise E3.8). You may also wish to try SymPy's solver, which can find them all at once. Give it a try, see Exercise E3.10.

E9.2 (a) Use the same method as in Exercise E9.1 to compute the eigenenergies of a single-well potential of width $(a-b)/2$, where a and b are the parameters for the double-well potential (9.6). Compare with the double-well energies in Figure 9.3 using the same parameters.

(b) Shrink the barrier width b by half, and calculate the wave functions with Program 9.2. What happens to the double hump?

Predict what will happen to the energy and wave function if the barrier width is increased, but the well width $(a-b)/2$ is kept the same. Repeat the calculation by doubling b. Discuss your results, and check degeneracy.

E9.3 (a) Verify the first derivative formula (9.60). Subtract the Taylor series $y(x \pm h)$ from Eqs. (2.9) and (2.10), keep terms up to h^3, and approximate y''' from Eq. (9.53).

(b) Carry out a von Neumann stability analysis (6.76) for Numerov's method (9.59), and show that it is stable regardless of the step size. Assume a simple function, for example, $f = c$.

(c) Experiment with instability in unidirectional integration. Pick a correct energy for the single-well potential, for example, the ground state energy from Table 9.1 or Exercise E9.1, integrate the Schrödinger equation from $-R$ to R (or where the wave function diverges badly) with Program 9.1 or 9.2. Plot the wave function. Reverse direction of integration. Discuss your observations.

E9.4 Fill in the steps leading from Eq. (9.12) to (9.13) in 1D FEM.

E9.5 Show that the expectation values of kinetic and potential energies

$$\langle \hat{T} \rangle = -\frac{1}{2}\int u^* \frac{d^2u}{dx^2}dx, \quad \langle \hat{V} \rangle = \int u^* V(x) u dx,$$

are given by Eq. (9.19) in terms of the respective $\hat{T}$ and $\hat{V}$ matrices, Eqs. (9.16a) and (9.16b).

E9.6 (a) Derive the eigenequation (9.24) for the Dirac δ molecule. Divide the space into three regions separated by the δ potentials, apply the boundary condition at $x = \pm\infty$, and finally match the wave function at $x = \pm\frac{a}{2}$, including the cusp condition (9.22).

(b) Solve Eq. (9.24) for the symmetric potential ($\alpha = \beta$) to express the solutions (9.25) in terms of the Lambert W function. Refer to Chapter 3 (Section 3.4) for an example on the general approach, for example, after Eq. (3.29). You can also solve it more quickly with SymPy, see Exercise E3.10.

E9.7 A delta-wall potential, shown in Figure 9.21, is composed of a delta potential $-\alpha\delta(x - \frac{a}{2})$ placed to the right of a hard wall at $x = 0$.

Show that there exists a bound state only if $\alpha a > 1$, the same condition for the existence of the excited state in the δ molecule (9.26), and that the energy is also the same (Eq. (9.25), k_-).

E9.8 Show that in the BEM, the Schrödinger equation (9.28) can be reduced to Eq. (9.17) with the matrix elements given by Eq. (9.29). Fill in the detail following similar projection methods from Eq. (S:8.24), or from Eq. (9.12) keeping the kinetic energy operator intact, that is, no integration by parts.

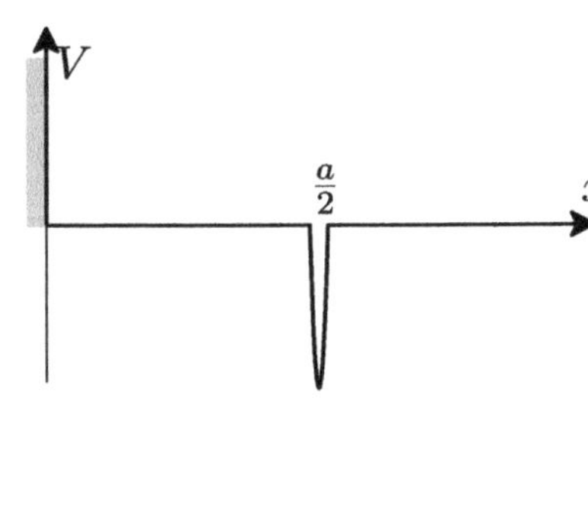

FIGURE 9.21 The delta-wall potential.

E9.9 Derive the potential matrix element V_{mn} (9.31) for the square well potential. Show that $V_{mn} = 0$ unless $m + n$ is even, in which case

$$V_{mn} = -\frac{2V_0}{\pi}\left[\frac{(-1)^{\frac{m-n}{2}}}{m-n}\sin\left(\frac{m-n}{2L}\pi a\right) - \frac{(-1)^{\frac{m+n}{2}}}{m+n}\sin\left(\frac{m+n}{2L}\pi a\right)\right].$$

E9.10 Show that the matrix elements for x, x^2, p, and p^2 in the SHO basis are

$$x_{mn} = \frac{a_0}{\sqrt{2}}\left(\sqrt{m_1}\,\delta_{m,n-1} + \sqrt{m}\,\delta_{m,n+1}\right),$$

$$p_{mn} = -i\frac{\hbar}{\sqrt{2}\,a_0}\left(\sqrt{m_1}\,\delta_{m,n-1} - \sqrt{m}\,\delta_{m,n+1}\right),$$

$$(x^2)_{mn} = \frac{a_0^2}{2}\left(\sqrt{m_1\,m_2}\,\delta_{m,n-2} + (2m+1)\delta_{mn} + \sqrt{m\,m_{-1}}\,\delta_{m,n+2}\right),$$

$$(p^2)_{mn} = -\frac{\hbar^2}{2a_0^2}\left(\sqrt{m_1\,m_2}\,\delta_{m,n-2} - (2m+1)\delta_{mn} + \sqrt{m\,m_{-1}}\,\delta_{m,n+2}\right),$$

where $m_j = m + j$. The easiest way is to express the position and momentum operators in terms of the raising and lowering operators as

$$x = \frac{a_0}{\sqrt{2}}(\hat{a} + \hat{a}^+), \quad p = -i\frac{\hbar}{\sqrt{2}\,a_0}(\hat{a} - \hat{a}^+),$$

where a_0 is the length scale given in Eq. (9.33). When acting on an eigenstate, the operators $\hat{a}$ and $\hat{a}^+$, respectively, lowers or raises the state by 1 [44],

$$\hat{a}u_n = \sqrt{n}\,u_{n-1}, \quad \hat{a}^+u_n = \sqrt{n+1}\,u_{n+1}.$$

E9.11 (a) Given the potential $V = \beta|x|$, use variable substitution to arrange the Schrödinger equation (9.1) in the form of Eq. (9.62), the differential equation for Airy functions. Show that the solutions are given by Eq. (9.61).

(b) Match the wave functions at $x = 0$ and obtain the eigenenergy equations (9.63) for even and odd states, respectively.

E9.12 Use the supplied mesh data file to calculate the eigenenergies and wave functions of the triangle quantum dot with Program 9.7. Compare the eigenenergies with the exact values (9.51), and plot the relative error for all states. Also plot the wave functions of the first 12 states like Figure 9.17 or Figure 9.20.

Projects

P9.1 (a) Run Program 9.1 with E starting below the bottom of the well, say $-2V_0 < E < -V_0$. Describe the shape of the wave function and explain why no bound state exists.

(b) Modify Program 9.1 to find the odd bound states. Instead of the derivative being zero at $x = 0$, the wave function itself is zero. Modify the conditional statement in the main loop to detect the sign change in the wave function. Also add a negative sign to the wave function for $x > 0$.

With the same parameters as in the program, how many odd states are there? Where do the energies fall in relation to the energies of even states?

(c) Modify the program so it can find both even and odd states. Double the width of the well, find the number of all bound states (both even and odd). Do the same, but double the depth. Which way is more effective at increasing the number of bound states? Briefly explain.

(d) Change the potential to the SHO potential $V = \frac{1}{2}x^2$, and find all states up to $E = 10$. Compare with Eq. (9.32) and comment on your results.

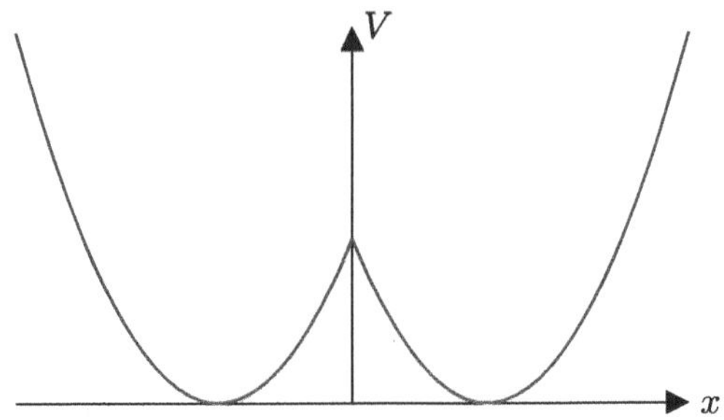

FIGURE 9.22 The distorted harmonic potential.

P9.2 Consider the distorted harmonic potential $V(x) = V_0 (|x| - x_0)^2$, graphed in Figure 9.22. If $x_0 = 0$, it is the SHO potential. Otherwise, the potential has a double well at $x = \pm x_0$, and approaches the SHO potential for large $|x| \gg x_0$.

(a) Let $V_0 = \frac{1}{2}$ and $x_0 = 0$. Use the shooting method (Program 9.2) to compute the first 10 to 20 eigenenergies. Plot and inspect a few wave functions. Compare the spectrum with Eq. (9.32). Are the states equidistant? Experiment with the parameters, including range, number of grids, and energy increment, and so on.

(b) Now let $x_0 = 1.2$. Repeat the above calculations. Discuss your results in relation to the SHO energies ($x_0 = 0$).

(c) Move the matching point M around, say just below or above x_0. Calculate the energy levels for three different M values. Do the energies change? Should they? Why?

P9.3 Modify Program 9.2 to use Numerov's method (9.59). Let the matching point be M. Integrate the Schrödinger equation upward or downward one step beyond M in both directions, so we can calculate the derivative using Eq. (9.60). See Program 9.5.

Profile your code to test the performance gain between Numerov's method and RK45 using the same step size for Project P9.2, for example. Optionally repeat the test for an SHO if the same accuracy is required.

P9.4 (a) Write a program to calculate the eigenenergies in the square well potential using the FDM method. Prepare the $\hat{T}$ and $\hat{V}$ matrices according to Eq. (9.10), and use the same step size h as the shooting method (Program 9.2).

(b) The square well is infinitely sharp, but numerically our grid is finite. This causes numerical error if the potential jumps from $-V_0$ to 0 at the walls. Rather than a steep rise, the result can be improved if we linearly interpolate the potential across the edges (see Eq. (3.51), Project P3.2). Modify your potential matrix so that the two values at the grid points immediately outside the well on each side is $-\frac{1}{2}V_0$. Repeat the calculation. Comment on the improvement of accuracy.

P9.5 (a) Evaluate the analytical matrix elements (9.18) for the square well potential. Replace the potential matrix in Program 9.3 by your analytical results. Repeat the calculation to reproduce the results in Table 9.1.

(b) The sharp edges of the square well affects the accuracy of the results. To improve it, linearly interpolate the potential between the two elements bracketing the edges at $\pm a/2$ as

$$V(x) = \begin{cases} 0, & \text{if } |x| > \frac{a}{2} + h, \\ -V_0, & \text{if } |x| < \frac{a}{2} - h, \\ -\frac{V_0}{2h}\left(\frac{a}{2} + h - |x|\right), & \text{otherwise,} \end{cases} \tag{9.52}$$

where h is the element size. The interpolated potential is identical to the actual potential outside the zones between the two immediate elements sandwiching the edges. In the zones, it varies linearly from $-V_0$ at $|x| = a/2 - h$ to 0 at $|x| = a/2 + h$. For best result, set the grid so the edges ($\pm a/2$) are also the nodes.

Use this potential in Program 9.3 and redo the calculation. Discuss your results.

P9.6 (a) Calculate the energy and wave function of the bound state of the Dirac δ atom. Assume $\alpha = 1$. Use the FEM code, Program 9.3, and modify the potential matrix so that the only nonzero matrix element is at the node where the δ potential is located. Plot the numerical and analytic wave functions in the same figure, normalize both at the center. Compare numerical and exact results for the energy and wave function.

(b) Check the quality of the wave function near the center. What are the numerical cusp values, $u'(0^{\pm})/u(0)$? How do they compare with the exact value?

(c) Calculate the expectation values of kinetic and potential energies from Eq. (9.19). See description of Program 9.3 on how to proceed.

(d) Compute the expectation values of $\langle x\rangle$, $\langle x^2\rangle$, $\langle p\rangle$, and $\langle p^2\rangle$. Use the matrix representations of the operators from Exercise S:E9.3 if available. If not, either derive them, or use numerical integration to find $\langle x\rangle$ and $\langle x^2\rangle$. Find the uncertainty $\Delta x\Delta p$, and verify that it satisfies the uncertainty principle.

(e)* Predict and sketch the wave function in momentum space. Calculate and plot the wave function using FFT. Estimate the widths $\Delta x'$ and $\Delta p'$ from their respective wave functions. Compare $\Delta x'\Delta p'$ with the result above.

(f)* Simulate the δ potential with the FDM. Represent the singularity as $V = -\frac{\alpha}{h}$ at $x = 0$ and zero elsewhere, where h is the grid spacing. Calculate the energy and cusp condition, and compare with the preceding FEM results.

P9.7 (a) Using the results from Exercise E9.9, modify Program 9.4 to compute the bound state energies for the square well potential with the box basis set. Also plot the corresponding wave functions.

(b) Calculate the eigenenergy of the Dirac δ-atom using the BEM. Compare with results from Project P9.6 for the same α. Change the number of states N in your basis, and determine the minimum N to achieve a comparable accuracy as the FEM.

P9.8 Study the energy structure of several model central potentials. Choose one potential, and use Program 9.5 as the base code.

(a) First, let us consider a potential that deviates slightly from the Coulomb potential in the power law,

$$V(r) = -\frac{Z}{r^{1+\epsilon}}.$$

Vary $\epsilon = 0.01$ to 0.1. Compute the energy-level diagram as shown in Figure 9.11. Compare with the hydrogen atom ($Z = 1$), and summarize the similarities and differences.

Note that all states are higher than hydrogen, except the $1s$ state. Explain why (a figure similar to Figure 9.12 should be helpful).

(b) Next, consider the screened Coulomb potential (Yukawa),

$$V(r) = -Z\frac{e^{-r/a}}{r}.$$

The parameter a is the screening length. This potential behaves like the Coulomb potential for $r \to 0$ but has a much shorter range. Let $a = 20$ and $Z = 1$. Calculate the energy-level diagram as in part (a). Find the energy difference between 2s and 2p states. Is the number of bound states finite?

Double $Z = 2$, repeat the calculation. Observe the difference with $Z = 1$.

(c) Last, consider the Hulthén potential,

$$V(r) = -\frac{Z}{a}\frac{e^{-r/a}}{1 - e^{-r/a}},$$

where Z is the nuclear charge, and a the screening length. What is the potential like in the limits $r \to 0$ and ∞? Based on your observations from earlier, sketch the energy-level diagram.

Choose $Z = 1$ and $a = 10$, and calculate the energy-level diagram. Find the largest l supported by the potential. Plot the wave function for the s states, and compare both with the hydrogen atom. Examine the number of nodes in the wave function so as not to miss any states.

Repeat the calculation for $Z = 2$, keep the same a.

(d)* Modify your program to compute the expectation values of kinetic and potential energies, respectively. Because the wave function is obtained on a regular grid, we can use Simpson's rule (Section 8.A.2) to carry out the numerical integral. For $\langle T \rangle$, do not compute the derivatives directly. Rather, use either the first derivative (9.60) in Eq. (8.18), or the second derivative from Eq. (9.8) with the integral in Exercise E9.5. In addition, properly scale your results by $C = \int u^* u dx$ because the wave function is not normalized.

Test your program on hydrogen atom to make sure it gives the correct results according to the virial theorem (9.40). For potentials given earlier, compute $\langle T \rangle$ and $\langle V \rangle$ for the lowest state in each l-series. Do they still obey the virial theorem? Comment on your results.

P9.9 (a) Calculate and plot the wave functions of the unit circle quantum dot (billiard, Section S:9.2.1) using the supplied mesh data file. Discuss the symmetry of the degenerate states.

(b) Generate a mesh for the circular billiard so it preserves rotational symmetry within the grid representation. One way to do this is start with the mesh grid of the hexagon or octagon, scale the points off the circle symmetrically about the radial they are on. Calculate the eigenenergies, and compare the degenerate states against Table S:9.1.

P9.10 (a) Generate a mesh for the hexagon using Program 9.9. Plot the kinetic energy matrix (or equivalently the overlap matrix) similar to Figure 7.12. You should see bands running along the diagonal initially, but a little later some bands break, jump away from the diagonal, and continue from there. This is caused by the discontinuity in numbering the nodes due to reflection.

(b) Modify the program so the nodes are consecutively numbered, or postprocess the mesh to achieve the same goal. Check the bandedness by plotting the matrix again.

(c)* Represent the new matrix in banded matrix format, see Figure 8.6 and Program 8.3. Solve the eigenvalue problem using the SciPy `eig_banded()` function.

Profile your programs and compare speeds between the standard and banded eigensolvers.

9.A NUMEROV'S METHOD

The Numerov method is useful for a second-order ODE of the following form that does not contain first derivatives

$$y'' + f(x)y = 0. \tag{9.53}$$

It turns out that we can apply a special technique to discretize it over a grid such that odd derivatives up to the fifth order can be canceled in the central difference scheme, giving us a method accurate to h^5, h being the grid size as usual.

To show this, we apply an operator to Eq. (9.53),

$$\left(1 + \frac{h^2}{12}\frac{d^2}{dx^2}\right)(y'' + fy) = y'' + fy + \frac{h^2}{12}\left[y'''' + (fy)''\right] = 0. \tag{9.54}$$

Using the Taylor series for $y(x \pm h)$ from Eqs. (2.9) and (2.10) and adding them up, we have,

$$y(x+h) + y(x-h) = 2y + h^2 y'' + \frac{h^4}{12} y'''' + O(h^6), \tag{9.55}$$

or in terms of the y''

$$y'' = \frac{y(x+h) - 2y + y(x-h)}{h^2} - \frac{h^2}{12} y'''' + O(h^4), \tag{9.56}$$

which is accurate to order h^4 (compare with Eq. (6.31b)).

Substituting Eq. (9.56) into (9.54), we see that the terms y'''' cancel, leaving us with

$$\frac{y(x+h) - 2y + y(x-h)}{h^2} + fy + \frac{h^2}{12}(fy)'' + O(h^4) = 0. \tag{9.57}$$

To approximate $(fy)''$, we can use Eq. (9.56) but keep only the first term so the error is still of the order h^4. Switching to the usual subscript notation, Eq. (9.57) becomes

$$\frac{y_{i+1} - 2y_i + y_{i-1}}{h^2} + f_i y_i + \frac{1}{12}(f_{i+1}y_{i+1} - 2f_i y_i + f_{i-1}y_{i-1}) + O(h^4) = 0. \tag{9.58}$$

Solving for y_{i+1}, we obtain Numerov's method

$$y_{i+1} = \left(1 + \frac{h^2}{12} f_{i+1}\right)^{-1} \left[2\left(1 - \frac{5h^2}{12} f_i\right) y_i - \left(1 + \frac{h^2}{12} f_{i-1}\right) y_{i-1}\right] + O(h^6). \tag{9.59}$$

Numerov's method is a three-term recursion relation. It is stable (Exercise E9.3) and has a local error of the order $O(h^6)$. This is the same order as the Runge-Kutta-Fehlberg method (RK45, Section 2.3.2), which requires six function calls. Numerov's method is more efficient with apparent three calls to $f(x)$, but effectively only one new call because the other two can be reused in subsequent iterations. However, it is not self-starting, as two seed values are necessary to get started. Program 9.5 contains an implementation of the method. We will include it in our ODE library `ode.py` from now on.

We can also obtain the first derivative to a higher order taking advantage of the special form Eq. (9.53). It can be shown that the central difference method leads to (Exercise E9.3)

$$y'_i = \frac{1}{2h}\left[\left(1 + \frac{h^2}{6} f_{i+1}\right) y_{i+1} - \left(1 + \frac{h^2}{6} f_{i-1}\right) y_{i-1}\right] + O(h^4). \tag{9.60}$$

The error is two orders smaller than the standard central difference formula (6.31a) and is the same as the standard five-point scheme. Equation (9.60) is useful for wave function matching in quantum problems.

We note that the standard Numerov method can be modified so it is applicable to ODEs containing first derivatives. The interested reader should check relevant literature if the need arises.

9.B THE LINEAR POTENTIAL AND AIRY FUNCTION

In the wedged linear potential $V = \beta|x|$, the wave function can be solved (see Exercise E9.11) in terms of a special function called the Airy function, $Ai(x)$,

$$u(x) = \begin{cases} C Ai(y - E/\beta x_0), & x \geq 0, \\ D Ai(-y - E/\beta x_0), & x < 0, \end{cases} \tag{9.61}$$

$$y = \frac{x}{x_0}, \quad x_0 = \left(\frac{\hbar^2}{2m_e \beta}\right)^{1/3}.$$

The Airy function is defined as a solution to the differential equation [10]

$$f''(x) - xf(x) = 0. \tag{9.62}$$

It has two solutions, the nondiverging regular solution $Ai(x)$ and a diverging solution $Bi(x)$. We are interested in the regular solution $Ai(x)$, which is graphed in Figure 9.23.

Because the potential is even, the eigenfunctions (9.61) have definite even or odd parities. As we have seen (e.g., Figure 9.3), even eigenfunctions must have a local maximum at the origin, $u'(0) = 0$, and odd ones must have a node, $u(0) = 0$. Accordingly, the eigenenergies must satisfy

$$\begin{aligned} Ai'(-E/\beta x_0) &= 0, \quad \text{(even)} \\ Ai(-E/\beta x_0) &= 0. \quad \text{(odd)} \end{aligned} \tag{9.63}$$

Let z_n and z'_n be the n-th zero of $Ai(x)$ and its derivative $Ai'(x)$, respectively. The eigenenergies can be found from Eq. (9.63) as

$$\begin{aligned} E_n &= -\beta x_0 z'_n, \quad \text{(even)} \\ E_n &= -\beta x_0 z_n. \quad \text{(odd)} \end{aligned} \tag{9.64}$$

To find the eigenenergies, we need to locate the zeros of the Airy function. The first several zeros are listed in Table 9.4.

If other zeros are needed, they may be generated from the following code using the SciPy special function library.

```
from scipy.special import ai_zeros
n = 10                              # number of zeros
zn, zpn, zbn, zbpn = ai_zeros(n)
print (zn, zpn)
```

On return, zn and zpn contain the zeros of $Ai(x)$ and $Ai'(x)$, respectively. The zeros of $Bi(x)$ and $Bi'(x)$ are in zbn and zbpn, which are of no interest to us here.

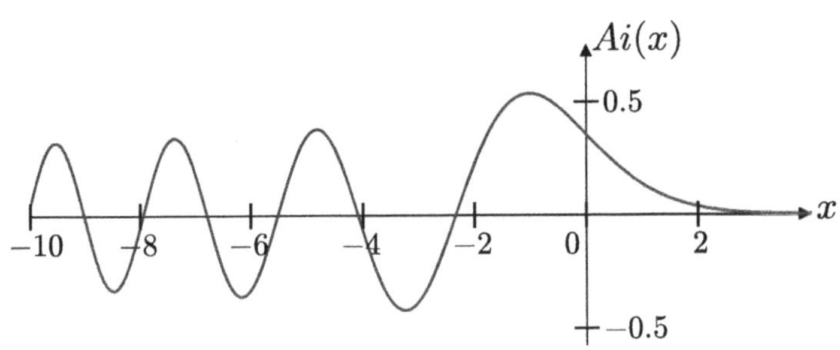

FIGURE 9.23 The Airy function.

Table 9.4 The First Six Zeros of Airy Function and Its Derivative.

n	1	2	3	4	5	6
z_n	−2.33811	−4.08795	−5.52056	−6.78671	−7.94413	−9.02265
z'_n	−1.01879	−3.24820	−4.82010	−6.16331	−7.37218	−8.48849

9.C PROGRAM LISTINGS AND DESCRIPTIONS

Program listing 9.1 Visual eigenstates in the square well (`squarewell.py`)

```
import numpy as np, visual as vp, ode, vpm
2
def V(x):                    # potential
4    return 0. if abs(x) > a/2. else -V0

6 def sch(psi, x):           # Schrodinger eqn
    return [psi[1], 2*(V(x)-E)*psi[0]]
8
# initialization and animation setup
10 a, V0 = 4.0, 4.                      # well width, depth
R, N = 4*a, 200                      # limit, intervals
12 xa = np.linspace(-R, R, 2*N+1)       # grid
h, z = xa[1]-xa[0], np.zeros(2*N+1) # step size
14 E, dE, dpsi, psix = -V0, 0.001, 1.0, np.zeros(2*N+1)

16 scene = vp.display(background=(.2,.5,1), range=1.5*a)
wf = vpm.line(xa, psix, z, vp.color.red, .05)
18 pot = vpm.line(xa, .5*np.vectorize(V)(xa), z, (1,1,1), .04) # pot. V
info = vp.label(pos=(0, -0.6*a, 0), box=False, height=20)
20
while (E < 0.0):
22    psi, x = [.0, .1], -R
    for i in range(N):                # WF for x <=0
24        psi = ode.RK45n(sch, psi, x, h)
        x += h
26        psix[i+1] = psi[0]
    psix[N+1:] = psix[N-1::-1]        # WF for x > 0 by reflection
28    if (dpsi*psi[1] < 0.):            # dpsi/dx changes sign
        info.text='Energy found, E=%5.4f' %(E-dE/2)
30        vpm.pause(scene)              # any key to continue
    else:
32        info.text='E=%5.3f' %(E)
    wf.move(xa, 2*psix/max(psix), z), vpm.wait(scene), vp.rate(2000)
34    dpsi = psi[1]                     # old dpsi/dx at E
    E += dE
```

The program assumes a symmetric potential well centered at the origin and calculates the energies of even states. The function `sch()` is explained in detail in Section 9.1.2. The range of integration is set to $[-R, R]$ where R should be large compared to the width of the well. The range is divided into $2N$ intervals ($2N + 1$ grid points). To display the potential, the NumPy `vectorize` function is used (line 18), which vectorizes a scalar function so it operates on an input array and returns an output array.

The program scans energy E from $-V_0$ to 0. At each E, we integrate the Schrödinger equation from $x = -R$ to the origin, starting with $\psi(-R) = 0$ (`psi[0]`) and a small, arbitrary value for the derivative $\psi'(-R)$ (`psi[1]`). The wave function for $x > 0$ is obtained by symmetry via slicing (line 27). We count backward from $N - 1$ to 0 to get the reflected wave function from $x = h$ to R.

For even states, the derivative will be zero at the origin if the energy is an eigenenergy. When the derivative changes sign, we assume a correct eigenenergy has been found. The program terminates when E becomes positive, and we have found all even eigenstates.

Program listing 9.2 Shooting method for double well (`qmshoot.py`)

```
import numpy as np, rootfinder as rtf, ode
import matplotlib.pyplot as plt

def V(x):                        # potential
    return 0 if (abs(x) > a/2. or abs(x) < b/2.) else -V0

def sch(psi, x):                 # Schrodinger eqn
    return [psi[1], 2*(V(x)-E)*psi[0]]

def intsch(psi, n, x, h):        # integrate Sch eqn for n steps
    psix = [psi[0]]
    for i in range(n):
        psi = ode.RK45n(sch, psi, x, h)
        x += h
        psix.append(psi[0])
    return psix, psi             # return WF and last point

def shoot(En):                   # calc diff of derivatives at given E
    global E                     # global E, needed in sch()
    E = En
    wfup, psiup = intsch([0., .1], N, xL, h)      # march upward
    wfdn, psidn = intsch([0., .1], N, xR, -h)     # march downward
    return psidn[0]*psiup[1] - psiup[0]*psidn[1]

a, b, V0 = 4.0, 1.0, 6.                # double well widths, depth
xL, xR, N = -4*a, 4*a, 500             # limits, intervals
xa = np.linspace(xL, xR, 2*N+1)        # grid
h, M = xa[1]-xa[0], 2                  # step size, M=matching point
E1, dE, j = -V0, 0.01, 1

plt.figure()
while (E1 < 0):      # find E, calc and plot wave function
    if (shoot(E1) * shoot(E1 + dE) < 0):       # bracket E
        E = rtf.bisect(shoot, E1, E1 + dE, 1.e-8)
        print ('Energy found: %.3f' %(E))
        wfup, psiup = intsch([0., .1], N+M, xL, h)      # compute WF
        wfdn, psidn = intsch([0., .1], N-M, xR, -h)
        psix = np.concatenate((wfup[:-1], wfdn[::-1]))  # combine WF
        scale = psiup[0]/psidn[0]
        psix[N+M:] *= scale                             # match WF
        ax = plt.subplot(2,2,j)
        ax.plot(xa, psix/max(psix))                     # plot WF
        ax.plot(xa, np.vectorize(V)(xa)/(2*V0))         # overlay V
        ax.set_xlim(-a,a)
        ax.text(2.2,0.7, '%.3f' %(E))
        if (j == 1 or j == 3): ax.set_ylabel(r'ψ')
        if (j == 3 or j == 4): ax.set_xlabel('x')
        if (j<4): j += 1               # 4 plots max
    E1 += dE
plt.show()
```

The potential function `V()` returns the double square well potential. The functions `sch()`, `intsch()`, and `shoot()` have been explained earlier in the text. The limits x_L and x_R are coupled to the width of the double well, which we take as the characteristic length of the system.

The main loop iterates through the trial energy in equal increments. When a root is bracketed, it is solved with the bisection method. Once a true eigenenergy is found, the wave function is obtained by integrating inward and matching at grid M, which should be off-center to avoid the node point for symmetric potentials. The wave function is combined by `np.concatenate` into one array `psix` for plotting. Note that the order for the downward wave function is reversed via slicing `wfdn[::-1]` before the combination, which is scaled (matched) according to Eq. (9.3) afterward. The layout of the plots are assumed to be 2×2. For other potentials with more bound states, it should be modified accordingly.

Program listing 9.3 Eigenenergies by FEM (`qmfem.py`)

```
import numpy as np
from scipy.sparse.linalg import eigsh

def V(x):                       # potential
    return 0. if (abs(x) >= a/2.) else -V0

def TB_mat(n):                              # fill in T and and B matrices
    Tm, B = np.diag([2.]*n), np.diag([4.]*n)
    Tm += np.diag([-1.]*(n-1),1) + np.diag([-1.]*(n-1),-1) # off diag
    B += np.diag([1.]*(n-1),1) + np.diag([1.]*(n-1),-1)
    return Tm/(2.*h), B*h/6.

def Vij(i, j):      # pot. matrix Vij over [x_i, x_{i+1}] by order-4 Gaussian
    x=np.array([0.3399810435848564, 0.8611363115940525]) # abscissa
    w=np.array([0.6521451548625463, 0.3478548451374545]) # weight
    phi = lambda i, x: 1.0 - abs(x-xa[i])/h         # tent function
    vV, hh = np.vectorize(V), h/2.0                  # vectorize V
    x1, x2 = xa[i] + hh - hh*x, xa[i] + hh + hh*x
    return hh*np.sum(w * (vV(x1) * phi(i, x1) * phi(j, x1) +
                       vV(x2) * phi(i, x2) * phi(j, x2)) )

def V_mat():        # fill potential matrix
    Vm = np.zeros((N-1,N-1))
    for i in range(N):        # for each element     # contribution to:
        if (i>0):  Vm[i-1,i-1] += Vij(i, i)          #    left node
        if (i < N-1): Vm[i,i] += Vij(i+1, i+1)       #    right node
        if (i>0 and i< N-1):
            Vm[i-1,i] += Vij(i, i+1)                 #    off-diagonals
            Vm[i,i-1] = Vm[i-1,i]                    #    symmetry
    return Vm

a, V0 = 4.0, 4.0                  # well width, depth
xL, xR, N = -4*a, 4*a, 600        # boundaries, num. of elements
xa = np.linspace(xL, xR, N+1)     # nodes
h = xa[1]-xa[0]                   # size of element

Tm, B = TB_mat(N-1)               # obtain T, B, V matrices
Vm = V_mat()

E, u = eigsh(Tm + Vm, 10, B, which='SA') # get lowest 10 states
print (E)
```

This is a general FEM program for 1D potentials. The routine `TB_mat()` prepares the kinetic energy and basis overlap matrices, both tridiagonal, according to Eqs. (9.16a) and (9.16c) using `np.diag` (see Program 6.3). The next couple of functions `Vij()` and `V_mat()` work together to compute the potential matrix (9.18) as explained in the text.

The main code initializes parameters such as the potential, space range, the number of finite elements, nodes, and so on. It then prepares the matrices for the kinetic energy `Tm`, overlap `B`, and the potential energy `Vm`. The eigenenergies and eigenvectors are obtained with the SciPy sparse matrix eigenvalue solver `eigsh` (line 40), which is more efficient than the standard `eigh` routine for large matrices. It works on symmetric real or complex Hermitian matrices. The eigenvalues are returned in `E` as a 1D array, and eigenvectors in `u` as a 2D array, in such a way that `u[:,i]`, a 1D array over the space grid, is the eigenvector corresponding to `E[i]`.

In many cases, we are interested in only a select group of eigenvalues, such as the low-lying bound states. It is also faster to solve only for a subset of the eigenvalues. We specify the number of eigenvalues (10 in this case) and the sorting option `'SA'` (lowest value first), so this combination of switches gives us the lowest 10 eigenstates. For larger dimensions, consider the banded eigenvalue solver `eig_banded` in SciPy for better convergence.

The eigenvector, that is, the wave function, returned by `eigsh` is properly normalized with `B` as the weight (9.20), meaning that the statement

```
np.dot(u[: ,i], np.dot(B, u[: ,i ]))
```

evaluates to 1 for any state i. Therefore, we can readily compute any average such as the kinetic or potential energies (9.19). For example, the kinetic energy of the first state (the ground state) is simply

```
ke = np.dot(u[:,0], np.dot(Tm, u[:,0]))
```

The function `np.dot` calculates the scalar product if the arguments are 1D arrays. If one of arguments is a 2D array, it is the same as matrix multiplication. This is the case for `dot(Tm, u[:,0])`, which returns a 1D array as a result of a square matrix `Tm` multiplied by a column matrix `u[:,0]`. A second `dot` operation computes the scalar product, a single number, from the resultant and the wave function, both 1D arrays.

Program listing 9.4 Eigenenergies by BEM (`bem.py`)

```
import numpy as np, integral as itg
from scipy.sparse.linalg import eigsh
from scipy.special import eval_hermite, gamma, ai_zeros

def V(x):                        # potential
    return beta*abs(x)

def uVu(x):                      # integrand u_m V u_n
    y, c = x/a0, a0*np.sqrt(np.pi*gamma(m+1)*gamma(n+1)*2**(m+n))
    return V(x)*eval_hermite(m,y)*eval_hermite(n,y)*np.exp(-y*y)/c

beta, omega, N = 1., 1., 20      # pot slope, nat freq, num. basis states
a0, x0 = 1/np.sqrt(omega), 1/(2*beta)**(1./3)  # length scale, x0
```

```
15 m, Vm = np.arange(N-2), np.zeros((N,N))
   mm = np.sqrt((m+1)*(m+2))                  # calc T matrix
17 Tm = np.diag(np.arange(1, N+N, 2,float))   # diagonal
   Tm -= np.diag(mm, 2) + np.diag(mm, -2)     # off-diagonal by +/-2
19 for m in range(N):                          # calc V matrix
       for n in range(m, N, 2):                # m+n is even every 2 steps
21         Vm[m, n] = 2*itg.gauss(uVu, 0., 10*a0) # use symm.
           if (m != n): Vm[n,m] = Vm[m,n]
23 Tm = Tm*omega/4.
   E, u = eigsh(Tm + Vm, 6, which='SA') # get lowest 6 states
25
   zn, zpn, zbn, zbpn = ai_zeros(3)           # find exact values
27 Ex = - beta*x0*np.insert(zpn, range(1, 4), zn)   # combine even/odd
   print (E, E/Ex)
```

For a given potential defined by the user, `V()`, the program calculates eigenenergies using the SHO basis set. It generates the kinetic energy matrix from Eq. (9.36). Because it is banded, we use `np.diag` to build the diagonal and off-diagonal (±2) bands. The potential matrix is computed using numerical integration. The integrand is defined in `uVu()`, which returns the product $u_m^* V u_n$ in Eq. (9.29). The basis functions u_n are evaluated using the Hermite polynomials from SciPy special functions library. You can also switch to the recurrence formula (9.35) to generate your own Hermite polynomials efficiently.

Note that the numerical integral is performed for positive x only and multiplied by 2 to account for the negative x by symmetry. This needs to be modified if the potential is not symmetric. Another caution is in order. If higher basis states are used, the integrand will be highly oscillatory, and the integral may become inaccurate. Make sure to check convergence, for example, by breaking the interval into two or varying the upper limit.

The eigenenergies obtained with `eigsh` are compared with exact values from Eq. (9.64) in terms of the zeros of the Airy function. The odd states are inserted after every even state (line 27).

Program listing 9.5 Hydrogen atom by Numerov's method (`hydrogen.py`)

```
import matplotlib.pyplot as plt
import numpy as np, rootfinder as rtf

def Veff(r):                        # effective potential
    return (L*(L+1)/(2*mass*r)-1)/r

def f(r):                           # Sch eqn in Numerov form
    return 2*mass*(E-Veff(r))

def numerov(f, u, n, x, h):         # Numerov integrator for u''+f(x)u = 0
    nodes, c = 0, h*h/12.           # given [u_0, u_1], return [u_0, u_1, ..., u_{n+1}]
    f0, f1 = 0., f(x+h)
    for i in range(n):
        x += h
        f2 = f(x+h)                 # Numerov method below, Eq. (9.59)
        u.append((2*(1-5*c*f1)*u[i+1] - (1+c*f0)*u[i])/(1+c*f2))
        f0, f1 = f1, f2
        if (u[-1]*u[-2] < 0.0): nodes += 1
    return u, nodes                 # return u, nodes
```

```
def shoot(En):
    global E                            # E needed in f(r)
    E, c, xm = En, (h*h)/6., xL + M*h
    wfup, nup = numerov(f, [0,.1], M, xL, h)
    wfdn, ndn = numerov(f, [0,.1], N-M, xR, -h)   # f' from Eq. (9.60)
    dup = ((1+c*f(xm+h))*wfup[-1] - (1+c*f(xm-h))*wfup[-3])/(h+h)
    ddn = ((1+c*f(xm+h))*wfdn[-3] - (1+c*f(xm-h))*wfdn[-1])/(h+h)
    return dup*wfdn[-2] - wfup[-2]*ddn

xL, xR, N = 0., 120., 2200              # limits, intervals
h, mass = (xR-xL)/N, 1.0                # step size, mass
Lmax, EL, M = 4, [], 100                # M = matching point

Estart, dE = -.5/np.arange(1, Lmax+1)**2-.1, 0.001    # ~ -1/2n^2
for L in range(Lmax):
    n, E1, Ea = L+1, Estart[L], []
    while (E1 < -4*dE):                 # sweep E range for each L
        E1 += dE
        if (shoot(E1)*shoot(E1 + dE) > 0): continue
        E = rtf.bisect(shoot, E1, E1 + dE, 1.e-8)
        Ea.append(E)
        wfup, nup = numerov(f, [0,.1], M-1, xL, h)      # calc wf
        wfdn, ndn = numerov(f, [0,.1], N-M-1, xR, -h)
        psix = np.concatenate((wfup[:-1], wfdn[::-1]))
        psix[M:] *= wfup[-1]/wfdn[-1]                   # match
        print ('nodes, n,l,E=', nup+ndn, n, L, E)
        n += 1
    EL.append(Ea)

plt.figure()                            # plot energy levels
for L in range(Lmax):
    for i in range(len(EL[L])):
        plt.plot([L-.3, L+.3], [EL[L][i]]*2, 'k-')
    plt.xlabel('l'), plt.ylabel('E')
    plt.ylim(-.51, 0), plt.xticks(range(Lmax))
plt.show()
```

The functions `Veff()` and `f()` calculate the effective potential and the Schrödinger equation (9.8), respectively. The routine `numerov()` is a standalone Numerov integrator. It receives $f(x)$ to be integrated as input, together with initial values $[u_0, u_1]$, number of steps n to advance, starting position x, and step size h. On return, the n new values plus the initial pair are contained in $u = [u_0, u_1, \cdots, u_{n+1}]$. The same module is included in the ODE library `ode.py` for convenience.

The module `shoot()` works the same way as the one in Program 9.2. It calls `numerov()` twice to integrate the Schrödinger equation inward from both ends, going one step past the matching point M in each direction (recall the number of steps counts from u_1 upward or u_{N-1} downward). This is so we can use the three-point formula (9.60) to compute the first derivative at M.

The main code sets parameters including the maximum angular momentum l_{max}. For each l, the main loop scans the energy range for eigenenergies. When one is found, the principal number n is increased. If no valid eigenenergies are skipped, $n - l - 1$ is equal to the number of nodes.

The energy levels are plotted at the end, grouped by l. The plotting statements for the wave function are similar to those in Program 9.2 and thus omitted.

For large n, l, the code can be modified to use the more efficient logarithmic grid. See Exercise S:E9.2.

Program listing 9.6 FEM library (`fem.py`)

```
import numpy as np

def abg(p1, p2, p3):            # return alpha, beta, gamma, area of element
    [x1,y1], [x2,y2], [x3,y3] = p1, p2, p3
    alfa = [x2*y3 - x3*y2, x3*y1 - x1*y3, x1*y2 - x2*y1]
    beta, gama = [y2-y3, y3-y1, y1-y2], [x3-x2, x1-x3, x2-x1]
    area = 0.5*(alfa[0] + alfa[1] + alfa[2])          # area of triangle
    return alfa, beta, gama, area

def overlap(i, j, p1, p2, p3):              # return ∫ φiφj dxdy, Eq. (9.49)
    a, b, c, area = abg(p1, p2, p3)         # over eltriangle and pts p1-p3
    X, Y, XY, X2, Y2 = 0., 0., 0., 0., 0.
    for [x, y] in [p1, p2, p3]:
        X, Y, XY, X2, Y2 = X+x, Y+y, XY+x*y, X2+x*x, Y2+y*y
    return ((a[i]*b[j]+b[i]*a[j])*X + (a[i]*c[j]+c[i]*a[j])*Y +
            (b[i]*b[j]*(X2+X*X) +(b[i]*c[j]+c[i]*b[j])*(XY+X*Y) +
             c[i]*c[j]*(Y2+Y*Y))/4 + 3*a[i]*a[j])/(12*area)

def A_mat(node, elm):       # fills matrix ∫ ∇φi · ∇φj dxdy, Eq. (7.19)
    A = np.zeros((len(node),len(node)))
    for e in elm:
        [x1,y1], [x2,y2], [x3,y3] = node[e[0]], node[e[1]], node[e[2]]
        a = 2*(x1*(y2-y3) + x2*(y3-y1) + x3*(y1-y2))      # 4A_e
        if (a<=0.0): print ('Warning: zero or negative elm area')
        beta, gama = [y2-y3, y3-y1, y1-y2], [x3-x2, x1-x3, x2-x1]
        for i in range(3):
            for j in range(i,3 ):                          # Eq. (7.27)
                A[e[i],e[j]] += (beta[i]*beta[j] + gama[i]*gama[j])/a
                if (i != j): A[e[j],e[i]] = A[e[i],e[j]]
    return A                    # A=twice KE, -1/2 < ∇^2 >

def B_mat(node, elm):       # overlap matrix, ∫ φiφj dxdy, Eq. (9.48)
    B = np.zeros((len(node),len(node)))
    for e in elm:
        p1, p2, p3 = node[e[0]], node[e[1]], node[e[2]]
        for i in range(3):
            for j in range(i,3 ):
                B[e[i],e[j]] += overlap(i, j, p1, p2, p3)
                if (i != j): B[e[j],e[i]] = B[e[i],e[j]]
    return B
```

This FEM library includes the necessary functions for FEM solutions of 2D quantum systems. Two functions are meant to be internal: `abg()` returning α, β, γ, and area of a basis function (7.11) at the three nodes of an element; and `overlap()` calculating the overlap integral (9.49) between two nodes.

The two external functions, `A_mat()` and `B_mat()`, compute the A and B matrices, respectively, as defined in Eqs. (9.48), (7.27), and (9.49). They iterate through the elements, calling the internal functions to compute the contributions from each element and assigning them to the appropriate matrix entry (see Figure 7.10 and associated discussion). After processing all the elements in this element-oriented approach, all contributions will have been accounted for, and the matrices A and B will be completely built.

Program listing 9.7 Quantum dot (`qmdot.py`)

```
import numpy as np, pickle, fileio, fem
import matplotlib.pyplot as plt
from scipy.sparse.linalg import eigsh
from mpl_toolkits.mplot3d import Axes3D

nt, n = 12, 0          # nt = tot num of states, n=state to plot
meshfile, eigenfile = 'meshdata.txt', 'eigendata.dat' # data files
node, elm, bp, ip = fileio.readmesh(meshfile)
print ('nodes/elements', len(bp), len(ip), len(elm))

try:
    file = open(eigenfile, 'r')             # if prev eigen data exists,
    E, u = pickle.load( file )              # read from file
except IOError:
    Tm = 0.5*fem.A_mat(node, elm)           # no eigendata, recalculate
    B = fem.B_mat(node, elm)
    Tm = np.delete(Tm, bp, axis=0)          # delete boundary rows
    Tm = np.delete(Tm, bp, axis=1)          # delete boundary cols
    B = np.delete(B, bp, axis=0)
    B = np.delete(B, bp, axis=1)
    E, u = eigsh(Tm, nt, B, which='SA') # solve
    file = open(eigenfile, 'w')             # file overwritten
    pickle.dump((E, u), file)               #
file.close()

print (E)                                   # print E, prep for wf
node, wf = np.asarray(node), u[:,n]
for i in bp: wf = np.insert(wf, i, 0.)      # add boundary values

plt.figure()                                # draw mesh
plt.subplot(111, aspect='equal')
plt.triplot(node[:,0], node[:,1], elm, 'o-', linewidth=1)
plt.xlabel('x', size=20), plt.ylabel('y', size=20)

fig = plt.figure()                          # plot wave function
ax = fig.add_subplot(111, projection='3d')
ax.plot_trisurf(node[:,0],node[:,1], wf, cmap=plt.cm.jet, linewidth=.2)
plt.axis('off')

plt.show()
```

This is a universal FEM program for quantum dot calculations. It requires the mesh data file (see Figure 9.24 and its associated mesh file format).

After reading in the mesh data, it tries to check the existence of the eigenvalue data file. If it exists, presumably from a previous calculation, it simply loads the data using the `pickle` file handler (line 13). This handler is very handy for outputting data to a file. It works with the dump function that can pickle, or package, any data and write it to a file. For our use, it is most convenient to form a tuple of eigenenergies and eigenfunctions (line 23) and dump it in one operation.

If the eigenvalue data file does not exist, indicated by the exception IOError, the program starts a new calculation. Matrices $\hat{T}$ and B are obtained from the FEM library, boundary nodes are deleted (same as in Program 7.3), and the generalized eigenvalue equation is solved using the sparse eigensolver from SciPy, `eigsh`. Because we usually do not want all states, we specify the number of states we do want, `nt`, which must be less than the rank of the matrix, or the number of internal nodes in our case. The results are dumped to the eigendata file, so it can be independently analyzed without repeating the calculation, which could be long for large problems. It is possible that `pickle` can run out of memory store the data set `(E, u)` for large matrices. If this is the case, either break up the set into smaller chunks, or store all eigenenergies and only a subset of eigenfunctions.

At the end, the program prints the eigenenergies and inserts the boundary values back into the wave function for graphing, assuming `bp` is already sorted. The wave function of a chosen state is graphed with the newer function `plot_trisurf(x,y,z)`. It interprets the x, y, and z values as 1D arrays defining the points of triangular patches. It fits our purpose well, and we do not have to regenerate the grid.

If the mesh is such that the system matrices are banded, the program can be modified to use a banded eigensolver, `eig_banded`. The program could then handle very large matrices efficiently. See Figure 8.6 and corresponding discussion for band matrix representation.

Program listing 9.8 Triangle-mesh plotting of wave function (`tripwf.py`)

```
import numpy as np, pickle, fileio
import matplotlib.pyplot as plt

meshfile, eigenfile = 'meshdata.txt', 'eigendata.dat' # data files
node, elm, bp, ip = fileio.readmesh(meshfile)
file = open(eigenfile, 'r')               # read pickled eigendata
E, u = pickle.load( file )
file . close ()

node, st = np.asarray(node), range(12)  # change st for other states
fig = plt.figure()
for n in range(len(st)):
    wf = u[:, st[n]]
    for i in bp:                          # bp should be sorted
        wf = np.insert(wf, i, 0.)         # add boundary values,
    ax = fig.add_subplot(4, 3, n+1, aspect='equal')   # 4x3 plots
    plt.tripcolor(node[:,0], node[:,1], wf, shading='gouraud')
    plt.title(repr(st[n]+1)), plt.axis('off')

plt.show()
```

This program plots the wave function over a triangular mesh using the Matplotlib function `tripcolor()`, which fits our need perfectly with shading. The code requires the mesh data and eigenvectors from Program 9.7 stored in respective files. The boundary points stored in `bp` must be sorted if meshes generated from other sources are used. As it is, the first 12 states are graphed, but we can change the state list `st` to graph any states desired.

Program listing 9.9 Hexagon mesh generator (`meshhex.py`)

```
1  import numpy as np, fileio
2  
3  def mesh(a, N):                      # generate Hexagon mesh
4      node, elm = [], []               # nodes, elements
5      bp, ip, h = [], [], a/N          # boundary and internal nodes, size
6      M = (N*(3*N+1))//2               # last node before y=0 (center row)
7      K = M + 2*N +1                   # last node at end of center row
8      ndn = lambda i, j: j*N + (j*(j+1))//2 + i   # node number at i,j
9  
10     for j in range(N+1):             # go up till y=0
11         x, y = - 0.5*(a + j*h), (j*h - a)*np.sqrt(3.0)/2.0
12         for i in range(N+1+j):
13             node.append([x + i*h, y])
14             if (j != N):
15                 elm.append([ndn(i,j), ndn(i+1,j+1),ndn(i,j+1)])
16                 if (i != N+j):
17                     elm.append([ndn(i,j), ndn(i+1,j),ndn(i+1,j+1)])
18             if (j == 0 or i == 0 or i == N+j): bp.append(ndn(i,j))
19             else: ip.append(ndn(i,j))
20 
21     node, elm = np.array(node), np.array(elm)   # get y>0 by reflection
22     flip = np.column_stack((node[:M][:,0], -node[:M][:,1])) # c-stack
23     node = np.concatenate((node, flip))
24 
25     flip = elm[:,[1, 0, 2]]           # swap nodes so they are ccw
26     flip [ flip<M] += K               # add offset, excluding center row
27     elm = np.concatenate((elm, flip))
28 
29     bp = np.concatenate((bp, K + np.array(bp[:-2])))
30     ip = np.concatenate((ip, K + np.array(ip[:-(N+N-1)])))
31     fileio .writemesh('meshHexagon.txt', node, elm.tolist(), bp, ip)
32     print ('Hex mesh: %d nodes (%d/%d bndry/intrnl), %d elements'
33            %(len(node), len(bp), len(ip), len(elm)))
34 
35 a, N = 1.0, 10        # side length, num intervals
36 mesh(a, N)
```

We can generate a mesh similar to Figure 9.18 with this program. It starts from the bottom row ($y_j = 0$), going from left to right, and up to the center row ($y = 0$). The nodes above the center row are obtained by reflection.

Below $y = 0$, the nodes are enumerated sequentially. Assuming each side divided into N intervals, the node number at row y_j and column x_i (counting from left) is given by $n_{ji} = jN + \frac{1}{2}j(j+1) + i$. This is calculated by `ndn` defined as a `lambda` function. Unless at the end of a row, two elements are formed at each node (i, j) in CCW order, the elements above and to the right (note that the row above is effectively left-shifted by one). This is similar to Program 7.3. The boundary and internal nodes are also recorded. The process stops after the center row. The number of nodes below the center row is $M = \frac{1}{2}N * (3N + 1)$.

For the other half above $y > 0$, we use reflection to copy the nodes below, reversing the sign of y coordinates. This is efficiently done using NumPy array slicing and column stacking (line 22), which puts the same x-column and the negative y-column side-by-side to make a 2D array. The flipped nodes are added (concatenated) to the nodes for $y \leq 0$. The elements are also copied over,

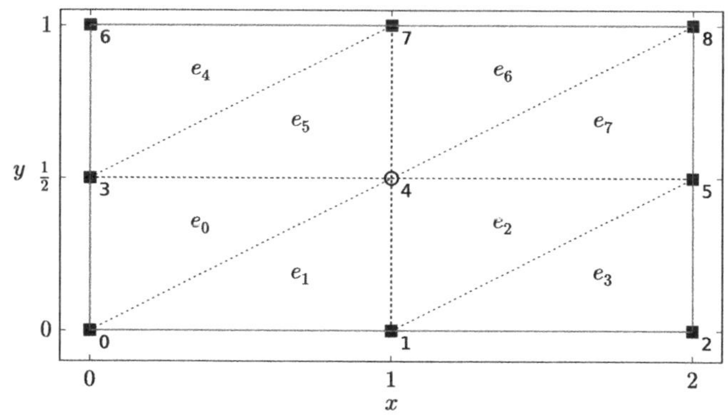

FIGURE 9.24 A sample mesh for a rectangular domain.

but nodes 0 and 1 are swapped so they are in CCW order (line 25). An offset is added to the copied nodes so their numbers start after the right-most node of the center row, $K = M + 2N + 1$. We used array mask, that is, truth array (line 26, see Section 1.D) to exclude the nodes on the center row. When finished, the total number of elements is $6N^2$.

Finally, the mesh data is written to a file containing, in order, the nodes, elements, boundary, and internal nodes. The `elm` array is converted to a list via `tolist` method for easy formating by the file writer. As an example, the format of the file for the mesh shown in Figure 9.24 is illustrated here, which can be read by `readmesh()`.

```
# this is a comment; blank lines start a new section
# each line can have an unlimited number of items, separated by commas

# nodes
[0.0, 0.0], [1.0, 0.0], [2.0, 0.0],
[0.0, 0.5], [1.0, 0.5], [2.0, 0.5],
[0.0, 1.0], [1.0, 1.0], [2.0, 1.0],

# elements
[0, 4, 3], [0, 1, 4], [1, 5, 4], [1, 2, 5],
[3, 7, 6], [3, 4, 7], [4, 8, 7], [4, 5, 8],

# boundary nodes
0, 1, 2, 3, 5, 6, 7, 8,

# internal nodes
4,
```

CHAPTER 10 Simple Random Problems

Many systems behave unpredictably and randomly, or at least seem to. Some are random due to a lack of information such as a coin toss or the golden autumn leaves fluttering and falling in the wind. Other systems, though perfectly deterministic and well defined, are still random because of their intrinsic, probabilistic nature, such as the radioactive particle decay and measurement of quantum systems.

We consider several simple random systems in this chapter to be familiar with random sampling and distributions needed in the next two chapters. First we introduce random numbers and model particle decay by random sampling. We also discuss random walks in 1D and 2D, as well as a stochastic model for Brownian motion, a simple but important phenomenon. Finally, we introduce Monte Carlo integration to evaluate multidimensional integrals including electrostatic potential energies.

10.1 RANDOM NUMBERS AND RADIOACTIVE DECAY

The first step to represent a random problem starts with the generation of random numbers. At initial glance, this appears paradoxical: On the one hand, the digital computer is as deterministic a machine as it gets, yet on the other hand we want it to produce random numbers. This is a dilemma. We cannot generate truly random numbers on the computer. We can, however, hope to generate a sequence of pseudorandom numbers that mimics the properties of a series of true random numbers.

The generation of pseudorandom numbers usually involves bits overflow in integer multiplication at the low level. For our purpose, Python has a random number library that provides several random distributions. For example, the `random()` function returns a uniform distribution in the interval $x \in [0, 1)$,

$$P(x)dx = \begin{cases} dx, & 0 \le x < 1, \\ 0, & \text{otherwise.} \end{cases} \tag{10.1}$$

We can repeat a random sequence by initializing it with the seed function `seed(n)`, where n is an identifier, typically an integer.

Truly random numbers have no correlation between them. Suppose we have a sequence of random numbers, $\{x\}$. Let us define a joint probability $P(x, x')$ as the probability of finding x' after x. No correlation means that the chance of finding x' must not depend on x before it. In other words,

$$P(x, x') = 1. \tag{10.2}$$

As a result, a correlation test between two random numbers is

$$\langle xx' \rangle = \int_0^1 \int_0^1 xx' P(x, x') dx dx' = \frac{1}{4}. \tag{10.3}$$

Similarly, the moment tests are defined as

$$\langle x^k \rangle = \int_0^1 x^k P(x) dx = \frac{1}{k+1}. \tag{10.4}$$

Table 10.1 Moments and Correlation of Pseudorandom Numbers Generated From `random()`.

Moment k	1	2	3	4	5	6
$\langle x^k \rangle$	0.49954	0.33262	0.24909	0.19896	0.16555	0.14170
Exact	0.5	0.33333	0.25	0.2	0.16666	0.14285
Correlation k	1	2	3	10	100	1000
$\langle x_i x_{i+k} \rangle$	0.24921	0.24986	0.24935	0.24988	0.24926	0.25019
Exact	0.25					

We can perform the tests numerically by generating a sequence of N random numbers $\{x_i\}$ and calculate the following averages

$$\langle x^k \rangle = \frac{1}{N}\sum_i x_i^k, \quad \text{(moment)}$$

$$\langle x_i x_{i+k} \rangle = \frac{1}{N}\sum_i x_i x_{i+k}. \quad \text{(correlation)} \tag{10.5}$$

Table 10.1 shows the results of moments and correlation of various order k. A million random numbers ($N = 10^6$) were drawn from `random()` in each case.[1] The values match their theoretical predictions to within the statistical uncertainty of $O(1/\sqrt{N})$.

Radioactive Nuclear Decay by Random Sampling

Radioactive nuclear decay is an intrinsically random process because it occurs via quantum transitions, which are probabilistic in nature as seen in Section 8.4. Let N be the number of nuclei alive at time t, and let $-\Delta N$ be the number of nuclei that will decay between t and $t + \Delta t$. Then, $-\Delta N$ will be proportional to the existing nuclei and the time step as

$$-\Delta N = \lambda N \Delta t,$$

$$-\frac{\Delta N}{N} = \lambda \Delta t \equiv p, \tag{10.6}$$

where λ is the decay constant related to the lifetime of the unstable nuclei as $\tau = 1/\lambda$ (Exercise E2.7).

We can interpret the LHS of Eq. (10.6) as the probability of decay in one step, given by p, a constant in terms of λ and Δt. To simulate the decay process, we draw a uniform random number x in $[0, 1)$ for each undecayed nucleus and compare it to p. If $x < p$, the particle decays; otherwise, the particle lives on. The code segment for each step is (import random as rnd)

```
deltaN = 0
for i in range(N):
    x = rnd.random()
    if (x < p):
        deltaN = deltaN+1
N = N - deltaN
```

We show in Figure 10.1 representative results for $p = 0.02$. The overall trend resembles an exponential decay law as expected. There are clear fluctuations in the number of decayed nuclei in each step (Figure 10.1, right). But their cumulative effect yields a relatively smooth curve of the

[1]NumPy also has a random library that is especially useful for generating arrays of random distributions. For example, see the use of `np.random.rand(N)` in Program 10.2.

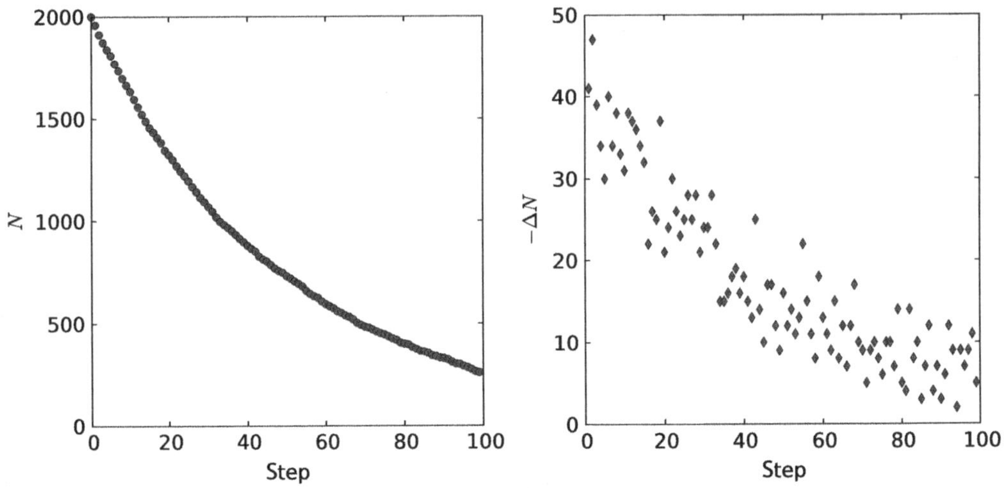

FIGURE 10.1 Nuclear decay by random sampling.

remaining particles (Figure 10.1, left), which is a Poisson distribution. The solution thus obtained by random sampling is an example of Monte Carlo simulations.

In our simulation, the step size is related to p and λ. In other words, the probability p implicitly determines the time scale of the decay process. If the rate of decay λ is large, Δt will be small and vice versa.

10.2 RANDOM WALK

Like radioactive decay, a random walk is another simple model defined in a straightforward way. A walker takes one unit step in one unit time, but the direction of each step is randomly distributed, uniformly or otherwise. Many practical applications are related to some features of random walks, including diffusive processes, Brownian motion (Section 10.3) and transport phenomena, and so on.

Random Walk in 1D

In one-dimensional random walks, there are only two possible directions, left or right. Let the unit step be l, so each step has a displacement of $\pm l$. If a walker starts from the origin, the position x, and its square x^2, after n steps are

$$x(n) = \sum_{i=1}^{n} s_i, \quad x^2(n) = \left(\sum_{i=1}^{n} s_i\right)^2, \tag{10.7}$$

where s_i is distributed according to

$$s_i = \begin{cases} +l, & \text{with probability } p, \\ -l, & \text{with probability } q, \end{cases} \tag{10.8}$$

subject to $p + q = 1$.

Figure 10.2 shows the results of ten separate walkers, each moving ten steps with $p = q = \frac{1}{2}$ and $l = 1$. The paths for different walkers show considerable variation (left panel). Because they move left or right with equal probabilities, the average position should be zero statistically. The solid line represents the average position of the walkers at each step number n. It fluctuates around zero with no clear trend. The average position squared (right panel) shows an upward trend, though there is fluctuation.

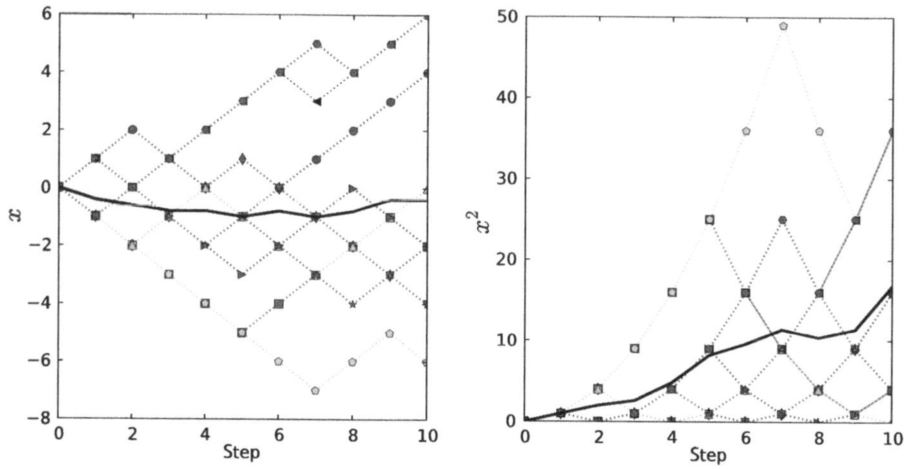

FIGURE 10.2 Random walks for 10 walkers moving 10 steps with equal probabilities $p = q = \frac{1}{2}$. The solid lines are the averages in each case.

Let N be the number of walkers, and $\langle x(n) \rangle$ and $\langle x^2(n) \rangle$ be the average values of $x(n)$ and $x^2(n)$ of the walkers after n steps, respectively. In the limit $N \gg 1$, the position follows a binomial distribution (Figure 10.3), and the distribution near the center resembles a Gaussian for large steps n (Exercise E10.3). The average values approach (see Exercise E10.4)

$$\langle x(n) \rangle = nl(p - q), \quad \langle x^2(n) \rangle = nl^2 \left(n - 4(n-1)pq\right). \tag{10.9}$$

As a measure of spread (dispersion), we can calculate the uncertainty as $\langle (\Delta x)^2 \rangle = \langle x^2(n) \rangle - \langle x(n) \rangle^2$. Given the expressions (10.9), we have

$$\langle (\Delta x)^2 \rangle = 4pql^2 n. \tag{10.10}$$

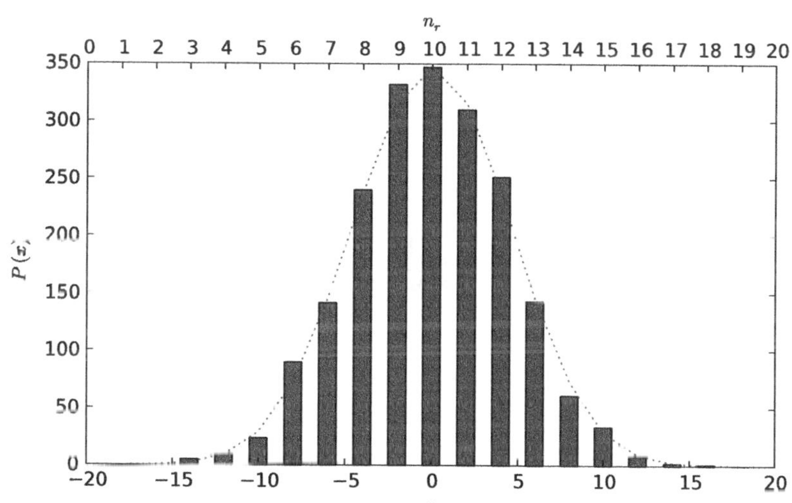

FIGURE 10.3 The distribution of position for 2000 walkers moving 20 steps each ($p = \frac{1}{2}$). The dotted line is the (unnormalized) binomial distribution. The upper scale n_r is the number of right steps.

The relationship is linear in step number n. It is an important result from random walks and is typical of random motion like Brownian motion. For continuous time, $t \propto n$, and the linear spread is $\sqrt{\langle(\Delta x)^2\rangle} = C\sqrt{t}$, C being some constant. For $p = q = \frac{1}{2}$, $\langle(\Delta x)^2\rangle = \langle x^2(n)\rangle$, so Figure 10.2 (right) shows the approximate linear trend. We can interpret the linear spread as the mean distance the walkers can travel in time t. To travel twice as far, we would have to wait four times as long.

Random Walk in 2D

We can generalize 1D random walks to 2D via the displacement vector $\vec{r} = x\hat{i} + y\hat{j}$. We need to determine x and y in some random fashion.

Unlike the 1D random walk, we have more freedom choosing the directions. There are many possibilities, including

- Random walk on a lattice. Choose with equal probability $\frac{1}{4}$ in the cardinal directions: left, right, up and down. Set the (x, y) pair accordingly as $(-1, 0)$, $(1, 0)$, $(0, 1)$, $(0, -1)$. One random number is needed. This seems to be the most direct way.
- Random directions. Choose a random angle θ in $[0, 2\pi]$, and set $x = l\cos\theta$ and $y = l\sin\theta$. Only one random number is needed.
- Variable step size. Choose x and y randomly in $[-l, l]$. Two random numbers are needed.

Figure 10.4 shows four walkers starting from the origin and moving with unit step size and uniform angular distribution (isotropic). The individual walks vary greatly as seen before in 1D random walks. We have to characterize an ensemble of walkers statistically. For instance, the average position and position squared are defined accordingly as ensemble averages,

$$\langle\vec{r}\rangle = \langle x\rangle\hat{i} + \langle y\rangle\hat{j}, \quad \langle r^2\rangle = \langle x^2\rangle + \langle y^2\rangle. \tag{10.11}$$

These averages will differ according to the chosen model. But, the general linear relationship (10.10) still holds, $\langle(\Delta r)^2\rangle \propto n$. We shall see an actual example with continuous time next in Brownian motion.

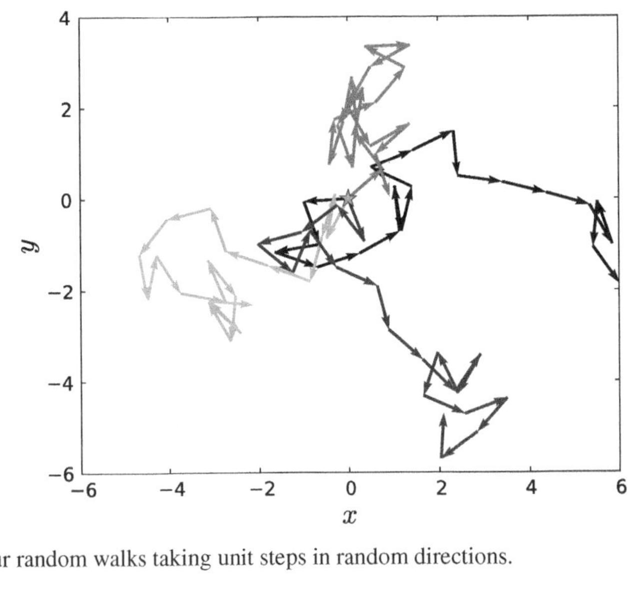

FIGURE 10.4 Four random walks taking unit steps in random directions.

10.3 BROWNIAN MOTION

Brownian motion refers to random motion of macroscopic particles such as fine dust grains in liquids observed under a microscope similar to Figure 10.4. It behaves in many respects like random walks, but is a real phenomenon. Brownian motion has great importance in physics and biophysical systems and often is the foundation toward understanding more complex phenomena in such systems. We model Brownian motion next.

10.3.1 A STOCHASTIC MODEL

Consider a particle suspended in a liquid and moving with velocity $\vec{v}$. Intuitively we would expect the particle to slow down due to drag (Section 3.2) and eventually to stop moving. But such particles observed in Brownian motion never stop moving. Rather, the motion carries on but has random fluctuations.

This means there has to be some random force, $F(t)$, besides drag. Because the speed is usually small, we can model the drag force as a viscous linear force (3.9), $-b_1\vec{v}$. The equation of motion including both forces is

$$m\frac{d\vec{v}}{dt} = -b_1\vec{v} + \vec{F}(t), \quad \text{(Langevin equation)} \tag{10.12}$$

where m is the mass, and b_1 the linear drag coefficient. This is known as the Langevin equation. It is just Newton's second law with random forces.

Both forces in Eq. (10.12) come from interactions with the same liquid, but why do we separate them? The reason has to do with different time scales. Roughly, we can think of the drag as representing some average force due to collective effects on a larger time scale and the random force due to collisions with individual atoms and molecules on a smaller time scale [76].

Clearly, $\vec{F}(t)$ must be rapidly fluctuating and very complex. Its precise form cannot be known, given the large number of atoms moving about. As a simplification, we model $\vec{F}(t)$ as a series of stochastic "kicks"

$$\vec{F}(t) = \sum_i \vec{F}_i\delta(t - t_i), \tag{10.13}$$

where $\vec{F}_i$ is the strength of the kick at t_i. Each kick is represented by a Dirac delta function $\delta(t - t_i)$ like in the kicked rotor (Figure S:5.1). The effect is to deliver an impulse to the particle after each kick. In general, both the strength $\vec{F}_i$ and the time t_i of the kicks are random.

Between kicks, there is only the drag force in Eq. (10.12), and the solutions can be generalized from the 1D results (3.12) and (3.13),

$$\vec{v} = \vec{v}_0\exp(-bt), \quad \vec{r} = \vec{r}_0 + \frac{\vec{v}_0}{b}[1 - \exp(-bt)], \tag{10.14}$$

where $b = b_1/m$, and $\vec{r}_0$ and $\vec{v}_0$ are the initial position and velocity. Note that $1/b$ has the dimension of time, so we can regard it as the larger time scale associated with drag stated earlier.

Let $\vec{v}_i$ be the velocity right before t_{i+1}. The velocity immediately after the kick at t_{i+1} is increased to

$$\vec{v}_{i+1} \stackrel{\epsilon\to 0}{=} \vec{v}_i + \int_{t_{i+1}-\epsilon}^{t_{i+1}+\epsilon}\left(-b\vec{v}_i + \frac{\vec{F}_{i+1}}{m}\delta(t - t_{i+1})\right)dt = \vec{v}_i + \frac{\vec{F}_{i+1}}{m}. \tag{10.15}$$

With the δ-kick model, Eq. (10.12) becomes a discrete linear map between kicks.

10.3.2 MODELING BROWNIAN MOTION

The model outlined earlier assumes arbitrary distributions for $\vec{F}_i$ and t_i. Now we just need to specify them to go forward. For simplicity, we assume $\vec{F}_i$ has a constant magnitude and random directions (this is not too crucial, see Project P10.2). For the distribution of t_i, it should be random, but we also expect that after a kick, the probability of not being kicked again should decrease with time. We choose to model the interval between kicks as an exponential distribution,

$$P(t) = \tau^{-1} e^{-t/\tau}. \tag{10.16}$$

This is a nonuniform distribution discussed in the next section (see also Section 10.B). In Eq. (10.16), τ is the average interval between kicks, $\tau = \int_0^\infty P(t)t\,dt$. It represents the smaller time scale of the rapidly fluctuating $\vec{F}(t)$. It should be small relative to $1/b$, that is, $\tau \ll 1/b$.

Here is our strategy for simulating Brownian motion. For each particle, sample the time interval t_i to the next kick according to Eq. (10.16), and randomly orient $\vec{F}_i$. Calculate the position and velocity from the kick-free motion (10.14) until t_i, then increment the velocity right after the kick from Eq. (10.15). Repeat the above steps.

Figure 10.5 shows snapshots from animation by Program 10.2 using this strategy. The time scales are $1/b = 10$ and $\tau = 2$ (arbitrary units). Note the different length scales between the two rows. All particles are "dropped" at the origin with zero initial velocity. In the beginning ($t \lesssim 10$), the size of the drop grows linearly with time, that is, $\sqrt{\langle r^2 \rangle} = Ct$. After that, the drop spreads at a slower square-root rate, $\sqrt{\langle r^2 \rangle} \propto \sqrt{t}$, the same as random walks (10.10).

The different power scalings are more clearly seen in Figure 10.6 showing the square of the distance (10.11), averaged over 1000 independent particles (an ensemble). We see two regimes: $t \lesssim t_1 = 10$ and $t \gtrsim t_2 = 30$, where the curve is roughly a straight line with different slopes (note the log-log scale). The slope is about 2 as expected below t_1, and almost 1 above t_2. Between t_1 and t_2, there is a transition region connecting the regimes. This different behavior at small and large time scales is predicted in a statistical model (see Eqs. (10.39) and (10.40), Section 10.A).

Physically, the small stochastic forces cause little effect over a short time period, and we expect the motion to be linear. We can see this from Eq. (10.14) as well, as $|\vec{r} - \vec{r}_0| \sim v_0 t$ for $bt \ll 1$. Over longer periods of time, the stochastic forces are more important because they provide the energy to the particle, so the motion resembles random walks with the same scaling (10.10).

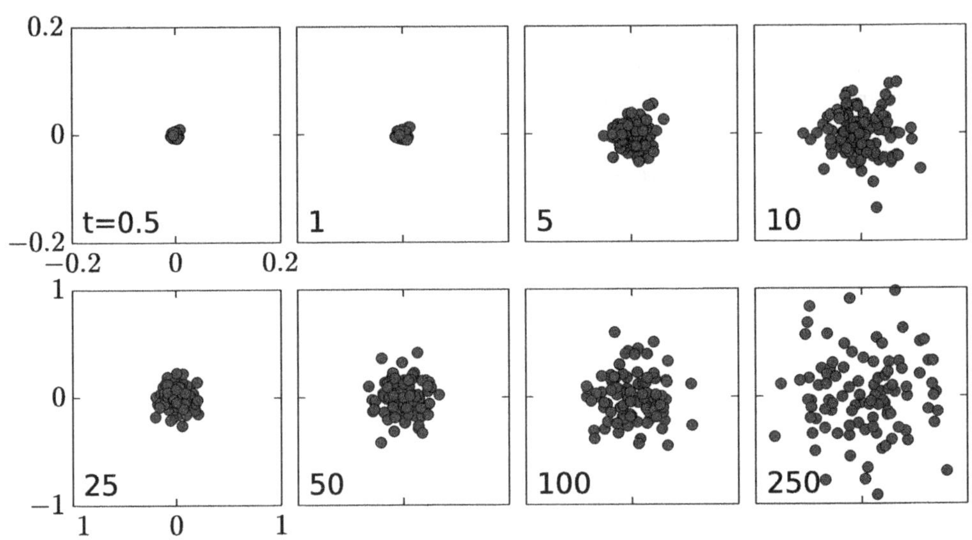

FIGURE 10.5 Snapshots of 2D Brownian motion.

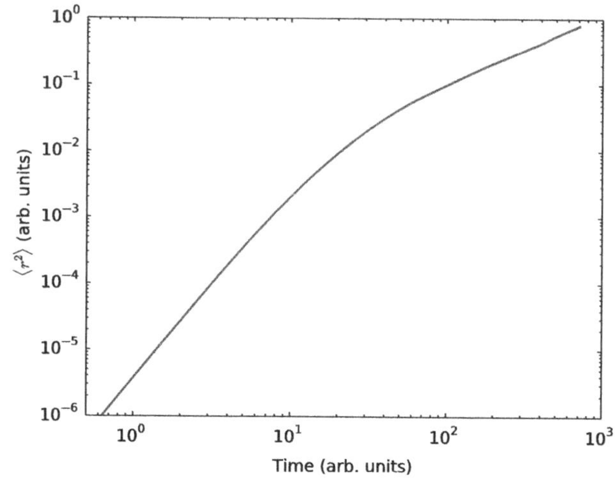

FIGURE 10.6 The average squared-distance of travel in Brownian motion.

We have determined the power laws for the distance of travel in Brownian motion. What about the constant C or the average speed? We suspect that they should depend on the strength and frequency of the kicks. Indeed they do, and they are related to temperature discussed in Chapter 11. Further investigation of Brownian motion is left to Project P10.2.

10.4 POTENTIAL ENERGY BY MONTE CARLO INTEGRATION

We have just described simulations of select simple problems with random numbers. It turns out that problem solving by random sampling is a powerful method and, for some complex problems, is the best method available. It falls into a category of the so-called Monte Carlo methods. Next we illustrate this method with an electrostatic problem.

Let us consider the electrostatic potential energy, U, between two charged spheres of radii r_1 and r_2, respectively, separated by a distance d between their centers on the x axis ($d > r_1 + r_2$, nonoverlapping). We can find U by

$$U = k \int_{V_1} \int_{V_2} \frac{\rho_1(x_1, y_1, z_1)\rho_2(x_2, y_2, z_2)\, dx_1 dy_1 dz_1 dx_2 dy_2 dz_2}{[(x_1 - x_2 - d)^2 + (y_1 - y_2)^2 + (z_1 - z_2)^2]^{1/2}}, \quad (10.17)$$

where k is the electrostatic constant in Eq. (7.1), ρ_1 and ρ_2 the charge densities in spheres 1 and 2, respectively. Equation (10.17) is a six-dimensional integral over the volumes of the spheres.

For arbitrary charge densities, Eq. (10.17) must be evaluated numerically (Project P10.3). We discussed grid-based numerical integration in Section 8.A. This approach becomes inefficient for higher multiple integrals like Eq. (10.17) because the number of grid points increases rapidly as a power of the dimensions. It can be expensive if this type of integrals must be calculated repeatedly during the course of a simulation.

Monte Carlo integration by random sampling can be useful in such cases. Its accuracy is independent of the dimensionality in principle. We discuss briefly this method next.

10.4.1 WEIGHTED AVERAGE METHOD

Let us consider the weighted average of a function $f(x)$ defined as

$$\langle f \rangle = \int_a^b f(x)P(x)dx, \quad (10.18)$$

where $P(x)$ is the weighting function. We assume it to be a normalized probability distribution, $\int_a^b P(x)dx = 1$.

We divide the space into small intervals as before (Figure 8.18) and convert Eq. (10.18) into a sum

$$\langle f \rangle \simeq \sum_j f(x_j)P(x_j)\Delta x. \tag{10.19}$$

If we randomly sample the points, $P(x_j)\Delta x$ tells us the probability with which the points should fall in the interval $\left[x_j, x_{j+1}\right]$. Let n_j be the number of points in that interval, and we have

$$P(x_j)\Delta x = \frac{n_j}{N}, \quad N = \sum_j n_j. \tag{10.20}$$

Substituting Eq. (10.20) into (10.19), we obtain

$$\langle f \rangle \simeq \frac{1}{N}\sum_j f(x_j)n_j = \frac{1}{N}\sum_{i=1}^{N} f(x_i). \quad (\{x_i\} \text{ distributed as } P(x)) \tag{10.21}$$

We have turned the group sum over j into an individual sum over i in Eq. (10.21). Once we have chosen a weighting function $P(x)$, we can compute the weighted average $\langle f \rangle$ and the integral (10.18). Let us choose a weighting function as

$$P(x) = \frac{1}{b-a}, \tag{10.22}$$

so that x is uniformly distributed in $[a, b]$. With this $P(x)$, Eq. (10.18) becomes

$$\langle f \rangle = \frac{1}{b-a}\int_a^b f(x)dx. \tag{10.23}$$

It follows that the integral is

$$I = \int_a^b f(x)dx = (b-a)\langle f \rangle_{mc}, \tag{10.24}$$

$$\langle f \rangle_{mc} = \frac{1}{N}\sum_{i=1}^{N} f(x_i). \quad (\{x_i\} \text{ uniformly distributed}) \tag{10.25}$$

The notation $\langle f \rangle_{mc}$ reminds us that the weighted average is to be calculated subject to Monte Carlo sampling, a uniform distribution in this case. We can generalize Eq. (10.25) to n-dimensional multiple integrals of the type

$$I = \int_{a_1}^{b_1} \cdots \int_{a_n}^{b_n} f(x_1, \ldots, x_n)dx_1 \cdots dx_n \simeq V\langle f \rangle_{mc}, \tag{10.26}$$

where V is the hypervolume over which the integration is to be done. Monte Carlo integration in this manner boils down to sampling the points uniformly within the volume of integration.

We apply Monte Carlo integration to the evaluation of the electrostatic potential energy U in Eq. (10.17). If we assume uniform charge densities, the result is known, $U = kQ_1Q_2/d$, where Q_1 and Q_2 are the charges on the spheres. This is the energy required to bring the two charges in from infinity. The Monte Carlo results are listed in Table 10.2, obtained from Program 10.3.

The main effort is to generate two points in a trial, one inside each sphere. This is done by `sample()`.

Table 10.2 The Electrostatic Potential Energy between Two Uniformly Charged Spheres of Radii $r_1 = 1$ and $r_2 = 2$ Separated by a Distance $d = 4$, in Units of kQ_1Q_2, by Monte Carlo Integration. The Exact Result is $\frac{1}{4}$.

N	100	200	400	1000	2000	4000	10000
MC	0.2387	0.2502	0.2464	0.2495	0.2502	0.2503	0.2503

```
def sample(r):        # return a random point inside a sphere
    while True:
        x, y, z = rnd.random(), rnd.random(), rnd.random()
        x, y, z = r*(x+x-1), r*(y+y-1), r*(z+z-1) # map to [-r, r]
        if (x*x + y*y + z*z <= r*r): break
    return x, y, z
```

The function samples points inside a sphere of radius r uniformly by the rejection method. A point at coordinates (x, y, z) is randomly selected within a cube of sides $2r$ containing the sphere. If the point is inside the sphere, it is selected and returned. Otherwise, it is rejected and a new point is generated until one is found inside. The main program simply evaluates the integrand at these points.

The Monte Carlo results are in satisfactory agreement with the exact value $\frac{1}{4}kQ_1Q_2$ for $d = 4$, even for N in the low hundreds. As N increases, the result stabilizes at the exact value. We expect the relative error to scale as $1/\sqrt{N}$, so at the largest $N = 10^4$ in Table 10.2, the error should be about 1%. We can easily control the relative error in Monte Carlo integration. Furthermore, we can improve the final result by cumulatively averaging different runs with proper weighting because they are independent.[2] Doing this with the seven runs in Table 10.2, we obtain 0.2501, the closest to the exact value yet. The overall convergence of the Monte Carlo approach is good with increasing N in this example. Its performance and simplicity cannot be matched by nested, grid-based integration.

10.4.2 IMPORTANCE SAMPLING

The integrand in the preceding example is monotonic and smooth. Like any numerical integration scheme, Monte Carlo integration suffers from ill-behaving integrands. The problem is particularly acute if the integrand is highly oscillatory or strongly peaked. Excessive oscillations cause large fluctuations in random sampling. Sharp peaks mean that most contributions to the integral come from small localized regions, making uniform sampling inefficient and often inadequate.

We can address the problem by the technique of importance sampling. Suppose we want to integrate $f(x)$ that is not smooth. We attempt to separate it into two parts as

$$\int_a^b f(x)dx = \int_a^b \frac{f(x)}{P(x)}P(x)dx = \int_a^b g(x)P(x)dx, \quad g(x) = \frac{f(x)}{P(x)}. \tag{10.27}$$

The new function $g(x)$ is assumed to be smoother than the original integrand $f(x)$. Like in Eq. (10.18), $P(x)$ is a probability distribution function, and Eq. (10.27) is the weighted average of $g(x)$ under $P(x)$. In other words, we have

$$\int_a^b f(x)dx = \langle g\rangle_{mc} = \frac{1}{N}\sum_i g(x_i) = \frac{1}{N}\sum_i \frac{f(x_i)}{P(x_i)}. \tag{10.28}$$

[2]This is true to the extent the pseudorandom numbers are uncorrelated, of course.

We stress that $\langle g \rangle_{mc}$ in Eq. (10.28) is calculated with x_i distributed according to probability $P(x)$. This is the basic idea of importance sampling. For a given integrand $f(x)$, we should choose $P(x)$ such that the modified integrand $g(x)$ becomes as smooth as possible. The distribution $P(x)$ guides us to sample the important regions. The integration volume does not appear explicitly in Eq. (10.28). It is implicitly taken into account by the sampling process. Of course, if we choose a uniform probability distribution as in Eq. (10.22), then Eq. (10.28) is reduced to (10.25).

As an example, let us evaluate the following integral

$$\int_a^b f(x)dx = \int_0^\infty \cos(x)e^{-x}dx = \frac{1}{2}. \tag{10.29}$$

The integrand is strongly peaked at the origin, where we expect most contributions to come from. At the same time, it also oscillates. But relatively speaking, the more pressing problem is to deal with the sharp exponential factor, not the cosine oscillation. Therefore, we choose an exponential probability function as $P(x) = \exp(-x)$, which is already normalized.

The modified integrand is $g(x) = f(x)/P(x) = \cos(x)$. The integral (10.29) becomes

$$\int_0^\infty \cos(x)e^{-x}dx = \int_0^\infty g(x)P(x)dx = \langle \cos(x) \rangle_{mc}. \tag{10.30}$$

The value $\langle \cos(x) \rangle_{mc}$ is to be calculated with the exponential probability distribution. The following code illustrates the procedure.

```
f, N = 0., 1000
for i in range(N):
    x = - np.log(rnd.random())      # exponential distribution
    f += np.cos(x)

print ('MC=', f/N)
```

One trial run gave a result 0.517, or about 3% error, consistent with the number of sampling points N. Without importance sampling, the error would be unacceptably large, even for a reasonably large N, as to make the result meaningless (see Exercise E10.7).

The key to the preceding code is the statement $x = -\ln(\text{random}())$, which transforms the logarithm of uniform random numbers in [0, 1) into a series satisfying the exponential distribution in $(0, \infty)$. It ensures that the small-x region, where the most contribution comes from, is sampled more frequently. We discuss this and general nonuniform sampling in Section 10.B.

CHAPTER SUMMARY

We discussed a few simple random problems, including radioactive particle decay by random sampling, and random walks in 1D and 2D. We constructed a stochastic model for Brownian motion and compared simulation results with a statistical treatment. We introduced Monte Carlo integration and importance sampling methods for obtaining nonuniform distributions.

The basic usage of the intrinsic Python random library was illustrated through the selected problems.

10.5 EXERCISES AND PROJECTS

Exercises

E10.1 (a) Test the uniformness of random number generation. Generate $N = 10^4$ samples, and plot the distribution as a histogram of 100 bins (see Program S:9.1).

(b) Write a program to calculate the moments and correlation listed in Table 10.1.

E10.2 (a) Suppose someone prefers white socks to red socks. On each shopping trip, he opens packages—sealed black boxes each containing a pair of white or red socks, until he finds a pair of white socks. Then he stops and must buy all the opened packages. After many such trips, will he have an oversupply of white socks relative to red ones? Prove it one way or the other by random sampling.

(b) The SOS game (stick or switch, similar to the Monty Hall game show *Let's Make a Deal*) is a game of chance and works as follows. The host hides a gold coin in one of four boxes. You pick a box first, and the host opens an empty box. You decide whether to stick with the box you picked, or switch to one of the two remaining boxes. Which way gives you a better chance of winning?

Simulate the SOS game by random sampling. Tabulate the results for N trials. Choose a large enough N so that you can draw definitive conclusions.

Assume on average half of the players switch and the other half stick. How much should the host charge the players per game to break even?

E10.3 (a) Generate random walk distributions analogous to Figure 10.3 for various p values, for example, 0.3, 0.6, 0.9, for example. Use enough walkers so the fluctuations are small. Show that as the maximum step number n_{max} increases, the distribution becomes sharper and more Gaussian-like.

(b) Two random walkers start at the same position. What is the probability that they will meet again after walking 20 steps each? Assume equal probability going left or right.

Simulate the process numerically. Use enough trials such that the event occurs about 100 times. How does the numerical result compare?

E10.4 Prove the relations (10.9). For $\langle x^2(n)\rangle$, group the terms as

$$x^2(n) = \sum_{i=1}^{n} s_i \sum_{j=1}^{n} s_j = \sum_{i=1}^{n} s_i^2 + \sum_{i\neq j}^{n} s_i s_j.$$

Compute the averages of the two terms on the right-hand side (RHS). The first term is independent of direction, and the double sum in the second term can be evaluated independently.

E10.5 Calculate the spread in 2D random walks

$$\langle(\Delta r)^2\rangle = \langle r^2\rangle - |\langle\vec{r}\rangle|^2,$$

for the following models (see Section 10.2): (a) random walks on a lattice; and (b) random directions.

For part (b), introduce $z = \sum z_j$, $z_j = x_j + iy_j = \exp(i\theta_j)$, and calculate $\langle r^2\rangle = \langle |z|^2\rangle$. Separate the equal and unequal indices in the sums as in Exercise E10.4.

E10.6 (a) Consider the n-fold integral

$$I(m,n) = \int_0^1 \cdots \int_0^1 (x_1 + x_2 + \cdots + x_n)^m dx_1 \cdots dx_n.$$

With the parametric identity

$$I(m,n) = \lim_{\lambda\to 0} \frac{\partial^m}{\partial\lambda^m} \int_0^1 \cdots \int_0^1 e^{\lambda(x_1+x_2+\cdots+x_n)} dx_1 \cdots dx_n,$$

show that

$$I(m,n) = \lim_{\lambda\to 0} \frac{\partial^m}{\partial\lambda^m} \left(\frac{e^\lambda - 1}{\lambda}\right)^n.$$

(b) Calculate $I(m, n)$ using Monte Carlo integration. First choose a fixed $m = 3$ and $n = 10$, and vary the number of sampling points from $N = 10$ to 10^6 by factors of 10. Compare the Monte Carlo and analytic results. Plot the results on a semilog N scale, and discuss convergence.

Next, fix N, for example, $N = 10^5$, and calculate the integral for $n = 10$ and $m = 0$ to 5. What do you expect for $n = 20$? Try it.

E10.7 Calculate the integral (10.29) by Monte Carlo integration by uniform sampling x in $[0, 8]$. Compare the accuracy with the exponential distribution by importance sampling for the same N (Section 10.4).

E10.8 (a) Show that the transform for the sine angular distribution is given by

$$p_\theta(\theta) = \frac{1}{2}\sin\theta, \quad \theta = \arccos(2x - 1), \quad 0 \le \theta < \pi,$$

and for the Lorentzian distribution by

$$p_y(y) = \frac{2}{\pi}\frac{a}{a^2 + y^2}, \quad y = a\tan\left(\frac{\pi}{2}x\right), \quad 0 \le y < \infty.$$

(b) Generate these distributions, plot them as histograms (see Program S:9.1). Use enough samplings so the fluctuations are small. Compare with the exact results.

(c) A Gaussian distribution can be generated using a pair of random variables as

$$y_1 = \sqrt{-2\ln x_1}\cos(2\pi x_2), \quad y_2 = \sqrt{-2\ln x_1}\sin(2\pi x_2).$$

This is known as the Box-Muller method. Because x_1 and x_2 are independent, the pair y_1 and y_2 are also independent, and both values can be used. Produce a histogram as shown earlier, and verify that it yields a Gaussian distribution. Find the standard deviation.

Projects

P10.1 Simulate 1D random walks. Construct a program that calculates individual path of a single walker, as well as the averages $\langle x(n)\rangle$ and $\langle x^2(n)\rangle$ for N walkers as a function of step number n.

Write a standalone subroutine `walknsteps()` as the workhorse. It should be similar to the function of the same name in Program 10.1. The function should assume arbitrary step size and probability p. At each step, draw a random number. If it is less than p, the walk is to the right, otherwise it is to the left. The main program should call `walknsteps` N times for a given maximum step number n_{max}. Compute the averages $\langle x(n)\rangle$ and $\langle x^2(n)\rangle$ after N walks.

(a) Assume unit step size $l = 1$, and choose $n_{max} \sim 50$ and $N \sim 100$. Vary p values, for example, 0.2, 0.5, 0.9, etc. Predict and sketch $\langle x(n)\rangle$ and $\langle x^2(n)$ in each case. Plot and discuss the actual results.

(b)* Modify `walknsteps()` to allow the step size l to vary between some minimum and maximum limits, $l \in [a, b]$. You can draw a random l in this range using `uniform(a,b)`, which is equivalent to $a + (b - a) \times r$, where r is a random number between $[0, 1]$. Set reasonable limits, for example, $[0, 2]$. Plot the results, and discuss differences and similarities with the preceding results.

P10.2 Let us investigate Brownian motion in this project. First explore Program 10.2 with animation on, change some parameters such as F, b, etc., but keep $\tau < 1/b$.

(a) Modify the program to calculate $\langle r^2\rangle$. You can disable animation. Add a function to the Brownian-motion class that returns $\langle r^2\rangle$ as a function of time. The average $\langle r^2\rangle$ at an instant can be obtained from

```
r2 = np.sum(self.r* self.r)/ self.N
```

The number of particles N should be 1000 or more to obtain good statistics. Append this to a list, and plot it as shown in Figure 10.6.

(b) Obtain $\langle v^2 \rangle$ in a similar way, and plot it as a function of time. Discuss the results. What is the temperature kT from the equipartition theorem (11.41) (set $m = 1$)?

(c) Reduce τ by a factor of two, and repeat the calculations for $\langle r^2 \rangle$ and $\langle v^2 \rangle$. Do the same with the kick strength. Discuss the results. Is the temperature higher or lower in each case? Why?

(d) Change the exponential distribution for t_i to a uniform distribution between 0 and 2τ, and repeat the calculations. Also, instead of a constant F, assume it follows a Gaussian distribution (see Eq. (S:10.4)). Sample it using `rnd.gauss()` in Python (see also Exercise E10.8), with a mean and standard deviation equal to the default F value. Discuss how these results differ from earlier ones.

(e)* Generalize the program to the 3D case. The position and velocity should now be $N \times 3$ arrays. The force $\vec{F}$ should be distributed uniformly in all directions. One way is, for a given magnitude F, to sample the angle φ uniformly in $[0, 2\pi]$, and then sample θ as a sine distribution in Exercise E10.8. Another method is to randomize $\vec{F}$ by an Euler rotation (see Eq. (S:12.125)). Compute $\langle r^2 \rangle$ and $\langle v^2 \rangle$ and compare with 2D results. Optionally visualize 3D Brownian motion using VPython.

P10.3 Solve the following problems by Monte Carlo integration. In each case, obtain the results to three digits of accuracy. Assume arbitrary units.

(a) Calculate the center of mass of a hemisphere of radius r with uniform mass density. Compare with the analytic result.

(b) Two spheres of radii r_1 and r_2 are separated by a distance d between the centers. Each carries a unit charge. The charge densities obey a linear relationship of the form $\rho_i = C_i(1 - r/r_i)$, with $i = 1$ and 2, and C_i are constants.

Assume $r_1 = 1$ and $r_2 = 2$. Determine the constants C_1 and C_2 analytically or numerically via Monte Carlo integration. Find the electrostatic potential energy between the spheres for $d = 4$. Discuss and compare your results with two uniformly charged spheres (Table 10.2).

(c) Consider the same setup as in part (b), but the charge densities have dipole angular distributions (Figure 7.24) $\rho_i = C_i \sin\theta_i$, where θ_i is the polar angle relative to central axis connecting the spheres. Predict the energy of this configuration relative to part (b), as well as to the uniformly charged spheres. Calculate the actual result, and compare the three systems.

P10.4 The hit-or-miss method can be used to find an area or an integral.

(a) Find the approximate value of π by this method. Imagine a unit circle is enclosed in a square board of side length 2, and we throw N darts randomly at the board. Afterward we inspect the board and count the number of hits inside the circle that is proportional to the area. We can calculate π as

$$\frac{\text{area of unit circle}}{\text{area of board}} = \frac{\pi}{4} \simeq \frac{\text{hits}}{\text{throws}}, \qquad \text{or } \pi = 4\,\frac{\text{hits}}{N}.$$

Simulate the dart game. For each throw, generate two random numbers x and y between $[-1, 1]$, and record the number of hits within the unit circle. Vary N, and calculate π. How many throws are needed to obtain π to two digits of accuracy consistently? How many throws to match the accuracy of 22/7? How about 355/113?[3]

(b) Design a Monte Carlo program to calculate the integral $\int_a^b f(x)dx$ by the hit-or-miss method (see Section 10.B). Test your code with $\int_0^\pi \sin x\, dx$, and compare the accuracy with the weighted average method.

[3] These were the ratios used to approximate π by the fifth-century Chinese mathematician Zu Chongzhi.

10.A STATISTICAL THEORY OF BROWNIAN MOTION

We briefly describe a statistical method for Brownian motion where the unknown force $\vec{F}(t)$ in the Langevin equation (10.12) is absorbed in the thermal properties of a system. We restrict ourselves to 1D motion.

Let us multiply both sides of Eq. (10.12) by x and divide by m, obtaining

$$x\frac{dv}{dt} = -bxv + \frac{1}{m}xF(t), \tag{10.31}$$

where $b = b_1/m$ as before. For convenience, we introduce $s = xv$, so that

$$s = x\frac{dx}{dt} = \frac{1}{2}\frac{dx^2}{dt}, \quad \frac{ds}{dt} = v^2 + x\frac{dv}{dt}. \tag{10.32}$$

Substituting s and $x\,dv/dt$ of Eq. (10.32) into the RHS and LHS of Eq. (10.31), respectively, we have

$$\frac{ds}{dt} = v^2 - bs + \frac{1}{m}xF. \tag{10.33}$$

To eliminate $F(t)$, let us take statistical (ensemble) averages on both sides of Eq. (10.33) to obtain

$$\frac{d\bar{s}}{dt} = \langle v^2\rangle - b\bar{s} + \frac{1}{m}\langle xF\rangle, \tag{10.34}$$

with $\bar{s} = \langle s\rangle$.

Because x and F are not correlated, the last term in Eq. (10.34) is zero because $\langle xF\rangle = \langle x\rangle\langle F\rangle = 0$. In thermal equilibrium, the first term can be obtained from the so-called equipartition theorem (11.41), which in 1D is

$$\frac{1}{2}m\langle v^2\rangle = \frac{1}{2}kT, \tag{10.35}$$

where k is the Boltzmann constant and T the temperature.

Using Eq. (10.35), we can simplify Eq. (10.34) to

$$\frac{d\bar{s}}{dt} = \frac{kT}{m} - b\bar{s}. \tag{10.36}$$

It yields the following solution, assuming initially $\bar{s} = 0$ at $t = 0$,

$$\bar{s} = \frac{kT}{bm}\left(1 - e^{-bt}\right). \tag{10.37}$$

Recalling $\bar{s} = \frac{1}{2}d\langle x^2\rangle/dt$ from Eq. (10.32), we can obtain $\langle x^2\rangle$ as

$$\langle x^2\rangle = 2\int_0^t \bar{s}\,dt = \frac{2kT}{bm}\left[t - \left(1 - e^{-bt}\right)/b\right], \tag{10.38}$$

subject to the initial condition $\langle x^2\rangle = 0$ at $t = 0$. This is the central result.

It seems that we have managed to eliminate the rapidly fluctuating force $F(t)$ in Eq. (10.38). But, its effect is hidden in temperature T. The strength (impulse) and frequency of $F(t)$ clearly depend on the temperature.

Still, the solution lets us obtain the limiting behavior of $\langle x^2\rangle$ at different time scales. For small $t \ll 1/b$, $\exp(-bt) \simeq 1 - bt + \frac{1}{2}b^2t^2$, and we have

$$\langle x^2\rangle \simeq \frac{kT}{m}t^2. \quad \text{(short time)} \tag{10.39}$$

For large $t \gg 1/b$, we can drop the second term in the bracket of Eq. (10.38) all together, so that

$$\langle x^2 \rangle \simeq \frac{2kT}{bm} t. \quad \text{(large time)} \tag{10.40}$$

This linear scaling is the same as in random walks (10.10) and other diffusive processes.

10.B NONUNIFORM DISTRIBUTIONS

The need for nonuniform probability distributions arises frequently in practical situations as demonstrated by several examples we have discussed so far. A number drawn from a nonuniform distribution is still random, but after many such drawings, the distribution becomes skewed. The general approach is to generate a nonuniform distribution from a uniform one.

Let us denote the nonuniform variable by y, and the uniform variable by x. We wish to generate a certain distribution $p_y(y)$ via a transformation of x. We will discuss two methods next.

Inverse Transform Method

In certain cases, it is possible to obtain a nonuniform distribution analytically via the inverse transform method. The basic premise is that there exists a transformation, say $y = G(x)$, which transforms the uniform distribution $p_x(x)$ to the nonuniform distribution $p_y(y)$.

Because probability is conserved, we have the following relationship

$$|p_y(y)dy| = |p_x(x)dx|. \tag{10.41}$$

The LHS of Eq. (10.41) gives the probability in the y-domain from y to $y + dy$. Because it is generated from x to $x + dx$ one-to-one by $G(x)$, it must be equal to the probability in the x-domain – the RHS of Eq. (10.41).

Because $p_x(x) = 1$ from Eq. (10.1), we can integrate Eq. (10.41) on both sides to obtain

$$\int p_y(y)dy = \int dx, \quad \text{or } F(y) \equiv \int p_y(y)dy, \quad F(y) = x. \tag{10.42}$$

In obtaining Eq. (10.42) we have dropped the || signs for convenience.

To the extent that the integral in Eq. (10.42) is analytically solvable for $F(y)$, and subsequently, the inverse of $F(y)$ exists, then we have found the transform

$$y = F^{-1}(x) = G(x), \tag{10.43}$$

where $G = F^{-1}$ is the inverse function of F.

As an example, let us obtain the exponential (Poisson) distribution

$$p_y(y) = \lambda^{-1} e^{-y/\lambda}. \tag{10.44}$$

Though this is also a standard distribution, we choose it for the simplicity. We find the integral from Eq. (10.42) as

$$F(y) = \left| \int \lambda^{-1} e^{-y/\lambda} dy \right| = e^{-y/\lambda}, \tag{10.45}$$

from which we can obtain the inverse transform

$$e^{-y/\lambda} = x, \quad y = -\lambda \ln x, \quad 0 < y < \infty. \tag{10.46}$$

This shows that the logarithm of a uniform distribution produces an exponential distribution. The two functions are just the inverse of each other in this case.

We have used this distribution in the exponential integral (10.29) in Section 10.4 with $\lambda = 1$, which controls the mean. Basically, the $-\ln x$ function compresses $x \sim 1$ toward $y \sim 0$ and stretches $x \sim 0$ toward $y \sim \infty$, so the resulting distribution is an exponential, like radioactive nuclear decay (Figure 10.1). More examples are given in Exercise E11.8.

Rejection Method

Often, it is not possible to find the inverse function analytically. For example, one of the most useful distributions, the Gaussian distribution, cannot be generated by the inverse transform method (though it is possible using two variables, see Exercise E10.8). Sometimes the distribution itself cannot be expressed as a simple, compact function. Rather, it may be in the form of discrete data like a lookup table. In such cases, we can use the rejection method, which always works.

The idea is similar to the hit-or-miss method (Project P10.4). Let H be greater or equal to the maximum value of $p_y(y)$ in the range $y \in [a, b]$. We put a box stretching from the lower-left corner $[a, 0]$ to the upper-right corner $[b, H]$. The sampling procedure is as follows (as usual, x is uniform in $[0, 1)$):

- Generate y randomly in $[a, b]$ by $y = a + (b - a)x$.
- Generate p randomly in $[0, H]$ by $p = Hx$.
- If $p < p_y(y)$, accept and return y.
- Otherwise, reject y, repeat the first step.

Each successful sampling requires at least two random numbers x. After many samplings, the number of y values accepted is proportional to $p_y(y)$, the desired distribution. The rejection method also requires the values of $p_y(y)$, in either functional or numerical form and its upper limit H. The closer H is to the actual maximum of $p_y(y)$, the more efficient the rejection method. The efficiency of the rejection method can be improved if a better comparison function closer to the actual $p_y(y)$, rather than the box, is suitably constructed.

10.C PROGRAM LISTINGS AND DESCRIPTIONS

Program listing 10.1 Random walk in 2D (`walk2d.py`)

```
import numpy as np, random as rnd
import matplotlib.pyplot as plt

def walknsteps(n):                              # walk n steps
    x, y = np.zeros(n+1), np.zeros(n+1)
    for i in range(n):
        phi = rnd.random()*2*np.pi              # random in [0, 2π]
        x[i+1] = x[i] + np.cos(phi)
        y[i+1] = y[i] + np.sin(phi)
    return x, y

nstep, N = 20, 4                                # steps, num of walkers
col = ['k', 'r', 'g', 'b', 'c', 'm']           # color codes
plt.subplot(111, aspect='equal')
```

```
for i in range(N):
    x, y = walknsteps(nstep)
    plt.quiver(x[:-1], y[:-1], x[1:]-x[:-1], y[1:]-y[:-1], scale=1,
               scale_units='xy', angles='xy', color=col[i%len(col)])
plt.plot([0], [0], 'y*', markersize=16)
plt.xlabel('x'), plt.ylabel('y')
plt.show()
```

The function `walknsteps()` simulates one walker taking n steps of unit length, in random directions between $[0, 2\pi]$, starting from the origin. The main program plots the paths of N walkers with `quiver()`, which connects the path by arrows having the same components as the steps.

Program listing 10.2 Brownian motion (`brownian.py`)

```
import numpy as np, random as rnd
import matplotlib.pyplot as plt
import matplotlib.animation as am

class BrownianMotion:                   # Brownian motion class
    def __init__(self, N=400, F=0.005, b=0.1, tau=2., h = 0.5):
        self.N, self.F, self.b, self.tau, self.h = N, F, b, tau, h
        self.r, self.v = np.zeros((N, 2)), np.zeros((N, 2))
        self.t = -np.log(np.random.rand(N))*tau   # initial kick times

    def move(self, r, v, dt):           # move between kicks
        edt = np.exp(-self.b*dt)
        return r + v*(1-edt)/self.b, v*edt

    def iterate(self):                  # advance one step
        r, v, t, h, F = self.r, self.v, self.t, self.h, self.F  # alias
        for i in range(self.N):
            if t[i] > h:                # no kick within current step
                dt = h                  # dt= time to end of step
            else:                       # suffers kicks before end of step
                tot, dt = 0., t[i]      # tot=time to last kick
                while t[i] <= h:
                    r[i], v[i] = self.move(r[i], v[i], dt)
                    phi = rnd.random()*2*np.pi           # apply kick
                    v[i] += [F*np.cos(phi), F*np.sin(phi)]
                    tot += dt
                    dt = -np.log(rnd.random())*self.tau     # sample dt
                    t[i] += dt          # next kick
                dt = h - tot            # dt= to end of current step
            r[i], v[i] = self.move(r[i], v[i], dt)  # no kick, just move
            t[i] -= h

def updatefig(*args):                       # update figure data
    bm.iterate()
    plot.set_data(bm.r[:,0], bm.r[:,1])     # update data
    return [plot]                           # return plot object

```

```
   bm = BrownianMotion()                       # create Brownian model
39 fig = plt.figure()
   plt.subplot(111, aspect='equal')
41 plot = plt.plot(bm.r[:,0], bm.r[:,1], 'o')[0]        # create plot object
   ani = am.FuncAnimation(fig, updatefig, interval=10, blit=True) # animate
43 plt.xlim(-1., 1.), plt.ylim(-1., 1.), plt.show()
```

Program 10.2 simulates Brownian motion according to the stochastic model in Section 10.3. We use object-oriented programming in this case because the model is self-contained, and there is little interaction with the main program. We define a Brownian-motion class consisting of an ensemble of N independent particles. The `__init__()` function is called when an object is created. In this case, it initializes the default parameters, sets the initial position and velocity vectors to zero, and samples the times of the initial kicks using the NumPy function `np.random.rand(N)` (line 9), which generates an array of N random numbers at once. The function `move()` calculates the position and velocity under the drag force only between kicks from Eq. (10.14).

The next module, `iterate()`, advances the solution forward by one step h. For each particle in the ensemble, we check whether it receives any kick within this step. That information is stored in `t[i]` (t_i), which indicates the time of its next kick from the start of the current step. If $t_i > h$, it means there is no kick within the current step. So Δt is set to h, the time to the end of the current step, and the particle will be moved according to Eq. (10.14) after the `if-else` statement.

On the other hand, if $t_i \le h$, there will be at least one kick. So we move the particle to obtain $\vec{r}_i$ and $\vec{v}_i$ just before t_i with `move()`. Then we sample a randomly oriented force $\vec{F}$ and increase the velocity by the impulse received (10.15). The variable `tot` keeps track of the time elapsed from the beginning of the step to the last kick. Now we sample from an exponential distribution (10.46) the interval to the next kick (line 27) and add it to t_i. If this t_i is still within the current step, we repeat the preceding treatment, until the next sampled kick is after the current step. We then set Δt to the end of the current step, which is the difference between the step size h and the time of the last kick `tot`.

After the `if-else` block, the particle is moved to the end of the current step (start of the next step), and t_i is adjusted accordingly.

The `updatefig()` function calls the `iterate()` method of a given object to update the x and y positions of the particles for animation. The plot object is returned as a list (line 36) required by the animation function.

The main program makes an instance of Brownian-motion objects `bm` with default parameters. It also creates a plot object for Matplotlib animation using `updatefig()` at regular intervals (10 ms) to update the figure as in Program 8.1. The `blit` option makes drawing faster.

Because `updatefig()` neither receives nor returns any data variable, the object-oriented approach is more convenient here because all data and computation (methods) are contained within the Brownian-motion object.

Program listing 10.3 Electrostatic energy (`mcint.py`)

```
import numpy as np, random as rnd

def sample(r):        # return a random point inside a sphere
    while True:
        x, y, z = rnd.random(), rnd.random(), rnd.random()
        x, y, z = r*(x+x-1), r*(y+y-1), r*(z+z-1) # map to [-r, r]
        if (x*x + y*y + z*z <= r*r): break
    return x, y, z
```

```

r1, r2, d = 1.0, 2.0, 4.0              # radii and separation, d>r1+r2

f, V, N = 0., (4*np.pi/3.)**2*(r1*r2)**3, 100
for i in range(N):
    x1, y1, z1 = sample(r1)
    x2, y2, z2 = sample(r2)
    f += 1./np.sqrt((x1-x2-d)**2+(y1-y2)*(y1-y2)+(z1-z2)*(z1-z2))

print ('MC=', V*f/N, 'exact=', V/d)
```

This program calculates the electrostatic potential energy between two spheres of uniform charge densities ($\rho_1 = \rho_2 = 1$, and electrostatic constant $k = 1$ in Eq. (10.17)). The function `sample()` returns a uniformly distributed point inside a sphere by the rejection method as explained in the text. The main program draws N points inside the spheres and computes the weighted average according to Eq. (10.17). The final result is multiplied by the combined volumes of the spheres.

CHAPTER 11 Thermal Systems

Up to the last chapter, we had built simulations from first principles, that is, calculations were done from Newton's laws or the Schrödinger equation. Thermal systems, however, require a different, statistical approach because they are made up of large numbers of atoms and molecules. Macroscopic thermal properties are determined by the microscopic interactions between these particles. Given that the number of particles is at least on the order of the Avogadro number $\sim 10^{23}$, we cannot hope to simulate thermal systems using first-principle calculations by tracking all the individual particles.[1] Nor would we want to. Even if their individual properties such as the energies or velocities were available, we would not be able to make sense out of that many particles without statistics. It turns out that we can model much smaller thermal systems and still simulate the statistical properties of large systems, provided we sample them correctly and be mindful of the limitations.

We begin with thermodynamics of equilibrium, exploring the role of energy sharing, entropy, and temperature as the driving forces toward equilibrium: the Boltzmann distribution. We then introduce the Metropolis algorithm and apply it to study 1D and 2D Ising models, carefully comparing numerical and analytic solutions where possible, including phase transitions. After fully assessing the Metropolis method, we extend it to the study of nonthermal systems such as the hanging tablecloth via simulated annealing. Finally, we investigate the thermalization of N-body atomic systems and kinetic gas theory by molecular dynamics, connecting first-principle calculations to thermal properties.

11.1 THERMODYNAMICS OF EQUILIBRIUM

The central idea in statistical thermodynamics is the Boltzmann distribution. It describes the energy distribution of a thermal system and is the most powerful tool we have in statistical mechanics. In this section we explore the equilibrium process and show, through numerical (Monte Carlo) experiments, that Boltzmann statistics arises naturally in thermal equilibrium as a matter of overwhelming probability.

11.1.1 TOWARD THERMAL EQUILIBRIUM

To understand probabilities of states, we begin with counting of states at the microscopic level. Imagine a thermal system made up of microscopic particles that can store energy internally in translational motion, rotation, or vibration. The specific mechanism is unimportant. For example, consider the Einstein solid model, one of the few analytically solvable problems. The system consists of N independent quantum simple harmonic oscillators (SHO). Each SHO has an infinite number of equidistant states (see Eq. (9.32)) and can store an unlimited quanta of energy. Suppose the system has a total energy of $q\epsilon$ with ϵ as the basic unit of energy. The energy can be shared in many ways among the N oscillators. A microstate refers to a particular distribution of energy.

At the macroscopic level, we are interested in the net, externally distinguishable effects, called macrostates. Consider two microstates in the Einstein solid example: one in which a single oscillator has all q units of energy and the rest $N-1$ oscillators have none, and another in which

[1]The amount of memory in all the computers in the world would not be nearly enough, short by a few orders of magnitude, to just store the positions of the particles in a can of air.

the first q oscillators have one unit of energy each and none for the rest. These two different microstates correspond to the same macrostate, a state of total energy q. In fact, all different microstates of an Einstein solid with a given energy is just one macrostate, irrespective of how it is distributed internally.

As there can be multiple microstates for a given macrostate, the correspondence between microstates and macrostates is not unique. The number of microstates in a given macrostate is called the multiplicity, denoted by Ω. The multiplicity, $\Omega(N, q)$, of the Einstein solid for a given N and q, can be found analytically [41, 82] (see Exercise E11.1).

To study equilibrium, let us consider an isolated system composed of two interacting solids, A and B, with N_A and N_B oscillators, respectively, sharing a total of q units of energy, $q_A + q_B = q$. We denote a macrostate n as a particular pair of numbers, $n \equiv (q_A, q_B)$, where $q_B = q - q_A$ would be fixed for a given q_A. The multiplicity of a macrostate n is equal to the product of multiplicities of the two solids,

$$\Omega(n) = \Omega_A(N_A, q_A) \times \Omega_B(N_B, q_B). \tag{11.1}$$

It tells us the number of possible ways (microstates) of arranging energies q_A and q_B within solids A and B, respectively.

To associate the multiplicity with a probability, we invoke the statistical postulate, which states that all accessible microstates are equally probable in an isolated system. This is a fundamental assumption in thermal physics.[2] With this postulate, we can assert that the probability of finding the system in macrostate n is

$$P(n) = \frac{\Omega(n)}{\Omega_T}, \quad \text{(statistical postulate)} \tag{11.2}$$

where $\Omega_T = \sum_n \Omega(n)$ is the total multiplicity.

We could calculate the probability distribution from Eq. (11.2) ("theory," Project P11.2), or we could simulate how this happens naturally ("experiment"). We choose the latter. To do so, we consider an asymmetric pair, $N_A \ll N_B$, a small solid A interacting with a large solid B. This simulates the common situation in which a system of interest is in thermal equilibrium with a large environment (the reservoir).

It is more convenient to define an Einstein-solid object via object-oriented programming rather than through a function as usual. This object will be used in more than one program. Furthermore, it can be expanded to two interacting solids. This object is defined as follows.

Program listing 11.1 Einstein solid class (`einsteinsolid.py`)

```
import random as rnd

class EinsteinSolid:                        # Einstein solid object
    def __init__(self, N=400, q=10):
        self.N = N
        self.cell = [q]*N                   # q units energy per cell

    def __add__(self, other):               # combine two solids
        self.N += other.N
        self.cell += other.cell
        return self
```

[2] We cannot prove it, but there are no other more plausible alternatives.

```
    def exchange(self, L=20):                     # iterate L times
        for i in range(L):
            take = rnd.randint(0, self.N-1) # random pair
            give = rnd.randint(0, self.N-1)
            while self.cell[take] == 0:     # find a nonzero-energy cell
                take = rnd.randint(0, self.N-1)
            self.cell[take] -= 1              # exchange energy
            self.cell[give] += 1
```

The function `__init__()` is called when an object is instantiated to initialize N oscillators and to store their energies in the attribute `cell` as a list.[3] The next function `__add__()` defines the addition operator so two solids can be combined such that if $C = A + B$, then $N_C = N_A + N_B$ and the cells (lists) are concatenated (Project P11.2). This is also called operator overloading. The double underscore marks a method as private to a class.

Because all allowed microstates are equally probable by the statistical postulate, we will let the oscillators exchange energy freely and randomly. This is carried out in `exchange()`, which makes a specified number of iterations through the oscillator list. In each iteration, it randomly picks a pair of oscillators for energy exchange using `randint(n1,n2)`, which returns a random integer between `n1` and `n2`, inclusive. After finding an oscillator with a nonzero energy, it takes one unit of energy from that oscillator and gives it to the other. As discussed later, this is equivalent to the interaction between a single oscillator and the rest.

Given an initial state, we allow many interactions to occur to reach equilibrium. We can count the number of cells with a given amount of energy to obtain the energy distribution. What is the distribution after equilibrium is established?

We first present results from Program 11.4 for a small Einstein-solid object (256 oscillators) in Figure 11.1, which shows the energy distribution in a 16×16 intensity image as a function of iterations elapsed between samplings. Darker cells correspond to higher energies. Not to play favoritism, we give every oscillator the same amount of energy in the initial microstate, indicated by the uniform intensity at the start (first frame). The visual effect clearly shows that, with increasing iterations, the number of white (empty) cells increases quickly, followed by ever smaller numbers of lighter to darker gray levels. We get the picture that the most probable states are at low energies.

To be more quantitative, we show the results from Program 11.5 in Figure 11.2 for a larger solid with $N = 1024$. It displays as histograms the relative fraction of oscillators having n units of energy between samplings. Because every oscillator is initialized to one unit of energy, we see a single peak at $n = 1$ in the first frame. As the number of iterations increases, the distribution widens, and the number of oscillators having zero energy grows rapidly, as observed in the intensity plot in Figure 11.1. After 1500 cumulative iterations from the start (the fifth frame), the zero-energy state is the most probable (to check if this might be due to the low initial energy, increase it and run Program 11.5). By the last three frames, the distribution has only small changes. It signals that the system has reached equilibrium, with what looks like an exponential distribution

$$P(n) = C \exp(-\alpha n). \tag{11.3}$$

Drawing smooth curves through the distribution, we find $\alpha \sim \ln 2$ gives the best visual fit. We will see why later (Section 11.1.4).

To confirm this, we plot the distribution in Figure 11.3, where $N(n)$ is the number of oscillators having n units of energy, this time on a semilog y scale. The data is from the last frame

[3] A Python list is faster than an ndarray for explicit looping (see Section 2.3.2).

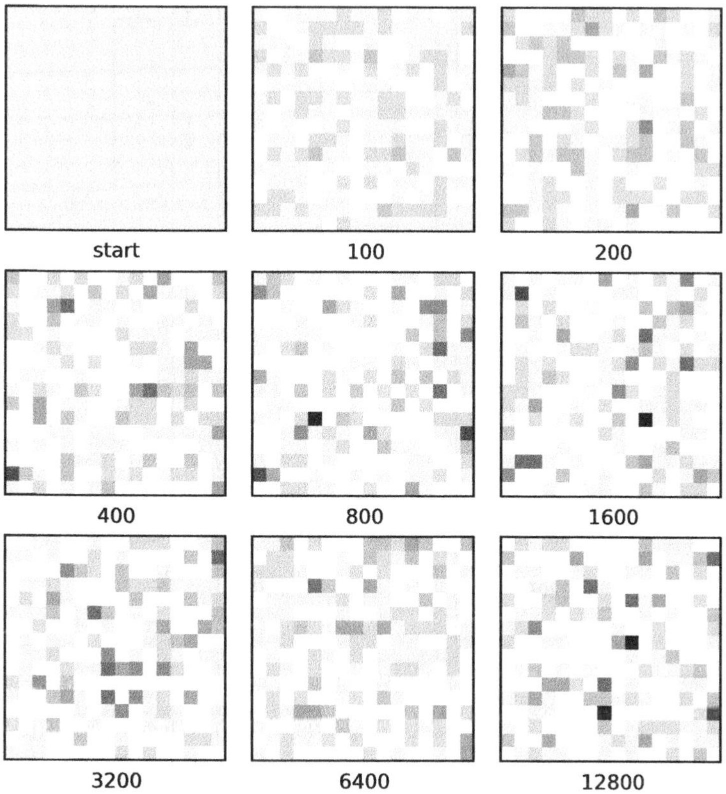

FIGURE 11.1 The energy distribution of an Einstein solid (256 oscillators) at given iterations (time). The darker the cell, the higher the energy.

of Figure 11.2. We can see a clear trend of the data points following closely the exponential function (11.3), which is a straight line on the semilog scale. In addition, we can also see the exponential tail at larger n, which is too small to see on a linear scale (Figure 11.2). The error bars are equal to the statistical fluctuation, $\sqrt{N(n)}$. They are produced from Program 11.5 with

```
plt.errorbar(range(len(bin)), bin, np.sqrt(bin), fmt=None)
```

In the simulation, we only know the total number of oscillators ($N = N_A + N_B = 1024$), but what are the implicit values of N_A and N_B? From the way we modeled the system, we can interpret the results as caused by the interaction between two solids, $N_A = 1$ and $N_B = 1023$. When we iterated the system, we picked two individual oscillators randomly to exchange energy. Because all oscillators are equivalent, each action is in effect an interaction between a single oscillator (A) and the rest (B). When we plot the histogram, we are effectively plotting the results of N iterations of a single oscillator.

This is the subtlety of the Monte Carlo experiment. We could have focused on a particular oscillator (say the first one) and exchanged energy between it and the rest of the system. Over the long run, we would still have the same cumulative distribution. However, it would be extremely inefficient and would take much longer to obtain results as accurate as these in Figure 11.3.

As we will see shortly (Section 11.1.4), the remarkable "experimental" results, obtained with a system so tiny as to be utterly insignificant compared to a normal thermal system, in fact

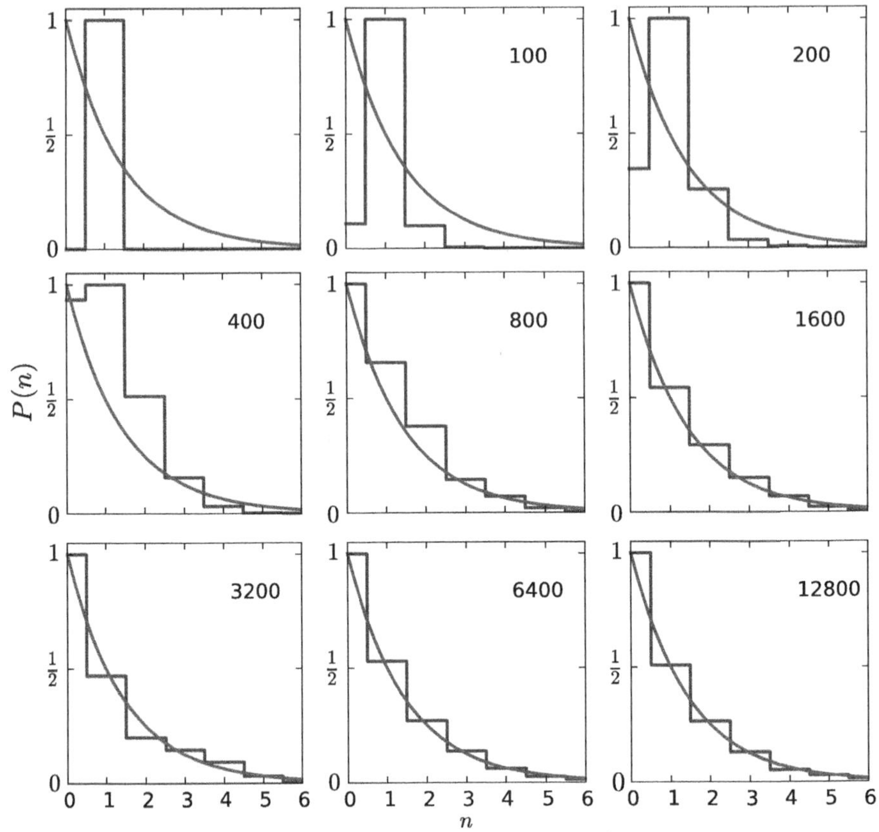

FIGURE 11.2 The energy distribution of an Einstein solid (1024 oscillators) at given iterations (time). The smooth curve is an exponential $e^{-\alpha n}$.

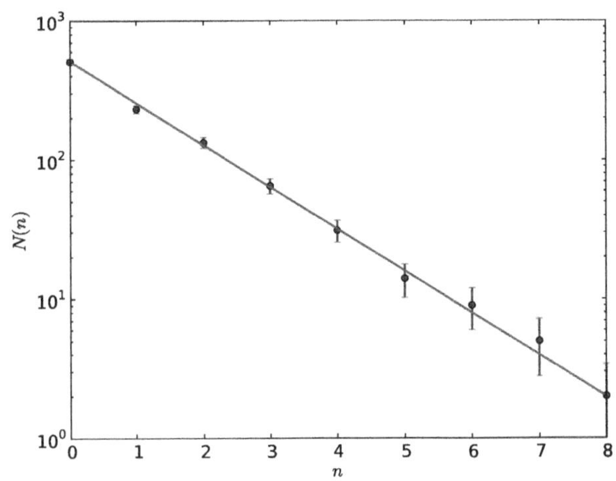

FIGURE 11.3 The energy distribution in equilibrium (last frame of Figure 11.2) on a semilog scale. The solid line is $N(0)e^{-\alpha n}$, $\alpha = \ln 2$.

do agree with Boltzmann statistics in actual systems. Although we used the Einstein solid model, the results are rather general because the exchange of energy is generic and does not assume any particular mechanism. It illustrates that we can efficiently simulate the behavior of much larger systems using a small but finite model system with correct sampling techniques and proper interpretation. Now that we have seen what equilibrium looks like, we try to understand why next.

11.1.2 ENTROPY

We have seen from the preceding that the thermal system marches from an ordered initial configuration toward somewhat disordered equilibrium configuration. The energy distribution follows a probabilistic course of action given by the exponential factor. The probability is extremely low that the system would spontaneously go back to the ordered initial state, even for our very tiny system of a few thousand oscillators. This all seems purely mathematical.

However, in thermal physics we deal with measurable physical quantities such as energy and heat, not just probability theory. How do we characterize the trend of thermal systems toward disorder? The answer is a physical quantity called entropy.

Entropy measures the degree of disorder, or more precisely, the number of ways a thermal system can be rearranged, namely the multiplicity Ω. Boltzmann defined entropy S as

$$S = k \ln \Omega, \tag{11.4}$$

where k is the Boltzmann constant. Because the multiplicity Ω is a very large number, the logarithm $\ln \Omega$ will be a large number. After multiplied by the small k, entropy S will be merely an ordinary number. More important, entropy thus defined is additive: The total entropy of a composite system is equal to the sum of individual entropies of its subsystems because the total multiplicity is equal to the product of multiplicities (11.1).

We can calculate the change of entropy in the preceding simulation of the Einstein solid. Because we know the probability distributions of an oscillator from Figure 11.2, it is more convenient to calculate entropy using an equivalent expression as [9] (see Section 11.A, Eq. (11.55))

$$S = -k \sum P_n \ln P_n, \tag{11.5}$$

where P_n is the probability of finding the system in state n.

The following function computes the entropy of one oscillator in the system.

Program listing 11.2 Entropy of an Einstein solid (`entropy.py`)

```
1 def entropy(cell):                       # entropy of Einstein solid
      N, n, nt, s = len(cell), 0, 0, 0.
3     while nt < N:                        # until all cells are counted
          cn = cell.count(n)               # num. of cells with En
5         n, nt = n + 1, nt + cn           # increase energy, cumul. cells
          p = cn/float(N)                  # probability
7         if (cn != 0): s -= p*np.log(p)   # entropy/k, Eq. (11.5)
      return s
```

It accepts the configuration (microstate) in `cell`, counts the number of oscillators with energy `n` (line 4), converts it to a probability, and adds it to the entropy, until all oscillators are accounted for.

We can put the preceding function in Program 11.5 (or make it a method of the Einstein-solid object in Program 11.1) to obtain the results in Figure 11.4, showing the entropy S of a single oscillator in a system of $N = 1024$ oscillators, in units of k to make its numerical

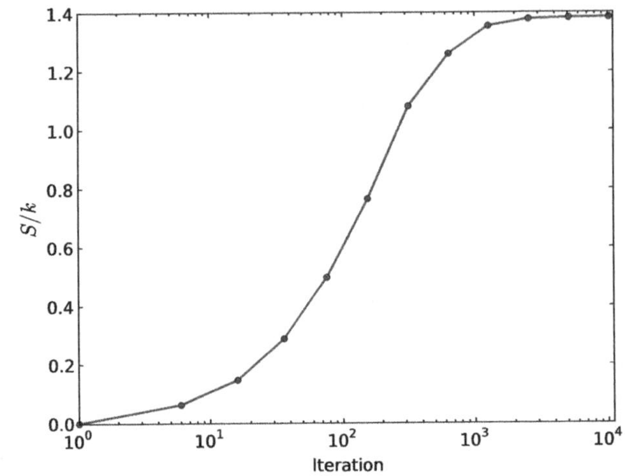

FIGURE 11.4 The entropy of an oscillator as a function iteration number.

value tidy. The total entropy of the system is N times larger because it is additive. Initially the entropy is zero because the initial state, where $P(1) = 1$ and $P(n \neq 1) = 0$, is perfectly ordered. As time increases, S grows rapidly, noting the semilog x scale. Near equilibrium at the end, S is maximum and stabilizes to a plateau. Numerical experiments lead us to this observation:

Entropy tends to increase.

This is the second law of thermodynamics [82]. Because entropy measures multiplicity, we can state the second law a little differently: A system in equilibrium will be found in a macrostate of greatest multiplicity (apart from fluctuations).

We can now understand why the most probable state is the ground state (lowest energy state) in terms of entropy. For a given amount of energy, the multiplicity of a macrostate as measured by the entropy is increased by a much greater factor if the energy is deposited in the large system B (reservoir) than in the small system A (microsystem). The energy makes many more microstates accessible in the reservoir than in the microsystem. Using our example $N_A = 1$, $N_B = 1023$ and $q = 1$, for instance, increasing the energy of system A by one unit has no effect on the number of microstates, it is always 1. But, there are 1023 possibilities, that is, additional microstates, of distributing one unit of energy in system B. Evidently, of all possibilities, an exponential distribution (i.e., Boltzmann distribution, Section 11.1.4) gives the maximum possible entropy for a system in equilibrium.

11.1.3 TEMPERATURE

We intuitively know that when two systems are in equilibrium, they have the same temperature. This leads us to expect that there is a connection between temperature and entropy reaching maximum in the preceding example.

Our small microsystem has very little effect on the temperature of the reservoir because the amount of energy exchanged is limited. We can gain more insight into the equilibrium process by changing the relative sizes of the interacting systems.

Figure 11.5 shows the changes of entropy for two Einstein solids of equal size, $N_A = N_B = 512$. We set up the systems so that initially solid A has 3 units of energy per oscillator, and solid B has 1. Both are in equilibrium before they are allowed to interact. Once interactions

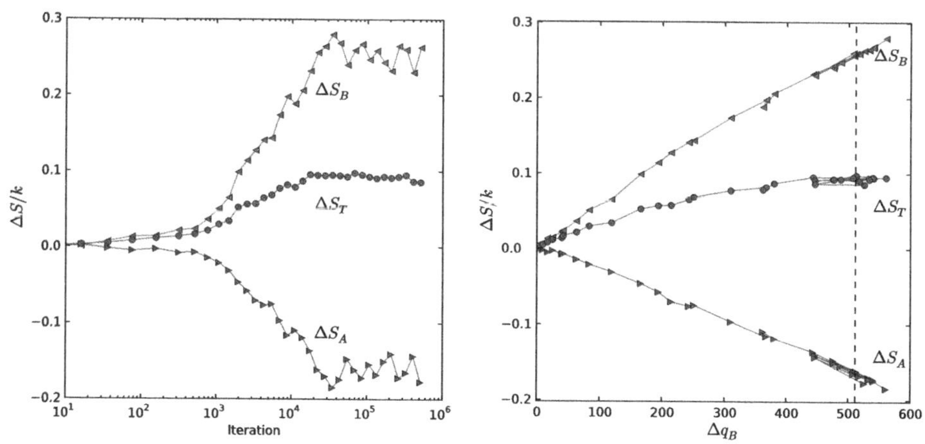

FIGURE 11.5 The changes of entropy per oscillator vs. iterations (left) and vs. energy exchange (right) between two equal-sized Einstein solids.

begin, the hotter solid A loses energy more readily than solid B. The change of entropy of A, ΔS_A, is negative and decreasing relative to the initial entropy. Conversely, the change of entropy of B, ΔS_B, is positive and increasing due to the absorbed energy $\Delta q_B > 0$ (Figure 11.5, right). But, the positive gain ΔS_B is more than the negative loss ΔS_A so the net change $\Delta S_T = \Delta S_A + \Delta S_B$ is positive.

Energy will continue to flow from solid A to solid B as long as the net change in entropy is positive. But as the energy transfer continues, the rate of entropy gain decreases. At the point of equilibrium ($\Delta q_B = 512$, dashed line in Figure 11.5, right), the net entropy stabilizes and fluctuates. We see the systems reach equilibrium, and their energies are equal $q_A = q_B$, apart from fluctuations.

The rate of entropy change is an important parameter. From Figure 11.5 (right) we can see that as solid B gains energy, the rate of ΔS_B decreases, whereas the rate of ΔS_A increases in magnitude, though the clarity of the trends is affected by fluctuations. A clearer picture is given in Figure 11.6, where the total multiplicity and changes of entropy are calculated from Eqs. (11.1) and (11.4) for the same parameters as in Figure 11.5 (Project P11.2). The scaled multiplicity shows a sharp peak near the equilibrium, which would become increasingly narrower for larger systems.

The values of entropy (Figure 11.6, bottom) are relative to a single solid at the beginning $q_B \sim 500$. The arrows indicate the rates of changes at the beginning and also at the equilibrium (dashed line). When the rates are equal in magnitude, the solids are in equilibrium. Because the total entropy is maximum at that point, we have the equilibrium condition

$$\frac{\partial S_T}{\partial q_B} = \frac{\partial S_B}{\partial q_B} + \frac{\partial S_A}{\partial q_B} = 0, \quad \text{or} \quad \frac{\partial S_B}{\partial q_B} = \frac{\partial S_A}{\partial q_A}. \tag{11.6}$$

We have used $q_B = q - q_A$ in the last term. Equation (11.6) is a statement on temperature, which is the "thing" that is the same for the solids or any systems in equilibrium. Considering that the temperature of a solid rises with increasing energy, we can therefore define the temperature as the inverse of the rate,

$$\frac{1}{T} = \frac{\partial S}{\partial E}. \tag{11.7}$$

We can see from the slope change in Figure 11.6 that the difference in temperature measures the tendency to give off energy. A body at a higher temperature tends to lose heat, and a body at

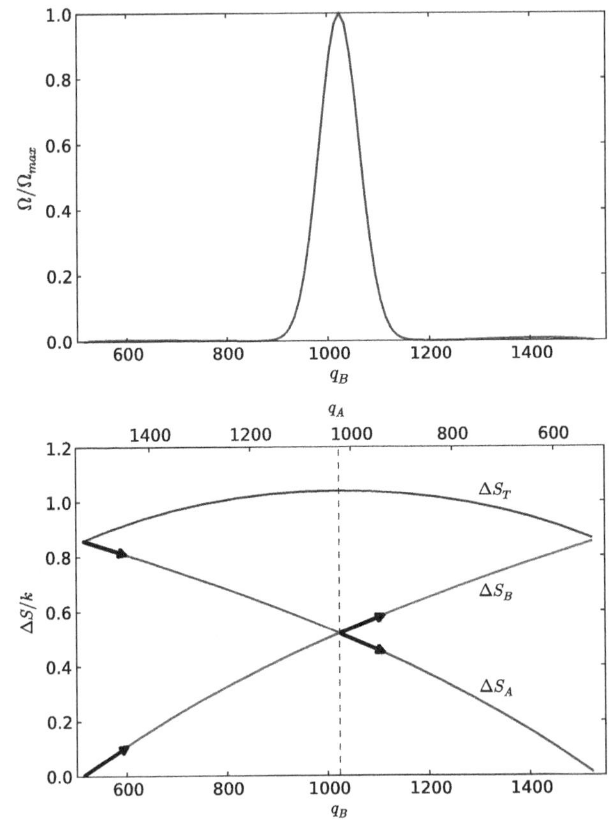

FIGURE 11.6 The multiplicity (top) and entropy (bottom) of two interacting solids $N_A = N_B = 512$, sharing a total energy $q = q_A + q_B = 2048$.

a lower temperature tends to gain heat. Restated in temperature, the equilibrium condition (11.6) is $T_A = T_B$.

Figure 11.7 shows the temperatures from our simulation. We calculate the temperature via the average energy from Eq. (11.13) rather than from the derivative relation (11.7).[4] The temperature of solid A steadily decreases whereas that of solid B increases. The solids reach an equilibrium temperature $kT \sim 2.5\epsilon$. The many data points around $q_B \sim 1024$ indicate the amount of fluctuation. We see the overall trends to be linear to a very good approximation. This is because the average energy $\langle E\rangle$ is high. For smaller $\langle E\rangle$, the lines would curve (Exercise E11.3).

The entropy in Einstein solids increases with energy (Figure 11.6), accompanied with increasing temperature. This is the normal behavior. However, it is not always the case. For systems such as a paramagnet that can store only a limited amount of energy, the entropy can decrease with increasing energy. The slope $\partial S/\partial E$ then is negative. According to Eq. (11.7), we have a negative temperature. What does it mean, though? We leave further exploration of such systems to Project P11.3.

[4]Numerically calculating derivatives of a fluctuating function is worse than undesirable and should be attempted only for educational purposes.

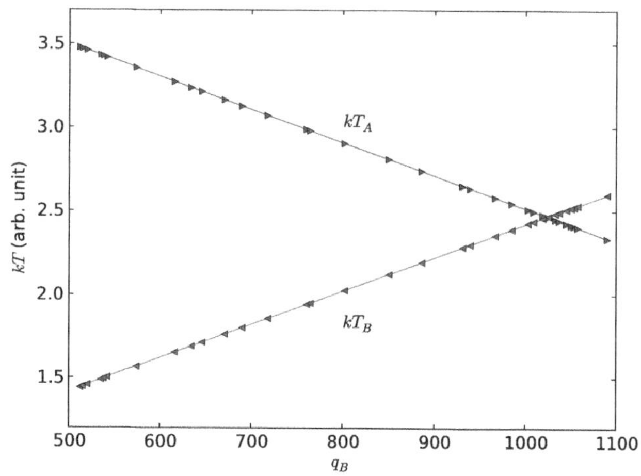

FIGURE 11.7 The temperatures of two interacting solids. The parameters are the same as in Figure 11.6.

11.1.4 BOLTZMANN DISTRIBUTION

Through numerical simulations, we have established that the energy distribution in equilibrium is an exponential factor (11.3), and temperature is related to entropy by Eq. (11.7).

We only have to relate α in Eq. (11.3) to temperature (see Section 11.A.1). Comparing Eq. (11.3) with the thermodynamic result (11.51) and noting that $E_n = n\epsilon$, we find $\alpha = \epsilon/kT$. So Eq. (11.3) becomes

$$P(E_n) = \frac{1}{Z}\exp(-E_n/kT). \tag{11.8}$$

This is the Boltzmann distribution. It describes the relative probabilities of states of a system in thermal equilibrium [82].

The $1/Z$ factor may be determined through normalization of probabilities $\sum_n P(E_n) = 1$, which yields the so-called partition function

$$Z = \sum_n \exp(-E_n/kT). \tag{11.9}$$

If we know the partition function Z, we can determine all properties of a thermal system in principle. For example, we can calculate the average energy as a weighted average from Eq. (11.8),

$$\langle E\rangle = \sum_n E_n P(E_n) = -\frac{\partial \ln Z}{\partial \beta}, \quad \beta \equiv \frac{1}{kT}. \tag{11.10}$$

Heat capacity is another important thermodynamic property. It is defined as

$$C(T) = \frac{\partial \langle E\rangle}{\partial T}. \tag{11.11}$$

The heat capacity measures the amount of energy required (or released) to raise (or lower) the temperature of a system by one degree. For numerical work, it is more convenient to use an alternative expression [41]

$$C(T) = \frac{\Delta^2 E}{kT^2}, \quad \Delta^2 E = \langle E^2\rangle - \langle E\rangle^2, \tag{11.12}$$

where $\Delta^2 E$ is the energy fluctuation (Exercise E11.2).

Many other quantities can be obtained this way. The exponential Boltzmann distribution (11.8) is central to thermodynamics.

As an illustration, we can finally determine why $\alpha = \ln 2$ in Figures 11.2 and 11.3. We expect α is related to temperature, which in turn should be related to the average energy of the system. For an SHO, the average energy is analytically available, and the temperature can be expressed as (see Exercise E11.3, ignoring zero point energy)

$$\alpha = \frac{\epsilon}{kT} = \ln\left(1 + \frac{\epsilon}{\langle E \rangle}\right), \tag{11.13}$$

where ϵ $(= \hbar\omega)$ is the unit of energy.

For the Einstein solid in our earlier example (Figure 11.2), the average energy is 1, $\langle E \rangle = \epsilon$, so $\alpha = \ln 2$. Equivalently, the temperature is $kT = \epsilon / \ln 2$. We have used this value for the exact Boltzmann distribution in Figures 11.2 and 11.3 that show very good agreement between numerical ("experimental") and analytical results.

11.2 THE ISING MODEL

The Einstein model considered earlier consists of independent oscillators exchanging energy with each other. Now we discuss a model in which the particles are interdependent, that is, coupled to each other. The simplest, yet nontrivial, model is the Ising model that can exhibit a range of thermodynamic properties. It is useful for understanding critical behaviors, including self-organization in magnetism such as ferromagnetism where magnetic dipole moments align themselves spontaneously in the same direction, giving rise to macroscopic domains of net magnetization. In terms of modeling, the simplicity of the Ising model is ideal for us to clearly highlight the important Metropolis algorithm whose basis rests on the Boltzmann distribution.

11.2.1 THE 1D ISING MODEL

We consider the 1D Ising model consisting of N identical spins (dipole moments), $\{s_i\}$, arranged in a chain as shown in Figure 11.8.

Each spin has two possible values: either up $s_i = 1$, or down $s_i = -1$. The interaction energy between any two spins is represented as

$$E_{i,j} = -\epsilon\, s_i s_j. \tag{11.14}$$

The parameter ϵ is the pair-wise energy (coupling). If $\epsilon > 0$, the interaction favors parallel spins, leading to ferromagnetism. For antiferromagnetism, $\epsilon < 0$, antiparallel spins are preferred. The specific value ϵ depends on the properties of the interaction between atoms, not the direct dipole–dipole interaction. The latter is too weak in comparison to the exchange energies arising from symmetries of the wave functions [76] (e.g., see Eq. (9.7) and Figure 9.3). For our purpose, we assume $\epsilon > 0$, treating it as the basic energy unit of a ferromagnetic system.

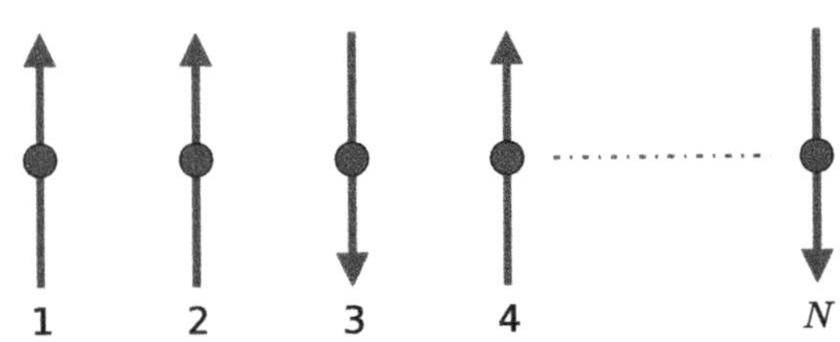

FIGURE 11.8 The one-dimensional Ising model.

In principle, all pairs interact with each other. To keep with simplicity, the Ising model takes into account only nearest neighbor interactions, neglecting all long-range interactions that are presumably weak. We can express the total energy of the 1D Ising model as the sum of neighboring pairs

$$E = \sum_i E_{i,i+1} = -\epsilon \sum_{i=1}^{N} s_i s_{i+1}. \tag{11.15}$$

Here, the last term involves s_{N+1}, which is undefined. We could just neglect that term, which is justified if N is large. However, a better way to deal with the situation without unnecessary complication is to use periodic boundary conditions,

$$s_{N+1} = s_1. \quad \text{(periodic BC)} \tag{11.16}$$

In effect, we assume that there were virtual identical copies of the chain to the left and to the right of the actual chain. This also helps to minimize finite size effects in Monte Carlo simulations. Another way to visualize Eq. (11.16) is the Ising chain arranged in a ring. The 1D Ising model is analytically solvable.

11.2.2 METROPOLIS SIMULATION

To simulate the dynamics of the Ising model, we need to put it in contact with a reservoir at temperature T, which we just introduced in Eq. (11.7). The reservoir acts as a large heat bath whose temperature remains unchanged while exchanging energy with the Ising system.

Qualitatively, for a given initial configuration and a specified T, the Ising chain will start exchanging energy by flipping the spins to reach equilibrium. If the energy of a microstate as given by Eq. (11.15) is too low, the system will absorb energy from the heat bath. Spins will tend to align antiparallel to each other so the energy of the system is increased. Conversely, if the energy of the microstate is too high, the system will give off energy to the heat bath. This causes the spins to flip toward parallel alignment.

These parallel and antiparallel spin flips will continue until, after sufficient time passes, the system reaches equilibrium. Then sporadic flips will still occur, but on average, the two types of flips balance each other out. The system will just fluctuate around equilibrium (see Project P11.4 and Figure 11.26 for an animated Ising model).

The question now is how to simulate this equilibrium process. A direct way is to use importance sampling (Section 10.4) to build up a set of microstates that obey the Boltzmann distribution. We would sample randomly a microstate $\{s_i\}$, record its energy E from Eq. (11.15), and accept it with a probability $\exp(-E/kT)$. The problem with the direct method is that the number of microstates is large even for a tiny system. For instance, if $N = 1000$, the number is already $2^N \sim 10^{300}$. Out of the uniformly sampled microstates, only a very small fraction will be near the equilibrium.[5] The direct approach would be highly inadequate, unless the process could be biased toward the equilibrium.

The Metropolis algorithm applies a proper bias to efficiently guide the system into equilibrium and keep it there [62]. It works as follows for the Ising model [41]. Starting from a given microstate, we pick a spin randomly and propose to flip it. Let ΔE be the energy difference caused by the flip. If the flip produces a lower energy, that is, $\Delta E < 0$, the flip is accepted and microstate changed. The reason is because lower energy configurations are always preferred according to the Boltzmann distribution. If the proposed flip produces a higher energy, $\Delta E > 0$, we do not discard

---

[5]A more serious technical barrier is the period of the pseudorandom number generator, which is 2^{19937} as currently implemented in Python. It means that we could simulate a system of no more than 10^4 spins by brute force.

the flip entirely. Instead, we accept it with a probability $\exp(-\Delta E/kT)$ according to Eq. (11.8). This is to account for thermal fluctuations, for otherwise the system would descend to the lowest possible state monotonically with no fluctuation whatsoever, a behavior contradictory to real systems.

This process is repeated many times. If we started with a configuration having an energy above its equilibrium value, the flips will lower the energy on average, and the Metropolis method will accept the flips. This drives the system toward lower energy configurations closer to equilibrium. The opposite happens if the initial configuration is below the equilibrium. Either way, once equilibrium is reached, the flips tend to raise or lower the energy equally, producing only fluctuations. In this pick-and-go manner, the Metropolis method samples microstates obeying the Boltzmann distribution.

Let us first see how the method works in a simulation and analyze it afterward. As stated earlier, we assume the Ising model is ferromagnetic, and without loss of generality, we can set $\epsilon = 1$. The temperature is measured in terms of the ratio kT/ϵ. The Metropolis sampling works as follows:

1. Set the initial microstate $\{s_i\}$. In principle, any random configuration will do. But at low temperatures $kT < 1$, the system takes longer to equilibrate. It is better in practice to start with a cold configuration, that is, more spins parallel to each other than antiparallel.

2. Pick a spin at random (say the i-th), and calculate the energy difference if the spin is to flip (a trial flip). Rather than calculating the energy anew, it is easier and more efficient to just find the energy difference due to the interaction with the nearest neighbors by

$$\Delta E = 2\epsilon\, s_i(s_{i-1} + s_{i+1}). \tag{11.17}$$

   Note, the value of s_i in Eq. (11.17) is the preflip value.

3. If $\Delta E < 0$, the trial flip is accepted, set $s_i \to -s_i$. Otherwise, draw a random number x. If $x < \exp(-\Delta E/kT)$, accept the flip, and set $s_i \to -s_i$; if not, reject the trial flip, and keep the microstate unchanged. This is the Metropolis algorithm.

4. Repeat steps 2 and 3 until equilibrium is established. As a rule of thumb, each spin should be given a sufficient number of chances to flip, say 10 to 100 (or more, depending on temperature).

Program 11.6 simulates the 1D Ising model with the Metropolis algorithm. The core module, `update()`, is listed next.

```
def update(N, spin, kT, E, M):       # Metropolis sampling
    i, flip = rnd.randint(0, N-1), 0
    dE = 2*spin[i]*(spin[i-1] + spin[(i+1)%N]) # Eq. (11.17), periodic bc
    if (dE < 0.0): flip=1             # flip if dE<0, else flip
    else:                             # according to exp(-dE/kT)
        p = np.exp(-dE/kT)
        if (rnd.random() < p): flip=1
    if (flip == 1):                   # accept
        E = E + dE
        M = M - 2*spin[i]
        spin[i] = -spin[i]
    return E, M
```

The Ising chain (N spins) is stored in the array `spin[]`. The parameters `kT`, `E`, and `M` are, respectively, the temperature, energy, and magnetization (11.18) of the current state. The module picks a random spin index between 0 and $N-1$ (inclusive) and calculates the change of energy ΔE from Eq. (11.17) if it is to be flipped. Note that the periodic boundary condition ($i=0$ or $N-1$) is automatically enforced via modulo operations. If the change ΔE is negative, the flip is accepted and the flag set accordingly. If not, it is accepted with a probability $\exp(-\Delta E/kT)$. At the end, the energy, magnetization and the spin itself are updated if the trial flip is accepted, `flip=1`.

Figure 11.9 shows the energies toward the equilibrium values of an Ising chain with $N=1000$ spins as a function of iterations (time). Three temperatures are used: low, $kT/\epsilon=0.5$ ($\epsilon=1$); intermediate, $kT=1$; and high, $kT=2$. We initialize the system to a configuration whose energy is below the high temperature but above the low and intermediate temperatures. In the beginning, the system rapidly exchanges energy with the heat bath and quickly approaches their respective equilibrium values from below or above. This is evidenced by the steep slope near the beginning.

In this specific run, the intermediate temperature case ($kT=1$) reaches equilibrium first, taking a bit less than one pass (one pass $=N$ iterations, each spin having one chance to flip on average). The quick equilibrium is helped by the initial energy closest to the analytic value for $kT=1$. The high temperature case ($kT=2$) rises to its equilibrium level around five passes. In both cases, the energy fluctuates around equilibrium as expected. After a relatively quick initial dive, the rate of energy loss in the low temperature case ($kT=0.5$) decreases substantially as the system becomes cooler. The hotter initial state tends to give off energy at a faster rate. But as it cools down, the temperature difference becomes smaller, resulting a slower rate. The system still has not reached equilibrium after ten passes. There is virtually no movement in the last two passes shown. The spins are more tightly aligned, and the chance of flipping to a lower energy is greatly reduced. In this case, only after more than 50 passes did the system reach equilibrium.

After equilibrium is achieved, we can start taking averages. Figure 11.10 displays the results for the average energy and magnetization as a function of temperature. The averages in

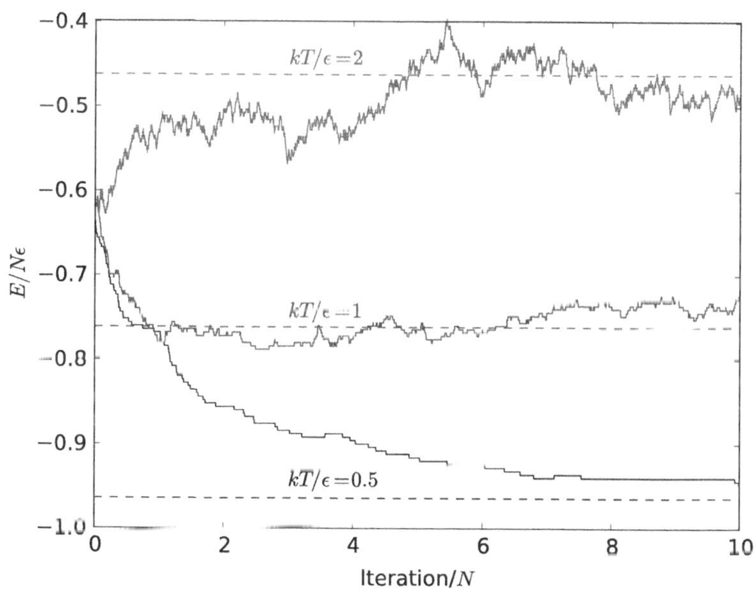

FIGURE 11.9 The energy of an Ising chain ($N=1000$) equilibrating at three temperatures. The dashed lines are the analytic values in equilibrium.

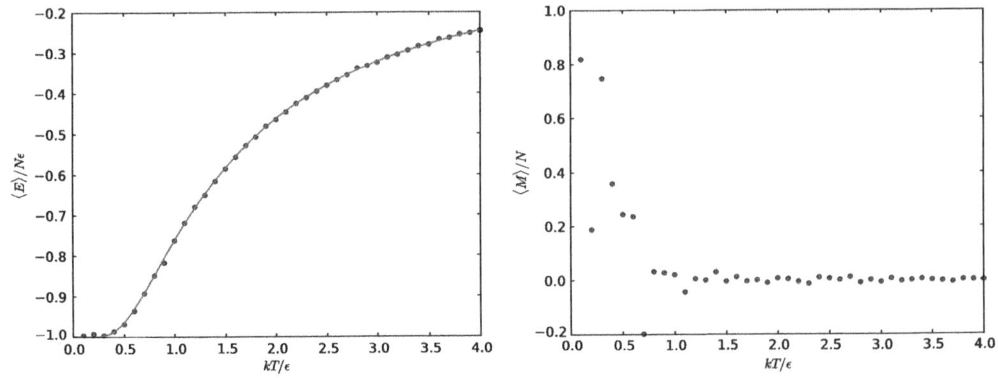

FIGURE 11.10 The equilibrium energy (left) and magnetization (right) per spin as a function of temperature. The solid line is the analytic result (11.19).

Monte Carlo Metropolis simulations are calculated as simple algebraic averages. For instance, the average energy and magnetization are

$$\langle E \rangle = \frac{1}{N_{mc}} \sum_{n=1}^{N_{mc}} E(n), \quad \langle M \rangle = \frac{1}{N_{mc}} \sum_{n=1}^{N_{mc}} M(n), \tag{11.18}$$

where $E(n)$ is the energy of microstate n (configuration $\{s_n\}$), computed from Eq. (11.15), and $M(n) = \sum_i s_i$ is the magnetization (net spin). The value N_{mc} is the total number of Monte Carlo samplings. Because we sample the microstates with the Metropolis algorithm, the importance-sampled microstates obey the Boltzmann distribution. As a result, Eq. (11.18) uses algebraic averages without the explicit weighting factor $\exp(-E/kT)$ as would be the case from Eq. (11.10) if the microstates were uniformly sampled. This is a another example of importance sampling (see Section 10.4.2).

The average energy per spin (Figure 11.10) is flat and very close to -1 at $T \to 0$, corresponding to all spins parallel to each other. As seen from Figure 11.9, we must take considerable care to make sure the Ising chain reaches equilibrium at low temperatures before taking averages. The energy grows with increasing temperature and should approach zero as $T \to \infty$, a state of totally random spin alignment.

The exact energy is given by [82] (Exercise E11.7)

$$\langle E \rangle = -N\epsilon \tanh(\beta\epsilon). \tag{11.19}$$

Overall, we see good agreement with analytic results for the average energy.

For magnetization, the agreement depends on temperature. Theoretically, magnetization is identically zero in the 1D Ising model, even at low temperature when all spins are aligned. The reason is twofold. First, absent any external magnetic field, there is no preferred direction for the spins. Second, in practically every microstate, a spin can spontaneously flip without changing the total energy. Sometimes this can lead to a domino effect, where the spins flip one after another, so the net spin is reversed. The average magnetization is therefore zero.

We see good agreement between numerical and analytic results for magnetization from intermediate to high temperatures (Figure 11.10). But for low temperature $kT < 1$, numerical data has considerable scatter, and the disagreement is large. Increasing the time for equilibrium and for sampling did not seem to noticeably improve the disagreement. On the one hand, most spins are aligned at low T, and there is a low probability of finding a spin with antiparallel neighbors where a flip requires no extra energy, effectively blocking the most favorable path. On the other hand, a positive amount of energy (11.17) is needed to flip a spin with parallel neighbors, and the probability of accepting the flip is prohibitively low due to the small exponential factor. For example, in such a flip, $\Delta E = 4$. If $kT = 0.5$, the probability is $\exp(-\Delta E/kT) = \exp(-8) \sim 3 \times 10^{-4}$. Basically, the spins are frozen in their alignment at low temperatures, and the time scale for the whole Ising chain to flip is very large. This also happens in actual thermal systems, which can get stuck in a state of finite entropy at low temperatures.

We show the heat capacity per spin, $C(T)/Nk$, in Figure 11.11. Numerically, it is calculated from energy fluctuation (11.12) (Project P11.4). The analytic result (solid line) is given by

$$C = Nk\frac{(\beta\epsilon)^2}{\cosh^2(\beta\epsilon)}. \tag{11.20}$$

Qualitatively, numerical and analytic results behave similarly. The heat capacity is small at low temperature. This is consistent with the average energy (Figure 11.10), which shows only small changes in energy when temperature is varied. At $T = 0$, $C(T) = 0$ analytically, a result often attributed to the third law of thermodynamics. At high temperature, the heat capacity again approaches zero as $C(T) \to 1/T^2$ because rising temperature is accompanied with little change in energy. In between we observe a maximum at $kT \sim 0.8$, corresponding to the change of curvature (bend) in the energy diagram (Figure 11.10).

Quantitatively, the agreement between numerical and analytic results is satisfactory from intermediate to high temperatures ($kT \geq 1$). At low temperature, the difference is significant, in line with similar observations for magnetization. For the same reasons as explained earlier, it is tricky to reduce the considerable scatter in the numerical results because of long relaxation times.

Finally, we show in Figure 11.12 the entropy as a function of temperature. To compute the entropy S in the simulation, we use the thermodynamic relationship (11.7)

$$\delta S = \frac{\delta E}{T}. \tag{11.21}$$

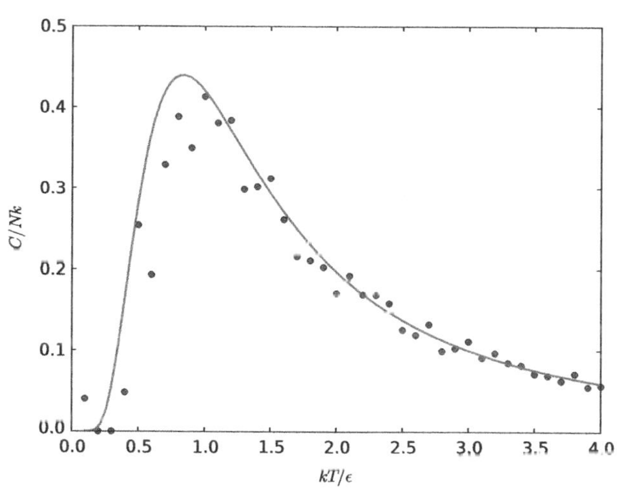

FIGURE 11.11 The heat capacity per spin as a function of temperature. The solid line is the analytic result.

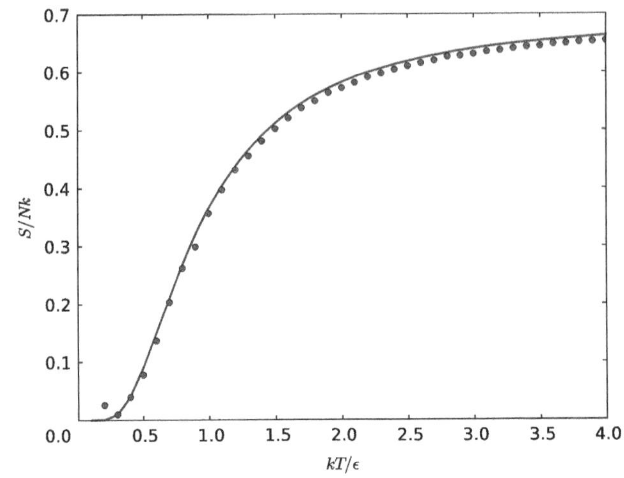

FIGURE 11.12 The entropy per spin as a function of temperature. The solid line is the analytic result.

We interpret δS as the change of entropy caused by the absorbed energy δE. The results in Figure 11.12 are the cumulative change of entropy, assuming $S = 0$ at $T = 0$. Equation (11.21) directly relates entropy to heat and temperature, all experimentally measurable.

The entropy starts from zero, increases quickly between $kT = 0.3$ and 1, and approaches the asymptotic limit ln 2, that is, random dipole orientations. The agreement is good between the numerical and analytic results. The latter is given by (Exercise E11.6)

$$S = Nk\Big[\ln(2\cosh(\beta\epsilon)) - \beta\epsilon\tanh(\beta\epsilon)\Big]. \tag{11.22}$$

Comparing entropy (Figure 11.12) with energy (Figure 11.10), the trends are very similar. On closer examination of Eq. (11.22), we see the first term dominates at higher temperatures, but has a weaker temperature dependence because of the logarithm. The second term, which is proportional to $\langle E\rangle/T$ per Eq. (11.19), dominates at intermediate temperatures. As a result, the entropy curve takes on the trend of the average energy at $kT \sim 1$, but it levels off faster at higher temperatures due to the $1/T$ factor.

Summarizing, all thermodynamic quantities presented are continuous and smooth, including energy, magnetization, heat capacity, and entropy. Therefore, the 1D Ising model does not exhibit phase transitions. Per our discussion on magnetization, a spin has only two neighbors in 1D, and the Ising chain can flip-flop freely without energy constraint. It makes critical phenomena impossible in 1D. However, it is possible in the 2D Ising model discussed shortly.

The Metropolis Algorithm Analyzed

We see from preceding discussions that the 1D Ising model is well described by simulations with the Metropolis algorithm in comparison with analytic results both qualitatively and quantitatively. We wish to analyze more closely the underlying mechanism.

Let E_1 and E_2 be the energies of two states in a trial transition (flip), and assume $E_1 < E_2$. Further, let $R(i \to f)$ denote the transition rate from state i to state f. According to the Metropolis algorithm, transitions from states $2 \to 1$ occur with unit probability, or $R(2 \to 1) = 1$, because it is going from a state of higher energy (E_2) to a lower one $\left(E_1\right)$. Conversely, transition $1 \to 2$ has a rate $R(1 \to 2) = \exp(-\Delta E/kT)$ with $\Delta E = E_2 - E_1$. In equilibrium, we expect the number of transitions to be equal in both directions $1 \Leftrightarrow 2$. Because that number is proportional to the product of the transition rate and the probability of finding the system in the initial state, we have

$$R(1 \to 2)P(E_1) = R(2 \to 1)P(E_2), \tag{11.23}$$

where $P(E_1)$ and $P(E_2)$ are the probabilities of finding the system in states 1 and 2, respectively. Substituting the Metropolis transition rates into Eq. (11.23) we obtain

$$\exp(-\Delta E/kT)P(E_1) = P(E_2), \quad \text{or} \quad \frac{P(E_1)}{P(E_2)} = \frac{\exp(-E_1/kT)}{\exp(-E_2/kT)}. \tag{11.24}$$

This is precisely the Boltzmann distribution (11.8).

Therefore, the Metropolis algorithm leads the system, sometimes slowly but surely, to its correct thermodynamic equilibrium. This algorithm is essential to thermodynamic simulations and can be adapted to problems that are nonthermodynamic in nature (see Section 11.3).

11.2.3 THE 2D ISING MODEL

The 1D Ising model offers many useful insights, but does not exhibit phase transitions. The 2D Ising model addresses this problem. It consists of spins arranged in rows and columns on a lattice, say an $N \times N$ square grid for a total of N^2 spins (Figure 11.13).

As in the 1D case, we still assume nearest neighbor interactions. Each spin now interacts with the four nearest neighbors: up, down, left, and right, doubling the number of nearest neighbors relative to its 1D sibling. The energy for a given microstate is

$$E = -\epsilon \sum_{(i,j)} s_i s_j, \tag{11.25}$$

where the sum is over all (unique) nearest neighboring pairs, (i,j).

The Monte Carlo simulation of the 2D Ising model runs parallel to that of the 1D Ising model. Let $s_{i,j}$ be the spin picked for a trial flip. The change of energy if the spin were to flip is

$$\Delta E = 2\epsilon s_{i,j}\left(s_{i-1,j} + s_{i+1,j} + s_{i,j-1} + s_{i,j+1}\right). \tag{11.26}$$

This is similar to Eq. (11.17) but with two more neighbors. If the picked spin is on the edge, we use the periodic boundary conditions (11.16) horizontally (column-wise) and vertically (row-wise).

The Metropolis algorithm can be applied in nearly the same manner as before, allowing for minor differences. The equivalent module for the 2D Ising model is given here.

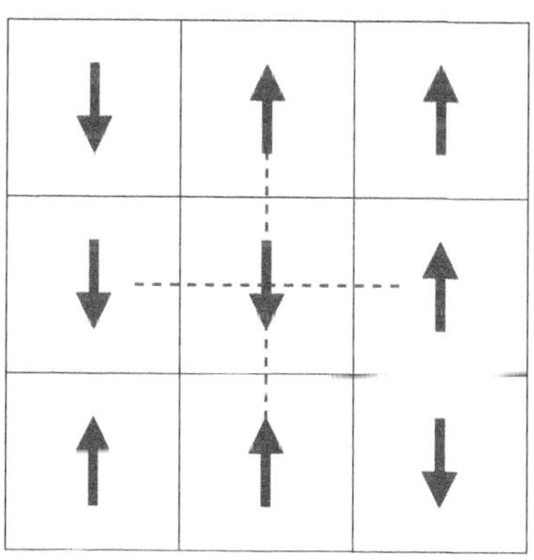

FIGURE 11.13 The 2D Ising model. Each spin interacts with the four nearest neighbors.

Program listing 11.3 Ising model in 2D (`metropolis2.py`)

```
def update2d(N, spin, kT, E, M):    # 2D Ising model, N x N lattice
    i, j, flip = rnd.randint(0, N-1), rnd.randint(0, N-1), 0
    dE = 2*spin[i][j]*( spin[i-1][j] + spin[(i+1)%N][j]
                   + spin[i][j-1] + spin[i][(j+1)%N] )
    if (dE < 0.0): flip=1            # flip if dE<0, else flip
    else:                            # according to exp(-dE/kT)
        p = np.exp(-dE/kT)
        if (rnd.random() < p): flip=1
    if (flip == 1):
        E = E + dE
        M = M - 2*spin[i][j]
        spin[i][j] = -spin[i][j]
    return E, M
```

Compared to the 1D case (Program 11.6), the only difference is that the spins are stored in a 2D $N \times N$ list. To select a spin, we draw a pair of random integers (i,j) representing the column and row indices, respectively. The change of energy is calculated as discussed earlier from Eq. (11.26). The rest is identical to the 1D case.

Next we discuss simulation results and leave the details of modeling to Project P11.5. We consider a system made up of 1024 spins arranged on a 32×32 square lattice. First, we show the spin alignment in Figure 11.14 at several temperatures. Starting from random initial states, the system is sampled after waiting sufficiently long so it has reached equilibrium. The domains may shift depending on when the sampling takes place, but the ratio of "up" to "down" spins should be constant within thermal fluctuations.

At the lowest temperature shown ($kT = 1$), all spins are aligned spontaneously (up). With increasing temperature, more down spins gradually appear. At the higher temperatures $kT \geq 2.8$, the numbers of up and down spins are roughly equal, a totally random distribution. However, something dramatic seems to occur between $kT = 2.2$ and 2.4, which upsets the gradual change in the up/down ratio. The abrupt change is a telltale sign of phase transitions [82].

Phase Transitions

To investigate further, we show the equilibrium energy and magnetization as a function of temperature in Figure 11.15. The energy is continuous and looks qualitatively similar to the 1D case. At low temperature, the energy per spin approaches -2ϵ, doubling the amount in the 1D Ising model. In 2D, a spin has four nearest neighbors. Because they are all aligned, the total energy is -4ϵ. Because this energy is mutual between each of the four pairs, the energy per spin is half the total. Another property, difficult to visually spot, is a discontinuity of the slope between $kT = 2.2$ and 2.4, the same interval observed earlier (Figure 11.14). Clearly, this is no coincidence.

The magnetization shows that the critical behavior is a phase transition occurring at the critical temperature $kT_c/\epsilon \simeq 2.2692$. Above T_c, the magnetization is identically zero. Below T_c, it rapidly rises to full, spontaneous alignment within a small temperature window.

The analytic results in Figure 11.15 are given in Section 11.B. The energy and magnetization per spin are [41]

$$\langle E\rangle/N\epsilon = -2\tanh 2\beta\epsilon - 2\frac{\sinh^2 2\beta\epsilon - 1}{\sinh 4\beta\epsilon}\left(\frac{2}{\pi}\mathsf{K}(m) - 1\right), \tag{11.27}$$

$$\langle M\rangle/N = \begin{cases} \left(1 - \sinh^{-4} 2\beta\epsilon\right)^{1/8}, & T < T_c, \\ 0, & T > T_c. \end{cases} \tag{11.28}$$

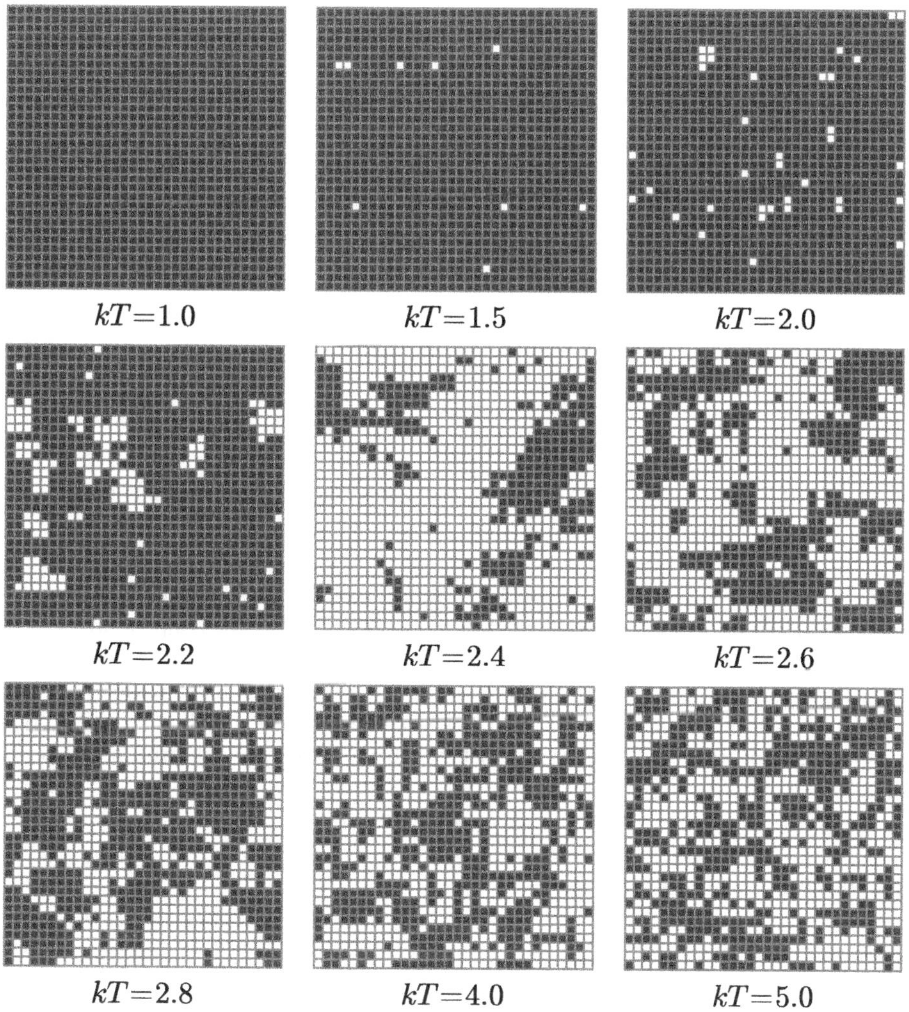

FIGURE 11.14 The Ising square lattice (32 × 32) in equilibrium. Filled squares (■) are "up" spins, and open squares (□) are "down" spins.

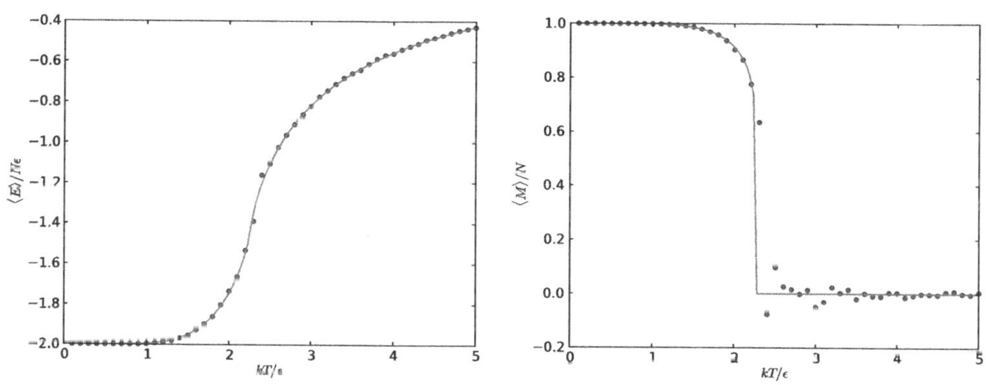

FIGURE 11.15 The energy and magnetization per spin as a function of temperature in the 2D Ising model. The solid lines are the analytic results. The phase transition occurs at $T_c = 2.2692\ \epsilon/k$.

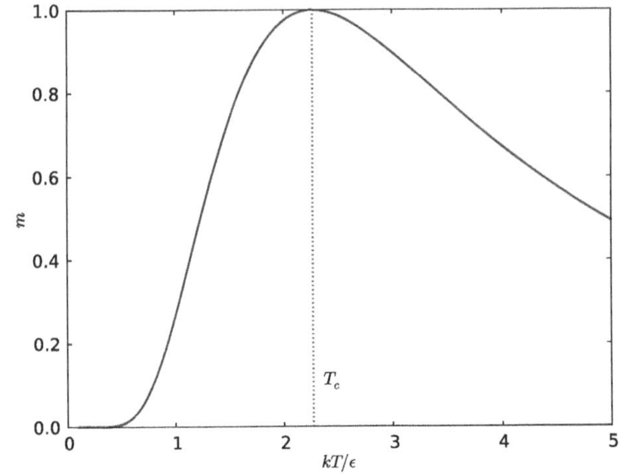

FIGURE 11.16 The parameter m (11.57) as a function of temperature. The value of m is 1 at $T_c = 2.2692$.

The function K(m) in Eq. (11.27) is the complete elliptic integral of the first kind (11.61), and m is a temperature-dependent dimensionless parameter (11.57). The temperature dependence of m is shown in Figure 11.16.

The value of m approaches 1 from both sides of T_c, the critical temperature. However, the function K(1) is infinite because of a logarithmic singularity at $m = 1$ (Figure 11.27). Because energy must remain finite, we require the factor in front of the second term in Eq. (11.27) to vanish at T_c,

$$\sinh(2\beta_c\epsilon) = \sinh(2\epsilon/kT_c) = 1, \tag{11.29}$$

yielding the critical temperature

$$kT_c/\epsilon = \frac{2}{\ln(1+\sqrt{2})} \simeq 2.2692. \tag{11.30}$$

The energy at the critical temperature is $\langle E(T_c)\rangle/N\epsilon = -\sqrt{2}$.

From the magnetization in Figure 11.15, we see the system undergoing a phase transition at the critical temperature T_c, known as the Curie temperature. Spin orientations are completely random just above T_c, whereas they spontaneously order themselves just below. Note, however, spins of the same orientation tend to "stick" together due to the ferromagnetic nature, forming pockets of various sizes like pieces on a go board (Figure 11.14). With increasing temperature, the sizes of the pockets become smaller.

Expanding Eq. (11.28) around T_c, we obtain

$$\langle M\rangle/N \sim \left(1 - \frac{T}{T_c}\right)^{1/8}, \quad T \lesssim T_c. \tag{11.31}$$

The power law scaling like Eq. (11.31) is a common feature in phase transitions. The critical exponent in this case is 1/8. Because the magnetization changes continuously from zero across T_c (even though its derivative is discontinuous), the phase transition may be classified as a continuous one.

Figure 11.17 displays the heat capacity as a function of temperature. The most prominent feature is the sharp spike at the critical temperature, another sign of possible phase transitions.

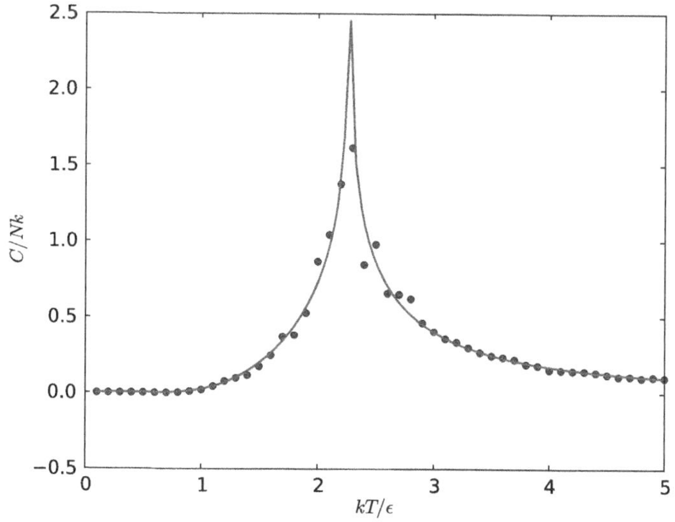

FIGURE 11.17 The heat capacity per spin as a function of temperature in the 2D Ising model. The solid line is the analytic result.

Unlike energy, there is no requirement that the heat capacity be finite. The critical behavior around T_c can be derived from the analytic expression (11.60),

$$C/Nk \sim A \ln \left|1 - \frac{T}{T_c}\right|. \tag{11.32}$$

This is a logarithmic discontinuity of the derivative of energy. The agreement between simulation and analytic results is good in regions away from the singularity around T_c. Near the singularity, the fluctuations in energy are large, resulting in larger scatter in the heat capacity.

We can now compare the 1D and 2D Ising models and understand why the 1D model does not show phase transitions. In 1D, we can reverse the whole spin chain without energy cost. The initial spin flip may require some extra energy, but it is given back at the end. This is always possible at finite temperatures. The exception is $T = 0$, when the spins are fully aligned. The Ising chain cannot reverse direction spontaneously without additional energy.

In the 2D model, a spontaneous full reversal of the lattice is not possible once the temperature is below some critical value when net magnetization becomes nonzero. On average, additional energies are required to get the process going due to the constraints of the domain walls and neighbors. For instance, consider one row of the spin lattice, the equivalent of a 1D spin chain. This row cannot flip spontaneously because of the interactions with the rows just above and below.

Finally, we discuss finite-size (or edge) effects. All our results have been obtained with a square lattice of 32×32, tiny compared to realistic systems. In most cases, our results are in good agreement with analytic results for $N \to \infty$. We conclude that finite-size effects are small except near phase transitions. Certainly, periodic boundary conditions help to minimize these effects. Nonetheless, we expect the accuracy to improve with larger systems in the region surrounding the phase transition.

In Figure 11.18, we show a larger Ising model containing 1048576 spins on a 1024×1024 square lattice at the critical temperature $T_c = 2.2692$. The initial microstate was hot with random spin directions. The snapshot was taken when equilibrium was well established after each spin had on average 3000 chances to flip, roughly 3×10^9 iterations. The energy of the sampled

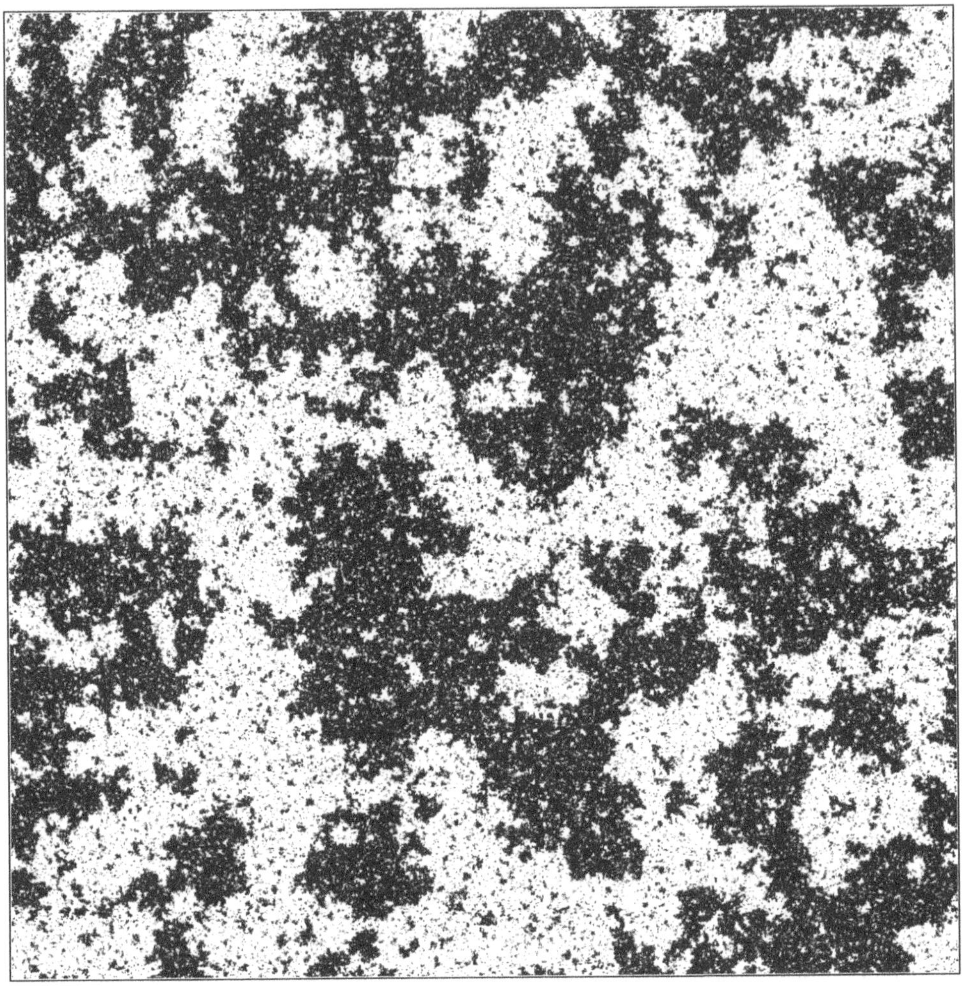

FIGURE 11.18
Snapshot of a 2D Ising model on a square lattice (1024×1024, 1048576 spins) in equilibrium at the critical temperature $T_c = 2.2692$.

state is -1.399, within $\sim$ 1% of the analytic value $-\sqrt{2}$. The error remains stable. Analytically, the magnetization should be zero, compared to the sampled value -9.7×10^{-4}. It means that only about 500 spins need to flip to achieve a perfect zero magnetization. Generally, the sampled magnetization has a higher fluctuation than energy, typically in the range ± 0.02.

We again see the tendency of parallel spins sticking together as the lattice is covered by large connected domains (continents), some spanning the full extent of the lattice. In the limit $N \to \infty$, the continents grow to infinity. But there are also smaller islands of varying sizes down to single sites, coexisting with the continents.

11.3 THERMAL RELAXATION BY SIMULATED ANNEALING

As the previous examples illustrate, a thermal system at high temperatures will typically be energetic and well mixed. As temperature decreases, it will become more ordered (homogeneous) and equilibrate toward a lower energy state. In the limit $T \to 0$, the system should approach the ground state, provided the temperature decreases gradually. This can take a long time.

This thermal relaxation occurs naturally in many processes such as annealing in materials. For example, heating a metal to higher temperatures and cooling it slowly can improve its structure and properties. Heating helps atoms to mix up with their neighbors, and gradual cooling ensures that they have enough time to find the lowest possible energy states and the most stable structure.

We can simulate the annealing process with the Metropolis method. There are many applications in which simulated annealing is useful and where the problems may not be thermodynamic in nature, including minimization problems in particular where finding the global minimum is essential. In these problems, we first find the equivalent parameters that play the role of energy and state (configuration), then apply the Metropolis method, starting from high to low temperature. At each temperature, sufficient time (Monte Carlo samplings) is allowed for equilibration. The temperature is then lowered, and the process is repeated. Hopefully, the system will be gradually brought to the lowest possible state at the lowest temperature. Next we discuss an example (more examples can be found in Section S:11.1).

The Falling Tablecloth Revisited

Let us revisit the relaxation of a free hanging tablecloth discussed in Section 6.8, where it was modeled as a square array of particles (Figure 6.20) connected by springs.

The relaxation problem was solved by integrating equations of motion in Section 6.8. If we treat it as a thermal process, it is essentially a minimization problem. Given a system of $N \times N$ particles of mass m on a square lattice of side L, suspended from the corners, and connected via springs of relaxed length l and spring constant k, what is the minimum energy of the system?

To solve this problem by simulated annealing, we randomly select a particle at grid (i, j). We then jiggle it slightly, causing a displacement of $\Delta\vec{r} = (\Delta x, \Delta y, \Delta z)$, and a new position vector of $\vec{r} = \vec{r}_{i,j} + \Delta\vec{r}$. We assume the (x, y) coordinates in the horizontal plane, and z in the vertical plane. The change of potential energy (see also Eq. (S:11.1)) for this proposed move is

$$\Delta E = \frac{1}{2}k \sum_n \left[(d'_n - l)^2 - (d_n - l)^2\right] + mg\Delta z, \tag{11.33}$$

where n denotes the nearest neighbors at $(i \pm 1, j)$ and $(i, j \pm 1)$, $d'_n = |\vec{r} - \vec{r}_n|$ is the new distance to the neighbor n, and d_n the original distance.

Having found ΔE, it is straightforward to implement the Metropolis method (Project P11.6). We show the results in Figure 11.19 for a $13 \times 13 = 169$ particle system. The parameters k, l, and m are the same as in Program 6.8.

We start from the flat configuration (Figure 11.19, upper left, $E = 0$) at $T = 10^{-3}$ (set $\epsilon = 1$ as usual). The next state (upper right) was taken after a few passes so the system is not in

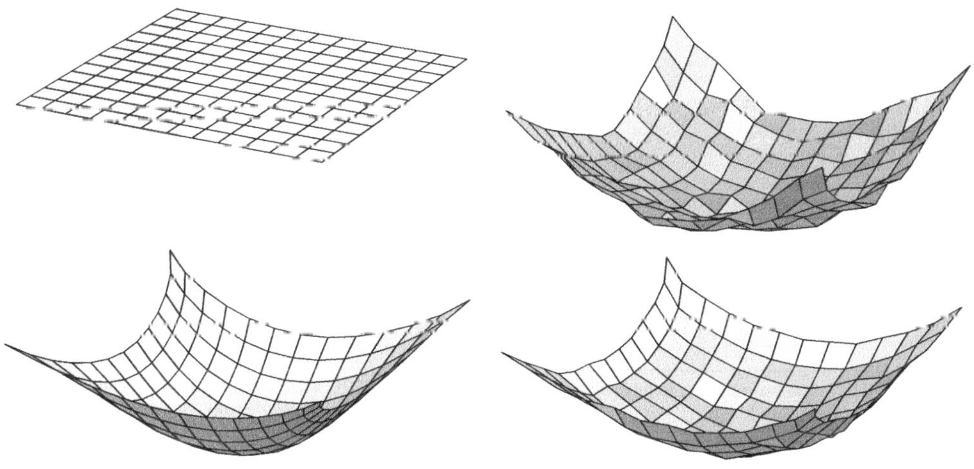

FIGURE 11.19 Snapshots of a hanging tablecloth in thermal relaxation. Starting from a flat surface (upper left), the temperature decreases in clockwise order, ending at $kT/\epsilon = 10^{-7}$ and a minimum energy -2.718ϵ.

equilibrium yet. Its energy is becoming negative, but the surface is rather rough. Temperature is decreased in clockwise order, and the next two states (lower right and left) are snapshots in equilibrium at $T = 5 \times 10^{-4}$ and 10^{-7}, respectively. The energy changes very little after the last state. The final energy is -2.718ϵ. Though the exact solution is unknown, it should be very close to the global minimum energy.

Compared to the direct integration method (Figure 6.21, Section 6.8), simulated annealing is simpler to implement, but is slower to converge to the global minimum. In direct integration, if the damping coefficient is chosen properly (at critical value), the system can relax to the global minimum quickly. There is no single parameter in simulated annealing to control this rate. It is the nature of thermal systems that changes occur very slowly at low temperatures. But simulated annealing is more efficient at exploring a larger parameter space and can be applied to solving very difficult problems such as the traveling salesman problem (Section S:11.1).

11.4 MOLECULAR DYNAMICS

We began this chapter by choosing statistical methods over first-principle calculations. Now we return to the latter for molecular dynamics, which simulates the motion of N-body systems (atoms or molecules) classically. Because the masses of atoms are much larger than the electron mass ($\sim$ 2000 times), the motion of an atom can be adequately described classically by Newton's laws at all but very low temperatures.

The reason for first-principle molecular dynamics simulations is not to replace statistical mechanics but to complement it. For instance, if we wish to study kinetic and thermal properties of gases such as Maxwell speed distributions (Section 11.4.4), we can use molecular dynamics to follow the motion of each atom classically and obtain thermal properties statistically.[6] Furthermore, many systems such as molecular protein or motor simulations involve thousands to millions of atoms in nonequilibrium motion, and molecular dynamics is a very useful tool. It can be used to study atomic and chemical structures or reactions because the internuclear degrees of freedom are effectively separated from the electronic degrees of freedom.

11.4.1 INTERACTION POTENTIALS AND UNITS

In a sense, molecular dynamics, an N-body problem, is an extension of the few-body problems discussed in Chapter 4. For instance, we can generalize the three-body equations of motion (4.42) to an N-body system as

$$\frac{d\vec{r}_i}{dt} = \vec{v}_i, \tag{11.34}$$

$$\frac{d\vec{v}_i}{dt} = \frac{1}{m_i}\sum_{j\neq i}\vec{f}_{ij}, \tag{11.35}$$

where $\vec{f}_{ij}$ is the force between atoms i and j. We typically consider systems of identical particles, so the mass $m_i = m$ will be the same for all N atoms.

Unlike gravity, the forces $\vec{f}_{ij}$ are Coulombic in nature, and there are no exact expressions for them due to many-body interactions (electrons and nuclei). For practical reasons, we use phenomenological forces that approximate the interactions between atoms, including the

[6]It is as if Richard Feynman was referring to molecular dynamics when he wrote [29] "what statement would contain the most information in the fewest words? I believe it is the *atomic hypothesis* that *all things are made of atoms—little particles that move around in perpetual motion, attracting each other when they are a little distance apart, but repelling upon being squeezed into one another*." feynmanlectures.caltech.edu/I_01.html

Morse potential (9.37) (Figure 9.10) and the Lennard-Jones potential (Project S:P9.3). The latter is a two-parameter empirical potential popular among chemists and commonly used in molecular dynamics simulations

$$V(r) = V_0 \left[\left(\frac{r_0}{r}\right)^{12} - 2\left(\frac{r_0}{r}\right)^{6} \right], \tag{11.36}$$

where V_0 is a positive constant and r_0 the equilibrium internuclear distance (Figure 11.20). The first term, $1/r^{12}$, describes the strong repulsive wall for $r \ll r_0$ where the electrons repel each other (Coulomb and Pauli exclusion), but the 12th power law is chosen more for convenience than for fundamental significance. The second term, $1/r^6$, describes the long-range attraction between neutral atoms due to dipole–dipole interactions (van der Waals force).[7] The switch-over between repulsive and attractive forces occurs at r_0, where the potential is minimum at $-V_0$.

The force on atom i by atom j is equal to the gradient $-\nabla V$,

$$\vec{f}_{ij} = 12\frac{V_0}{r_0}\left[\left(\frac{r_0}{r_{ij}}\right)^{6} - 1\right]\left(\frac{r_0}{r_{ij}}\right)^{8}\frac{\vec{r}_{ij}}{r_0}, \tag{11.37}$$

where $\vec{r}_{ij} = \vec{r}_i - \vec{r}_j$ is the relative position vector between atoms i and j, and $r_{ij} = |\vec{r}_i - \vec{r}_j|$ is the magnitude. As Figure 11.20 shows, the force behaves qualitatively similar to the potential, a sharply rising $(1/r^{13})$ repulsive force for small r and a rapidly decreasing $(1/r^7)$ attractive force for large r. In between, the force is most negative at $r = (13/7)^{1/6} r_0 \sim 1.11 r_0$, slightly larger than the equilibrium distance, with the value $f \sim -2.69 V_0/r_0$. This force is effective within a small radius $r/r_0 \lesssim 2$ and is negligible for larger values $r/r_0 \gtrsim 3$. It is a short-ranged force. We will use Eq. (11.37) in the equations of motion (11.35).

Like before, we need to choose a convenient unit system for the problem at hand. We could use atomic units (Table 8.1), but it is unwieldy as the mass would be unnaturally large and velocities and energies very small. Because we will be using the Lennard-Jones potential, we shall define energy in units of V_0, length in units of r_0 as listed in Table 11.1.

Like the unit system for restricted three-body problems (Table 4.3), Table 11.1 is also a floating unit system whose absolute scale is determined by the Lennard-Jones parameters and the mass of atoms. In this sense, simulation results with this unit system are universal. It is particularly

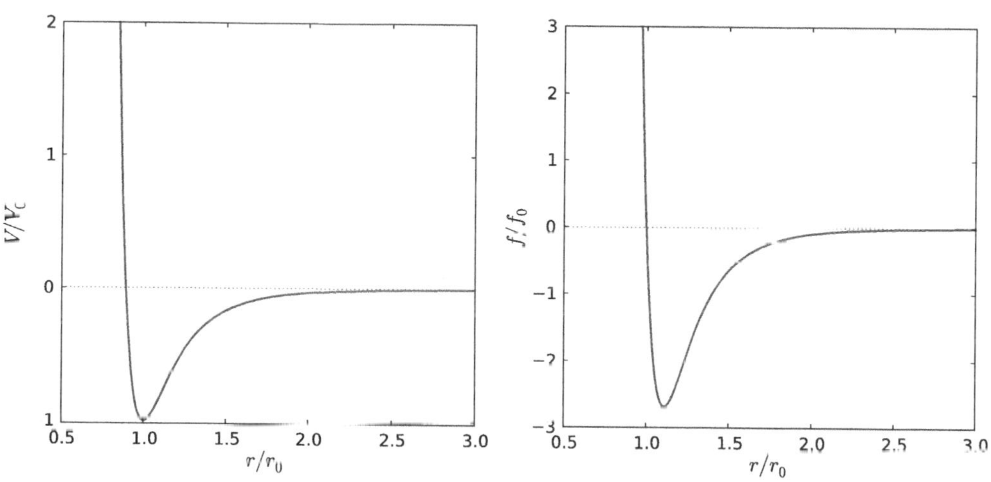

FIGURE 11.20 The Lennard-Jones potential and force.

[7] Naturally, the Lennard-Jones potential is also given the moniker of "6–12" potential.

Table 11.1 Unit System for Molecular Dynamics Based on the Lennard-Jones Potential.

Quantity	Energy	Length	Mass	Time	Velocity	Force
Units	V_0	r_0	m	$r_0(m/V_0)^{1/2}$	$(V_0/m)^{1/2}$	V_0/r_0

tidy now for numerical work, for we can just drop V_0, r_0 and m from Eqs. (11.37) and (11.35). For example, the parameters for argon are $V_0 = 0.01$ eV, $r_0 = 3.9$ Å, and $m = 40$ amu (6.7×10^{-26} kg). The other derived units are: velocity, $v_0 = 150$ m/s; time, $t_0 = 2.5 \times 10^{-12}$ s; temperature $T_0 = 120$ K; and pressure, $P_0 = 2.7 \times 10^7$ Pa = 270 atm.

11.4.2 PERIODIC BOUNDARY CONDITIONS

There is just one more problem before we start writing the program, and it is this: what boundary conditions to use? In actual thermal systems, there are around 10^{23} particles. Collisions among the particles determine the thermal properties of the systems. Boundary effect are negligible, for we know that air in a ball, a bottle, or a box behaves the same way. We cannot possibly keep track of that many particles. However, we have seen that it is possible to simulate behaviors of real thermal systems with much smaller model systems if we are careful with boundary conditions.

Suppose we set up the system in a cube, for instance, and let the atoms interact and move about. Some atoms will eventually collide with the walls. To prevent them from escaping, we could calculate the point of collision and reflect the atom off the wall elastically, as was done in the stadium billiard problem (Section S:5.1). For a few hundred or thousand particles, collisions with the boundaries will be much more frequent than collisions between the particles themselves. So the boundaries will artificially affect the behaviors, contrary to how real systems behave.

It turns out that, again, we can call on periodic boundary conditions to come to the rescue. We imagine that identical cubes containing identical but virtual systems surround our central cube [42]. For every atom moving in our system, there is an atom with exactly the same position (relative) and velocity in every other virtual cube (Figure 11.21). If an atom escapes our cube, say from the right side (Figure 11.21, atom ❶), another one (①) enters from the left side to replace it. This way, we do not have to worry about collisions with the walls because there are no collisions, and we never lose atoms. When a particle exits the cube on one side, it is "teleported" to the opposite side (①↶❶, Figure 11.21) without altering its velocity.

But teleporting atoms this way creates another problem. As Figure 11.21 illustrates, the distance between atoms ❶ and ❷ changes abruptly when the former is teleported, from r to r'.

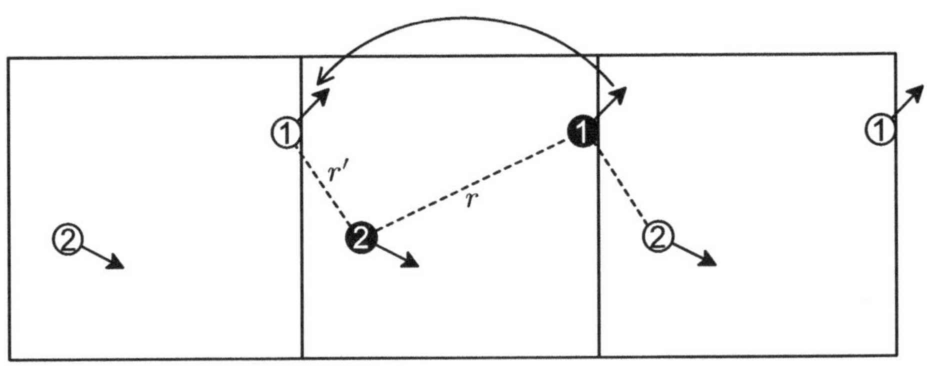

FIGURE 11.21 Periodic boundary conditions in molecular dynamics simulations. The central real box is surrounded by virtual copies all around.

This causes a discontinuous and unphysical jump in the potential energy and the force. Worse, the teleported atom can exert such a strong, and sudden, force on the atoms already in the close vicinity of its new position that some can be ejected from the system in just a few steps due to numerical error, causing the system to disintegrate.

A careful look of Figure 11.21 between the real atom ❶ and the virtual atom ② (in the right box) reveals that if we use the distance r' instead of r, we can avoid the discontinuity in the force. It suggests that we should use the smaller of the two horizontal separations, $x_{❶} - x_{❷}$ and $x_{❶} - x_{②}$, between the atoms. Physically, this means that an atom always finds another atom, be it a real atom or a virtual clone, that is closer along a given axis to interact with. Let us call this close-neighbor interaction.

Examination of Figure 11.21 shows that, within close-neighbor interaction, the largest horizontal or vertical separation between two atoms (real or virtual) is equal to half the width of the box. For example, if we shift ❶ horizontally between ❷ and ②, the largest separation occurs when ❶ is exactly in the middle of the two atoms.

Let L be the width of the box, and $\Delta x = x_1 - x_2$. If $|\Delta x| \leq \frac{1}{2}L$, we accept it as is. But if $|\Delta x| > \frac{1}{2}L$, we require the following correction

$$\Delta x \to \begin{cases} \Delta x - L, & \text{if } \Delta x > \frac{1}{2}L, \\ \Delta x + L, & \text{if } \Delta x < -\frac{1}{2}L. \end{cases} \qquad \text{(periodic BC)} \tag{11.38}$$

Similar relationships hold for y and z directions.

With the simple and elegant periodic boundary conditions (11.38) as described earlier, we solve both the boundary-wall collision problems and the abrupt jumps in interaction energies and forces. There is one drawback, however. At exactly $|\Delta x| = \frac{1}{2}L$, the force is ambiguous. Using atom ❶ as the example (Figure 11.21), the force would change sign depending whether it interacts with ❷ to the left or with the virtual clone, ②, to the right. We justify this situation by arguing that, for a box size $L \gg r_0$, the forces between atoms separated at $\frac{1}{2}L$ are small as to be negligible. For the Lennard-Jones potential (Figure 11.20), this is true to a very good degree of approximation because the force is short-ranged and is negligible for $r \gtrsim 3$ as stated earlier (Figure 11.20). Take $L = 10$, for instance, the magnitude of the force at $\frac{1}{2}L$ is $\sim 10^{-4}$. In this case, we expect that the periodic boundary conditions should work well.

Of course, if there were significant spurious effects arising from the use of Eq. (11.38), we would need to amend our approach. Fortunately, Eq. (11.38) works well for molecular dynamics simulations using potentials that decrease rapidly for large r as does the Lennard-Jones potential.

11.4.3 SIMULATION AND VISUALIZATION

We can now write a molecular dynamics program employing periodic boundary conditions. Taken from the complete Program 11.7, the heart of the simulation computing the core N-body dynamics is given next.

```
def nbody(id, r, v, t):                  # N-body MD
    if (id == 0):                        # velocity
        return v
    a = np.zeros((N,3))                  # acceleration
    for i in range(N):
        rij = r[i]-r[i+1:]               # rij for all j>i
        rij[rij > HL] -= L               # periodic bc
        rij[rij < -HL] += L
        r2 = np.sum(rij*rij, axis=1)     # |rij|^2
```

```
10          r6 = r2*r2*r2
11          for k in [0,1,2]:                    # L-J force in x,y,z
12              fij = 12.*(1. - r6)*rij[:,k]/(r6*r6*r2)
13              a[i,k] += np.sum(fij)
14              a[i+1:,k] -= fij                  # 3rd law
15      return a
```

The module `nbody()` returns the equations of motion (11.34) and (11.35) in the required format for leapfrog integration. The variables `r` and `v` are arrays of vectors, effected by 3-element NumPy arrays, containing the positions and velocities of the atoms. If `id=0`, velocities are returned. Otherwise, acceleration is calculated in an optimized manner with NumPy arrays.

For a given atom i, we obtain the forces, $\vec{f}_{ij}$, between it and all higher-numbered atoms $j = i + 1$ to N (of course $j \neq i$, an atom does not exert forces on itself). The forces $\vec{f}_{ji} = -\vec{f}_{ij}$ are found by Newton's third law, reducing computation by half. The cumulative effect is that for atom i we only need to find $\vec{f}_{ij}$ for $j > i$ because the other forces ($j < i$) had been found already in prior iterations. For instance, when $i = 1$, forces $\vec{f}_{12}$, $\vec{f}_{13}$, and so on, will be calculated. When $i = 2$, forces $\vec{f}_{21}$, $\vec{f}_{23}$, and so on, are needed, but $\vec{f}_{21} = -\vec{f}_{12}$ was already available and recorded and need not be recalculated.

Normally, one would need an explicit inner loop over j, but we avoid it for the sake of speed. We first find the relative position vectors (line 6) $\vec{r}_{ij} = \vec{r}_i - \vec{r}_j$ between i and all $j > i$, simultaneously. Here, $\vec{r}_i$ is broadcast into a $(N - i - 1) \times 3$ array (shape of `r[i+1:]`), equivalent to itself duplicated $N - i - 1$ times, and then $\vec{r}_j$ for $j > i$ is subtracted from it via element-wise operations. Next, the components of $\vec{r}_{ij}$ are checked for periodic boundary conditions (11.38) to ensure close-neighbor interactions. This is accomplished with truth array indexing (see Section 1.D, Section 4.4.2), so that separations greater than $L/2$ in any direction (e.g., $|x_{ij}| > L/2$) are mapped according to Eq. (11.38), again all at once. The distances squared between the atoms are obtained by summing $|r_{ij}|^2$ over the three directions via `np.sum` along the second axis, that is, across columns (line 9). It is slightly faster to compute r^6 and r^{14} from r^2 using multiplications rather than powers (`r**n`) which is slower for low `n`.

Finally, each component of the net force on atom i due to atoms $j > i$ is summed up, and the same component on atom j due to i is obtained via Newton's third law. On exit, accelerations are returned as an $N \times 3$ array of vectors.

Figure 11.22 shows two snapshots of an animated molecular dynamics simulation. We used $N = 16$ atoms, initially randomly distributed in a cube of side length $L = 10$. The initial velocities were set randomly between $[-1, 1]$. We can see from the trails that close-by atoms interact strongly (sharp turns), and distant ones interact weakly (nearly straight line trajectories). In particular, the dark-colored (red) atom (Figure 11.22, left) suffers a strong force near the beginning, being close to several atoms to the right. It makes a sharp turn toward the left wall. After a short time, it exits the cube and reenters from the right side by the periodic boundary conditions. For the duration shown between the two frames, this is the only atom escaping the cube. But several atoms near the top and back in the second frame are just about to exit the cube. When they do, the same periodic boundary conditions will be applied, and they will continue to interact with other atoms without collisions with boundary walls.

11.4.4 MAXWELL DISTRIBUTIONS AND KINETIC PROPERTIES

We can obtain a slew of properties by analyzing relevant data. We show only results and leave the work to Project P11.7. All results are in the floating-scale units (Table 11.1). Figure 11.23 displays the velocity $P(v_x)$ (similar for v_y and v_z) and speed $P(v)$ distributions. We divide the

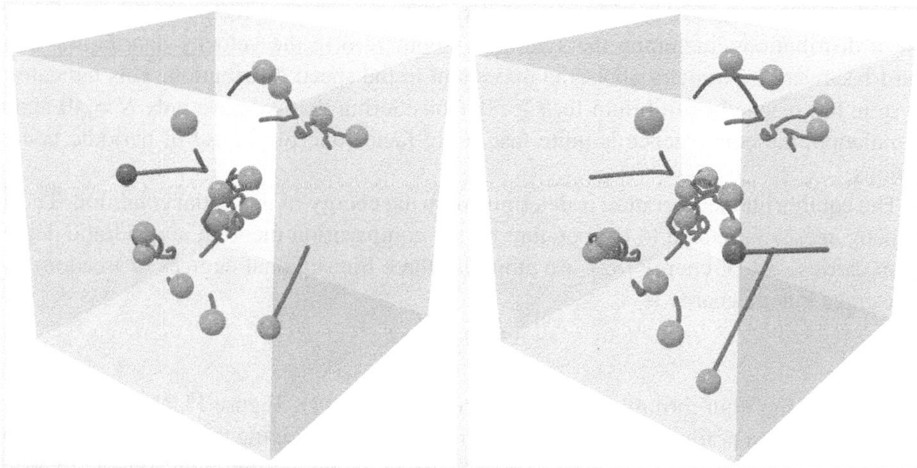

FIGURE 11.22 Animated molecular dynamics simulation. The left-most particle (red/dark color) in the left frame exited the cube and is wrapped around by periodic boundary conditions (right frame).

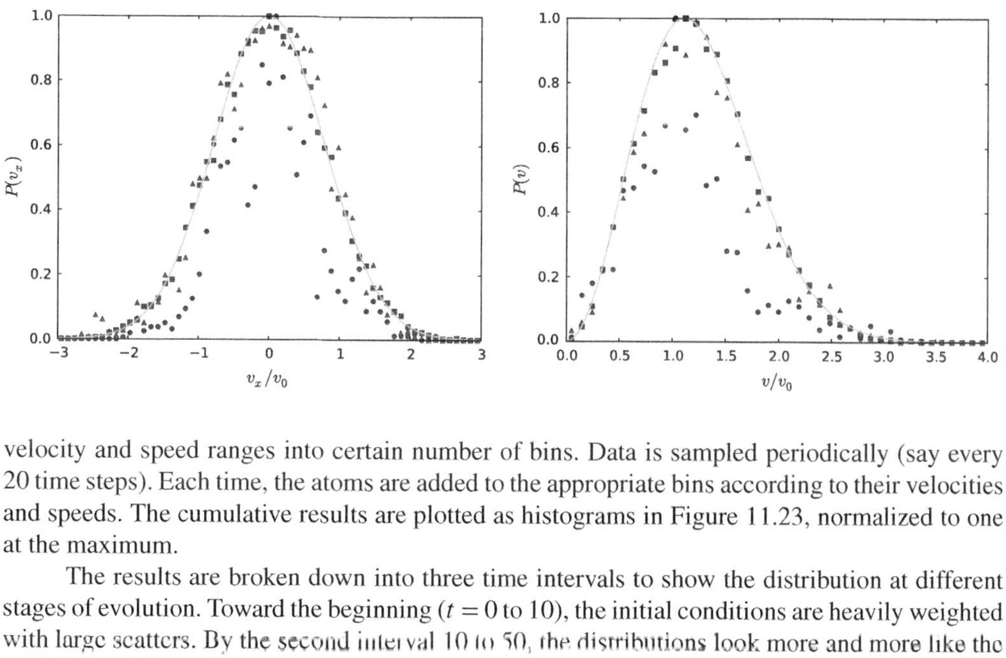

FIGURE 11.23 Velocity (left) and speed (right) distributions for $N = 40$ atoms. Data are taken in different time intervals: ●, $t = 0$ to 10; ▲, 10 to 50; ■, 50 to 400. The solid curves are the Maxwell distributions.

velocity and speed ranges into certain number of bins. Data is sampled periodically (say every 20 time steps). Each time, the atoms are added to the appropriate bins according to their velocities and speeds. The cumulative results are plotted as histograms in Figure 11.23, normalized to one at the maximum.

The results are broken down into three time intervals to show the distribution at different stages of evolution. Toward the beginning ($t = 0$ to 10), the initial conditions are heavily weighted with large scatters. By the second interval 10 to 50, the distributions look more and more like the equilibrium Maxwell velocity distribution [82]

$$P(v_x) = \left(\frac{m}{2\pi kT}\right)^{1/2} \exp\left(-\frac{mv_x^2}{2kT}\right), \tag{11.39}$$

and the speed distribution

$$P(v) = 4\pi \left(\frac{m}{2\pi kT}\right)^{3/2} v^2 \exp\left(-\frac{mv^2}{2kT}\right). \tag{11.40}$$

Roughly speaking, Eqs. (11.39) and (11.40) are a result of the Boltzmann distribution (11.8), with $E_x = \frac{1}{2}mv_x^2$ and $E = \frac{1}{2}mv^2$, respectively. The prefactor $4\pi v^2$ in Eq. (11.40) comes from the volume element $d^3v = 4\pi v^2 dv$.

In the last interval (50 to 400), the results are in very good agreement with the respective Maxwell distributions, including the symmetry about zero in the velocity distribution and the forward-backward asymmetry about the maximum in the speed distribution. This indicates that the system has reached equilibrium for $t \gtrsim 50$. Considering that we have only $N = 40$ atoms in the simulation, the convergence is quite fast, a big factor due to the use of periodic boundary conditions.

The equilibrium temperature is determined by the energy of the initial condition. The average kinetic energy is related to temperature by the equipartition theorem: a quadratic degree of freedom carries $\frac{1}{2}kT$ of energy [82]. An atom has three translational degrees of freedom in 3D, so its average kinetic energy is

$$\frac{1}{2}m\langle v^2\rangle = \frac{3}{2}kT. \tag{11.41}$$

We can obtain the equilibrium temperature from Eq. (11.41). Figure 11.24 shows the speed squared as a function of time. It is calculated periodically by averaging the speed squared v^2 of all atoms at once. The average value of v^2 rises steadily in the beginning, indicating an oversupply of potential energy in the initial condition. It quickly reaches a plateau around $\langle v^2\rangle \sim 1.85$ at $t \sim 40$, consistent with the equilibration time observed from Figure 11.23. Using this value, the temperature from Eq. (11.41) is $kT = \langle v^2\rangle/3 = 0.62$. We have used this temperature in the Maxwell velocity and speed distributions in Figure 11.23.

There is considerable fluctuation in v^2. The main cause is due to the small number of atoms. During motion, some atoms can "bunch" up, creating local concentrations that have higher or lower than normal potential energies. As a result, kinetic energies rise or fall accordingly, leading to the observed fluctuations.

We can also obtain pressure from our simulation (Figure 11.25), even though using periodic boundary conditions we have managed to avoid any atom-wall collisions, which generates pressure in real systems. Pressure is calculated as follows. When an atom escapes the cube, say from the side, we record its velocity v_x before teleporting it. We assume that, as in a real system without periodic boundary conditions, the atom would be spectrally reflected from the wall, reversing the x velocity to $-v_x$. The change of momentum (impulse) would be $\Delta p_i = -2mv_x$.

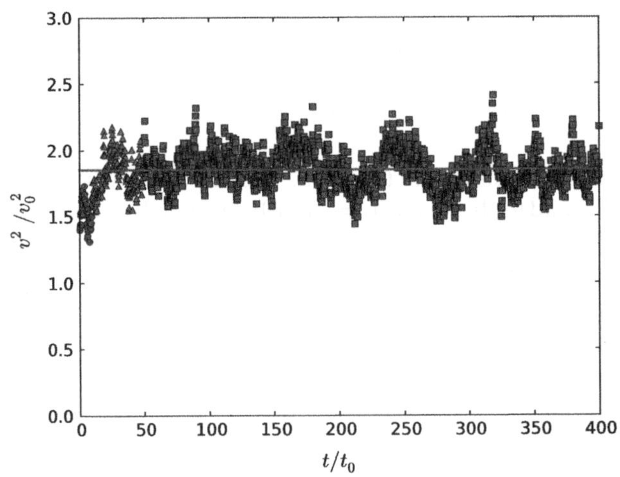

FIGURE 11.24 The average of speed squared as a function of time. The system is the same as in Figure 11.23.

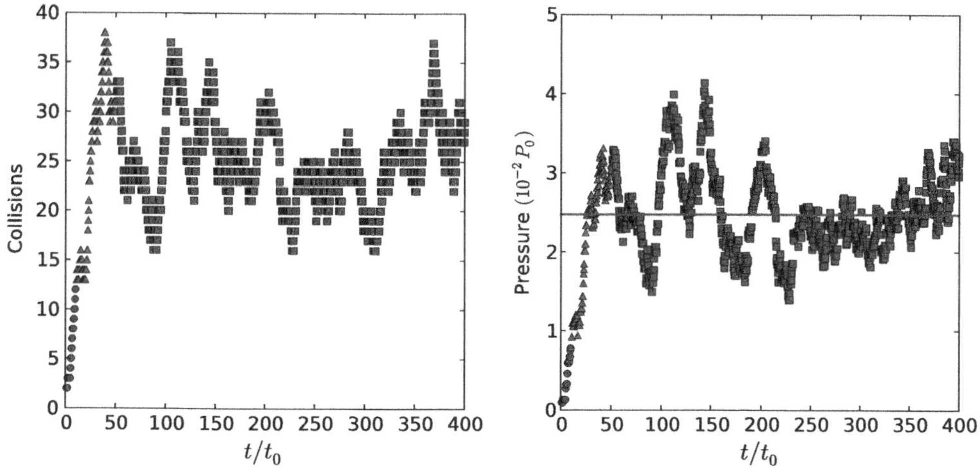

FIGURE 11.25 The number of collisions (left) and the average pressure (right) as a function of time. The system is the same as in Figure 11.23. The solid curve shows the expected pressure for an ideal gas at $kT = 0.62$.

We keep cumulating the impulse within a given time interval Δt (e.g., 20 time steps). The force exerted on the wall would be $F = \sum_i \Delta p_i/\Delta t$, which yields the pressure, $P = F/L^2$.

Figure 11.25 shows the number of collisions per $\Delta t = 0.2$ (left) and the pressure (right) obtained in the manner described. For this simulation, there are on average about 30 atoms escaping one side of the cube in each interval (about 1.5 atoms per step), a relatively low number. The fluctuations are again mainly due to the atoms bunching up as discussed earlier. The pressure generally follows the trend in the number of collisions as expected, but it is modified by the momentum carried by the atom. We also show the pressure for an ideal gas, $PV = NkT$, assuming the equilibrium temperature found earlier. There is good agreement between numerical results and the ideal gas law. We conclude that although our gas is not ideal because of short-range interactions instead of point-like hard spheres, the system is sufficiently dilute that the ideal gas law is approximately valid.

However, the actual equation of state for an interacting gas deviates from the ideal gas law. According to the theory of weakly interacting gases, the pressure can be expanded as

$$P/kT = n + B_2(T)n^2 + B_3(T)n^3 + ..., \tag{11.42}$$

where $n = N/V$ is the particle density. The constants B_2 and B_3, and so on, are called the virial coefficients [76], which are zero for ideal gases. For low-density gases, the leading correction is the second term, $B_2(T)n^2$. The second virial coefficient B_2 is given by

$$B_2(T) = -2\pi \int_0^\infty (e^{-V/kT} - 1)r^2\, dr. \tag{11.43}$$

If we take V to be the Lennard-Jones potential (11.36), the average potential energy should be negative at low temperatures, and we expect the integrand in Eq. (11.43) to be positive, yielding a negative B_2. At high temperatures, higher (positive) values of the potential energy near the hard wall become important. The value B_2 should be positive because the integrand will be negative ($e^{-V/kT} < 1$ if $V > 0$). With regard to the results in Figure 11.25, the particle density is $n = 4 \times 10^{-2}$. With n this small and only one temperature, we cannot clearly extract the effect of the correction term. Further investigation of the second virial coefficient is left to Project P11.7.

CHAPTER SUMMARY

In the preceding sections we discussed equilibrium thermodynamics and Monte Carlo simulations of thermal systems. With the Einstein solid model, we presented simulations that showed the Boltzmann distributions as a natural outcome of energy exchange maximizing the multiplicity, that is, probability. This enabled us to develop the concept of entropy and temperature and to construct algorithms to calculate them. We showed that the total entropy of interacting solids increases and drives the systems toward equilibrium.

We introduced the Metropolis algorithm in the discussion of 1D and 2D Ising models. It showed that we can simulate statistical properties of large systems using much smaller thermal systems. We did a careful examination of numerical and analytic results, including phase transitions absent in 1D but present in 2D Ising models. Simulated annealing, that is, Monte Carlo sampling with the Metropolis algorithm, was introduced to study nonthermal problems such as the hanging tablecloth by treating it as a thermal relaxation process.

Thermalization of N-body systems was studied with molecular dynamics simulations, showing the importance of periodic boundary conditions to avoid the artifacts of particle-wall collisions. At the same time, we used the method of virtual particle-wall collisions to extract pressure and to study the kinetic properties of weakly interacting gases.

We have used extensively the capabilities of Matplotlib and NumPy libraries for visualizing data including computation and presentation of various distributions. We also find that the combination of the leapfrog method and vector operations simulated by NumPy arrays is adequate for medium-scale N-body simulations.

11.5 EXERCISES AND PROJECTS

Exercises

E11.1 (a) Consider a chain of N spins, each can be up or down. A microstate refers to any ordering of the spins, and a macrostate refers to the number of "up" spins (see e.g., Ref. [82]). How many microstates are there? How many macrostates?

For a given macrostate $N_\uparrow$, show that the multiplicity is

$$\Omega(N, N_\uparrow) = \frac{N!}{N_\uparrow!(N - N_\uparrow)!} = \binom{N}{N_\uparrow}. \tag{11.44}$$

This is a binomial coefficient $\binom{m}{n}$, which reads "m choose n."

(b) For an Einstein solid with N SHOs and q units of energy, show that the multiplicity is

$$\Omega(N, q) = \frac{(N - 1 + q)!}{(N - 1)!\, q!} = \binom{N - 1 + q}{q}. \tag{11.45}$$

It gives the number of ways of distributing q quanta among N oscillators, like putting q eggs in N baskets.

One method is to convert the problem into one similar to the spin chain shown earlier. Let q be the eggs (O, "up spins"), and the SHOs be the baskets ($\boxed{n}$, "down spins"). Consider the number of eggs to the left of a basket as the units of energy in that SHO, with no eggs allowed to the right of the last SHO. For instance, the following layout illustrates three units of energy in a system of four SHOs:

O $\boxed{1}$ $\boxed{2}$ O O $\boxed{3}$ $\boxed{4}$

We interpret this configuration as follows: the first SHO has one unit of energy, the second none, the third two, and the last none. The effective length of the "spin" chain is $N - 1 + q$ because the last SHO does not move.

E11.2 (a) Verify the relationship (11.10) between the average energy and the partition function.

(b) Derive the heat capacity (11.12). First rewrite Eq. (11.11) in terms of $\partial\langle E\rangle/\partial\beta$, $\beta = 1/kT$, and show that

$$\frac{\partial\langle E\rangle}{\partial\beta} = \underbrace{-\frac{1}{Z^2}\frac{\partial Z}{\partial\beta}\sum E_n e^{-\beta E_n}}_{\langle E\rangle^2} - \underbrace{\frac{1}{Z}\sum E_n^2 e^{-\beta E_n}}_{\langle E^2\rangle}$$

$$= \langle E\rangle^2 - \langle E^2\rangle.$$

E11.3 (a) Show that the partition function and the average energy for an SHO are

$$\frac{1}{Z} = \exp\left(\frac{\epsilon}{2kT}\right) - \exp\left(-\frac{\epsilon}{2kT}\right), \quad \langle E\rangle = \frac{1}{2}\epsilon + \frac{\epsilon}{\exp(\epsilon/kT) - 1}.$$

Assume $E_n = \left(n + \frac{1}{2}\right)\epsilon$, $n = 0, 1, 2, \ldots$, and $\epsilon = \hbar\omega$ is a unit of energy.

The first term in $\langle E\rangle$ is the zero point energy at $T = 0$. Argue that it represents an energy shift and can be safely ignored in practical calculations. Obtain the low and high T limits.

(b) Calculate the heat capacity per oscillator from Eq. (11.11). Explain the limiting behavior for low and high T.

E11.4 (a) For $N \gg 1$, Stirling's approximation can be used to compute $N!$ as

$$N! \simeq N^N e^{-N}\sqrt{2\pi N}.$$

Compare the exact values with Stirling's approximation for $N = 1$, 5, 10, 50, and 100. Calculate the absolute and relative error in each case.

(b) A related result is the logarithm of $N!$,

$$\ln N! \simeq N\ln N - N.$$

Calculate the absolute and relative error for $N = 10^n$, $n = 0$ to 6.

E11.5 Working with large numbers in thermodynamics, numerical problems such as over- or underflow can occur in intermediate steps even though the final results are reasonable and well defined. A useful technique is to use logarithm for the intermediate calculations and convert back in the final step. For example, with the identity $x = \exp(\ln x)$, we can calculate $m!/n! = \exp(\ln m! - \ln n!)$ by computing the exponent first and the exponential last. The logarithm of the Gamma function can be found from scipy.special, `gammaln()`.

(a) Let $N! = 10^x$. Find x for $N = 1000$.

(b) Evaluate $2^{-200}\frac{200!}{95!105!}$ exactly, and compare with the value using Stirling's approximation for the factorials.

E11.6 Consider a paramagnetic system with $N = N_\uparrow + N_\downarrow$ dipoles. Each dipole can be in one of two states, $E_\mp = \mp\epsilon$, depending on whether it is aligned (↑) or antialigned (↓) with the external magnetic field B. The energy unit is $\epsilon = \mu B$, with μ being the magnetic dipole moment. The total energy is $E/\epsilon = -N_\uparrow + N_\downarrow = N - 2N_\uparrow$. For all the quantities that follow, sketch their temperature dependence, and derive the low and high T limits.

(a) Write down the partition function for a single dipole, and show that the average energy per dipole is

$$\langle\epsilon\rangle = -\epsilon\frac{\exp(\beta\epsilon) - \exp(-\beta\epsilon)}{\exp(\beta\epsilon) + \exp(-\beta\epsilon)} = -\epsilon\tanh(\beta\epsilon).$$

The total energy is then $\langle E\rangle = N\langle\epsilon\rangle = -N\epsilon\tanh(\beta\epsilon)$.

(b) Calculate the entropy of the paramagnetic system. Express the number of "up" dipoles $N_\uparrow$ in terms of energy $q = \langle E\rangle/\epsilon$ from above. Using the multiplicity (11.44) from Exercise E11.1, show that for large N the entropy is given by

$$S = \frac{1}{2}Nk\left[2\ln 2 - \ln(1-x^2) + x\ln\frac{1-x}{1+x}\right], \quad x = \frac{q}{N}.$$

Rewrite it in terms of temperature to show the equivalent expression (11.22).

(c) The magnetization is the average dipole moment. Show that for a single dipole, the magnetization is

$$\langle m\rangle = \mu P(E_-) + \mu P(E_+) = \mu\tanh(\beta\epsilon).$$

The total magnetization is $\langle M\rangle = N\langle m\rangle$. Compare the low T limit with Curie's law, $M \propto 1/T$.

(d) Derive the heat capacity

$$C = Nk\frac{(\beta\epsilon)^2}{\cosh^2(\beta\epsilon)}.$$

E11.7 Consider the 1D Ising model consisting of N dipoles. The partition function can be written as

$$Z = \sum_{s_1\ldots s_N}\prod_{i=1}^{N} e^{\beta\epsilon s_i s_{i+1}}.$$

Assuming periodic boundary conditions, $s_{N+1} = s_1$, show that

$$Z = \left[2\cosh(\beta\epsilon)\right]^N.$$

Prove that the properties of the 1D Ising system are identical to the paramagnetic system (Exercise E11.6).

E11.8 (a) Derive the energy (11.59) and heat capacity (11.60) of the 2D Ising model.

(b) Calculate and graph the entropy of the 2D Ising model as a function of temperature. Use Eq. (11.56) with the partition function (11.58). It may be convenient to expand the logarithm inside the integral as $\ln(1+x) \sim x$ and use the elliptic integral (11.62). Comment on your result around the critical temperature T_c.

Projects

P11.1 Explore the Einstein solid model.

(a) Make a table listing all possible microstates of an Einstein solid for $N = 3$ and $q = 4$. Verify that the total number agrees with the multiplicity from Eq. (11.45).

(b) Assuming $N = 400$, graph the entropy S/k as a function of energy $q = 0$ to 500 from Eq. (11.4). To avoid overflow, either use the techniques of Exercise E11.5, Stirling's approximation (Exercise E11.4), or the combination function `comb` from `scipy.misc`.

(c) Modify Program 11.5 to calculate the entropy of an Einstein solid in equilibrium. Use the same N as part (b). Vary q in the range 0 to 500 at equal increments (say 20). Initialize the solid for each q, equilibrate the system, and record the stabilized value of the entropy (the plateau part of Figure 11.4). To reduce fluctuation, sample entropy multiple times after equilibrium, and take the average. Graph the results and compare with the preceding analytic results.

(d)* Choose a q (say 200), plot the progress toward equilibrium as intensity images similar to Figure 11.1. Produce the energy distribution histograms, determine the temperature, and compare with the Boltzmann distribution.

P11.2 Consider an isolated system consisting of two interacting Einstein solids with N_A and N_B oscillators, respectively, and sharing $q = q_A + q_B$ units of energy. You can use SciPy's combination function (Project P11.1) for the first two parts.

(a) Let $N_A = 6$, $N_B = 4$, and $q = 10$. Calculate the multiplicity as a function of q_A from Eqs. (11.1) and (11.45), and plot the results. What is the probability of finding the most probable macrostate?

(b) Let $N_A = 600$, $N_B = 400$, and $q = 300$. What is the most probable macrostate? Plot the multiplicity distribution as a function of q_A, normalized to 1 at the maximum. Also plot the entropy analogous to Figure 11.6.

(c) Simulate the interaction process of the solids. Create a program similar to Program 11.5. Initialize the solids with $q_A = 100$ and $q_B = 200$, equilibrate each well before interactions can begin. Monitor the entropy or the Boltzmann distribution to ensure equilibrium is reached. Calculate the initial temperatures, and the final equilibrium temperature.

Allow the systems to interact by combining them into one solid as

```
C = A + B            # combine solids
C.exchange(L)        # energy exchange
```

This enables both inter- and intrasolid interactions. At any instant, solid A is in the first N_A elements (`C.cell[:Na]`) and solid B in the next N_B elements (`C.cell[Na:]`).

From the start of the interaction, calculate the changes of entropy and temperature in each solid. Sample the systems at intervals $I_{n+1} = fI_n$ (geometric distribution) between iterations, starting with $I_1 = 2$. First set $f = 2$, and once I_n is larger than some value, say 100, reduce $f \sim 1.2$ or so. This speeds up the simulation in later rounds without sacrificing details in the early stages. Append the results in arrays, including the energies q_A and q_B. Stop sampling once equilibrium has been well established.

Plot ΔS_A, ΔS_B, and ΔS_T vs. q_A on the linear scale and vs. the (cumulative) iteration number on the semilog x scale.

(d) Calculate the heat capacity of solid A (or B) from Eq. (11.12). Plot the numerical results as a function of temperature. Compare with analytic results (Exercise E11.3).

P11.3 Study the two-state paramagnetic system outlined in Exercise E11.6, which also contains the analytic results needed in this project.

(a) First consider a single system with $N = 10000$ dipoles. Graph the entropy and temperature as a function of $q = [-N, N]$ using the analytic expression. Explain the general shapes of the plots.

Next, divide the system into two interacting subsystems A and B with $N_A = N_B = N/2$ sharing a total of $q = 0$ units of energy. Plot the multiplicity and entropy of the system as a function of q_A from analytic formulas.

(b) Write a program to simulate the interaction between A and B. Initialize the subsystems with $q_A = -2000$, $q_B = 1000$. This determines the number of "up" ($N_\uparrow$) and "down" ($N_\downarrow$) dipoles within each subsystem. Let the dipoles interact with each other.

In each iteration, pick a pair of dipoles randomly, and swap their orientations. If they are from the same subsystem, do nothing. If they are from different subsystems, adjust the dipoles and energies appropriately. For instance, if an up dipole in A is swapped with a down dipole from B, reduce $N_{A\uparrow}$ by one and raise $N_{A\downarrow}$ by one, and do the opposite for $N_{B\uparrow}$ and $N_{B\downarrow}$. At the same time, increase q_A by two and decrease q_B by two. Sample the subsystems at regular iteration intervals to calculate entropies from Eq. (11.5), where P_n is now the probability of a dipole pointing up or down. Also calculate temperatures and heat capacities of the subsystems.

Record the data points until equilibrium has been well established.

Plot the results as a function of the change of energy, Δq_A. What are the trends of the changes of entropy? Are the temperatures of either subsystem negative? If so, what does negative temperature mean? Discuss what happened to the temperature of subsystem B when its energy q_B crossed zero and became negative. Is there a singularity in entropy, heat capacity, or temperature?

P11.4 Explore the standard and modified 1D Ising model including varying the number and strength of interacting neighbors.

(a) For the standard Ising model, calculate the heat capacity and entropy as a function of temperature. Modify Program 11.6 to compute the fluctuation of energy and heat capacity from Eq. (11.12). For the entropy, use Eq. (11.21). Let S_1 be the entropy at T_1, and δE be the difference of energy between T_1 and T_2. Compute the entropy at T_2 as $S_2 \simeq S_1 + \delta E/\overline{T}$ with $\overline{T} = (T_1 + T_2)/2$. Assume $S(T = 0) = 0$. Watch fluctuations in your results, particularly in heat capacity. You may need many passes to keep the data scatter small.

Plot both the heat capacity and entropy along with analytical results. Compare your results with Figures 11.11 and 11.12.

(b) An improved method for calculating entropy is to integrate the energy-temperature curve (Figure 11.10). We already have the energy at each temperature. Use the trapezoid rule (Section 8.A, Eq. (8.69)) to obtain $S(T)$. Compare with the preceeding results.

Try a different way using Eq. (11.5). Assume P_n is the probability of a spin pointing up or down, same as in the paramagnetic system (Exercise E11.6). Plot this result together with the above results. Does this method work? Why?

(c) Test a modified 4-neighbor 1D Ising model. Increase the number of interacting neighbors to include the nearest and the next nearest neighbors. The change of energy of a spin flip is

$$\Delta E = 2\epsilon\, s_i(rs_{i-2} + s_{i-1} + s_{i+1} + rs_{i+2}).$$

The constant r is a reduction factor simulating weaker interactions between next nearest neighbors. If $r = 0$, we recover the standard Ising model.

First set $r = 1$. Modify the program used in part (a) to calculate the energy, magnetization, heat capacity, and entropy as a function of temperature between [0.1, 5] in steps of 0.1 or less. Discuss your results and note any differences with the standard Ising model.

Based on your observations, predict how the results will change if r is decreased. Run the simulation with $r = 0.5$. Discuss your findings.

(d)* Use VPython to animate the 1D Ising chain. Represent the spins as arrows (see Program 7.9), and choose $N \sim 50$ to 100. Start with an initial microstate, animate spin flips toward equilibrium at a given temperature. Integrate keyboard detection in the program to change the temperature, for example, "up" or "down" key for increasing or decreasing the temperature by a certain amount. A sample scene is shown in Figure 11.26. Note like spins tend to be clustered.

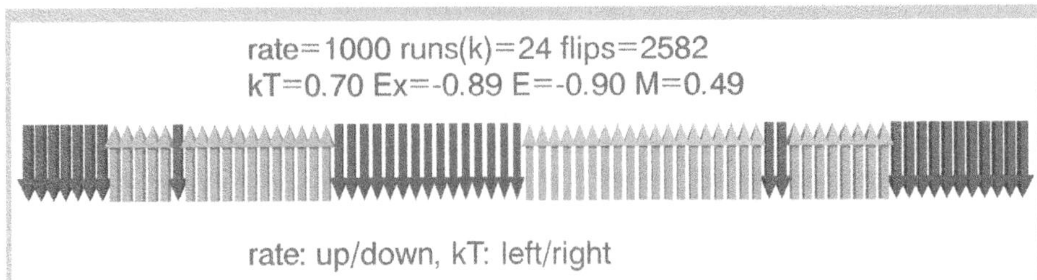

FIGURE 11.26 Animated Ising model.

P11.5 Investigate the properties of the 2D Ising model.

(a) Write a program to simulate the Ising model on an $N \times N$ lattice. Use Program 11.6 or the code developed from Project P11.4 as a template, implement the Metropolis algorithm in Section 11.2.3 into your program. Test it and reproduce the results shown in Figures 11.15 and 11.17.

(b) Set $N = 32$. Calculate the entropy of the system from your simulation as a function of temperature. Use the method discussed in Project P11.4 (the integral method). Compare with analytic results (see Exercise E11.8). What is the reason for the high numerical values near T_c?

(c) Study hysteresis of the 2D Ising model in an external magnetic field. Let the magnetic interaction energy be $-\eta s_i$, so the energy of the system is modified from Eq. (11.25)

$$E = -\epsilon \sum_{(i,j)} s_i s_j - \eta \sum_{i,j} s_{i,j}.$$

The parameter $\eta = \mu B$ is the strength of the magnetic interaction. Accordingly, we add an extra term, $2\eta s_{i,j}$, to ΔE in Eq. (11.26).

Assume $\epsilon = 1$. Calculate the magnetization as a function of temperature at $\eta = \pm 1, \pm\frac{1}{2}$, and 0. Discuss your results.

Calculate and plot the magnetization as a function of η at several temperatures, from $T = 1.5$ to 2.5 in steps 0.2. For a given temperature, plot the magnetization for increasing $\eta = -1 \to +1$ in small steps (≤ 0.02); then do the same but reverse the field $\eta = +1 \to -1$. At each η value, make 10 passes to obtain the magnetization.

Discuss the hysteresis loops. What are the differences below and above the critical temperature? Why does hysteresis occur? Does it change if you vary the number of passes between changes of η?

(d)* Explore the ferromagnetic domain on a larger system, $N = 1024$ or larger. Plot the equilibrium results (after 1000 passes) similar to Figure 11.18 at $T = 2$, 2.2692, and 2.5 (use `imshow`, see Program 7.4). This will be computation intensive, so consider using speed boost such as Program S:11.4. Or, write your program such that you can dump the data to a file with `pickle` after a number of passes, and reload it to continue. Observe the qualitative changes of the domain across the critical temperature. Present a few unique plots, and discuss what you think is interesting.

P11.6 Simulate the hanging tablecloth problem, and find the minimum energy by thermal relaxation.

(a) Write a program based on Program 6.8. Design a Metropolis updater using Eq. (11.33) as the change of energy. Convergence may be faster by sampling Δx, Δy, and Δz individually rather than all simultaneously in each iteration. Furthermore, because the relaxation is driven mostly by gravity, the vertical z direction should be sampled more frequently. Pick a particle to move, excluding the four corner particles. Sample x and y with probabilities $\frac{1}{4}$ each, and sample z with $\frac{1}{2}$, illustrated here.

```
dx, dy, dz = 0., 0., 0.
t = rnd.random()
if (t < .25):
    dx = rnd.gauss(0., sd)
elif (t < .5):
    dy = rnd.gauss(0., sd)
else:
    dz = rnd.gauss(0., sd*2)
```

In the preceding code, we also double the standard deviation in the z direction to reflect larger movements in that direction.

Use the same parameters for k, l, and m as in Program 6.8. Set the standard deviation $\sigma = \frac{1}{4}l$.

Run the simulation. Start at $T = 10^{-3}$ and a flat configuration, let the system reach equilibrium, then decrease T by a factor of 2, and repeat. Check the final results against those in Section 11.3 or Section 6.8. Increase the animation rate `vp.rate` to gain execution speed. How do you check whether the system has reached equilibrium at a given T?

(b) Disable animation, and automate the process of finding the minimum energy for a given $N \times N$ system. Calculate the minimum for $N = 3$ to 20 (9 to 400 particles). Plot the minimum energy per particle, excluding the particles held fixed at the corners, as a function of temperature. Explain the results.

P11.7 Use molecular dynamics simulations to investigate the behavior of an interacting gas.

(a) Run Program 11.7 with different parameters to get a feel on the program execution and speed. For example, vary $N = 40$ to 100, step size $h = 0.001$ to 0.01, and so on. Fine-tune the code for optimal performance, observing the points raised next.

Add a function that calculates the total energy of the system. Plot the relative error (see Project P4.1). How well is the energy conserved?

If you run the program multiple times, it is likely that you will observe the system collapse for some initial conditions where the atoms too close to one another can fly off. You can either rerun the program or avoid it by following the prescription given in Program 11.7.

Another issue is speed. Consider using compiled code in Programs S:11.3 and S:11.4. Push the upper limits of N and h while preserving accuracy, by monitoring the total energy for instance. Because the force is negligible for $r \gtrsim 3$ (Figure 11.20), modify `nbody()` such that it is evaluated only if $r^2 < 9$ (`r2`). For the following tasks, use animation sparingly or just turn it off.

(b) Calculate the velocity and speed distributions. Write a module that evolves the system and samples periodically the distributions to build cumulative histograms. For instance, the following code can be used for this purpose.

```
vxbin = np.zeros(nvbin, dtype=int)
speedbin = np.zeros(nsbin, dtype=int)
......
v2 = np.sum(v*v, axis=1)
speed = np.sqrt(v2)
speedbin += np.histogram(speed, nsbin, (0., smax))[0]
vxbin += np.histogram(v[:,0], nvbin, (-vmax, vmax))[0]
```

The variables `vxbin` and `speedbin` are integer ndarrays for storing the cumulative histograms of x velocity and speed, respectively. The NumPy function `histogram` accepts an input array, the number of bins and the range (a pair of numbers), and returns two arrays: the counts and the bin edges. We only need the first. For the speed distribution, we calculate the square of speed first using the `np.sum` function along axis 1 (i.e., $v_x^2 + v_y^2 + v_z^2$) for all particles at once and then take the square root by element-wise operations to obtain speed (similar to the calculation of distances in Program 11.7).

Compute the histogram periodically (e.g., every 20 steps), and cumulate the counts separately in time windows $t = 0$ to 10, 10 to 50, and 50 to 200. In the last time window, compute the average kinetic energy and the corresponding temperature. Plot the results similar to Figure 11.23, and compare with the Maxwell velocity and speed distributions.

Optionally change the initial condition to a higher equilibrium temperature, for example, set the range of the initial velocities to $[-2, 2]$. Repeat the simulation. Beware that you may need to adjust the step size or find a new initial condition to maintain stable integration.

(c) Calculate pressure on one side of the cube. Track the atoms exiting the side, find the impulse over regular time intervals to compute the pressure as described following Figure 11.25. Plot the pressure as a function of time, and compare the equilibrium pressure with the ideal gas value.

(d)* Extract numerically the second virial coefficient B_2 of Eq. (11.42). To do so, vary the number of atoms $N = 100$ to 10000 (adjust the upper limit according to your computing power, and consider speed boost mentioned earlier). Note that as the particle density increases, it is crucial to start the atoms somewhat uniformly in space to avoid large forces destroying the system and to use small step sizes to keep numerical error low. Lowering the initial speeds also helps.

At each N, calculate the pressure as described earlier. Obtain at least five data points. Plot P/nkT vs. n. The linear slope is equal to B_2. Compare B_2 with theoretical results in Exercise S:E11.2. The effect is more pronounced at low temperatures. Change the initial condition corresponding to half the temperature, and repeat the simulation.

11.A BOLTZMANN FACTOR AND ENTROPY

11.A.1 BOLTZMANN FACTOR

Consider an isolated system consisting of a microsystem (e.g., an oscillator) in contact with a reservoir. To keep it simple, we assume the microsystem is nondegenerate, that is, there is only one microstate for a given energy. In other words, the multiplicity of the microsystem is 1. Suppose the microsystem is found in state E_n. The multiplicity for any macrostate is $1 \times \Omega_n = \Omega_n$, where Ω_n is the multiplicity of the reservoir with energy $E_R = E - E_n$, E being the total energy.

We can write Ω_n according to Eq. (11.4) as

$$\Omega_n = e^{S_R(n)/k}, \tag{11.46}$$

where $S_R(n) \equiv S_R(E - E_n)$ is the entropy of the reservoir.

The probabilities of finding the microsystem in states E_n and E_m are proportional to the multiplicities of the macrostates, so their ratio is

$$\frac{P_n}{P_m} = \frac{\Omega_n}{\Omega_m} = e^{[S_R(n)-S_R(m)]/k}. \tag{11.47}$$

Because $\Delta E = E_n - E_m$ is small compared to the total energy E, we can approximate the exponent in Eq. (11.47) by a Taylor series,

$$S_R(m) = S_R(E - E_n + \Delta E) = S_R(n) + \frac{\partial S_R(n)}{\partial E_R}\Delta E + O[(\Delta E)^2]. \tag{11.48}$$

Relating the second term to temperature from Eq. (11.7) and ignoring the second-order correction, we simplify Eq. (11.48) to

$$S_R(m) = S_R(n) + \Delta E/T. \tag{11.49}$$

Substitution of this $S_R(m)$ into Eq. (11.47) gives

$$\frac{P_n}{P_m} = e^{-\Delta E/kT} = \frac{e^{-E_n/kT}}{e^{-E_m/kT}}. \tag{11.50}$$

Equation (11.50) shows that the probability of a microsystem with energy E_n is given by the Boltzmann factor,

$$P_n = Ce^{-E_n/kT}, \tag{11.51}$$

where C is a normalization constant. It is the same as the numerical result (11.3), of course.

If the microsystem is degenerate, we can generalize Eq. (11.51) to include degeneracy as $P_n = Cd_n \exp(-E_n/kT)$, where d_n is the number of states with E_n.

11.A.2 ENTROPY AND PROBABILITIES

We are now interested in the entropy of the microsystem. Let S, S_R, and S_T be the entropies of the microsystem, the reservoir, and the total (isolated) system, respectively. Because entropy is additive, we have

$$S + S_R = S_T. \tag{11.52}$$

Let P_n be the probability that the microsystem will be found in the state n. It is the same as the probability that the total system will be found in the macrostate associated with n because the microsystem is assumed to be nondegenerate.

Statistically, the entropy of the reservoir is given by the weighted average

$$S_R = \sum_n P_n S_R(n) = k \sum_n P_n \ln \Omega_n, \tag{11.53}$$

where we have substituted $S_R(n) = k \ln \Omega_n$ according to Eq. (11.4). The multiplicity of the reservoir Ω_n is related to the total multiplicity of the isolated system Ω_T as $\Omega_n = P_n \Omega_T$ by Eq. (11.2). Substituting it into Eq. (11.53), we obtain

$$S_R = k \sum_n P_n(\ln P_n + \ln \Omega_T) = k \sum_n P_n \ln P_n + k \ln \Omega_T, \tag{11.54}$$

where we have used the normalization $\sum_n P_n = 1$ to obtain the second term. We recognize that $S_T = k \ln \Omega_T$ is the total entropy of the isolated system.

Substituting Eq. (11.54) into (11.52) and eliminating S_T on both sides, we obtain the entropy S of the microsystem of interest as

$$S = -k \sum_n P_n \ln P_n. \tag{11.55}$$

This expression gives the entropy in terms of probabilities. It can be extended to nonthermal systems such as information analysis. In that case, the Boltzmann constant k is often set to 1. Equation (11.55) shows directly that entropy is related to disorder. The more possibilities, the higher the entropy. For instance, suppose there are N possible events. If only one event occurs with unit probability, the entropy is $S = 0$. But, if all events can occur with equal probability $1/N$, the entropy is $S = k \ln N$.

Another useful expression for entropy can be obtained if the partition function Z (11.9) for a thermal system is known. Putting $P_n = \exp(-E_n/kT)/Z$ into $\ln P_n$ in Eq. (11.55) and using Eq. (11.10), the entropy can be rewritten as

$$S = \frac{E - F}{T}, \quad F = -kT \ln Z, \tag{11.56}$$

where E is the average energy, and F is called the Helmholtz free energy.

11.B EXACT SOLUTIONS OF THE 2D ISING MODEL

The derivation of the exact solution of the 2D Ising model, first worked out by Onsager, is very involved. Here, we only quote the results [41].

Let m be a dimensionless parameter

$$m = \frac{4 \sinh^2 2\beta\epsilon}{\cosh^4 2\beta\epsilon}. \tag{11.57}$$

It varies between 0 and 1 and is graphed in Figure 11.16.

The partition function can be written as

$$Z = Z_1^N,$$
$$\ln Z_1 = \ln\left(\sqrt{2}\cosh 2\beta\epsilon\right) + \frac{1}{\pi}\int_0^{\pi/2} \ln\left(1 + \left(1 - m\sin^2\varphi\right)^{1/2}\right) d\varphi. \tag{11.58}$$

From Eq. (11.10), it can be shown that the energy per spin is (Exercise E11.8)

$$\langle E\rangle/N\epsilon = -2\tanh 2\beta\epsilon - 2\frac{\sinh^2 2\beta\epsilon - 1}{\sinh 4\beta\epsilon}\left(\frac{2}{\pi}\mathsf{K}(m) - 1\right), \tag{11.59}$$

and the heat capacity per spin is, from Eq. (11.11),

$$C/Nk = \frac{1}{\pi}(2\beta\epsilon\coth 2\beta\epsilon)^2\left[\left(1 + \cosh^{-4} 2\beta\epsilon - \frac{m}{4}\right)\mathsf{K}(m) - \mathsf{E}(m) - \frac{\pi}{2}\cosh^{-2} 2\beta\epsilon\right]. \tag{11.60}$$

In Eqs. (11.59) and (11.60), the functions $\mathsf{K}(m)$ and $\mathsf{E}(m)$ are the complete elliptic integrals of the first and second kind (Sec. 11.12 of Ref. [10]), respectively,

$$\mathsf{K}(m) = \int_0^{\pi/2} \frac{1}{\sqrt{1 - m\sin^2\varphi}}\, d\varphi, \tag{11.61}$$

$$\mathsf{E}(m) = \int_0^{\pi/2} \sqrt{1 - m\sin^2\varphi}\, d\varphi. \tag{11.62}$$

These integrals first appeared in Chapter 4, Eq. (4.30). They are graphed in Figure 11.27. The function $\mathsf{K}(m)$ has a logarithmic singularity at $m = 1$.

We can find the elliptic integrals as defined in Eqs. (11.61) and (11.62) in the SciPy special function library called `ellipk` and `ellipe`. If other sources are used, beware about the argument

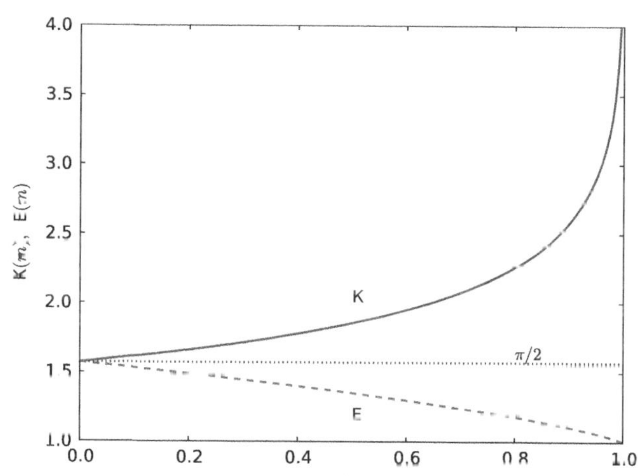

FIGURE 11.27 The complete elliptic integrals of the first and second kind. The first kind $\mathsf{K}(m)$ diverges logarithmically as $m \to 1$.

m to the elliptic integrals. Some sources use m^2 in the definitions of Eqs. (11.61) and (11.62). In such cases, give $\sqrt{m}$ as the argument.

The magnetization per spin can be obtained separately as

$$\langle M \rangle / N = \begin{cases} \left(1 - \sinh^{-4} 2\beta\epsilon\right)^{1/8}, & T < T_c, \\ 0, & T > T_c. \end{cases} \tag{11.63}$$

The preceding results are valid for both ferromagnetic ($\epsilon > 0$) and antiferromagnetic ($\epsilon < 0$) systems. For the latter, the magnetization (11.63) should be interpreted as the staggered magnetization (see Project S:P11.1).

11.C PROGRAM LISTINGS AND DESCRIPTIONS

Program listing 11.4 Equilibrium energy sharing (`equishare.py`)

```
from einsteinsolid import EinsteinSolid
import matplotlib.pyplot as plt, numpy as np
import matplotlib.animation as am

def updateimg(*args):                       # args[0] = frame
    L = 20 if args[0]<400 else 200          # slower init rate
    solid.exchange(L)
    plot.set_data(np.reshape(solid.cell, (K,K))) # update image
    return [plot]                           # return line object in a list

K = 20                                      # grid dimension
solid = EinsteinSolid(N = K*K, q=10)  # 10 units of energy/cell
fig = plt.figure()
img = np.reshape(solid.cell, (K,K))         # shape to KxK image
plot = plt.imshow(img, interpolation='none', vmin=0, vmax=50)
plt.colorbar(plot)

anim = am.FuncAnimation(fig, updateimg, interval=1, blit=True) # animate
plt.show()
```

The process of thermal equilibrium energy sharing is visualized on a square lattice composed of $K \times K$ interacting oscillators.

The main program creates an Einstein-solid object with 10 units of energy per oscillator. The energy distribution is represented by `imshow()` from Matplotlib. Like in Program 10.2, we use the Matplotlib animation function to update the figure with the supplied function `updateimg()`. It calls the `exchange()` method for a number of iterations, which randomly exchanges energy between the oscillators as explained in Program 11.1, then updates image data.

Program listing 11.5 Boltzmann distribution (`boltzmann.py`)

```
from einsteinsolid import EinsteinSolid
import matplotlib.pyplot as plt, numpy as np

solid = EinsteinSolid(N=1024, q=1)       # initialize solid
L, M, kT = 100, 6, 1./np.log(2.)         # L = base iteration num.
E, bin = np.linspace(0., M, 100), np.zeros(M+1)
fig = plt.figure()
for i in range(9):                       # 3x3 subplots
    ax = fig.add_subplot(3, 3, i+1)
```

```
    for n in range(M+1):
        bin[n] = solid.cell.count(n)     # count n
    plt.step(range(len(bin)), bin/max(bin), where='mid')
    plt.plot(E, np.exp(-E/kT))
    if (i == 3): plt.ylabel('P(n)', fontsize=14)
    if (i == 7): plt.xlabel('n', fontsize=14)
    if (i <= 5):
        plt.xticks(range(M+1), ['' for j in range(M+1)])
    if (i > 0): plt.text(M-2, 0.8, repr(L*2**(i-1)))
    solid.exchange(L*2**i)                 # double L each time
plt.show()
```

This program simulates the equilibrium distribution of an N-oscillator Einstein solid. It uses the same `exchange()` method as in Program 11.4. The average energy per oscillator is 1. This determines the temperature according to Eq. (11.13). The main loop plots the energy distribution as histograms between iterations. The number of oscillators with a given amount of energy, n, is conveniently tallied by the list count method `cell.count(n)`, which returns the number of times `n` appears in `cell`. The histograms are made with the step plotter `plt.step()` (see Figure S:9.1), normalized to 1 at the maximum and centered at the midpoints. To reduce crowdedness, the x tick labels for the first two rows are omitted.

Program listing 11.6 Ising model in 1D (`ising1d.py`)

```
import random as rnd, numpy as np
import matplotlib.pyplot as plt

def initialize (N):  # set initial spins
    p, spin, E, M = 0.5, [1]*N, 0., 0.       # p = order para.
    for i in range(1, N):
        if (rnd.random() < p): spin[i]=-1
        E = E - spin[i-1]*spin[i]
        M = M + spin[i]
    return spin, E - spin[N-1]*spin[0], M+spin[0]

def update(N, spin, kT, E, M):         # Metropolis sampling
    i, flip = rnd.randint(0, N-1), 0
    dE = 2*spin[i]*(spin[i-1] + spin[(i+1)%N]) # periodic bc
    if (dE < 0.0): flip=1              # flip if dE<0, else flip
    else:                              # according to exp(-dE/kT)
        p = np.exp(-dE/kT)
        if (rnd.random() < p): flip=1
    if (flip == 1):                    # accept
        E = E + dE
        M = M - 2*spin[i]
        spin[i] = -spin[i]
    return E, M

N, passes = 1000, 10
iter, Nmc = passes*N, passes*N
T, Eavg, Mavg = [], [], []
for i in range(1,41):                  # temperature loop
    kT = 0.1*i                         # kT = reservoir temperature
    spin, E, M = initialize (N)
```

```
    for k in range(iter):              # let it equilibrate
        E, M = update(N, spin, kT, E, M)

    E1, M1 = 0., 0.
    for k in range(Nmc):               # take averages
        E, M = update(N, spin, kT, E, M)
        E1, M1 = E1 + E, M1 + M
    E1, M1 = E1/Nmc, M1/Nmc
    T.append(kT), Eavg.append(E1/N), Mavg.append(M1/N)

plt.figure()
plt.plot(T, Eavg, 'o', T, -np.tanh(1./np.array(T)))
plt.xlabel('kT/ε'), plt.ylabel(r'⟨E⟩/Nε')
plt.figure()
plt.plot(T, Mavg,'o')
plt.xlabel('kT/ε'), plt.ylabel(r'⟨M⟩/N')
plt.show()
```

Program 11.6 simulates the 1D Ising model using the Metropolis algorithm (Section 11.2.2). The module `initialize()` initializes the spin chain stored in a list. Plain lists are faster to manipulate in explicit (serial) iterations than NumPy arrays. The p-value is an order parameter that specifies the probability of down spins. The net spin is given by $1-2p$. A p-value different from $\frac{1}{2}$ may be preferred in some situations such as at low temperatures to reduce equilibration time.

The Metropolis algorithm is programmed in `update()`, explained in the text. The main code steps through temperature T. At a given T, it initializes the spin chain, waits for it to equilibrate, and samples the system N_{mc} times to obtain the average energy and magnetization. Note that the cumulative energy and magnetization, `E1` and `M1`, are defined to be floats to avoid potential inaccuracies by integer division (Python 2.xx). Though this is not an issue in Python 3.xx, which defaults to floating point divisions, it is still good practice to be precise (Section 1.3.1). At low temperatures, the number of passes to reach equilibrium will need to be increased for accurate sampling afterward (Figure 11.9). The averages are plotted at the end.

Program listing 11.7 *N*-body molecular dynamics (`md.py`)

```
import ode, vpm, random as rnd
import numpy as np, visual as vp

def nbody(id, r, v, t):                  # N-body MD
    if (id == 0):                        # velocity
        return v
    a = np.zeros((N,3))                  # acceleration
    for i in range(N):
        rij = r[i]-r[i+1:]               # rij for all j>i
        rij[rij > HL] -= L               # periodic bc
        rij[rij < -HL] += L
        r2 = np.sum(rij*rij, axis=1)     # |rij|^2
        r6 = r2*r2*r2
        for k in [0,1,2]:                # L-J force in x,y,z
            fij = 12.*(1. - r6)*rij[:,k]/(r6*r6*r2)
            a[i,k] += np.sum(fij)
            a[i+1:,k] -= fij             # 3rd law
    return a
```

```
19
   L, N = 10.0, 32                                # cube size, num. atoms
21 atoms, HL, t, h = [], L/2., 0., 0.002
   r, v = np.zeros((N,3)), np.zeros((N,3))
23
   scene = vp.display(background=(.2,.5,1), center=(L/2, L/3, L/2))
25 vp.box(pos=(HL,HL,HL), length=L, height=L, width=L, opacity=0.3)
   for i in range(N):                             # initial pos, vel
27     for k in range(3):
           r[i,k] = L*rnd.random()
29         v[i,k] = 1-2*rnd.random()
       atoms.append(vp.sphere(pos=r[i], radius=0.04*L, color=(1,0,1)))
31 v -= np.sum(v, axis=0)/N                       # center of mass frame

33 while (1):
       vpm.wait(scene), vp.rate(1000)
35     r, v = ode.leapfrog(nbody, r, v, t, h)
       r[r > L]  -= L                             # periodic bc
37     r[r < 0.] += L
       for i in range(N): atoms[i].pos = r[i]     # move atoms
```

We can simulate full 3D molecular dynamics with Program 11.7. The N-body dynamics is modeled in `nbody()`, which is written in the required format for leapfrog integration and is explained in Section 11.4. The main code initializes the parameters such as the cube size and number of atoms, and so on. The position and velocity vectors are stored in the variables `r` and `v`, which are created as $N \times 3$ arrays. After randomly initializing positions within the cube and velocity components between $[-1, 1]$, the center of mass velocity $\vec{v}_{CM}$ is computed by summing over the three components of all atoms separately, via `np.sum` along the first axis, that is, the particle index. To eliminate the net drift caused by the center of mass motion, we subtract the average $\vec{v}_{CM}/N$ from the velocities. Here (line 31), $\vec{v}_{CM}/N$ is broadcast into an $N \times 3$ array before element-wise operations are carried out.

Animation elements include a transparent cube and a list of atoms, `atoms`. In the infinite loop, the leapfrog integrator is called to evolve the N-body system. After each step, the atoms are checked whether they have escaped the cube. We do so with the same truth array indexing as in `nbody()`, so the position in any direction outside the cube (e.g., $x < 0$ or $x > L$) is mapped inside by periodic conditions. Finally, the `atoms` list is updated to move the atoms.

Because the initial positions are randomly selected, there are situations when atoms are placed too close to each other. This can cause a very large force due to strong repulsion $1/r^{13}$ of the Lennard-Jones potential. Unless the time step h is very small, numerical error can destroy the system, often ejecting the atoms into a distant oblivion. If this happens, just rerun the simulation. Note that this rarely happens once the simulation starts normally. To guard against this situation, we can require a minimum initial distance between the atoms, seed the random number generator for a stable initial condition (Section 10.1, Program S:11.4), or use the N-body leapfrog method with time transformation (see Program S:12.1).

Our code is fairly optimized and efficient. It is adequate for N-body simulations up to perhaps $N \lesssim 1000$. For larger simulations, it would be too slow, with most time spent computing the forces. In this case, we can use compiled codes for `nbody()` in Fortran via F2Py, or in C/C++ via Cython or Weave (see Section 1.3.2). We describe an F2Py modification in Programs S:11.3 and S:11.4, which shows a dramatic increase in speed by a factor of about 100.

CHAPTER 12 Classical and Quantum Scattering

Most problems we have discussed so far, from harmonic oscillations to planet motion, are bound systems. There are unbound systems in our physical world, such as light scattering from water droplets making rainbows, as well as proton–proton collisions leading to the recent discovery of Higgs boson in the world's largest atom smasher. These are scattering problems. To study them, we need a fundamentally new approach to account for the fact that particles come in from infinity and go out to infinity.

Much of what we know about the fundamental forces (e.g., Coulomb, strong, and weak) comes from particle scattering studies. In a typical scattering setup, a beam of particles (projectiles), prepared in well-defined initial states, is sent to a target region for collisions with the constituents. After the beam passes through, the outgoing particles will be scattered relative to their incoming direction, possibly accompanied by other fragments or secondary emissions. By detecting the scattered particles, on the one hand, we can infer the underlying forces that caused the deflection or scattering. This was how the atomic nucleus was discovered in Rutherford's alpha-particle and gold-foil experiment. On the other hand, if the interaction forces are presumed to be well known, scattering studies can test and improve our understanding of collision dynamics by comparing experimental observations with theoretical models and predictions. The study of scattering is vital not only to understanding the fundamental interactions, but also to many practical applications, including radiation and transport physics (Section S:11.2), fusion energy research, and the scattering and absorption of radiation by the atmosphere.

We start by introducing the concept of scattering cross sections and numerical determination of deflection functions. After studying general potential scattering, we present classical analysis of rainbow and glory scattering. Quantum mechanically, we define the scattering amplitude and relate it to the cross section in analogy to scattering of waves. Then, the scattering (continuum) wave functions and amplitudes are calculated in terms of phase shifts from partial wave analysis.

12.1 SCATTERING AND CROSS SECTIONS

In Chapter 4 we discussed bound (planetary) motion in central fields for negative energies $E < 0$. Because scattering is a natural extension from bound to unbound motion for positive energies (4.11), we begin our study of classical scattering from central field motion for $E > 0$.

12.1.1 DEFLECTION FUNCTION AND SCATTERING ANGLE

Figure 12.1 (top) illustrates the schematics of the scattering process. The incoming particle comes in from far left (infinity), where the force is negligible, with a given kinetic energy E and velocity $\vec{v}$ toward the target (scattering center). Taking the incident direction to be horizontal, the initial vertical distance b from the target is called the impact parameter. The impact parameter may not be available experimentally, but we can specify it theoretically. After passing through the target region, the outgoing particle is scattered to a deflection angle Θ at infinity, where the force again becomes negligible.

Each value of b produces a well-defined deflection angle Θ. But in general we cannot say that a given Θ corresponds to a unique b. There are cases where different values of b lead to the same Θ as we will see shortly. The goal of scattering studies is to determine the relationship between the deflection angle $\Theta(b)$ and the impact parameter b. The function $\Theta(b)$ is called the deflection function.

Given a collision energy E, we can find the incident speed v and angular momentum L as

$$v = \sqrt{2E/m}, \quad L = mvb. \tag{12.1}$$

Because the angular momentum L depends on b, a typical scattering experiment samples outcomes of different angular momenta even though the energy is fixed. In other words, angular momentum may be conserved in each individual collision, but the overall outcome has contributions from different angular momenta.

The deflection angle Θ for a given b is defined as the angle of rotation of the velocity vector from the incident direction to the final outgoing direction (Figure 12.1, top). To find Θ, it is easier to work with the polar angle of the radial vector. Consider the radial bisector as the straight line from the target to the point of closest approach $(r = r_{min})$ bisecting the trajectory. Let α be the angle swept by the radial vector from the radial bisector to infinity ($r = \infty$). From symmetry, this is also equal to the angle swept by the radial vector in the incoming direction. Accordingly, the deflection angle can be written as

$$\Theta = \pi - 2\alpha. \tag{12.2}$$

The derivation for Θ and α is given in Section 12.A, and the result is

$$\Theta(b) = \pi - 2\int_{r_{min}}^{\infty} \frac{b\,dr}{r^2\sqrt{1 - \frac{V(r)}{E} - \frac{b^2}{r^2}}}, \tag{12.3}$$

where $V(r)$ is the interaction potential. The parameter r_{min} is the distance of closest approach where the denominator in Eq. (12.3) is zero, defined in Eq. (12.48). For certain potentials, the deflection function can be calculated analytically [40]. A well-known example is Rutherford

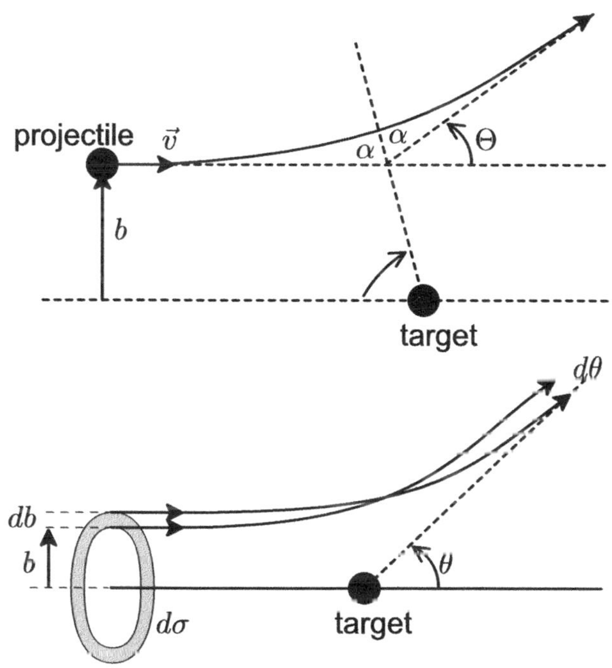

FIGURE 12.1 Schematics of particle scattering from a central field (top). All particles entering the shaded ring of cross section $d\sigma = 2\pi b\,db$ are scattered to the angle between θ and $\theta + d\theta$ (bottom).

scattering with the Coulomb potential (12.52). For most potentials, the integral in Eq. (12.3) can only be evaluated numerically.

We make a distinction between the deflection angle Θ and the scattering angle θ. The scattering angle θ is the physically observed angle after the particle leaves the scattering region, and it is restricted to $0 \le \theta \le \pi$. The deflection angle Θ can be positive or negative (attractive potentials) and is not limited to a fixed range. For instance, in some attractive potentials, the particle could be orbiting quite a few times before leaving the interaction region, resulting in a deflection angle exceeding $-n\pi$, n being a positive integer. However, the deflection angle is not directly observable. It is related to the observable scattering angle through

$$\theta = \Theta \quad (\text{if } \Theta > 0),$$
$$\theta = |2n\pi + \Theta|, \; n = \text{Int}\left[\frac{\pi - \Theta}{2\pi}\right] \quad (\text{if } \Theta < 0). \tag{12.4}$$

For negative Θ, Eq. (12.4) maps it to the observable θ range and accounts for the possibility of orbiting ($\Theta < -2\pi$, Figure 12.8).

12.1.2 CROSS SECTIONS

To measure the scattering events, we imagine placing particle detectors far outside the target region. The outcomes as measured experimentally consist of the number of particles (dN) recorded by the detector subtending a solid angle, $d\Omega$. That number is proportional to the beam flux I, defined as the number of incident particles per unit area. It is useful to introduce a quantity independent of I as

$$\sigma(\Omega) = \frac{\text{number of particles scattered into } \theta \text{ and } \theta + d\theta}{(\text{particle flux } I) \cdot d\Omega} = \frac{1}{I}\frac{dN}{d\Omega}. \tag{12.5}$$

The value $\sigma(\Omega)$ gives the likelihood of a particle being scattered to the angle θ, but has the dimensions of an area. Therefore, $\sigma(\Omega)$ is called the differential scattering cross section, or simply the cross section. It is the single, most useful quantity in scattering.

As Figure 12.1 (bottom) shows, all particles (dN) entering the ring in the impact parameter range b to $b + db$ are scattered to the angle between θ and $\theta + d\theta$. The area of the ring is $d\sigma = 2\pi b\, db$, so $dN = I d\sigma = I 2\pi b\, db$. Because scattering in central fields is independent of the azimuthal angle φ, the solid angle element is $d\Omega = 2\pi \sin\theta d\theta$, and $\sigma(\Omega) = \sigma(\theta)$. Substituting dN and $d\Omega$ into Eq. (12.5), we obtain

$$\sigma(\theta) = \frac{b}{\sin\theta}\left|\frac{db}{d\theta}\right| \equiv \frac{d\sigma}{d\Omega}. \tag{12.6}$$

We take the absolute value of the derivative $|db/d\theta|$ in Eq. (12.6) to ensure that $\sigma(\theta)$ is positive. The classical interpretation of $d\sigma$ is that of the cross-sectional area around the scattering center (Figure 12.1, bottom). Note that the flux I drops out of $\sigma(\theta)$.

The cross section $\sigma(\theta)$ can be measured experimentally (12.5) and calculated theoretically (12.6). Because the latter requires the interaction force as input, comparison between experiment and theory can reveal the nature of that interaction.

Theoretically, because Θ is not necessarily bounded or single-valued, it is possible that different deflection angles contribute to the same scattering angle θ (see Eq. (12.4)). To be general, we can extend the definition of the cross section (12.6) to include this possibility as

$$\sigma(\theta) = \frac{1}{\sin\theta}\sum_i b_i \left|\frac{db}{d\Theta}\right|_i, \tag{12.7}$$

where the sum is over all values of Θ and b that yield the same observable scattering angle θ. We shall see an example of this later on.

The total cross section, σ_t, can be obtained from Eq. (12.6) by

$$\sigma_t = \int \sigma(\theta) d\Omega = \pi b_{max}^2. \quad (b_{max} = \text{maximum impact parameter}) \tag{12.8}$$

In classical mechanics, the total elastic cross section is infinite if the potential has an infinite range (long-ranged), no matter how small or how rapidly the force decreases at large distances. This is because there is always a small deflection at any large but finite b (at least theoretically). In quantum mechanics, however, the total cross section is finite if the potential decreases faster than $1/r$ (see Section 12.4.5).

12.1.3 RUTHERFORD SCATTERING

Rutherford scattering is an example of scattering by long-ranged potentials, the Coulomb interaction in this case, $V = k/r$. The trajectories of scattering motion are described by Eq. (12.50) (Section 12.A). They are plotted in Figure 12.2.

For repulsive potentials, the trajectories curve to the same side of the incoming particle, and for attractive potentials they curve to the opposite side. Accordingly, the deflection angles have the same sign as the potential. In either case, the scattering angle is the absolute value of the deflection angle $\theta = |\Theta|$ (no orbiting). At small impact parameters, the scattering angle approaches π due to the singularity at the origin (hardcore potentials). For repulsive potentials, the particle is pushed back, and for attractive potentials, it is turned around the target. As the impact parameter increases, the scattering angle decreases as expected. We can see graphically (Figure 12.2) that the scattering angle for a given impact parameter is the same regardless of the sign of the Coulomb potential.[1]

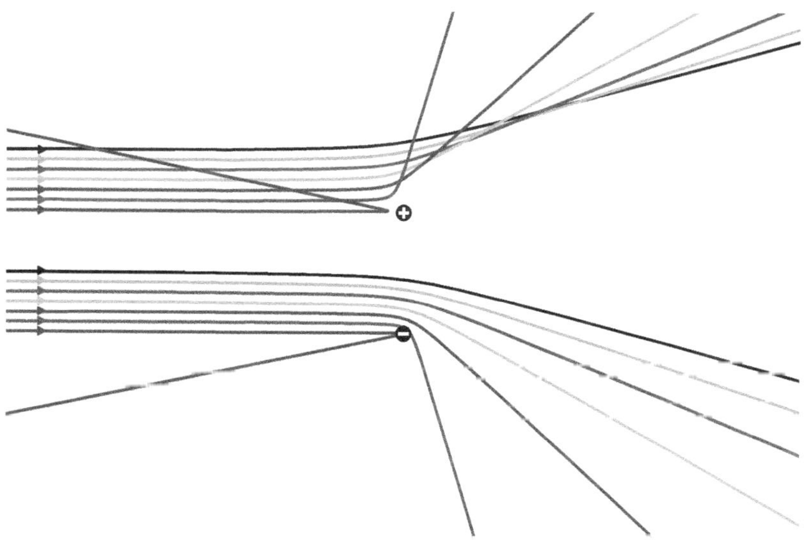

FIGURE 12.2 Scattering from repulsive (top) and attractive (bottom) potentials.

[1]This fact is an exception rather than the rule. It is related to the special behavior (symmetry) of the Coulomb potential, which is also responsible for the Runge-Lenz vector (4.20) being a constant of motion in $1/r$ fields. In general, the scattering angle depends on the sign of noncoulombic potentials.

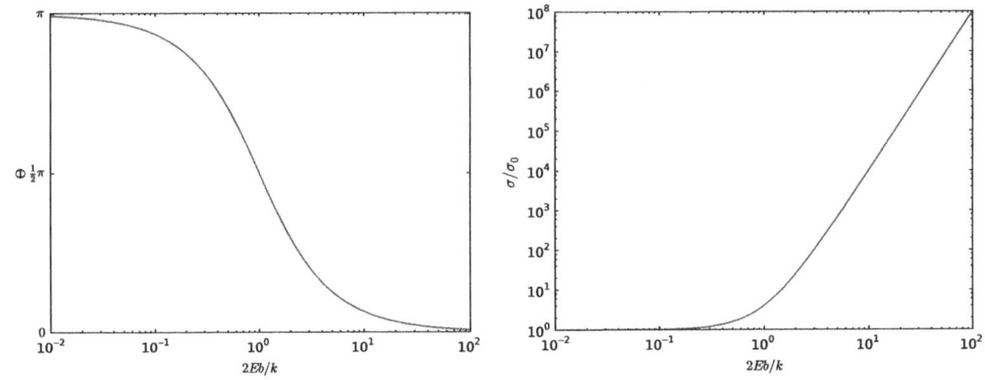

FIGURE 12.3 Universal deflection function (left) and cross section (right) for Rutherford scattering.

The deflection function is given in Eqs. (12.52) and (12.54). Differentiating $db/d\Theta$ from Eq. (12.54), we obtain the cross section (12.7) as

$$\sigma(\theta) = \frac{k^2}{16E^2}\frac{1}{\sin^4\frac{\theta}{2}}. \tag{12.9}$$

This is the celebrated Rutherford cross section. It is the same for repulsive or attractive potentials because of the k^2. Moreover, it is also identical in both classical and quantum mechanics.

We plot the deflection function and cross section in Figure 12.3, displaying both as a function of the dimensionless variable $\eta = 2Eb/k$. Consequently, the plots are universal. For instance, if energy E is doubled, the impact parameter b should be halved (or k doubled) to obtain the same value for Θ or σ. The net effect is to shift the curves to the left by the same factor. For small values of $\eta \to 0$ (low energy, small impact parameter, or strong potential), the deflection angle is near π. In the limit of large $\eta \gg 1$, it approaches zero.

The cross section (Figure 12.3) is scaled by $\sigma_0 = k^2/16E^2$, the value at $\eta = 0$. Because it is plotted in terms of η rather than the conventional scattering angle θ, the Rutherford cross section can be displayed as a universal curve. We can relate η and θ graphically via the deflection function (Figure 12.3, left), or analytically via Eq. (12.54). We use semilog and log-log plots to show the details over the full scales of η and σ.

The Rutherford cross section diverges at large impact parameters as b^4, or equivalently as θ^{-4} at small angles. This gives rise to an infinite total cross section. At small impact parameters (large scattering angle $\sim \pi$), the cross section is finite. In the original Rutherford experiment, alpha particles were sent to gold foils. From the number of back-scattered alpha particles, Rutherford concluded that there had to be a point-like charge at the scattering center, the nucleus. Without a hardcore potential (singularity), large-angle scattering would not be possible, as evidenced next.

12.2 RAINBOW AND GLORY SCATTERING

Unlike the Coulomb potential, for the majority of potentials encountered in actual scattering, the integral in Eq. (12.3) cannot be carried out analytically or elegantly. It is easier to obtain the deflection function by numerical integration. Next we discuss two examples, rainbow and glory scattering from softcore potentials.

12.2.1 RAINBOW SCATTERING

Consider the plum-pudding potential given by Eq. (S:7.6) (Project S:P7.2),

$$V(r) = \begin{cases} \dfrac{Z}{2R}\left[\dfrac{r^2}{R^2} - 3\right], & \text{if } r \le R, \\ -\dfrac{Z}{r}, & \text{if } r > R. \end{cases} \quad \text{(Plum pudding model)} \tag{12.10}$$

It describes the potential between an electron and a uniformly charged sphere of charge Z and radius R (in atomic units [a.u.], which is implicitly assumed thereafter). The "plum" potential is a softcore potential because the force is finite (zero in this case) at the center. Outside the sphere, it is the normal Coulomb potential (Figure 12.4). The force is linear in r inside the sphere and $1/r^2$ outside. The potential is attractive when the nuclear charge Z is positive and repulsive if Z is negative.

The deflection functions for the repulsive plum-pudding potential ($Z = -1$ and $R = 1$) are shown in Figure 12.5. They are computed from Program 12.1, which calculates the integral (12.49) in $u = 1/r$ space. At the lower energy $E = 1.2$, the scattering angle spans the full range $[0, \pi]$. The particle does not have sufficient energy to penetrate the repulsive core at small impact parameters and is deflected backward including $\theta = \pi$ (Figure 12.6, top).

At the higher energy $E = 2$, the particle is energetic enough to move straight through the softcore potential, and the scattering angle is zero at $b = 0$ (Figure 12.6, bottom). This cannot happen if the potential is singular as discussed earlier regarding Rutherford scattering (Figure 12.2). With increasing b, the scattering angle increases at first due to the rising repulsive force (Figure 12.4). As b further increases, the effect of the stronger force is neutralized by the reduced time spent in the charged sphere, indicated by the shrinking portion of trajectories inside the circle in Figure 12.6 (bottom). Consequently, the scattering angle reaches a maximum value near $b \sim 0.4$ and decreases thereafter. The maximum scattering angle, $\sim \pi/4$ in this case, is called the rainbow angle. For a given angle below the rainbow angle, there are two impact parameters contributing to the cross section, indicated by two crossings on the deflection function (Figure 12.5, dotted line).

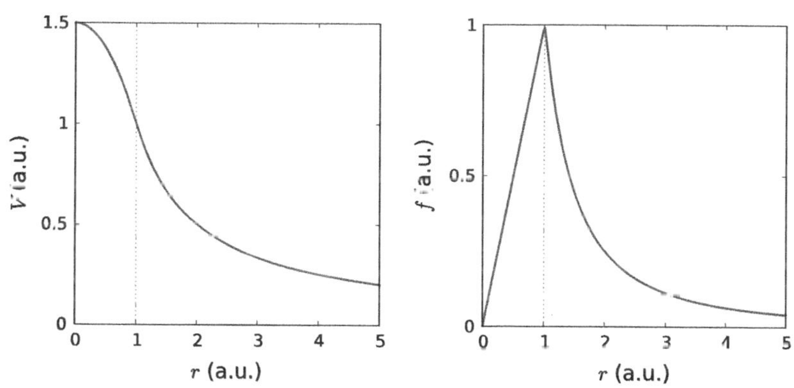

FIGURE 12.4 The repulsive plum-pudding potential (left) and force (right). The radius of the sphere is 1 (dotted line).

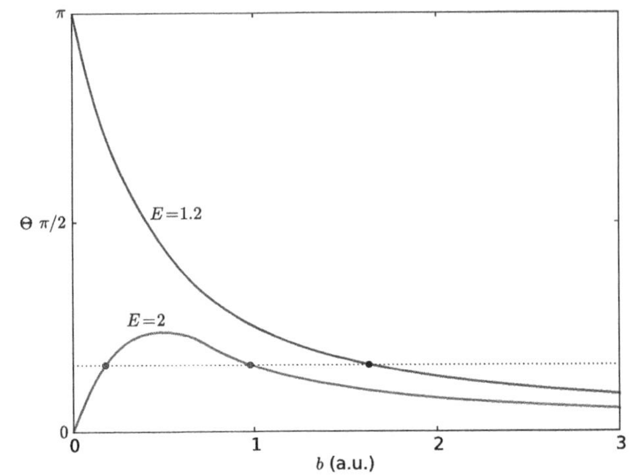

FIGURE 12.5 Deflection functions of the plum potential at two energies. The dotted line shows that there is only one impact parameter contributing at $E = 1.2$ but there are two at $E = 2$.

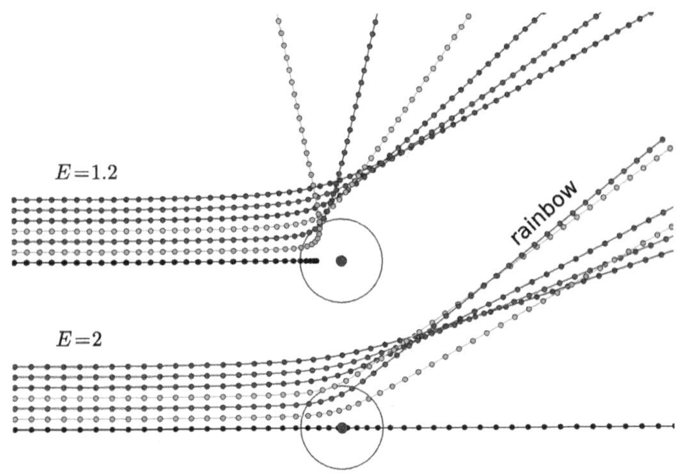

FIGURE 12.6 Scattering from the plum potential at two energies. The circle represents the size of the plum sphere.

At large impact parameters, both deflection functions behave like Rutherford scattering because the potential is Coulombic outside the sphere.

Computing Cross Sections

Knowing the deflection functions, we can obtain the cross sections that are directly observable experimentally from Eq. (12.7). A direct but less accurate method is to compute the numerical derivative $d\Theta/db$ of the deflection function from Eq. (2.13) as

$$\frac{d\Theta}{db} \simeq \frac{\Theta(b+\Delta b) - \Theta(b-\Delta b)}{2\Delta b}. \tag{12.11}$$

Equation (12.11) is useful if Θ is known at discrete impact parameters b.

To obtain the cross section at arbitrary b values, the derivative $d\Theta/db = (db/d\Theta)^{-1}$ would need to be evaluated. It can be obtained from Eq. (12.49) as (Exercise E12.2)

$$\frac{d\Theta}{db} = -2\int_0^{u_{min}} \frac{du}{\sqrt{f}} - 4b^2 \int_0^{u_{min}} \frac{2uf' - u^2 f''}{\sqrt{f}f'^2}\, du, \tag{12.12}$$

where $f = 1 - V(1/u)/E - b^2u^2$, and f' and f'' are the first and second derivatives with respect to u. Equation (12.12) can be accurately calculated and has the advantage of being on equal footing as the deflection function (12.3) (see Project P12.4). Finally, we can also obtain the cross section directly by virtual experiments (see Section S:12.B, Project S:P12.2).

Here we will use the direct method (12.11) for the results (Program 12.1). Figure 12.7 shows the cross sections for the plum potential. At the lower energy (Figure 12.7, left), the cross section behaves like Rutherford scattering from small angles to just below $\pi/2$, where a small plateau develops. The energy is such that the particle can enter the outer layers of the core, but is eventually deflected because the energy is still below the peak of the potential. We can see a small shoulder in the deflection function responsible for this effect (a slight change in curvature, barely visible in Figure 12.5 near $\frac{1}{5} \lesssim b \lesssim \frac{1}{2}$). Beyond the plateau, large scattering angles are predominantly due to scattering at the core.

At the higher energy (Figure 12.7, right), the cross section is nonzero in a limited range only, from zero up to the rainbow angle. At every angle, there are two contributing impact parameters. They converge to the rainbow angle where the cross section is infinite. It is caused by the singularity in the derivative $db/d\Theta = \infty$ at the rainbow angle (Figure 12.5). This is known as rainbow scattering (see Exercise E12.3). The terminology comes from the analogy to the optical phenomenon of rainbows formed by light rays incident on water droplets.[2] The primary rainbow, observed at the angle $\sim 42°$ with the back against the incident light, is due to a reflection between two refractions inside the droplet (Figure 12.20). This angle occurs at an impact parameter just shy of the radius of the droplet, gathering light from a ring and giving rise to enhanced light intensity, the rainbow [8, 69]. The enhancement of cross sections by rainbow scattering is also present in quantum mechanics (Figure 12.19) and has been observed experimentally.

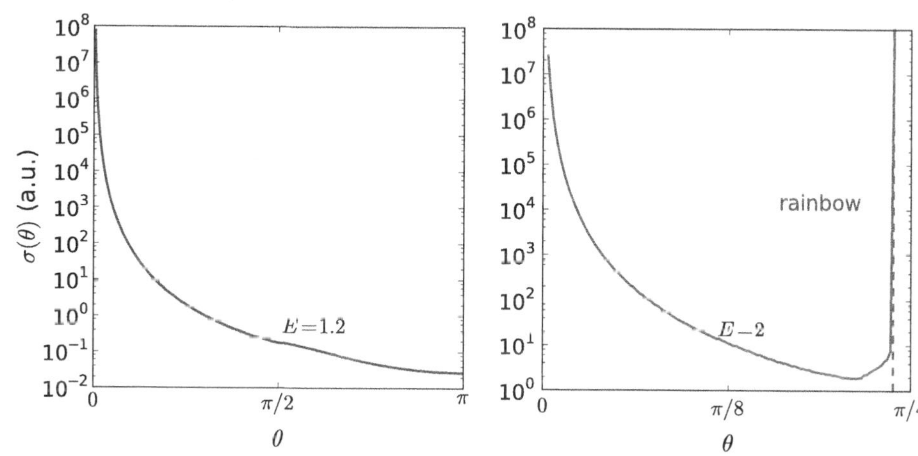

FIGURE 12.7 The cross sections for the plum potential at two energies. The cross section diverges at the rainbow angle ($E = 2$, right).

[2]A rainbow in the sky looks slightly different to observers at different locations, so everyone sees their own rainbow. You can easily make your own rainbow by holding a mist-emitting hose facing away from the afternoon sun.

12.2.2 GLORY SCATTERING

For attractive potentials, the deflection angle is negative. As alluded to earlier, in some cases the particle could be deflected by more than $-\pi$. We discuss a representative case involving the screened Coulomb potential (Yukawa),

$$V = -Z\frac{e^{-r/a}}{r}, \tag{12.13}$$

where a is the screening length, and Z the nuclear charge as usual. The potential is widely used to model ion–atom interactions (Section S:11.2.3). It is a short-range potential because of the exponential reduction relative to the unscreened Coulomb potential, which is long-ranged. As a result, scattering at intermediate energies or above is predominantly at forward angles. At low energies, however, several interesting behaviors emerge. We discuss this case next. The results are obtained by modifying Program 12.1 for the Yukawa potential. Calculations are left to Project P12.4.

Figure 12.8 shows the deflection and the scattering angles as a function of impact parameters at energy $E = 0.1$. Starting from large b when the deflection angle is zero, the value of Θ steadily decreases (more negative) for decreasing b, until the onset of a precipitous drop around $b \sim 3.5$. The deflection angle reaches $-\pi$ at $b \sim 3$, crosses it, and falls rapidly past to the lowest point $\Theta \sim -2.7\pi$, giving rise to a rainbow. It rebounds back to $-\pi$ at $b \sim 0$ (hardcore potential) from below.

When Θ crosses $-\pi$, it produces an infinite cross section (12.6) due to $\sin\theta \to 0$ but $db/d\Theta$ being finite. The is known as the glory scattering.[3] The infinite cross section is due to the vanishing solid-angle element $d\Omega$. Physically, it means channeling a finite flux into an infinitely small area, so the cross section is greatly enhanced.

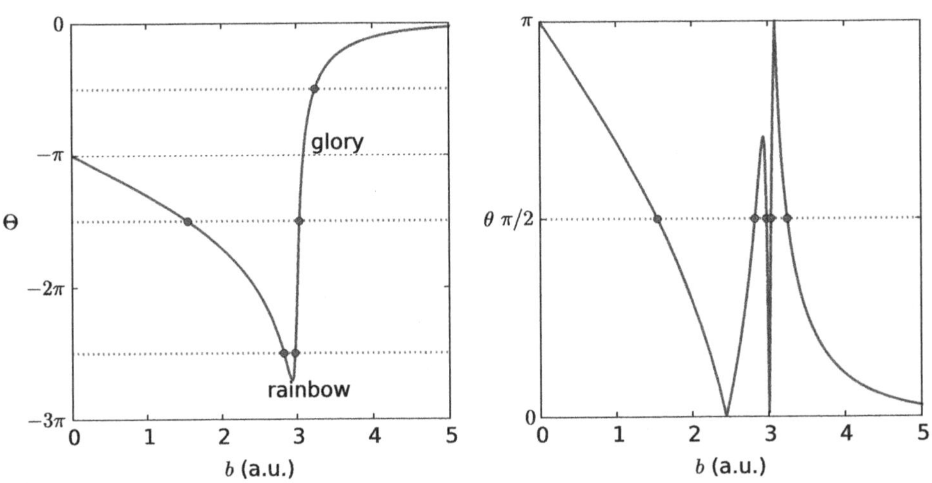

FIGURE 12.8 The deflection (left) and scattering (right) angles for the Yukawa potential ($Z = 2$, $a = 1$) at $E = 0.1$. Five impact parameters contribute to the scattering angle $\frac{1}{2}\pi$, marked as (•) on the curves.

[3]Like the rainbow, the glory has its origin in sunlight back-scattered by water droplets. Basically, light rays entering the sides of a droplet are first refracted critically, then reflected from the back to the other side, and refracted again, finally emerging from the droplet in the opposite direction to the incident light [13]. The net effect is a colored ring around the shadow of an observer's head on the cloud. It is rarer to observe the glory than the rainbow.

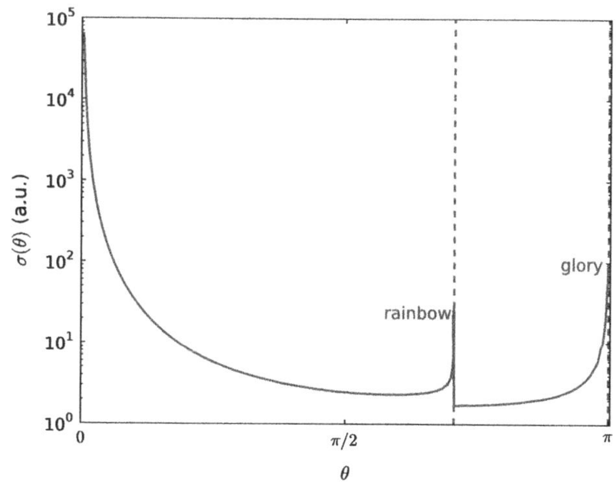

FIGURE 12.9 The cross sections for the Yukawa potential with the same parameters as in Figure 12.8.

Immediately following the glory, a rainbow occurs as the deflection angle has reached the minimum. This is not the minimum for the scattering angle θ, however. The latter can be obtained from Eq. (12.4), shown side by side in Figure 12.8. The sharp tips are the result of mapping Θ into the range $[0, \pi]$. We can see that the deflection and the scattering angles can be quite different. For instance, for the scattering angle $\theta = \frac{1}{2}\pi$, there are three Θ values contributing, $-\frac{1}{2}\pi$, $-\frac{3}{2}\pi$, and $-\frac{5}{2}\pi$. In terms of the impact parameter, there are five b values contributing to the cross section for all scattering angles θ below the rainbow angle ($\sim 0.7\pi$). The five impact parameters for $\theta = \frac{1}{2}\pi$ are marked by the crossings in Figure 12.8. Between the rainbow angle and π, we find three contributing impact parameters, one from small b, and two from both sides of the glory.

The cross section shown in Figure 12.9 is dominated by small-angle scattering, the same as in Rutherford scattering. The smooth background is punctuated by two singularities, the rainbow and the glory, which theoretically are infinite above the background. The cross section drops to a lower background value immediately past the rainbow angle because the number of contributing impact parameters has decreased from five to three as discussed earlier. Because of the glory, the cross section does not remain finite as θ approaches π as would be expected otherwise for hardcore potentials.

Like the rainbow, the glory scattering has also been observed in electron-atom scattering experiments and predicted by quantum scattering discussed next.

12.3 QUANTUM SCATTERING AMPLITUDE

Classical scattering discussed earlier breaks down at the atomic scale. To correctly describe scattering phenomena of atomic particles, we again turn to quantum mechanics as we did previously (Chapters 8 and 9). Full quantum scattering theory is itself a rich and expansive topic [51]. We will only briefly summarize the key elements of this theory within a limited scope for our purpose. We use atomic units (Table 8.1) unless otherwise noted.

The basic premise in quantum scattering is the same as classical scattering of waves such as sound waves. There are differences, of course, that in quantum mechanics, we are dealing with matter waves that obey the Schrödinger equation (8.2) instead of classical wave equations (6.62). Nonetheless, we can freely borrow ideas from classical wave scattering to construct a qualitatively correct picture of quantum scattering.

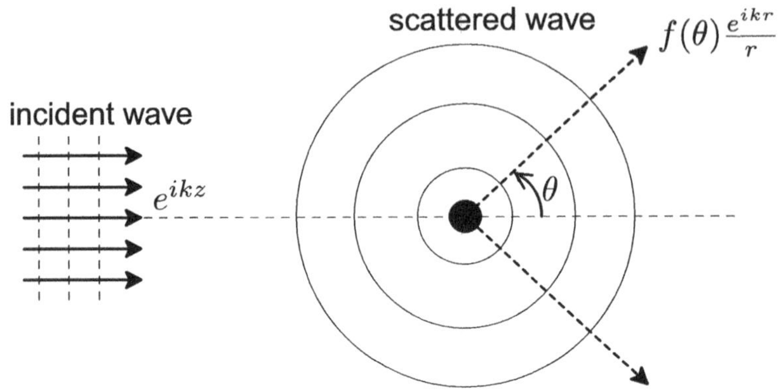

FIGURE 12.10 Schematics of quantum scattering from a central field.

Figure 12.10 shows the schematics of wave scattering from a potential field. The incident wave is dispersed into scattered waves on encountering the scattering center. We can write the total wave function as a superposition of the incident wave $\psi_{\rm in}$ and the scattered wave $\psi_{\rm sc}$,

$$\psi = \psi_{\rm in} + \psi_{\rm sc}. \tag{12.14}$$

We take the incident wave as an idealized plane wave, $\psi_{\rm in} = e^{ikz}$ (Figure 8.1). Far from the scattering center, the scattered wave is an outgoing spherical wave, its magnitude modulated by an angle-dependent factor $f(\theta)$. We need to find the scattering solutions satisfying the boundary condition of outgoing spherical waves,

$$\psi \overset{r\to\infty}{\longrightarrow} e^{ikz} + f(\theta)\frac{e^{ikr}}{r}, \tag{12.15}$$

where $k = \sqrt{2mE}/\hbar$ is the wave vector, $E > 0$ the energy, and m the mass of the particle. The factor $f(\theta)$ is called the scattering amplitude. Equation (12.15) makes it clear that the amplitude $f(\theta)$ has the dimension of length. In contrast to bound states, scattering wave functions (12.15) for continuum states do not vanish at infinity.

The amplitude of the incident plane wave is chosen to be one, which sets the overall particle flux and does not affect the cross sections. The incident flux is (Exercise E8.5)

$$I = -i\frac{\hbar}{2m}\left(\psi^*_{\rm in}\frac{\partial\psi_{\rm in}}{\partial z} - \psi_{\rm in}\frac{\partial\psi^*_{\rm in}}{\partial z}\right) = \frac{\hbar k}{m}. \tag{12.16}$$

Similarly, we can find the outgoing radial flux, far from the scattering center $r \to \infty$, from the asymptotic spherical waves (12.15) to be

$$I_r = \frac{\hbar k}{m}\frac{|f(\theta)|^2}{r^2} = I\frac{|f(\theta)|^2}{r^2}. \tag{12.17}$$

Suppose a particle detector is placed at r and θ outside the interaction region (Figure 12.1). Let ds and $d\Omega$ be the surface area and the solid angle element subtended by the detector, respectively. They are related by $ds = r^2 d\Omega$. The number of particles going into the detector is

$$dN = I_r ds = I_r r^2 d\Omega = I|f(\theta)|^2 d\Omega. \tag{12.18}$$

We have used Eq. (12.17) to eliminate I_r in Eq. (12.18). Substituting dN into Eq. (12.5), the product $Id\Omega$ cancels, and we obtain the scattering cross section

$$\sigma(\Omega) = |f(\theta)|^2. \tag{12.19}$$

The calculation of cross sections in quantum scattering boils down to finding the scattering amplitude $f(\theta)$, which is to be extracted from the scattering wave function ψ in Eq. (12.14) (see also Eq. (S:12.10)). In essence, the scattering amplitude is to quantum scattering as the deflection function is to classical scattering.

12.4 PARTIAL WAVES

The partial wave method is an effective way to calculate the scattering amplitude $f(\theta)$ just defined. In this method, we decompose the incident waves into components, or partial waves, of fixed angular momenta. This is in exact analogy to the analysis of classical scattering in terms of the impact parameter. The commonality exists because angular momentum is conserved in central field potentials. In quantum scattering, each partial wave can be treated separately to find the phase shift, which determines the scattering amplitudes and cross sections.

12.4.1 PHASE SHIFT

Before discussing scattering in 3D space, let us introduce phase shifts in scattering from a 1D potential with a hard wall at $x = 0$, $V(0) = \infty$. Figure 12.11 (top) shows a square well potential, but it can be any short-ranged potential decaying faster than $x^{-1-\epsilon}$, $\epsilon > 0$, as $x \to \infty$.[4]

We assume an incident plane wave e^{-ikx} traveling from right to left. The scattered wave can be quite complex in the interaction region, but in the asymptotic region where the potential vanishes, it emerges again as a plane wave, moving to the right and phase-shifted by 2δ, $-e^{2i\delta}e^{ikx}$. According to Eq. (12.14), we have the scattering wave function in the asymptotic region as

$$\psi \xrightarrow{x\to\infty} e^{-ikx} - e^{2i\delta}e^{ikx}. \tag{12.20}$$

The negative sign in the scattered wave comes from collision with the hard wall, similar to inverted waves on a string (Figure 6.16, Section 6.5.5). The magnitude of the phase factor is one, $|e^{2i\delta}| = 1$, because the incident flux must be equal to the outgoing flux. Therefore, the phase shift δ is real. The factor of two is included for convenience to account for the fact that

$$\psi \xrightarrow{x\to\infty} -2ie^{i\delta}\sin(kx + \delta). \tag{12.21}$$

We can now interpret the phase shift in real sine waves. Figure 12.11 (bottom) shows two cases, a positive phase shift corresponding to a sine wave pulled back, and a negative one to a sine wave pushed forward (see Figure 12.17 for an actual case). Attractive potentials produce positive phase shifts, and repulsive ones produce negative phase shifts. Another explanation for the factor of two is that the wave is shifted once on the way in and once on the way out.

Regardless, the phase shift provides information on the potential. It is analogous to probing a dark well by sending waves into it and inferring the structure of the well from the phase shift of the returned waves.

[4] Potentials decaying at or slower than x^{-1} such as the Coulomb potential are long-ranged. Particles are never free in long-ranged potentials, suffering distortions even at infinity (see Exercise S:E12.7). As a result, scattering in these potentials does not have well-defined phase shifts. The same effect causes an infinite number of bound states in the Coulomb potential (Chapter 9, Section 9.5).

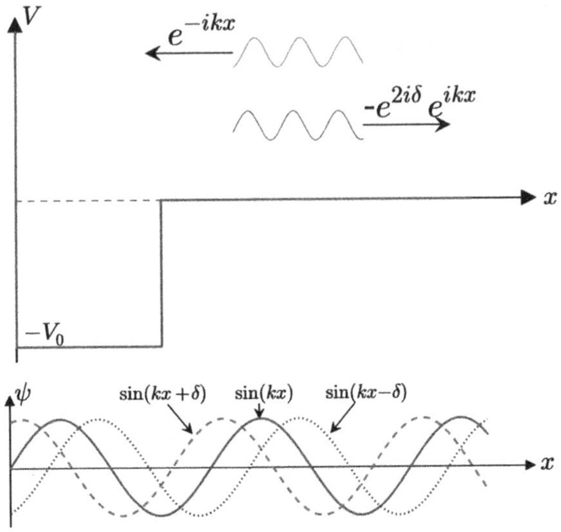

FIGURE 12.11 Scattering in 1D (top) and the phase shifts (bottom, $\delta > 0$).

Figure 12.12 shows an example of phase shifts from the delta-wall potential, a delta potential $-\alpha\delta(x - a/2)$ placed to the right of a hard wall (Figure 9.21). We can find the phase shift analytically (Exercise E12.6),

$$\delta = \tan^{-1}\left(\frac{2\beta\sin^2\frac{ka}{2}}{1-\beta\sin ka}\right), \quad \beta = \frac{m\alpha}{\hbar^2 k}. \tag{12.22}$$

The results in Figure 12.12 confirm that the phase shift is positive for attractive potentials ($\alpha > 0$) and negative for repulsive potentials ($\alpha < 0$). The phase shifts in this example vanish exactly whenever $ka/2 = n\pi$, n being an integer. This is the condition when standing waves are formed between the wall and the delta potential. When this happens, the scattered waves are not shifted at all (other than being reflected). This is an extreme case of resonance.

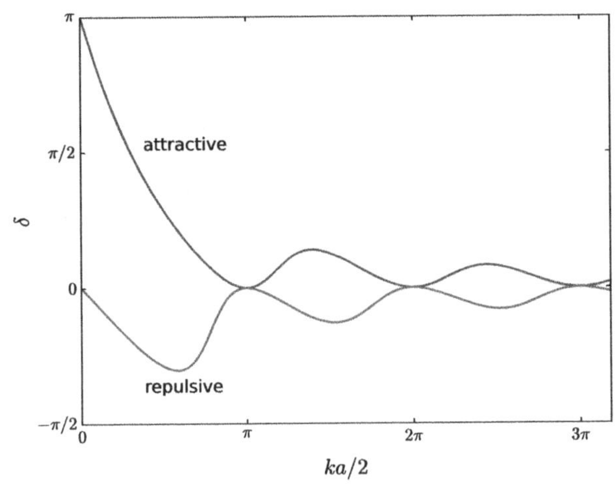

FIGURE 12.12 Phase shift from the delta-wall potential (Figure 9.21).

At low energies, Figure 12.12 shows that the phase shift approaches zero for the repulsive potential. But for the attractive potential, it approaches π. The latter is due to an interesting but general property known as Levinson's theorem [60], which states that the phase shift at $k = 0$ is equal to

$$\delta(k \to 0) = n_b \pi, \quad \text{(Levinson's theorem)} \tag{12.23}$$

where n_b is the number of bound states supported by the potential.

The parameters used in Figure 12.12 are such that there is one bound state (Exercise E9.7). To ensure the correct calculation of the phase shift, we require that δ is continuous and $\delta \to 0$ as $k \to \infty$ (usually). In plotting Figure 12.12, we have made corrections to account for the phase shift jumping by π due to the multivaluedness of the inverse tangent function. Only then does $\delta(k = 0)$ approach the correct limit of Levinson's theorem (12.23). Further explorations can be found in Exercise E12.6.

12.4.2 PARTIAL WAVE EXPANSION

The preceding concept of phase shifts in 1D is equally applicable to 3D. We just need to separate the radial and angular variables in the 3D Schrödinger equation as $\psi = R(r)P(\theta)$ (Section 9.5), so we obtain an effective 1D radial equation. Furthermore, the angular solution is given by $P_l(\cos\theta)$, the well-known Legendre polynomials (Section 8.A.3), independent of the potential (so long it is central). The order parameter l corresponds to a discrete angular momentum state. As such, any wave function can be written as a superposition $\sum A_l R_l P_l$.

For instance, the incident plane wave e^{ikz}, a free-particle solution ($V = 0$, see Eq. (S.12.2)), can be written as [4]

$$e^{ikr\cos\theta} = \sum_{l=0}^{\infty} (2l+1) i^l j_l(kr) P_l(\cos\theta), \tag{12.24}$$

where $j_l(kr)$ is the spherical Bessel function of the first kind (see Section 12.B). For nonzero potentials, we seek solutions of the form [80]

$$\psi = \sum_{l=0}^{\infty} (2l+1) i^l R_l(k,r) P_l(\cos\theta). \tag{12.25}$$

The radial wave function $R_l(k,r)$ satisfies the radial Schrödinger equation (9.38) for scattering states $E > 0$,

$$\frac{1}{r^2}\frac{d}{dr}\left(r^2 \frac{dR_l}{dr}\right) + \left[k^2 - \frac{2m_e}{\hbar^2}\left(V(r) + \frac{l(l+1)\hbar^2}{2m_e r^2}\right)\right] R_l = 0. \tag{12.26}$$

The free-particle radial wave functions, $j_l(kr)$, are solutions to Eq. (12.26) for $V = 0$ (see Eq. (12.55)). They have the asymptotic behavior [4]

$$j_l(kr) \xrightarrow{r\to\infty} -\frac{e^{-i(kr - l\pi/2)} - e^{i(kr - l\pi/2)}}{2ikr} = \frac{\sin(kr - \frac{l\pi}{2})}{kr}. \tag{12.27}$$

Like the 1D case (12.20), we recognize $e^{\mp i(kr - l\pi/2)}/r$ in Eq. (12.27) as the incoming and outgoing spherical waves, respectively. This fact readily lends itself to the interpretation that the incoming wave is scattered into an outgoing wave with zero phase shifts because $V = 0$.

When the interactions are not zero, we expect the asymptotic behavior of $R_l(k,r)$ to be (see Section 12.B, Eqs. (12.66a) to (12.66c))

$$R_l \xrightarrow{r\to\infty} e^{i\delta_l} \cos\delta_l \, j_l(kr) - e^{i\delta_l} \sin\delta_l \, n_l(kr) \tag{12.28a}$$

$$= -\frac{1}{2ikr}\left[e^{-i(kr-l\pi/2)} - e^{2i\delta_l}e^{i(kr-l\pi/2)}\right] \tag{12.28b}$$

$$= e^{i\delta_l}\frac{\sin(kr - \frac{l\pi}{2} + \delta_l)}{kr}, \tag{12.28c}$$

in full analogy to Eqs. (12.20) and (12.21). In Eq. (12.28a), $n_l(kr)$ is the spherical Bessel function of the second kind. Comparing Eq. (12.28b) with (12.27), we conclude that the outgoing wave is phase-shifted by $2\delta_l$, yet to be determined.

Substituting Eqs. (12.24) and (12.25) into (12.14), we can express the scattered wave as

$$\psi_{sc} = \psi - e^{ikr\cos\theta} = \sum_{l=0}^{\infty}(2l+1)i^l\left[R_l(k,r) - j_l(kr)\right]P_l(\cos\theta). \tag{12.29}$$

With Eqs. (12.27) and (12.28b), we can obtain the asymptotic behavior of the scattered wave,

$$\psi_{sc} \xrightarrow{r\to\infty} \frac{1}{2ik}\sum_{l=0}^{\infty}(2l+1)i^l P_l(\cos\theta)(e^{2i\delta_l} - 1)e^{-il\pi/2}\frac{e^{ikr}}{r}, \tag{12.30}$$

where the incoming waves cancel out.

Equation (12.30) is a pure outgoing wave, $f(\theta)\frac{e^{ikr}}{r}$. We identify $f(\theta)$ as the scattering amplitude from Eq. (12.15),

$$f(\theta) = \frac{1}{2ik}\sum_{l=0}^{\infty}(2l+1)i^l P_l(\cos\theta)(e^{2i\delta_l} - 1)e^{-il\pi/2}$$

$$= \frac{1}{k}\sum_{l=0}^{\infty}(2l+1)e^{i\delta_l}\sin\delta_l P_l(\cos\theta), \tag{12.31}$$

where we have used $e^{-il\pi/2} = (-i)^l$. The scattering amplitude is complex, so the cross section is

$$\sigma(\theta) = |f(\theta)|^2 = \frac{1}{k^2}\left|\sum_{l=0}^{\infty}(2l+1)e^{i\delta_l}\sin\delta_l P_l(\cos\theta)\right|^2. \tag{12.32}$$

We can obtain the total cross section by integrating Eq. (12.32) (Exercise S:E12.5)

$$\sigma_t = \int \sigma(\theta)\,d\Omega = \frac{4\pi}{k^2}\sum_{l=0}^{\infty}(2l+1)\sin^2\delta_l. \tag{12.33}$$

12.4.3 EXTRACTION OF PHASE SHIFTS

The preceding discussion shows that the central quantity is the phase shift δ_l. Knowing the asymptotic form of $R_l(k,r)$, we can extract the phase shift from Eqs. (12.28a) to (12.28c). Numerically, it is more accurate to use Eq. (12.28a) in the potential free region, where kr is just large enough that the solutions are well described by a combination of $j_l(kr)$ and $n_l(kr)$, but not so large that the asymptotic forms (12.27) and (12.60) are sufficiently satisfied.

Suppose the matching radius is $r = a$, beyond which the potential is negligible. Differentiating Eq. (12.28a) on both sides with respect to r, we obtain the logarithmic condition

$$\frac{R_l'(k,a)}{R_l(k,a)} = k\frac{\cos\delta_l\, j_l'(ka) - \sin\delta_l\, n_l'(ka)}{\cos\delta_l\, j_l(ka) - \sin\delta_l\, n_l(ka)}, \tag{12.34}$$

where $j_l'(x)$ and $n_l'(x)$ are derivatives with respect to x, evaluated at $x = ka$. Rearranging Eq. (12.34), we obtain the phase shift as

$$\tan\delta_l = \frac{\beta j_l(ka) - k j_l'(ka)}{\beta\, n_l(ka) - k\, n_l'(ka)}, \qquad \beta = \frac{R_l'(k,a)}{R_l(k,a)}. \tag{12.35}$$

The phase shift δ_l depends on l and k, but should be independent of the matching radius a. The latter can be a useful numerical check.

Like bound states in central field potentials (Section 9.5, Eq. (9.39)), it is easier to work with $u_l(k,r) = rR_l(k,r)$, satisfying

$$\frac{d^2 u_l}{dr^2} + \left[k^2 - \frac{2m_e}{\hbar^2} V_{eff}(r)\right] u_l = 0, \tag{12.36}$$

$$V_{eff}(r) = V(r) + \frac{l(l+1)\hbar^2}{2m_e r^2}.$$

In terms of u_l, we can express the phase shift as

$$\tan\delta_l = \frac{(\gamma a - 1) j_l(ka) - ka\, j_l'(ka)}{(\gamma a - 1) n_l(ka) - ka\, n_l'(ka)}, \qquad \gamma = \frac{u_l'(k,a)}{u_l(k,a)}. \tag{12.37}$$

Given the logarithmic derivatives of the radial wave functions β or γ, analytically or numerically, phase shifts can be calculated from Eqs. (12.35) or (12.37). They can also be obtained from the phase shift integral (S.12.85) (see Section S:12.A).

12.4.4 SCATTERING BY A HARD SPHERE

As an example, consider scattering from a hard sphere,

$$V(r) = \begin{cases} \infty, & \text{if } r \le a, \\ & \qquad\qquad \text{(hard sphere)} \\ 0, & \text{if } r > a. \end{cases} \tag{12.38}$$

The wave function vanishes inside the sphere at $r \le a$. Outside the sphere $r > a$, the radial wave function is a solution to the free-particle radial equation (12.55)

$$R_l(k,r) = e^{i\delta_l}(\cos\delta_l\, j_l(kr) - \sin\delta_l\, n_l(kr)). \tag{12.39}$$

We can obtain the phase shifts by the continuity condition, $R_l(k,a) = 0$. It leads to the analytic result

$$\tan\delta_l = \frac{j_l(ka)}{n_l(ka)}. \tag{12.40}$$

The same result can be derived from Eq. (12.35) by setting $\beta \to \infty$.

Figure 12.13 shows the phase shift as a function of wave vector (ka) for the first several partial waves $l = 0$ to 4 (this and the following results can be obtained as described in Project P12.5). The values of $j_l(ka)$ and $n_l(ka)$ are obtained from SciPy special functions (see Program 12.2 for examples). All phase shifts start from zero at $ka = 0$, and are negative (repulsive potential). At low energies, the predominant partial wave is the s-wave ($l = 0$), the lowest angular momentum state. As energy increases, higher partial waves start to contribute more to scattering.

The phase shifts at large energies $ka \gg 1$ approach $\delta_l = l\pi/2 - ka$, linearly decreasing with ka. This behavior is atypical, as we expect the phase shift to approach zero as $ka \to \infty$ for finite

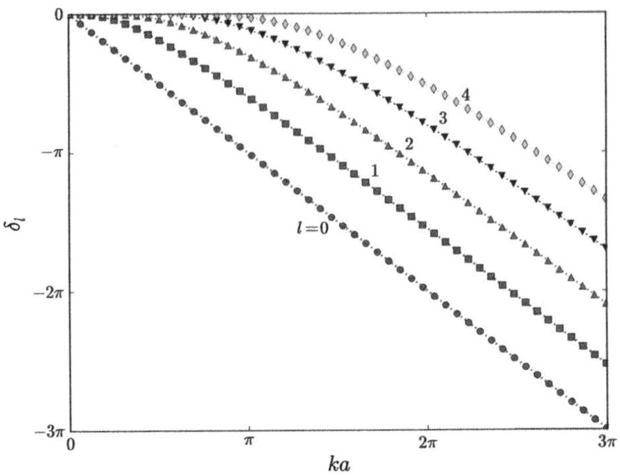

FIGURE 12.13 Phase shifts for scattering from a hard sphere.

potentials. For ideal hard spheres, however, no matter how high the energy is, a particle will never penetrate the sphere. The wave function remains zero at $r = a$ (Figure 12.14) while the wavelength decreases, causing the magnitude of the phase shifts to increase without bound.

The wave functions squared are shown in Figure 12.14 for two energies $ka = 0.01$ and 2 in the left and right columns, respectively. They are computed from Eq. (12.39). The incident wave ψ_{in} moves horizontally from left to right. At each energy, the total wave function ($|r\psi|^2$, Eq. (12.25)) is shown above the scattered wave function ($|r\psi_{\text{sc}}|^2$, Eq. (12.29)). We plot $|r\psi|^2$ instead of $|\psi|^2$ to emphasize the details in the asymptotic region. On the sphere, the full wave function vanishes as required, and the scattered wave function is maximum. Because the incident waves are not zero on the sphere, the scattered waves must be properly phase-shifted such that they cancel each other, $\psi_{\text{sc}} = -\psi_{\text{in}}$ at $r = a$.

At the lower energy, both the total and the scattered waves are nearly independent of directions, showing a high degree of isotropy. Here, the de Broglie wavelength is long compared to the sphere or any structure of the potential, so the scattering amplitude $f(\theta)$ has little angular dependence.

At the higher energy, the total wave peaks right in front of the sphere, and there is a dark shadow immediately behind. The overall shape illustrates the scattering of plane waves by a hard sphere, resembling sheets of paper bending under stress. There must be strong destructive interference between the incident and the scattered waves to create the shadow. Indeed, the scattered waves are strongest in the shadow behind the sphere, and are predominantly focused in the forward direction. Because of this, we expect that the scattering amplitude $f(\theta)$ must have a strong anisotropy. We also expect the focus to be increasingly narrower at higher energies.

The cross sections, both differential and total, are shown in Figure 12.15. At the lowest energy ($ka = 0.01$), the differential cross section is flat, that is, isotropic. As we have seen from Figure 12.13, low-energy scattering is dominated by the s-wave and hence is nondirectional because $P_0(\cos\theta) = 1$ in Eq. (12.32). This is consistent with the isotropy in the wave functions (Figure 12.14).

The differential cross section peaks at increasingly forward angles as energy increases, over a smooth background extending to the backward scattering angles up to 180°. The background value happens to coincide with the classical differential cross section, which remains constant over the entire angular range, irrespective of energy (Exercise E12.4). This is no coincidence, of course. At higher energies, or equivalently at large momentum transfers, the effective de Broglie

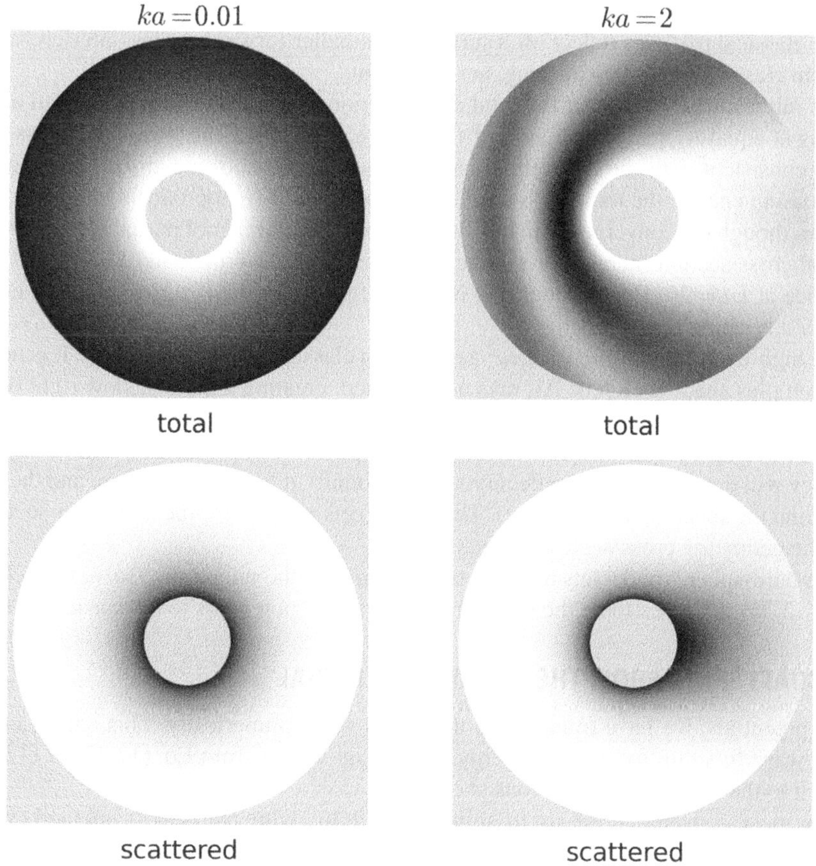

FIGURE 12.14 The total and scattered waves, $|r\psi|^2$ and $|r\psi_{sc}|^2$, shown in the top and the bottom rows, respectively, for scattering from a hard sphere at two energies, $ka = 0.01$ (left column) and 2 (right column).

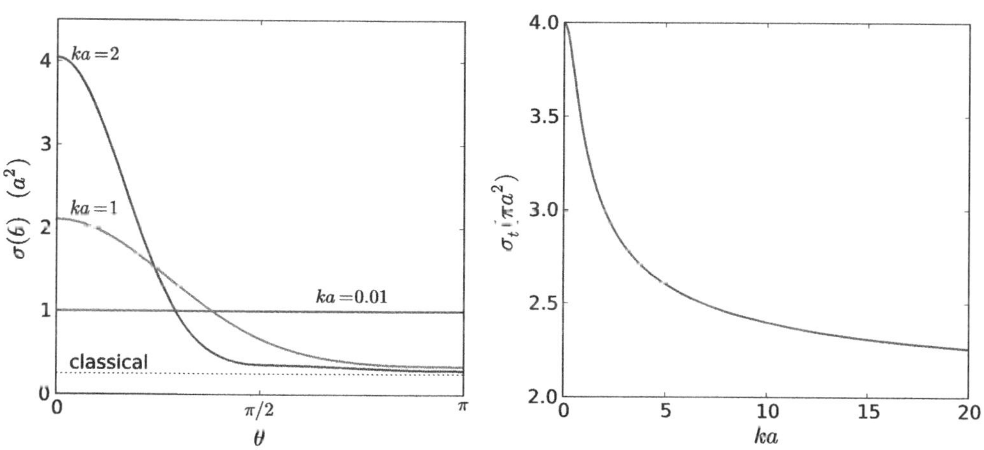

FIGURE 12.15 The differential (left) and total cross sections (right) for scattering from a hard sphere. The dotted line indicates the classical cross section.

wavelength becomes very small. The waves behave more like geometric rays scattering in the same way classical particles reflect off a surface. At smaller forward angles, interference effects, absent from classical particle scattering, are important.

It is interesting to note that the total cross section (Figure 12.15) is maximum at $ka \to 0$, $4\pi a^2$. This is equal to the surface area of the sphere, four times the geometric cross-sectional area. It is consistent with the differential cross section being isotropic at $\sigma(\theta) = a^2$ for $ka \sim 0$. With increasing energy, the total cross section decreases steadily, approaching a limit $\sigma_t \to 2\pi a^2$ as $ka \to \infty$, though the convergence to this limit is slow. The number of partial waves contributing to the total cross section increases for large ka. Numerically, we included $l = 0$ to 40 to obtain convergence at $ka = 20$. Even though the phase shifts (Figure 12.13) grow without bound, the cross section remains well defined.

The high-energy limit of $\sigma_t = 2\pi a^2$ agrees with classical optical scattering. It is interpreted as reflection plus shadow effects. All rays are reflected, creating a dark shadow right behind the sphere. However, the space far behind the sphere is still filled with waves (Figure 12.14), so this can be construed as waves scattered from behind the sphere. If detectors are placed away from the sphere, they will detect the actual reflections (background) in front of the sphere and the scattered waves behind the sphere (shadow effect). The two effects account for the total cross section equal to twice the geometric cross-sectional area.

This hard-sphere model lets us extract phase shifts without having to solve the Schrödinger equation (12.36). It ceases to be the case for other realistic potentials to be considered next.

12.4.5 SCATTERING FROM THE YUKAWA POTENTIAL

For most potentials, we have to calculate the phase shifts numerically. This entails computing the radial wave function for a given energy and partial wave l from Eq. (12.26) or (12.36), and matching it to the free-particle solutions (12.34).

It is most convenient for us to solve Eq. (12.36) with Numerov's method, which we have used for bound states in central field potentials such as the hydrogen atom (Section 9.5). For scattering states, the problem is even easier to handle. Recall that for bound states the solutions diverge in the upward direction of integration as $r \to \infty$ (Section 9.1.1). We do not have this difficulty for scattering (continuum) states because scattering wave functions oscillate rather than decay at large values of r. It is not necessary to integrate Eq. (12.36) upward and downward and match in the middle.

The procedure is very similar to Program 9.5. We only need to start from the origin with $u(0) = 0$, and integrate Eq. (12.36) upward until the field-free region, $r \geq a$, is reached. We apply the outgoing boundary condition to obtain the correct wave function (12.28a), hence the phase shift (12.37). The full modified program is given in Program 12.2. The following code illustrates the core function.

```
def wf(M, xm):                      # find w.f. and deriv at xm
    c = (h*h)/6.
    wfup, nup = ode.numerov(f, [0,.1], M, xL, h)     # 1 step past xm
    dup = ((1+c*f(xm+h))*wfup[-1] - (1+c*f(xm-h))*wfup[-3])/(h+h)
    return wfup, dup/wfup[-2]
......
jl, dj = sph_jn(Lmax, k*a)          # (j_l, j_l ') tuple
nl, dn = sph_yn(Lmax, k*a)          # (n_l, n_l ')
```

The module `wf()` integrates the radial equation upward for M steps using Numerov's method from the ODE library. Its function is similar to `shoot()` in Program 9.5. The energy E and angular momentum l are properly set before `wf()` is called. It returns the wave function on the grid, as well as $\gamma = u_l'/u_l$. The latter is used to calculate the phase shift from Eq. (12.37), which requires the spherical Bessel functions and their derivatives at ka, namely, j_l, n_l, j_l', and n_l'. Like the Hermite polynomials (Program 9.4), we can obtain these values from the SciPy special functions library as `sph_jn` and `sph_yn`. Each returns a pair of arrays with indices from 0 to L_{max}, containing the function values and the derivatives. Once the phase shifts have been calculated, we can obtain the cross sections and normalized wave functions.

We first test Program 12.2 against a known case of scattering from the square spherical barrier,

$$V(r) = \begin{cases} V_0, & \text{if } r \le a, \\ 0, & \text{if } r > a. \end{cases} \tag{12.41}$$

The solutions are analytically available in terms of spherical Bessel functions (see Exercise E12.7). But, only for $l = 0$ can the phase shift be easily calculated analytically. For $l > 0$, we still need to rely on numerical evaluation of these functions.

We calculate the phase shifts from Program 12.2, shown in Figure 12.16. When compared to the analytic results (Project P12.6), they are identical within numerical accuracy. The spherical barrier does not support bound states. All phase shifts are negative, approach zero as $k \to 0$ with zero slope, except $l = 0$ with a finite slope. The latter is similar to the s-wave for the hard sphere, with different slopes. For a finite V_0, the waves can enter the barrier, however slightly at low energies, leading to a less steep slope. As $V_0 \to \infty$, the result for the finite barrier will approach that for a hard sphere.

With increasing energies, the phase shifts reach some minima and rise up sharply, followed by plateaus and smaller oscillations. The sharp edges are well-known features of resonances in the effective potential (see Section S:8.1.2, Figure S:8.5 for the 1D analog). At high energies, the phase shifts vanish.

We now turn our attention to scattering from the Yukawa potential (12.13). The nuclear charge and screening length are $Z = 2$ and $a = 1$, respectively. Note that a is not the matching radius, which is set to 10 in this case, with a grid size $h = 0.05$.

Figure 12.17 shows the first two radial wave functions $u_l(kr)$ at the energy $E = 1$. They start from zero and show faster oscillations than free-particle solutions in the inner region.[5]

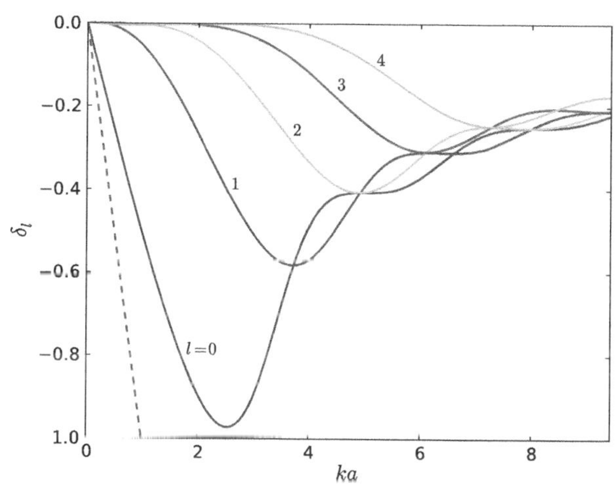

FIGURE 12.16 Phase shifts by a square spherical barrier, $a = 1$ and $V_0 = 2$. The dashed line is the δ_0 result by a hard sphere.

[5]For $l \ge 1$, true oscillations begin only after entering the classically allowed region beyond the turning point (Figure S:12.1).

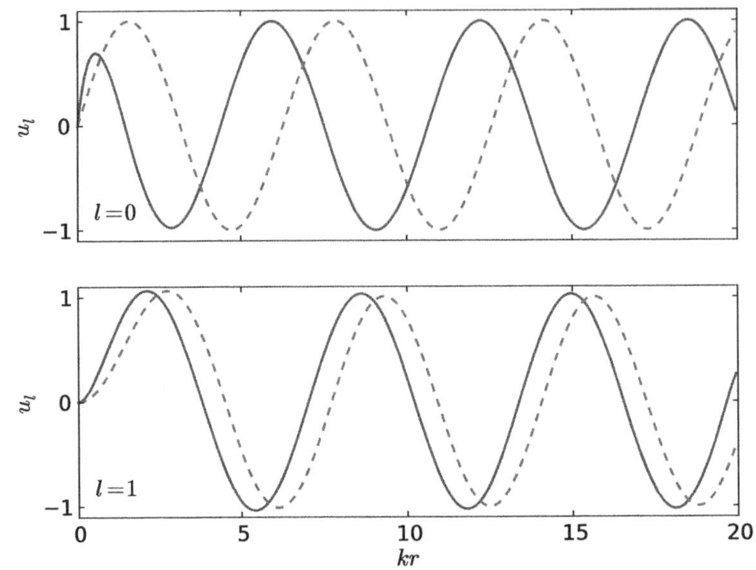

FIGURE 12.17 Radial wave functions u_l for $l = 0$ and 1 for the Yukawa potential with $Z = 2$ and $a = 1$. The energy is $E = 1$, or $k = \sqrt{2}$. The dashed lines are the field-free solutions $j_l(kr)$.

For direct comparison, the peak values are normalized to be equal. The faster oscillations, or shorter wavelengths, cause the scattered waves to accumulate phase at a faster rate than the unscattered free waves, giving rise to positive phase shifts. Conversely, if the potential was repulsive, the scattered waves would be suppressed (larger wavelengths or decaying waves), leading to negative phase shifts. Summarizing these observations, we have for the sign of phase shifts,

$$\delta_l = \begin{cases} \text{positive if } V < 0, \\ \text{negative if } V > 0. \end{cases} \tag{12.42}$$

For potentials with mixed signs, the sign of the phase shift is indeterminate, depending on the average effects of the interaction.

For the Yukawa potential, which decreases exponentially, the field-free region is established quickly at $r \sim 3$. As a result, starting at $kr \gtrsim 5$, the scattered waves behave just like the phase-shifted free-particle waves. In this case ($k = \sqrt{2}$), the phase shifts are $\delta_0 \sim 1.9$ and $\delta_1 \sim 0.7$, as can be estimated visually from Figure 12.17.

We show in Figure 12.18 the phase shifts as a function of l for fixed energies and as a function of energy for fixed l values. At low energies, only low partial waves are important. Higher partial waves become significant for increasing energies. To show this more clearly, we have multiplied the phase shifts by the factor $2l + 1$, the multiplicity of level degeneracy of partial wave l (Figure 12.18, right).

For a given energy, we can estimate the typical range of significant partial waves as $l_{max} \sim k\langle r\rangle$, where $\langle r\rangle$ is the range of the potential. For the Yukawa potential, it is on the order of a, the screening length. If the energy is high, the value of l_{max} can be large to achieve convergence. In that case, perturbative methods such as the Born approximation (Section S:12.3.2) may be more appropriate.

As $k \to 0$, only s-wave phase shift remains, approaching $\delta_0 = \pi$. According to Levinson's theorem (12.23), we expect that there is one bound state for $l = 0$. Indeed, this is the case for the potential parameters used. For $l \neq 0$, no bound states exist, as indicated by the vanishing phase shifts for $\delta_{l\neq 0} = 0$ as $k \to 0$.

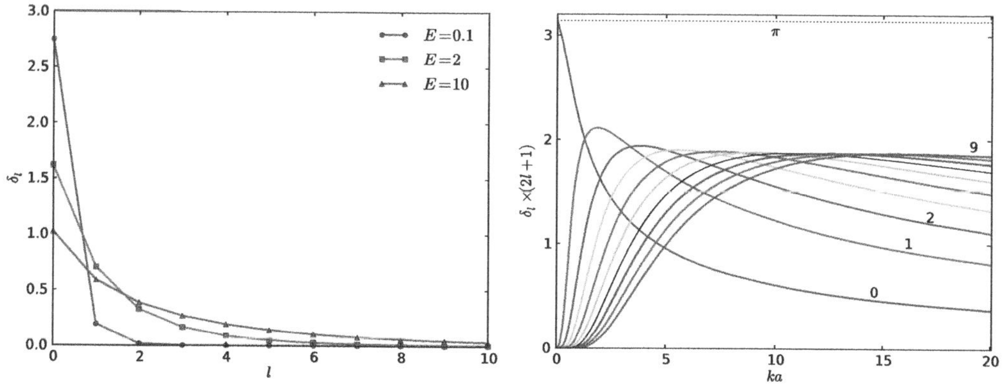

FIGURE 12.18 Phase shifts vs. partial waves l (left) and vs. wave vector ka (right). The latter is scaled by $2l + 1$, the multiplicity of degeneracy. The potential parameters are the same as Figure 12.17.

We present in Figure 12.19 the cross sections (12.32) at low and intermediate energies. The classical (Figure 12.9) and Born (Eq. (S:12.13)) results are also plotted for comparison. The latter is valid from intermediate to high energies, so only $E = 2$ results are shown.

At low energies, the quantum mechanical results show a minimum at $\theta \sim 50°$. It is due to interference of the s- and p-waves ($l = 0$ and 1), which are the predominant partial waves at $E = 0.1$ (Figure 12.18). For stronger potentials, the minimum can develop into a narrow, sharp dip that has been observed experimentally.

The differences between classical and quantal results are very large in forward scattering angles up to 90°. We expect the classical results to fail at low energies in general because the de Broglie wavelength is large, and the scattering is highly quantum mechanical. In the forward angles, in particular, we expect the disagreement to be even worse because any potential without a cutoff radius will cause a classical particle to deflect, no matter how large the impact parameter is. This causes the total cross section to always diverge as stated earlier (Section 12.1.2). Quantum mechanically, however, the uncertainty principle means that we can resolve $\Delta\theta$ only to within some finite interval limited by the spread of angular momentum, which is finite. As a result, quantal total cross sections are finite for short-range potentials.

Toward large angles, agreement starts to emerge between quantal and classical results. The classical rainbow and glory singularities manifest as enhanced but smooth quantal cross sections, which, in fact, become maximum at $\theta = 180°$.

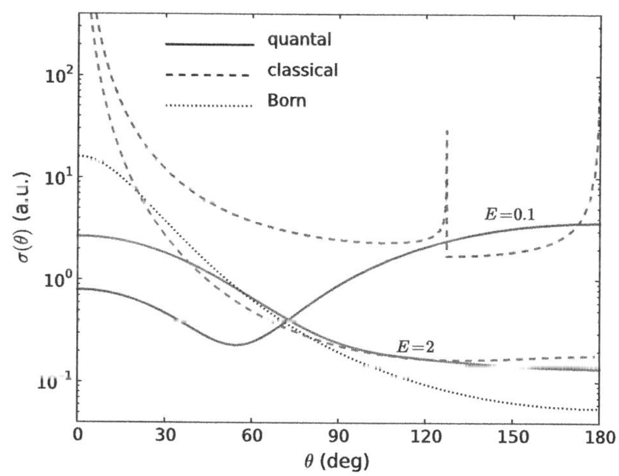

FIGURE 12.19 Cross sections at two energies by the partial wave method. The classical and Born ($E = 2$ only) results are also plotted. The potential parameters are the same as Figure 12.17.

At the intermediate energy $E = 2$, the quantal results start to favor forward scattering. The agreement with the classical results is much better for $\theta > 30°$. There is still some discrepancy at backward angles (150° to 180°), mainly due to a handful of partial waves interfering with each other. The differences disappear at higher energies (not shown) when a large number of partial waves contribute, and interference effects are insignificant.

The Born results are not terribly accurate, but not grossly inaccurate either. It is often useful as a quick check on the order-of-magnitude estimate at this energy. Like the classical results, it becomes accurate at higher energies (Section S:12.3.2).

CHAPTER SUMMARY

We discussed potential scattering in both classical and quantum mechanics and compared the two where appropriate.

The direct method for classical potential scattering is to calculate the deflection function on a grid and obtain the cross section by numerical differentiation. We can also conduct numerical experiments to determine the cross section. We studied rainbow and glory scattering in some detail.

Quantum mechanically, the partial wave analysis is the easiest way to compute the cross sections via phase shifts. Used with Numerov's method, the radial Schrödinger equation can be efficiently solved for scattering states just as it could for bound states (Chapter 9). At low energies, only the s-wave is important, which also contains bound state information (Levinson's theorem). Quantum scattering cross sections show enhancement, rather than singularities, at the classical rainbow and glory angles.

In the calculation of phase shifts and cross sections, we made extensive use of the SciPy special function library, including spherical Bessel functions and Legendre polynomials.

12.5 EXERCISES AND PROJECTS

Exercises

E12.1 (a) From Eq. (12.48), show that the distance of closest approach in Rutherford scattering ($V = k/r$) is given by

$$r_{min} = \frac{2Eb^2}{k(\pm e - 1)}.$$

The $\pm$ signs correspond to repulsive ($k > 0$) and attractive ($k < 0$) potentials, respectively.

(b) Integrate Eq. (12.49) directly to derive the deflection function (12.52) for the Coulomb potential. Optionally try your luck with SymPy's symbolic integration (Project P2.7).

E12.2 Derive the derivative $d\Theta/db$ (12.12) from Eq. (12.49). Note that the upper limit u_{min} depends on the impact parameter b. Carefully handle the diverging terms arising from the differentiation of the upper limit and the denominator.

E12.3 The primary rainbow refers to light rays back-scattered from water droplets. Figure 12.20 illustrates how the rainbow is formed.

The incident light ray entering the droplet undergoes a refraction, reflection, and another refraction, finally exiting as the rainbow ray at a scattering angle θ. Derive a formula for θ using Snell's law in terms of the impact parameter b, droplet radius R, and refractive index n.

Plot θ vs. b, assuming $n = 1.33$ and $R = 1$. Compare the plot with the deflection function for the attractive plum potential (Project P12.2) if available, or the repulsive potential (Figure 12.5).

Show that $\theta_m \leq \theta \leq 180°$. Find θ_m and the value of b where the minimum angle occurs. Optionally, draw the paths of rainbow rays over select b values between $[0, R]$.

E12.4 (a) Derive the deflection function for scattering from a hard sphere (12.38). Sketch the Θ-b curve.

(b) Show that the differential cross section is $\sigma(\theta) = \frac{1}{4}a^2$, independent of energy or the scattering angle, and the total cross section is therefore $\sigma_t = \pi a^2$. Explain both results.

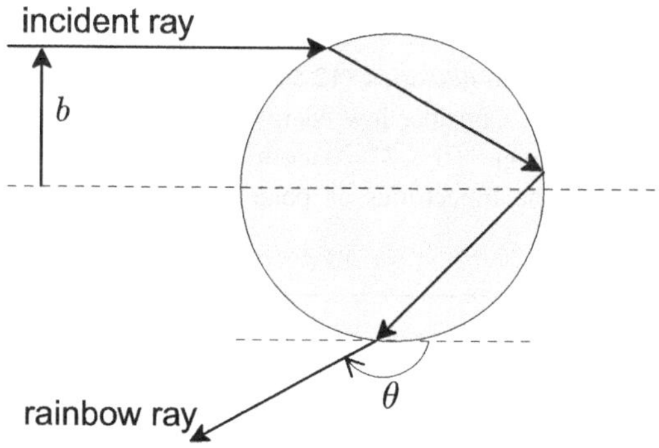

FIGURE 12.20 Schematics of optical rainbow scattering.

E12.5 Similar to the incident flux (12.16), the radial flux is given by

$$I_r = -i\frac{\hbar}{2m}\left(\psi_r^* \frac{\partial \psi_r}{\partial r} - \psi_r \frac{\partial \psi_r^*}{\partial r}\right),$$

where $\psi_r = f(\theta)e^{ikr}/r$ is the asymptotic outgoing spherical waves (12.15). Calculate I_r and prove that it is given by Eq. (12.17).

E12.6 (a) Show that the phase shift of the delta-wall potential (Figure 9.21, Exercise E9.7) is given by Eq. (12.22). Assuming $m = 1$ and $\alpha = 1$, calculate and plot the phase shifts for $a = \frac{1}{2}$, 1, and 2 (all in a.u.). Make sure the curves are continuous and $\delta \to 0$ as $ka \gg 1$.

(b) Calculate the phase shift from the square well potential of width a and depth V_0 (Figure S:8.4), and show that it is given by

$$\delta = -ka + \tan^{-1}\left(\frac{k}{q}\tan(qa)\right),$$

where $k = \sqrt{2mE}/\hbar$ and $q = \sqrt{2m(E+V_0)}/\hbar$.

Plot the phase shift as a function of $k \in [0, 10]$ for $m = 1$, $V_0 = 1$ and $a = 1$. Do the same with the same parameters as in Figure S:8.5 ($a = 3.8$). Discuss your results in relation to the transmission and reflection coefficients there.

Optionally explore the phase shift at small $k \sim 0$ by increasing the parameters a and V_0. Verify that $\delta(k \sim 0) = n\pi$, n being the number of bound states according to Levinson's theorem. Check n against the case shown in Table 9.1.

E12.7 (a) Show that for the square spherical potential well, $V = -V_0$ for $r \le a$, and $V = 0$ for $r > a$, the β value in Eq. (12.35) is given by

$$\beta = \frac{qj_l'(qa)}{j_l(qa)}, \quad q = \frac{\sqrt{2m(E+V_0)}}{\hbar}.$$

Prove that for $l = 0$, with the β value above, Eq. (12.35) is identical to the 1D result of Exercise E12.6.

(b)* For the square spherical potential barrier, $V_0 < 0$, show that the β value is the same as above for $E > |V_0|$, and for $E < |V_0|$, it is

$$\beta = \frac{qi_l'(qa)}{i_l(qa)}, \quad q = \frac{\sqrt{2m|E+V_0|}}{\hbar},$$

where $i_l(x) = (-i)^l j_l(ix)$ is the modified spherical Bessel function of the first kind.

Projects

P12.1 (a) Explore Rutherford scattering numerically. Plot the trajectories (12.50) for different impact parameters as in Figure 12.2. Assume atomic units. Consider low energy $E/k = \frac{1}{2}$ and intermediate energy $E/k = 2$ regimes. Vary b in the range $[10^{-2}, 2]$. At each b, find the distance of closest approach r_{min} (Exercise E12.1). Plot the trajectories as polar plots using `polar` projection as

```
plt.subplot(111, polar=True)
plt.plot(theta, r)
```

or convert (r, θ) to (x, y) plots. In either case, note that the angle $-\alpha < \theta < \alpha$ in Eq. (12.50) is relative to the radial bisector (Figure 12.1).

(b) Calculate the deflection function from Eq. (12.47). First graph the integrand for a given E and b. Note that it has a singularity at r_{min}, but it is an integrable singularity. Estimate what the upper limit should be for the integral to converge.

Evaluate the integral numerically with an open-end integrator such as Gaussian integration. Incrementally increase the upper limit until the integral converges to at least three digits. Break up the interval as necessary to maintain accuracy. Compute the deflection function at low and intermediate energies for b in $[10^{-2}, 10^{2}]$. Plot the results on a semilog scale together with the analytic result.

Accelerate convergence by isolating the integrable singularity (using a technique similar to Eq. (8.79) or Project S:P11.7). Make a variable substitution $r - r_{min} = x^2$, and repeat the evaluation. Explain the reason for the improvement in the results.

Apply the preceding techniques to u-space (12.49). Assess the advantage or disadvantage relative to r-space (12.47).

P12.2 Study scattering from the attractive plum-pudding potential (12.10). Use the parameters $R = 1$ and $Z = -1$.

(a) Make predictions about the deflection function and the cross sections. Sketch them for low and intermediate energies.

Run Program 12.1 to compute the deflection function and the cross sections at $E = 0.5$ and 2, respectively. Discuss the results and compare with the repulsive potential (Figures 12.5 and 12.7). Record the rainbow angles and impact parameters. What parts of the results are the same for the deflection function? Cross section?

(b) Based on the deflection function, sketch the particle trajectories. Write a program to compute and plot the trajectories. Use the leapfrog solver to integrate the equations of motion, and set the initial condition to $\vec{r} = (-a, b)$ and $\vec{v} = (v, 0)$, where a is large compared to the range of the potential (say $a = 10$), b the impact parameter, and v according to the collision energy (12.1). Divide b evenly between $[0, b_{max}]$, for example, $b_{max} = 2$, in roughly 10 intervals. Stop the integration when the particle leaves the interaction region, for instance when $r \geq 1.5a$.

Plot the trajectories for all impact parameters for a given energy. Discuss how they relate to the rainbow angles and impact parameters.

P12.3 Consider classical scattering from the square spherical well $V = -V_0$ for $r \leq a$, and $V = 0$ for $r > a$.

(a) Sketch a typical trajectory for $b < a$ and $V_0 > 0$. Also sketch the deflection function as a function of b. Explain your reasoning.

Modify Program 12.1 to calculate the deflection function Θ and the cross section $\sigma(\theta)$. Use $a = 1$ and $V_0 = 1$, and plot $\Theta(b)$ for $b \leq a$ and $\sigma(\theta)$ up to the maximum scattering angles for two energies $E = 0.5$ and 1. Take precaution to treat the discontinuity in the potential at

$r = a$, for example, by breaking the integration range into two parts on either side of a and properly shifted for the integration variable. The plots should be sufficiently smooth. Comment on your results. Can you tell if $\sigma(\theta)$ is finite as $\theta \to 0$?

(b) Show that the scattering is equivalent to an optical refraction problem if the refractive index of the sphere is defined as

$$n = \sqrt{1 + \frac{V_0}{E}}.$$

Calculate the deflection function from Snell's law, and compare with the numerical results from part (a). Because you have the exact $\Theta(b)$, calculate the cross section analytically from $\Theta(b)$ (evaluated numerically, of course).

Discuss the source of discrepancies, particularly at small scattering angles, between the cross sections obtained by finite difference and by analytic differentiation. How can the differences be reduced?

(c)* Let $V_0 = -1$, so the scattering is from a spherical barrier. Repeat the preceding parts, but for energies $E = 0.5$, 1.2, and 2. Compare the results for the lowest energy with scattering from the hard sphere, Exercise E12.4.

At higher energies $E > |V_0|$, investigate the existence of critical scattering angles where, according to part (b), the index of refraction n is less than 1 when $V_0 < 0$. Explore the dependence of the critical angle on the impact parameter and energy. What is the physical explanation of the critical angle in the case of scattering?

P12.4 Carry out the calculations of classical scattering for the Yukawa potential (12.13), assuming parameters $Z = 2$, $a = 1$, and $E = 0.1$.

(a) Compute the deflection function Θ for $b \in [10^{-2}, 10]$ using Program 12.1 or codes developed from preceding projects. Plot the results (Figure 12.8). Ensure the data points are dense enough to obtain an overall smooth curve, especially around the sharp dip at the rainbow angle. Map the deflection angle to the scattering angle θ according to Eq. (12.4), and plot it alongside Θ.

(b) Obtain the cross sections. It may be easier to search for all contributing impact parameters from the mapped scattering angle using Eq. (12.11). But, watch for artificial jumps in the angle due to the mapping (e.g., at $b \sim 2.5$ and 3), and resolve them appropriately to ensure continuity of the derivatives (12.11). Plot the results.

Depending on the spacing of data points for the impact parameter, your results may not be smooth in the neighborhood of the rainbow and glory angles. If so, increase the density of data points in the neighborhood until the curve is satisfactorily smooth.

(c)* Accurately calculate the cross section from Eq. (12.12). For a given scattering angle θ, we need to know the contributing impact parameters. To do so, identify all Θ values corresponding to a given θ from the deflection function Θ (already available from earlier), and bracket the contributing b values. Use the bisection root solver to find the exact b values from Eq. (12.3) (see Project S:P11.7 for solving equations involving integrals). Plug these values into Eqs. (12.12) and (12.7) to obtain the cross sections. Plot the results and compare with the direct but crude approximations above.

P12.5 Numerically study quantum scattering from the ideal hard sphere (12.38), $V = \infty$ for $r \leq a$, and $V = 0$ for $r > a$. Assume $a = 1$.

(a) Calculate the phase shifts δ_l from Eq. (12.40) for $l \leq l_{max} = 20$ as a function of k from 0 to 10 in equal steps of 0.1 or smaller. Obtain j_l and n_l values from the SciPy special function library as demonstrated in Program 12.2. Adjust δ_l by adding or subtracting π so that they are continuous and approach $\delta_l(k \to 0) = 0$. Plot the results as in Figure 12.13.

Compute phase shifts by integrating the Schrödinger equation with Program 12.2. Modify the starting point to begin from $r = a$. Integrate to a short distance away, say $2a$, and match the wave functions. Determine the phase shifts, and compare with the analytic results from earlier.

(b) First calculate the differential cross sections at several k values from low to high energies, for example, $k = 10^{-2}$, 1 and 10. Plot the results as in Figure 12.15.

Next, calculate the total cross section $\sigma_t(k)$ as a function of k. Extend the upper end of k to 10^2 or higher. Make sure to adjust l_{max} for larger k. Plot the shifted results, $\sigma_t/\pi - 2$, on a log-log scale. How do you know if the l_{max} is adequate? For large $k \gg 1$, you should see the curve approaching a straight line. Determine the slope. What is the meaning of the slope? Explain.

(c) Generate and plot the scattering wave functions as shown in Figure 12.14. Divide the radius between $[a, r_{max}]$ into m intervals, and θ between $[0, 2\pi]$ into n intervals (though the maximum θ is π, it is easier to cover the full angular range without the need for reflection). Form an $m \times n$ mesh over (r, θ) and subsequently a grid over (x, y) (see Program 7.6 for an example).

On the (r, θ) mesh, calculate the full wave function ψ in Eq. (12.25) using the radial wave function (12.39). At each grid point, you must perform the sum over l. Do the same for the scattered wave function ψ_{sc} by subtracting $j_l(kr)$ from the radial wave function (12.39). Once the wave functions are generated, plot $|r\psi|^2$ and $|r\psi_{sc}|^2$ over the (x, y) grid as filled contours with

```
plt.contourf(x, y, wf, 256, cmap=plt.cm.jet)
```

This creates a 256-level contour plot. Plot the wave functions for k in the low, intermediate, and high energies, for example, $k = 10^{-2}$, 1, and 5. Adjust the grid sizes so the plots are smooth.

(d)* Calculate and plot the scattered radial flux $r^2 I_r(\theta)$ (Exercise E12.5). Plot it as a vector field with the `plt.quiver()` function (see Program 7.5) over the same grid and with the same k values as above.

P12.6 Investigate quantum scattering from the square spherical potential well or barrier, $V = -V_0$ for $r \le a$, and $V = 0$ for $r > a$, considered in Project P12.3 for classical scattering.

(a) First assume a potential well, $V_0 = 1$ and $a = 1$. Replace the potential in Program 12.2, and calculate the phase shifts δ_l for $l \le l_{max}$ as a function of $k \in [10^{-2}, 10]$ incremented by a constant factor (say 1.1). Set l_{max} such that higher partial waves than l_{max} are negligible at the largest k. Ensure that the jumps in π are accounted for. Plot the phase shifts as a function of k.

Calculate the differential cross section at energies $E = 0.5$ and 1. Compare with the classical results from Project P12.3 if available.

(b) Due to the sharp cutoff of the potential, accuracy of results will suffer unless the discontinuity is properly treated. To do so, linearly interpolate the potential between the two immediate grid points sandwiching the radius a according to Eq. (9.52) (Project P9.4).

With the same grid size h, calculate the phase shifts. Compute the relative errors of the results with and without the interpolation from the exact results (Exercise E12.7). Plot the relative errors for the first several partial waves as a function of k. Explain the improvement.

(c) Check that Levinson's theorem (12.23) works correctly. Compute the phase shifts for $l = 0$, 1, and 2 as a function of k. Use the same parameters for V_0 and a as in Table 9.1. For δ_0, the number of bound states n_b is the same in 1D and 3D. For $l \ne 0$, use Program 9.5 to find n_b. Show that δ_l approaches the correct limit as $k \to 0$.

(d) Repeat the preceding investigation for the square barrier, $V_0 = -1$.

P12.7 Comparatively study quantum scattering from the Yukawa potential (12.13) against several other potentials including,

(i) the Gaussian potential, $V(r) = -\frac{Z}{a}e^{-r^2/a^2}$;

(ii) the Hulthén potential, $V(r) = -\frac{Z}{a}\frac{e^{-r/a}}{1-e^{-r/a}}$;

(iii) the Buckingham polarization potential, $V(r) = -\frac{Za^3}{(r^2+a^2)^2}$.

(a) Plot (or sketch to scale) these potentials for the same parameters $Z = 2$ and $a = 1$. Compare the shapes of the potential. Predict which potential will have the greatest effect on scattering relative to the Yukawa potential.

(b) Using Program 12.2 or a similar one you have developed, calculate the phase shifts at $E = 2$ for the potential predicted earlier. Plot the results along with those for the Yukawa potential on the same graph as a function of l. Also calculate and plot the differential cross sections. Discuss the similarities and differences in the results. Do they support your prediction?

(c) Calculate the total cross sections for each of the potentials as a function of energy from $E = 0.1$ to 20. Comment on the results. Which potential has the largest effect? The least effect? Explain.

P12.8 The cross section for electron scattering from light atoms, represented by weak Yukawa potentials (Eq. (12.13), $Z \sim 1$), is relatively featureless. For heavier atoms, however, some interesting effects can occur, including the so-called Ramsauer-Townsend effect. Investigate this effect in this project.

Let $Z = 10$ and $a = \frac{1}{4}$. Calculate the phase shifts as a function of $k = [10^{-2}, 10]$ for $l = 0$ to 5. Compute and plot the total cross section as a function of k. You should see a graph like Figure 12.21. Note the peak at $k \sim 3$ and the quick descend into the valley at $k \sim 1$. The Ramsauer-Townsend effect refers to the sharp decrease of the cross section at low energies. Why does this happen?

To explore this effect, plot the phase shifts. Below $k \lesssim 5$, are higher partial waves $l = 2$ to 5 negligible? Confirm by plotting the partial cross sections

$$\sigma_l = \frac{4\pi}{k^2}(2l+1)\sin^2\delta_l.$$

Now consider only the first two partial waves $l = 0$ and 1. How does δ_0 change when k decreases from 3 to 1? How does σ_0 behave? How does that explain the rapid drop in the cross section around $k \sim 2$? What is the role of σ_1?

Repeat the preceding calculation for $a = 0.15$. Does the Ramsauer-Townsend effect occur? Does your explanation apply? How about for $Z = 5$ and $a = \frac{1}{4}$?

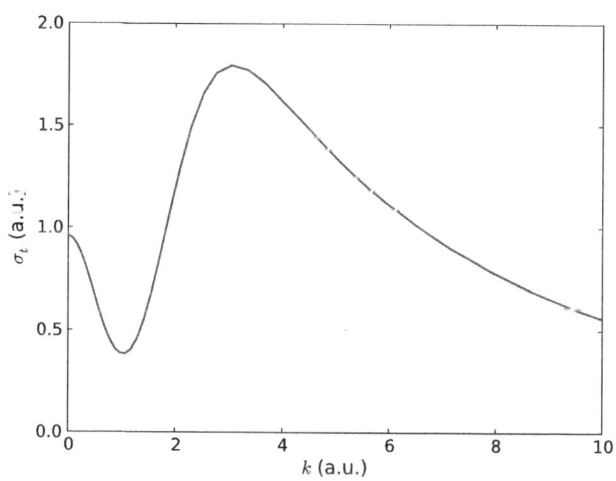

FIGURE 12.21 The Ramsauer-Townsend effect (Project P12.8).

12.A DERIVATION OF THE DEFLECTION FUNCTION

To find the polar angle in terms of the radial vector, we first express their time derivatives $\dot{\theta}$ and $\dot{r}$ from Eqs. (4.6) and (4.8),

$$\dot{\theta} = \frac{L}{mr^2}, \quad \dot{r} = \sqrt{\frac{2}{m}\left(E - V(r) - \frac{L^2}{2mr^2}\right)}. \tag{12.43}$$

We can derive the relationship between θ and r by eliminating time from Eq. (12.43) with

$$d\theta = \frac{d\theta}{dr}dr = \frac{d\theta/dt}{dr/dt}dr = \frac{\dot{\theta}}{\dot{r}}dr. \tag{12.44}$$

Substituting Eq. (12.43) into (12.44) and integrating both sides, we obtain

$$\theta(r) - \theta_0 = \int_{r_0}^{r} \frac{\frac{L}{mr^2}}{\sqrt{\frac{2}{m}\left(E - V(r) - \frac{L^2}{2mr^2}\right)}} dr. \tag{12.45}$$

Because α in Eq. (12.2) is the angle swept by the radial vector from r_{min} to ∞ (Figure 12.1), we can obtain α by replacing r_0 and r with r_{min} and ∞ in Eq. (12.45), respectively,

$$\alpha = \int_{r_{min}}^{\infty} \frac{\frac{L}{mr^2}}{\sqrt{\frac{2}{m}\left(E - V(r) - \frac{L^2}{2mr^2}\right)}} dr. \tag{12.46}$$

Here, r_{min} is the turning point where $\dot{r} = 0$.

Using Eq. (12.1) to simplify α explicitly in terms of the impact parameter b, the resulting expression for the deflection function from Eq. (12.2) is

$$\Theta(b) = \pi - 2\int_{r_{min}}^{\infty} \frac{b\,dr}{r^2\sqrt{1 - \frac{V(r)}{E} - \frac{b^2}{r^2}}}, \tag{12.47}$$

where the distance of closest approach (turning point) satisfies

$$1 - \frac{V(r_{min})}{E} - \frac{b^2}{r_{min}^2} = 0. \tag{12.48}$$

Sometimes it is useful to convert the integral of an infinite range into a finite range by $u = 1/r$. This gives

$$\Theta(b) = \pi - 2\int_{0}^{u_{min}} \frac{b\,du}{\sqrt{1 - \frac{V(1/u)}{E} - b^2u^2}}. \tag{12.49}$$

For Coulomb potentials $V(r) = k/r$, the integral (12.47) can be carried out analytically. However, we had already described the result for the orbit of the Kepler problem, which is hyperbolic for positive energies, Eq. (4.11). Combining Eqs. (4.13) and (4.14), we have

$$r = \frac{\pm 1 - e}{\pm 1 - e\cos\theta} r_{min}, \tag{12.50}$$

where $e > 1$ is the eccentricity (4.10). The $\pm$ signs correspond to repulsive and attractive potentials, respectively. Equation (12.50) is equivalent to setting the reference point $\theta = 0$ at the distance of closest approach. Therefore, the value of α is equal to θ when $r = \infty$ in Eq. (12.50) (Figure 12.1, top), which means that the denominator is zero,

$$\cos\alpha = \pm\frac{1}{e}. \tag{12.51}$$

From Eq. (12.2), we have $\alpha = \frac{\pi}{2} - \frac{\Theta}{2}$, so it follows from Eq. (12.51) that

$$\sin\frac{\Theta}{2} = \pm\frac{1}{e}, \quad \text{or } e = \pm 1/\sin\frac{\Theta}{2}. \tag{12.52}$$

Because the eccentricity e is positive, the deflection angle Θ must be negative for attractive potentials.

Equation (12.52) gives Θ as an implicit function of the impact parameter and energy through the eccentricity e. To make it explicit, we use Eq. (4.10) and the relations (12.1) to obtain

$$e = \sqrt{1 + \frac{4E^2}{k^2}b^2}, \quad b^2 = \frac{k^2}{4E^2}(e^2 - 1). \tag{12.53}$$

Substituting Eq. (12.52) into (12.53), we find

$$b = \frac{k}{2E}\cot\frac{\Theta}{2}. \tag{12.54}$$

We see that as $b \to 0$, $\Theta \to \pi$, and as $b \to \infty$, $\Theta \to 0$.

12.B PARTIAL WAVE ANALYSIS

Phase shifts are determined by the behavior of the radial wave function in the asymptotic region where the potential is negligible. If $V = 0$, the radial equation (12.26) can be simplified to

$$\frac{1}{r^2}\frac{d}{dr}\left(r^2\frac{dR_l}{dr}\right) + \left[k^2 - \frac{l(l+1)}{r^2}\right]R_l = 0. \quad \text{(free particle)} \tag{12.55}$$

It admits two linearly independent solutions, $j_l(kr)$ and $n_l(kr)$, the spherical Bessel functions of the first and second kind, respectively.

The first several spherical Bessel functions are ($x = kr$) [10]

$$j_0(x) = \frac{\sin x}{x}, \quad j_1(x) = \frac{\sin x}{x^2} - \frac{\cos x}{x},$$
$$n_0(x) = -\frac{\cos x}{x}, \quad n_1(x) = -\frac{\cos x}{x^2} - \frac{\sin x}{x}. \tag{12.56}$$

Higher orders of j_l and n_l can be obtained via the recurrence relation

$$j_{l+1}(x) - \frac{2l+1}{x}j_l(x) + j_{l-1}(x) = 0. \tag{12.57}$$

The same holds true for $n_l(x)$.[6]

[6]The recurrence relation (12.57) is stable in the downward direction for j_l and in the upward direction for n_l (see Exercise S:E12.6).

For small $x \sim 0$, $j_l(x)$ and $n_l(x)$ approach the limits

$$j_l(x) \to \frac{x^l}{(2l+1)!!}, \qquad n_l(x) \to -\frac{(2l-1)!!}{x^{l+1}}. \tag{12.58}$$

For large $x \to \infty$, the asymptotic behaviors are

$$j_l(kr) \overset{r\to\infty}{\longrightarrow} \frac{\sin(kr - \frac{l\pi}{2})}{kr} = \frac{e^{i(kr-l\pi/2)} - e^{-i(kr-l\pi/2)}}{2ikr}, \tag{12.59}$$

$$n_l(kr) \overset{r\to\infty}{\longrightarrow} -\frac{\cos(kr - \frac{l\pi}{2})}{kr} = -\frac{e^{i(kr-l\pi/2)} + e^{-i(kr-l\pi/2)}}{2kr}. \tag{12.60}$$

The function $j_l(kr)$ is regular in the entire range $r \in [0, \infty)$ and is always a solution to Eq. (12.55), as discussed earlier relative to the incident wave (12.24). But, $n_l(kr)$ diverges at the origin and is therefore a valid solution only if the origin is excluded. This is the case in the asymptotic region of Eq. (12.26), where the general solution $R_l(k, r)$ can be written as

$$\begin{aligned} R_l(k,r) &\overset{r\to\infty}{\longrightarrow} A_l j_l(kr) + B_l n_l(kr) \\ &= -\frac{1}{2ikr}\left[(A_l + iB_l)e^{-i(kr-l\pi/2)} - (A_l - iB_l)e^{i(kr-l\pi/2)}\right]. \end{aligned} \tag{12.61}$$

We have used the asymptotic behavior (12.59) and (12.60) in Eq. (12.61), where the first and second terms are the incoming and outgoing spherical waves, respectively.

In comparison to the incident wave (12.27), we require the incoming wave be unmodified,

$$A_l + iB_l = 1, \qquad \text{or } A_l - iB_l = 2A_l - 1, \tag{12.62}$$

so that Eq. (12.61) becomes

$$R_l(k,r) \overset{r\to\infty}{\longrightarrow} -\frac{1}{2ikr}\left[e^{-i(kr-l\pi/2)} - (2A_l - 1)e^{i(kr-l\pi/2)}\right]. \tag{12.63}$$

Analogous to the 1D case (12.20), we also require that the incident flux be equal to the outgoing flux,

$$|2A_l - 1| = 1, \qquad \text{or } 2A_l - 1 = e^{2i\delta_l}, \tag{12.64}$$

where δ_l is a real number, to be identified as the phase shift.

Solving Eqs. (12.62) and (12.64), we obtain the coefficients A_l and B_l as

$$A_l = e^{i\delta_l}\cos\delta_l, \qquad B_l = -e^{i\delta_l}\sin\delta_l. \tag{12.65}$$

Finally, substituting Eq. (12.65) into (12.61) and (12.63), we obtain the asymptotic limit of the radial wave function in terms of δ_l,

$$R_l(k,r) \overset{r\to\infty}{\longrightarrow} e^{i\delta_l}\cos\delta_l\, j_l(kr) - e^{i\delta_l}\sin\delta_l\, n_l(kr) \tag{12.66a}$$

$$= \frac{1}{2ikr}e^{i\delta_l}\left[e^{i(kr-l\pi/2+\delta_l)} - e^{-i(kr-l\pi/2+\delta_l)}\right] \tag{12.66b}$$

$$= e^{i\delta_l}\frac{\sin(kr - \frac{l\pi}{2} + \delta_l)}{kr}. \tag{12.66c}$$

We can see from Eq. (12.66c) that the meaning of δ_l is a phase shift.

Program listing 12.1 Deflection function (`deflect.py`)

```python
import matplotlib.pyplot as plt
import numpy as np, rootfinder as rtf, integral as itg

def V(r):            # plum potential
    return Z*(3-r*r/(R*R))/(R+R) if (r<R) else Z/r

def fu(u):           # f(u), turning point eqn
    return 1 - V(1./u)/E - b*b*u*u

def fx(x):           # integrand, called by gauss
    u = umin - x*x
    return 2*x*b/np.sqrt(fu(u))

def xection(theta, ba, da):     # cross section
    cs = 0.0                    # ba=impact para, da=deflection angle
    for i in range(len(ba)-1):
        if ((theta-da[i])*(theta-da[i+1]) < 0.):   # theta bracketed
            db = ba[i+1] - ba[i]
            cs += (ba[i] + db/2.)*abs(db/(da[i+1]-da[i]))
    return cs/np.sin(theta)

Z, R = 1.0, 1.0      # nuclear charge Z, radius of plum potential
E, b, bmax = 1.2, 0.01, 20.              # energy, initial b, bmax
eps, tiny =1.E-14, 1.E-5                 # rel error, u-limit
ba, theta = [], []                       # impact para, deflection
while (b <= bmax):
    umin = rtf.bisect(fu, tiny, 1./b, eps)   # find turning pt
    alpha = itg.gauss(fx, 0., np.sqrt(umin))
    ba.append(b), theta.append(np.pi - 2*alpha)
    b *= 1.02

plt.figure(), plt.plot(ba, theta)        # plot deflection function
plt.xlabel('$b$ (a.u.)'), plt.ylabel('$\Theta$', rotation=0)
plt.yticks([0, np.pi/2, np.pi], ['$0$','$\pi/2$', '$\pi$'])
plt.xlim(0,3)

cs, sa = [], np.linspace(0.01, np.pi-.01, 500)  # scattering angle
for x in sa:                             # calc, plot cross section
    cs.append(xection(x, ba, theta))
plt.figure(), plt.plot(sa, cs)
plt.xlabel(r'$\theta$'), plt.ylabel(r'$\sigma(\theta)$')
plt.xticks([0, np.pi/2, np.pi], ['$0$','$\pi/2$', '$\pi$'])
plt.xlim(0, np.pi), plt.semilogy()

plt.show()
```

Given a potential $V(r)$, Program 12.1 calculates the deflection function and the differential cross sections, Eqs. (12.3) and (12.7), respectively. The function `fu()` returns the equation defining the turning points (12.48) in $u = 1/r$ space. The other function, `fx()`, computes the integrand (12.49), but with a change of variable $u = u_{min} - x^2$ to improve convergence (see Project P12.1 and Project S:P11.7).

The main code sets the relevant parameters and iterates through the impact parameter range. In each iteration, it finds the turning point u_{min}. The bisection method is used to solve Eq. (12.48). Because the plum potential is assumed to be positive, the turning point is such that $b < r_{min} < \infty$, or equivalently $0 < u_{min} < 1/b$. A small value `tiny` is used for zero to avoid overflow. Having obtained u_{min}, the α value is calculated from Eq. (12.49) by Gaussian integration, using the variable substitution found earlier.

The impact parameter is increased by a constant factor in each iteration, so we can step through a large range without wasting time. This is required if we are interested in small angle scattering. The rest of the code plots the deflection function and the cross section.

The cross section is calculated by the module `xection()` from the deflection function with the approximation (12.11), requiring the impact parameter array ba and the corresponding deflection angle array da. For a given scattering angle θ, it searches through the deflection angles that bracket θ (line 17). If found, $\left|b\frac{db}{d\theta}\right|$ is approximated at the midpoint. The cumulative contributions from all impact parameters (12.7) are included.

Program listing 12.2 Partial wave method (`qmscatt.py`)

```
from scipy.special import sph_jn, sph_yn, lpn
import matplotlib.pyplot as plt, numpy as np, ode

def V(r):                          # Yukawa potential
    Z, sa = 2., 1.0                # nuclear charge, screening length
    return -Z*np.exp(-r/sa)/r

def f(r):                          # Sch eqn in Numerov form
    return 2*(E - V(r)) - L*(L+1)/(r*r)

def wf(M, xm):                     # find w.f. and deriv at xm
    c = (h*h)/6.
    wfup, nup = ode.numerov(f, [0,.1], M, xL, h)     # 1 step past xm
    dup = ((1+c*f(xm+h))*wfup[-1] - (1+c*f(xm-h))*wfup[-3])/(h+h)
    return wfup, dup/wfup[-2]

xL, a, M = 0., 10., 200                    # limits, matching point
h, Lmax, E =(a-xL)/M, 15, 2.               # step size, max L, energy

k, ps = np.sqrt(2*E), np.zeros(Lmax+1) # wave vector, phase shift
jl, dj = sph_jn(Lmax, k*a)                 # (j_l, j_l ') tuple
nl, dn = sph_yn(Lmax, k*a)                 # (n_l, n_l ')

for L in range(Lmax+1):
    u, g = wf(M, a)                        # g= u'/u
    x = np.arctan(((g*a-1)*jl[L] - k*a*dj[L])/    # phase shift Eq. (12.37)
                  ((g*a-1)*nl[L] - k*a*dn[L]))
    while (x < 0.0): x += np.pi            # handle jumps by pi
    ps[L] = x

theta, sigma = np.linspace(0., np.pi, 100), []
cos, La = np.cos(theta), np.arange(1,2*Lmax+2,2)
for x in cos:                               # calc cross section
    pl = lpn(Lmax, x)[0]                    # Legendre polynomial
    fl = La*np.exp(1j*ps)*np.sin(ps)*pl     # amplitude
    sigma.append(np.abs(np.sum(fl))**2/(k*k))
```

```
37
   plt.figure()                                   # plot phase shift vs L
39 plt.plot(range(Lmax+1), ps, '-o')
   plt.xlabel('$l$'), plt.ylabel(r'$\delta_l$', fontsize=16)
41
   plt.figure()
43 plt.plot(theta*180/np.pi, sigma)               # plot cross sections
   xtck = [0, 30, 60, 90, 120, 150, 180]
45 plt.xticks(xtck, [repr(i) for i in xtck])      # custom ticks
   plt.xlabel(r'$\theta$ (deg)')
47 plt.ylabel(r'$\sigma(\theta)$ (a.u.)'), plt.semilogy()

49 plt.show()
```

Phase shifts, cross sections, and normalized wave functions may be obtained from Program 12.2. It efficiently solves quantum scattering problems from central field potentials by partial wave expansion with Numerov's method. It is based on Program 9.5.

It starts by importing the spherical Bessel functions and other packages. The function `V(r)` calculates the Yukawa potential, and `f()` returns the corresponding function $f(r)$ in Eq. (12.36) dependent on the energy E and the effective potential. The code up to line 22 had been explained in the main text, except for the two lines in the middle initializing parameters including the matching radius a, the number of intervals M, step size h, maximum partial waves l_{max}, and scattering energy E. Depending on how quickly the potential decays, or how high the energy is, it may be necessary to adjust a, M (h) to ensure accuracy. Generally speaking, the higher the energy, the more rapidly the wave function oscillates, and therefore a larger M should be used.

The next L-loop calls `wf(M,a)` integrating the radial equation for a given l to obtain the wave function $u_0, u_1, ..., u_{M+1}$. Note that the derivative is evaluated at $u'_M = u'(a)$, however. The phase shift δ_l is calculated from Eq. (12.37) with the returned γ (variable g). For attractive potentials, line 28 ensures that the phase shift is positive due to jumps by π in the arctan function. All phase shifts are stored in the array `ps`.

The cross section is calculated next. For each angle θ, the SciPy special function `lpn()` is called to calculate the Legendre polynomial $P_l(\cos\theta)$. Like the spherical Bessel functions, `lpn()` also returns a pair of arrays, the function values and the derivatives, for all orders $0 - l_{max}$. We only need the function values, so we select the first array in line 34. Next (line 35), the scattering amplitude is calculated from Eq. (12.31) for all partial waves at once, and the cross section from Eq. (12.32). At higher energies, l_{max} should be sufficiently large to include all contributions.

The remaining code plots the phase shifts and the cross sections. We use customized tick marks (line 45) for the latter.

List of Programs

We have built many programs in this book. The following table summarizes these programs and dependencies. Standard libraries such as Matplotlib (2D), NumPy, and SciPy are omitted. Library abbreviations are: fft—fast Fourier transform (`fft.py`, Chapter 5); fem—finite element method (`fem.py`, Chapter 9); fileio—file input/output (`fileio.py`, Chapter 9); itg—numerical integration (`integral.py`, Chapter 8); ode—ordinary differential equation (`ode.py`, Chapter 2); rtf—root finders (`rootfinder.py`, Chapter 3); and vpm—VPython Modules (`vpm.py`, Chapter 6). In addition, we also note the use of 3D and animation libraries: Axes3D—Matplotlib 3D plotting; am—Matplotlib animation; and vp—VPython.

Page numbers with the "S:" prefix are entries from the Digital Online Companion.

Program	**Description**	**Dependence**
3body, 139	Three-body motion	ode, rtf, vp
balltoss, 58	Ideal projectile motion	vp
baseball, 89	Flight of baseball	ode, vp
bdipole, 268	Magnetic dipole field	vp
bem, 354	Basis expansion method	itg
bisect, 88	Bisection root finder	
boltzmann, 424	Boltzmann distribution	
bouncing_ball, 27	Bouncing ball	vp
brownian, 379	Brownian motion	am
coupled, S:112	Quantum transitions	ode, vp, vpm
ctmc, S:263	Classical ion-atom collisions	ode, rtf, vp, vpm
deflect, 459	Deflection function	itg, rtf
dipole, 270	Dipole distribution	Axes3D
earth, 93	Planetary motion	ode, vp
edipole, 269	Electric dipole radiation	vp
eigh, 194	Eigenvalue problem	
einsteinsolid, 383	Einstein solid class	
entropy, 387	Entropy of an Einstein solid	
equishare, 424	Thermal energy sharing	am
fem, 357	FEM library	
fft, S:44	Fast Fourier transform	
fractal, 182	Mandelbrot fractal	
freefall_euler, 39	Free fall with modular Euler	
freefall_plot, 30	Free fall with inline Euler	
gauss_elim, S:57	Gauss elimination	
gaussxw, 312	Gauss abscissa and weight	
hockey, 261	Electric field hockey	ode, vp
hydrogen, 355	Atomic structure	rtg
integral, 311	Numerical integration	

Continued on next page

Continued from previous page

Program	Description	Dependence
ising1d, 425	Ising model	
lambertw, 69	Lambert *W* function	
laplace, 262	Relaxation solutions	Axes3D, vp
laplacefem, 264	Laplace equation by FEM	Axes3D
laplacerbf, 266	Meshfree method for PDEs	
leapfrog, 45	Leapfrog method	
leapfrog_tt, 103	Time-transformed leapfrog	
leapfrog_ttN, S:224	*N*-body transformed leapfrog	
life, S:149	Game of life	
logisdiff, 179	Difference of logistic map	
logisticmap, 146	Logistic map iterates	
longwire, 269	Magnetic fields of a long wire	vp
lyapunov, 180	Lyapunov exponent	
mcint, 380	Monte Carlo integration	
md, 426	Molecular dynamics	ode, vp, vpm
mdf2py, S:192	Molecular dynamics F2Py	ode, nbodyf
mercury, 135	Precession of Mercury	ode, vp
meshhex, 360	FEM meshes of a hexagon	fileio
metropolis2, 400	Ising model in 2D	
motion, 7	1D motion and graphing	
nbody, S:266	*N*-body Coulomb interaction	
nbodyf, S:189	*N*-body interaction, Fortran	
newton, 89	Newton's root finder	
nns, S:135	Energy spectrum unfolding	
nonlindro, 180	Nonlinear pendulum	ode
odewrap, 55	SciPy ODE wrapper	
perioddbl, 150	Bifurcation diagram	
plane, 270	Plane electromagnetic wave	vp, vpm
poincare, 181	Poincaré map	ode
projectile, 62	Motion with linear drag	
qmdot, 358	Quantum dot	Axes3D, fem, fileio
qmfem, 353	Eigenenergies by FEM	
qmplane, 307	Quantum plane wave	am
qmscatt, 460	Central field scattering	ode
qmshoot, 352	Quantum shooting method	ode, rtf
qmwaves2d, 310	Quantum waves in 2D	vp, vpm
r3body, 142	Restricted 3-body motion	ode, vp
r3body_veff, 140	Lagrange points	Axes3D
range, 66	Range with linear drag	rtf
relax, S:59	Relaxation of a string	Axes3D, ode
relaxtd, S:187	Thermal relaxation	
rk2, 35	Runge-Kutta order 2	
rk4, 36	Runge-Kutta order 4	
rk45n, 55	Runge-Kutta-Fehlberg	
rk4n, 54	Nonvector RK4	
rvfit, 136	Radial velocity modeling	

Program	Description	Dependence
sdlf, 307	Quantum wavepacket motion	ode, vp, vpm
sho, 184	Simple harmonic oscillator	vp
sho_lf, 46	Oscillator with leapfrog	ode
shoot, 82	Shooting method	ode, rtf
slinky, S:60	Motion of a slinky	ode, vp, vpm
splitop, 309	Quantum free fall	vp, vpm
squarewell, 351	Visualizing eigenstates	ode, vp, vpm
stringfdm, 197	Displacement of a string	
stringfun, 222	Fundamental frequencies	
tablecloth, 224	Falling tablecloth	ode, vp, vpm
transport, S:188	Particle transport	
triatom, 222	Triatomic vibration	vp
tripwf, 359	Triangular mesh plots	fileio
vpm, S:61	VPython modules	vp
walk2d, 378	Random walk in 2D	
waves, 208	Waves on a string	vp, vpm
waves2d, 223	Waves on a membrane	vp, vpm

Bibliography

1. M. Abramowitz and I. A. Stegun. *Handbook of mathematical functions: with formulas, graphs, and mathematical tables.* (Dover, New York), 1970.
2. R. K. Adair. The physics of baseball. *Physics Today*, **48**(5): 26–31, (1995).
3. R. K. Adair. *The physics of baseball.* (Perennial, New York), 2002.
4. G. B. Arfken and H. J. Weber. *Mathematical methods for physicists.* (Academic Press, New York), 6th edition, 2005.
5. T. Asai, K. Seo, O. Kobayashi, and R. Sakashita. Fundamental aerodynamics of the soccer ball. *Sports Eng.*, **10**:101–110, (2007).
6. N. W. Ashcroft and N. D. Mermin. *Solid state physics.* (Saunders College, New York), 1976.
7. E. B. Becker, G. F. Carey, and J. T. Oden. *Finite elements: an introduction.* (Prentice Hall, Englewood Cliffs, NJ), 1981.
8. C. A. Bennett. *Principles of physical optics.* (Wiley, New York), 2008.
9. S. J. Blundell and K. M. Blundell. *Concepts in thermal physics.* (Oxford University Press, New York), 2010.
10. M. L. Boas. *Mathematical methods in the physical sciences.* (Wiley, New York), 3rd edition, 2006.
11. B. H. Bransden and C. J. Joachain. *Physics of atoms and molecules.* (Prentice Hall, New York), 2002.
12. K. Briggs. A precise calculation of the Feigenbaum constants. *Math. Comp.*, **57**:435–439, (1991).
13. H. C. Bryant and N. Jarmie. The glory. *Sci. Am.*, **231**:60–71, July (1974).
14. R. P. Butler, J. T. Wright, G. W. Marcy, D. A. Fischer, S. S. Vogt, C. G. Tinney, H. R. A. Jones, B. D. Carter, J. A. Johnson, C. McCarthy, and A. J. Penny. Catalog of nearby exoplanets. *Astrophys. J.*, **646**:505–522, (2006).
15. A. Chenciner and R. Montgomery. A remarkable periodic solution of the three-body problem in the case of equal masses. *Ann. of Math.*, **152**:881–901, (2000).
16. G. M. Clemence. The relativity effect in planetary motions. *Rev. Mod. Phys.*, **19**:361–364, (1947).
17. C. J. Cohen and E. C. Hubbard. Libration of the close approaches of Pluto to Neptune. *Astrono. J.*, **70**:10–13, (1965).
18. J. W. Cooley and J. W. Tukey. An algorithm for the machine calculation of complex Fourier series. *Math. Comp.*, **19**: 297–301, (1965).
19. S. N. Coppersmith. A simpler derivation of Feigenbaum's renormalization group equation for the period-doubling bifurcation sequence. *Am. J. Phys.*, **67**:52–54, (1999).
20. R. M. Corless, G. H. Gonnet, D. E. G. Hare, D. J. Jeffrey, and D. E. Knuth. On the Lambert *W* function. *Adv. Comp. Math.*, **5**:329–359, (1996).
21. J. Crank and E. Nicolson. A practical method for numerical evaluation of solutions of partial differential equations of the heat-conduction type. *Adv. Comp. Math.*, **6**:207–226, (1996).
22. R. Cross. *Physics of baseball and softball.* (Springer, New York), 2011.
23. C. R. Davis and H. M. Schey. Eigenvalues in quantum mechanics: a computer-generated film. *Am. J. Phys.*, **40**: 1502–1508, (1972).
24. P. DeVries and J. Hasbun. *A first course in computational physics.* (Jones & Bartlett, Sudbury, MA), 2010.
25. E. Doolittle. The secular variations of the elements of the orbits of the four inner planets computed for the epoch 1850.0 G.M.T. *Trans. Am. Phil. Soc.*, **22**:37–189, (1912).
26. P. Fallahi, A. C. Bleszynski, R. M. Westervelt, J. Huang, J. Walls, and E. J. Heller. Imaging a single-electron quantum dot. *Nano Lett.*, **5**:223–226, (2005).
27. G. E. Fasshauer. *Meshfree approximation methods with MATLAB.* (World Scientific, Hackensack, NJ), 2007.
28. J. Feigenbaum. Quantitative universality for a class of nonlinear transformations. *J. Stat. Phys.*, **19**:25–52, (1978).
29. R. Feynman. *The Feynman Lectures on Physics*, volume I. (Addison-Wesley, New York), 1970. feynmanlectures .caltech.edu.
30. Debra A. Fischer, R. Paul Butler, Geoffrey W. Marcy, Steven S. Vogt, and Gregory W. Henry. A sub-Saturn mass planet orbiting HD 3651. *Astrophys. J.*, **590**:1081–1087, (2003).
31. G. R. Fowles and G. L. Cassiday. *Analytical mechanics.* (Thomson Brooks/Cole, Belmont, CA), 7th edition, 2004.
32. C. Frohlich. Aerodynamic drag crisis and its possible effect on the flight of baseballs. *Am. J. Phys.*, **52**:325–334, (1984).
33. T. Gay. *The physics of football.* (HarperCollins, New York), 2005.
34. N. Giordano and H. Nakanishi. *Computational physics.* (Benjamin Cummings, New York), 2nd edition, 2005.
35. J. Gleick. *Chaos: making a new science.* (Penguin, New York), 1988.

36. J. E. Goff and M. J. Carré. Trajectory analysis of a soccer ball. *Am. J. Phys.*, **77**:1020–1027, (2009).
37. A. Goldberg, H. M. Schey, and J. L. Schwartz. Computer-generated motion pictures of one-dimensional quantum-mechanical transmission and reflection phenomena. *Am. J. Phys.*, **35**:177–186, (1967).
38. J. Golde, J. Shertzer, and P. Oxley. Finite element solution of Laplaces equation for ion-atom chambers. *Am. J. Phys.*, **77**:81–86, (2009).
39. H. Goldstein. Prehistory of the Runge-Lenz vector. *Am. J. Phys.*, **43**:737–738, (1975).
40. H. Goldstein, C. Poole, and J. Safko. *Classical mechanics.* (Addison-Wesley, New York), 2002.
41. H. Gould and J. Tobochnik. *Statistical and thermal physics with computer applications.* (Princeton University Press, Princeton, NJ), 2010.
42. H. Gould, J. Tobochnik, and W. Christian. *An introduction to computer simulation methods: Applications to physical systems.* (Addison-Wesley, New York), 3rd edition, 2007.
43. D. J. Griffiths. *Introduction to electrodynamics.* (Prentice Hall, Upper Saddle River, NJ), 1999.
44. D. J. Griffiths. *Introduction to quantum mechanics.* (Prentice Hall, Upper Saddle River, NJ), 2005.
45. P. Harrison. *Quantum wells, wires and dots: theoretical and computational physics of semiconductor nanostructures.* (Wiley, New York), 3rd edition, 2010.
46. B. Hayes. Why *W*? *Am. Scientist*, **93**:104–108, (2005).
47. W. H. Heintz. Determination of the Runge-Lenz vector. *Am. J. Phys.*, **42**:1078–1082, (1974).
48. P. Hellings. *Astrophysics with a PC.* (Willmann-Bell Inc., Richmond, VA), 1994. An introduction to computational astrophysics.
49. D. Hestenes. *New foundations for classical mechanics.* (Kluwer Pub., New York), 1999. Ch. 8, Celestial mechanics.
50. J. D. Jackson. *Classical electrodynamics.* (Wiley, New York), 1975.
51. C. J. Joachain. *Quantum collision theory.* (North-Holland, Amsterdam), 1983.
52. E. J. Kansa. Multiquadrics—a scattered data approximation scheme with applications to computational fluid dynamics II: Solutions to parabolic, hyperbolic, and elliptic partial differential equations. *Comp. Math. Appl.*, **19**:147–161, (1990).
53. R. Landau, M. Páez, and C. Bordeianu. *Computational physics: Problem solving with computers.* (Wiley, New York), 2nd edition, 2007.
54. H. P. Langtangen. *A primer on scientific programming with Python.* (Springer, New York), 3rd edition, 2012.
55. E. Larsson and B. Fornberg. A numerical study of some radial basis function based solution methods for elliptic PDEs. *Comp. Math. Appl.*, **46**:891–902, (2003).
56. S. M. Lea. *Mathematics for physicists.* (Thomson Brooks/Cole, Belmont, CA), 2004.
57. R. L. Liboff and J. Greenberg. The hexagon quantum billiard. *J. Stat. Phys.*, **105**:389–402, (2001).
58. J. J. Lissauer. Chaotic motion in the Solar System. *Rev. Mod. Phys.*, **71**(3):835–845, (1999).
59. E. N. Lorenz. Deterministic nonperiodic flow. *J. Atmo. Sci.*, **20**:130–141, (1963).
60. G. D. Mahan. *Quantum mechanics in a nutshell.* (Princeton University Press, Princeton, NJ), 2009.
61. B. Marmaras and J. Wang. Simulation and visualization of few-body systems and the differential precession of Mercury. *Computing in Sci. & Eng.*, **16**:42–50, (2014).
62. N. Metropolis, A. W. Rosenbluth, M. N. Rosenbluth, A. H. Teller, and E. Teller. Equation of state calculations by fast computing machines. *J. Chem. Phys.*, **21**:1087–1092, (1953).
63. S. Mikkola and S. Aarseth. A time-transformed leapfrog scheme. Integration of few-body systems with large mass ratios. *Celes. Mech. Dyn. Astrono.*, **84**:343–354, (2002).
64. M. H. Mittleman. *Introduction to the theory of laser-atom interactions.* (Plenum, New York), 1993.
65. A. Morbidelli. Modern integrations of solar system dynamics. *Annu. Rev. Earth Planet. Sci.*, **30**:89–112, (2002).
66. NASA exoplanet archive. http://exoplanetarchive.ipac.caltech.edu/.
67. A. M. Nathan. The effect of spin on the flight of a baseball. *Am. J. Phys.*, **76**:119–124, (2008).
68. M. Newman. *Computational physics.* (CreateSpace Independent Publishing), 2012.
69. H. M. Nussenzveig. The theory of the rainbow. *Sci. Am.*, **236**:116–127, (1977).
70. T. Pang. *An introduction to computational physics.* (Cambridge Univ. Press, Cambridge), 2nd edition, 2006.
71. W. H. Press, B. P. Flannery, S. A. Teukolsky, and W. T. Vetterling. *Numerical recipes: the art of scientific computing.* (Cambridge Univ. Press, Cambridge), 1992.
72. M. P. Price and W. F. Rush. Nonrelativistic contribution to Mercury's perihelion precession. *Am. J. Phys.*, **47**:531–534, (1979).
73. E. Rabe. Determination and survey of periodic Trojan orbits in the restricted problem of three bodies. *Astrono. J.*, **66**:500–513, (1961).
74. I. I. Rabi. Space quantization in a gyrating magnetic field. *Phys. Rev.*, **51**:652–654, (1961).
75. J. N. Reddy. *An introduction to the finite element method.* (McGraw-Hill, New York), 1993.
76. F. Reif. *Fundamentals of statistical and thermal physics.* (Waveland Press, Long Grove, IL), 2009.
77. S. M. Reimann and M. Manninen. Electronic structure of quantum dots. *Rev. Mod. Phys.*, **74**:1283–1342, (2002).
78. M. J. Romanelli. Runge-Kutta methods for the solution of ordinary differential equations. In A. Ralston and H. S. Wiff, editors, *Mathematical methods for digital computers*, pages 110–120. (Wiley, New York), 1960.
79. A. E. Roy. *Orbital motion.* (IOP Publishing, Bristol), 2005.

80. J. J. Sakurai and J. Napolitano. *Modern quantum mechanics*. (Addison-Wesley, New York), 2011.
81. G. S. Sawicki, M. Hubbarda, and W. J. Stronge. How to hit home runs: Optimum baseball bat swing parameters for maximum range trajectories. *Am. J. Phys.*, **71**:1152–1162, (2003).
82. D. V. Schroeder. *An introduction to thermal physics*. (Addison-Wesley, New York), 1999.
83. R. Seydel. Determinant factors of the table tennis game: measurement and simulation of ball-flying curves. *Int. J. Tab. Tenn. Sci.*, **1**:1–7, (1992).
84. C. Simó. New families of solutions in N-body problems. *Proceedings of the ECM 2000, Barcelona (July 10–14)*, (2000).
85. M. G. Stewart. Precession of the perihelion of Mercury's orbit. *Am. J. Phys.*, **73**:730–734, (2005).
86. S. H. Strogatz. *Nonlinear dynamics and chaos*. (Westview Press, Cambridge, MA), 1994.
87. D. F. Styer. Quantum revivals versus classical periodicity in the infinite square well. *Am. J. Phys.*, **69**:56–62, (2001).
88. G. J. Sussman and J. Wisdom. Numerical evidence that the motion of Pluto is chaotic. *Science*, **241**:433–437, (1988).
89. J. Thijssen. *Computational physics*. (Cambridge University Press, Cambridge), 2nd edition, 2007.
90. S. R. Valluri, D. J. Jeffrey, and R. M. Corless. Some applications of the Lambert W function to physics. *Can. J. Phys.*, **78**(9):823–831, (2000).
91. M. C. Vargas, D. A. Huerta, and V. Sosa. Chaos control: The problem of a bouncing ball revisited. *Am. J. Phys.*, **77**:857–861, (2009).
92. L. Verlet. Computer "experiments" on classical fluids. I. Thermodynamical properties of Lennard-Jones molecules. *Phys. Rev.*, **159**:98–103, (1967).
93. J. Wang, J. Burgdörfer, and A. Bárány. Ionization spectrum for ion-atom collisions with short-ranged potentials in one and three dimensions. *Phys. Rev. A*, **43**:4036–4039, (1991).
94. R. Warburton and J. Wang. Analysis of asymptotic projectile motion with air resistance using the Lambert W function. *Am. J. Phys.*, **72**:1404–1407, (2004).
95. P. R. Weissman and G. W. Wetherill. Periodic Trojan-type orbits in the Earth-Sun system. *Astrono. J*, **79**:404–412, (1974).
96. H. Wendland. *Scattered data approximation*. (Cambridge University Press, Cambridge), 2005.
97. J. Wesson. *The science of soccer*. (IOP Publishing, Bristol), 2002.
98. F. Yamamoto, Y. Tsuji, and G. Chen. Basic theory and experiment for the simulation of ball trajectory. *Int. J. Tab. Tenn. Sci.*, **3**:1–15, (1996).

Index

Page numbers with the "S:" prefix are entries from the Digital Online Companion.

Printed and bound by CPI Group (UK) Ltd, Croydon, CR0 4YY

17/09/2026

14964251-0001